GRAVITATIONAL-WAVE ASTRONOMY

Gravitational-Wave Astronomy

Exploring the Dark Side of the Universe

Nils Andersson

Mathematical Sciences and STAG Research Centre,
University of Southampton, Southampton, UK

OXFORD

UNIVERSITY PRESS

Great Clarendon Street, Oxford, OX2 6DP,
United Kingdom

Oxford University Press is a department of the University of Oxford.
It furthers the University's objective of excellence in research, scholarship,
and education by publishing worldwide. Oxford is a registered trade mark of
Oxford University Press in the UK and in certain other countries

First Edition published in 2020

Impression: 2

Published in the United States of America by Oxford University Press
198 Madison Avenue, New York, NY 10016, United States of America

British Library Cataloguing in Publication Data
Data available

Library of Congress Control Number: 2019945425

ISBN 978–0–19–856803–2

DOI: 10.1093/oso/9780198568032.001.0001

Printed and bound by
CPI Group (UK) Ltd, Croydon, CR0 4YY

Preface

Whenever you are writing a book, people are bound to ask: 'What kind of book is it? Who is it for?' These questions are reasonable, but the answers may not be that obvious. You may, for example, have embarked on the project simply because it seemed like a good idea at the time. So, with this in mind, what kind of a book is this? Having lived with it for longer than I care to figure out, I still find it difficult to give a clear answer. It is much easier to explain what it is not. This book is not an exhaustive review of gravitational-wave astronomy. At least not in the sense that it provides a 'complete' reference list and a detailed account of the historical developments of the ideas and the scope of the field. It is much more 'subjective' than that. This may be frustrating to colleagues that have contributed to the developments over the last several decades, but the reality is that I had to make choices. It was simply not manageable to peek into (and report back on) every nook and cranny, no matter how fascinating this might have been. Instead, I have tried to provide an entry point to the vast (and rapidly growing!) literature on the different aspects of gravitational waves and related astrophysics.

In essence, I have tried to build a bridge across different areas of physics that have fascinated me for a long time. On the one hand, we have gravity—with Einstein's warped spacetime providing an astonishing example of what the human mind is capable of. On the other hand, there is the extraordinary range of astrophysics and cosmology that comes into play when we try to understand the gravitational-wave sky. And finally, we need to consider the sublime technology that was developed to catch these faint whispers from the distant Universe. This book maps out a journey through this complex landscape—introducing a combination of overlapping areas of research, many of which require their separate books for a fair treatment. The different chapters (especially in the second part) are intended to narrow the gap between a basic understanding and current research. An important part of this involves introducing the relevant language—making the involved concepts less 'mysterious'.

The book is intended to work as a platform, sufficiently low that anyone with an interest in gravitational waves can scramble onto it, but at the same time high enough that it connects with current research—and exciting discoveries that are happening right now. It may only be an introduction, but I think it has potential... If you are an astronomer and you want a basic understanding of this new window to the Universe, including a brief (relatively self-contained) glimpse at Einstein's theory, then this book may work for you. Similarly, if you spend most of your time analysing data from gravitational-wave detectors and you would like a better picture of what you are looking for (and perhaps why theorists find it so difficult to make firm predictions) then other parts of the book could work for you. Finally, there is a connection to nuclear physics—which is natural, since gravitational-wave signals from neutron stars may help constrain our ideas for matter at extreme densities. Relevant aspects are addressed at various places in the book, which may help nuclear and particle physicists appreciate how their work fits into the bigger picture. Whichever direction you are coming from, and regardless of where you are going, this book may be of interest to you.

In terms of teaching, the scope of the book is likely too vast for a single undergraduate or masters-level course. But the material is flexible. The first part introduces the key ideas, following a general overview chapter and including a brief reminder of Einstein's theory. This part can be taught as a (fairly) self-contained undergraduate one semester course. In fact, the material is based on a course we have had on the books for over a decade. So I know it works. Depending on the background and interest of the students, I would select topics from the second (much longer) part of the book to connect with the actual state of the art. The chapters are written to work as 'set pieces' with core material that can be adapted to specific lectures and additional material that provide context and depth. At least that's the way I like to think about it. Some of the chapters have been road-tested at summer schools and other events so I am confident they work. The one thing that is missing in terms of teaching material is exercises. However, it is quite easy to identify steps that need filling in and to come up with questions that go beyond the material, so this should not be a major issue.

Before we embark on the journey, it is useful to make a few comments on notation and conventions. Throughout the book I have chosen to work with a spacetime metric with signature +2. There is one exception: The discussion of the Newman–Penrose formalism used to discuss the dynamics of spinning black holes. I have adopted the convention that spacetime indices are given by letters from the beginning of the alphabet, $a, b, c, ...$, while spatial indices start with $i, j, k, ...$. Many text books use Greek letters for the former. Repeated indices (spacetime or spatial) indicate summation.

With these formalities out of the way, let's get started.

Contents

Part 2 The dark side of the universe

A long time ago, in a galaxy far away, the two black holes edged closer. Dancing around each other in a nearly perfect circle. Drawn together by gravity, through the emission of gravitational waves. Faint ripples encoded the change in gravity over eons. In the last few moments the motion grew frantic. A storm of warped space and time raged as the two objects came together. An energy equal to the obliteration of several suns was released in a fraction of a second. Then it was over. All that remained was a single black hole. And empty space.

The signal moved unchanged over the vast distances of space until, after more than a billion years, it reached the Earth. When the signal was created, this insignificant blue planet hosted single cell organisms. When the signal arrived, there was an advanced civilization. A civilization curious about the Universe. A civilization with technology to catch the elusive spacetime whisper. Their advanced detectors registered a disturbance.

This was the beginning.

1

Opening the window

1.1 The beginning

The first direct detection of gravitational waves was announced to the world on the 11th of February 2016 with a triumphant 'We did it!'. The signal, which had been picked up by the two LIGO detectors on the 14th of September 2015, matched the predictions from numerical simulations of the merger of a pair of black holes with masses $36M_\odot$ and $29M_\odot$, forming a larger black hole with mass $62M_\odot$ (Abbott *et al.*, 2016*b*). The missing mass—the equivalent of about 3 solar masses—had been radiated as gravitational waves. This extraordinary event, which only lasted a fraction of a second, was the most powerful astronomical event ever observed. It was the beginning of a new kind of astronomy.

The breakthrough detection came nearly a century after Einstein's prediction that changes in gravity should propagate as waves (Einstein, 1916). It was an extraordinary moment of success, following decades of technology development, political wrangling to secure funding, and several false starts. It was a moment of glory, rewarding an enormous amount of patient and hard work from a lot of people.

The LIGO project was initiated in the early 1990s Abramovici *et al.* (1992) and the first generation of kilometre-scale gravitational-wave interferometers reached their initial design sensitivity in a broad frequency window in November 2005 (during the fifth science run, S5). More than one year's worth of quality data was taken during the following science run (S6) in 2009–10. Many research papers were written, but no signals were found. After a couple of years' downtime to improve the technology, the first 'observing run' (O1) of the advanced interferometers started in September 2015. The immediate detection of the black-hole signal led to a collective sigh of relief. It had been a long journey.

The first detection brought the promise of gravitational-wave astronomy into sharp focus. It was much more than a confirmation that gravitational waves exist and that we can catch them. We learned that there are double black-hole systems in the Universe and that they merge due to the emission of gravitational radiation. The observed signal agreed with the predictions from general relativity, showing the expected inspiral, merger, and ringdown phases seen in numerical supercomputer simulations (Chapter 19). It was the first test of Einstein's theory in a dynamical, strong-field setting. The signal allowed us to identify more massive black holes than so far found in X-ray binaries, and it also provided interesting constraints on the spin of the individual black holes.

Gravitational-Wave Astronomy: Exploring the Dark Side of the Universe. Nils Andersson, Oxford University Press (2020).
© Nils Andersson. DOI: 10.1093/oso/9780198568032.001.0001

The underlying theory may be complex, but the observed signal was simple. It swept upwards in amplitude and frequency from 30 to 250 Hz in a perfect example of the anticipated chirp (see the time-frequency plots in the lower panels of Figure 1.1). At its peak, the gravitational-wave strain, $h \approx 10^{-21}$, corresponded to a luminosity equivalent to emitting the mass-energy of about 200 suns in a second. The event took place 1.3 billion light years from the Earth (Abbott *et al.*, 2016*b*). In terms of the Universe, it was ancient history.

Binary signals, like GW150914, carry unique information on the masses and spins of the sources. In the case of neutron stars, the gravitational waves also encode the internal structure, which depends on the state of matter at extreme densities. In essence, gravitational-wave observations have the potential to probe many fundamental physics issues. Given the weakly interacting nature of gravitational waves, the information they carry provides an important complement to electromagnetic observations. In fact, they

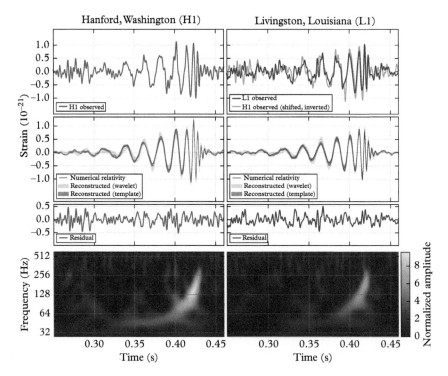

Figure 1.1 *The first gravitational-wave signal (GW150914) observed by the LIGO Hanford (H1, left) and Livingston (L1, right) interferometers. The top row shows how the gravitational-wave strain varied with time in the two detectors (with a direct comparison after a time shift of 10 ms corresponding to the travel time—at the speed of light—between the two instruments). The middle row compares the signal to results from numerical relativity simulations, showing inspiral, merger, and ringdown of two coalescing black holes. The bottom row gives a time-frequency representation of the gravitational-wave strain, again showing the signal frequency and strength increasing with time. (Reproduced from Abbott et al. (2016b), Creative Commons Attribution 3.0 License.)*

shed light on aspects that cannot be probed by traditional means, like the internal dynamics of a supernova explosion or quantum fluctuations in the very early Universe just after the Big Bang. In order to understand the wide range of possibilities, we need to explore the mechanisms that generate gravitational waves in the first place. We need to be able to predict the character of the signals and consider the challenges associated with detecting them. As this involves many complex questions, and it is important to appreciate the context, we need to start from the beginning.

1.2   A new kind of astronomy

With his theory of general relativity, Einstein revolutionized our view of space and time (Einstein, 1915). By explaining gravity in terms of the geometry of a combined spacetime he provided a fresh perspective on the Universe. This led to the introduction of exciting concepts that have become part of mainstream culture. Most notably, *black holes*, formed when massive stars die, and the *Big Bang*, the explosion which gave birth to the Universe some 14 billion years ago. Moreover, Einstein's general relativity is a *dynamic* theory of gravity, where space and time are flexible concepts. The theory predicts that changes in gravity propagate as waves, ripples in spacetime moving at the speed of light. These *gravitational waves* are elusive. For decades they caused debate and controversy[1] and, until recently, attempts to detect them proved futile.

It is not really surprising that the detection of gravitational waves proved such a challenge. Early generations of instruments may have been remarkably sensitive—from an everyday life point of view—but they would still only have been able to catch unique events in our own Galaxy and its immediate neighbourhood and such events are rare. Take supernova explosions, which occur only a few times per century in a typical galaxy, as an example. Population modelling and our understanding of stellar evolution tell us that we need to reach further out into the Universe if we want to detect such events. Exactly how far, we do not know at this point. It is relatively easy to work out the energy that must be released in order for a given source to be detectable, but very difficult to provide a reliable model of the complex physics associated with most gravitational-wave scenarios. Yet, it is clear that we will always be dealing with faint signals. This is in sharp contrast with mainstream astronomy, where observations are traditionally made at large signal-to-noise ratios.

As the sensitivity of the available detectors improved—gradually—we learned valuable lessons. It is fairly easy to identify 'milestone' results leading up to the breakthrough in 2015. For example, the initial LIGO–Virgo detectors were sensitive enough that they would have been able to catch a gravitational-wave burst from a Milky Way supernova, should one have occurred during the series of science runs (Abadie *et al.*, 2012). The absence of detections hardly challenged our view of the Universe, but it was nevertheless an important step. The fact that the gravitational-wave contribution to the spin-down of

---

[1] A meeting at Chapel Hill in January 1957 is often seen as the turning point. In particular, Richard Feynman famously provided a 'sticky bead' argument to demonstrate that gravitational waves must carry energy.

the Crab Pulsar—a neutron star born in a supernova recorded by Chinese astronomers in 1054—can be constrained to be less than a fraction of a percent of the observed rate (Abbott *et al.*, 2008*a*) may only be mildly interesting from the astrophysics point of view, but it was nevertheless a milestone achievement as it constrained the asymmetry of a distant astronomical object in a way that could not be done by other means.

Gravitational-wave astronomy is a fascinating area that involves a range of complex issues, from the development of detector technology to data-handling techniques and theory modelling. In order to progress, we need to improve on all these aspects. As we celebrate the first successful detections, it is useful to keep in mind the effort behind the success. Over decades, generations of scientists turned an impressive engineering project into an astronomical observatory. This was a spectacular achievement, but we are far from done. Future observing runs will probe a much larger volume of space. We will have more, better quality, data. Conservative population synthesis models suggest that we will detect many inspiralling compact binaries (consisting of black holes and/or neutron stars) every year. Given that such 'bread and butter' binary signals are well understood (and depend very little on the composition of the binary companions) and the data analysis algorithms are (more or less) developed, this should allow us to probe the parameters of such systems, shedding light on the cosmic compact binary population and the relevant formation channels.

The wider range of gravitational-wave sources put more emphasis on the involved physics and high-quality modelling of relevant astrophysical scenarios. Inevitably, this requires an exchange of expertise with mainstream astronomers. For a long time the emphasis was on detector development and data analysis strategies. As we establish this new area of astronomy, we need rapid change. We need to address challenging modelling problems. Many relevant gravitational-wave scenarios involve extreme physics that cannot be tested in the laboratory and precision searches require an understanding beyond 'order of magnitude' precision.

The future is, of course, bright. Once third-generation detectors, like the Einstein Telescope (Punturo *et al.*, 2010; Sathyaprakash *et al.*, 2012) or the Cosmic Explorer (Abbott *et al.*, 2017*c*), come on-line we will firmly be in the era of gravitational-wave astronomy. These instruments will improve the broadband sensitivity by another order of magnitude, reaching another factor of 1,000 in volume of space. This may seem remote, given that such detectors are still at the design stage, but we need to consider their promise now. We are talking about 'big science' and we need to understand its potential in order to argue the case for building such hugely expensive instruments. It is relevant to ask what we can hope to achieve with an Einstein Telescope, but not (necessarily) with Advanced LIGO. How much better can we do with (roughly) an order of magnitude improvement in sensitivity? Are there situations where this improvement is needed to see the signals in the first place, or is it a matter of doing better astrophysics by getting improved statistics and more precise parameter extraction? There are many interesting and complicated issues to consider.

Perhaps in contrast, it is straightforward to argue the case for a space-based detector, like the LISA project which is expected to launch in 2034 to address the European

Space Agency's science theme of the Gravitational Universe (Amaro-Seoane *et al.*, 2017). Sensitive to low-frequency gravitational waves, LISA is perfectly tuned to typical astronomical timescales (hours to minutes). If the instrument works as planned—and there is no reason to think that it should not, given the impressive results from the LISA Pathfinder (Armano *et al.*, 2018)—detection is guaranteed. In fact, many known binary systems can be used to verify that the detector is working as intended. The challenges that the LISA project faces are different. Given the number of, in principle, detectable binaries in the Galaxy, the data analyst may suffer an embarrassment of riches. The science may (to some extent) be confusion limited. However, the fact that LISA is sensitive to signals from supermassive black holes (either merging or capturing smaller objects) throughout the Universe makes it an extremely exciting mission.

On a timescale of 20 years or so we should have a network of high-precision instruments searching the skies for gravitational-wave signals over a range of up to eight decades in frequency; see Figure 1.2. These detectors will provide us with unprecedented insights into the dark side of the Universe, and allow us to probe much exciting physics. Further improvements in data quality may allow us to extract the gravitational-wave component in the cosmic microwave background. In addition, ultra-low-frequency gravitational waves are likely to have been detected by pulsar timing arrays. In parallel, we can expect to see breakthroughs in related areas of physics. Following the detection of the Higgs boson by the Large Hadron Collider, the colliders probe higher energies and may eventually find evidence for supersymmetry. Experiments aimed at detecting dark

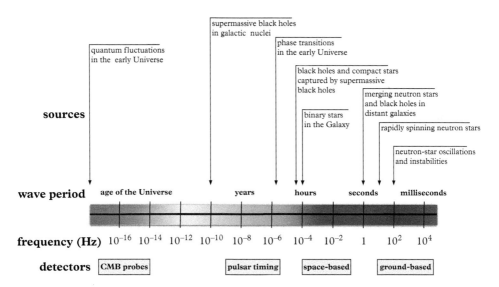

Figure 1.2 *The spectrum of anticipated gravitational-wave sources and the different methods that may be used to detect them, across more than 20 decades in frequency. The physical timescales range from the age of the Universe to a fraction of a millisecond.*

matter signals may provide indisputable data. We ought to have a better understanding of dark energy, e.g. a constraint on the cosmic 'equation of state'. These developments will stimulate theorists as well as experimenters, leading to dramatic improvements in our understanding of the Universe in which we live.

1.3   Audio not video

Most of the information we have about the Universe was gleaned from electromagnetic signals; from beautiful high-resolution images from the Hubble Space Telescope to X-ray timing with the Rossi X-ray Timing Explorer (RXTE) and spectra from Chandra, from pulsar timing with radio dishes to the cosmic microwave data from the Wilkinson Microwave Anisotropy Probe (WMAP) and the Planck experiment, the Sloane Digital Sky Survey, and so on. In the past 50 years we have learned that the Universe is a violent place where stars explode and galaxies collide. There are massive black holes at the centre of most galaxies, and their evolution (through accretion or mergers) may be closely linked to the formation of large-scale structures in the first place. The amount of information we have gathered is truly awesome. Yet, our current Universe is no less mysterious than that of the early 1960s. As we improve our understanding, there are surprises and new questions. At the present time, specific questions concern the dynamics of black holes and their role in evolutionary scenarios, and the state of matter under the extreme conditions in a neutron star core. The big puzzles concern dark energy and (obviously) the still uncomfortable marriage between gravity and physics at the quantum scale.

The gravitational-wave effort should be viewed from this perspective. It is natural to start by comparing and contrasting signals carried by gravity and electromagnetic ones. From the theory point of view, there is a close analogy between electromagnetic and gravitational waves. However, one must not push this too far. The two problems are conceptually rather different. Electromagnetic radiation corresponds to oscillations of electric and magnetic fields propagating *in* a given spacetime, while gravitational waves are oscillations *of* the spacetime itself. In order to identify a gravitational wave one must identify an oscillating contribution to spacetime, varying on a lengthscale much smaller than that of the 'background' curvature (which we experience as our everyday gravity). This distinction can be confusing. Other differences hint at the promises and challenges of gravitational-wave astronomy:

(i) While electromagnetic waves are radiated when individual particles are accelerated, gravitational waves are due to asymmetric bulk motion of matter. In essence, the incoherent electromagnetic radiation generated by many particles carry information about the thermodynamics of the source. Gravitational radiation probes large-scale dynamics.

(ii) The electromagnetic waves that reach our telescopes will have been scattered many times since their generation. In contrast, gravitational waves couple weakly to matter and arrive at the Earth in pristine condition. They carry key information about violent processes that otherwise remain hidden, e.g. associated with the heart of a supernova core collapse or merging black holes. Of course, the waves also interact weakly with our detectors, making their detection a challenge.

(iii) Mainstream astronomy is based on deep imaging of small fields of view, while gravitational-wave detectors cover virtually the entire sky. A consequence of this is that the ability to pinpoint a source in the sky is not particularly good. On the other hand, any source in the sky will in principle be detectable, not just ones towards which we aim the detector (which we cannot do anyway!). This could lead to difficulties if the sources are plentiful, which may be a problem for space-based instruments like LISA.

(iv) Electromagnetic radiation typically has a wavelength much smaller than the size of the emitter. Meanwhile, the wavelength of a gravitational wave is usually comparable to or larger than the size of the radiating source. Hence, gravitational waves can not be used for 'imaging'. Gravitational-wave astronomy is more like listening to the radio than watching television. It may be a matter of taste, but let us not forget that radio offers quality entertainment...

The bottom line is that, gravitational waves carry information about the most violent phenomena in the Universe; information that is complementary to (in fact, very difficult to obtain from) electromagnetic data.

1.4   On the back of an envelope

Without (at this point) getting immersed in technical detail, let us outline the key ideas involved in modelling gravitational-wave sources and at the same time take the opportunity to get a rough idea of the strength and character of typical astrophysical signals. As we will derive the key results later—after developing the required tools—this also provides us with an idea of the road ahead.

We start by noting that, since gravitational-wave signals tend to be weak, it is often sufficient to work at the level of linear perturbations of a given spacetime. In essence, one makes a distinction between a (known) background spacetime and a deviation that lives in this spacetime. In terms of the metric g_{ab}, which provides distance measurements in the curved spacetime, we then have

$$g_{ab} = g_{ab}^{B} + h_{ab},\tag{1.1}$$

where g_{ab}^{B} is some known background metric and $|h_{ab}|$ is suitably small. The metric is, of course, a tensor and each index runs from 0 to 3 to represent the four dimensions of spacetime. It must also satisfy Einstein's field equations, essentially a set of 10 coupled nonlinear partial differential equations. Massaging these equations (by choosing a particularly useful set of 'coordinates') one can show that h_{ab} satisfies a wave equation. Changes in the gravitational field propagate as waves, travelling at the speed of light.

If we consider the effect that the waves of gravity have on matter, we find that they are transverse and have two possible polarizations. They act like a tidal force, which means that they change all distances by the same ratio. If we consider two 'free masses' a distance L apart, then the gravitational-wave induced strain $h \sim |h_{ab}|$ leads to a change ΔL such that

$$h \approx \frac{\Delta L}{L}. \tag{1.2}$$

This allows us to quantify the effect that a passing wave will have on a detector (see Figure 1.3 for a simple thought experiment). Of course, to do this we need an estimate of the typical magnitude of h. We get this from a well-known formula that relates the gravitational-wave luminosity (the energy radiated per unit time) to the strain h

$$\frac{c^3}{16\pi G}|\dot{h}|^2 = \frac{1}{4\pi d^2}\dot{E}, \tag{1.3}$$

where G is Newton's gravitational constant, c is the speed of light, d is the distance to the source, and the dots represent time derivatives. This relation is exact for the weak waves that bathe the Earth.

Suppose we characterize a given event by a timescale τ and assume that the signal is monochromatic, with frequency f. Then we can use $\dot{E} \approx E/\tau$ and $\dot{h} \approx 2\pi f h$. Introducing the relevant scales in the problem, we find that

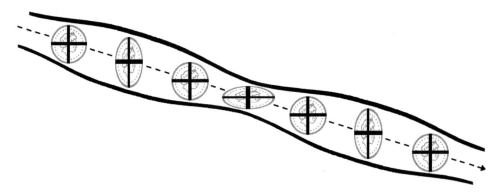

Figure 1.3 *In order to illustrate the effect that a gravitational wave has on matter, let us consider a simple thought experiment. Paint a cross on a coin and place it on a table. Then wait until a gravitational wave passes through the coin. Somewhat simplistically, the gravitational wave will alternately stretch and squeeze the coin (as shown in the illustration) and we should be able to monitor how the cross changes shape as a result. Of course, the impact of gravitational waves from astrophysical sources is far too minuscule to be detected this way. Nevertheless, the principle behind this experiment is the same as that used in bar detectors. (Illustration by O. Dean.)*

$$h \approx 5 \times 10^{-22} \left(\frac{E}{10^{-3} M_\odot c^2} \right)^{1/2} \left(\frac{\tau}{1 \text{ ms}} \right)^{-1/2} \left(\frac{f}{1 \text{ kHz}} \right)^{-1} \left(\frac{d}{15 \text{ Mpc}} \right)^{-1}, \tag{1.4}$$

where we have scaled the distance to the Virgo cluster, the nearest supercluster of galaxies about 15 Mpc away from us. This kind of scaling is necessary to ensure a reasonable event rate for many astrophysical scenarios. For example, at this distance one would expect to see several supernovae per year, which means that one can hope to catch the birth of a few neutron stars/black holes during one year of observation. We have taken the energy radiated to be a thousandth of the energy equivalent of the mass of the Sun ($M_\odot c^2$), which would represent a very powerful event, and assumed that the typical timescale of the dynamics that generated the gravitational waves is a millisecond.

We learn that the effect the waves will have on a terrestrial detector is minuscule. They would stretch a one metre ruler by a puny 10^{-22} m, much less than the diameter of the nuclei that make up the atoms of the ruler. This highlights the severe challenge associated with detecting this kind of signal. Fortunately, we can do better. We can define an 'effective amplitude' that reflects the fact that detailed knowledge of the signal can be used to dig deeper into the detector noise. A typical example is based on the use of matched filtering (see Chapter 8), for which the effective amplitude improves roughly as the square root of the number of observed signal cycles, $\mathcal{N}$. This is a good approximation when $\mathcal{N}$ is large, so the estimate will be reliable for persistent sources (like a slowly evolving source) but obviously less so for short bursts associated with explosive events. Anyway, using $\mathcal{N} \approx f\tau$ we arrive at

$$h_c = \sqrt{\mathcal{N}} h \approx 5 \times 10^{-22} \left(\frac{E}{10^{-3} M_\odot c^2} \right)^{1/2} \left(\frac{f}{1 \text{ kHz}} \right)^{-1/2} \left(\frac{d}{15 \text{ Mpc}} \right)^{-1}. \tag{1.5}$$

This relation shows us that the effective gravitational-wave strain, essentially the 'detectability' of the signal, depends only on the radiated energy and the characteristic frequency. This allows us to assess the relevance of a range of proposed sources without having to work out the detailed signals.

To make progress we need a better idea of the typical frequencies associated with different classes of sources. Luckily, this is straightforward. We only have to note that the dynamical frequency of any self-bound system with mass M and radius R can be approximated by

$$f \approx \frac{1}{2\pi} \left(\frac{GM}{R^3} \right)^{1/2}. \tag{1.6}$$

Given this, the natural frequency of a (non-rotating) black hole (for which $R = 2GM/c^2$) should be

$$f_{\text{BH}} \approx 10^4 \left(\frac{M_\odot}{M} \right) \text{ Hz}, \tag{1.7}$$

immediately suggesting that medium-sized black holes, with masses in the range $10 - 100 M_\odot$, should be prime sources for ground-based interferometers since the "sweet spot" of these detectors tends to be located around 100 Hz; see Figure 1.4. Basically, these instruments are perfectly tuned to events like GW150914. We also see that neutron stars, with a typical mass of $1.4 M_\odot$ compressed inside a radius of 10 km or so, would be expected to radiate at

$$f_{NS} \approx 2 \text{ kHz.} \tag{1.8}$$

In other words, they require detectors that are sensitive at high frequencies. This is a key science target for future ground-based instruments.

Compact object binaries—involving black holes, neutron stars, or white dwarfs—provide particularly promising sources. One reason for this is that the signal strength is calibrated by the masses, so it is fairly easy to assess the detectability. We have already seen that the signal from a pair of black holes with mass of order $10 M_\odot$ are within reach of today's ground-based detectors. A simple scaling argument then tells us that supermassive black-hole binaries—resulting from galaxy mergers—radiate in the LISA frequency band. In fact, the frequency range of the space-based interferometer (down to 10^{-4} Hz) is a good match to the timescale of many known astronomical systems. Different classes of galactic binary systems radiate gravitationally in the LISA band and should lead to detectable signals. The most common such systems are (i) binary white dwarfs, (ii) binaries comprising an accreting white dwarf and a helium donor star, and (iii) low-mass X-ray binaries. There may be more than a billion galactic binaries in the LISA range. Finally, pulsar timing arrays allow us to probe ultra-low frequencies (nano-Hz) for signals from truly gigantic black holes.

These back-of-the-envelope estimates provide a sketch of the gravitational-wave sky. They do not tell us the whole truth but serve to motivate more detailed thinking. Unfortunately, the next step tends to be difficult, either involving poorly understood physics (as in the case of neutron stars) or complex nonlinear dynamics (as for black-hole collisions), or both (as in the case of neutron star mergers and supernova core collapse). These requirements have led to the development of numerical relativity as a high-powered tool for astrophysical simulations (see Chapters 19 and 20). At the same time, a range of issues bridging nuclear physics, particle physics and quantum field theory, low-temperature physics, and hydrodynamics relevant for neutron stars are being investigated. Fundamental physics associated with the early Universe and the dark matter/energy models in modern cosmology is also under vigorous scrutiny.

1.5   Binary inspiral and merger

Before we turn to the detailed theory, let us sketch a set of problems that provide interesting modelling challenges. These problems (obviously) do not provide a complete list in any sense. Rather, they have been selected to illustrate particular aspects and provide an idea of the bigger picture.

It is natural to start with inspiralling binaries. Long before the first detection of gravitational waves, compact binaries provided convincing—although indirect—support for Einstein's theory. Detailed monitoring of the famous Binary Pulsar PSR B1913+16, discovered in data from the Arecibo radio telescope in 1974 (Hulse and Taylor, 1975), and the more recently found (and more relativistic) Double Pulsar PSR J0737-3039 (Lyne *et al.*, 2004), provides clear evidence of orbits decaying at a rate that agrees with the predictions of general relativity.

However, in contrast to Newtonian gravity, the two-body problem remains 'unsolved' in general relativity. Given the lack of suitable solutions to the Einstein field equations, significant effort has gone into developing approximations and numerical approaches to the problem. For the inspiral phase of a binary system, the post-Newtonian expansion (essentially a low-velocity expansion; see Chapter 11) is particularly useful. Within the post-Newtonian scheme, the leading order radiation effects are described by the so-called quadrupole formula, according to which the gravitational-wave strain follows from the second time derivative of the source's quadrupole moment

$$Q^{jk} = \int \rho x^j x^k dV,$$ (1.9)

where x^i is the position vector and ρ is the mass density. If we consider the simple situation of two (effectively) point masses with mass M separated by a distance a (see Chapter 5 for a detailed discussion), then we see that

$$Q \sim Ma^2.$$ (1.10)

The gravitational-wave strain follows from

$$h_{jk} = \frac{2G}{dc^4} \frac{d^2 Q_{jk}}{dt^2} \rightarrow h \sim \frac{Ma^2 f^2}{d} \sim \frac{M^2}{da},$$ (1.11)

where d is the distance to the source and we have used the frequency for a Keplerian orbit $f \sim M/a^3$.

As the system emits gravitational waves, it loses energy and the orbit shrinks. From (1.3) we see that

$$\dot{E} \sim d^2 \dot{h}^2 f^2 \sim \frac{M^5}{a^5}.$$ (1.12)

Balancing the rate of energy loss to the orbital energy, $E \sim M^2/a$, we arrive at an evolutionary timescale for the decay

$$t_D \sim \frac{a^4}{M^3}.$$ (1.13)

Assuming that this timescale is shorter than the observation time (which obviously means that we are considering the final stages before merger), we can use (1.6) to estimate the effective amplitude of the binary signal

$$h_c \approx \sqrt{f t_D}\, h \sim \frac{M}{d} \left(\frac{a}{M}\right)^{1/4}. \tag{1.14}$$

This estimate shows that, even though the raw signal gets stronger as the frequency *chirps* up towards its cut-off value at plunge and merger, the detectability decreases as the orbit shrinks. Hence, we need to make sure our detectors are sensitive at low frequencies, where the binary system spends more time. This is, however, problematic due to gravity gradient noise (earthquakes, human activity, clouds, you name it...) below a few hertz.

Ground-based detectors should (eventually) be able to track a neutron star binary as it evolves all the way from a few hertz through to coalescence, radiating around 10^4 cycles in the process. In principle, one can enhance the detectability by roughly a factor of 100 if one can follow the signal through the entire evolution (without losing a single cycle of the wave-train). This influences detector design. It also motivates the development of high-order post-Newtonian approximations to the waveforms (especially the phase), as well as fine-tuned signal analysis algorithms (see Chapter 8).

The estimated inspiral time, t_D, tells us that any binary system which is observable from the ground will coalesce within hours. Statistics based on the known radio pulsar population (see Chapter 9) then tells us that these events should happen less regularly than once every 10^5 yrs in our Galaxy. Hence, we need to detect events from a volume of space containing at least 10^6 galaxies in order to see a few such mergers every year. Translating this into distance, using our understanding of the mass distribution in the Universe, we learn that a detector must be sensitive enough to see coalescing binaries beyond a few hundred megaparsec in order for the event rate to be reasonable.

This explains why, first of all, it would have been surprising to find a binary signal in initial LIGO data. Given even the most optimistic rate estimate from population synthesis models, such events would be extremely rare in the observable volume of space. The situation is drastically different given Advanced LIGO level sensitivity so it is (perhaps) not surprising that the first detection came immediately after the detector upgrade in 2015. It may have been a surprise that the first observed signal came from merging black holes, but given our ignorance about the black-hole binary population (with some models suggesting they should not form at all and others stating they would be plentiful) this may be overstating the case.

Third-generation detectors may still be required if we want to consider population statistics. They are also likely to be needed if we want to study the final stages of neutron-star binary inspiral, including the merger. This is a very interesting phase of the evolution since the merger will lead to the formation of a hot compact remnant with violent dynamics (see Chapter 20), generating a relatively high-frequency gravitational-wave signal that should be rich in information. In particular, it may tell us whether a massive neutron star or a black hole was formed. As neutron stars are magnetized, the merger may also trigger a gamma-ray burst. As a rough rule-of-thumb, if the inspiral phase is observable with Advanced LIGO then the Einstein Telescope should be able to detect the merger (Andersson *et al.*, 2011). In other words, the development of third-generation detectors is likely to be essential if we want to study these events. In parallel, we need to improve our models of the merger phase. This requires nonlinear simulations in full general relativity, accounting for the complex physics associated with a high-density/extremely hot remnant. Realistic models need to include magnetic fields and account for energy loss due to neutrino emission. Progress is being made towards this goal, but many challenges remain.

The situation is quite different for black-hole binaries. The last decade has seen a breakthrough in numerical relativity, to the point where the problem of inspiralling and merging black holes is considered 'solved' (see Chapter 19). This progress was of immense importance as it provided experimenters with reliable templates that could be used to develop optimal data analysis strategies, impressively demonstrated in the case of GW150914. Since a black-hole binary should be more massive than one comprising neutron stars, it will lead to a stronger gravitational-wave signal. This means that we expect to see more distant black-hole binaries, ultimately reaching out to cosmological distances with instruments like the Einstein Telescope and LISA.

Detailed calculations show that, in the case of unequal masses the leading order signal depends only on the so-called 'chirp-mass'; the combination $\mathcal{M} = \mu^{3/5} M^{2/5}$ where μ is the reduced mass and M the total mass. If one observes the decay of the orbit as well as the gravitational-wave amplitude, then one can infer the chirp mass and the distance to the source. This means that coalescing binaries are 'standard sirens' which may be used to infer the Hubble constant and other cosmological parameters (see Chapter 22). By extracting higher-order post-Newtonian terms one can hope to infer the individual masses, the spins and maybe also put constraints on the graviton mass (the speed of the waves compared to the speed of light).

In the case of LISA, one would expect to be able to observe mergers of supermassive black holes with very high signal-to-noise ratio (several 1,000s; see Figure 1.4). This means that one may be able to see such events no matter where they occur in the Universe. The information gained from such observations will shed light on the evolution of these gigantic black holes, via a sequence of mergers or accretion, and improve our understanding of the development of large-scale structures in the Universe.

Another key problem for LISA concerns the capture of smaller bodies by large black holes. A space-based detector should be able to detect many such events. Their detailed signature provides information about the nature of the spacetime in the vicinity of the black hole. To model these systems is, however, far from trivial, in particular since the orbits may be highly eccentric. The main challenge concerns the calculation of the effects of radiation reaction on the smaller body (see Chapter 16). In addition to accounting for the gravitational self-force and the radiation reaction, one must develop a computationally efficient scheme for modelling actual orbits. This is not easy, but at least we know what the key issues are.

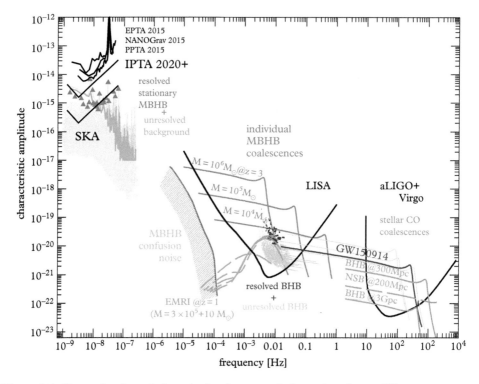

Figure 1.4 *Comparing the typical gravitational-wave strain for various classes of binary systems to the sensitivity of current and future detectors (showing dimensionless characteristic strain amplitude as function of frequency). The current ground-based detectors, Advanced LIGO and Virgo, are sensitive above 10 Hz. Third generation instruments (like the Einstein Telescope, not shown) are expected to improve the sensitivity by an order of magnitude across a similar frequency range. The space-based interferometer LISA (Amaro-Seoane et al., 2017) will be a supreme instrument for detecting signals from massive black holes in the range $10^4 - 10^7 M_\odot$. There is also expected to be a population of extreme mass-ratio inspirals arising from the gravitational capture of smaller objects by supermassive black holes. The most massive black-hole binaries in the Universe, radiating at nanohertz frequencies, may be detectable via pulsar timing arrays. Inspiralling black holes with mass $10^9 - 10^{10} M_\odot$ lead to to a stochastic low-frequency background with a few individual, loud sources. (Figure provided by A. Sesana.)*

1.6   Supernovae

At this point a word of caution is in order. Our expectations are not always brought out by more detailed modelling. Sometimes the devil is in the detail and our intuition falters. For example, one might expect apparently powerful events like supernova explosions and the ensuing gravitational collapse to lead to very strong gravitational-wave signals. This was, indeed, first thought to be the case (Thorne, 1979). However, the outcome depends entirely on the asymmetry of the collapse process. That this is the case is clear from (1.11). The difference between the initial and the final state does not matter. It is the route the system takes—how it evolves—that determines the strength of the gravitational-wave signal. Unfortunately, numerical simulations suggest that the level of radiation from core collapse events is low. Typical results suggest that an energy equivalent to $\sim 10^{-7} M_\odot c^2$ (or less!) will be radiated (see Chapter 20).

Combining this anticipated level of energy release with the typical dynamical timescale for a collapsing compact core, around a millisecond (frequency $\sim$1 kHz), we see from Eq. (1.5) that the gravitational-wave amplitude may be on the order of $h_c \sim 10^{-22}$ for a source in the Virgo cluster. This estimate (which accords reasonably well with full numerical simulations) suggests that these sources are unlikely to be detectable beyond the local group of galaxies. This would make observable events rare. It is expected that only three to four supernovae will go off every century in a typical galaxy so we would be very lucky to see one in our Galaxy given only a decade or so of observation. However, a single stand-out event could provide great insight into supernova physics. The gravitational waves carry unique information and the detailed signature may allow us to distinguish between different proposed explosion mechanisms. While the optical signal emerges hours, and the neutrino burst several seconds after the collapse, the gravitational waves are generated during the collapse itself. As a result, they carry a clean signature of the collapse dynamics. This information may be impossible to extract in any other way.

1.7   Spinning neutron stars

When the dust settles from the supernova event we are left with a compact remnant, either a neutron star or a black hole. As we have already seen, both sets of objects are relevant to the gravitational-wave physicist. Black holes involve extreme spacetime curvature, and allow us to probe strong field aspects of Einstein's theory, while neutron stars are cosmic laboratories of exciting physics that cannot be tested by terrestrial experiments. With a mass of more than that of the Sun compressed inside a radius of about 10 km, their density reaches beyond that of the atomic nucleus ($\sim 10^{14}$ g/cm^3). We already have a wealth of data from radio, X-ray, and gamma-ray observations, providing evidence of an incredibly rich phenomenology. We know that neutron stars appear in many different guises, from radio pulsars and magnetars to accreting millisecond pulsars, radio transients, and intermittent pulsars. Our models for these systems remain somewhat basic, despite 40 years of observations and attempts to understand the physics of the pulsar emission mechanism, glitches, and the evolution of accreting systems.

Importantly, neutron stars can radiate gravitationally through a number of different mechanisms. Relevant scenarios include the binary inspiral and merger that we have already discussed, rotating stars with deformed elastic crusts (Chapter 14), various modes of oscillation, and a range of associated instabilities (Chapter 13). Modelling these different scenarios is not easy since the physics of neutron stars is far from well known. To make progress we must combine supranuclear physics (the elusive equation of state) with magnetohydrodynamics, the crust elasticity, a description of superfluids/superconductors, and potentially also exotic phases of matter involving a deconfined quark-gluon plasma or hyperons (see Chapter 12). Moreover, in order to be quantitatively accurate, the models have to account for relativistic gravity.

Much effort has been invested in understanding the rich spectrum of oscillations of 'realistic' neutron star models (see Chapters 13 and 18). This is natural since such oscillations may be excited to a relevant level at different stages in a neutron star's life. Gravitational waves from a pulsating neutron star could provide an excellent probe of the star's properties, and may allow us to infer the mass and radius with good precision. This would help constrain the supranuclear equation of state. The most promising scenarios involve unstable oscillations. As an unstable pulsation mode grows, it may reach a sufficiently large amplitude that the emerging gravitational waves can be detected. In the past couple of decades the inertial r-modes have been under particular scrutiny, following the realization that they are particularly prone to a gravitational-wave-induced instability (see Chapter 15). This is an active area of current research.

As soon as a newly born neutron star cools below roughly 10^{10} K (within a few minutes) its outer layers begin to crystallize; freezing to form the neutron star crust. The crust is not very rigid—in fact, it is rather like jelly—but it can still sustain shear stresses, something that a fluid is unable to do. Asymmetries in the crust, expected to arise due to its evolutionary history, will slowly leak rotational energy away from a spinning neutron star. Such sources would be the gravitational-wave analogue of radio pulsars, radiating at twice the star's spin frequency (see Chapter 6). On the one hand, rotating neutron stars will emit low amplitude waves, but on the other hand, they radiate continuously for a long time. This means that observers can carry out targeted searches for known radio and X-ray pulsars, with frequency and position provided from radio and X-ray data (see Chapter 14).

It is straightforward to estimate the signal strength for this kind of source. We are essentially dealing with a rotating bar. Expressing the associated asymmetry in terms of an ellipticity ϵ, we find that

$$h \approx 8 \times 10^{-28} \left(\frac{\epsilon}{10^{-6}}\right)\left(\frac{f}{100 \text{ Hz}}\right)^2 \left(\frac{10 \text{ kpc}}{d}\right), \qquad (1.15)$$

where we have used the fact that the gravitational-wave frequency f is twice the rotation frequency (and assumed a canonical $M = 1.4 M_\odot$ and $R = 10$ km neutron star). The source distance has been scaled for objects in our Galaxy, since this is all we can hope to detect, anyway. This signal is far too weak to be detected directly, but the effective amplitude increases (roughly) as the square root of the number of detected cycles. Accounting for this and assuming an observation time of one year, we need

$$\epsilon > 2.5 \times 10^{-6} \left(\frac{100 \text{ Hz}}{f}\right)^{5/2} \left(\frac{d}{10 \text{ kpc}}\right)\left(\frac{h_c}{10^{-22}}\right). \qquad (1.16)$$

Combined with the expected detector sensitivities, i.e. some idea of the achievable h_c, this estimate allows us to assess whether a given deformation is likely to be detectable.

However, we also need some idea of the level of asymmetry one would expect a neutron star to have. This is a complicated physics problem. The answer depends not only on the properties of the star, but also on its evolutionary history. To a large extent, modelling has focused on establishing what the largest possible neutron star 'mountain' would be. Detailed models suggests that $\epsilon < 2 \times 10^{-5}(\sigma_{\mathrm{br}}/0.1)$ and recent molecular dynamics simulations suggest that the breaking strain is $\sigma_{\mathrm{br}} \approx 0.1$ (much larger than originally anticipated). In comparison to terrestrial materials, which have $\sigma_{\mathrm{br}} \approx 10^{-4} - 10^{-2}$, the neutron star crust appears to be super-strong! But the induced asymmetries are still tiny.

Observations of targeted radio pulsars have provided interesting results, even in the absence of a detection. As an example, one month of data from the third and fourth LIGO science runs was used to set the constraint $\epsilon < 7 \times 10^{-7}$ for PSR J2124-3358 (Abbott *et al.*, 2007). This tells us that this, relatively fast-spinning, 4.9 ms period pulsar is far from maximally deformed. It is also quite easy to estimate how the result should improve in the future since the effective amplitude of a periodic signal increases as the square root of the observation time. For example, Advanced LIGO, with more than an order of magnitude better sensitivity, should be able to reach $\epsilon < 10^{-8}$ for this pulsar (still assuming a one-year integration). The Einstein Telescope may push the limit as far as $\epsilon < 10^{-9}$. At this point, the deformation of the star would be constrained to the micron level—an astonishing level of symmetry. One might expect a signal to be detected before this level is reached, but we do not know this for sure. The main issue concerns the cause of the star's deformation. Why would the neutron star be deformed in the first place? This is an urgent question that needs to be addressed by theorists.

As far as evolutionary scenarios are concerned, accreting neutron stars in low-mass X-ray binaries have attracted the most attention. This is natural for a number of reasons. First of all, the currently observed spin distribution in these systems seems consistent with the presence of a mechanism that halts the spin-up due to accretion well before the neutron star reaches the break-up limit. Gravitational-wave emission could provide a balancing torque (Chapter 6). Using a simple accretion torque model one finds that the required deformation is smaller than the allowed upper limit. We need

$$\epsilon \approx 4.5 \times 10^{-8} \left(\frac{\dot{M}}{10^{-9} M_{\odot}/\mathrm{yr}} \right)^{1/2} \left(\frac{300 \ \mathrm{Hz}}{\nu_s} \right)^{5/2}, \tag{1.17}$$

where $\dot{M}$ is the mass accretion rate and ν_s is the star's spin frequency. It is fairly easy to see why asymmetries would develop in an accreting system, as the matter flow should be channelled along the star's magnetic field. However, accreting systems are messy, and we do not understand the accretion torque very well. Hence, reality could be quite different from our rough estimate. Another issue concerns the need to integrate the signal for a long time to build up the signal-to-noise ratio. Given the somewhat erratic behaviour of accreting neutron stars it is not clear that we will be able to track such systems (and integrate coherently) for long enough.

1.8   Fundamental physics

Arguably, the most fundamental observation we can hope to make with future detectors would be a cosmological background from the Big Bang. Asymmetries in the very early Universe would be amplified by the expansion, resulting in a broad gravitational-wave spectrum in the present Universe (see Chapter 22). The slope of this spectrum and possible peaks provide information about masses of particles, energies of relevant phase transitions, and perhaps the sizes of extra dimensions. Detection of such stochastic signals is rather different from the problems we have considered so far. In particular, it requires cross-correlation of the output from several detectors. This presents a new set of challenges.

Constraints on the energy density of the stochastic gravitational-wave background (normalized to the critical energy density of the Universe; see Chapter 4) confronts some of the most extreme theoretical suggestions but the standard scenario remains very safe. We may have to live with this situation for some time. Gravitational waves associated with the standard inflationary scenario will be out of reach for all planned ground-based instruments and LISA. However, LISA's frequency band ($\sim$mHz) represents gravitational waves that had the horizon size at the electroweak phase transition. If this transition were first order, then there could be a detectable stochastic background (Chapter 22). Another relevant possibility involves cosmic strings, which emit gravitational waves with a characteristic signature. These waves may be detectable even if they do not make a significant contribution to the mass budget of the Universe. The best search window for such signals, essentially free of 'local' gravitational-wave sources, is around 0.1–1 Hz. This would require a LISA follow-up space mission. Any such venture is not going to take place soon.

Future detectors may nevertheless make significant contributions to our understanding of fundamental physics. As an example of this, consider the fact that we expect to be able to infer distances to coalescing binary systems, providing a distance scale of the Universe in a precise calibration-free measurement (see Chapter 22). This idea will be pursued with ground-based detectors in the first place, but their reach is obviously not as impressive as that of LISA. In fact, LISA will have excellent sensitivity to massive black hole mergers at redshift $z = 1$ and would be able to detect $10^4 M_\odot$ systems out to $z = 20$. If some of these merger events have an observable electromagnetic counterpart this would provide us with redshift information, and as a consequence we may be able to use LISA data to constrain the dark energy equation of state. In other words, LISA may also be a dark energy mission. This is an interesting idea, providing yet another example of the interdisciplinary nature of gravitational-wave physics.

1.9   Many different messengers

Bringing the different aspects of the discussion together, gravitational-wave astronomy promises to shed light on the 'dark side' of the Universe. Because of their strong gravity, black holes and neutron stars are ideal sources and we hope to probe the extreme physics associated with them. The potential for this is clear, but in order to detect the signals and extract as much information as possible we need to improve our theoretical models.

The binary black-hole problem may have been 'solved' by the progress in numerical relativity, but for neutron star binaries many issues remain. We need to work out precisely when finite size effects begin to affect the evolution (see Chapter 21). We need to consider tidal resonances and ask to what extent they affect the late stages of inspiral. For hot merger remnants, we need to refine our nonlinear simulations. The simulations must use 'realistic' equations of state, and consider composition, heat/neutrino cooling, and magnetic fields with as few 'cheats' as possible. Similar issues arise for supernova core-collapse simulations, which set the current standard for including realistic physics (Chapter 20). In parallel, we need to improve our understanding of neutron star oscillations and instabilities. This effort should aim at accounting for as much of the interior physics as possible. We need a clearer phenomenological understanding of pulsar glitches, accreting neutron stars, and magnetar flares. These are ambitious targets, but there is no reason why we should not make progress on them. Observations (gravitational and electromagnetic) will help us understand aspects of extreme physics that seem mysterious to us today.

The future will see significant improvements in the various observational channels, and we should expect great progress in our understanding of the Universe. Gravitational-wave physics, with its promise to provide information that is complementary to electromagnetic observations, has an important part to play in this enterprise. The precision modelling and sophisticated data analysis tools that are essential for gravitational-wave experiments should be valuable also for X-ray and radio astronomy. Evidence for this is clear from the emergence of numerical relativity simulations as a reliable tool for studying violent astrophysical phenomena and the use of sophisticated data analysis techniques being adapted to detect gamma-ray pulsars in Fermi data (Clark *et al.*, 2018). These developments may have been motivated by gravitational-wave physics, but they are finding applications in a wider context. One would expect these kinds of synergies to continue to develop, particularly as many gravitational-wave sources may have electromagnetic counterparts. The most exciting results may come from a combination of gravitational-wave, electromagnetic, and neutrino channels—true multimessenger astronomy.

1.10   The golden binary

On the 17th of August 2017 we arrived in the future. The two LIGO detectors picked up the gradually evolving signature of a binary neutron star system, a signal lasting over a minute before diving into the detector noise (Abbott *et al.*, 2017j). The signal was fainter in the Virgo instrument, which had actively joined the detector network at the beginning of the month. This suggested that the source must be in one of Virgo's blind spots, which helped the observers constrain the location in the sky (with an uncertainty of about 30 degrees). A trigger message was sent to electromagnetic observers around the globe. They responded—and struck gold.

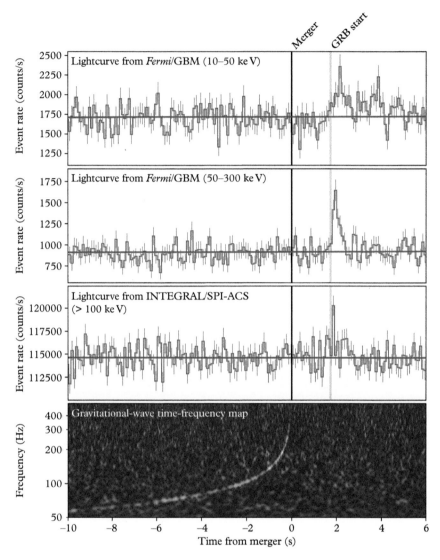

Figure 1.5 *The joint multi-messenger detection of GW170817 and the short gamma-ray burst GRB 170817A. Top: the light curve from the Fermi Gamma-ray Burst Monitor (GBM) between 10 and 50 keV. Second: the same, but in the energy range 50–300 keV. Third: the light curve from Integral with the energy range starting approximately at 100 keV and with a high energy limit of least 80 MeV. Bottom: the time-frequency map of GW170817 obtained by coherently combining LIGO-Hanford and LIGO-Livingston data. (Reproduced from Abbott et al. (2017j), Creative Commons Attribution 3.0 License.)*

The GW170817 event was detected across the electromagnetic spectrum, from gamma rays to X-rays, radio and optical. First to join the party were the gamma-ray instruments Fermi and Integral (Abbott *et al.*, 2017*j*; Troja *et al.*, 2017). They picked up a gamma-ray flash (GRB 170817A) emitted just two seconds after the neutron star merger; see Figure 1.5. The signal was faint, but the features were generally consistent with those of short gamma-ray bursts, long suspected to have their origin in neutron star mergers. GW170817 was unique in that the event occurred more than 10 times closer to the Earth (130 million light years away) than any previously known short gamma-ray burst. This makes it much easier to study, but the signal was weak. Perhaps we did not look right down the barrel of the outflowing jet (Chapter 21).

The event was caught by traditional optical telescopes (Cowperthwaite *et al.*, 2017), confirming predictions that neutron star mergers lead to an outflow of matter that radiates in a characteristic way. By tracking the evolution of the light signal as it faded away over several days, peaking in the infrared, astronomers identified the glow emitted as neutrons and protons combine to form heavy elements (Metzger and Berger, 2012). This extraordinary event told us that all the gold, platinum and uranium in the Universe come from merging neutron stars. GW170817 was a multi-sensory treasure trove of information.

Part 1

From theory to experiment

2

A brief survey of general relativity

In order to understand the nature of gravitational waves and model relevant sources we, first of all, need to have a working understanding of general relativity. There are many excellent textbooks that cover Einstein's theory in much more detail than we will be able to here,[1] but it nevertheless makes sense to provide a brief survey, to remind us of the main ideas and introduce the various tools we will need. This involves motivating why it makes sense to consider gravity in terms of a curved spacetime, discussing what this implies for the motion of both light and massive objects and finally explaining the reasoning behind the Einstein field equations, which govern the gravitational field. As we outline the theory, it becomes clear that issues associated with the measurement of distance are fundamental. They are also intimately linked to the origin and nature of gravitational waves.

The theory of special relativity, which deals with bodies in relative uniform motion, is based on the simple postulate that the speed of light, c, is a universal constant, taking the same the value according to all (inertial) observers. From this seemingly innocuous starting point, Einstein revolutionized our view of space and time. His 1905 theory tells us that time and space are flexible; time runs slow on a moving clock and moving rods appear shorter than they are at rest. These effects follow immediately from the fact that the theory in invariant under the Lorentz transformation, which relates time and space measurements in two coordinate systems, one moving relative to (in this case, away from) another one (see Figure 2.1 for an illustration). Using primes for the moving coordinates, we have

$$t' = \gamma(t - vx), \qquad x' = \gamma(x - vt), \qquad y' = y, \qquad z' = z, \qquad (2.1)$$

where v is the (constant) relative velocity (assumed to be in the x-direction) and

$$\gamma = (1 - v^2)^{-1/2} \qquad (2.2)$$

---

[1] Every practitioner of relativity has his/her own favourite books on the subject. Different aspects are covered particularly well by particular authors. It may also be a matter of taste. A useful starting point, at the undergraduate level, would be either Schutz (2009) or Hartle (2003), while more advanced topics are covered by Poisson and Will (2014) and Thorne and Blandford (2017). The overview in this chapter is (clearly) incomplete and the reader would be well advised to dig deeper—it will not be a waste of time.

Gravitational-Wave Astronomy: Exploring the Dark Side of the Universe. Nils Andersson, Oxford University Press (2020).
© Nils Andersson. DOI: 10.1093/oso/9780198568032.001.0001

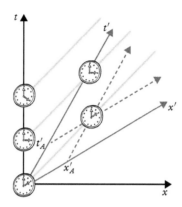

Figure 2.1 *In Einstein's relativity, different spacetime events and observers carry their own clocks to measure time (and rulers to measure space distance). Since the speed of light is a universal constant, the different clocks can be synchronized using light signals. The idea is indicated in this illustration which shows the relation between the two sets of coordinates used in (2.1), with the primed coordinate system moving away from the unprimed one. An observer at x'_A will synchronize his/her clock with another observer at $x = 0$ in such a way that $t = 0$ corresponds to t'_A.*

(in units where the speed of light is unity[2]). The key lesson is that we have to consider time and space together.

However, special relativity does not (really) deal with acceleration. It also does not consider gravity. After completing the theory, Einstein spent the next decade trying to extend the framework to handle both acceleration and gravity. His vision was a geometric theory—where the curvature of spacetime encodes gravity—inspired by a simple thought experiment.

Imagine asking a person in a lift (without windows, obviously) to carry out an experiment to establish if the lift is moving. He/she decides to do this by dropping a ball. If the lift is at rest in a gravitational field the ball will accelerate towards the floor due to gravity. But the experiment would have the same outcome if the lift were accelerating upwards in absence of gravity. Once the ball is dropped it will be floating free, but the floor of the lift accelerates towards it. By arranging the rates of acceleration to be the

---

[2] Setting the speed of light $c = 1$ simply means that we are measuring distance in (say) light-seconds. Later we will work in *geometric units*, where the gravitational constant G is also set to unity. This is convenient, but it may lead to confusion when we try to put actual numbers in. And it gets worse if we submit to the temptation to also set the Boltzmann and Planck constants equal to 1 when we weave in thermodynamics and quantum aspects. If we are not careful, we end up in a right mess.

same in the two cases one can ensure that the observer in the lift is confused (as long as the rocket engine is very quiet, of course).

This is known as the *equivalence principle*. We cannot tell the difference between gravity and acceleration. They are two sides of the same coin. This, in turn, suggests that we should be able to describe gravity as a non-inertial effect. The idea may seem simple, but the development of general relativity still took the best part of a decade. Einstein had to learn the mathematics required to describe motion in curved spaces (tensor calculus/differential geometry) and he needed to verify that his theory led back to Newton's gravity in the appropriate limit. The final theory from November 1915 remains one of the greatest achievements of modern physics—a geometric description of gravity that has (so far) passed all tests with flying colours.

2.1 A simple thought experiment

Following in Einstein's footsteps (even though the shoes may be a bit too big for us!) we can describe the theory in terms of a simple thought experiment.

Let us extend the lift experiment in such a way that we are able to deduce the presence of a gravitational field. To do this we note that falling bodies move towards the centre of gravity, which means that the relative distance between two falling objects should change. Extending the setting of the experiment to deal with this problem, we let our experimenter drop two objects in such a way that they initially fall along parallel trajectories. For the purpose of this thought experiment we will assume that the two objects are 'ideal' in the sense that their motion is only affected by gravity. That is, they are able to fall unimpeded through the matter of the Earth.

Since gravity attracts towards the centre of the Earth, we expect the trajectories of the two objects to cross eventually. In Newtonian physics, this happens because of the universal gravitational attraction. As a first step towards understanding Einstein's explanation, compare the trajectories of the bodies in the thought experiment to two great circles on the surface of the Earth (assumed to be a sphere), as in Figure 2.2. Two great circles that start out orthogonal to the equator must cross at the north and south poles, because of the curvature of the surface. This motivates an intuitive explanation of the crossing of the trajectories of our falling objects in terms of the curvature of the spacetime in which they are moving.

Before we develop the mathematical machinery we need to describe this problem, it is worth pointing out that the two ways to understand gravity differ significantly in philosophy. While Newton's theory *describes* the gravitational attraction, Einstein's general relativity *explains* the origin of gravity in terms of the curved spacetime. In the first case an inverse square law is imposed at the outset. In the second, gravity follows as a consequence of the geometry of spacetime.

Let us now develop—step by step—the tools we need to describe the thought experiment. By comparing the Newtonian result to a framework based on a curved spacetime, we will be able to derive Einstein's field equations.

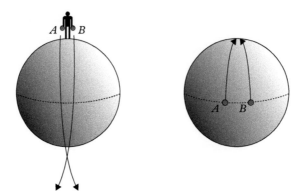

Figure 2.2 *The left image illustrates a thought experiment where two ideal objects A and B (only influenced by gravity) are dropped along initially parallel trajectories from the surface of the Earth. The right image shows the close analogy with great circles on a sphere. Even though two great circles A and B start out parallel at the equator, they cross at the poles.*

2.2   The tidal tensor

We begin with the Newtonian description of the problem. Introducing a vector $\boldsymbol{\xi}$ that describes the separation between the two falling objects, we must initially have

$$\frac{d\boldsymbol{\xi}}{dt} = 0, \tag{2.3}$$

since the trajectories start out parallel. However, the second derivative will not vanish. Symbolically, we can write the relative acceleration as

$$\frac{d^2\boldsymbol{\xi}}{dt^2} = -\mathcal{E}(\_,\boldsymbol{\xi}), \tag{2.4}$$

defining the *tidal tensor* $\mathcal{E}$. When this object acts on the displacement vector $\boldsymbol{\xi}$ we end up with another vector, encoding the tidal acceleration.

Tensors play a central role in much of the following, so we need to understand what they are and how to work with them. There are different ways to approach this, but we will choose a route that is both elegant and conceptually simple. The key idea is to think of a tensor as a geometric object, representing a *linear* function of a number of vectors. Somewhat simplistically, we think of a tensor as a machine that turns a set of vectors into a single number. Each tensor has a number of 'slots' that can be occupied by vectors. The number of slots is called the *rank* of the tensor. If all slots are filled, the result is a real number; if one slot is left empty the outcome is a vector; and so on. The tidal tensor $\mathcal{E}$ has rank 2 (note that one slot is left empty in (2.4)), and the object on the right-hand side of (2.4) is a vector. A tensor is completely specified once we know what real number we get when all the slots are filled. These ideas may seem somewhat abstract at this point, but they should become clear as we start calculating.

This advantage of this way of thinking about tensors is that it allows us to understand the geometric nature of many physical relations. This is useful since we expect nature to obey the *principle of relativity* which essentially says that true physical laws should not depend on our choice of coordinate system or frame of reference.

Of course, in terms of actually working things out (and designing experiments!) we often want to express tensors in terms of their components, using a basis associated with a given set of coordinates. Since it helps us understand the concept of a tensor, consider the familiar fact that any vector can be expressed in terms of a set of orthonormal basis vectors $\hat{e}_j$ (where the index $j = 1 - 3$ simply labels the vectors) as

$$\boldsymbol{\xi} = \sum_j \xi^j \hat{e}_j = \xi^j \hat{e}_j. \tag{2.5}$$

The second identity introduces Einstein's summation convention: repeated indices are implicitly summed over.[3] As the basis vectors are orthogonal and normalized to unit length we have

$$\hat{e}_i \cdot \hat{e}_j = \delta_{ij}, \tag{2.6}$$

where δ_{ij} is the Kronecker delta ($= 1$ if $i = j$ and 0 otherwise). Given this, the individual vector components follow from the scalar product

$$\boldsymbol{\xi} \cdot \hat{e}_j = \xi^k \hat{e}_k \cdot \hat{e}_j = \xi^k \delta_{jk} = \xi_j. \tag{2.7}$$

Equivalently, in the spirit of (2.4), we can think of the vector components as given by

$$\xi^j = \boldsymbol{\xi} \left(\hat{e}_j \right). \tag{2.8}$$

Returning to the tidal tensor, we then have

$$\mathcal{E}^{jk} = \boldsymbol{\mathcal{E}}(\hat{e}_j, \hat{e}_k), \tag{2.9}$$

and since tensors are linear, we can write this as

$$\boldsymbol{\mathcal{E}} = \mathcal{E}^{jk}(\hat{e}_j \otimes \hat{e}_k). \tag{2.10}$$

---

[3] We have also (sneakily) introduced a distinction between upstairs and downstairs indices. If we are dealing with Cartesian tensors, this distinction is not so important, but we will soon see that it is crucial in a curved spacetime. Hence, it makes sense to introduce the convention from the beginning. In the following, the Einstein summation convention applies only to repeated up-down indices. Repeated down-down indices (say) render an expression meaningless.

This relation introduces a new symbol, $\otimes$ (representing a *tensor product*), but there is no reason to run in terror at the sight of this. It is fairly easy to understand. First of all, $\mathcal{E}_{jk}$ is the number we get if we insert the basis vectors $\hat{e}_j$ and $\hat{e}_k$ in the two slots of $\hat{e}_j \otimes \hat{e}_k$. In terms of two (general) vectors $\boldsymbol{u}$ and $\boldsymbol{v}$, the tensor product is then defined as

$$\hat{e}_j \otimes \hat{e}_k(\boldsymbol{u},\boldsymbol{v}) = (\hat{e}_j \cdot \boldsymbol{u})(\hat{e}_k \cdot \boldsymbol{v}). \tag{2.11}$$

The first scalar product projects out the part of $\boldsymbol{u}$ which is parallel to $\hat{e}_j$, while the second projects out the part of $\boldsymbol{v}$ parallel to $\hat{e}_k$. It is then easy to see that (2.9) follows if we take $\boldsymbol{u}$ and $\boldsymbol{v}$ to be the basis vectors.

Let us now return to the fundamental equation of tidal gravity. In practical calculations, one would often focus on the individual components of a given expression. Leaving out the basis vectors, the geometrical relation (2.4) takes the form[4]

$$\frac{d^2\xi^i}{dt^2} = -\mathcal{E}^i{}_j \xi^j. \tag{2.12}$$

We need to relate this expression to the gravitational potential Φ. In Newtonian gravity, the acceleration is proportional to the gradient of the gravitational potential. Letting the trajectory of a falling object be described in terms of coordinates x^i we have

$$\frac{d^2 x^i}{dt^2} = -\delta^{ij}\frac{\partial \Phi}{\partial x^j} = -\delta^{ij}\partial_j \Phi \tag{2.13}$$

(where we have introduced a useful short-hand notation for the partial derivative) for each of the two particles. With the vector separating the particles defined as (see Figure 2.3)

$$\xi^j = x_A^j - x_B^j, \tag{2.14}$$

we have

$$\frac{d^2\xi^i}{dt^2} = \text{expand in Taylor series} \approx -\delta^{ij}\left(\frac{\partial^2\Phi}{\partial x^j \partial x^k}\right)\xi^k, \tag{2.15}$$

as long as the trajectories are suitably close. Comparing this result to (2.12) we identify

$$\mathcal{E}^i{}_k = \delta^{ij}\left(\frac{\partial^2\Phi}{\partial x^j \partial x^k}\right). \tag{2.16}$$

---

[4] We are using a simplified mathematical machinery here. When a metric is present one can simplify the notation to 'avoid' row vectors and work only with column vectors. In the language of differential geometry, we can get rid of one forms and work solely with vectors. In the context of gravitational waves, which is our main focus, we can get by with this simplified approach.

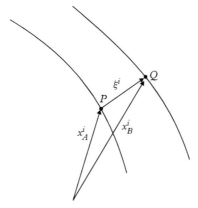

Figure 2.3 *The trajectories of two (ideal) objects A and B (only influenced by gravity) falling along initially parallel trajectories.*

This completes the Newtonian description of the thought experiment. Once we know the gravitational potential—which is determined by the mass distribution in the usual way—we can work out the relative motion of the two particles.

2.3 Introducing the metric

We now turn to the more challenging task of describing the thought experiment in terms of a curved spacetime. To do this, we need a new set of tools. This is the price we have to pay as soon as we work in a curved space. We need to think carefully about concepts we would normally take for granted. However, there is still no need to panic. Many of the ideas are intuitive and the final computational machinery is just as easy to use as the one we are accustomed to.

The principle of relativity is the key to the mathematics we need. If the physics is to be the same regardless of the frame of reference, then the mathematical expressions should not depend on whatever coordinates we decide to use. This naturally leads us to thinking of spacetime as made up of *events* $\mathcal{P}$ (say), which can be defined without resorting to a specific coordinate system. The idea is that different observers should always agree that something specific happened (like 'a star exploded in that particular galaxy'), even though they may associate the event with different coordinate locations and times (as they may be using their own rulers and clocks). Of course, given a set of coordinates x^a (where we add the time coordinate x^0 to the usual space coordinates x^i so the index a runs from 0 to 3), there should be a one-to-one mapping between points and particular coordinate values. It is, however, important to develop the physical concepts in a coordinate independent way.

To stress this point, consider two spacetime events $\mathcal{P}$ and $\mathcal{Q}$. We can easily define the vector $\Delta \boldsymbol{x}$ that separates the corresponding points in spacetime. These concepts do not rely on coordinates for their definition. It should also be clear that, even though different observers may disagree on the 'coordinate location' of the points $\mathcal{P}$ and $\mathcal{Q}$, they should be able to confirm that they are discussing the same events and the same separation vector. In fact, the principle of relativity forces the *interval* between the two events

$$ds^2 = (\Delta s)^2 = (\Delta \boldsymbol{x})^2 \tag{2.17}$$

to be the same in all frames of reference. This is an important observation. However, in order to make sense of this, we need a way of quantifying the concept of squared length. The object we need is the *metric tensor*, which encodes the geometry of spacetime and provides us with the means to measure distance.

Given a scalar product

$$\boldsymbol{A} \cdot \boldsymbol{B} = \text{ real number}, \tag{2.18}$$

we define the metric tensor $\boldsymbol{g}$ in such a way that

$$\boldsymbol{g}(\boldsymbol{A}, \boldsymbol{B}) = \boldsymbol{A} \cdot \boldsymbol{B}. \tag{2.19}$$

Since we have already introduced the concept of the squared length of a vector as equal to the associated spacetime interval ds^2 we can now compute the dot product between any two vectors. To do this, we note that

$$\boldsymbol{A} \cdot \boldsymbol{B} = \frac{1}{4}[(\boldsymbol{A}+\boldsymbol{B}) \cdot (\boldsymbol{A}+\boldsymbol{B}) - (\boldsymbol{A}-\boldsymbol{B}) \cdot (\boldsymbol{A}-\boldsymbol{B})]. \tag{2.20}$$

Moreover, we see that

$$\boldsymbol{A} \cdot \boldsymbol{B} = \boldsymbol{B} \cdot \boldsymbol{A}, \tag{2.21}$$

which implies that the metric tensor will be symmetric in its two slots. When written in component form it is symmetric in its two indices

$$g_{ab} = \boldsymbol{g}(\hat{\boldsymbol{e}}_a, \hat{\boldsymbol{e}}_b) \quad \longrightarrow \quad g_{ab} = g_{ba}. \tag{2.22}$$

In four spacetime dimensions the metric has 10 components.

It is worth noting that we can always isolate the symmetric part of a given tensor. For example, the symmetric part of a tensor, A_{ab}, follows from

$$A_{(ab)} = \frac{1}{2}(A_{ab} + A_{ba}), \tag{2.23}$$

and in the case of the metric we have $g_{(ab)} = g_{ab}$. Similarly, the anti-symmetric part is given by

$$A_{[ab]} = \frac{1}{2}(A_{ab} - A_{ba}), \tag{2.24}$$

and it is clear that $g_{[ab]} = 0$.

When discussing motion in spacetime, it is useful to consider a specific observer. We need someone to carry out our measurements. One can imagine this observer carrying a clock, the rate of which is associated with an infinitesimal vector $d\boldsymbol{P}$. This allows us to ask, 'How fast does an ideal clock tick?', as we move along with the observer. If the answer is $d\tau$, we have

$$\text{(length of } d\boldsymbol{P})^2 = ds^2 = -d\tau^2 < 0, \tag{2.25}$$

and $d\tau > 0$ is called the *proper time* interval. It represents the time recorded on the co-moving clock. Intervals such that ds^2 is negative are called *timelike* because they represent curves along which one can imagine a physical observer (limited by the speed of light) carrying a clock. In contrast, a *spacelike* vector $d\boldsymbol{Q}$ leads to

$$\text{(length of } d\boldsymbol{Q})^2 = ds^2 > 0, \tag{2.26}$$

where ds measures the spatial distance in the particular reference frame in which the vector is at constant time. Such intervals cannot be connected by observers restricted by the speed-of-light speed limit. Finally, light signals are *null*, meaning that $ds^2 = 0$. Centered on any particular point in spacetime, null intervals define the *light cone* which limits the movement of massive particles and defines their possible past and attainable future. The idea is illustrated in Figure 2.4.

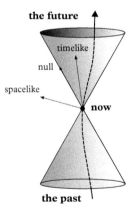

Figure 2.4 *As no signals can travel faster than light, actual clocks are confined to remain inside the light cone associated with an event (now). This constrains both the past history of the event and future events it may communicate with. Spacetime intervals A^a (say) that may be associated with an observer carrying a clock are called timelike and satisfy $g_{ab}A^aA^b < 1$. In contrast, spacelike intervals, for which $g_{ab}A^aA^b > 1$, would have to involve superluminal motion. Finally, intervals associated with light signals (called null) are such that $g_{ab}A^aA^b = 0$.*

2.4   The four-velocity

In order to discuss motion in more detail, it is useful to revisit the concept of an *inertial frame*. An inertial frame can be viewed as a lattice of rods and clocks that moves in such a way that it is not affected by any forces. It also does not rotate with respect to distant observers. We let the rods form an orthogonal lattice with uniform length intervals that can be used to set up orthonormal Cartesian coordinates. The clocks are synchronized by light pulses, as in Figure 2.1, and measure time in a uniform way. Given such an inertial system we have a natural coordinate system for spacetime, $x^a = \{t, x, y, z\}$ (say). The coordinates associated with a certain event $\mathcal{P}$ is then given by the location $\{x, y, z\}$ in the lattice of rods, and the time t measured by the clock at that coordinate location. In a curved spacetime we cannot construct a global inertial frame. Instead, we are forced to consider *local* inertial frames, which are relevant only in a small region of spacetime.

The *world line* of a particle is the sequence of events $\mathcal{P}(\tau)$, where τ is the *proper time* measured by an ideal clock carried along by an imagined observer riding along with the particle (as in Figure 2.5). The world line's tangent vector $\boldsymbol{u}$ is called the four-velocity. Mathematically, we can define the four-velocity through the standard limiting procedure for derivatives

$$\boldsymbol{u} = \frac{d\mathcal{P}}{d\tau} = \lim_{\Delta\tau \to 0} \frac{\mathcal{P}(\tau + \Delta\tau) - \mathcal{P}(\tau)}{\Delta\tau}. \tag{2.27}$$

In practice, one would think of the world line as a trajectory $x^a(\tau)$, leading to

$$u^a = \frac{dx^a}{d\tau}. \tag{2.28}$$

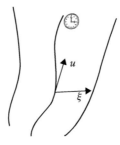

Figure 2.5 *An illustration of the four-velocity $\boldsymbol{u}$ as the tangent to a given worldline, with proper time measured on a co-moving clock. As long as no forces act on the particle it follows a geodesic. The separation $\boldsymbol{\xi}$ to a neighbouring geodesic plays a key role in the discussion of geodesic deviation.*

Anyway, it is easy to show that we must have

$$\boldsymbol{u}^2 = \frac{d\mathcal{P}}{d\tau} \cdot \frac{d\mathcal{P}}{d\tau} = \frac{d\mathcal{P} \cdot d\mathcal{P}}{(d\tau)^2} = -1. \tag{2.29}$$

This follows since $d\mathcal{P} \cdot d\mathcal{P}$ is the squared length of the interval $d\mathcal{P}$, which is equal to ds^2, while $d\tau^2$ is (by definition) minus this interval.

It is worth stressing the following (perhaps trivial) point: Even though the difference in the numerator of the expression on the right-hand side of (2.27) becomes small as $\Delta\tau \to 0$, the four-velocity $\boldsymbol{u}$ itself may nevertheless be 'large' (since we divide by $\Delta\tau$). This means that we should not expect to be able to 'lay the tangent vector down' in the curved space. The situation is similar for a calculation carried out on a spherical surface; see Figure 2.6. In order for the vector to follow the space we need to bend it in a nonlinear manner. But this is not allowed because tensor analysis is linear. Hence, all vectors, like $\boldsymbol{u}$, live in the flat *tangent space* associated with each individual spacetime point $\mathcal{P}$.

Now consider a freely falling object. That the object is freely falling means that it should follow a straight line in a local inertial frame (a frame carried along by the particle and in which it remains at the origin). In this frame we should have

$$\boldsymbol{u} = \text{constant}. \tag{2.30}$$

That the four-velocity is constant in this frame means the worldline of the object is 'as straight as possible'. Thinking about this for a moment, we see that it boils down to the requirement that the derivative of $\boldsymbol{u}$ along itself must vanish. In terms of a 'suitably defined' spacetime derivative ∇ and inspired by the directional derivative from vector calculus, one would expect to end up with a frame-independent relation:

$$(\boldsymbol{u} \cdot \nabla)\boldsymbol{u} = \nabla_{\boldsymbol{u}}\boldsymbol{u} = 0. \tag{2.31}$$

This is the *geodesic equation,* which describes the motion of freely falling objects in any chosen coordinate system. However, we have moved a little bit too fast. While we

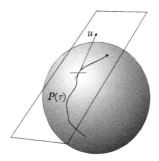

Figure 2.6 *An illustration of the fact that the four-velocity* $\boldsymbol{u}$ *lives in the tangent space associated with each individual spacetime point* $\mathcal{P}$ *along the worldline of a particle.*

understand the meaning of this kind of expression for three-dimensional vectors in flat space, we have not yet discussed derivatives in a curved space. We need to take a couple of steps back in order to move on.

Consider a curve $\mathcal{P}(\zeta)$ parameterized in terms of a suitable parameter ζ (which one might take to be the proper time τ in the case of a massive particle). The curve has tangent vector

$$\frac{d\mathcal{P}}{d\zeta} = \lim_{\Delta\zeta\to 0} \frac{\mathcal{P}(\zeta+\Delta\zeta) - \mathcal{P}(\zeta)}{\Delta\zeta}. \tag{2.32}$$

These are all coordinate independent concepts. Now assume that we choose coordinates $\{x^a\}$ and that all points in spacetime are described by $\mathcal{P}(x^a)$. Then we can build a *coordinate basis* for vectors through the partial derivatives

$$\boldsymbol{e}_a = \frac{\partial\mathcal{P}}{\partial x^a}. \tag{2.33}$$

This is important. For any chosen set of coordinates we have a natural basis in terms of which we can express tensor components. Given this basis, we define the tangent vector as

$$\boldsymbol{t} = t^a\boldsymbol{e}_a = \frac{d\mathcal{P}}{d\zeta} = \frac{dx^a}{d\zeta}\frac{\partial\mathcal{P}}{\partial x^a} = \frac{dx^a}{d\zeta}\boldsymbol{e}_a. \tag{2.34}$$

Note that standard differential calculus works even though, strictly speaking, $\mathcal{P}$ is a set of points. Thus, we identify

$$t^a = \frac{dx^a}{d\zeta}. \tag{2.35}$$

Contrast this to the directional derivative of a scalar field $\psi(\mathcal{P})$. As in standard vector calculus, we define the derivative of ψ along $\boldsymbol{t}$ as

$$\nabla_t\psi = \boldsymbol{t}\cdot\nabla\psi = t^a\frac{\partial\psi}{\partial x^a}. \tag{2.36}$$

In essence, we have

$$\nabla_t = t^a\frac{\partial}{\partial x^a} = \frac{dx^a}{d\zeta}\frac{\partial}{\partial x^a} = \frac{d}{d\zeta}, \tag{2.37}$$

evaluated along the curve $\mathcal{P}$, we then have

$$\nabla_t\psi = \frac{d}{d\zeta}\psi[\mathcal{P}(\zeta)]. \tag{2.38}$$

This highlights the correspondence between vectors and directional derivatives. In fact, from the mathematical point of view, we can define a vector to be a directional derivative. From a practical point of view, we can take

$$t = \nabla_t = \frac{d\mathcal{P}}{d\zeta} = \frac{d}{d\zeta},$$

(2.39)

to have the same meaning. Similarly, we have for the coordinate basis

$$e_a = \frac{\partial \mathcal{P}}{\partial x^a} = \frac{\partial}{\partial x^a}.$$

(2.40)

It is important to note that the natural basis is orthogonal but not normalized. The latter is only the case in flat space.

Let us now ask what happens if we transform from one set of coordinates to another, say from x^a to x'^a. From our definitions we have (thinking of the unprimed coordinates as functions of the primed ones and using the chain rule)

$$t = t^a e_a = t^a \frac{\partial}{\partial x^a} = t^a \frac{\partial x'^b}{\partial x^a} \frac{\partial}{\partial x'^b} = t^a \frac{\partial x'^b}{\partial x^a} e'_b = t'^b e'_b.$$

(2.41)

We see that the components must transform in such a way that

$$t'^b = \frac{\partial x'^b}{\partial x^a} t^a.$$

(2.42)

This rule defines a *contravariant* object; what we usually call a vector.

Meanwhile, in the natural basis the components of the metric tensor follow from

$$g_{ab} = g(e_a, e_b) = e_a \cdot e_b.$$

(2.43)

Even though we often do not 'buy much' by using the full machinery of differential geometry (and dealing with one-forms and dual spaces) when we discuss applications of general relativity (basically because we have a metric theory) it is useful to introduce the concept of a *dual basis*. Suppose we have determined a basis $\{e_a\}$ at the point $\mathcal{P}$. Then we can define another basis (the dual) $\{e^b\}$ through

$$e^b \cdot e_a = g(e^b, e_a) = \delta^b_a.$$

(2.44)

This leads us to introduce *covariant* objects, like the dual vector (more commonly referred to as a one form)

$$B = B_a e^a,$$

(2.45)

where B_a are the components in the dual basis.

As we have already pointed out, it is often useful to think of tensor expressions as abstract (geometric) relations where the 'slots have labels' rather than as a value in a particular coordinate basis (even though there obviously is a direct correspondence between the two pictures). For example, for the dot product we have

$$\boldsymbol{g}(\boldsymbol{A},\boldsymbol{B}) = \boldsymbol{A} \cdot \boldsymbol{B} = A^a B^b \boldsymbol{e}_a \cdot \boldsymbol{e}_b = A^a B^b g_{ab}. \tag{2.46}$$

Here we can interpret g_{ab} as the geometric object $\boldsymbol{g}(-,-)$ where we have 'labeled' the two slots a and b. The repeated indices in $A^a B^b g_{ab}$ mean that we have inserted the vector $\boldsymbol{A}$ in the first slot of $\boldsymbol{g}$ and $\boldsymbol{B}$ in the second. That is, we are dealing with the real number $\boldsymbol{g}(\boldsymbol{A},\boldsymbol{B})$. Making use of the dual basis, we alternatively have

$$\boldsymbol{g}(\boldsymbol{A},\boldsymbol{B}) = \boldsymbol{A} \cdot \boldsymbol{B} = A^a B_b \boldsymbol{e}_a \cdot \boldsymbol{e}^b = A^a B_a. \tag{2.47}$$

This has to be the same number as before. Comparing the two expressions, we see that

$$B_a = g_{ab} B^b. \tag{2.48}$$

This is a very important result. It shows that we can use the metric to *lower* a given tensor index.

Similarly, we can define the inverse metric as

$$g^{ab} = \boldsymbol{g}(\boldsymbol{e}^a, \boldsymbol{e}^b) = \boldsymbol{e}^a \cdot \boldsymbol{e}^b. \tag{2.49}$$

Not surprisingly, this object can be used to *raise* component indices. Finally, the mixed metric follows from

$$g^a{}_b = \boldsymbol{g}(\boldsymbol{e}^a, \boldsymbol{e}_b) = \boldsymbol{e}^a \cdot \boldsymbol{e}_b = \delta^a{}_b, \tag{2.50}$$

or, equivalently,

$$g^{ab} g_{bc} = \delta^a{}_c. \tag{2.51}$$

We can also use the dot product to figure out how a covariant object transforms if we change coordinates. Basically, we have

$$\boldsymbol{A} \cdot \boldsymbol{B} = A_a B^a = A_a \frac{\partial x^a}{\partial x'^b} B'^b = A'_b B'^b, \tag{2.52}$$

where we have used (2.42) (exchanging primed and unprimed variables). We see that

$$A'_b = \frac{\partial x^a}{\partial x'^b} A_a. \tag{2.53}$$

This is the transformation rule for a covariant object.

The dot product illustrates the concept of tensor *contraction*. Essentially, contraction is a process that lowers the rank of a tensor by two. In component expressions a contraction is represented by a sum over repeated indices. When carrying out actual calculations one can freely change the labels of such repeated indices (often called *dummy indices*).

2.5 The covariant derivative

Since physical laws tend to be described in terms of (partial) differential equations, we must add differentiation of tensors to our geometrical armour. Remember that, strictly speaking, we still do not fully understand the meaning of the geodesic equation (2.31).

Formally, the introduction of derivatives poses no problem. We can define derivatives through the usual limiting process. Indeed, in the case of the derivative of a scalar field ψ we see from (2.36) that

$$\nabla = e^a \frac{\partial}{\partial x^a}, \tag{2.54}$$

and it follows that (if $\boldsymbol{u}$ is the four-velocity)

$$\boldsymbol{u} \cdot \nabla \psi = \left(u^a e_a\right) \cdot \left(e^b \frac{\partial \psi}{\partial x^b}\right) = u^a \frac{\partial \psi}{\partial x^a} = \frac{d\psi}{d\tau}. \tag{2.55}$$

The question is what happens for tensors. In principle, given a tensor $\boldsymbol{A}$ (which should be interpreted as a *tensor field* $\boldsymbol{A} = \boldsymbol{A}[\mathcal{P}(\zeta)]$) we can define the rate of change along a given world line $\mathcal{P}$ as

$$\nabla_t \boldsymbol{A} = \lim_{\Delta\zeta \to 0} \frac{\boldsymbol{A}[\mathcal{P}(\zeta + \Delta\zeta)] - \boldsymbol{A}[\mathcal{P}(\zeta)]}{\Delta\zeta}, \tag{2.56}$$

where t is the tangent vector from before. However, there is a slight problem here. We have seen that tensors are linear objects that live in the tangent space associated with each spacetime point; see Figure 2.6. The two tensors in the expression on the right-hand side of Eq. (2.56) represent different events and hence belong to different tangent spaces. In order to work out the derivative, we need to be able to transport one of the vectors in such a way that we take the difference in the same tangent space. In doing this we need to account for any stretching and twisting of the basis vectors due to the curvature of spacetime. We need to identify the *connection* between the tangent spaces.

This may sound like a tricky problem, but the procedure turns out to be straightforward. In flat space we can *parallel transport* a vector by keeping it aligned with itself while holding its length fixed. This idea is easily extended to tensors. We simply keep all the components fixed in an orthonormal coordinate basis. However, in a curved spacetime we do not have the luxury of such a basis, apart from at the local level. Of course, in constructing derivatives we are dealing with infinitesimal differentials. One can convince oneself that the difference between the tangent spaces (the deviation from a local inertial frame at (say) $\mathcal{P}(\zeta)$) occurs at second order in the distance $\Delta\zeta$. This means that we can introduce coordinates sufficiently close to the flat space ones that we can define parallel transport in the usual way.

The *gradient* of a tensor is then defined as

$$\nabla_t A(\_) = \nabla A(\_, t). \tag{2.57}$$

That is, if A is a tensor of rank n, then its gradient has rank $n+1$. Alternatively, we write this as

$$\nabla A = \nabla_{e_a} A = \nabla_a (A^b e_b) e^a = (\nabla_a A^b)(e_b \otimes e^a). \tag{2.58}$$

This is the *covariant derivative*.

In the following we will use either ∇_a or a semi-colon followed by the relevant spacetime index to represent the covariant derivative in component expressions. Partial derivatives will be given by ∂_a or a comma followed by the index. The latter notation may be confusing at first, but as it is commonly used it is important to be comfortable with it. For example, the *divergence* follows as the contraction

$$\nabla \cdot A = \nabla_a A^a = A^a_{\ ;a}. \tag{2.59}$$

The question is, how do we work out the coefficients of these expressions? First of all, it makes sense that the covariant derivative should reduce to the standard gradient for scalars. After all, there is no issue about tangent spaces for scalar quantities. However, for tensors, it is easy to show that the partial derivative does not respect the expected transformation properties. Partial differentiation of vectors is not a tensor operation. The solution to the problem follows once we realize that the induced change in any given basis vector e_a must be a linear combination of all the basis vectors. That is, we should have (again with u the four-velocity, in order to make contact with (2.31))

$$
\begin{aligned}
(u \cdot \nabla)u &= \left(u^b e_b\right) \cdot \left(e^c \partial_c\right)\left(u^a e_a\right) \\
&= u^b \partial_b \left(u^a e_a\right) = u^b \left[e_a(\partial_b u^a) + u^a(\partial_b e_a)\right] \\
&= u^b \left(\partial_b u^a + \Gamma^a_{cb} u^c\right) e_a = (u^b \nabla_b u^a) e_a,
\end{aligned}
\tag{2.60}
$$

where we have defined

$$\partial_b e_a = \Gamma^c_{ab} e_c, \tag{2.61}$$

and we identify the components of the covariant derivative as

$$\nabla_b u^a = u^a_{\ ;b} = \partial_b u^a + \Gamma^a_{cb} u^c. \tag{2.62}$$

The first term measures the rate of change of the components of $\boldsymbol{u}$ while the second term arises because of the change in the basis vectors. The quantities Γ^c_{ab} are known as the *connection coefficients*. For a given metric (in a coordinate basis) they are often called the *Christoffel symbols*.

Next, the fact that $\boldsymbol{e}^a \cdot \boldsymbol{e}_b = \delta^a_b$ leads to

$$\partial_b \boldsymbol{e}^a = -\Gamma^a_{cb} \boldsymbol{e}^c, \tag{2.63}$$

and it follows that

$$\nabla_a A_b = A_{b;a} = \partial_a A_b - \Gamma^c_{ba} A_c. \tag{2.64}$$

For higher rank tensors we have to account for the change in each basis vector, which means that all indices must be corrected for in the same way as in these examples.

Finally, we need to be able to calculate the connection coefficients. One can think of different ways of doing this. One way would be to define the derivative in such a way that the metric is effectively constant with respect to it. This is the choice that is made in general relativity. It is sometimes described as the covariant derivative being 'compatible' with the metric. We then require the derivative to be such that

$$\nabla \boldsymbol{g} = 0 \quad \text{or} \quad \nabla_a g_{bc} = 0. \tag{2.65}$$

From our definition of the covariant derivative, it follows that

$$\nabla_a g_{bc} = \partial_a g_{bc} - g_{dc}\Gamma^d_{ab} - g_{bd}\Gamma^d_{ac} = 0. \tag{2.66}$$

Given this relation we can cyclically permutate the free indices to get three relations, combine these, and use the fact that the connection coefficients are symmetric in the last two indices[5] to show that

$$\Gamma^a_{bc} = \frac{1}{2}g^{ad}(\partial_c g_{bd} + \partial_b g_{cd} - \partial_d g_{bc}). \tag{2.67}$$

2.6 The geodesic equation

We have arrived at a point where, once we have introduced coordinates, we can work out the corresponding coordinate basis and build the metric to get a notion of distance. We also have a derivative that allows us to discuss how quantities vary along a given curve. We are ready to make sense of the geodesic equation.

---

[5] Caution: This is an assumption of general relativity. Other theories may have non-vanishing torsion, in which case this symmetry does not hold.

A freely falling particle will follow the straightest possible path in spacetime. This means that the four-velocity $\boldsymbol{u}$ satisfies the geodesic equation

$$\nabla_{\boldsymbol{u}}\boldsymbol{u} = (\boldsymbol{u} \cdot \nabla)\boldsymbol{u} = 0, \tag{2.68}$$

or

$$u^a \nabla_a u^b = (\partial_a u^b + \Gamma^b_{ca} u^c) u^a = 0. \tag{2.69}$$

In essence, the vector $\boldsymbol{u}$ is parallel transported along itself. In a given set of coordinates, $x^a(\tau)$ (where τ could be proper time, in the case of a material object), the first term can be written

$$u^a \partial_a u^b = \frac{\partial u^b}{\partial x^a}\frac{dx^a}{d\tau} = \frac{du^b}{d\tau} = \frac{d}{d\tau}\left(\frac{dx^b}{d\tau}\right), \tag{2.70}$$

while, by definition,

$$u^a u^c = \frac{dx^a}{d\tau}\frac{dx^c}{d\tau}. \tag{2.71}$$

Hence, the geodesic equation becomes

$$\frac{d^2 x^b}{d\tau^2} + \Gamma^b_{ca}\frac{dx^c}{d\tau}\frac{dx^a}{d\tau} = 0. \tag{2.72}$$

These four coupled ordinary differential equations determine the trajectory for given initial conditions.

Let us digress for a moment to consider an issue closely related to the motion of an object: the existence (or not) of conserved quantities. In general, a conserved quantity is associated with the problem having some underlying symmetry. As we will now show, this leads to the existence of a so-called Killing vector, an object that satisfies

$$\nabla_a k_b + \nabla_b k_a = 0. \tag{2.73}$$

It is straightforward to show that, for any vector that satisfies this equation, we have

$$\frac{d}{d\tau}(k_a u^a) = u^b \nabla_b (k_a u^a) = 0. \tag{2.74}$$

In effect

$$\boldsymbol{k} \cdot \boldsymbol{u} = k_a u^a \tag{2.75}$$

is constant along a geodesic. We have a constant of motion. It is also easy to demonstrate that this conserved quantity is associated with a spacetime symmetry. Such a symmetry typically results in there existing a set of coordinates such that the metric does not depend on one of them. Let us take the time coordinate as an example. Suppose we have a Killing vector associated with x^0 in some set of coordinates. That is, let

$$k^a = \delta_0^a \qquad \longrightarrow \qquad k_a = g_{ab}k^b = g_{a0}. \tag{2.76}$$

Making use of the covariant derivative (2.64) and the explicit expression for the Christoffel symbols from (2.67) we see that

$$\nabla_a k_b + \nabla_b k_a = \partial_0 g_{ab} = 0. \tag{2.77}$$

In other words, the existence of this particular Killing vector implies that the metric does not depend on the x^0 coordinate, and vice versa. Later, when we consider particle motion in black-hole spacetimes (see Chapter 10), we will see that this particular symmetry leads to the conservation of energy.

2.7 Curvature

Having developed the notion of tensor derivatives, we are primed to discuss the concept of curved spaces in more detail. To do this, we need to introduce the notion of a *commutator*. Given two vectors A and B (or identically, directional derivatives) the commutator is defined by

$$[A,B]\psi = [\nabla_A, \nabla_B]\psi = (\nabla_A\nabla_B - \nabla_B\nabla_A)\psi$$
$$= \left(A^a\frac{\partial B^b}{\partial x^a} - B^a\frac{\partial A^b}{\partial x^a} \right)\psi = \nabla_C\psi, \tag{2.78}$$

where ψ is a scalar field. Geometrically, we can understand this in the following way. Consider two vectors which start at the same point but aim in different directions. Let us call them A and B. Then parallel transport one of the vectors (A) until its tail is at the tip of the other vector (B), and vice versa. In flat space, we know that this would lead to the tips of the transported vectors touching. In a curved space the answer may be different. As we move the vectors to different points they will change a little. $A^a\partial B^b/\partial x^a$ measures the amount by which the vector B changes as one moves it along A, and so on. The commutator measures the 'failure' of the tips of the two transported vectors to meet at the end of the exercise, encoding the spacetime curvature.

Having introduced the commutator, it is natural to highlight the different types of bases one might want to use for practical calculations. As we have already seen, given a set of coordinates $x^a(\mathcal{P})$ one can construct a *coordinate basis*, whose basis vectors are given by

$$\boldsymbol{e}_a = \frac{\partial \mathcal{P}}{\partial x^a} = \frac{\partial}{\partial x^a} \qquad \longrightarrow \qquad [\boldsymbol{e}_a, \boldsymbol{e}_b] = \left[\frac{\partial}{\partial x^a}, \frac{\partial}{\partial x^b}\right] = 0. \tag{2.79}$$

Obviously, a *non-coordinate basis* must then be such that

$$[\hat{\boldsymbol{e}}_a, \hat{\boldsymbol{e}}_b] = c_{ab}{}^c \hat{\boldsymbol{e}}_c \neq 0, \tag{2.80}$$

where $c_{ab}{}^c$ are called the commutation coefficients. It is important to appreciate the difference between these two sets. For example, when taking partial derivatives one must know (i) what one is differentiating with respect to, and also (ii) what is held fixed. Perhaps confusingly, one may want to use both kinds of basis in a practical calculation. Typically, one would use an orthonormal (non-coordinate) basis, where the basis vectors have unit length, when discussing physical interpretations (the experimenter's view). At the same time, many calculations are easier to carry out in a coordinate basis (the theorist's view). We will see examples of both strategies later.

Having come quite far since we set out, we now embark on the final leg of the journey. Equipped with the appropriate tools we are able to discuss the spacetime curvature. Particularly important for this discussion will be the *Riemann tensor*, which is a measure of the failure of second derivatives to commute. Given a vector field $\boldsymbol{A}$ we define the Riemann tensor through

$$(\nabla_c \nabla_d - \nabla_d \nabla_c) A^a = R^a{}_{bcd} A^b. \tag{2.81}$$

This shows that the Riemann tensor must be anti-symmetric in its last two indices (associated with the derivatives). Although it is less trivial, one can also prove that it must be anti-symmetric in the first two indices, and symmetric if the two pairs of indices are interchanged. In addition one can prove that

$$\nabla_e R_{abcd} + \nabla_c R_{abde} + \nabla_d R_{abec} = 0. \tag{2.82}$$

These are the *Bianchi identities*. Combining all this information, one can work out that, in four spacetime dimensions the Riemann tensor has 20 independent components. Generally, these components can be calculated from

$$R^a{}_{bcd} = \partial_c \Gamma^a_{bd} - \partial_d \Gamma^a_{bc} + \Gamma^a_{ce} \Gamma^e_{bd} - \Gamma^a_{de} \Gamma^e_{bc}. \tag{2.83}$$

This expression follows (admittedly after a little bit of work) from the definition of the covariant derivative. In terms of a prescribed metric we get, in a local inertial frame (where we can ignore first, but not second derivatives of the metric),

$$R_{abcd} = \frac{1}{2} (\partial_c \partial_b g_{ad} + \partial_d \partial_a g_{bc} - \partial_d \partial_b g_{ac} - \partial_c \partial_a g_{bd}). \tag{2.84}$$

This result will be important to us later.

Given the Riemann tensor, we can contract on the first and third indices to obtain the *Ricci tensor*

$$R_{ab} = g^{cd} R_{dacb} = R^c{}_{acb}. \tag{2.85}$$

The 10 independent components of this object follow from

$$R_{ab} = \partial_c \Gamma^c{}_{ab} - \partial_b \Gamma^c{}_{ac} + \Gamma^c{}_{ab} \Gamma^e{}_{ce} - \Gamma^c{}_{ae} \Gamma^e{}_{bc}. \tag{2.86}$$

Contracting one more time we arrive at the *Ricci scalar*

$$R = g^{ab} R_{ab} = R^a{}_a. \tag{2.87}$$

Given these definitions we note that, if we carry out the Ricci contraction on the Bianchi identities (2.82), we find that the divergence of the so-called *Einstein tensor*

$$G_{ab} = R_{ab} - \frac{1}{2} g_{ab} R \tag{2.88}$$

vanishes. That is, we have

$$\nabla \cdot \boldsymbol{G} = 0 \qquad \text{or} \qquad \nabla_a G^{ab} = 0. \tag{2.89}$$

The importance of this result will soon become apparent.

2.8   A little bit of matter

So far we have focused on how the motion of an object is affected by the spacetime curvature. However, our aim is to design a theory of gravity, where the presence of matter generates the gravitational field that causes the spacetime to bend in the first place. In order to do this we need to introduce one more tensor: the stress-energy tensor. This is a rank 2 object that encodes all energy, momentum, and stresses associated with matter. As we also want to account for electromagnetism, elasticity, and other matter properties, this can be a very complicated object. We will consider some of the complications later when we discuss neutron star physics (Chapter 12), but for the purposes of the present discussion we will take a more simplistic view. Let us ask what the simplest tensor we can build out of the ingredients we already have available may be. This may seem more of a mathematics question than one of physics, but it turns out that this approach leads us directly to two widely used matter models; the dust model (representing a gas of non-interacting and hence pressure-less particles) that is used in cosmology and the perfect fluid that is used to build stellar models.

First, consider a set of moving particles, each with its own individual worldline (as in Figure 2.5). Averaging over the particles (on some length scale) the congruence of worldlines provides us with a four-velocity (field) u^a. With this as our only building block, there is only one rank 2 object we can construct:

$$T^{ab} = \varepsilon u^a u^b. \tag{2.90}$$

This model represents what is called 'dust' and ε is the corresponding energy density. By contracting twice with the four-velocity, we have

$$\varepsilon = u_a u_b T^{ab}. \tag{2.91}$$

This model is very simple, yet it forms the basis for relativistic cosmology (Chapter 4). The basic idea is that individual stars/galaxies can be viewed as non-interacting (dust) particles and on a suitably large scale the average (2.90) makes sense.

We obviously want to be able to deal with more complicated models. The natural next step would be to ask what happens if we build the stress-energy tensor out of both the four-velocity and the spacetime metric. Then we have

$$T^{ab} = A u^a u^b + p g^{ab}. \tag{2.92}$$

As before, the energy density follows from contractions with the four-velocity. We now have

$$\varepsilon = A - p, \tag{2.93}$$

which means that

$$T^{ab} = (p + \varepsilon) u^a u^b + p g^{ab} = \varepsilon u^a u^b + p \perp^{ab}. \tag{2.94}$$

This model is called the perfect fluid. It does not allow for shear stresses, but it allows for particle interactions that give rise to the (isotropic) pressure p. In Chapter 4 we will see how this model leads to the familiar equations for fluid dynamics.

Before we move on, let us make two remarks. First of all, we have (sneakily) introduced a new object, $\perp^{ab}$, in (2.94). This is known as the projection. To see why this name is appropriate, note that

$$u_a \perp^{ab} = 0. \tag{2.95}$$

The projection isolates components orthogonal to the flow associated with u^a. This will prove useful later.

Secondly, it is worth considering the Newtonian limit. Special relativity tells us that energy and mass are equivalent. As a result, the energy density generally takes the form (for a fluid made up of a single particle species)

$$\varepsilon = mnc^2 + \epsilon(n), \tag{2.96}$$

where n is the number density of particles, m is the rest mass of each particle, and ϵ is the internal energy due to interactions between the particles. Formally, the Newtonian limit involves letting $c \to \infty$ (see Chapter 4), which means that the rest mass contribution dominates and the leading part of the stress-energy tensor is (returning to geometric units)

$$T^{00} \approx mn = \rho. \tag{2.97}$$

We will make use of this result shortly.

2.9   Geodesic deviation and Einstein's equations

We now have the tools we need to attempt a relativistic description of the thought experiment from the beginning of the chapter. Recall that we are interested in the paths of two freely falling particles; see Figures 2.2 and 2.5. Let us try to describe the motion of the two particles A and B in the thought experiment, assuming that the two trajectories start out parallel. To do this, we again introduce a vector $\boldsymbol{\xi}$ to measure the separation between the two trajectories. Assuming that this vector is purely spatial according to the trajectory of A, which we also assign to measure time (such that the corresponding four-velocity only has a time-component), we trivially have

$$\boldsymbol{u} \cdot \boldsymbol{\xi} = u^a \xi_a = 0. \tag{2.98}$$

The second derivative of $\boldsymbol{\xi}$ will be affected by the spacetime curvature. In order to quantify this effect, we assume that the geodesics are labelled by a parameter λ in such a way that

$$\boldsymbol{u} = \left(\frac{\partial}{\partial \tau} \right)_\lambda \quad \text{and} \quad \boldsymbol{\xi} = \left(\frac{\partial}{\partial \lambda} \right)_\tau. \tag{2.99}$$

Then it is easy to show that

$$[\boldsymbol{u}, \boldsymbol{\xi}] = \nabla_{\boldsymbol{u}} \boldsymbol{\xi} - \nabla_{\boldsymbol{\xi}} \boldsymbol{u} = 0, \tag{2.100}$$

or

$$u^a \nabla_a \xi^b - \xi^a \nabla_a u^b = 0 \tag{2.101}$$

This means that we get

$$\nabla_{\boldsymbol{u}} \nabla_{\boldsymbol{u}} \boldsymbol{\xi} = \nabla_{\boldsymbol{u}} \nabla_{\boldsymbol{\xi}} \boldsymbol{u} = (\nabla_{\boldsymbol{u}} \nabla_{\boldsymbol{\xi}} - \nabla_{\boldsymbol{\xi}} \nabla_{\boldsymbol{u}}) \boldsymbol{u}, \tag{2.102}$$

where we have made use of the geodesic equation. In other words, the relative acceleration is caused by the failure of the double gradients to commute. In terms of components (and making use of (2.101)), we have

$$u^c \nabla_c (u^b \nabla_b \xi^a) = u^c \xi^b (\nabla_c \nabla_b - \nabla_b \nabla_c) u^a = -R^a{}_{dbc} u^d \xi^b u^c, \tag{2.103}$$

where we have used the definition of the Riemann tensor, (2.81). This is the equation of *geodesic deviation*.

As we want to make contact with the Newtonian description of our falling-body thought experiment, it is useful to introduce (in analogy with the result for scalar fields) a total time derivative such that

$$\frac{D}{D\tau} = u^a \nabla_a, \tag{2.104}$$

which means that (2.103) becomes

$$\frac{D^2 \xi^a}{D\tau^2} = -R^a{}_{dbc} u^d \xi^b u^c. \tag{2.105}$$

Now that we have an expression for the relative acceleration caused by the spacetime curvature, we can complete the description of the thought experiment. In the local inertial frame associated with particle A we have (since the particle remains at rest in this frame)

$$u^0 = 1, \quad u^j = 0, \tag{2.106}$$

$$\xi^0 = 0, \quad \xi^j \neq 0. \tag{2.107}$$

Also replacing the proper time τ with the coordinate time in this frame t, we have

$$\frac{\partial^2 \xi^j}{\partial t^2} = -R^j{}_{abc} u^a \xi^b u^c = -R^j{}_{0b0} \xi^b = -R^j{}_{0k0} \xi^k. \tag{2.108}$$

Comparing this to the Newtonian equation for tidal gravity (2.12), we identify

$$R^j{}_{0k0} = \mathcal{E}^j{}_k = \delta^{jl} \left(\frac{\partial^2 \Phi}{\partial x^l \partial x^k} \right). \tag{2.109}$$

This provides a constraint that any curved spacetime theory must satisfy in the Newtonian limit (weak gravitational fields, low velocities etcetera). In essence, the Riemann tensor represents a generalized tidal field.

In the Newtonian case we also have the Poisson equation for the gravitational potential

$$\nabla^2 \Phi = 4\pi G \rho = \delta^{jk} \partial_j \partial_k \Phi = \mathcal{E}^j{}_j, \tag{2.110}$$

where G is Newton's gravitational constant and ρ is the mass density. This relation provides a constraint on the trace of the tidal tensor. We must have (in geometric units)

$$\mathcal{E}^j{}_j = 4\pi\rho. \tag{2.111}$$

One might then expect the field equations of general relativity to look something like

$$R^j{}_{0j0} = 4\pi\rho. \tag{2.112}$$

However, this is not an acceptable solution. As it is not a coordinate independent statement it does not have the form expected of the 'true physics'. We need to have spacetime indices, rather than spatial ones. However, noting that

$$R^0{}_{000} = 0, \tag{2.113}$$

we have

$$R_{00} = R^a{}_{0a0} = R^j{}_{0j0} = 4\pi\rho, \tag{2.114}$$

which takes us part of the way (as we are now summing from 0 to 3). Inspired by these relations, let us *guess* that the required form of the field equations might be

$$R_{ab} = 4\pi T_{ab}, \tag{2.115}$$

where we have used the fact that $T_{00} \approx \rho$ in the weak-field limit. This would seem a natural extension of the Newtonian result. Unfortunately, it is wrong...

The reason why the above expression is not (quite) what we want is the underlying coordinate freedom of the spacetime problem. In essence, Eq. (2.115) corresponds to 10 equations for 10 unknown metric coefficients. In normal circumstances this is exactly what one would need for a well-posed mathematical problem. However, the geometric description allows us to freely choose the coordinate system, i.e. specify the four functions $x^a(\mathcal{P})$. We can use this freedom to make four of the metric functions g_{ab} anything we want. So the problem we have posed is overdetermined. We need to formulate the field equations in such a way that we end up with six independent equations.

With the benefit of more than a century's hindsight, the fix is simple. We need to introduce a constraint to remove four degrees of freedom. We can do this by working with the Einstein tensor G_{ab}, which is required to satisfy the four equations (2.89). Thus, we are led to

$$G_{ab} = R_{ab} - \frac{1}{2}g_{ab}R = 8\pi T_{ab}, \tag{2.116}$$

where the factor of 8π is determined by taking the Newtonian limit. If we reinstate G and c we have the final result[6]

$$G_{ab} = R_{ab} - \frac{1}{2}g_{ab}R = \frac{8\pi G}{c^4}T_{ab}. \qquad (2.117)$$

These are the *Einstein field equations*. They describe how the presence of matter at a point affects the average spacetime curvature in its neighbourhood, and how the curvature, in turn, influences the motion of the matter.

The theory is now consistent because the four constraints from (2.89) reduce the problem to six independent equations, as required. The model brings us back to Newtonian gravity in the appropriate limit. In addition, it leads to a number of testable predictions. We will consider such tests in Chapter 10.

Before we move on, it is useful to note that (2.89) and (2.117) imply that

$$\nabla \cdot \boldsymbol{T} = 0, \quad \text{i.e.} \quad \nabla_a T^{ab} = 0. \qquad (2.118)$$

These four equations represent the conservation of energy and momentum[7] and can be used to work out the equations of motions for any given matter model. We will discuss this further (from different perpectives) in Chapters 4 and 20.

We have retraced the steps Einstein took in developing general relativity, from the original idea of the equivalence principle in 1907 to the final version of the field equations almost a decade later (Einstein, 1915). It is remarkable to note that he did not yet have the final form of the equations at the beginning of the frantic month of November 1915. His thinking may still have been in flux, but he nevertheless arrived at a theory that has passed every single test it has been subjected to—an extraordinary achievement.

As he completed the theory, Einstein knew that the field equations were remarkably complicated. He even suggested that they may never be solved, at least not in closed form. In this case, he was wrong. We will discuss this in Chapter 4. First, we will explore another prediction of the theory—the existence of gravitational waves.

---

[6] The correct power of c can be figured out from the fact that the left-hand side represents spacetime curvature so has dimension of inverse length squared, while the stress-energy tensor has the dimension of an energy density.

[7] It is worth pointing out that (2.118) is not a 'true' conservation law. The actual argument assumes the presence of a Killing vector—a k_a satisfying (2.73)—and the flux $j^b = k_a T^{ab}$ such that

$$\nabla_b j^b = k_a \nabla_b T^{ab} = 0. \qquad (2.119)$$

Flux conservation requires (2.118) to be satisfied, but this is only part of the argument.

3

Gravitational waves

General relativity explains gravity in terms of a dynamical spacetime, where changes in the gravitational field propagate as waves. These waves are generated by the bulk acceleration of matter and they travel at the speed of light. We want to understand the nature of these gravitational waves,[1] explain how they are generated, and discuss how they move through spacetime. These are all important steps.

In order to model astrophysical sources we need to understand what gravitational waves are and how they are generated. We need to establish how they carry energy from the source to our detectors, what effect this has on the source, and how a wave passing through our instruments affects them. We need to know to what extent the waves are altered as they travel through the Universe, e.g. via gravitational lensing or through interaction with the interstellar medium. If the waves are of cosmological origin we also need to weigh in the fact that the Universe evolves (see Chapter 22).

As the combined problem is challenging, it makes sense to build our understanding step by (careful) step. Hence, we begin by considering different ways to analyse slightly perturbed spacetimes, starting from weak waves travelling in an otherwise flat spacetime and then extending the analysis to the general case of a curved background. As a key part of the development, we consider the origin of the waves and the effect they have on a given source. This leads us to the so-called *quadrupole formula*, which provides a very useful estimate of the signal. Along the way, we also learn how one can derive a stress–energy tensor for the waves.

Let us begin by asking why there should be gravitational waves in the first place. In a sense this brings us back to the speed of light providing a universal limit for signal propagation in Einstein's theory, but one may still ask why we end up with waves. To show that this is natural, consider a weak disturbance propagating in an otherwise flat spacetime. Working in flat space we can construct a global inertial reference frame with coordinates $\{x^a\} = \{t, x, y, z\}$. We also have the Minkowski metric

---

[1] There are a number of books dedicated to gravitational waves (although nowhere near as many as there are general relativity texts), for example, Maggiore (2007) and Creighton and Anderson (2011). These texts predate the first detection but they are nevertheless useful complements to this book. In particular, the latter provides a focus on data analysis strategies. A more recent text, Maggiore (2018) includes a detailed discussion of the detections to date and also provides an exhaustive discussion of relevant cosmology aspects.

Gravitational-Wave Astronomy: Exploring the Dark Side of the Universe. Nils Andersson, Oxford University Press (2020).
© Nils Andersson. DOI: 10.1093/oso/9780198568032.001.0001

$$g_{ab} = \eta_{ab} = \text{diag}(-1, 1, 1, 1), \tag{3.1}$$

and all derivatives become partial. This setting is comfortably familiar. Now, any deviation from the flat background will lead to curvature which can be expressed in terms of the Riemann tensor, R_{abcd}. Without too much effort one can show that the individual components of the Riemann tensor satisfy

$$\Box R_{abcd} = \eta^{ef} \partial_e \partial_f R_{abcd} = 0, \tag{3.2}$$

where (in flat space, Cartesian coordinates, and geometric units)

$$\Box = -\frac{\partial}{\partial t^2} + \frac{\partial}{\partial x^2} + \frac{\partial}{\partial y^2} + \frac{\partial}{\partial z^2}. \tag{3.3}$$

In other words, the Riemann tensor components satisfy a wave equation. Moreover, the wave operator is the same as in electromagnetism, so the gravitational waves travel at the speed of light.

3.1   Weak waves in an otherwise flat spacetime

Since gravity is a tidal interaction, the presence of a gravitational wave can never be detected by a local experiment. This follows immediately from the fact that we have the freedom to choose inertial coordiclnates associated with any observer. We can always find a frame such that spacetime is flat in the neighbourhood of some given point. In order to begin to understand the nature of gravitational waves we need to revisit the setting of the thought experiment that led us to Einstein's equations in the first place (see Chapter 2). Thus, we consider the influence of a wave on two test particles, A and B, initially taken to be at rest. As before, we arrange the setting to be such that the particles are separated by a purely spatial vector ξ^j. A gravitational wave will 'push the particles around' with respect to each other, which means that our reference frame can no longer be globally inertial.

To discuss the motion of the particles let us use the local inertial frame in which particle A remains at rest at the origin. This situation is illustrated in Figure 3.1. In this frame we have

$$g_{ab} = \eta_{ab} + O(|x|^2/R^2), \tag{3.4}$$

where $R^2 \sim 1/|R_{abcd}|$ is a measure of the radius of spacetime curvature induced by the waves.

Recalling the analysis that motivated the field equations of general relativity, we realize that we need to make use of the equation of geodesic deviation

$$\nabla_u \nabla_u \xi = -R(\_, u, \xi, u). \tag{3.5}$$

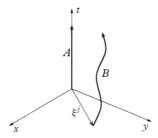

Figure 3.1 *A schematic illustration of the effect a gravitational wave has on two particles, A and B, as viewed in a local inertial frame moving along with particle A.*

In our chosen frame we have $u^0 = 1$ and $u^j = 0$. In essence, the coordinate time t of A is identical to the proper time (because the metric is flat in the vicinity of the origin of our coordinate system). Thus, we get

$$\frac{\partial^2 \xi^j}{\partial t^2} = -R^j{}_{0k0}\xi^k.$$

(3.6)

Now assume that

$$\xi^j = x_0^j + \delta x^j,$$

(3.7)

where x_0^j represents the unperturbed location of B and δx^j describes the wave-induced change. The latter is obviously small for weak waves. This leads to

$$\frac{\partial^2 \delta x^j}{\partial t^2} \approx -R^j{}_{0k0}x_0^k.$$

(3.8)

One can prove that, for weak gravitational waves and in a local inertial frame, all components of the Riemann tensor can be determined from R_{j0k0}. There are, in fact, only two independent components. Intuitively, one might expect R_{j0k0} to have six independent components, but symmetries reduce this to two degrees of freedom. It is customary to define the gravitational-wave field h_{jk} through

$$R_{j0k0} = -\frac{1}{2}\frac{\partial^2 h_{jk}}{\partial t^2},$$

(3.9)

where the factor of $1/2$ is a convention. We now get

$$\frac{\partial^2 \delta x^j}{\partial t^2} \approx \frac{1}{2}\frac{\partial^2 h^j{}_k}{\partial t^2}x_0^k,$$

(3.10)

which integrates to (if we assume that the particles are initially at rest)

$$\delta x^j = \frac{1}{2} h^j{}_k x_0^k. \tag{3.11}$$

This means that, if the two particles are initially a distance L apart, then a passing gravitational wave ($h_{jk} \sim h \sin \omega t$) will lead to a change in the separation ΔL such that

$$h \approx \frac{\Delta L}{L}. \tag{3.12}$$

This example describes exactly the situation one deals with in discussing gravitational-wave detection with interferometers (see Chapter 7). Of course, it is a simplification. For example, in the case of a ground-based detector one must also account for effects due to the Earth's gravity. Nevertheless, the analysis remains essentially the same. Most importantly, as long as the wavelength of the wave is much larger than the size of the detector, Eq. (3.4) is an adequate approximation.

3.2   Effect on matter

In order to investigate the properties of the (frame-dependent) gravitational-wave field, we may orient our coordinate system in such a way that the waves travel in the z-direction. Then we have $h_{jk} = h_{jk}(t - z)$, since the waves propagate at the speed of light (and we use geometric units). Furthermore, we have only two independent components. It turns out (we will show this later) that one can always find a coordinate system where h_{ij} is such that

$$h_{xx}^{TT} = -h_{yy}^{TT}, \tag{3.13}$$

$$h_{xy}^{TT} = h_{yx}^{TT}, \tag{3.14}$$

and all other components vanish. The TT subscript indicates that h_{jk} is *transverse* and *traceless*. It is conventional to introduce the notation

$$h_+ = h_{xx}^{TT} = -h_{yy}^{TT}, \tag{3.15}$$

$$h_\times = h_{xy}^{TT} = h_{yx}^{TT}, \tag{3.16}$$

where the two polarizations are referred to as 'h-plus' and 'h-cross'.

We can now use (3.11) to figure out the effect a wave with (say) pure plus polarization h_+ has on matter. Consider a particle initially located at (x_0, y_0) and let $h_\times = 0$ to find that

$$\delta x = \frac{1}{2}h_{xx}^{\mathrm{TT}}x_0 + \frac{1}{2}h_{xy}^{\mathrm{TT}}y_0 = \frac{1}{2}h_+x_0, \tag{3.17}$$

$$\delta y = \frac{1}{2}h_{xy}^{\mathrm{TT}}x_0 + \frac{1}{2}h_{yy}^{\mathrm{TT}}y_0 = -\frac{1}{2}h_+y_0. \tag{3.18}$$

From this we see that, during a wave-induced oscillation, $h_+ \sim h\sin\omega t$ first induces a stretch in the x-direction accompanied by a squeeze along the y-axis. One half-cycle later, the stretch is in the y-direction, with the squeeze is along the x-axis. Similarly, we readily show that the influence of $h_\times$ is such that

$$\delta x = \frac{1}{2}h_\times y_0, \tag{3.19}$$

$$\delta y = \frac{1}{2}h_\times x_0. \tag{3.20}$$

The effect is the same as that of h_+ but rotated by 45°. The characteristic stretch and squeeze that a gravitational wave induces on a ring of test particles is illustrated in Figure 3.2.

We naturally have to exercise some caution at this point. We carried out our analysis in a particular reference frame. In essence, we determined how particle B moves due to a gravitational wave in a coordinate system where particle A remains at the origin. How do things change if we use some other set of coordinates? In order to see under what conditions our derivation remains valid we need to examine (3.4) a bit closer. Suppose

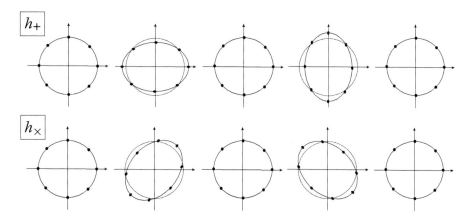

Figure 3.2 *A schematic illustration of the effect that the two gravitational-wave polarizations h_+ and $h_\times$ have on a ring of test particles (the unperturbed case is represented by the dashed circle). As shown in the top panel, h_+ induces a stretch in the x-direction, accompanied by a squeeze along the y-axis. One half-cycle later, the stretch is in the y-direction, while the squeeze is along the x-axis. The effect of $h_\times$ is analogous but rotated by 45°, as seen in the bottom panels. (The illustration is, of course, a vast exaggeration of the true magnitude of the motion.)*

we carry out an experiment using a 'detector' of size L. That is, we take $|x_0^j| \sim L$ in our calculation. Considering gravitational waves with reduced wavelength $\lambda = \lambda/2\pi$ we then have

$$\frac{1}{R^2} \sim |R_{abdc}| \sim |R_{j0k0}| \sim |\ddot{h}_{jk}| \sim \frac{h_+}{\lambda^2}. \tag{3.21}$$

Hence, Eq. (3.4) becomes

$$g_{ab} = \eta_{ab} + O\left(h_+ L^2/\lambda^2\right). \tag{3.22}$$

This shows that, as long as $L \ll \lambda$, we have

$$g_{ab} = \eta_{ab} + h_{ab}, \qquad \text{with} \qquad h_{ab} \sim h_+ \left(\frac{L}{\lambda}\right)^2 \ll h_+. \tag{3.23}$$

This proves the statement we made earlier. As long as the system is small compared to the gravitational-wave wavelength we do not need to worry about the curved background when carrying out a measurement. We will make this notion more precise later.

If we compare the wavelength of a gravitational wave with frequency of order 100 Hz (where ground-based interferometers are the most sensitive), i.e. $\lambda \approx 500$ km, we see that with interferometer arms a few kilometers long we are safely in the regime where the calculation is valid. As far as the detector is concerned we are (essentially) in flat space and in order to understand the basic principles of detection one never really has to calculate anything in full general relativity!

3.3   The wave equation

At this point it is natural to consider the equations that govern gravitational waves, and how they connect with a general matter source. As before (see Eq. (3.4)), let us consider a metric that differs only slightly from the Minkowski case. That is,

$$g_{ab} = \eta_{ab} + \epsilon h_{ab}, \tag{3.24}$$

where ϵ is a small (dimensionless) parameter introduced to make the smallness of the metric deviation explicit. By carrying out calculations accurate to linear order in ϵ, i.e. neglecting all quadratic and higher order terms, one can show that

$$g^{ab} = \eta^{ab} - \epsilon h^{ab}. \tag{3.25}$$

In essence, at this level of approximation one may think of h_{ab} as a field living in the fixed background space and indices can be raised and lowered with the flat metric. For example, we have

$$h^a_{\ b} = \eta^{ac} h_{cb}. \tag{3.26}$$

From (2.84) we find that the Riemann tensor, to order ϵ, is given by

$$R_{abcd} = \frac{1}{2}\epsilon\left(\partial_c\partial_b h_{ad} - \partial_c\partial_a h_{bd} - \partial_d\partial_b h_{ac} + \partial_d\partial_a h_{bc}\right). \tag{3.27}$$

Contracting the first and third index of this expression—using the flat metric as we neglect higher order terms in ϵ—we find that the Ricci tensor takes the form

$$R_{ab} = -\frac{1}{2}\epsilon\left(\Box h_{ab} + \partial_a\partial_b h - \partial_c\partial_b h^c_{\ a} - \partial_a\partial_c h^c_{\ b}\right), \tag{3.28}$$

where we have introduced the trace $h = h^a_{\ a}$ and the wave operator $\Box = \eta^{ab}\partial_b\partial_a$, as before. Another contraction leads to the Ricci scalar

$$R = -\epsilon\left(\Box h - \partial_c\partial_d h^{cd}\right). \tag{3.29}$$

These equations are still quite complicated. However, we have not yet made use of the coordinate freedom in Einstein's theory. It makes sense to ask whether we can find a set of coordinates in which the relations simplify. As we will now demonstrate, this is straightforward. First of all, we note that an infinitesimal coordinate transformation,

$$x^a \to x'^a = x^a + \epsilon\xi^a, \tag{3.30}$$

affects the metric g_{ab} in such a way that

$$h_{ab} \to h'_{ab} = h_{ab} - \partial_b\xi_a - \partial_a\xi_b. \tag{3.31}$$

In principle, this means that we can adjust some of the components of h_{ab} by choosing coordinates in a clever way. Secondly, we introduce a new variable (inspired by the form of the Einstein tensor (2.88))

$$\bar{h}_{ab} = h_{ab} - \frac{1}{2}\eta_{ab}h. \tag{3.32}$$

Note that this operation reverses the sign of the trace

$$\bar{h} = \bar{h}^a_{\ a} = -h^a_{\ a} = -h. \tag{3.33}$$

Under the coordinate transformation, this new variable changes in such a way that

$$\bar{h}_{ab} \to \bar{h}'_{ab} = \bar{h}_{ab} - \partial_b\xi_a - \partial_a\xi_b + \eta_{ab}\partial_c\xi^c. \tag{3.34}$$

Combining the results we find that the linearized Einstein tensor can be written

$$G_{ab} = R_{ab} - \frac{1}{2}\eta_{ab}R = -\frac{1}{2}\epsilon\left(\Box\bar{h}_{ab} + \eta_{ab}\partial_c\partial_d\bar{h}^{cd} - \partial_c\partial_b\bar{h}^c{}_a - \partial_c\partial_a\bar{h}^c{}_b\right). \tag{3.35}$$

Let us now specify the coordinates in such a way that

$$\partial_a\bar{h}^a{}_b = 0. \tag{3.36}$$

These four conditions determine what is known as *Lorenz gauge*.[2] With this choice, the problem simplifies to

$$G_{ab} = -\frac{1}{2}\epsilon\Box\bar{h}_{ab} = 8\pi T_{ab}, \tag{3.37}$$

where the stress-energy tensor enters at order ϵ since we assumed that the background spacetime was flat. This makes the example somewhat contrived, but we will soon deal with that issue.

We have arrived at the final result (reinstating G and c and setting $\epsilon = 1$ since it was only used for bookkeeping)

$$\Box\bar{h}_{ab} = -\frac{16\pi G}{c^4}T_{ab}. \tag{3.38}$$

The linear metric perturbations satisfy a wave equation, as expected. This particular form of the equation only describes gravitational waves in weak gravity, but we will soon see that the analysis can be extended to a more general setting.

3.4   Transverse-traceless (TT) gauge

Before we move on, let us digress on the choice of coordinates in the derivation of the wave equation. The procedure we used leads to coordinates that are 'as close to inertial as possible'. To see this, first, note that the Lorenz gauge remains unchanged for any coordinate transformation such that

$$\Box\xi^a = 0. \tag{3.39}$$

This should not come as a great surprise since (3.36) is a condition on the divergence of the gravitational field. It leaves four 'integration constants' unspecified. This means that we can impose four further conditions on h_{ab}. It is natural to take the vanishing of the trace $h^a{}_a$ as the first of these. Of course, if we do this then we have $\bar{h}_{ab} = h_{ab}$.

---

[2] Also referred to as *harmonic coordinates*. Generally, the word 'gauge' is used to describe a chosen set of coordinates or, equivalently, constraints imposed on the metric.

One can also prove that, if we orient the coordinate system in such a way that the waves propagate in the z-direction and impose as our remaining three conditions

$$h_{00} = h_{j0} = h_{0j} = 0 \tag{3.40}$$

(note that the z-components vanish automatically because of the transverse nature of the waves), we have the final result

$$h_{jk} = h_{jk}^{\text{TT}}. \tag{3.41}$$

This is known as transverse-traceless (or simply TT) gauge. Since the gravitational-wave field takes a very simple form in TT-gauge it is often used in source modelling and discussions of wave propagation.

If we want to understand the behaviour of a gravitational-wave detector that is large compared to the wavelength, we need to abandon the 'flat space description' and use the TT-gauge. In the case of a space-based interferometer like LISA, a typical gravitational-wave signal will have frequency 10^{-2} Hz, i.e. $\lambdabar \approx 10^6$ km, which is similar to the planned interferometer armlength. Hence, we cannot use the local inertial frame argument to discuss the effect that gravitational waves have on the detector.

Since we will be using the TT-gauge extensively in the following, it is worth discussing it in a bit more detail. First note that we have

$$g_{ab} = \eta_{ab} + h_{ab}^{\text{TT}}. \tag{3.42}$$

Given this, consider the geodesic equation for a single particle, i.e.

$$\frac{d^2 x^a}{d\tau^2} + \Gamma_{bc}^a \frac{dx^b}{d\tau} \frac{dx^c}{d\tau} = 0. \tag{3.43}$$

If the particle is initially at rest we have

$$\frac{d^2 x^j}{d\tau^2} = -\Gamma_{00}^j \left(\frac{dx^0}{d\tau} \right)^2 = -\frac{1}{2}(2\partial_0 h_0^j - \partial^j h_{00}) = 0 \tag{3.44}$$

(where $\partial^j = \eta^{jk} \partial_k$). After integration this leads to

$$\frac{dx^j}{d\tau} = 0. \tag{3.45}$$

In other words, we have $x^j = $ constant. This shows that the TT-gauge represents coordinates which move with the particles—the system is as 'inertial as possible'.

If we again consider the two test particles A and B we now have the situation shown in Figure 3.3 (where we have oriented the coordinate system in such a way that the two

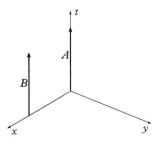

Figure 3.3 *A schematic illustration of the effect that a gravitational wave has on two test particles, A and B, when the problem is considered in TT-gauge.*

particles are both on the *x*-axis). At first sight this result may seem odd. How can the TT picture be compatible with the local inertial frame description from Figure 3.1? The answer follows if we consider the proper distance between the particles:

$$(\Delta s)^2 = g_{xx}(\Delta x)^2. \tag{3.46}$$

In the local inertial frame associated with particle *A* we have

$$g_{xx} = \eta_{xx} = 1, \tag{3.47}$$

and if we take the initial separation to be $x_0 = L$, we get

$$(\Delta s)^2 = (\Delta x)^2 = \left(L + \frac{1}{2}h_+L\right)^2 \rightarrow \Delta s \approx \left(1 + \frac{h_+}{2}\right)L. \tag{3.48}$$

Meanwhile, in TT-gauge we find that

$$(\Delta s)^2 = (1 + h_{xx}^{TT})(\Delta x)^2 = (1 + h_{xx}^{TT})L^2 \rightarrow \Delta s \approx \left(1 + \frac{h_+}{2}\right)L. \tag{3.49}$$

So the two pictures are consistent, after all. We just need to keep in mind that it is the distance in spacetime that matters, not the spatial separation in some particular set of coordinates.

Finally, suppose we have managed to find a solution to the wave equation (3.38). Because of the assumptions made in the derivation, this solution will obviously be in Lorenz gauge. However, this means that the result will in general not be transverse-traceless. However, it is often useful to work in the TT gauge. So how easy is it to transform a given result into this gauge? We will consider two options. The first approach is pragmatic, completely sacrificing mathematical rigour. Suppose that we are in a gauge where the waves propagate at the speed of light. Assume that wavefronts are nearly planar, i.e. that we are far away from the source, and orient the coordinate system in such a way that the waves move in the z-direction. Then one can simply:

1. Throw away all the time-space and time-time components,
2. Drop all components with z indices,
3. Keep the remaining off-diagonal terms unchanged, and
4. Remove the trace from the diagonal components.

These steps bring the result into the TT-gauge.

A more formal procedure (which essentially proves why the 'ad hoc' approach actually works) makes use of the projection operator. The aim is to first project out the transverse part of $\bar{h}_{ab}$ and then remove the trace. This can be done via the (spatial) projection

$$\perp_j^k = \delta_j^k - n_j n^k,\tag{3.50}$$

where $\boldsymbol{n} = \boldsymbol{x}/r$ is a unit vector, and the waves move away from the source in the direction of $\boldsymbol{n}$. Clearly, we then have

$$\perp_j^k n^j = 0.\tag{3.51}$$

In order to project out the transverse part of $\bar{h}_{jk}$ we need to act on each of the two indices. Also making sure to remove the trace, we have

$$h_{jk}^{\mathrm{TT}} = \left(\perp_j^l \perp_k^m - \frac{1}{2}\perp_{jk}\perp^{lm}\right)h_{lm}.\tag{3.52}$$

3.5 The quadrupole formula

Having derived the wave equation for gravitational waves (3.38) and considered the implications of the particular gauge in which the equation was shown to hold, we now turn to the link between the motion of the source and the waves. The aim is to outline a framework that allows us to estimate the rate of emission associated with proposed phenomena. In a very practical sense, this involves working out an approximate solution to (3.38). As we will demonstrate, this can be done in terms of a formal expansion away from Newtonian gravity, leading to the so-called *quadrupole formula*. Traditionally, this is the first port of call for anyone trying to estimate the gravitational-wave emission for a given astrophysical scenario. We will derive the key result, and consider the implications for various relevant problems later.

We know that a weak gravity source leads to gravitational waves governed by the wave equation (again, in geometric units)

$$\Box \bar{h}_{ab} = -16\pi \, T_{ab}. \tag{3.53}$$

As we are working in flat space, we can obtain a formal solution by means of the standard (retarded) Green's function for the wave equation. This leads to the integral equation

$$\bar{h}_{ab}(t, \boldsymbol{x}) = 4 \int \frac{T_{ab}(\boldsymbol{x}', t' = t - |\boldsymbol{x} - \boldsymbol{x}'|)}{|\boldsymbol{x} - \boldsymbol{x}'|} d^3 x. \tag{3.54}$$

Moreover, following the procedure outlined in the previous section, we readily translate the result into the TT-gauge:

$$\bar{h}_{jk}^{\mathrm{TT}}(t, \boldsymbol{x}) = \left\{ 4 \int \frac{T_{jk}(\boldsymbol{x}', t' = t - |\boldsymbol{x} - \boldsymbol{x}'|)}{|\boldsymbol{x} - \boldsymbol{x}'|} d^3 x \right\}^{\mathrm{TT}}, \tag{3.55}$$

where we recall that we have $h_{ab}^{\mathrm{TT}} = \bar{h}_{ab}^{\mathrm{TT}}$.

At first sight, this may not appear to have taken us any closer to an actual solution of the problem. But—as a matter of fact—it has! The trick is to use the integral form of the solution and iterate over a succession of approximations, until we reach the desired precision. This forms the basis of the post-Newtonian approximation scheme that we discuss in more detail in Chapter 11. In order to get a feeling for the nature of the solutions we obtain this way, and the steps involved, we focus on the leading order approximation here.

We start by manipulating (3.55). First, assume that we are dealing with a slow-motion source, i.e. that there is a characteristic velocity v such that $v \ll c \to \lambdabar \gg L$. Then the source resides deep within its own 'near zone' (we will discuss this concept in more detail later), $|\boldsymbol{x} - \boldsymbol{x}'| \approx r =$ the distance to the centre of the source, and the waves do not change much as they move across the source (since it is smaller than $\lambdabar$). In essence, we have

$$t' = t - |\boldsymbol{x} - \boldsymbol{x}'| \approx t - r, \tag{3.56}$$

so

$$\bar{h}_{jk}^{\mathrm{TT}}(t, \boldsymbol{x}) = \left\{ \frac{4}{r} \int T_{jk}(\boldsymbol{x}', t' = t - r) d^3 x \right\}^{\mathrm{TT}}. \tag{3.57}$$

Next, we use the equations of motion for the matter

$$\partial_b T^{ab} = 0, \tag{3.58}$$

where the use of partial derivatives is warranted since we are neglecting the self-gravity of the source. This leads to conservation laws for energy and momentum (we explore this in more detail in Chapter 4)

$$\partial_t T^{00} + \partial_j T^{0j} = 0 \quad \text{energy conservation,} \tag{3.59}$$

$$\partial_t T^{j0} + \partial_k T^{jk} = 0 \quad \text{momentum conservation,} \tag{3.60}$$

where we have used $x^0 = ct$ (with $c = 1$ throughout the calculation). After some algebra, using the fact that we can work in a global inertial frame, we find that

$$\partial_t^2 T^{00} x^j x^k = \partial_l \partial_m (T^{lm} x^j x^k) - 2\partial_l (T^{lj} x^k + T^{lk} x^j) + 2T^{jk}. \tag{3.61}$$

This relation is extremely useful because it enables us to deduce the components of the stress–energy tensor from various divergences. It is a handy trick since, if we perform a volume integration, the various divergence terms vanish (as long as we can ignore surface terms), and we are left with

$$2 \int T^{jk} d^3x = \int \partial_t^2 T^{00} x^j x^k d^3x. \tag{3.62}$$

Hence, (3.57) reduces to

$$\begin{aligned}
\bar{h}^{\text{TT}jk}(t, \boldsymbol{x}) &= \left\{ \frac{2}{r} \int T^{00}_{,00} x'^{j} x'^{k} d^3x' \right\}^{\text{TT}}_{t'=t-r} \\
&= \left\{ \frac{2}{r} \partial_t^2 \int T^{00} x'^{j} x'^{k} d^3x' \right\}^{\text{TT}}_{t'=t-r}.
\end{aligned} \tag{3.63}$$

If we now define the *mass quadrupole moment* of the source as

$$M_{jk} = \int \rho x_j x_k d^3x, \tag{3.64}$$

and use the fact that $T^{00} = \rho$ in any nearly Newtonian situation, we see that we have

$$\bar{h}^{\text{TT}}_{jk} = \frac{2}{r} \ddot{M}^{\text{TT}}_{jk}. \tag{3.65}$$

It is customary to express the final result in terms of the *reduced quadrupole moment*, defined by

$$\mathcal{I}_{jk} = \int \rho \left(x_j x_k - \frac{1}{3} r^2 \delta_{jk} \right) d^3x. \tag{3.66}$$

Simply noting that $M_{jk}^{\mathrm{TT}} = \mathcal{I}_{jk}^{\mathrm{TT}}$ (since the trace is removed by the TT operation), and reinstating G and c, the final result can be written

$$\bar{h}_{jk}^{\mathrm{TT}} = \frac{2G}{rc^4}\ddot{\mathcal{I}}_{jk}^{\mathrm{TT}}(t-r). \tag{3.67}$$

This is the *quadrupole formula*. As we will see later, it forms the basis for many useful gravitational-wave estimates.

Perhaps the most important insight we gain from this discussion is that the amount of gravitational waves that a source emits depends on how the involved masses accelerate. The difference between initial and final state (say in terms of gravitational binding energy) makes no difference. In gravitational-wave physics the destination is less important than the route you take to reach it.

3.6   The energy carried by gravitational waves

It is obviously important to estimate the rate at which energy is carried away from a system. However, it is not at all trivial to quantify the energy associated with gravitational waves. In fact, the issue was quite contentious (Kennefick, 2007), until it was finally resolved in the late 1950s.

On dimensional grounds one would expect the energy per unit volume to be proportional to the square of the time derivative of the gravitational-wave field; recall (1.3). As we are using geometric units, we should have

$$\left(\frac{dE}{d^3x}\right)_{\mathrm{GW}} \sim \frac{\mathrm{g}}{\mathrm{cm}^3} \sim \frac{1}{\mathrm{cm}^2} \sim \frac{1}{\mathrm{s}^2} \sim \dot{h}^2. \tag{3.68}$$

Given this, one ought to be able to formulate a *gravitational-wave stress-energy tensor* that is quadratic in the wave amplitude. However, there is a problem. According to the equivalence principle one can always find a local inertial frame (carried along by an observer) in which the gravitational field vanishes. Does this mean that gravitational waves do not carry energy, or even worse, that they do not even exist? Not at all; it is simply an indication that we cannot localize the effect of the wave, or the energy it carries. This should not come as a great surprise. We have already seen that gravitational waves represent tidal effects that can only be measured through the relative influence on two (or more) bodies.

The trick to defining a stress-energy tensor for gravitational waves is to average over several wavelengths. To see how this works, let us extend the deviation away from flat space to higher orders. As before, consider $g_{ab} = \eta_{ab} + h_{ab}$ but now let

$$h_{ab} = \epsilon h_{ab}^{(1)} + \epsilon^2 h_{ab}^{(2)} + \mathcal{O}\left(\epsilon^3\right), \tag{3.69}$$

where ϵ is a bookkeeping parameter (that we will set to unity later), and $h_{ab}^{(1)}$ is the linear wave from before. If we plug this expression into the Einstein equations in vacuum we get

$$G_{ab} = \overset{(0)}{G}_{ab}(\eta_{cd}) + \overset{(1)}{G}_{ab}(h_{cd}) + \overset{(2)}{G}_{ab}(h_{cd}) + \mathcal{O}(\epsilon^3) = 0. \tag{3.70}$$

The notation is a bit clumsy, but we want to distinguish between the expansion in powers of ϵ and the different contributions to the Einstein tensor (the label on top of the quantities), where the second term on the right-hand side collects all linear terms in h_{cd}, while the third term contains all quadratic ones. Now, the first term vanishes because background spacetime is flat and we already know that the linear term leads to a wave equation for $h_{ab}^{(1)}$. At second order in ϵ we have two contributions. These lead to

$$\overset{(1)}{G}_{ab}(\overset{(2)}{h}_{cd}) = -\overset{(2)}{G}_{ab}(\overset{(1)}{h}_{cd}) = \frac{8\pi G}{c^4} t_{ab}. \tag{3.71}$$

In this equation, the right-hand side effectively acts in the same way as a matter source with stress–energy tensor t_{ab}. The waves at order ϵ generate curvature at order ϵ^2. We see that t_{ab} is symmetric and if h_{ab} satisfies the wave equation, then $\partial^a t_{ab} = 0$. Hence, it may be tempting to interpret t_{ab} as the gravitational-wave stress-energy tensor, but we have to be a little bit more careful than that. One problem is that t_{ab} is not gauge invariant. Indeed, we would not expect it to be since we have already argued that there can be no local notion of energy for the gravitational field. However, if we average t_{ab} over a small volume, rather than evaluating it at a point, then we do get a meaningful measure. And it turns out to be gauge-invariant, as well. In view of this, we define

$$T_{ab}^{\text{GW}} = \langle t_{ab} \rangle = -\frac{c^4}{8\pi G} \left\langle \overset{(2)}{G}_{ab} \right\rangle = -\frac{c^4}{8\pi G} \left\langle \overset{(2)}{R}_{ab} - \frac{1}{2}\eta_{ab}\overset{(2)}{R} \right\rangle, \tag{3.72}$$

where the angle brackets indicate averaging over several wavelengths. At this point we can abandon the messy notation, as we only need to keep track of the linear wave contribution. Working out the quadratic contribution to the Ricci tensor we get

$$\begin{aligned}
\overset{(2)}{R}_{ab} = \frac{1}{2}\Bigg[&\frac{1}{2}\partial_a h_{cd}\partial_b h^{cd} + h^{cd}\left(\partial_a\partial_b h_{cd} - \partial_b\partial_d h_{ca} - \partial_a\partial_d h_{cb} + \partial_c\partial_d h_{ab}\right) \\
&+ \partial^d h_b^c\left(\partial_d h_{ca} - \partial_c h_{da}\right) + \partial_d h^{cd}\left(\partial_c h_{ab} - \partial_b h_{ca} - \partial_a h_{cb}\right) \\
&+ \frac{1}{2}\partial^c h\left(\partial_a h_{cb} + \partial_b h_{ca} - \partial_c h_{ab}\right)\Bigg].
\end{aligned} \tag{3.73}$$

This expression may look complicated but it simplifies dramatically once we perform the averaging. Crucially, this step allows us to 'integrate by parts'.[3] As an example, once we use this trick the second term in the bracket becomes

$$h^{cd}\partial_a\partial_b h_{cd} \longrightarrow -\partial_a h_{cd}\partial_b h^{cd} + \text{discarded surface terms}. \qquad (3.74)$$

This term obviously combines with the first one. Remarkably, this is all we are left with once we impose the condition for TT-gauge. That (3.73) collapses completely is easy to see. We first of all have $\bar{h} = -h = 0$, which removes the last few terms in the expression. Most of the other terms vanish because of the gauge condition (which now holds also for h_{ab}), although in some cases we need to integrate by parts to see this. Finally, we have

$$\partial^d h_b{}^c \partial_d h_{ca} \longrightarrow h_b{}^c \Box h_{ca} = 0. \qquad (3.75)$$

The same arguments show that the contribution to (3.72) from the Ricci scalar also vanishes.

At the end of the day, we are left with a neat result (usually referred to as the Isaacson stress-energy tensor (Isaacson, 1968b))

$$T_{ab}^{GW} = \frac{c^4}{32\pi G} \left\langle \partial_a \bar{h}_{cd}\partial_b \bar{h}^{cd} \right\rangle. \qquad (3.76)$$

Noting that the time components of the wave field can be set to 0, we arrive at the commonly used form for the gravitational-wave stress-energy tensor:

$$T_{ab}^{GW} = \frac{c^4}{32\pi G} \left\langle \partial_a h_{ij}^{TT}\partial_b h^{TTij} \right\rangle. \qquad (3.77)$$

This formula localizes the gravitational-wave energy to within a wavelength, but it cannot provide more specific information. For example, we cannot say whether the energy is associated with a wave peak or a wave trough. This leads to obvious trouble if we want to discuss the backreaction of the waves on a compact source. Consider, for example, a neutron star which radiates at (say) 1 kHz. The wavelength would then be of order 50 km, and since the radius of the star is about 10 km we cannot meaningfully associate the emission of the wave with a specific point in the star.

---

[3] As we are integrating over a volume with a boundary we can use integration by parts for spatial derivatives ∂_i and discard all surface terms (assuming periodic boundary conditions as would be relevant if we average over one or several wave cycles). Moreover, we know that the time dependence of h_{ab} is in terms of retarded time. If we, as an example, assume the waves are propagating in the z-direction, then we have $h_{ab}(x_0 - z)$ with $x_0 = ct$. This means that $\partial_0 h_{ab} = -\partial_z h_{ab}$. In other words, we can replace a time derivative by a spatial one, integrate by parts, and then reinstate the time-derivative. The upshot of this argument is that time and space derivatives are on equal footing in the average even though we are only working out a spatial integral.

Suppose we want to work out the energy in terms of the two wave polarizations. One can readily show that the above expression leads to

$$T_{ab}^{\text{GW}} = \frac{1}{16\pi} \langle \partial_a h_+ \partial_b h_+ + \partial_a h_\times \partial_b h_\times \rangle . \tag{3.78}$$

Let us assume that we are interested in a wave moving in the z-direction (in a local coordinate system such that the z-direction points away from the source). The energy flux then follows from (recall the comment on the derivatives from the previous footnote)

$$T_{0z}^{\text{GW}} = -T_{00}^{\text{GW}} = -\frac{1}{16\pi} \langle \dot{h}_+^2 + \dot{h}_\times^2 \rangle . \tag{3.79}$$

Assuming that the wave is monochromatic with frequency ω, the energy flux is

$$\mathcal{F} = -T_{0z}^{\text{GW}} = \frac{\omega^2}{16\pi} \langle h_+^2 + h_\times^2 \rangle . \tag{3.80}$$

Finally taking $h_+ \sim h_\times \sim h \sin \omega (t - z)$ and using $\langle \sin^2 \omega (t - z) \rangle = 1/2$, we arrive at

$$\dot{E} = \frac{h^2 \omega^2}{16\pi} , \tag{3.81}$$

and if we (finally) integrate over a sphere with radius d, we get

$$\dot{E} = \frac{\omega^2 d^2}{4} h^2 \rightarrow |\dot{h}|^2 = \frac{4G}{c^3 d^2} \dot{E} . \tag{3.82}$$

This is the flux formula we used to obtain our first gravitational-wave estimates in Chapter 1. After a considerable amount of work, we have managed to derive what may have seemed like a fairly simple result at the time. In fact, you could argue that the result follows more or less directly from dimensional analysis. The detailed arguments are needed to provide the correct numerical prefactor.

3.7   The radiation reaction force

Up to this point, we have discussed the energy loss in terms of the radiation that reaches a distant observer. We never required information about the effect that the emission of the waves might have on the source. We are, in fact, fortunate to be able to discuss the problem in this somewhat indirect way, because it is far from easy to account for the gravitational-wave back reaction. This is more or less obvious since (i) gravitational waves typically have wavelengths larger than the size of the radiating system, and (ii) we can only localize the energy carried by the waves to one wavelength or so.

Despite these conceptual difficulties we need to be able to model the effect the waves have on a source. This would certainly be the case if we wanted to describe a system that evolves as the radiation is emitted, i.e. in situations where averaging over one wavelength may not be enough. Typical astrophysical systems where we need this kind of modelling would be a supernova core collapse leading to the formation of a black hole or a neutron star, an eccentric orbit inspiral of a small body into a rapidly rotating black hole, and the growth of unstable oscillations of a rotating neutron star.

We can make progress on this tricky problem by expressing the radiation reaction acting on a body in terms of a local (potential force), F^i. To work this out, we first of all need to connect our result for the gravitational-wave energy with the quadrupole formula. Again taking the local direction of wave propagation to be the z-direction, the energy that flows through an area element dA of a sphere with radius r is

$$\frac{dE}{dtdA} = T_{0z}^{\mathrm{GW}} = -\frac{1}{c}T_{00}^{\mathrm{GW}} = -\frac{c^3}{32\pi G}\left\langle \dot{h}_{ij}^{\mathrm{TT}}\dot{h}^{\mathrm{TT}ij}\right\rangle = -\frac{G}{8\pi r^2 c^5}\left\langle \dddot{I}_{ij}^{\mathrm{TT}}\dddot{I}^{\mathrm{TT}ij}\right\rangle. \qquad (3.83)$$

The energy radiated into a solid angle $d\Omega$ is then

$$\frac{dE}{dtd\Omega} = -\frac{G}{8\pi c^5}\left\langle \dddot{I}_{ij}^{\mathrm{TT}}\dddot{I}^{\mathrm{TT}ij}\right\rangle. \qquad (3.84)$$

In order to integrate this over the sphere we need to account for the fact that

$$I_{ij}^{\mathrm{TT}} = \left(\perp_{ik}\perp_{jl} - \frac{1}{2}\perp_{ij}\perp_{kl}\right)I^{kl}, \qquad (3.85)$$

where we recall that $\perp_{ij} = \delta_{ij} - n_i n_j$ with n^i a unit vector aligned with the wave direction. It is also worth noting that, since I_{ij} is traceless, we have $\perp_{ij} I^{ij} = -n_i n_j I^{ij}$. We now have

$$\frac{dE}{dtdA} = -\frac{G}{4\pi c^5}\left\langle \frac{1}{2}\dddot{I}_{ij}\dddot{I}^{ij} - n_i n_j \dddot{I}_{ik}\dddot{I}^j_k + \frac{1}{4}n_i n_j n_k n_l \dddot{I}^{ij}\dddot{I}^{kl}\right\rangle. \qquad (3.86)$$

In order to integrate over the angles we need

$$\int d\Omega = 4\pi, \qquad (3.87)$$

$$\int n_i n_j d\Omega = \frac{4\pi}{3}\delta_{ij}, \qquad (3.88)$$

and

$$\int n_i n_j n_k n_l d\Omega = \frac{4\pi}{15}\left(\delta_{ij}\delta_{kl} + \delta_{ik}\delta_{jl} + \delta_{il}\delta_{jk}\right). \qquad (3.89)$$

Putting this all together we arrive at the rate at which energy is lost by the system

$$\mathcal{F} = -\frac{dE^{GW}}{dt} = \frac{G}{5c^5}\left\langle \dddot{\mathcal{I}}_{ij}\dddot{\mathcal{I}}^{ij}\right\rangle. \tag{3.90}$$

We want to turn this result into a force F^i such that

$$\int F_i v^i \, dt = \frac{G}{5c^5}\int \dddot{\mathcal{I}}_{ij}\dddot{\mathcal{I}}^{ij} \, dt, \tag{3.91}$$

where v^i represents the velocity of (say) a local fluid element or an individual particle. Integrating by parts (twice) and neglecting boundary terms, this leads to

$$\int F_i v^i \, dt = \frac{G}{5c^5}\int \dot{\mathcal{I}}^{ij}\mathcal{I}^{(5)}_{ij} \, dt, \tag{3.92}$$

where the (5) indicates five time derivatives. If we consider a small localized body with mass m, then

$$\dot{\mathcal{I}}^{ij} = m\frac{d}{dt}\left(x^i x^j - \frac{1}{3}r^2\delta^{ij}\right) = m(v^i x^j + x^i v^j) \tag{3.93}$$

(assuming that the distance r^2 is constant). Making use of this (and the symmetry of $\mathcal{I}^{ij}$) we get

$$\int F_i v^i \, dt = \frac{2G}{5c^5}\int m v^i x^j \mathcal{I}^{(5)}_{ij} \, dt, \tag{3.94}$$

from which we read off the desired result

$$F_i = \frac{2G}{5c^5} m x^j \mathcal{I}^{(5)}_{ij}. \tag{3.95}$$

Taking one final step, we can express the result in terms of a radiation-reaction potential (Burke and Thorne, 1970)

$$\phi^R = \frac{G}{5c^5}\mathcal{I}^{(5)}_{jk} x^j x^k, \tag{3.96}$$

such that

$$F_i = m\partial_i\phi^R. \tag{3.97}$$

This is the expression we need.

It is important to keep in mind that, while it is consistent with the rate of energy loss we predict from the quadrupole formula, the force we have identified only (really) works in an averaged sense.

3.8   The radiated angular momentum

Given a workable expression for the force gravitational waves exert on a source (albeit only in an averaged sense), it is a straightforward exercise to work out the rate at which angular momentum is radiated. We only have to recall that the torque exerted by a force is given by

$$\tau^i = \epsilon^{ijk} x_j F_k.\tag{3.98}$$

In our case, we have the instantaneous torque

$$\tau^i = \frac{2G}{5c^5} m \epsilon^{ijk} x_j x^l \mathcal{I}_{kl}^{(5)}.\tag{3.99}$$

However, in the TT-gauge we have (again, for a small body like a point particle or an individual fluid element)

$$m x_j x_k = \mathcal{I}_{jk},\tag{3.100}$$

so, in fact, we have

$$\tau^i = \frac{2G}{5c^5} \epsilon^{ijk} \mathcal{I}_j^l \mathcal{I}_{kl}^{(5)}.\tag{3.101}$$

Integrating this expression by parts twice (ignoring boundary terms, as before) we arrive at

$$\tau^i = \frac{2G}{5c^5} \epsilon^{ijk} \overset{..}{\mathcal{I}}_j^l \overset{...}{\mathcal{I}}_{kl}.\tag{3.102}$$

From this we learn that the rate at which gravitational waves carry angular momentum away from a source is given by

$$\mathcal{G}^i = -\left\langle \tau^i \right\rangle = -\frac{2G}{5c^5} \left\langle \epsilon^{ijk} \overset{..}{\mathcal{I}}_j^l \overset{...}{\mathcal{I}}_{kl} \right\rangle.\tag{3.103}$$

We will make use of this result later.

3.9   A stab at perturbation theory

So far we have assumed that the background gravitational field is weak enough that it can be well approximated by flat space. This is, obviously, an enormous restriction on the physics. Fortunately, it is easy to do better. Moreover, we can show that all the concepts we have introduced extend in a natural way to a more general curved spacetime setting.

As a first step towards generalizing the description, we consider gravitational waves that propagate through the Universe, and assume that the wavelength of the waves is such that $\lambda \ll R$, with R the radius of background curvature of spacetime. To isolate the waves, we need to make a distinction between this *background* curvature and that associated with the gravitational waves. We can do this by averaging over several gravitational wavelengths. This provides us with a measure of the background curvature

$$g_{ab}^{B} = \langle g_{ab} \rangle, \tag{3.104}$$

where g_{ab}^{B} is a solution to the likewise averaged Einstein equations.

Now write the full metric as

$$g_{ab} = g_{ab}^{B} + h_{ab}, \tag{3.105}$$

where h_{ab} represents a short wavelength ripple on the slowly varying (now curved) background (the separation is similar to that of the small dimples on the curved surface of an orange, say). With this decomposition, and adding the assumption[4] that the waves are weak $|h_{ab}| \ll 1$, we can derive the linearized Einstein equations

$$\underbrace{h_{ab|c}{}^{|c} + g_{ab}^{B} h^{cd}{}_{|cd} - 2h_{c(a}{}^{|c}{}_{|b)}}_{\sim 1/\lambda^2} + \underbrace{2R_{cadb}^{B} h^{cd} - 2R_{c(a}^{B} h_{b)}^{c}}_{\sim 1/R^2} = -16\pi T_{ab}, \tag{3.106}$$

where $|$ denotes the covariant derivative (replacing the semi-colon) with respect to the defined background metric g_{ab}^{B} (defined in such a way that $g_{ab|c}^{B} = 0$). Also recall that the parentheses indicate the symmetric part of a tensor; see Chapter 2. In this case, this means that

$$R_{c(a}^{B} h_{b)}^{c} = \frac{1}{2} \left[R_{ca}^{B} h_{b}^{c} + R_{cb}^{B} h_{a}^{c} \right]. \tag{3.107}$$

Finally, the stress-energy tensor contribution on the right-hand side of (3.106) should be taken to represent the wave-generating motion of any matter source. That is, it contains only the part of the complete T_{ab} that is due to asymmetric matter motion.

---

[4] This requirement is not as trivial as it may seem. The magnitude of a tensor component depends on the chosen coordinates. We take $|h_{ab}| \ll 1$ to mean that there exists a coordinate system in which the assumption holds in a suitably large spacetime region.

Since we have assumed that $\lambdabar \ll R$ the above relation immediately simplifies to

$$h_{ab|c}{}^{|c} + g^B_{ab} h^{cd}{}_{|cd} - 2h_{c(a}{}^{|c}{}_{|b)} = -16\pi T_{ab}. \tag{3.108}$$

We can simplify this expression further by introducing the trace-reversed metric perturbation

$$\bar{h}_{ab} = h_{ab} - \frac{1}{2} g^B_{ab} h^c{}_c, \tag{3.109}$$

and working in Lorenz gauge, which is now defined by

$$\bar{h}_{ab}{}^{|b} = 0. \tag{3.110}$$

Thus, we arrive at the simple (and familiar) result

$$\Box \bar{h}_{ab} = -16\pi T_{ab}. \tag{3.111}$$

This shows that one can, quite generally, think of gravitational waves as a field living in a given curved background spacetime. The one key difference from the wave equation we derived earlier is that the wave operator is no longer that of flat space. Instead, we have

$$\Box = g^{Bab} \nabla_a \nabla_b, \tag{3.112}$$

where ∇_a is the covariant derivative associated with g^B_{ab}.

The previous results for weak fields follow immediately from the above equations, but the general framework is clearly more powerful. Of course, there is an obvious question: What is the appropriate choice of background metric g^B_{ab}? The answer depends on the physics under consideration. In the first instance, it makes sense to explore how the use of symmetries allows us to build suitable solutions to Einstein's field equations. We are led to consider black holes, relativistic stars, and (on a much larger scale) the evolution of the entire cosmos.

4

From black holes to stars and the Universe at large

The scope of gravitational-wave astronomy should be apparent from the discussion of the key aspects of the theory. Moreover, it is clear what we have to do if we want to model specific gravitational-wave sources. We have seen that the waves represent small scale variations of a smooth background spacetime and we have connected the origin of the waves to asymmetric motion of a matter source. However, these results are extremely general. In order to make sure we do not take on more than we can handle, we need to focus on more specific situations. This leads us to two issues. First of all, we can make the notion of a given background spacetime more precise by considering exact solutions to the Einstein field equations. Secondly, we can take a closer look at the relation between the stress-energy tensor and the motion of matter. The first issue involves introducing symmetries to simplify the mathematics to the point where we can solve the equations. The second issue is a bit messier as it involves introducing a matter description, which by necessity includes some understanding of the relevant 'microphysics' (particle interactions, thermodynamics, and so on). Nevertheless, we can make progress on modelling problems on a vast range of scales, from single astrophysical objects, like black holes and stars, to the entire cosmos.

4.1 The Schwarzschild solution

Given the complexity of the equations of general relativity, Einstein did not think it would be possible to find exact solutions. However, much to his surprise, the first such solution was discovered by Karl Schwarzschild only two months after the publication of the theory (see Schwarzschild (1999) for a translated version). Today we have a large number of analytical solutions, and there are systematic ways of using computer algebra to generate new ones (Stephani *et al.*, 2009). Unfortunately, very few of these solutions are of astrophysical interest. In fact, Schwarzschild's original solution remains one of the most interesting as it describes the exterior of both non-rotating black holes and stars.

Any attempt to find an exact solution to the relativistic field equations relies on simplifications, typically associated with assumed symmetries. For example, in order to

Gravitational-Wave Astronomy: Exploring the Dark Side of the Universe. Nils Andersson, Oxford University Press (2020).
© Nils Andersson. DOI: 10.1093/oso/9780198568032.001.0001

be in equilibrium an object must be either static or stationary. Intuitively, we know what these words mean. The difference between the two cases should become clear under time inversion. In the static case—when there is no motion—the solution looks the same, but a stationary configuration looks different if we 'run the film backwards' (a rotating star is seen to spin the other way).

In order to make these ideas precise, we define a stationary spacetime to be one for which there exists a set of coordinates x^a such that

$$\frac{\partial g_{ab}}{\partial x^0} \overset{*}{=} 0, \tag{4.1}$$

where x^0 is a time-like coordinate, and the asterisk indicates that the statement is true only in this particular coordinate system. A static spacetime requires the additional assumption that the line element

$$ds^2 = g_{ab}dx^a dx^b \tag{4.2}$$

is invariant under the change $x^0 \rightarrow -x^0$ (time reversal). This indicates that the metric must be such that

$$g_{0j} \overset{*}{=} 0, \quad j = 1 - 3. \tag{4.3}$$

If we, in addition, require the solution to be spherically symmetric, then there should exist coordinates $\{t, r, \theta, \varphi\}$ such that the $t = $ constant, $r = $ constant surfaces have the geometry of a sphere. Without loss of generality, one can show that such a metric can be written

$$ds^2 = -e^\nu dt^2 + e^\lambda dr^2 + r^2 (d\theta^2 + \sin^2\theta\, d\varphi^2). \tag{4.4}$$

Let us see if we can find a solution to the vacuum Einstein equations for a metric of this form. That is, we are trying to solve (2.117) in the absence of matter. We can read off the non-zero metric components from the line element and use the results to work out the Einstein tensor. After some algebra we arrive at three independent equations,

$$G_{00} = \frac{e^{\nu-\lambda}}{r^2}\left(r\lambda' + e^\lambda - 1\right) = 0, \tag{4.5}$$

$$G_{01} = \frac{\dot\lambda}{r} = 0, \tag{4.6}$$

$$G_{11} = \frac{1}{r^2}\left(r\nu' - e^\lambda + 1\right) = 0, \tag{4.7}$$

where a dot represents a time-derivative and a prime is a derivative with respect to r. The second of these equations shows that λ must be a function only of r, which means that we can integrate the first equation to get

$$e^{\lambda} = \left(1 - \frac{2GM}{c^2 r}\right)^{-1},$$ (4.8)

where we have (suggestively) introduced the integration constant $-2GM/c^2$. Combining the first and the third equations, we see that

$$\lambda' + v' = 0 \quad \longrightarrow \quad v(t, r) = -\lambda(r) + h(t).$$ (4.9)

Finally, introducing a new time coordinate by making the change $e^{h(t)/2} dt \rightarrow dt$ we arrive at the line element

$$ds^2 = -\left(1 - \frac{2GM}{c^2 r}\right) dt^2 + \left(1 - \frac{2GM}{c^2 r}\right)^{-1} dr^2 + r^2 (d\theta^2 + \sin^2 \theta \, d\varphi^2).$$ (4.10)

This is the Schwarzschild solution. It represents a non-rotating black hole with mass M. At first sight, the solution appears to be singular at the *event horizon*, $r = 2GM/c^2$ (the so-called Schwarzschild radius). However, one can show that this is a coordinate artefact (Kruskal, 1960). It is possible to find coordinates which are perfectly regular at the horizon. One particular such alternative—*isotropic coordinates*—follows from the replacement (in geometric units)

$$r = r' \left(1 + \frac{M}{2r'}\right)^2.$$ (4.11)

This leads to

$$ds^2 = -\left(\frac{1 - M/2r'}{1 + M/2r'}\right)^2 dt^2 + \left(1 + \frac{M}{2r'}\right)^4 \left[dr'^2 + r'^2 \left(d\theta^2 + \sin^2 \theta \, d\varphi^2\right)\right],$$ (4.12)

which is only singular at the origin (this is a real singularity).

The Schwarzschild solution (4.10) is also relevant in the exterior of a non-spinning star. However, in order to describe the star's interior we need to account for the presence of matter on the right-hand side of the Einstein equations. We need to explore relativistic fluid dynamics.

4.2 Relativistic fluids

In order to model stars we need to understand how fluids move in a curved spacetime. We need to describe how the matter affects the geometry and, in turn, how spacetime feeds back on the motion. To illustrate this problem, let us assume that the matter is represented by the perfect fluid stress–energy tensor

$$T^{ab} = (p + \varepsilon) u^a u^b + p g^{ab},$$ (4.13)

where ε is the energy density, p is the isotropic pressure, and u^a is the four-velocity associated with the (averaged) flow of the fluid elements. The four-velocity has a double role as it is also taken to be the observer that measures the fluid properties. In order to describe the motion of the fluid we need four equations, representing the conservation of energy and momentum. These follow from

$$\nabla_a T^{ab} = 0, \tag{4.14}$$

which leads to

$$(p+\varepsilon)\, u^a \nabla_a u^b + u^b \nabla_a \left[(p+\varepsilon)\, u^a\right] + g^{ab} \nabla_a p = 0. \tag{4.15}$$

It is useful to decompose this expression into a component aligned with the flow and an orthogonal part. The component along the four-velocity (contract the equation with u_b) becomes

$$u^a \nabla_a \varepsilon + (p+\varepsilon)\, \nabla_a u^a = 0. \tag{4.16}$$

In order to interpret this result we need a little bit of thermodynamics. Assuming that we are dealing with a fluid composed of a single-particle species, we have $\varepsilon = \varepsilon(n)$, where n is the particle number density. The associated chemical potential (essentially the energy cost associated with adding a single particle to the system; see Andersson and Comer (2007))

$$\mu = \frac{d\varepsilon}{dn} \tag{4.17}$$

then satisfies the thermodynamics relation

$$p + \varepsilon = n\mu. \tag{4.18}$$

Making use of these relations, we find that (4.16) leads to

$$\nabla_a \left(n u^a\right) = 0. \tag{4.19}$$

This is simply the statement that the particle flux is conserved. The model does not account for particle creation/destruction.

Next we use the projection $\perp_{ab} = g_{ab} + u_a u_b$ to obtain the component orthogonal to the four-velocity. This leads to

$$(p+\varepsilon)\, \dot{u}_c + \perp^a_c \nabla_a p = 0, \tag{4.20}$$

where the four-acceleration is given by

$$\dot{u}^a = \frac{du^a}{d\tau} = u^b \nabla_b u^a. \tag{4.21}$$

These are the relativistic Euler equations. They simply show that the fluid accelerates (deviates from geodesic motion) as a result of pressure gradients. The intuition is exactly the same as in Newtonian physics.

4.3 How to build a star

Let us now show how we build a non-rotating star. Since the system is static, the left-hand side of the Einstein equations will remain as in the Schwarzschild problem. We only need to provide the right-hand side. That is, we need the four-velocity. In the static case, when there is no fluid motion, the only non-trivial component is u^0 and we have

$$-1 = u^a u_a = g_{ab} u^a u^b = g_{00} \left(u^0 \right)^2 = -e^\nu \left(u^0 \right)^2 \quad \longrightarrow \quad u^0 = e^{\nu/2}. \tag{4.22}$$

For a perfect fluid, we then have, using the metric inferred from (4.4),

$$T_{00} = \varepsilon e^\nu, \tag{4.23}$$
$$T_{11} = p e^\lambda. \tag{4.24}$$

Combining these results with the Einstein tensor components from the previous section we have, first of all,

$$G_{00} = 8\pi T_{00} \quad \longrightarrow \quad \frac{1}{r^2} \frac{d}{dr} \left[r \left(1 - e^{-\lambda} \right) \right] = 8\pi\varepsilon. \tag{4.25}$$

Inspired by the Schwarzschild result, we define

$$m(r) = \frac{r}{2} \left(1 - e^\lambda \right), \tag{4.26}$$

which means that

$$m' = 4\pi\varepsilon r^2 \quad \longrightarrow \quad m = 4\pi \int \varepsilon r^2 dr. \tag{4.27}$$

In effect, $m(r)$ represents the mass-energy contained inside radius r. Next we have

$$G_{11} = 8\pi T_{11} \quad \longrightarrow \quad \nu' = \frac{2e^\lambda}{r^2} \left(m + 4\pi p r^3 \right). \tag{4.28}$$

These equations are known as the Tolman–Oppenheimer–Volkoff equations (Tolman, 1939; Oppenheimer and Volkoff, 1939) and, once they are supplemented by an equation of state $p(\varepsilon)$—or, equivalently, $\varepsilon(n)$—they allow us to build models of relativistic stars. It is also useful to note that the radial component of the fluid momentum equation (4.20) leads to

$$p' = -\frac{1}{2}(p+\varepsilon)v'. \tag{4.29}$$

This is the relativistic version of a familiar result from Newtonian gravity—the equation for hydrostatic equilibrium. As this is an important connection, it is worth considering it in more detail.

4.4 The Newtonian limit

It is often useful to build intuition about possible gravitational-wave sources in the simplified context of Newtonian gravity, making use of the quadrupole formula. However, while doing this one must keep an eye on the simplifications assumed in the analysis.

Let us consider the weak-field limit of the relativistic fluid equations. As in Chapter 3, we can consider the gravitational field as a small deviation away from flat space. In fact, even though we are not interested in the dynamical aspects of the problem, it is useful to take the derived wave equation (3.38) as our starting point. Noting that

$$|T_{00}| \approx \varepsilon \gg |T_{0i}| \gg |T_{ij}| \approx p \tag{4.30}$$

in the weak-field regime, and that $\varepsilon \approx \rho = mn$, we focus on

$$\Box \bar{h}_{00} = -16\pi\, T_{00} \approx -16\pi\rho. \tag{4.31}$$

We also have

$$\Box \bar{h} = -\Box h = -16\pi\, T \approx 16\pi\rho, \tag{4.32}$$

so it follows that

$$\Box h_{00} \approx -8\pi\rho. \tag{4.33}$$

For low velocities, $v \ll c$, this simplifies further. If $\partial_t \sim v\partial_j$ then

$$\partial_t \ll \partial_j \quad \longrightarrow \quad \Box \approx \nabla^2 \tag{4.34}$$

(where ∇^2 is the familiar three-dimensional Laplacian), and we arrive at

$$\nabla^2 h_{00} \approx -8\pi G\rho \qquad (4.35)$$

(where we have reinstated the gravitational constant, for convenience). Comparing this to the Poisson equation for the gravitational potential Φ in Newton's theory

$$\nabla^2 \Phi = 4\pi G\rho, \qquad (4.36)$$

we identify

$$h_{00} \approx -2\Phi, \qquad (4.37)$$

and find that (to leading order) we have the line element

$$ds^2 = -c^2 d\tau^2 = -c^2 \left(1 + \frac{2\Phi}{c^2}\right) dt^2 + \eta_{ij} dx^i dx^j, \qquad (4.38)$$

where x^i ($i = 1 - 3$) are Cartesian coordinates and η_{ij} is the flat three-dimensional metric.

In essence, the Newtonian limit consists of writing the general relativistic field equations to leading order in an expansion in powers of the speed of light c. The Newtonian equations are obtained in the limit where $c \to \infty$.

Let us now see what this implies for the equations of fluid dynamics. With τ the proper time measured along a fluid element's worldline, the curve it traces out can be written

$$x^a(\tau) = \{ct(\tau), x^i(\tau)\}. \qquad (4.39)$$

In order to work out the four-velocity,

$$u^a = \frac{dx^a}{d\tau}, \qquad (4.40)$$

we note that (4.38) leads to

$$d\tau^2 = \left(1 + \frac{2\Phi}{c^2} - \frac{g_{ij}v^i v^j}{c^2}\right) dt^2, \qquad (4.41)$$

with $v^i = dx^i/dt$, the usual Newtonian three-velocity of the fluid. Since the velocity is assumed to be small, in the sense that

$$\frac{|v^i|}{c} \ll 1, \qquad (4.42)$$

this leads to

$$\frac{dt}{d\tau} \approx 1 - \frac{\Phi}{c^2} + \frac{v^2}{2c^2},\tag{4.43}$$

where $v^2 = \eta_{ij} v^i v^j$, and

$$u^0 = \frac{dx^0}{d\tau} = c\frac{dt}{d\tau} \approx c\left(1 - \frac{\Phi}{c^2} + \frac{v^2}{2c^2}\right).\tag{4.44}$$

It is also easy to see that

$$u^i = \frac{dx^i}{d\tau} = v^i \frac{dt}{d\tau} \approx v^i.\tag{4.45}$$

In order to obtain the covariant components we use the metric (which is obviously diagonal). Thus, we find that

$$u_0 = g_{00} u^0 = -c\left(1 + \frac{2\Phi}{c^2}\right)\left(1 - \frac{\Phi}{c^2} + \frac{v^2}{2c^2}\right) \approx -c\left(1 + \frac{\Phi}{c^2} + \frac{v^2}{2c^2}\right),\tag{4.46}$$

and

$$u_i = v_i.\tag{4.47}$$

Note that these equations lead to

$$u^a u_a = -c^2\left(1 - \frac{\Phi}{c^2} + \frac{v^2}{2c^2}\right)\left(1 + \frac{\Phi}{c^2} + \frac{v^2}{2c^2}\right) + v^2 \approx -c^2.\tag{4.48}$$

We can now work out the Newtonian limit for the conserved particle flux

$$\nabla_a(nu^a) = 0 \longrightarrow \frac{1}{c}\partial_t\left(nu^0\right) + \nabla_i\left(nv^i\right) = 0$$

$$\longrightarrow \partial_t n + \nabla_i\left(nv^i\right) = \mathcal{O}\left(c^{-1}\right).\tag{4.49}$$

To leading order we retain the standard result

$$\partial_t n + \nabla_i\left(nv^i\right) = 0.\tag{4.50}$$

If we introduce the mass density $\rho = mn$, with m the mass per particle, we recover the continuity equation, usually taken to imply mass conservation,

$$\partial_t \rho + \nabla_i \left(\rho v^i \right) = 0, \tag{4.51}$$

If we reintroduce the basis vectors, we have

$$\partial_t \rho + \nabla \cdot \left(\rho \boldsymbol{v} \right) = 0. \tag{4.52}$$

In order to work out the corresponding limit of the Euler equations, we need the curvature contributions to the covariant derivative. However, from the definition (2.67) and the weak-field metric, we see that only g_{00} gives a nonvanishing contribution. Moreover, it is clear that

$$\Gamma^a_{bc} = \mathcal{O}(1/c^2), \tag{4.53}$$

which is why we did not need to worry about this in the case of the flux conservation. The curvature contributes at higher orders.

Explicitly, we now have

$$u^a \nabla_a u^b = u^a \partial_a u^b + \Gamma^b_{ca} u^a u^c = \frac{1}{c} u^0 \partial_t u^b + u^i \partial_i u^b + \Gamma^b_{ca} u^a u^c. \tag{4.54}$$

We only need the spatial components, so we set $b = j$ to get

$$
\begin{aligned}
u^a \nabla_a u^j &= \frac{1}{c} u^0 \partial_t u^j + u^i \partial_i u^j + \Gamma^j_{ca} u^a u^c \\
&= \partial_t v^j + v^i \partial_i v^j + c^2 \Gamma^j_{00} + \text{higher order terms} \\
&= \partial_t v^j + v^i \partial_i v^j + \frac{1}{2} \eta^{jk} \partial_k \left(\frac{2\Phi}{c^2} \right) \\
&= \partial_t v^j + v^i \partial_i v^j + \eta^{jk} \partial_k \Phi.
\end{aligned}
\tag{4.55}
$$

Finally, we need the pressure contribution. For this we note that the projection becomes

$$\perp^{ab} = g^{ab} + \frac{1}{c^2} u^a u^b, \tag{4.56}$$

in order to be dimensionally consistent. We also need $\varepsilon \gg p$. This means that we have

$$\perp^{ba} \nabla_a p \quad \longrightarrow \quad \eta^{jk} \partial_k p, \tag{4.57}$$

and we (finally) arrive at the Euler equations

$$\partial_t v^j + v^i \partial_i v^j = -\eta^{jk}\left(\frac{1}{\rho}\partial_k p + \partial_k \Phi\right),$$
(4.58)

or

$$(\partial_t + \boldsymbol{v} \cdot \nabla)\boldsymbol{v} + \frac{1}{\rho}\nabla p + \nabla\Phi = 0.$$
(4.59)

By setting $v^i = 0$ we arrive at the equation for hydrostatic equilibrium

$$\partial_j p = -\rho \partial_j \Phi,$$
(4.60)

a result that we could have obtained by taking the weak-field limit of (4.29).

4.5 Modelling the Universe

Finally, let us turn to what might be the mightiest task problem of all: to model the entire Universe. Quite naturally, we have to simplify things in a dramatic fashion before we can even begin to contemplate this problem.

We are obviously not going to try to track every single speck of dust, star, or galaxy. Instead, we focus on the gross properties and the large-scale evolution. Imagine zooming out until you reach a scale where all granularity has been smeared out. At this point the problem is closely related to that of fluid dynamics. It makes sense to—at least in the first instance—work at this averaged level and assume that the Universe is (spatially) homogeneous and isotropic. There are no preferred directions and things look pretty much the same everywhere. One can then introduce a 'fibration' of spacetime[1] associated with a four-velocity u^a and a corresponding cosmological time t.

The assumed spatial symmetry implies the Robertson–Walker line element[2]

$$ds^2 = -dt^2 + [R(t)]^2\left[\frac{dr^2}{1 - kr^2} + r^2\left(d\theta^2 + \sin^2\theta\, d\varphi^2\right)\right].$$
(4.61)

In this expression, we recognize the geometry of the two-sphere—which is natural since the assumed isotropy implies spherical symmetry with respect to every point in spacetime. The function $R(t)$, known as the *scale factor*, represents the 'size' of the Universe. Finally, the curvature constant k is $-1, 0$, or 1, depending on whether the

---

[1] It is useful to compare this approach to the 3+1 foliation used in numerical relativity; see Chapter 20.
[2] More detailed introductions to relativistic cosmology are provided by Carroll (2004) and Weinberg (2008). A less mathematical introduction is provided by Liddle (2003), while a detailed analysis of many relativistic cosmology scenarios is provided by Ellis and van Elst (1999).

Universe is open, flat or closed. As we will see later (Chapter 22), current observations favour a flat Universe. It is easy to see that, if we set $k = 0$ in (4.61) the spatial part is a flat (3D) geometry scaled by an overall (conformal) factor $R^2(t)$.

Assuming that the matter in the Universe can be described as a perfect fluid, we can work out the Einstein equations for the geometry in (4.61) and the stress-energy tensor from (2.94). This leads to the Friedmann equations

$$\left(\frac{\dot{R}}{R}\right)^2 = \frac{1}{3}(8\pi\varepsilon + \Lambda) - \frac{k}{R^2}, \tag{4.62}$$

where the dot represents a derivative with respect to t, and

$$\frac{\ddot{R}}{R} = -\frac{4\pi}{3}(\varepsilon + 3p) + \frac{\Lambda}{3}. \tag{4.63}$$

In these equations we have accounted for a *cosmological constant*, which arises if we add a term Λg_{ab} to the left-hand side of the Einstein equations. The original motivation for introducing this term was that the second equation could only lead to a static Universe if $\varepsilon + 3p = 0$. That is, if the pressure vanishes (as in the favoured dust model), then the energy density must vanish as well. The introduction of the cosmological constant resolves this conundrum. However, following the discovery that distant galaxies move away from us—the Universe is expanding (Hubble, 1929)—the use of Λ fell out of favour. Einstein is said to have referred to it as a blunder. However, it has since been reintroduced as it appears to be required to explain observations. The modern argument for its presence is that it represents the vacuum energy, although the calculation of this quantity from first principles remains a vexing issue.

We can combine the two Friedmann equations to get

$$\dot{\varepsilon} + 3(p + \varepsilon)\frac{\dot{R}}{R} = 0. \tag{4.64}$$

If we want to progress, we need to provide an equation of state that relates p and ε. In cosmology, it is usually assumed that this relation is linear. That is, we have

$$p = w\varepsilon, \tag{4.65}$$

where w is constant. This leads to

$$\varepsilon \sim R^{-3(1+w)}, \tag{4.66}$$

and we can identify three distinct cases. First we may consider a universe consisting of dust, i.e. take $w = 0$. Then $\varepsilon \sim R^{-3}$, which simply states that the density decreases as the volume of the universe increases. The next option is to assume a universe dominated by radiation, in which case $w = 1/3$. This leads to $\varepsilon \sim R^{-4}$; i.e. the radiation energy decays faster as the universe expands. Finally, if the cosmological constant (read: the vacuum energy) dominates, then $w = -1$ and $\varepsilon \sim$ constant. In reality, all three components contribute, but the simple scalings suggest that an expanding universe will go through different eras where each term dominates in turn. The hot early Universe was dominated by radiation. At some point there was a phase-transition to our current matter-dominated era. Finally, if the Universe continues to expand there will be a late epoch when matter becomes so dilute that the cosmological constant takes over.

In order to illustrate the range of possible behaviour, it is worth solving the Friedmann equations in two simplified scenarios. Let us first assume that the Universe is flat ($k = 0$). Then we get

$$\dot{R}^2 = \frac{8\pi \varepsilon_0}{3} R^{-3(1+w)} \quad \longrightarrow \quad R(t) = R_0 \left(\frac{t}{t_0}\right)^{2/(3+3w)}, \tag{4.67}$$

where ε_0 is the current density, R_0 is the current size of the Universe, and

$$t_0 = \frac{2}{3(1+w)} \frac{1}{H_0} \tag{4.68}$$

is its 'age'. The quantity H_0 is the current value of the Hubble expansion parameter

$$H = \frac{\dot{R}}{R}, \tag{4.69}$$

one of the main 'observables' in cosmology (see Chapter 22). If matter dominates, we have $w = 0$, which leads to the Einstein–deSitter model in which

$$R(t) = R_0 \left(\frac{t}{t_0}\right)^{2/3} \quad \text{and} \quad t_0 = \frac{2}{3H_0}. \tag{4.70}$$

Until quite recently, this simple Universe—that expands uniformly forever and ends in a 'big chill'—was the favoured model.

As an alternative, consider the case where the cosmological constant term dominates. That is, let $w = -1$. This leads to

$$\dot{R}^2 = \frac{8\pi \varepsilon_0}{3} R^2 \quad \text{or} \quad \dot{R} = H_0 R \quad \longrightarrow \quad R(t) = R_0 e^{H(t-t_0)}. \tag{4.71}$$

This shows that, in a flat Universe, the cosmological constant may drive an exponential expansion. Our Universe is thought to have undergone such an era of rapid expansion in its early stages. This is known as *inflation*.

To conclude this brief initial consideration of cosmology, and prepare the ground for a discussion of gravitational waves in a cosmological setting (in Chapter 22), it is useful to introduce the critical density, ρ_c, as

$$\rho_c = \frac{3H^2}{8\pi},$$ (4.72)

along with the density parameter, Ω, such that

$$\Omega = \frac{\varepsilon}{\rho_c}.$$ (4.73)

These definitions allow us to rewrite the first Friedmann equation (4.62) as (ignoring the cosmological constant for now)

$$\Omega - 1 = \frac{k}{H^2 R^2}.$$ (4.74)

This simple relation shows that the sign of k depends on Ω (hence the notion of a critical density). If $\varepsilon > \rho_c$ then $k > 0$ and the Universe is closed. There is enough energy that the cosmic expansion will reverse at some point. Meanwhile, a flat Universe ($k = 0$) must be fine-tuned to have $\Omega = 1$ and, finally, if there is not enough energy, so that $\varepsilon < \rho_c$ then the expansion can proceed indefinitely. It is straightforward to extend this argument to account for additional energy contributions. In particular, the notion of a density ρ_{gw} associated with gravitational waves will prove useful when we consider stochastic cosmological backgrounds .

4.6   Was Einstein right?

After more than a century, general relativity remains a cornerstone of modern physics. The theory has been battered by a variety tests but always passes intact (Will, 2005). This is (obviously!) remarkable. Yet, we know that Einstein's theory must be wrong. We just do not know exactly how, or at what level, the theory breaks.

There are good reasons to believe that general relativity requires modifications in strong gravitational fields. We also know that it does not mesh with quantum theory—it is an entirely classical theory, after all—so it must also break down at small scales (high energies). As an example, high-energy corrections may help us avoid the formation of the singularities that seem inevitable when we consider black-hole interiors. Ultimately, we need a quantum theory of gravity to provide the answer. At the moment we do not have such a theory, but even incomplete constructions (at the moment, string theory (the monograph by Green *et al.* (1987) is largely outdated—see Polchinski (2005) for a modern alternative—but it still provides a motivation for the theory) and loop quantum gravity (Ashtekar and Pullin, 2017) make predictions and it makes sense to ask to what extent we can test them. At the very least, we should learn how to formulate meaningful questions.

Our understanding of the large scales of cosmology also raises a number of issues. What is the origin of the mysterious dark energy? Why is the value of the cosmological constant so remarkably small, compared to our expectations? Again, the answers may require modification(s) of Einstein's theory...

Any effort to modify general relativity poses interesting challenges. On the one hand, we do not seem to have much freedom to move. Assuming Lorentz invariance and a massless spin-2 particle (the graviton, which represents gravitational waves from the particle physics perspective; see Maggiore (2007)), Einstein's theory is unique. If we want to add physics we need to add degrees of freedom, but the changes must be compatible with the slew of tests we have already carried out. We are up against laboratory experiments, a range of solar system tests, and precise observations of binary pulsars (see Chapter 10). On the other hand, we have an embarrassment of riches. If we let imagination roam free, then we can devise a plethora of new theories, deviating from Einstein's theory in their own special ways. This makes the problem difficult, because we need to avoid confusion and contradictions. We need some kind of 'taxonomy' of alternative gravity theories, a dictionary that allows us to keep careful track of what we are testing (Berti *et al.*, 2015).

To start with, we can try to list the aspects of the theory we are challenging. If we do this, we typically end up with four main categories:

- Adding fields: This is, perhaps, the most famous (and best tested) approach to tweaking general relativity. One may, for example, couple gravity to a dynamical scalar field (Brans and Dicke, 1961). Some versions of this category of theories have multiple extra (possibly vector or tensor) fields. The fields may be passive or contribute to the dynamics, for example, leading to additional gravitational-wave polarizations.

- Introducing extra dimensions: The presence of dimensions beyond the familiar three space dimensions is a requirement of string theory. These dimensions may be 'compactified' but this does not mean they do not affect experiments. There are a number of different alternatives, including models with a single extra space dimension—brane cosmology, where our everyday reality is restricted to 3+1 dimensions while gravity acts in one additional dimension (Randall and Sundrum, 1999).

- Violating the equivalence principle: The introduction of the Einstein tensor G_{ab} (Chapter 2) ensured that the equations of general relativity are well posed. At closer inspection, this was dictated by the requirement that the stress-energy tensor must be divergence free, which follows from the (weak) equivalence principle. One can imagine theories that go beyond this (minimal) coupling between matter and geometry. However, the wriggle room in this direction is limited because the (weak) equivalence principle has been tested to very high precision.

- Breaking Lorentz invariance: This is a key ingredient already in special relativity and it is well tested by particle physics (as it is a key part of the standard model), but one may still imagine breaking this symmetry at high energies. This opens the door to alternative theories. The invariance under diffeomorphisms dictate that the graviton is massless—gravitational waves propagate at the speed of light. If this symmetry is broken, then gravity signals should move slower, and we can try to use experiments to test this.

Turning to the issue of possible additional gravitational-wave polarizations, we recall Figure 3.2 where there was no stretch or squeeze in the z-direction. This is a prediction of general relativity. It follows from the equivalence principle, which demands that gravity is a metric theory such that (i) spacetime is endowed with a symmetric metric g_{ab}, (ii) freely falling bodies move along geodesics of this metric, and (iii) according to an inertial observer, all nongravitational physics remain as in special relativity. However, as we have just mentioned, we can formulate more general metric theories of gravity. A famous example is the theory proposed by Brans and Dicke (1961), which couples the metric to a scalar field. Such scalar–tensor theories would be natural in the context of string theory or higher-dimensional models. In the most general case, all six components of R_{j0k0} may be expressed as wave polarizations (see Will (1999) for a light-touch discussion). Three of these would be transverse to the direction of propagation. Two are the quadrupole deformations we have already discussed. The third represents an axisymmetric 'breathing' mode. The remaining three polarizations are longitudinal. One induces stretching in the direction of propagation, and the remaining two are quadrupolar modes in each of the two orthogonal planes containing the direction of propagation. Out of these possibilities, scalar–tensor theory allows the standard transverse quadrupole modes and the scalar breathing mode. Given a suitable array of detectors one would hope to be able to infer the actual polarizations present in a detected wave, and hence put constraints on the theory of gravity. If one were to observe additional polarizations then general relativity would fall. Conversely, absence of the breathing mode would limit the parameters of scalar–tensor gravity.

In terms of observations, it is difficult to constrain the polarization of any incoming gravitational waves with the two LIGO instruments. As they have similar orientations— a design decision aimed at maximizing the chance of detections in the first place—little information about polarizations can be obtained from these two detectors alone. The situation changes with the addition of Virgo. With data from the three instruments, one can project the wave amplitude onto the detectors to constrain the polarization. The first such test was carried out for the black-hole merger GW170814 (Abbott *et al.*, 2017*h*), basically confirming that the signal was consistent with Einstein's theory. A similar test on signals from rotating deformed neutron stars also returned a null result (Abbott *et al.*, 2018*b*).

When it comes to the possible existence of extra dimensions, one may search for evidence that gravitational waves and photons experience different number of spacetime dimensions. If this were the case, the gravitational waves may 'leak' into the extra dimensions, effectively reducing their amplitude compared to expectations. This will, in turn, introduce a systematic error in the inferred distance. Assuming that electromagnetic signals are not similarly affected, one can use counterpart observations to constrain the extra-dimensional leakage. This idea has been tested for the neutron star merger GW170817 (Pardo *et al.*, 2018). The results indicate that we live in the expected 3+1 spacetime dimensions.

The first set of detections has also allowed us to test the notion that the graviton has mass. The idea is quite simple. If gravity is associated with a massive field, then long wavelength gravitational waves will travel more slowly than short wavelength ones. This introduces a distortion of the binary inspiral signal (it is 'squashed' as high-frequency waves try to catch up with the lower frequency ones emitted earlier), which can be constrained using matched filtering (Will, 1998). The effect is typically quantified in terms of the Compton wavelength of the graviton

$$\lambda_g = \frac{h}{m_g c}. \tag{4.75}$$

Combining data from the three events GW150914, GW151226, and GW170104 one arrives at a constraint (Abbott *et al.*, 2017*g*)

$$\lambda_g > 1.6 \times 10^{13} \text{ km}, \tag{4.76}$$

which improves bounds set in the 1980s using solar-system data by about a factor of 6. However, an update on the latter leads to a new bound about one order of magnitude more stringent than the gravitational-wave one (Will, 2018). The situation should improve with more detections, but progress may be slow. For ground-based interferometers, the bound scales roughly as (Will, 2018)

$$\lambda_g \sim S_0^{-1/4} f_0^{-1/3} \mathcal{M}^{11/12}, \tag{4.77}$$

where S_0 and f_0 represent the noise 'floor' at the frequency of the sensitivity sweet spot of the detector, respectively, and $\mathcal{M}$ is the 'chirp mass'. It is worth noting that the result is largely independent of the distance. Basically, the decrease in parameter estimation accuracy cancels the cumulative effect of the massive graviton on the signal propagation. The bound increases roughly linearly with the chirp mass, but binaries significantly more massive than those seen so far will merge before entering the detectors' sensitive band.

   In addition to being 'viable'—in the sense of passing all available observational tests—any suggestion for an alternative theory should—in order to be practically useful—lead to distinct predictions in regimes that have not yet been probed. As we have seen, the gravitational-wave events observed so far constrain a variety of mechanisms associated with the generation and propagation of gravitational waves (see Yunes *et al.* (2016) for an in-depth discussion). In particular, we are beginning to probe strong-field gravity, involving the dynamics of black holes and neutron stars. The gravitational-wave signature associated with different scenarios involving such compact objects comes to the fore. Simply speaking, we can try to establish to what extent the 'black holes' we observe are the Kerr black holes (Chapter 17) predicted in Einstein's theory, or if they are somehow different. Similarly, we can probe the properties of neutron stars, but (as we will discuss in Chapter 12) this problem is complicated by the fact that we do not fully understand the involved nuclear physics. A key point is that we need to be able to determine what a 'black hole' (say) would look like in our new theory, and we need to quantify how the answer differs from what we find in general relativity. This is far from trivial, but there has been some progress on understanding (in particular) higher dimensional black-hole solutions. Still, we need to explore the nonlinear merger regime in any relevant modified gravity scenario, and this may be far from easy.

There are many tricky issues and we are not yet equipped to explore them in detail. Once we have further developed the theory—to the point where we appreciate the possibilities—we will return to some of the relevant points. Given our scope, we will be particularly interested in using gravitational-wave observations to test to what extent Einstein was right.

5

Binary inspiral

Equipped with the quadrupole formula from Chapter 3 and some understanding of the context, we are ready to turn our attention to the gravitational waves associated with specific astrophysical systems. This will highlight an entirely different set of questions that need to be resolved if we want our models to be useful. In particular, it will become apparent that we need an accurate representation of the physics of the source. In order to illustrate some of the relevant issues, we will first consider the gradual inspiral of a compact binary system, driven by the loss of orbital energy as gravitational waves are emitted. This is the archetypal gravitational-wave source—and it was the first to be detected, as well!

Gravitational waves are emitted as stars or black holes orbit each other and as a result the binary separation decreases. In Newtonian gravity the two-body problem is conservative and—as we will see—easily solved. In general relativity, the emission of gravitational waves, leading to a shrinking orbit, complicates the problem immensely and it can no longer be solved in closed form. Given the lack of suitable analytical solutions, considerable effort has gone into developing approximations and numerical approaches to the problem. In the first instance, we will try to get an idea of the gravitational-wave signal by making use of the quadrupole formula. We will turn to numerical simulations later (in Chapter 19).

5.1 Basic celestial mechanics

As we plan to explore the binary problem in the full glory of general relativity, and it is useful to appreciate similarities and differences, we will start by going back to basics. This allows us to introduce the concepts we need to generalize. Thus, we consider two bodies moving around each other due to their mutual gravitational interaction, as in Figure 5.1. Let the masses of the two bodies be M_1 and M_2 and their locations (with respect to an arbitrary origin) be given by the two vectors r_1 and r_2. That is, the separation between the two masses is $r = r_1 - r_2$ and the relative distance is $r = |r|$. Because of Newton's third law, the forces acting on each body are equal and opposite:

$$F_1 = -F_2 = -G\frac{M_1 M_2}{r^2}\hat{r}. \tag{5.1}$$

Gravitational-Wave Astronomy: Exploring the Dark Side of the Universe. Nils Andersson, Oxford University Press (2020).
© Nils Andersson. DOI: 10.1093/oso/9780198568032.001.0001

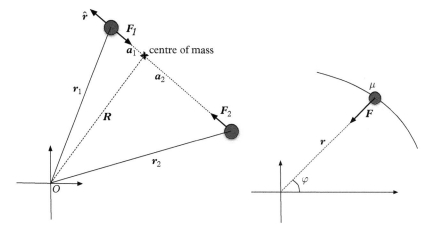

Figure 5.1 *Left: The two-body problem, where two bodies of mass M_1 and M_2 orbit around one another due to their mutual gravitational interaction. Right: The effective one-body picture, where a fiducial body of mass equal to the reduced mass μ orbits around a different origin (note that there is no body at this location).*

Each force (obviously) accelerates the corresponding mass according to Newton's second law.

In order to work out the orbital motion, it is useful to reformulate the situation as an effective one-body problem. To see why this makes sense, note that

$$\ddot{\boldsymbol{r}} = \ddot{\boldsymbol{r}}_1 - \ddot{\boldsymbol{r}}_2 = \frac{\boldsymbol{F}_1}{M_1} - \frac{\boldsymbol{F}_2}{M_2} = \left(\frac{1}{M_1} + \frac{1}{M_2}\right)\boldsymbol{F}_1 = \frac{1}{\mu}\boldsymbol{F}_1, \qquad (5.2)$$

where the dots represent time derivatives and we have introduced the reduced mass

$$\mu = \frac{M_1 M_2}{M_1 + M_2}. \qquad (5.3)$$

We see that μ can be interpreted as the mass of a body with position $\boldsymbol{r}$ moving under the influence of a central force $-F_1\hat{\boldsymbol{r}}$. Defining the total mass $M = M_1 + M_2$, we have

$$\ddot{\boldsymbol{r}} = -\frac{GM}{r^3}\boldsymbol{r}. \qquad (5.4)$$

The two-body system has two constants of motion, the energy E and angular momentum L of the orbit. Since the velocity of the reduced body is $\boldsymbol{v} = \dot{\boldsymbol{r}}$, the energy is given by

$$E = \frac{1}{2}\mu v^2 - G\frac{\mu M}{r}. \qquad (5.5)$$

If we assume that the system is oriented as in Figure 5.1, i.e. that the body rotates around the z-axis of a Cartesian coordinate system centred on the fiducial origin for the reduced body, then it is natural to let the angle to the x-axis be φ. Working in polar coordinates in the orbital plane, the angular momentum is then given by

$$L = \mu r^2 \dot{\varphi}, \tag{5.6}$$

and

$$v^2 = \dot{r}^2 + r^2 \dot{\varphi}^2 \tag{5.7}$$

leads to

$$E = \frac{1}{2}\mu \dot{r}^2 + \frac{1}{2}\frac{L^2}{\mu r^2} - \frac{G\mu M}{r} = \text{constant.} \tag{5.8}$$

Rewriting this using the mass and angular momentum per unit mass, $\tilde{E} = E/\mu$ and $\tilde{L} = L/\mu$, respectively, we have

$$\dot{r}^2 = 2\tilde{E} - \frac{\tilde{L}^2}{r^2} + \frac{2GM}{r} = 2\tilde{E} - V(r), \tag{5.9}$$

where $V(r)$ is an effective potential that governs the binary motion. This is a powerful concept, readily extended to bodies moving in a curved spacetime. In fact, we will contrast the present Newtonian results with their relativistic counterparts in Chapter 10.

It is easy to use (5.9) to infer the key properties of the orbital motion. First of all, we can take the time derivative of the relation (keeping in mind that $\tilde{E}$ and $\tilde{L}$ are both constant) to get

$$\ddot{r} - \left(\frac{\tilde{L}^2}{r^3} - \frac{GM}{r^2} \right) = 0. \tag{5.10}$$

From this we learn that circular orbits must be such that

$$r = \frac{\tilde{L}^2}{GM} = \frac{L^2}{G\mu M_1 M_2} \equiv p, \tag{5.11}$$

which defines the so-called *semilatus rectum p*. Quite intuitively, it takes a certain angular momentum L to keep the system in an orbit with radius r.

Assuming instead that the orbit is elliptic, we learn from (5.9) that the motion must be such that $\tilde{E} \geq V/2$. Each orbit then has two turning points, at which $\dot{r} = 0$. From (5.9) we see that these are given by the roots to

$$r^2 + \frac{GM}{\tilde{E}}r - \frac{\tilde{L}^2}{2\tilde{E}} = 0. \tag{5.12}$$

Introducing

$$e = \left[1 + \frac{2\tilde{L}^2\tilde{E}}{(GM)^2}\right]^{1/2}, \tag{5.13}$$

we have

$$r = -\frac{GM}{2\tilde{E}}(1 \pm e). \tag{5.14}$$

Moreover, noting that (5.11) leads to

$$\frac{p}{1 - e^2} = -\frac{GM}{2\tilde{E}}, \tag{5.15}$$

we see that the two turning points are

$$r = \frac{p}{1 - e^2}(1 \pm e). \tag{5.16}$$

This result is useful as it allows us to parameterize the motion in terms of the semilatus rectum and the *eccentricity*, e.

In describing an orbit it is also worth noting that the point of closest approach, the periastron r_p, and the furthest distance, the apastron r_a, are such that

$$p = \frac{2r_a r_p}{r_a + r_p}, \tag{5.17}$$

and

$$e = \frac{r_a - r_p}{r_a + r_p}. \tag{5.18}$$

Finally, it is useful to introduce the orbital separation

$$a = \frac{1}{2}(r_p + r_a) = \frac{p}{1 - e^2}, \tag{5.19}$$

such that

$$\tilde{E} = -\frac{GM(1 - e^2)}{2p} = -\frac{GM}{2a}. \tag{5.20}$$

Let us now determine the motion of the system. This problem becomes straightforward if we first introduce a new variable

$$u = u(\varphi) = \frac{1}{r} \quad \longrightarrow \quad \dot{r} = -\tilde{L}\frac{du}{d\varphi}. \tag{5.21}$$

Using this in (5.9) (and taking a derivative with respect to φ) we get

$$\frac{d^2 u}{d\varphi^2} + u = \frac{GM}{\tilde{L}^2}. \tag{5.22}$$

Not surprisingly (since the orbit is periodic), we have arrived at the equation for a harmonic oscillator. The solution can be written (after setting suitable initial data)

$$u = \frac{GM}{\tilde{L}^2}(1 - e\cos\varphi) \quad \longrightarrow \quad r = \frac{p}{1 - e\cos\varphi}. \tag{5.23}$$

As expected, the orbit is an ellipse. This suggests an easy way to work out the orbital period. Noting that an angular increment $d\varphi$ corresponds to the motion sweeping out an area segment $dA = r^2 d\varphi/2$ we can integrate over a complete orbit to get

$$\dot{A} = \frac{\tilde{L}}{2} \quad \longrightarrow \quad A = \int dA = \frac{\tilde{L}}{2}\int dt = \frac{\tilde{L}}{2}P = \pi ab, \tag{5.24}$$

where the semi-major axis is $b = a(1 - e^2)^{1/2}$. It follows that the orbital period, $P = 2\pi/\Omega$, with Ω is the angular frequency, is given by

$$P = 2\pi\left(\frac{a^3}{GM}\right)^{1/2} \quad \longrightarrow \quad \Omega^2 = \frac{GM}{a^3}. \tag{5.25}$$

This is, of course, no surprise. We have simply derived Kepler's law.

Having characterized the orbital motion, we want to use the quadrupole formula to work out the rate of gravitational-wave emission. To do this, we need to know the motion of each individual body rather than the reduced mass. Fortunately, this is easy to work out. From Figure 5.1 we have

$$\boldsymbol{r} = \boldsymbol{r}_1 - \boldsymbol{r}_2 = \boldsymbol{a}_1 - \boldsymbol{a}_2 = \boldsymbol{a}, \tag{5.26}$$

and making use of the centre of mass

$$\boldsymbol{R} = \frac{M_1\boldsymbol{r}_1 + M_2\boldsymbol{r}_2}{M}, \tag{5.27}$$

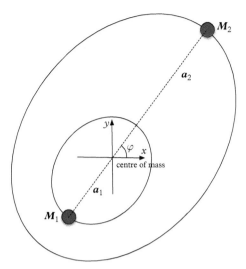

Figure 5.2 *A schematic illustration of the orbital motion of a binary system.*

we see that

$$a_1 = r_1 - R = \frac{\mu}{M_1} a, \tag{5.28}$$

and

$$a_2 = r_2 - R = -\frac{\mu}{M_2} a. \tag{5.29}$$

Each body moves around the centre of mass in the same way that the reduced body moves about its (fiducial) centre. The individual distances are simply shortened by the factor μ/M_A $(A = 1, 2)$. Since the reduced mass moves along an ellipse, the individual bodies in the binary also move along ellipses, as shown in Figure 5.2.

5.2 Circular orbits

Let us now focus on the simplest case, that of circular orbits. Then the two distances to the centre of mass, a_1 and a_2, remain fixed. Provided the stars are far enough apart that we can ignore the tidal interaction (see Chapter 21), we can treat the two bodies as point masses, which means that the integrals required to work out the mass multipole moments we need for the quadrupole formula are trivial. Working in a Cartesian coordinate system, as shown in Figure 5.2, with the z-axis associated with the rotation, we have the reduced quadrupole moment from (3.66)

$$\mathcal{I}_{xx} = (M_1 a_1^2 + M_2 a_2^2)\left[\cos^2 \varphi - \frac{1}{3}\right] = \frac{\mu a^2}{2}\cos 2\varphi + \text{constant}, \qquad (5.30)$$

where $\varphi = \Omega t$. We do not need to worry about the constant piece since we will take time derivatives to arrive at the gravitational-wave field (see, for example, (3.67)). The other time-varying contributions are

$$\mathcal{I}_{xx} = -\mathcal{I}_{yy} = \frac{\mu a^2}{2}\cos 2\varphi, \qquad (5.31)$$

$$\mathcal{I}_{xy} = \mathcal{I}_{yx} = \frac{\mu a^2}{2}\sin 2\varphi. \qquad (5.32)$$

Combining these results with (3.90), we see that the gravitational-wave luminosity is (where the angle brackets indicate an average over at least one orbit)

$$\mathcal{F} = \frac{G}{5c^5}\left(\frac{\mu a^2}{2}\right)^2 (2\Omega)^6 \left(2\sin^2 2\varphi + 2\cos^2 2\varphi\right) = \frac{32}{5}\frac{G}{c^5}\mu^2 a^4 \Omega^6. \qquad (5.33)$$

Using the orbital frequency from Kepler's law (5.25) and introducing the so-called *chirp mass*

$$\mathcal{M} = \mu^{3/5} M^{2/5}, \qquad (5.34)$$

we have

$$\mathcal{F} = \frac{32\mathcal{M}\Omega}{5c^5}(G\mathcal{M}\Omega)^{7/3}. \qquad (5.35)$$

This result provides the rate at which gravitational waves carry energy away from the system. Later, when we discuss higher order post-Newtonian corrections (see Chapter 11), it is useful to understand how the luminosity scales with the characteristic velocity of the system. Hence, we note that the orbital velocity is $v = a\Omega$, in terms of which (5.33) becomes

$$\mathcal{F} = \frac{32}{5}\frac{c^5}{G}\eta^2 \left(\frac{v}{c}\right)^{10}, \qquad (5.36)$$

where we have used the symmetric mass ratio $\eta = \mu/M$.

In order to predict how fast the orbit shrinks as a result of the gravitational-wave emission, we also need the energy from (5.20). Again, making use of the chirp mass, we have

$$E = -\frac{\mathcal{M}}{2}(G\mathcal{M}\Omega)^{2/3}. \qquad (5.37)$$

Working out the back reaction on the binary orbit, we only need to keep in mind that the gravitational waves carry energy away from the system. As long as the evolution is slow and gradual, the change in the system's energy is given by $\dot{E} = -\mathcal{F}$, so the evolution takes place on a characteristic timescale

$$t_D = \frac{E}{\dot{E}} = -\frac{E}{\mathcal{F}} \sim \mathcal{M}^{-5/3} \Omega^{-8/3}. \tag{5.38}$$

Note that, for an equal mass system, this agrees with the back-of-the-envelope estimate from Chapter 1

$$t_D \sim \frac{a^4}{M^3}. \tag{5.39}$$

The evolution of the relevant orbital parameters follows from

$$\frac{\dot{P}}{P} = -\frac{\dot{\Omega}}{\Omega} = \frac{3}{2}\frac{\dot{a}}{a} = -\frac{3}{2}\frac{\dot{E}}{E} = \frac{3}{2}\frac{\mathcal{F}}{E}. \tag{5.40}$$

For example, the orbital period evolves according to

$$\frac{\dot{P}}{P} = -\frac{96}{5c^5}(G\mathcal{M}\Omega)^{5/3}\,\Omega. \tag{5.41}$$

Notably, the rate of orbital decay only depends on the chirp mass and the orbital frequency. The estimate (5.41) should be valid as long as the evolution is slow compared to the orbital velocity

$$|\dot{a}| \ll \Omega a. \tag{5.42}$$

However, it is easy to see that this condition ought to be satisfied. We have

$$|\dot{a}| = \frac{2}{3}\frac{\dot{\Omega}}{\Omega}a = (\Omega a)(\Omega t_D)^{-1}, \tag{5.43}$$

where

$$\Omega t_D = \frac{5}{64}\eta^{-1/5}\left(\frac{v}{c}\right)^{-5} \gg 1. \tag{5.44}$$

Determining the gravitational-wave amplitude is a little bit more complicated since we need to introduce a specific observer; see Eq. (3.67). In essence, any given observer will only see waves due to transverse motion according to his/her location relative to the source. As an example, let the observer sit a distance d away from the source on the y-axis and work out h_{xx}^{TT}. Then we use $\boldsymbol{n} = \boldsymbol{e}_y$ and the TT-projection from (3.52) to get

$$\mathcal{E}_{xx}^{TT} = \perp_x^l \perp_x^m \mathcal{E}_{lm} - \frac{1}{2} \perp_{xx} \left(\perp^{lm} \mathcal{E}_{lm} \right) = \frac{1}{2} \mathcal{E}_{xx}. \tag{5.45}$$

This leads to

$$h_{xx}^{TT} = -\frac{2\mu a^2 \Omega^2}{d} \cos 2\Omega (t - r) = -\frac{2\mathcal{M}^{5/3}\Omega^{2/3}}{d} \cos 2\Omega (t - r). \tag{5.46}$$

The gravitational waves emerge at twice the orbital frequency, which makes sense if we consider the symmetry of the situation. Moreover, we see that the chirp mass, $\mathcal{M}$, plays a central role. In fact, this is the only combination of the two masses that can be inferred from an observed gravitational-wave signal at this level of approximation. We also see that if one can observe both the shrinkage of the orbit and the gravitational-wave field (h_{xx}^{TT}, say) then one should be able to infer both the chirp mass and the distance to the source, d. This means that coalescing binaries can act as 'standard candles' which may be used to measure distance in the Universe. That is, we can hope to infer the Hubble constant H_0 (see Chapter 22). By extracting higher order post-Newtonian terms we should also be able to infer the individual masses, the spins and perhaps even put constraints on the mass of the graviton (see Chapter 4).

5.3   The Binary Pulsar

The first double neutron star system—the Binary Pulsar PSR B1913+16—was discovered in 1974 by Russell Hulse and Joseph Taylor in data from the Arecibo radio telescope (Hulse and Taylor, 1975). This system allowed the first test of Einstein's theory in the (moderately) strong field regime. The discovery (and the subsequent analysis of the timing data) is celebrated because it provided the first quantitative (albeit indirect) confirmation of the existence of gravitational waves. Given the historical significance, it is natural to use the Binary Pulsar as a benchmark for our estimates from the quadrupole formula.

If we combine the observed parameters from Table 5.1 with the calculated change in the orbital period from (5.41), then we predict that

$$\dot{P} \approx -2 \times 10^{-13} \text{ s/s}. \tag{5.47}$$

However, the precision of pulsar radio timing allows a very accurate tracking of the system's evolution. The observed change in the orbit is

$$\dot{P} \approx -2.3 \times 10^{-12} \text{ s/s}. \tag{5.48}$$

Our estimate appears to be wrong by an order of magnitude. This does not look good. However, we used the results for circular orbits. In reality the orbit of the Binary Pulsar is eccentric, see Table 5.1, and this makes a difference. In principle, one would expect the

Table 5.1 *Key observed parameters for the Binary Pulsar PSR B1913+16. (Data from Weisberg et al. (2010).)*

Pulsar mass (M_1)	$1.441 M_\odot$
Pulsar spin period	59.02999792988 ms
Companion mass (M_2)	$1.387 M_\odot$
Total mass (M)	$2.828379 M_\odot$
Reduced mass (μ)	$0.707 M_\odot$
Chirp mass ($\mathcal{M}$)	$1.23 M_\odot$
Semimajor axis (a)	1,950,100 km
Orbital period (P)	7.75 hours (27,900 s)
Eccentricity (e)	0.617
Distance (d)	6,400 pc (1.97×10^{17} km)

gravitational-wave emission from an eccentric system to differ from our estimate because the signal is stronger during the part of the orbit when the stars are closer and weaker when they are further apart. To make progress on the comparison, we need to quantify this effect.

As a slight aside before we move on, it is interesting to note that the power radiated as gravitational waves by the Binary Pulsar is about 7×10^{23} W—enough to power quite a few light bulbs. As a result, the orbit shrinks almost 4 m each year. The system will eventually merge, but we have to wait another 300 million years.

5.4   Eccentic orbits

It is relatively straightforward to extend our analysis of the gravitational-wave emission to eccentric binary systems (Peters and Mathews, 1963). After all, we have already worked out the general orbital motion. But the problem is messier than the circular-orbit one. Without working out all the details, let us illustrate why this is the case. The good news is that the expressions for the mass multipoles remain exactly as before. We have

$$M_{xx} = \mu r^2 \cos^2 \varphi, \tag{5.49}$$
$$M_{xy} = M_{yx} = \mu r^2 \sin \varphi \cos \varphi, \tag{5.50}$$

and

$$M_{yy} = \mu r^2 \sin^2 \varphi. \tag{5.51}$$

For circular orbits, we had $r = a = $ constant so it was easy to identify the time-varying parts of these moments. In the general case, this is not quite so straightforward. Now we have

$$r = a\frac{1 - e^2}{1 - e\cos\varphi}, \tag{5.52}$$

so the time dependence of the multipole moments will be more complicated. We need to use

$$\dot\varphi = \frac{\tilde{L}}{r^2} = \left(\frac{GM}{a^3}\right)^{1/2}\left(1 - e^2\right)^{-3/2}(1 - e\cos\varphi)^2. \tag{5.53}$$

After some algebra, we arrive at

$$\mathcal{F} = \frac{8}{15}\frac{\Omega(GM\Omega)^{7/3}}{c^5}\frac{1}{a^5\left(1 - e^2\right)^5}\left\langle(1 - e\cos\varphi)^4\left[12(1 - e\cos\varphi)^2 + e^2\sin^2\varphi\right]\right\rangle. \tag{5.54}$$

In order to work out the required average, we integrate over one complete orbit. This leads to the final result

$$\mathcal{F} = f(e)\mathcal{F}_0, \tag{5.55}$$

where $\mathcal{F}_0$ follows from (5.36) and

$$f(e) = \frac{1}{(1 - e^2)^{7/2}}\left[1 + \frac{73}{24}e^2 + \frac{37}{96}e^4\right]. \tag{5.56}$$

This can be quite a large factor. In the case of the Binary Pulsar we have $e = 0.617$ (see Table 5.1), which leads to $f(e) \approx 11.8$. Accounting for the ellipticity thus brings the theoretical estimate much closer to the observed result. In fact, after more than 40 years of timing of PSR 1913+16, the observations agree with the predictions of general relativity to within a fraction of a percent; see Figure 5.3—a remarkable confirmation of Einstein's theory.

In the case of circular orbits, we only needed the gravitational-wave luminosity in order to work out the orbital evolution. For eccentric orbits we need more information since the eccentricity evolves as the orbit shrinks. As we will see, the orbit becomes more circular with time.

If we want to account for this effect, we need the rate at which gravitational waves carry angular momentum away from the system. From (5.13) we have

$$\dot{e} = -\left(\frac{1 - e^2}{e}\right)\left(\frac{\dot{E}}{2E} + \frac{\dot{L}}{L}\right) = \left(\frac{1 - e^2}{e}\right)\left(\frac{\mathcal{F}}{2E} + \frac{\mathcal{G}}{L}\right). \tag{5.57}$$

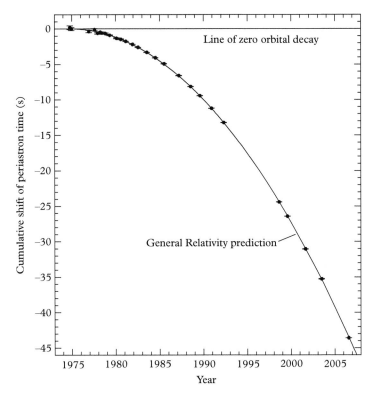

Figure 5.3 *Orbital decay caused by the loss of energy to gravitational radiation from the Binary Pulsar B1913+16. The parabola shows the shift of periastron time relative to an unchanging orbit, as predicted by general relativity. Data points represent measurements, with error bars mostly too small to see. (Reproduced from Weisberg et al. (2010) with permission by the AAS.)*

In this expression, $\mathcal{G}$ is the rate at which the system radiates angular momentum, which follows from (3.103). With the setup as in Figure 5.2 we get (in the orbital plane and after using integration by parts for two of the terms)

$$\mathcal{G} = \frac{4G}{5c^5}\langle\dddot{\mathcal{I}}_{xy}(\dddot{\mathcal{I}}_{xx} - \dddot{\mathcal{I}}_{yy})\rangle = \frac{4G}{5c^5}\langle\dddot{M}_{xy}(\dddot{M}_{xx} - \dddot{M}_{yy})\rangle. \tag{5.58}$$

This leads to

$$\mathcal{G} = -\frac{32}{5}\frac{\mathcal{M}(G\mathcal{M}\Omega)^{7/3}}{c^5}\frac{1}{(1-e^2)^2}\left(1 + \frac{7e^2}{8}\right). \tag{5.59}$$

Combining this result with the expression for the energy flux, we arrive at

$$\dot{e} = -\frac{304}{15}\frac{(G\mathcal{M}\Omega)^{5/3}\Omega}{c^5}\frac{e}{(1-e^2)^{5/2}}\left(1+\frac{121}{304}e^2\right),$$ (5.60)

which leads to

$$\frac{da}{de} = \frac{192}{304}\frac{a}{e(1-e^2)}\left(1+\frac{73}{24}e^2+\frac{37}{96}e^4\right)\left(1+\frac{121}{304}e^2\right)^{-1}.$$ (5.61)

This can be integrated to give

$$a = C\frac{e^{12/19}}{1-e^2}\left(1+\frac{121}{304}e^2\right)^{870/2299},$$ (5.62)

where C is constant for any given system.

Let us consider what this implies for a system like the Binary Pulsar. To get an idea, we simplify the expression assuming that $e^2 \ll 1$. Then we have

$$e \approx Da^{19/12},$$ (5.63)

where the constant factor is fixed by the current orbital parameters, $D \approx 2 \times 10^{-5}$. Extrapolating into the distant future, we see that when this system has evolved to the point where it enters the sensitivity band of a ground-based detector (say, at a frequency of 10 Hz), the ellipticity has decreased to $e \approx 4 \times 10^{-6}$. For all intents and purposes the orbit will be circular.

The main lesson from this exercise is that the emission of gravitational waves drives eccentric systems towards circular orbits on a timescale much shorter than that of the inspiral.

5.5   The orbital evolution

With confidence in our results for the gravitational-wave emission, let us make a few additional observations. Since we have established that the orbit tends to circularize, it makes sense to assume that $e \approx 0$. We can then use our estimates to write down an equation for the rate of change of the binary separation. Solving this equation we find that

$$a = a_0\left(1-\frac{t}{t_m}\right)^{1/4},$$ (5.64)

where a_0 is the initial separation and $t_m = t_D/4$. Parameterizing this timescale we have

$$t_m \approx 30 \left(\frac{f}{10 \text{ Hz}}\right)^{-8/3} \left(\frac{10 M_\odot}{\mathcal{M}}\right)^{5/3} \text{ s,} \tag{5.65}$$

where the gravitational-wave frequency is twice the orbital frequency, $f = \Omega/\pi$. This gives us a more precise idea of the time that remains until merger (even though, in reality, the merger will not correspond to $a = 0$, exactly). A sample of typical values for t_m for the types of binaries expected to be the most relevant for ground-based detectors is given in Table 5.2. From the results we see that all binaries likely to be observed from the ground will merge within hours from the time when they first enter the detector bandwidth.

In order to appreciate another key aspect of the problem, let us work out the energy radiated as the system evolves. Taking the orbital energy as our starting point, we have

$$\frac{dE}{df} = \frac{\mathcal{M}}{3} (\pi G \mathcal{M})^{2/3} f^{-1/3}. \tag{5.66}$$

We see that the energy released from the point when the signal first enters the detector bandwidth, f_{min} (at time t_0), up to the frequency that signals the end of the inspiral, f_{max} (at t_1), follows from

$$\Delta E = \frac{\pi^{2/3}}{2G} (G\mathcal{M})^{5/3} \left(f_{\text{max}}^{2/3} - f_{\text{min}}^{2/3}\right). \tag{5.67}$$

If we assume that $f_{\text{max}} \gg f_{\text{min}}$, then

$$\Delta E \approx 4 \times 10^{-2} M_\odot \left(\frac{\mathcal{M}}{1.2 M_\odot}\right)^{5/3} \left(\frac{f_{\text{max}}}{1 \text{ kHz}}\right)^{2/3}. \tag{5.68}$$

Table 5.2 *Estimated time until merger ($t_m = t_D/4$) for various classes of binary systems that may be observed by ground-based detectors, the most common of which are expected to be double neutron star (each with mass about $1.4 M_\odot$) systems, double (here $10 M_\odot$ each) black-hole systems, and mixed black hole–neutron star binaries. f_{min} is the gravitational-wave frequency of the signal when the system first enters the detector's sensitivity band.*

f_{min}	$1.4 - 1.4 \ M_\odot$	$1.4 - 10 \ M_\odot$	$10 - 10 \ M_\odot$
1 Hz	6 days	1 day	4.8 hrs
10 Hz	17 min	4 min	38 s
100 Hz	2 s	0.5 s	0.08 s

As we will discuss in Chapter 10, the inspiral will end when the system reaches the innermost stable circular orbit (Bardeen *et al.*, 1972). Approximating this with the result for a small body moving in the Schwarzschild spacetime (see Chapter 10)—an assumption which may not be particularly accurate—we have

$$a_{\text{isco}} \approx \frac{6GM}{c^2} \quad \longrightarrow \quad f_{\text{max}} \approx \frac{c^3}{6^{3/2}\pi GM}. \tag{5.69}$$

Hence, we arrive at an estimate for the radiated energy

$$\Delta E \approx \frac{1}{12}\mu c^2, \tag{5.70}$$

where μ is the reduced mass of the system, as before. This is a lot of energy, but the signal will still be weak. Moreover, binary inspirals are rare in a typical galaxy. To detect them we need to search a volume of space that contains many galaxies. In order to make this statement precise, and get an idea of how sensitive a detector will have to be, we need to know what the event rate may be. We will consider this question in Chapter 9.

Let us wrap up the discussion by highlighting a complicating aspect. We have seen that the frequency evolves according to

$$\dot{f} = \frac{96\pi^{8/3}}{5}(G\mathcal{M})^{5/3}f^{11/3}. \tag{5.71}$$

We can use this relation to work out the number of gravitational-wave cycles emitted during the inspiral

$$\mathcal{N} = \int_{t_0}^{t_1} f\,dt = \int_{f_{\text{min}}}^{f_{\text{max}}} \frac{f}{\dot{f}}\,df = \frac{1}{32\pi^{8/3}}\frac{c^5}{(G\mathcal{M})^{5/3}}\left(f_{\text{min}}^{-5/3} - f_{\text{max}}^{-5/3}\right), \tag{5.72}$$

or (if $f_{\text{max}} \gg f_{\text{min}}$)

$$\mathcal{N} \approx 1.6 \times 10^4 \left(\frac{10\text{ Hz}}{f_{\text{min}}}\right)^{5/3}\left(\frac{1.2M_\odot}{\mathcal{M}}\right)^{5/3}. \tag{5.73}$$

When we consider the problem of identifying weak signals in a noisy data stream (see Chapter 8) we learn that the loss of a single cycle leads to a significant drop in the signal-to-noise ratio. In essence, we need very accurate source models. For binary inspirals, the leading order post-Newtonian results we have used so far will not suffice. We need to do much better. This means going to higher orders in the post-Newtonian expansion, a technically challenging problem which we will return to in Chapter 11.

6

Spinning stars and cosmic recycling

As a step towards problems where the detailed physics have a more decisive impact, let us consider gravitational waves from rotating neutron stars. A spinning star radiates gravitationally if its shape deviates from perfect symmetry. Colloquially, the associated (quadrupole) deformations are often referred to as 'mountains', even though their actual height may not be particularly impressive. Stars also have internal dynamics, often represented in terms of a set of normal modes of oscillations, that may be exited during the star's life. These two aspects force us to account for the physics of the stellar interior (see Chapter 12). Neutron star mountains may be associated with the star's crust—the outer kilometre or so—where, moving towards the centre, increasingly neutron-rich nuclei form a Coulomb lattice with elastic properties. The crust may also sustain shear modes of oscillation, but due to the relatively low density of the crust region these may not be very important for gravitational-wave physics. From our point of view oscillations involving the high-density core will be more relevant. The relevant modes of oscillation, and their damping, then depend on the supranuclear physics, e.g. the matter composition, encoded in the equation of state. We consider this problem in Chapters 13 and 18. For the moment we ignore any internal dynamics and assume that the star can be considered a rigid body.

As soon as a newly born neutron star cools below 10^{10} K or so (within seconds to minutes after its birth) the outer layers will begin to crystallize, forming the neutron star crust. Even though the crust is not very rigid, it can sustain shear stresses, something that a fluid is unable to do. This means that neutron stars may have long-lived deformations which generate a continuous gravitational-wave signal. The signal will inevitably be weak, but there are many neutron stars in the Galaxy and some of them are quite close to us. Moreover, the effective amplitude increases (roughly) as the square-root of the observation time (see Chapter 8), so an observing run lasting (say) a year would provide a significant enhancement of the raw signal.

6.1 Rotating deformed stars

In order to understand the neutron-star mountain problem, we need to recall some results from classical mechanics. Even though neutron stars are not rigid—in a rigid body the

Gravitational-Wave Astronomy: Exploring the Dark Side of the Universe. Nils Andersson, Oxford University Press (2020).
© Nils Andersson. DOI: 10.1093/oso/9780198568032.001.0001

distance between the different particles does not change, while a neutron star is expected to have a fluid core and an elastic crust, both of which may sustain oscillations—the rigid-body problem provides a useful introduction to the main concepts.

To describe the motion, we use two coordinate systems: an inertial system with coordinates x^i and (orthonormal) basis vectors e_i, with $i, j, k, \ldots = 1 - 3$, and a moving system with coordinates $x^{\hat{i}}$ and basis vectors $e_{\hat{i}}$ with $\hat{i}, \hat{j}, \hat{k}, \ldots$ also running from 1 to 3. This latter system is fixed to the body. We will occasionally use standard Cartesian coordinates x, y, and z for the inertial system. For simplicity, we let these coordinates be centred at the body's centre of mass and neglect any translational motion.

When the body rotates with angular velocity Ω^i we then have

$$v = \Omega \times r, \quad \text{or} \quad v^i = \epsilon^{ijk} \Omega_j x_k, \tag{6.1}$$

where x^i are the components of the position vector of a given particle in the body, such that $x_i x^i = r^2$ gives the distance from the centre of the star. Using this in the expression for the kinetic energy, we get

$$E = \frac{1}{2} \int \rho v^2 dV = \frac{1}{2} \int \rho (\Omega \times r)^2 dV = \frac{1}{2} \int \rho [\Omega^2 r^2 - (\Omega^i x_i)^2] dV, \tag{6.2}$$

where the integral is over the volume of the star, V. Alternatively, since the angular velocity must be the same for all parts of the body,

$$E = \frac{1}{2} \Omega^i \Omega^j \int \rho [r^2 \delta_{ij} - x_i x_j] dV = \frac{1}{2} I_{ij} \Omega^i \Omega^j, \tag{6.3}$$

which defines the moment of inertia tensor I_{ij}.

Similarly, the definition of the angular momentum of a spinning body leads to

$$\mathcal{J}^i = \epsilon^{ijk} \int \rho x_j v_k d^3 x = \epsilon^{ijk} \epsilon_{klm} \int \rho x_j \Omega^l x^m dV = \Omega^j \int \rho (\delta_{ij} r^2 - x_i x_j) dV = I_{ij} \Omega^j. \tag{6.4}$$

It is worth noting that, when we work out the gravitational-wave signal, we need the reduced quadrupole moment. This follows from, see Eq. (3.66),

$$I_{jk} = -\mathcal{I}_{jk} + \frac{2}{3} \delta_{jk} \int \rho r^2 d^3 x, \tag{6.5}$$

where the second term is constant.

It is also worth pointing out that I_{ij} is additive. The total moment of inertia is the sum of contributions from the various parts of the body. This is an important insight, useful whenever we want to consider an object made up of several distinct components (like a neutron star with a rigid crust and a fluid core; see Chapter 12). We also see that the moment of inertia tensor only depends on the density distribution and the shape of the object.

Since I_{ij} is symmetric, one would expect to be able to diagonalize it by making a clever choice of coordinates. The corresponding axes, which we will take as the system $e_{\hat{i}}$, defines the principal axes of the body. It is natural to refer to this as the 'body frame'. We denote the corresponding moments of inertia by $I_{\hat{i}}$. In this system the kinetic energy takes the very simple form

$$E = \frac{1}{2}(I_1\Omega_1^2 + I_2\Omega_2^2 + I_3\Omega_3^2),\tag{6.6}$$

and

$$\mathcal{J}_{\hat{i}} = I_{\hat{i}\hat{j}}\Omega^{\hat{j}}.\tag{6.7}$$

For future reference, it is worth noting that these relations imply

$$\dot{E} = \Omega^{\hat{i}}\dot{\mathcal{J}}_{\hat{i}},\tag{6.8}$$

where the dots represent time derivatives, provided the body-frame moments of inertia are fixed (i.e. the body is truly rigid).

In general, it may be difficult to determine the principal axes, but the problem simplifies if the body has symmetries. Then the centre of mass must be located on any axis of symmetry and the principal axes must share the symmetry of the body.

Without specifying the shape of the body, let us move on and consider the gravitational-wave luminosity. From the quadrupole expression (3.90) we know that we need to work out

$$\frac{dE}{dt} = \frac{G}{5c^5}\left\langle \dddot{\mathcal{F}}_{jk}\dddot{\mathcal{F}}^{jk} \right\rangle,\tag{6.9}$$

where it follows from (6.5) that

$$\dddot{\mathcal{F}}_{jk} = -\dddot{I}_{jk},\tag{6.10}$$

and we recall that the angular brackets indicate time-averaging over several periods. We now note that it is much easier to work in the body frame than in the inertial frame, and it is wise to take advantage of this. In fact, since $I_{\hat{i}\hat{j}}$ is constant in the body frame, the time derivatives we need are encoded in the transformation between the two systems. In general, we need

$$I_{jk} = R_j^{\hat{j}}R_k^{\hat{k}}I_{\hat{j}\hat{k}},\tag{6.11}$$

where the transformation follows from combining $x^i = R^i_j \hat{x}^j$ with (6.1) (keeping in mind that the transformation to the inertial frame involves rotating the axes in the opposite direction to the body's rotation in the inertial frame). We will make use of this when we consider the general problem of a wobbling star in Chapter 14. For now we will settle for a discussion of the simpler case of rigid body rotation.

Consider a spinning star endowed with a small asymmetry. Assume that the star has principal moments of inertia I_1, I_2 and I_3 and that it rotates around the $z = e_3 = e_{\hat{3}}$ axis, as in Figure 6.1. Provided that $I_2 \neq I_1 = I_3$ the star is not axisymmetric and will radiate gravitationally.

In this case the tranformation we need in order to estimate the gravitational-wave luminosity from (6.9) is the standard rotation matrix

$$R = \begin{pmatrix} \cos\varphi & \sin\varphi & 0 \\ -\sin\varphi & \cos\varphi & 0 \\ 0 & 0 & 1 \end{pmatrix}, \tag{6.12}$$

and (6.11) is simply

$$I_{\text{inertial}} = R^T I_{\text{body}} R. \tag{6.13}$$

If we define $\Delta = I_1 - I_2$, then it follows that

$$I_{xx} = -I_{yy} = \frac{1}{2}\Delta\cos 2\varphi, \tag{6.14}$$

and

$$I_{xy} = I_{yx} = \frac{1}{2}\Delta\sin 2\varphi, \tag{6.15}$$

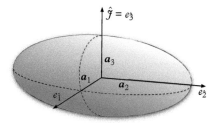

Figure 6.1 *A schematic illustration of an asymmetric rotating neutron star. The body frame is described by axes $e_{\hat{i}}$ and the star rotates around the inertial $z = e_3$-axis, associated with the (conserved) angular momentum $\mathcal{J}^i$.*

where φ is the angle between the e_1-axis and the inertial x-axis, such that $\varphi = \Omega t$ with Ω the angular frequency (and we recall that we are only interested in contributions that vary in time). If we assume that the rotation is steady and Ω is constant, then

$$\frac{dE}{dt} = \frac{32G}{5c^5}\Delta^2\Omega^6. \qquad (6.16)$$

In order to relate this result to an actual star, we need the moments of inertia. For simplicity, we consider the case of a uniform density ellipsoid. We let the three axes of the ellipsoid, along which the principal axes lie, have length a_1, a_2, and a_3, respectively. To work out the corresponding moments of inertia we need to integrate over the volume. This exercise is easier if we note that the change of variables $x_1 = a_1\xi_1$ (and similar for x_2 and x_3), turns the equation for an ellipsoidal surface

$$\frac{x_1^2}{a_1^2} + \frac{x_2^2}{a_2^2} + \frac{x_3^2}{a_3^2} = 1, \qquad (6.17)$$

into that of the unit sphere

$$\xi_1^2 + \xi_2^2 + \xi_3^2 = 1. \qquad (6.18)$$

We then find that, if we consider rotation about the $\hat{e}_1$-axis, the relevant moment of inertia is

$$I_1 = \rho \int (x_2^2 + x_3^2)dx_1\,dx_2\,dx_3$$
$$= \rho a_1 a_2 a_3 \int (a_2^2\xi_2^2 + a_3^2\xi_3^2)d\xi_1\,d\xi_2\,d\xi_3 = \frac{1}{5}M(a_2^2 + a_3^2), \qquad (6.19)$$

where M is the star's mass and we have used the volume of the ellipsoid: $4\pi a_1 a_2 a_3/3$. The other moments of inertial follow from a cyclic permutation of the indices.

Assuming that $a_2 \neq a_3 = a_1$, as in Figure 6.1, we have

$$\Delta = \frac{1}{5}M(a_2 - a_1)(a_1 + a_2). \qquad (6.20)$$

We now define the ellipticity

$$\epsilon = \frac{a_2 - a_1}{(a_1 + a_2)/2}, \qquad (6.21)$$

and relate the result to a spherical star with the same volume, which would have radius

$$R^3 = a_1^2 a_2. \qquad (6.22)$$

This leads to

$$\Delta = \frac{2\epsilon MR^2}{5} = \epsilon I_0, \tag{6.23}$$

where $I_0 = 2MR^2/5$ is the moment of inertia of a uniform density sphere with mass M and radius R. Thus, we have the final result for the gravitational-wave luminosity

$$\frac{dE}{dt} \approx \frac{32G}{5c^5} \epsilon^2 I_0^2 \Omega^6. \tag{6.24}$$

It is worth noting that we have assumed that the star is stretched into the shape of an American football. We will later distinguish this prolate configuration from the oblate case, where the star is squashed to look more like a disc. Effectively, this involves changing the sign of Δ. However, it is already clear from our calculation that the sign of Δ has no effect on the gravitational-wave luminosity.

The simple model problem we have considered, essentially assuming that we can describe the star as a rotating bar, raises several tricky questions. The two most obvious ones may be: How large can we reasonably expect the ellipticity ϵ to be? This is a difficult problem as it involves issues—like the breaking strain—that are poorly understood even for terrestrial materials and evolutionary considerations—which should determine why the deformation developed in the first place. We will return to these problems in Chapter 12. A second set of questions concern observations: In particular, how do we confront our estimates with actual data?

6.2 The Crab Pulsar

Even though we know very little about the actual asymmetries of astrophysical neutron stars, we can get important hints from radio pulsar observations. It is, for example, straightforward to compare the predicted gravitational-wave spin-down rate to observations. From (6.24) we see that the emitted gravitational waves lead to a change in the spin period (recall that the energy is drained from the star's rotation)

$$\frac{\dot{P}}{P} = -\frac{\dot{E}}{2E} = -\frac{32G}{5c^5} \epsilon^2 I_0 \Omega^4, \tag{6.25}$$

where $P = 2\pi/\Omega$ is the spin period and we have taken the rotational energy to be $E = I_0 \Omega^2/2$.

As an indicative example, let us consider the famous Crab Pulsar PSR B0531+21. When the Crab Pulsar was discovered—shortly after the first discovery of pulsars (Hewish *et al.*, 1968)—it was the first pulsar to be associated with a supernova remnant. Impressively, the association with a historical supernova from 1054 is confirmed by old

Chinese records.[1] The Crab Pulsar is interesting in many ways, but for now all we need to know is that it currently spins with a period $P = 33$ ms while the observed spin-down rate is

$$\dot{P}_{obs} \approx 4.2 \times 10^{-13} \text{s/s}. \tag{6.26}$$

If we want to compare these observations to (6.24), then we need to know the star's mass and radius. Unfortunately, we do not. However, if we assume 'canonical' neutron star parameters, a mass of $1.4M_\odot$ and radius $R = 10$ km (see Chapter 12), then we find that the pulsar should slow down at a rate

$$\dot{P}_{gw} \approx 8 \times 10^{-7} \epsilon^2. \tag{6.27}$$

Comparing the results we see that we need $\epsilon \approx 7 \times 10^{-4}$ in order to explain the observations. This estimate, which is often quoted as an upper limit on the possible gravitational-wave strength, would correspond to a deformation of about 10 m on the surface of the star.

So far, this looks reasonable. Our estimate suggests that the Crab Pulsar might be an interesting gravitational-wave source, so let us consider the detectability of these waves. We can readily use the energy balance argument from (3.82) to estimate the strength of the emerging gravitational waves. This leads to

$$h \approx 8 \times 10^{-28} \left(\frac{\epsilon}{10^{-6}} \right) \left(\frac{f}{100 \text{ Hz}} \right)^2 \left(\frac{10 \text{ kpc}}{d} \right), \tag{6.28}$$

where we have used the fact that the gravitational-wave frequency, $f = \Omega/\pi$, is twice the star's rotation frequency. Noting that the Crab Pulsar is about 2 kpc away from us, we would have

$$h \approx 10^{-24} \quad \text{at} f \approx 60 \text{ Hz}. \tag{6.29}$$

This is too weak to be detected directly by any present or, indeed, future detector. However, as we have already mentioned, the effective amplitude h_c increases (roughly) as the square-root of the number of detected cycles. If one could observe the system for an entire year, then this buys a large factor, and we would have

$$h_c \approx 4 \times 10^{-20}. \tag{6.30}$$

---

[1] ... and petroglyphs by the Anasazi in the US southwest.

This estimate is often referred to as the Crab Pulsar *spin-down limit*. In general, we see that (6.25) implies that

$$\epsilon_{sd} = \left[\frac{5c^5 P^3 \dot{P}}{32(2\pi)^4 I_0} \right]^{1/2} \tag{6.31}$$

provides an upper limit on the allowed ellipticity of any observed system (see Figure 14.4).

However, nature is unlikely to be this generous. We have assumed that the pulsar spins down entirely due to gravitational-wave emission. It is easy to argue that this is optimistic. For a handful of relatively young pulsars, including the Crab, observations provide the second derivative of the spin. This allows us to work out the so-called braking index, a dimensionless quantity defined as

$$\dot{\Omega} = \text{constant} \times \Omega^n \rightarrow n = \frac{\Omega \ddot{\Omega}}{\dot{\Omega}^2}. \tag{6.32}$$

From our gravitational-wave formulas we would expect $n_{gw} = 5$, while electromagnetic dipole radiation would lead to $n_{em} = 3$ (see Espinoza *et al.* (2011*b*) and Chapter 9). The observed result for the Crab Pulsar is $n \approx 2.51$, clearly closer to electromagnetic result. This is a strong indication that gravitational radiation cannot be the main spin-down mechanism for this pulsar. The argument does, of course, not exclude a small gravitational-wave component. We will return to this problem in Chapter 14.

6.3   Contact binaries

One can imagine taking the discussion in different directions at this point. We could, for example, dig deeper into neutron star physics and consider the crust elasticity that is needed to support the mountains we have discussed. We might consider other observed pulsars to see if we can find systems that are more promising than the Crab. We will, indeed, discuss these problems later (see Chapter 14). Right now, we will embark on what may (at first) seem like a detour.

So far we have considered problems where the gravitational-wave emission is readily understood from the quadrupole formula. These examples provide useful and important illustrations, but they do not tell the full story. Our calculations only apply to idealized situation, like well-separated binaries or isolated spinning neutron stars. Reality is more complicated. The late stage of binary inspiral likely requires a detailed understanding of the neutron-star interior and we need to consider nonlinear aspects of relativity. Similarly, there are many relevant scenarios where neutron stars interact with their environment, and where this interaction impacts on the star's evolution.

As a first step in this direction, let us consider close binary systems where matter may flow from one partner to the other. We will discuss how this mass transfer affects the orbital evolution and how the accretion of matter may spin up the receiving star. These problems are important and, as we will see, complex. They provide a first insight into why the modelling of binary evolution is difficult. This also adds relevant historical context. The Binary Pulsar PSR B1913+16 provided impressive quantitative (albeit indirect) evidence that binary systems emit gravitational waves at the rate predicted by Einstein's theory, but this was not the first indication that the theory is correct. The first such evidence came from systems undergoing mass transfer (Faulkner, 1971).

In order to see how gravitational-wave emission affects systems undergoing mass transfer, let us start by considering two bodies in circular Keplerian orbit (as in Chapter 5), but now ask how a small particle in the mutual gravitational potential would move. Any gas flow between the stars should simply be governed by the equations of fluid dynamics. If we translate the Newtonian equations from Chapter 4 into a frame that rotates along with the two stars, accounting for both centrifugal effects and the Coriolis force, we have

$$(\partial_t + \boldsymbol{v} \cdot \nabla)\boldsymbol{v} = -\frac{1}{\rho}\nabla p - \nabla \Phi_R - 2\boldsymbol{\Omega} \times \boldsymbol{v}. \tag{6.33}$$

Here, ρ and p are the density and pressure of the gas, respectively, and $\boldsymbol{v}$ is the velocity. The last term in the equation is the Coriolis force. The so-called Roche potential, Φ_R, is given by

$$\Phi_R = -\frac{GM_1}{|\boldsymbol{r}-\boldsymbol{r}_1|} - \frac{GM_1}{|\boldsymbol{r}-\boldsymbol{r}_2|} - \frac{1}{2}(\boldsymbol{\Omega} \times \boldsymbol{r})^2, \tag{6.34}$$

where the last term is the centrifugal force. The inner equipotential lines for this kind of potential are sketched in Figure 6.2. Far away from the system, the potential remains that of a single point mass at the centre of mass, but the situation becomes more complicated—and interesting—as we zoom in. Close to each star any gas flow would be dominated by the pull of the closest body. This means that there will be a critical surface that joins the two stars. This surface will have the shape of an eight. This defines the *Roche lobe* associated with each of the stars and also the (inner) Lagrange point L_1,

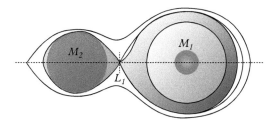

Figure 6.2 *An illustration of the Roche lobes that surround two stars in close orbit and the associated mass transfer from the secondary to the primary at the Lagrange point L_1. The figure also shows the accretion disk that forms around the primary.*

at which the two lobes intersect. Near this point it is easier for orbiting material to enter the Roche lobe of the other star than it is for it to escape the critical surface.

In general, one would expect the tidal interaction to lead to the stars rotation becoming synchronized with the orbital rotation. In such a system, the Roche lobes represent the maximum volume that each star can fill while remaining in hydrostatic equilibrium. If either star swells up to fill its Roche lobe at some point, mass will be transferred to the companion. Matter will simply fall from the secondary (the lobe-filling star) onto the primary. This transfer will be stable as long as the secondary continues to fill its Roche lobe. Such systems are called semi-detached binaries. In detached binaries both stars are much smaller than their Roche lobes, while both stars fill their Roche lobes in contact binaries. These phases of matter exhange are important for stellar evolution and impact on, for example, the formation rate for stellar mass black holes (discussed in Chapter 9).

It is important to consider the stability of the mass transfer. If we (for simplicity) assume that the total mass is conserved, then the mass exchange affects the system's angular momentum in such a way that

$$\frac{\dot{a}}{a} = \frac{2\dot{J}}{J} - \frac{2\dot{M}_2}{M_2}(1-q), \tag{6.35}$$

where $q = M_2/M_1$ is the mass ratio. Let us first consider the conservative case where $\dot{J} = 0$ and assume that $M_2 < M_1$. We see that, if $\dot{M}_2 < 0$, then $\dot{a} > 0$. As mass is transferred to the more massive companion, more matter is moved closer to the centre of mass, which means that the mass M_2 must move out to conserve angular momentum. Mass transferred to the lighter partner would decrease the orbit. We can approximate the mean radius of the secondary's Roche lobe as (Paczyński, 1971)

$$R_{\mathrm{L}} \approx \frac{2a}{3^{4/3}} \left(\frac{q}{1+q}\right)^{1/3}, \tag{6.36}$$

which means that

$$\frac{\dot{R}_{\mathrm{L}}}{R_{\mathrm{L}}} = \frac{\dot{a}}{a} + \frac{\dot{M}_2}{3M_2}, \tag{6.37}$$

and the Roche lobe is also affected by the change in mass ratio and separation. Combining this with our earlier result we find that

$$\frac{\dot{R}_{\mathrm{L}}}{R_{\mathrm{L}}} = \frac{2\dot{J}}{J} - \frac{2\dot{M}_2}{M_2}\left(\frac{5}{6} - q\right). \tag{6.38}$$

In the conservative case, when $\dot{J} = 0$, mass transfer from the less massive star cannot be stable if $q < 5/6$. The Roche radius R_{L} will simply expand and end the process. In contrast, if $q > 5/6$, then R_{L} will shrink onto the star and accelerate the transfer. This will lead to a violent evolution until q falls below $5/6$ and we return to the first case. The key insight from this argument is that we cannot have stable mass transfer unless the system loses angular momentum .

However, since many observed systems shine in X-rays, stable mass transfer must take place. Gravitational waves are thought to play a key role in facilitating this (Faulkner, 1971). The idea is simple. The stars gradually spiral together until the secondary fills its Roche lobe. At that point the mass transfer will become self-sustained as long as gravitational waves remove enough angular momentum to stop the Roche lobe from growing faster than the secondary's radius expands (a rate which obviously depends on the matter equation of state).

It is also relevant to ask what happens to the matter that crosses the Lagrange point. So far we assumed that it simply accretes directly onto the primary. However, this may not be the case. From the point of view of the primary, the Lagrange point acts like a rotating nozzle spraying material. Unless the binary period is long, this nozzle will rotate so fast that the material will be injected with considerable angular momentum. If we take L to be the distance from the centre of the primary to L_1, then the matter would have specific angular momentum

$$j = L^2 \Omega. \tag{6.39}$$

Once the matter enters the primary's Roche lobe its motion will be dictated by that star's gravitational field. This means that matter will orbit at a distance determined by the amount of angular momentum it has. This leads to an orbital frequency

$$\Omega_g = \left(\frac{GM_2}{R_c^3} \right)^{1/2}, \tag{6.40}$$

where the distance R_c is obtained by balancing the angular momentum

$$R_c^2 \Omega_g = L^2 \Omega. \tag{6.41}$$

If we assume Keplerian orbits, we have

$$R_c = a(1+q) \left(\frac{L}{a} \right)^4. \tag{6.42}$$

This is called the circularization radius. If R_c is smaller than the radius of the primary, then the matter will crash (obliquely) onto the surface. If, on the other hand, R_c is greater than the star's radius, the matter cannot accrete until it loses angular momentum. Instead, the matter forms an accretion disk around the primary, as indicated in Figure 6.2. Over a longer timescale viscosity moves angular momentum outwards, which allows the matter to creep closer to the primary and eventually accrete onto it. The angular momentum stored in the disk also enters in the previous mass transfer argument, making the problem more complicated.

6.4 Cosmic recycling

Let us return to the problem of gravitational waves from rotating neutron stars. The discussion of contact binaries allows us to broaden the discussion to include a wider set of astrophysical system. In particular, we can now consider accreting neutron stars in X-ray binaries.

Neutron stars may be born with a range of spin frequencies, basically depending on the formation process and the detailed dynamics of the collapsing core following the supernova explosion. Simple conservation of angular momentum arguments would suggest that a newly born neutron star ought to spin fast, but one can easily come up with counterarguments involving efficient coupling to the extended envelope during the collapse (Spruit and Phinney, 1998). Observations provide some support for these arguments. If we trace the spin-evolution of young pulsars back in time we typically find that they were not spinning near the break-up limit, above which the centrifugal force causes mass shedding (see Chapter 12). In the case of the Crab Pulsar one finds that it would have been born with a period ~ 19 ms, while the mass-shedding limit corresponds to a period of ~ 1 ms.

The fastest known pulsars are all old. They also have weak magnetic fields, which may suggest that the field decays as the stars age. The most rapidly rotating millisecond radio pulsar is PSR J1748-2446ad with spin frequency $\nu_s = 716$ Hz. In the standard picture, such systems form through cosmic recycling (Alpar *et al.*, 1982; Radhakrishnan and Srinivasan, 1982). The neutron star gains its angular momentum during a period of accretion from a (low-mass) binary partner, following the mass transfer scenario we already discussed. The idea is supported by the general observation that approximately 80% of millisecond pulsars are in binary systems, compared to only 1% of slower pulsars (see Figure 9.3). There is also a growing body of evidence connecting fast-spinning neutron stars with low-mass companions to the radio millisecond pulsars. The 1998 discovery of the first accreting millisecond X-ray pulsar SAX J1808-3658 (see Table 6.1 and Wijnands and van der Klis (1998)) added important support for the recycling paradigm and we now have evidence of transitional systems which appear to be in the process of moving from one population to the other (Archibald *et al.*, 2010).

The low-mass X-ray binaries could be interesting gravitational-wave sources. All observed systems appear to rotate far below the mass-shedding limit for a typical neutron star (see Figure 6.3). Could it be that nature imposes a speed limit on spinning stars? If so, what is the mechanism that enforces this? The answer may involve gravitational waves (Papaloizou and Pringle, 1978; Wagoner, 1984; Chakrabarty *et al.*, 2003).

In order to work out whether it is reasonable to assume that the spin-up of accreting neutrons stars may be stalled because of gravitational radiation, we need to consider two issues. First, we have to understand the angular momentum transferred as matter accretes onto the star. Secondly, we need to argue that the accreting star develops

Table 6.1 *Observational parameters for the accreting millisecond X-ray pulsar SAX J1808-3658.*

Spin frequency (Hz)	401
Rate of spin evolution (Hz/s)	2×10^{-13}
Peak luminosity (erg/s)	$\sim 10^{36}$
Distance (kpc)	2.5
Orbital period (hours)	2.1
Companion mass ($M_\odot$)	0.05–0.1

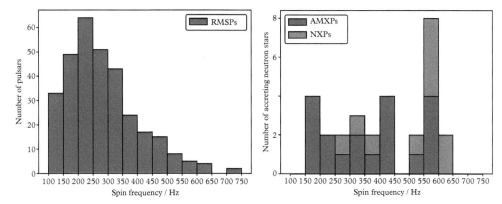

Figure 6.3 *The observed distribution of fast-spinning neutron stars. Left: the millisecond radio pulsars, based on data from http://www.atnf.csiro.au/research/pulsar/psrcat/, with PSR J1748-2446ad being the fastest known system ($\nu_s = 716$ Hz). Right: the accreting neutron stars in low-mass X-ray binaries (both accretion powered, AMXP, and nuclear powered, NXP), based on data from Patruno et al. (2017). The fastest known accreting neutron star is 4U 1608-522 (at $\nu_s = 620$ Hz). (Figures provided by F. Gittins.)*

a significant asymmetry. We will leave aside the latter problem for the moment, and simply take it as given that the star is deformed and emits gravitational radiation at twice the spin frequency. We can then work out how large the deformation needs to be for the gravitational-wave emission to balance the accretion torque. In the simplest case (see Chapter 14 for a more detailed model), we can approximate the accretion torque as the angular momentum change associated with orbiting matter falling onto the star's equator. Effectively assuming that the accretion disk reaches the star's surface, we then have

$$\dot{J} = \dot{M}\sqrt{GMR}. \tag{6.43}$$

We balance this with the loss due to gravitational-wave emission (6.24), using

$$\dot{E} = \Omega \dot{J},$$

(6.44)

to relate the energy and angular momentum losses. Combining this with (6.24) (assuming a neutron star with mass $1.4 M_\odot$ and radius 10 km) we arrive at an estimate for the deformation required to balance accretion spin-up:

$$\epsilon \approx 10^{-8} \left(\frac{\dot{M}}{10^{-9} M_\odot/\text{yr}} \right)^{1/2} \left(\frac{500 \text{ Hz}}{\nu_s} \right)^{5/2}.$$

(6.45)

This estimated deformation seems small enough to make it plausible that gravitational-wave spin-down could play a role in these systems.

The predicted ellipticity is certainly much smaller than that required to 'explain' the spin-down of the Crab Pulsar. In that case we learned that our estimate was optimistic. Confronting our new estimate with observational data for accreting systems is much more difficult. If we want to move beyond the back-of-the envelope understanding, we need to consider a number of complicating aspects. The connection with observations typically links the observed X-ray luminosity to the mass accretion rate. This is done by assuming that the gravitational potential energy released by the infalling matter is radiated as X-rays. This leads to

$$L_x \approx \frac{GM\dot{M}}{R}.$$

(6.46)

Observations then give us a direct estimate for the required deformation

$$\epsilon \approx 3 \times 10^{-9} \left(\frac{L_x}{10^{36} \text{ erg/s}} \right)^{1/2} \left(\frac{500 \text{ Hz}}{\nu_s} \right)^{5/2}.$$

(6.47)

However, most of the observed systems are transients with variable, and often low, mass transfer rates. For example, SAX J1808-3658 has an average X-ray luminosity in quiescence of about 10^{31} erg/s. Meanwhile, during X-ray outbursts the luminosity rises to 10^{35} erg/s on a timescale of days. The system seems to undergo bursts lasting for roughly three weeks every two years or so. Using a single luminosity to infer the mass accretion rate is clearly an oversimplification, but at least we have an idea of where to start.

We may, for example, approximate the maximum accretion rate by balancing the pressure due to spherically infalling gas to that of the emerging radiation. This leads to the Eddington limit

$$\dot{M}_\text{Edd} \approx 1.5 \times 10^{-8} \left(\frac{R}{10 \text{ km}} \right) M_\odot/\text{yr},$$

(6.48)

which would correspond to an X-ray luminosity

$$L_x \approx 1.8 \times 10^{38} \left(\frac{M}{1.4 M_\odot} \right) \left(\frac{\dot{M}}{\dot{M}_{\mathrm{Edd}}} \right) \text{ erg/s.} \tag{6.49}$$

These estimates give us an idea of what the strongest gravitational-wave sources may be. An often-quoted example is Sco X1, one of the brightest X-ray sources in the sky. In this case we have $L_x \approx 2 \times 10^{38}$ erg/s (Bildsten, 1998). This suggests that the neutron star must accrete near the Eddington limit, which leads to

$$\epsilon \approx 4 \times 10^{-8} \left(\frac{500 \text{ Hz}}{\nu_s} \right)^{5/2}. \tag{6.50}$$

Unfortunately, the spin frequency of the Sco X1 neutron star is not known. In order to search for it one has to consider a range of frequencies and this is computationally expensive (see Chapter 14).

The second aspect we need to consider is the detailed accretion torque. Our estimate (6.43) is too naive. The accretion problem is messy and it has proved difficult to use observations to constrain the theory. In fact, there does not appear to be an immediate association between the X-ray luminosity (the rate of accretion) and the rate of change of the spin frequency. For example, the neutron star in SAX J1808-3658 was seen to spin up during outbursts in 1998 and 2000 (with $\dot{\nu}_s \approx 2 - 3 \times 10^{-13}$ Hz/s), but it was undergoing constant spin-down during the following outburst in 2002. The problem is further complicated by the fact that one should account for fluctuations in the luminosity, on a timescale of hours to days, during quiescence.

Nevertheless, we can do better than (6.43) by noting that, for a magnetized star, the accretion disk may not reach all the way to the surface. At some point, the matter flow will be dominated by the magnetic field and funnelled onto the star's surface along the field lines. This effect may not be overwhelming for weakly magnetized neutron stars in low-mass X-ray binaries, but we need to quantify it if we want to improve our understanding. We will consider this problem in Chapter 14.

6.5 Spin–orbit evolution

As the physics involved tends to be complicated, one often has to settle for phenomenological models. This may seem far removed from the elegance and precision of general relativity, but it may be the best that we can hope for. And one should not underestimate the importance of simple, qualitative arguments. They may, after all, hint at directions where more effort should be invested.

As an example of an indicative scenario—essentially how we can use estimated timescales to explore the behaviour of a system—let us consider the spin–orbit evolution of an accreting neutron star (borrowing results from Ho *et al.* (2011*b*)). This ties together the different concepts we have discussed. We have considered how mass transfer affects the binary evolution, and accounted for the angular momentum associated with gravitational waves from the orbit. We have also discussed the spin-up of the primary star as matter and angular momentum from the secondary are accreted onto it. We have suggested that observational data hints at the need for an additional mechanism for angular momentum loss, potentially associated with gravitational waves from the primary. In this case we have additional angular momentum loss and it is interesting to consider how this impacts on the orbital evolution.

Let us consider an accreting neutron star with mass $M_1 = m_x M_\odot$ (using $m_x = 1.4 M_\odot$ as the canonical example) and a binary companion with mass $M_2 = m_c M_\odot$. Then we can use (6.36) to work out when the donor star fills its Roche lobe. Taking the companion to be a main sequence star, $m_c \approx R_c/R_\odot$, where $R_\odot$ is the solar radius. Using (6.36) and Kepler's law, we then have

$$a = 0.8 R_\odot (m_x + m_c)^{1/3} \left(\frac{P_{\rm orb}}{2\,{\rm hr}} \right)^{2/3}, \tag{6.51}$$

where $P_{\rm orb}$ is the orbital period. We also have a relationship between the mass of the companion and the orbital period

$$m_c = 0.23 \hat{m}_c \left(\frac{P_{\rm orb}}{2\,{\rm hr}} \right), \tag{6.52}$$

where $\hat{m}_c \equiv m_c/m_c^{\rm MS}$ and $m_c^{\rm MS}$ is the mass of a main-sequence star that just fills its Roche lobe at $P_{\rm orb}$. The factor of $\hat{m}_c$ (< 1) is due to the fact that the companion star in a binary has a larger radius for its given mass than an isolated star. This encodes an uncertainty in the (less evolved, since we only consider $m_c < m_x$) evolutionary state of the companion.

Now consider mechanisms that would lead to a change in the angular momentum of the orbit or the neutron-star spin, and thus cause a system to move in the v_s–$P_{\rm orb}$-plane. We have seen that mass transfer can cause an increase in the size of the orbit, while orbital angular momentum loss due to magnetic braking or gravitational-wave emission causes the orbit to decrease. The orbital period which separates expansion and decay is estimated to be > 0.5 days, while magnetic braking is dominant for orbital periods longer than one hour. A typical (see Chapter 9) estimate for the magnetic braking torque leads to a timescale for orbital decay

$$t_{\rm decay} \approx 2.1 \times 10^8 \hat{m}_c^{-4/3} m_x^{2/3} \left(\frac{P_{\rm orb}}{2\,{\rm hr}} \right)^{-2/3} {\rm yr}, \tag{6.53}$$

and a mass transfer rate (in units of $10^{-11} M_\odot\,{\rm yr}^{-1}$)

$$\dot{M}_{-11} = 80 \hat{m}_c^{7/3} m_x^{-2/3} \left(\frac{P_{\rm orb}}{2\,{\rm hr}} \right)^{5/3}. \tag{6.54}$$

Note that the mass accretion rate increases with the orbital period. It is also worth noting that, at very short orbital periods, $P_{\rm orb} \lesssim 4$ hr, gravitational radiation becomes important and some accreting millisecond pulsar companions are degenerate stars. These systems may be evolving to longer orbital periods, but this should happen slowly because the mass transfer and angular momentum loss rates are low. For example, SAX J1804-3658 has a degenerate companion and the orbital period is seen to increase on a timescale $P_{\rm orb}/\dot{P}_{\rm orb} \approx 6 \times 10^7$ yr.

As we have seen, accretion of angular momentum from the companion can spin up the neutron star. The timescale for spin-up is (again; see Chapter 14)

$$t_{\rm su} \approx 1.5 \times 10^9 \, B_8^{-2/7} \dot{M}_{-11}^{-6/7} m_{\rm x}^{-3/7} \left(\frac{\nu_s}{100 \text{ Hz}}\right) \text{ yr}, \tag{6.55}$$

where B_8 is the neutron star magnetic field (in units of 10^8 G). Using the mass accretion rate from (6.54), we see that

$$t_{\rm su} \approx 3.4 \times 10^7 \, B_8^{-2/7} \hat{m}_{\rm c}^{-2} m_{\rm x}^{1/7} \left(\frac{P_{\rm orb}}{2 \text{ hr}}\right)^{-10/7} \left(\frac{\nu_s}{100 \text{ Hz}}\right) \text{ yr}. \tag{6.56}$$

Meanwhile, a deformed neutron star spins down due to gravitational waves. From (6.25) we have the spin-down timescale

$$t_{\rm sd} \approx 2.9 \times 10^{10} \, m_{\rm x}^{-2} \epsilon_{-8}^{-2} \left(\frac{\nu_s}{100 \text{ Hz}}\right)^{-4} \text{ yr}, \tag{6.57}$$

where $\epsilon_{-8} = \epsilon/10^{-8}$ is the star's ellipticity. We ignore spin-down by electromagnetic dipole radiation, which only becomes dominant at

$$B_8 \gtrsim 10\epsilon_{-8} \left(\frac{\nu_s}{100 \text{ Hz}}\right). \tag{6.58}$$

The evolution of the star's spin, ν_s, and the orbital period, $P_{\rm orb}$, are (primarily) determined by the process with the shortest timescale. As an example, a comparison of the different estimates is provided in Figure 6.4. For the ranges displayed, we identify three regions representing; spin-up from the accretion torque, spin-down from gravitational radiation, and orbital decay from magnetic braking. It is natural to focus on the transitions between the different regimes. First of all, an accreting neutron star moves towards higher ν_s if the pulsar rotation rate is below both ν_1 (from $t_{\rm su} < t_{\rm sd}$) and ν_3 (from $t_{\rm su} < t_{\rm decay}$), where ν_1 is given by

$$\nu_1 \approx 330 \, \epsilon_{-8}^{-2/5} B_8^{2/35} \hat{m}_{\rm c}^{2/5} \left(\frac{P_{\rm orb}}{2 \text{ hr}}\right)^{2/7} \text{ Hz}, \tag{6.59}$$

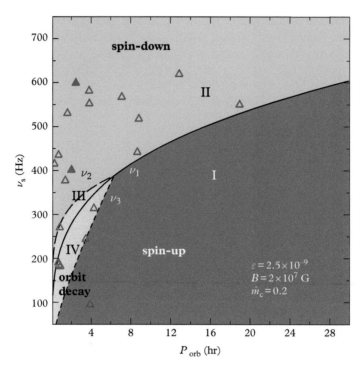

Figure 6.4 *Estimated regimes of spin–orbit evolution for accreting millisecond pulsars. A system in the spin-up/spin-down regime (I/II) moves to higher/lower spin frequencies, as well as shorter orbital periods on a longer timescale. Systems in the orbit decay regime (III/IV) move to shorter orbital periods, as well as to lower/higher spin frequencies on a longer timescale. The lines (labeled ν_1, ν_2, ν_3) separating the regimes are given by Eqs. (6.59)–(6.61). Open triangles denote accreting millisecond pulsars, while the solid triangles are the two specific systems SAX J1808.4-3658 at $(P_{orb}, \nu_s)=$(2 hr, 401 Hz) and IGR J00291+5934 at (2.5 hr, 599 Hz), which exhibit short-term spin-up and long-term spin-down. (Reproduced from Ho et al. (2011b) with permission by the AAS.)*

while ν_3 follows from

$$\nu_3 \approx 750\, B_8^{2/7}\, \hat{m}_c^{2/3} \left(\frac{P_{\mathrm{orb}}}{2\ \mathrm{hr}}\right)^{16/21} \mathrm{Hz}. \qquad (6.60)$$

This defines spin-up region (I) in Figure 6.4.

Meanwhile, if the rotation rate is above both ν_1 and ν_2 (from $t_{\mathrm{sd}} < t_{\mathrm{decay}}$), where

$$\nu_2 \approx 270\, \epsilon_{-8}^{-1/2}\, \hat{m}_c^{1/3} \left(\frac{P_{\mathrm{orb}}}{2\ \mathrm{hr}}\right)^{1/6} \mathrm{Hz}, \qquad (6.61)$$

then the star moves towards lower ν_s. This is the spin-down regime (II) in Figure 6.4.

Finally, a decrease in orbital period occurs when the rotation rate is below ν_2 but above ν_3. Note that deviations from low values of $\hat{m}_c$ (up to a factor of order unity, e.g. due to degenerate companions) have the strongest effect on ν_3, but do not qualitatively change the picture.

Though (obviously) dependent on the various parameters, the timescale of the dominant process in each regime bears out the observed population. In the spin-up region (long P_{orb}, low ν_s) we have

$$t_{\text{su}} \sim 7 \times 10^6 \hat{m}_c^{-2} \left(\frac{P_{\text{orb}}}{10 \text{ hr}} \right)^{-10/7} \left(\frac{\nu_s}{200 \text{ Hz}} \right) \text{ yr.} \tag{6.62}$$

In the spin-down region (high ν_s), the evolution is independent of the orbital period and occurs on a timescale

$$t_{\text{sd}} \sim 2 \times 10^9 \left(\frac{\epsilon}{10^{-9}} \right)^{-2} \left(\frac{\nu_s}{500 \text{ Hz}} \right)^{-4} \text{ yr.} \tag{6.63}$$

Finally, in the orbit decay region (short P_{orb}), the evolution is independent of the spin frequency and occurs on a timescale

$$t_{\text{decay}} \sim 2 \times 10^8 \hat{m}_c^{-4/3} \left(\frac{P_{\text{orb}}}{3 \text{ hr}} \right)^{-2/3} \text{ yr.} \tag{6.64}$$

The main lesson is that we can try to explain the observed population of fast-spinning accreting neutron stars in terms of the evolution of ν_s and P_{orb}, depending on the process (magnetic braking, mass accretion, and gravitational radiation) with the shortest timescale. A system born at low ν_s and long P_{orb} very quickly spins up to high ν_s. On a much longer timescale, these fast-spinning sources slow down, while their orbits shrink. The first phase happens on a timescale of a million years, whereas the second may last a thousand times longer. Once mass accretion stops, there is no longer a spin-up torque. The binary then contains a rotation-powered millisecond pulsar that can move into the (former) spin-up region in Figure 6.4 by spinning down (as a result of gravitational-wave emission or electromagnetic radiation) or by expanding its orbit.

This model is, obviously, only a qualitative sketch, but it accords reasonably well with observations. It is, for example, instructive to compare the estimates to data for the two accreting millisecond pulsars SAX J1808-3658 and IGR J00291+5934 (Galloway *et al.*, 2005). These systems show an overall spin-down with timescales $\nu_s/|\dot{\nu}_s| \approx 2 \times 10^{10}$ and 5×10^9 yr, respectively (interupted by short spin-ups during outbursts with timescales 10^8 yr and 2×10^7 yr, respectively). Given this, one would expect both systems to reside in the spin-down region (II). Adjusting the parameters of the model accordingly, we end up with the results in Figure 6.4. Here is it worth noting a lack of observed slow spin–long orbital period systems. The discovery of a system at the relatively low ν_s (100 Hz $\lesssim \nu_s \lesssim$ 400 Hz) and long P_{orb} would clearly challenge the scenario (or at least question the parameters assumed in Figure 6.4). Nevertheless, it is suggestive that the quadrupole deformation required for the gravitational-wave emission is $\epsilon \gtrsim 10^{-9}$. For smaller deformations, spin-down by gravitational radiation becomes irrelevant—see eqs. (6.59) and (6.61)—and all observed systems would be in spin-up or orbit expansion/decay, contrary to what is seen. It is interesting to note that the estimated ellipticity is well below the theoretical maximum (see Chapter 12) and also (although much less so) below the current limit set by gravitational-wave searches for signals from known pulsars (see Chapter 14).

7

Catching the wave

With an understanding of the basic theory and some astrophysics perspective, we can turn our attention to the experimental challenge. We already have an idea of the effect that a passing gravitational wave will have on matter and we know it will be tiny. Given that the coupling is proportional to $G/c^5 \approx 3 \times 10^{-53}$ s^3/kg m^2 we need to compensate by involving large masses and/or high velocities. Still, we realistically need to measure relative displacements of order 10^{-21}. Common sense would suggest this is impossible (Saulson, 2000). After all, even if we design a kilometre-size instrument we need to resolve a change in distance about 1,000 times smaller than the size of the atomic nucleus. Luckily, experimenters do not always listen to reason.

The question is: How do we turn the notion of independent test particles, as in Figure 3.2, into a practical proposition? We clearly cannot reproduce this setup in the laboratory, but perhaps we can get close to it. In principle, any object can serve as a gravitational-wave detector, but it is not enough to absorb the energy from the passing wave. We need to amplify the signal to be able to measure it. One possibility would be to make use of resonances. Consider as a simply toy detector two masses joined by a spring (ignoring the mass of the spring). A gravitational wave will push the masses together, then apart, and so on. The spring tries to counter the effect. In general, the motion of the masses would be tiny but if the wave were to have frequency close to one of the natural frequencies of the system (representing its free oscillations), the associated resonance would amplify the effect. This simple idea motivated the first attempt to build a gravitational-wave detector, by Joseph Weber in the 1960s. In practice, Weber (and those following in his footsteps) used massive cylindrical metal bars to represent the resonant system. Later generations of instruments were cooled to a low temperature to counteract thermal fluctuations (Visco and Votano, 2000; Fafone, 2004) .

Another possibility would be to replace the spring connecting the masses with a pendulum suspending each mass, monitoring their movement with laser light. By making sure that the natural frequencies associated with the suspension lie outside the frequency range of interest, and using interferometry to monitor the motion of the test masses, such instruments can be made extremely sensitive. In fact, the use of laser interferometry— pioneered by Rai Weiss at MIT and Ron Drever in Glasgow, and after decades of

Gravitational-Wave Astronomy: Exploring the Dark Side of the Universe. Nils Andersson, Oxford University Press (2020).
© Nils Andersson. DOI: 10.1093/oso/9780198568032.001.0001

developments leading to the Advanced LIGO instruments[1]—was the decisive move towards successful detection.

A third option would be to implement the idea envisaged in Figure 3.2 and actually use freely floating test masses. This is obviously not possible on the ground, but it can be achieved in space. This is the idea behind space-borne detectors like LISA (Amaro-Seoane *et al.*, 2017).

The idea behind the different kinds of detectors may be simple, but the engineering challenge associated with reaching the required sensitivity is still immense. Not the least since any detector is limited by a number of noise sources. In fact, in order to understand the behaviour of any given detector, it is crucial to have a good handle on the limiting noise sources.

7.1   Resonant mass detectors

Suppose that a gravitational wave propagating along the z axis, with pure plus polarization amplitude h_+, and frequency ω, passes by our proposed mass-spring toy detector, now aligned with the x-direction (say). Let us try to calculate the amplitude of the oscillations induced in the instrument and the amount of energy absorbed by it. As outlined in Chapter 3, the tidal force induced on the detector follows from the equation for geodesic deviation, and as a result the masses move according to

$$\ddot{\xi} + \dot{\xi}/\tau + \omega_0^2 \xi = -\frac{1}{2}\omega^2 L h_+ e^{i\omega t}, \tag{7.1}$$

where ω_0 is the natural vibration frequency of the detector, τ is the damping time of the oscillator due to friction, L is the separation between the two masses, and ξ is the relative change in the distance between them. The passing gravitational wave provides a driving force for the oscillator, and the solution is simply

$$\xi = h_+ \frac{L e^{i\omega t}}{2} \frac{\omega^2}{\omega_0^2 - \omega^2 + i\omega/\tau}. \tag{7.2}$$

If the frequency of the impinging wave is near the natural frequency of the oscillator (so that we are close to resonance), the detector is excited into large-amplitude motion and rings like a bell. Actually, for $\omega = \omega_0$ we have the maximum amplitude

$$\xi_{\text{max}} = \frac{1}{2}\omega_0 \tau h_+ L. \tag{7.3}$$

---

[1]   The historical developments are nicely summarized in Saulson (2005), while an in-depth description of the technology can be found in Saulson (2017) and Bond *et al.* (2017). The approach to space-borne instruments is reviewed by Tinto and Dhurandhar (2014).

Since the size of the detector, L, and the amplitude of the gravitational waves are fixed, large-amplitude motion can be achieved only by increasing the quality factor $Q = \omega_0 \tau$ of the instrument. Moreover, in practice the frequency of the detector is fixed by its size so the only improvement we can make is to choose materials with long relaxation times.

In reality, the sensitivity of a resonant bar detector depends on many other parameters. Even if we assume perfect isolation from external noise sources (acoustic, seismic, electromagnetic), we are limited by thermal noise. In order to detect a signal, the energy deposited by the gravitational wave every τ seconds must be larger than the energy $k_B T$ due to thermal fluctuations. This leads to an expression for the minimum detectable energy flux of gravitational waves, which, in turn, leads to a minimum detectable strain

$$ h_{\min} \leq \frac{1}{\omega_0 L Q} \left(\frac{15 k_B T}{M} \right)^{1/2} , \tag{7.4} $$

where M is the mass of the detector. In the case of Weber's first detector (Weber, 1967), with $M = 1{,}410$ kg, length $L = 1.5$ m, resonant frequency $\omega_0 = 1{,}660$ Hz and quality factor $Q = 2 \times 10^5$ at room temperature, the smallest detectable strain would be of order 10^{-20}. However, this estimate applies only to gravitational waves whose duration is at least as long as the damping time of the bar's vibration and whose frequency perfectly matches the resonance of the detector. For short bursts or periodic signals off resonance the sensitivity would decrease by several orders of magnitude. You would have to be very lucky to detect gravitational waves with an instrument operating at this level.[2]

More modern bar detectors (Visco and Votano, 2000; Fafone, 2004) are complicated devices, consisting of a solid metallic cylinder weighing a few tons suspended in vacuum by a cable wrapped under the centre of gravity. This suspension system protects the antenna from external mechanical shocks. The whole system is cooled down to temperatures of a few kelvin (or even millikelvin). To monitor the vibrations of the bar, piezoelectric transducers (or more modern capacitive ones) are attached to the bar. These converted the bar's mechanical energy into electrical energy. The signal is amplified by an ultra-low-frequency amplifier, via a device called a SQUID (Superconducting QUantum Interference Device), before becoming available for analysis.

Throughout much of the 1990s, a number of such devices were in (nearly) continuous operation at several institutions around the world. At best, they achieved sensitivities of a few times 10^{-21}, but never provided conclusive evidence for gravitational waves. At the given level of sensitivity, they would have had a reasonable chance of detecting a signal from a supernova explosion in our Galaxy, but they would still have had to be lucky, because such events are rare (a few per century).

More advanced ideas include the construction of spherical resonant detectors (see Lobo (2000) for the principles), the advantages of which would be a high mass and a

---

[2] There were claimed events, but they were relatively easy to dismiss with astrophysics arguments. For example, the energy associated with the coincidences reported in Weber (1969) would be at the level of the Sun having exploded.

broader sensitivity range (a bandwidth of up to 100–200 Hz). Such a detector would also be omnidirectional (as five modes of the sphere could be excited). This would make it possible, in the case of detection, to obtain information about the polarization of the wave and the direction to the source.

However impressive the technology, a resonant mass detector is ultimately a narrow-band instrument, typically operating at a frequency $500 - 1,500$ Hz with a bandwidth of a few tens of hertz. Many astronomical sources are expected to radiate at far lower frequencies and (arguably) the most important sources, inspiralling compact binaries, produce broadband signals. The chance of detecting such gravitational waves is far greater with a detector that operates at lower frequencies and has wide bandwidth. It was perhaps inevitable that laser interferometers would eventually become the technology of choice.

7.2   Gravitational waves and light beams

There is a (seemingly simple) question everyone that ever tried to explain gravitational-wave interferometers to a general audience will be familiar with. In the description of the freely floating test masses in Chapter 3, we described the effect of gravitational waves as stretching and squeezing distances. How come, when we try to measure that distance, the ruler we use is not stretched and squeezed, as well? To some extent, we have already seen the answer. We can, indeed, set up the problem in such a way that the individual masses remain at the same coordinate location (this is the TT gauge, illustrated in Figure 3.3). The reason we can nevertheless measure the effect is that we are measuring spacetime distance, not just spatial distance. We have already sketched this argument in Chapter 3, but it is worth considering it in more detail.

Understanding the effect gravitational waves have on null and time-like geodesics is an essential step towards appreciating how different detectors work, and it is also necessary if we want to figure out new methods of detection. For example, laser interferometric detectors are based on the principle that the round trip travel time of laser beams in the interferometer arms is affected by a passing gravitational wave. Similarly, arrival times of radio pulses from a millisecond pulsar will be modulated by a gravitational wave passing by the Earth (see Chapter 22). It is, therefore, instructive to explore how gravitational waves interact with beams of light in general.

We already know from the equivalence principle that it is impossible to distinguish between non-inertial reference frames and gravitational fields in a local neighbourhood of spacetime. No experiment restricted to small time- and length-scales can differentiate between accelerated reference frames and gravitational fields. For instance, as Einstein himself taught us, no local experiment would detect the presence of the Earth's gravity in a freely falling lift. One way to infer the presence of such a field is to compare the frequency of a standard source of light as the light beam propagates from one point to another. As light 'climbs' up a gravitational potential it gets redshifted. Another way to detect gravitational fields is to watch two nearby freely falling test masses, as in the thought experiment from Chapter 2. After some time, the two masses will be seen to approach each other. The effect of gravitational waves on light beams and free test masses is not different from the effect of gravity itself.

Let us recall the form of the metric for weak gravitational waves and consider the associated symmetries. As discussed in Chapter 3, the metric of a plane gravitational wave travelling in the z direction is given by

$$ds^2 = -c^2 dt^2 + [1 + h_+(t - z/c)] dx^2 + [1 - h_+(t - z/c)] dy^2$$
$$+ 2h_\times(t - z/c)dx\,dy + dz^2, \tag{7.5}$$

where $\{t, x, y, z\}$ are the flat spacetime coordinates, and h_+ and $h_\times$ are the amplitudes of the plus and cross polarizations of the wave. Note that the metric coefficients are independent of the coordinates x and y. This independence is due to the symmetry of the problem. As discussed in Chapter 2, whenever the metric is independent of a certain coordinate there will be an associated Killing vector. The above form of the metric explicitly implies that there are at least two such Killing vectors. Moreover, from the fact that the metric depends on t and z only through the combination $t - z/c$ (it is propagating wave) we expect that there should be a third Killing vector. The existence of this vector can be most easily seen by transforming to a new set of (advanced and retarded) coordinates $\{\xi, x, y, \chi\}$, defined by

$$\xi \equiv ct - z, \quad \chi \equiv ct + z, \tag{7.6}$$

leaving x and y unchanged. These are called null coordinates, as lines of constant ξ and χ define null rays. In this new coordinate system the metric takes the form

$$ds^2 = -d\xi\,d\chi + [1 + h_+(\xi)]dx^2 + [1 - h_+(\xi)]dy^2 + 2h_\times(\xi)dx\,dy. \tag{7.7}$$

We see that, in null coordinates the metric is independent of the coordinates χ, x and y. Therefore, there are three Killing vectors, given by

$$k_1 = \frac{\partial}{\partial\chi}, \quad k_2 = \frac{\partial}{\partial x}, \quad \text{and} \quad k_3 = \frac{\partial}{\partial y}. \tag{7.8}$$

The components of these Killing vectors are

$$k_1^{'a} = (0, 0, 0, 1), \quad k_2^{'a} = (0, 1, 0, 0), \quad k_3^{'a} = (0, 0, 1, 0), \tag{7.9}$$

where the primes indicate that the components refer to the $\{\xi, x, y, \chi\}$ coordinates. However, the calculation of the effect of gravitational waves on light beams is simpler if we work with the $\{t, x, y, z\}$ coordinates. The components of the Killing vectors in these coordinates can be determined using the standard transformation law for vectors (or simply noting that $k_1^a = \partial x^a/\partial\chi$, etc.). Thus, we have

$$k_1^a = \frac{1}{2}(1,0,0,1), \quad k_2^a = (0,1,0,0), \quad k_3^a = (0,0,1,0). \tag{7.10}$$

Following the demonstration in Chapter 2 (see Eq. (2.73)), we then have a conserved quantity ϕ_i associated with each k_i^a ($i = 1-3$) along the spacetime trajectory of a test mass or light, given by

$$\phi_i = g_{ab} k_i^a V^b, \tag{7.11}$$

where V^a is the four-velocity (for massive objects) or the tangent to the trajectory (for light rays).

Let us now consider the effect that a passing gravitational wave has on light (like the laser beam in a detector). The path of light in a gravitational field is described by null geodesics. Consider a beam of light in the field of a gravitational wave, for the sake of simplicity assuming that the gravitational wave is propagating in the z direction. Let us suppose a beam of light is sent from an emitter, at the origin of the chosen coordinate system, to a receiver a distance L away from the emitter, as in Figure 7.1. The directions of propagation of the light beam and gravitational wave define a plane, which we take to be the $x-y$ plane. Furthermore, let us assume that the wave consists of only the plus polarization, i.e. $h_+ \neq 0$ and $h_\times = 0$. In this case the metric simplifies to:

$$ds^2 = -c^2 dt^2 + [1 + h_+(t - z/c)] dx^2 + [1 - h_+(t - z/c)] dy^2 + dz^2. \tag{7.12}$$

Even though we have assumed a specific polarization and direction of propagation of the waves, the final result will be covariant so remains valid in general (as long as the

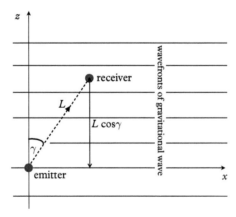

Figure 7.1 *A light beam (dashed line), making an angle γ with the z axis and travelling in the $x-z$ plane, is sent from an emitter to a receiver located a distance L away. The frequency of light at the receiver is Doppler modulated, relative to the emitter, as it travels in the field of a gravitational wave moving in the z direction (with wave fronts parallel to the x-axis). (Illustration by B. Sathyaprakash.)*

wavelength of the wave is far greater than the distance between the emitter and the receiver).

Let the light beam make an angle γ with the z-axis. In flat spacetime such a light beam will be described by a null vector

$$U^a = v\,(c, c\sin\gamma, 0, c\cos\gamma), \tag{7.13}$$

where v is the frequency of the light. We want to calculate the time it takes for a light beam to traverse from the emitter to the receiver and back in the presence of gravitational waves. In the absence of waves the time interval Δt_R between consecutive light pulses arriving at the receiver will be the same as the time interval $\Delta t_E = v^{-1}$ between consecutive pulses at the emitter. A gravitational wave will alter the time of flight of the light pulses. We can work out the change in travel time from the Doppler shift in the frequency of the light beam. That is, we focus on the change in the frequency of light as it moves from emitter to receiver.

As discussed in Chapter 2, the vector U^a parallel transported along itself defines the light ray and, in flat spacetime, both the frequency v and the angle γ remain fixed as the beam propagates. In a curved spacetime, however, both will change. Let us denote by v_E and v_R the frequencies, by γ_E and γ_R, the angles and by V_E^a and V_R^a the null vectors, at the emitter and receiver, respectively, in the gravitational-wave spacetime. To linear order in the metric perturbation h_{ab}, a null vector V^a in the perturbed spacetime is related to the flat spacetime null vector U^a by

$$V^a = U^a - \frac{1}{2}\eta^{ab}h_{bc}U^c. \tag{7.14}$$

Using this relation it is easy to see that

$$V_E^a = v_E\left[c, c\sin\gamma_E\left(1 - \tfrac{1}{2}h_+(t_E)\right), 0, c\cos\gamma_E\right], \tag{7.15}$$

$$V_R^a = v_R\left[c, c\sin\gamma_R\left(1 - \tfrac{1}{2}h_+(t_R - L\cos\gamma_R/c)\right), 0, c\cos\gamma_R\right], \tag{7.16}$$

where t_E and $t_R = t_E + L/c$ are the times when the beam leaves the emitter and is received at the receiver, respectively. Note that the gravitational field at the receiver is not evaluated at time t_R but at an earlier time $t_R - L\cos\gamma_R/c$. The gravitational-wave phase does not quite catch up with the phase of the light beam because they travel in different directions.

The null vectors at the emitter and receiver are related by parallel transport. We could parallel transport V_E^a from the emitter to the receiver and compare the final vector to V_R^a to arrive at a relation between v_E and v_R. However, we can readily derive such a relation by making use of the conserved quantities ϕ_i. We know that ϕ_i at the emitter must be the same as at the receiver. Let us first note that, due to our choice of geometry, ϕ_3 is identically zero. Meanwhile, for $i = 1, 2$ we have

$$\phi_{1E} = \phi_{1R} \quad \Rightarrow \quad v_E(1 - \cos\gamma_E) = v_R(1 - \cos\gamma_R),$$

$$\phi_{2E} = \phi_{2R} \quad \Rightarrow \quad v_E\left[1 + \tfrac{1}{2}h_+(t_E)\right]\sin\gamma_E = v_R\left[1 + \tfrac{1}{2}h_+(t_R - L\cos\gamma/c)\right]\sin\gamma_R.$$

We can eliminate γ_R from these equations and solve for the Doppler shift of the beam caused by the wave. Keeping only terms of linear order in h_+ we find that

$$\frac{v_R - v_E}{v_E} = \frac{1 + \cos\gamma}{2}\left[h_+(t) - h_+(t + L(1 - \cos\gamma)/c)\right], \qquad (7.17)$$

where t refers to the time at the emitter and for γ one can use either γ_E or γ_R, leaving the Doppler shift unchanged at linear order in the metric perturbation.

When the beam travels in the same direction as the wave (i.e., when $\gamma = 0$ or π) there is no Doppler modulation and $v_R = v_E$. When $\gamma = 0$, the light beam is 'riding' on the gravitational wave and so the frequency of light does not change as it traverses from emitter to receiver. When $\gamma = \pi$, the amplitude of the effect is zero (although the change in phase of the gravitational wave is the greatest). For all other angles, the Doppler modulation of the light beam can be used to detect the gravitational wave.

To see this, we can write (7.17) in terms of time intervals $\Delta t_E = v_E^{-1}$ and $\Delta t_R = v_R^{-1}$. To linear order in h_+, we have

$$\frac{\Delta t_R}{\Delta t_E} = 1 + \frac{1 + \cos\gamma}{2}\left[h_+(t + L(1 - \cos\gamma)/c) - h_+(t)\right]. \qquad (7.18)$$

Given this, one can detect a passing gravitational wave by comparing the rate at which light pulses arrive at a distant point as compared to the rate at which they were sent. This is essentially the way pulsar timing arrays work (see Chapter 22). The largest fractional difference in the ticks is $|1 - \Delta t_R/\Delta t_E| \simeq h$ so the sensitivity of a one-path system as a detector would be limited by how stable the clocks are. Since the best atomic clocks have a stability of 10^{-16}, the maximum detectable amplitude would be $h \sim 10^{-16}$ (see Figure 1.4). Pulsar timing arrays improve on this by using data from several stable pulsars. The effective gravitational-wave amplitude increases (roughly) as the square root of the number of pulsars used.

An alternative to the above strategy would be to compare the rate at which light pulses return to the emitter after being reflected by the receiver. Starting from (7.18), we can deduce that the beam reflected by the receiver arrives at the emitter at intervals $\Delta t_{\rm ret}$ given by:

$$\frac{\Delta t_{\rm ret}}{\Delta t_R} = 1 + \frac{1 + \cos(\gamma + \pi)}{2}\left[h_+(t + 2L/c) - h_+(t + L(1 - \cos\gamma)/c)\right]. \qquad (7.19)$$

Here we have made use of the fact that the return beam makes an angle $\gamma + \pi$ with the z-axis and when the beam returns to the emitter the phase of the gravitational wave at the emitter corresponds to time $t + 2L/c$. Multiplying (7.18) and (7.19) and keeping terms linear in h_+ we get

$$\frac{\Delta t_{\text{ret}}}{\Delta t_E} = 1 + \frac{1}{2}\left[(1 - \cos\gamma)h_+(t + 2L/c) + 2\cos\gamma\, h_+(t + L(1 - \cos\gamma)/c)\right.$$
$$\left. - (1 + \cos\gamma)h_+(t)\right]. \tag{7.20}$$

This provides the time interval between consecutive light pulses as they travel to the receiver and are reflected back to the emitter.

In an interferometric detector, the frequency of a standard laser source acts like a clock. However, the frequency of even the best lasers is not very stable and a one-arm detector would be severely limited by the frequency fluctuations. Interferometry avoids this problem by comparing the round trip travel time of light beams in two perpendicular directions; the round trip travel time in one arm of the interferometer is used as a reference clock against which the round trip travel time along the other arm is compared. This is the basic principle behind interferometric gravitational-wave detectors.

7.3 Advanced interferometers

Returning to the issue of designing a sensitive gravitational-wave detector, we know from our estimates that we can improve a resonant-mass instrument by increasing the dimensions. We can simply increase the length L of the instrument. Of course, at some point this becomes problematic. Instead, we can develop the strategy we have just outlined and try to measure the relative change in distance between two well-separated masses, e.g. by monitoring the separation via a laser beam that bounces back and forth between them. As we have already mentioned, we can compare the time of flight of laser beams in the two perpendicular arms of a Michelson interferometer. This kind of instrument has a long history in testing aspects of relativity going back to Albert Michelson designing and performing experiments (already in 1881) to test the presence of the luminiferous aether. The famous null result of the Michelson and Morley experiment in detecting any such medium ultimately led to the formulation of special relativity (Michelson and Morley, 1966). The same idea is now being used, more than a century later, to measure tiny variations in gravity arising from violent cosmic events.

A sketch of a typical gravitational-wave interferometer is provided in Figure 7.2. The simplest design consists of a source of monochromatic light (a laser beam) that is sent along two perpendicular arms with the use of a 50–50 beam splitter—a half-silvered mirror that passes half of the light and reflects the other half. Highly reflective mirrors at the end of the two arms reflect the light beams back towards the beam splitter. Beams that show no net difference in path length interfere constructively upon reaching the beam splitter and exit the interferometer towards the laser source and no light is transmitted towards the photo diode. That is, the photo diode is set on a dark fringe when the two arms are of equal length.[3] A passing gravitational wave changes the path lengths, which

---

[3] More precisely, the photo diode will be on a dark fringe if the arm lengths differ by a multiple of the wavelength and will be on the main, or central, dark fringe when the arms are of equal length.

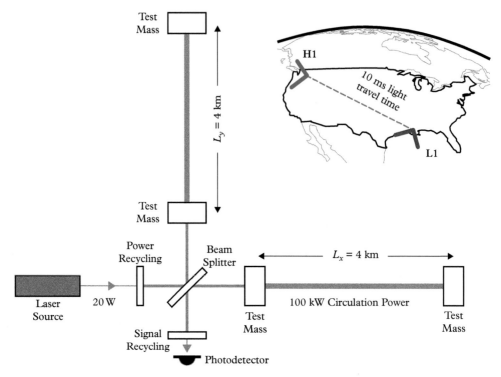

Figure 7.2 *Schematic illustration of the key elements of an advanced gravitational-wave interferometer. The light from a laser source passes through a beam splitter that equally divides the light along two perpendicular arms. Having been reflected by mirrors at the end of the arms, the returning beams interfere and the output is recorded on a photo diode. The design also includes an optical cavity that reflects the laser light back and forth many times in each arm, enhancing the effect of a gravitational wave on the phase of the laser light; a power recycling mirror that increases the power of the laser in the interferometer as a whole and a signal recycling mirror that further optimizes the signal extracted at the photodetector. The inset on the right shows the locations and orientations of the two LIGO instruments and indicates the light travel time between them (about 10 ms). (Adapted from Abbott et al. (2016b), Creative Commons Attribution 3.0 License.)*

causes the fringe to shift. The associated change in light intensity at the photo diode can be directly linked to the gravitational-wave amplitude.

Recall, from Chapter 3, that the relative change in interferometer arm length is related to the strain amplitude through $\Delta L/L \approx h$. That is, an instrument with a longer arm length is sensitive to a weaker signal. A difference ΔL in arm length induces a phase shift $\Delta \phi = 2\pi \Delta L/\lambda$ in the two light beams, or a fringe shift of $\Delta \phi/2\pi = \Delta L/\lambda$, where λ is the wavelength of laser.

What magnitude of fringe shift can we expect from astronomical gravitational-wave sources? As we have argued, the strongest sources are expected to have strain amplitude $h \sim 10^{-21}$ and so the change in length in a 1-km arm is $\Delta L \sim 10^{-18}$ m. For a Nd:Yag infrared laser with wavelength $\lambda = 1,064$ nm the fringes then shift by a tiny fraction of $\Delta L/\lambda \sim 10^{-12}$. Differential length changes of this magnitude can be caused by unwanted noise sources, some of which are technical (and so might require new designs and technology) in origin and others fundamental (and difficult or impossible to circumvent). Novel interferometer designs have made it possible to address some of the main causes of noise.

Interferometers cannot detect arbitrary length changes because of fundamental noise that mimics the effect of gravitational waves. For example, even when the arms are of equal length the photo diode could receive some light because photons from the laser obey Poisson statistics and arrive at random times. This fluctuation in photon number, which is called shot noise, limits the sensitivity of the interferometer. In an instrument that uses a laser of wavelength λ, the smallest change that can be measured is limited to

$$\Delta L_{\text{shot}} = \lambda/(2\pi\sqrt{N}), \tag{7.21}$$

where N is the number of available photons. The energy of each photon is $2\pi\hbar c/\lambda$ ($\hbar$ being the reduced Planck constant) so the number of photons contained in a laser of power P is

$$N = P\tau\lambda/(2\pi\hbar c), \tag{7.22}$$

where τ is the time duration over which the measurement is made. Assuming that we collect photons over a time $\tau = 1$ (in suitable units), we arrive at

$$N = P\lambda/(2\pi\hbar cf). \tag{7.23}$$

It follows that the sensitivity that can be achieved due to the shot noise limit is

$$h_{\text{shot}} \geq \frac{\Delta L_{\text{shot}}}{L} = \frac{1}{L}\sqrt{\frac{\hbar c\lambda f}{2\pi P}}. \tag{7.24}$$

Interferometeric detectors require continuous lasers for operation and the best (current) lasers have powers $P \sim 100$ W. This would limit the sensitivity to strain amplitudes $h \sim 10^{-18}$ at $f \sim 1$ kHz. It is also worth noting that the sensitivity scales as $f^{1/2}$, so it is more difficult to detect high-frequency signals.

Photon shot noise can be reduced by using two clever designs, both of which involve adding extra mirrors to the instrument. First of all, we can add a mirror at the interferometer output. This *power recycling mirror* does exactly what it says. Remember that when no gravitational waves are passing by, the light beam exits the interferometer. In early designs this light was wasted by diverting it away from the interferometer. The power recycling mirror sends this light back into the interferometer *in phase with the input laser.* The maximum power P then depends on the reflectivity R, or transmittance $(1-R)$, of the recycling mirror and the mirrors at the end of the two arms. The power keeps building up until transmission losses add up to the input laser power P_{in}, that is $P(1 - R) = P_{in}$, so the input laser power is amplified by a factor $1/(1 - R)$. Current mirror coating technology has achieved power losses of few parts per million leading to amplification $1/(1 - R) \sim 10^5$ in arm cavity laser power.

We can also add extra mirrors after the beam-splitter in each arm. These *Fabry–Perot mirrors* convert the two arms into cavities, effectively increasing the path length of the beams. The storage time can, however, not be arbitrarily large. Recall that the interferometer mirrors oscillate at the same frequency as that of incident gravitational waves. So the differential length change that builds up during the first half of a gravitational-wave period will be undone by the next half. So the storage should not be larger than half the gravitational wave period, or $1/2f$. This corresponds to $n = c/(2f)L$ or ~ 100 bounces for $L = 1\,km$ and $f \sim 1\,kHz$.

Power recycling and Fabry–Perot cavities together improve the strain sensitivity by a factor $\sqrt{(1 - R)/n}$, so the limiting sensitivity in this case will be:

$$h_{FP} \geq \frac{\Delta L_{FP}}{L} = \frac{1}{L}\sqrt{\frac{\hbar c f \lambda (1 - R)}{2\pi n P}}. \tag{7.25}$$

Taking $(1 - R) = 10^{-5}$ and $n = 100$, we have $h_{FP} \sim 2 \times 10^{-22}$, which sets the design sensitivity of the current generation of advanced interferometers (Abramovici *et al.*, 1992; Aasi *et al.*, 2015a; Martynov *et al.*, 2016). For the purpose of data analysis (see Chapter 8), what is relevant is the spectral density of noise fluctuations caused by shot noise. This is usually expressed in units of strain per $\sqrt{Hz}$ and can be estimated to be $\tilde{h}_{FP} = h_{FP}/\sqrt{f}$. At 1 kHz, the expected spectral sensitivity is then $\tilde{h}_{FP} \sim 10^{-23}\,Hz^{-1/2}$.

It is a bit more complicated to estimate the effect of introducing a signal recycling mirror. Gravitational waves incident on the interferometer can be thought to create side bands in the laser light of frequencies $\nu \pm f$ where ν and f are the laser and gravitational-wave frequencies, respectively. The signal recycling mirror is used to selectively resonate one of these side bands. Relative to the signal recycling mirror the rest of the interferometer can be thought of as an effective mirror, with which the signal recycling mirror forms a cavity. By carefully adjusting the position of the signal recycling mirror relative to the beam splitter it is possible to improve the sensitivity at specific frequencies. In essence, what happens is that the photons used in the measurement are made to resonate at a specific gravitational-wave frequency, f, so there will be gain in sensitivity around this frequency at the expense of loss at other frequencies. The reflectivity of the signal recycling mirror determines the improvement in sensitivity and the bandwidth: the smaller the chosen transmittance, the greater the bandwidth.

The sensitivity of all detectors on the ground is severely limited below about 10 Hz because of fluctuations in gravity gradients caused by motion of the ground. Human activity of various kinds and environmental factors (variations in the density of air, clouds, tumble-weed, and so on) are extremely difficult to control.

7.4 An international network

At the time of writing, three kilometre-scale laser interferometeric detectors have been built and operated on the ground—in addition to the two LIGO instruments, the advanced Virgo detector (Acernese *et al.*, 2015) started taking data in August 2017. The discussion of future instruments is well underway, exploring the potential of similar technology and developing the plans for the space-based LISA detector (Amaro-Seoane *et al.*, 2017), which is scheduled to fly in 2034. Ground-based interferometers operate from a few hertz up to about 10 kHz, with the best sensitivity in the range from 20 Hz to 2 kHz. Given that a compact binary of mass M has the greatest luminosity just prior to coalescence, when the gravitational-wave frequency is $f \sim 200(M/20\,M_\odot)\,\mathrm{Hz}$, these ground-based detectors are essentially sensitive to stellar-mass sources.

The first detection in 2015 was the culmination of decades of technology development. The work on interferometric detectors began with prototypes at MIT (10-m arms) led by Rai Weiss, CalTech (40-m) pioneered by Ron Drever, the University of Glasgow (10-m) led by Jim Hough, and the Max-Planck-Institut für Quantenoptik in Munich (30-m) under Albrecht Rüdiger. The instruments were gradually improved as many complex issues were resolved. As early as 1989, coincident operation of the Glasgow and Munich prototypes for about 100 hours provided a demonstration that the technology to build and operate detectors together over a long baseline was becoming available (Nicholson *et al.*, 1996).

Cutting a long story short, the CalTech 40-m prototype work eventually led to the foundation of the Laser Interferometer Gravitational-wave Observatory (LIGO; see Abramovici *et al.* (1992)), which has grown to a scientific collaboration involving over 1,000 scientists at over 100 institutions worldwide. The two LIGO installations have 4 km long arms and are built in Livingston, Louisiana and Hanford, Washington. The detector sensitivity reached the initial design target; see Figure 7.3, by the time of the fifth science run (S5) in 2005 and over a year's worth of data was taken during S6 (from July 2009 to October 2010).

At the same time, a French–Italian–Hungarian–Dutch collaboration built the Virgo detector, with 3-km arms, outside Pisa. The sensitivity of the Virgo detector is similar to that of the LIGO ones. The British–German GEO600 instrument outside Hannover is a 600-m arm interferometer. It is less sensitive than LIGO and Virgo, but has served as an important test bed for new technologies like monolithic silica suspensions, signal recycling, squeezed light, and so on. A 300-m interferometer called TAMA was also operated in Tokyo between 2001–2005.

The initial detector era saw impressive strides in technology development. The data placed 'interesting' upper limits for some predicted sources, but there were (obviously) no detections. This changed as soon as the technology upgrade of the LIGO detectors was completed in the summer of 2015 (see Figure 7.3). The first detection came on the

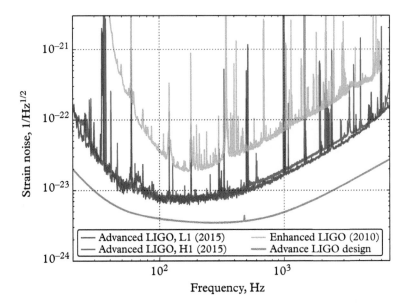

Figure 7.3 *The strain sensitivity, in units of strain per $\sqrt{Hz}$, for the LIGO Livingston detector (L1) and the LIGO Hanford detector (H1) during the O1 observing run. Also shown is the noise level for the Advanced LIGO design (grey curve) and the sensitivity during the final data collection run (S6) of the initial detectors. (Reproduced from Martynov et al. (2016).)*

14th of September, pretty much the moment the detectors were switched on. The first Observing run (O1) lasted from September 2015 to January 2016, when the detectors were again taken offline for fine tuning.

The LIGO instruments resumed data taking at the end of November 2016 and the advanced Virgo detector joined the venture during the final month of this second observing run (O2), in August 2017. This was a very exciting month, with several black-hole events—including GW170814 (Abbott *et al.*, 2017*h*), which provided a clear demonstration of the value of adding a detector to the international network. As can be seen in Figure 7.4 the sky localization improved by about a factor of 10 compared to the two LIGO detectors on their own. This improvement was, in fact, crucial for the spectacular neutron star binary merger GW170817 (Abbott *et al.*, 2017*f*). The inferred position in the sky allowed electromagnetic observers to point their telescopes in the right direction, leading to the identification of counterpart signals across the spectrum (see Chapter 21).

The ultimate aim of the advanced detector development is to reach strain sensitivity a factor of 10 better (a survey volume a factor 1,000 larger) than the initial instruments (see Figure 7.3). The detectors are expected to reach this target in the next few years. In parallel, Japan is building a new 3-km arm, underground instrument, called KAGRA, in the Kamioka mines near the city of Toyashima (Akutsu *et al.*, 2017). This is an exciting

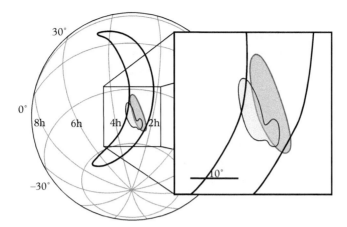

Figure 7.4 *An illustration of the 90% probability localizations for GW170814, indicating the improvement associated with adding the Virgo detector to the two LIGO instruments. The large banana shaped curve represents a sky area of 1160 square degrees, obtained from the data in the two LIGO detectors. The green area shows the improvement after adding Virgo, leading to an area of about 100 square degrees. The purple map shows the final localization after a full parameter estimation analysis, an area of about 60 square degrees. (Reproduced from Abbott et al. (2017h), Creative Commons Attribution 4.0 License.)*

project because it allows a glimpse of future technology. In particular, an underground detector should experience greatly reduced gravity gradients. The KAGRA detector will also deploy cryogenic mirrors to beat the thermal noise. Further developments of this kind of technology may be crucial for third-generation ground-based detectors, like the (aptly named) Einstein Telescope (Punturo *et al.*, 2010).

Initial LIGO had a third interferometer with 2-km arms at the Hanford installation, operating in the same vacuum tube as the 4-km instrument. It was realized that the scientific potential of the interferometer network would vastly improve if this interferometer were to be installed at a different site, such as Australia or India, far away from the other locations. The network of LIGO, Virgo, and KAGRA consisting of four interferometers are all located roughly in the same plane. As a result, they are not able to resolve some of the degeneracies in the parameter space of sources like coalescing binaries. Specifically, there is a strong degeneracy between the distance to the source and the inclination angle ι of the orbital plane with respect to the radial vector pointing in the direction of the source from a detector (see Figure 7.5). The inclination angle is also strongly correlated with the polarization angle ψ (see Eqs. (7.26)–(7.27)). A detector in the southern hemisphere, or India, would partly break this degeneracy. More importantly, the largest light travel time between detectors increases by 60% to $\sim 40\,\mathrm{ms}$ as opposed to $\sim 25\,\mathrm{ms}$ for the three-detector network. This improves the localization of sources by a factor 5 or so. Given this promise it was decided to ship the third Advanced LIGO detector to India. The LIGO-India detector is expected to join the other instruments in the mid 2020s.

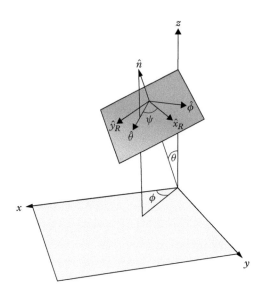

Figure 7.5 *The geocentric coordinate system where the origin is assumed to be at the centre of the Earth, with the x-axis pointing along zero degree longitude and z-axis pointing along the zenith. This coordinate system is used to define the antenna pattern functions of gravitational-wave detectors. (Illustration by B. Sathyaprakash)*

7.5   The antenna pattern

Interferometric detectors are essentially omnidirectional (quadrupole) antennas. Unlike optical telescopes which can only observe a small field-of-view at any given time, they have good sensitivity over a large fraction of the sky. The antenna pattern characterizes the sensitivity of a gravitational-wave detector to different directions.

We can use the results from Section 7.2 to understand an interferometer's antenna pattern. First of all, we need to introduce a system of coordinates in which to specify sky position. We take this to be the geocentric coordinate system sketched in Figure 7.5. An interferometer located at the north pole in this coordinate system will have its arms along x- and y-axes and the z-axis points in the direction of the zenith. Gravitational waves coming from a sky position (θ, ϕ) propagate in the negative radial direction, $-\hat{\boldsymbol{n}}$, and the wavefronts lie in the *radiation plane* normal to $\hat{\boldsymbol{n}}$. $\hat{\boldsymbol{\theta}}$ and $\hat{\boldsymbol{\phi}}$ are unit vectors in the direction of increasing θ and ϕ in the radiation plane.

Recall that the metric perturbation h_{ab} takes a particularly simple form in the transverse-traceless gauge: $h_{xx}^{\mathrm{TT}} = -h_{yy}^{\mathrm{TT}} = h_+$, $h_{xy}^{\mathrm{TT}} = h_\times$, with all other components vanishing. The coordinate system that refers to this gauge is defined by unit vectors $(\hat{\boldsymbol{x}}_R, \hat{\boldsymbol{y}}_R, \hat{\boldsymbol{z}}_R)$, with $\hat{\boldsymbol{z}}_R = -\hat{\boldsymbol{n}}$, and we assume that the vector $\hat{\boldsymbol{x}}_R$ makes an angle ψ with the

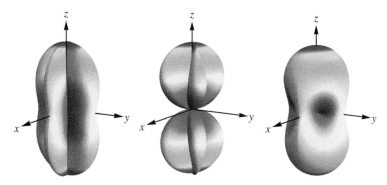

Figure 7.6 *Plots of the antenna pattern functions F_+ (left) and $F_\times$ (middle), showing the sensitivity of an interferometer to plus- and cross-polarised waves coming from a given direction in the sky. We also show the total response function $\sqrt{F_+^2 + F_\times^2}$ (right) which is independent of the polarization angle ψ. The interferometer arms are assumed to be in the $x - y$ plane. Red indicates directions where the interferometer's sensitivity is null (blind spots), while blue corresponds to directions (orthogonal to the plane containing the interferometer arms) in which the interferometer has the greatest sensitivity. (Illustration provided by B. Sathyaprakash.)*

vector $\hat{\boldsymbol{\theta}}$. That is, $\hat{\boldsymbol{\theta}} \cdot \hat{\boldsymbol{x}}_R = \cos \psi$. In this coordinate system the plus and cross antenna pattern functions of a hypothetical interferometer located at the centre of the Earth are:

$$F_+ = \frac{1}{2}\left(1 + \cos^2 \theta\right) \cos 2\phi \cos 2\psi - \cos \theta \sin 2\phi \sin 2\psi, \qquad (7.26)$$

$$F_\times = \frac{1}{2}\left(1 + \cos^2 \theta\right) \cos 2\phi \sin 2\psi + \cos \theta \sin 2\phi \cos 2\psi. \qquad (7.27)$$

These results are shown in Figure 7.6. In general, the antenna patterns are functions of three angles: the direction in the sky (θ, ϕ) and the polarization angle ψ.

By operating a network of detectors one can add sensitivity in directions where a single instrument would be blind (or perhaps, rather, deaf). This notion is demonstrated by Figure 7.7 which gives an idea of the response of the currently planned worldwide detector network to different polarisations. The figure shows

$$F^2 \equiv \sum_A (F_+^A)^2 + (F_\times^A)^2, \qquad (7.28)$$

where the sum is over the different detectors included. Basically, a network of four gravitational-wave detectors, LIGO-Hanford, LIGO-India, LIGO-Livingston and Virgo (HILV), would have a decent sky coverage, achieving good sky localization of binary inspiral sources. Once the KAGRA detector in Japan is added, blind spots are essentially removed.

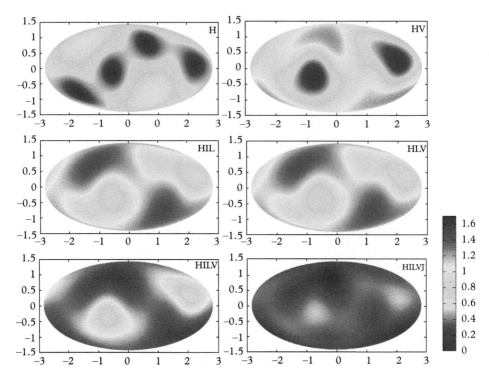

Figure 7.7 *Mollweide plots (equal area projections commonly used for global maps) of the antenna patterns of single and multiple detectors for different combinations of current and upcoming interferometers. Starting from the top left: Hanford (H), H-Virgo (HV), H-India-Livingston (HIL), HLV, HILV, HIL-Japan-V (HILJV) networks. The network of five detectors consisting of HILJV has roughly uniform sensitivity over the whole sky. (Illustration provided by B. Sathyaprakash.)*

7.6   The road to the future

As we enter the era of gravitational-wave astronomy, it is natural to take a look into the crystal ball and try to predict where this new field may take us. This is an important exercise because any new generation of gravitational-wave detectors will require huge investment. We are talking about billion-dollar facilities, akin to the Large Hadron Collider at CERN. It is debatable if single countries can bear the cost of such an experiment. Moreover, if we want to widen the search and look for low-frequency gravitational waves, as expected from massive black holes in distant galaxies, then we will need a detector in space. Such missions are costly and on top of that they require a significant lead time. This is perhaps the main lesson from the drawn-out development of the LISA project, which was first proposed in the late 1980s (Faller *et al.*, 1985) and which is expected to launch in the 2030s (Amaro-Seoane *et al.*, 2017).

7.6.1 The view from the ground

If we want to build more sensitive detectors, then it is useful to look back at how we arrived at the present instruments—how the different noise sources were eventually beaten down to the level where detections became possible. This is a complicated story, but we basically learn that the natural steps towards better instruments involve more powerful lasers, longer arm lengths, and a colder environment. It is also important to understand the fundamental limitations. For laser interferometers this leads us to the so-called quantum limit.

We have already seen that the photon shot noise limits the sensitivity at high frequencies. This limit arises due to the quantum nature of light. Expressing the effect in terms of the displacement noise spectrum, we have

$$S_{\text{shot}}^{\text{x}} = \frac{\hbar c \lambda}{2\pi P}, \tag{7.29}$$

where P is the laser power and λ is the wavelength of the laser light. This leads to the previous estimate for the dimensionless strain (7.24) since

$$h_{\text{shot}} = \sqrt{\frac{S_{\text{shot}}^{\text{x}} f}{L^2}}. \tag{7.30}$$

However, it is useful to work directly with the noise spectrum since we can simply add different contributions to the overall noise. We also see that $S_{\text{shot}}^{\text{x}}$ is independent of the gravitational-wave frequency—we are dealing with a white noise contribution. Anyway, from this estimate it is clear that we can suppress this noise by increasing the laser power. However, we cannot push this too far. As we increase the power, the radiation pressure on the suspended mirrors increases. This leads to a low-frequency noise which can be estimated as

$$S_{\text{rp}}^{\text{x}} = \frac{P}{\lambda c} \frac{\hbar}{2\pi^3 m^2 f^4}. \tag{7.31}$$

This noise increases linearly with the laser power so we pay a penalty for trying to improve the instrument. After adding the two noise sources, we see that we are limited by

$$\frac{\hbar c \lambda}{2\pi P} + \frac{P}{\lambda c} \frac{\hbar}{2\pi^3 m^2 f^4} \geq \frac{\hbar}{\pi^2 m f^2}. \tag{7.32}$$

Notably, the lower limit (which follows from working out when the two contributions are equal) is independent of the laser power. It represents what is known as the standard quantum limit

$$S_{\text{ql}}^{\text{x}} = \frac{\hbar}{\pi^2 m f^2} \quad \longrightarrow \quad h_{\text{ql}} \approx \frac{1}{\pi L} \left(\frac{\hbar}{m f} \right)^{1/2}. \tag{7.33}$$

This is basically as a trade-off between the shot noise and radiation-pressure noise. As we increase or decrease the optical power, the power-independent lower bound of the total spectrum will trace over S_{ql}^{x}.

The rules of quantum mechanics are strict, but it is possible to cheat. For example, recall that the uncertainty principle limits the product of two conjugate variables. You can imagine improving the sensitivity to one of these variables while sacrificing the other. This idea leads to the notion of squeezed light, which allows experimenters to go beyond the quantum limit. Squeezing of light has already been implemented and tested in the GEO600 detector (Willke *et al.*, 2007) and is one of the key steps towards future instruments.

The discussion of future instruments is gathering pace, with several design studies considering the next generation of detectors (see Figure 7.8). A common target for these designs is to enable the detection of binary signals from cosmological distances. The aim is to study populations, rather than individual events. The first steps towards this ambitious goal will involve upgrades of the existing installations (in the first instance, leading to a detector configurations referred to as LIGO-A^{+}). The LIGO Voyager configuration pushes the technology to the limit, still within the original infrastructure (Abbott *et al.*, 2017c). On a longer timescale, one would envisage instruments with a

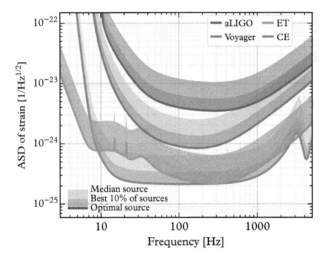

Figure 7.8 *Possible sensitivity of future ground-based detectors (amplitude spectral density, ASD) for monochromatic sources distributed isotropically in sky position, inclination, and polarization. The solid lines indicate the effective strain sensitivity for an optimally oriented source. The coloured bands denote the effective sensitivity for the best 10% of sources, and the median source. The blue curve shows the Advanced LIGO design sensitivity. The orange curve below it represents the so-called Voyager design, which uses the current infrastructure but pushes the technology to the limit. The capability of third-generation instruments, with longer arm lengths, is indicated by the predicted sensitivity for the Einstein Telescope (ET, green) and the LIGO Cosmic Explorer (CE, pink). (Reproduced from Hall and Evans (2019).)*

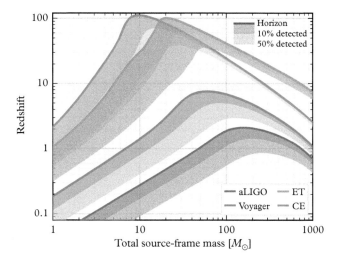

Figure 7.9 *Indicative detector horizon—the redshift beyond which none of the sources are detected—of selected second- and third-generation detectors for equal-mass, non-spinning binaries, shown as a function of total source-frame mass. The binaries are distributed isotropically in sky location and inclination angle. The shaded bands show the redshifts at which 10% and 50% of the sources at that distance are detected (a source is assumed to be detected if it leads to a matched-filter signal-to-noise ratio ≥ 8). The blue curve shows the Advanced LIGO design sensitivity. The orange curve below it represents the so-called Voyager design, which uses the current infrastructure but pushes the technology to the limit. The capability of third-generation instruments, with longer arm lengths, is indicated by the predicted sensitivity for the Einstein Telescope (ET, green) and the LIGO Cosmic Explorer (CE, pink). (Reproduced from Hall and Evans (2019)).*

longer baseline. In Europe, a conceptual design study for an underground detector, called the Einstein Telescope (ET), has been completed (Punturo *et al.*, 2010). The proposed design has a triangular layout with 10-km arms and operates three broadband detectors at a single site by using each arm of the triangle twice. The overall ET strain sensitivity is 10 times better than the current advanced detectors. A similarly ambitious project discussed in the USA is known as the Cosmic Explorer (Abbott *et al.*, 2017c). In this case the design is L-shaped, as in LIGO, but the arm-length is a stunning 40 km. At the moment, these are simply discussion documents aimed at focusing the mind. There may be a long way to actual constructions, but these are very exciting projects. With a suggested detector horizon reaching a redshift of several for solar-mass binaries, third-generation interferometers should be able to detect signals from all related mergers in the Universe.

7.6.2   Going into space

The sensitivity of ground-based detectors below a few hertz will always be limited by gravity gradient noise that arises as a result of variations in the surface density of the

Earth due to seismic waves, variations in the density of air caused by wind and other environmental factors, and, more generally, noise due to human activity. Some of these noise sources can be reduced by building a detector deep underground (as in the case of KAGRA) where the density of air and anthropogenic noise will cease to be problems and the effect of seismic waves is suppressed.

One can try to avoid low-frequency noise entirely by placing the detector in space. Plans for such a mission have been under development since the early 1990s, under the leadership of Karsten Danzmann (Amaro-Seoane *et al.*, 2017). It has been a long journey, but space missions require a very long lead time. Finally, in June 2017, the Laser Interferometer Space Antenna (LISA) was selected by the European Space Agency's steering committee for the L3 mission launch slot. It is now due to fly in 2034. LISA will have a set of three free flying spacecraft. Separated from each other by 2.5 million km (in the current design; see Figure 7.10), the spacecraft will fly in a triangular formation, trailing the Earth's heliocentric orbit by about 20°. The instrument will be sensitive to sources in the frequency range $0.01 - 100$ mHz, meaning that it can probe radiation from supermassive black-hole binaries from the far end of the Universe. It would essentially make it possible to trace the history of large black holes across all stages of galaxy evolution. A range of anticipated LISA sources is summarized in Figure 7.11.

An instrument like LISA has its own design challenges. For example, due to diffraction losses it is not feasible to reflect the laser beams back and forth between the spacecraft, as is done on the ground. Instead, each spacecraft will have its own laser. These lasers

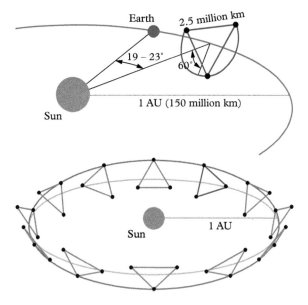

Figure 7.10 *Illustration of the planned LISA orbit, with the triangular interferometer configuration lagging around 20° behind the Earth as it travels around the Sun. (Reproduced with permission from Amaro-Seoane et al. (2017).)*

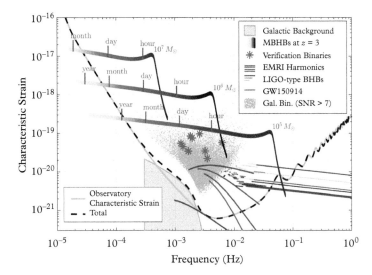

Figure 7.11 *Predicted sources in the frequency range of LISA, compared to the sensitivity of the proposed three-arm configuration. The figure shows the dimensionless characteristic strain amplitude for three tracks of equal mass black-hole binaries, located at a redshift of z = 3 with total intrinsic masses 10^7, 10^6, and $10^5 M_\odot$. The remaining time until plunge is indicated along each track. Simultaneously evolving harmonics of an extreme-mass-ratio inspiral source at z = 1.2 are also shown, as are the tracks of a number of stellar origin black-hole binaries of the type discovered by LIGO (in particular GW150914). Several thousand galactic binaries will be resolved after one year of observation. Some of the known systems will, in fact, serve as verification signals. Millions of other binaries result in a confusion-limited signal, with a detected amplitude that is modulated by the motion of the constellation over the year (the grey-shaded area). (Reproduced with permission from Amaro-Seoane et al. (2017).)*

will be phase locked, achieving the same kind of phase coherence as LIGO does with mirrors. The configuration then functions as three, partially independent and partially redundant, gravitational-wave interferometers (Tinto and Dhurandhar, 2014). The individual test masses must be placed in as drag-free an environment as possible. The relevant technology has been (spectacularly) demonstrated in flight. The results obtained by the LISA Pathfinder mission (basically a mini-version of the LISA setup, flying from December 2015 through March 2016) shows that the required sub-femto-g/$\sqrt{\text{Hz}}$ spurious acceleration requirement can be met (Armano *et al.*, 2018); see Figure 7.12. This was a spectacular result. At the time of the measurements, the LISA Pathfinder test masses were in one of the stillest places in the Universe. The experiment was more sensitive than the weight of a virus.

The main LISA noise sources are relatively easy to understand. At low frequencies, below 3 mHz, the instrument is limited by acceleration noise. This aspect was tested by the Pathfinder mission. At higher frequencies the photon shot noise determines LISA's sensitivity. The sensitivity curve then steepens at $f \sim 3 \times 10^{-2}$ Hz because at higher frequencies the gravitational-wave period is shorter than the round-trip light travel time

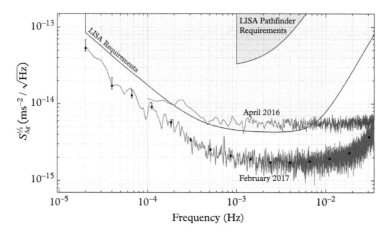

Figure 7.12 *The level of parasitic differential acceleration of the LISA Pathfinder test masses as a function of frequency. Data refer to a roughly 13-day-long run and the red, noisy line is the amplitude spectral density estimated with the standard periodogram technique averaging over 10, 50% overlapping periodograms each 2×10^5 s long. The blue noisy line represents an earlier, less sensitive, measurement. The results are compared to the LISA requirements. Fulfilling requirements implies that the noise must be below the corresponding shaded area at all frequencies. (Reproduced from Armano et al. (2018), Creative Commons Attribution 4.0 License.)*

in each arm. LISA's sensitivity is roughly the same as that of LIGO, but at a frequency 10^5 times lower. Since the gravitational-wave energy flux scales as $\mathcal{F} \sim f^2 h^2$, this corresponds to an energy sensitivity that is 10^{10} times better than LIGO.

7.7 Doppler tracking

So far we have discussed different detector designs, where the primary purpose of the experiment is to catch gravitational waves. We may also make progress using technology developed for a different purpose. This would typically involve a variation on the 'single-arm' light-travel experiment from Figure 7.1. As an example of this, the Doppler delay of signals between the Earth-based communication stations and a spacecraft provides a gravitational-wave detector (Armstrong, 2006). A radio signal of frequency f_0 is transmitted to a spacecraft and then coherently transported back to Earth, where it is received and the frequency is measured with a highly stable clock (typically a hydrogen maser). The relative change f/f_0 is monitored as a function of time. A gravitational-wave propagating through the Solar System would cause a small perturbation in the frequency ratio, proportional to the amplitude of the gravitational waves.

With a baseline of 1–10 AU, we can use this technique to search for gravitational waves in the millihertz regime, and as atomic clocks are extremely stable, it is possible to achieve sensitivities of order $h \sim 10^{-15} - 10^{-13}$. Noise sources for this experiment can be divided into two broad classes: (i) instrumental and (ii) related to propagation. At the high-frequency end of the band accessible to Doppler tracking, thermal noise dominates over all other noise sources (typically at about 0.1 Hz). Among the other sources of instrumental noise (transmitter and receiver, mechanical stability of the antenna, stability of the spacecraft, etc.), clock noise has been shown to be the most important source of frequency fluctuations. The propagation noise is due to fluctuations in the index of refraction of the troposphere, ionosphere, and the interplanetary solar plasma.

Both NASA and ESA have performed this kind of measurements, with increasing sensitivity due to advances in technology, since the 1980s (Armstrong, 2006). The first opportunities came with Voyager in 1980, for which a burst search was carried out. Similar proof-of-principle studies for a stochastic background, periodic waves, and chirps were carried out on data from Pioneer 10 and 11 and Ulysses. This work motivated innovations in signal analysis, setting the scene for the Galileo and, in particular, Cassini missions. Cassini was launched on a mission to Saturn in 1997 and there were several data-taking campaigns in 2001–3. Opportunities to continue this effort are provided by the Jupiter mission Juno and, in the near future, the Mercury orbiter BepiColombo.

7.8 Pulsar timing arrays

A population of highly stable millisecond pulsars, with timing accuracies of $\sim 10\,\mathrm{ns}$ over several years, could serve as an array of clocks whose regular ticks would be coherently modulated by gravitational waves passing the Earth. Given this, there is a worldwide effort to observe stable millisecond pulsars and exploit them for detecting gravitational waves.

Pulsar timing arrays can be thought of as nature's own interferometers, with galactic-scale arm lengths. They are sensitive to much lower frequencies than ground-based instruments. Typically, precise timing of an array of pulsars could detect the nanohertz gravitational waves one would expect from merging supermassive black-hole binaries with masses in the range $10^9 - 10^{10} M_{\odot}$. The array should also be sensitive to stochastic gravitational waves.

As we will discuss in Chapter 22, the strategy behind pulsar timing arrays is intimately linked to the problem of extracting the signal. The detector design is up to nature. The observers job is 'simply' to identify, and continue to track, the largest possible set of extremely stable pulsars. There are three current efforts in this direction, operating under the joint umbrella of the International Pulsar Timing Array (IPTA) (Manchester *et al.*, 2013): NANOgrav in the USA, the Parkes Pulsar Timing Array (PPTA) in Australia, and the European Pulsar Timing Array (EPTA). The current sensitivity of the experiments is close to the level where one may expect a detection. This is already exciting and the future seems bright. Once it comes on-stream in the next decade, the Square-Kilometre Array (SKA) is expected to lead to significant further improvements (Stappers *et al.*, 2018).

8

Mining the data

Gravitational-wave detectors are complex instruments, with (more or less) 'individual personalities' requiring intricate and specialized control systems. Moreover, they are wide-band instruments sensitive to (more or less) all sky directions. A single detector monitors sources over a large fraction of the sky and a network of instruments may achieve near isotropic sensitivity. Signals from a variety of sources—transients, continuous waves, binary inspiral and merger signals, stochastic backgrounds—may be present in the same data set. In addition, the background noise is likely to have considerable non-stationarity (short duration glitches caused by electronics, suspension systems, stray magnetic fields acting on the electronics, etc.), which means that it cannot simply be described as a simple Gaussian background. Disentangling different sources and determining their properties is a serious challenge. However, over the past decades many different search techniques have been developed and considerable progress has been made on addressing the key problems.

In order to complete our first pass through the main issues facing a gravitational-wave astronomer, we will now introduce the basic principles behind signal detection and parameter estimation.[1] The main aim is to develop the ideas to the point where we can assess the detectability of any proposed source. The main focus will be on standard matched filtering. To an expert, this may seem somewhat old fashioned, but it is natural to take this 'frequentist' approach as it does not assume any prior knowledge of statistics. However, both the detection and parameter extraction problems ultimately involve the maximization of the likelihood that a signal is present in the data and that the signal parameters take certain values. Hence, it is equally natural that modern analysis discussions tend to be based on a Bayesian strategy. Given this, we also provide a brief introduction of Bayesian statistics and how it can be applied to the gravitational-wave analysis problem.

Before turning to the details it is useful and instructive to underline why any data analysis strategy for gravitational waves must be different from the conventional approach in astronomy.

---

[1] The interested reader will find a more exhaustive discussion in the book by Creighton and Anderson (2011). Other useful starting points are the reviews by Thorne in Hawking and Israel (1989) and Sathyaprakash and Schutz (2009).

Gravitational-Wave Astronomy: Exploring the Dark Side of the Universe. Nils Andersson, Oxford University Press (2020).
© Nils Andersson. DOI: 10.1093/oso/9780198568032.001.0001

(i) Gravitational-wave antennas are omnidirectional, with a response better than 50% of the root mean square over 75% of the sky (see Chapter 7). As we do not have the ability to 'point' the instrument, we have to carry out all-sky searches.

(ii) Interferometers are typically broadband, covering a frequency range of (up to) three orders of magnitude. This is an advantage, as it helps track sources whose frequency evolves, but it means that we need to search over a wide range of frequencies.

(iii) General relativity predicts that gravitational radiation has two independent polarizations (h_+ and $h_\times$; see Chapter 3). Measuring the polarization is of fundamental importance (as there are alternative theories of gravity with more than two polarizations; see Chapter 4) and may have astrophysical implications too (it would, for example, be one way to resolve the mass-inclination degeneracy of binary systems). A polarization measurement requires a network of detectors, which means that we need analysis algorithms that work with data from multiple antennas.

(iv) Astrophysical gravitational waves are detected coherently, by following the phase of the radiation, rather than the energy (as in standard astronomical observations). The signal-to-noise ratio is built up by coherent superposition of many wave cycles emitted by a source. The phase evolution contains more information than the amplitude and the signal structure encodes the underlying physics. Nevertheless, tracking a signal's phase means that searches will have to be made over a vast region of parameter space for each source, placing severe demands both on the theoretical understanding of the emitted waveforms and the data analysis algorithms.

(v) Gravitational-wave detection is computationally intensive. Advanced detectors will collect data continuously for several years at the rate of several megabytes per second (the sampling rate of LIGO is 16,384 Hz). About 1% of this data is signal data; the rest is housekeeping information that monitors the operation of the instrument. The large parameter space requires that the signal data be filtered many times for different searches, and this puts huge demands on computing hardware and algorithms.

Despite these challenges, data analysis for broadband detectors has been developed since the mid-1980s (see Thorne's contribution to Hawking and Israel (1989)). Much of the theory is now well understood, but implementation strategies depend on available computer resources, data volumes, astrophysical knowledge, and source modelling, and so are under constant development.

8.1 Random noise

As we turn to the actual data analysis problem, it is natural to begin by considering the issue of random noise. Our real interest may be in astrophysical gravitational-wave signals, but it is crucial that we understand the detector noise. After all, we need to devise clever schemes for digging weak signals out of the noisy data stream. As a first step, it is useful to consider how we characterize the noise in a given detector.

Noise is a random process. As the detector output is sampled at some finite rate, the noise is given by some discrete time series $n_i(t)$ (say). However, we will nevertheless consider the problem in the continuous limit. That is, we take the noise to be some function of time $n(t)$. The extension to the discrete case is straightforward, e.g. by changing the Fourier transforms below to their discrete versions.

Because the noise is random we cannot know exactly which realization we are dealing with at any give time. We have to consider the problem from a statistical point of view.

Let us take p_n to be the probability (density) for a given n at time t. We then have the expectation value

$$\langle n \rangle = \int n p_n(n) dn \tag{8.1}$$

(where the notation should not be confused with the averaging over wave-periods from the previous chapters). If we assume that the noise is stationary[2] we can use the time average

$$\langle n \rangle = \lim_{T \to \infty} \frac{1}{T} \int_{-T/2}^{T/2} n(t) dt. \tag{8.2}$$

For stationary Gaussian processes the mean value $\langle n \rangle = 0$, but the corresponding power spectrum will be nontrivial. The power associated with the noise follows by integrating n^2 over some time T and averaging, so we have

$$\langle n^2 \rangle = \lim_{T \to \infty} \frac{1}{T} \int_{-T/2}^{T/2} n^2 dt. \tag{8.3}$$

Assuming that the noise vanishes outside the time window we are considering, i.e. using,

$$n(t) = \begin{cases} n_T, & -T/2 \le 0 \le T/2, \\ 0, & \text{elsewhere,} \end{cases} \tag{8.4}$$

we have

$$\langle n^2 \rangle = \lim_{T \to \infty} \frac{1}{T} \int_{-\infty}^{\infty} n_T^2 dt = \lim_{T \to \infty} \frac{2}{T} \int_0^{\infty} |\tilde{n}_T(f)|^2 df \equiv \int_0^{\infty} S_n(f) df, \tag{8.5}$$

---

[2] The assumption that the noise is stationary boils down to assuming that (in absence of signals) the detector output remains (statistically) unchanged. The noise would have the same properties on Tuesday as it had on, say, Monday. This property is desirable, but it is always going to be an idealization.

where we have used the Fourier transform

$$\tilde{n}(f) = \int_{-\infty}^{\infty} n(t)e^{2\pi ift}\, dt \tag{8.6}$$

with inverse

$$n(t) = \int_{-\infty}^{\infty} \tilde{n}(f)e^{-2\pi ift}\, df. \tag{8.7}$$

We have also used the fact that the noise is real, so that $\tilde{n}(-f) = \tilde{n}^*(f)$, where the asterisk represents complex conjugation, and applied Parseval's theorem. The result (8.5) defines the (one-sided) power spectral density, $S_n(f)$, which provides the all-important measure of the strength of the noise.

Before we proceed, it is worth noting that, the spectral noise density is (twice) the Fourier transform of the auto-correlation[3] of the noise. With

$$R_n(\tau) = \langle n(t)n(t+\tau)\rangle, \tag{8.8}$$

we have

$$S_n(f) = 2\int_{-\infty}^{\infty} R_n(\tau)e^{2\pi if\tau}\, d\tau. \tag{8.9}$$

8.2   Matched filtering and the optimal signal-to-noise ratio

We now want to see what happens when we add in a signal. As a first step towards understanding this problem, it makes sense to introduce matched filtering—a well-established data analysis technique that efficiently searches for a signal of known shape buried in noise. The method involves correlating the output of a detector with a waveform, known as a template or filter. Given a signal $h(t)$ buried in noise $n(t)$, the task is to find an 'optimal' template $K(t)$ that produces, on average, the best possible signal-to-noise ratio.

Intuitively, it is clear what we expect the matched filtering to achieve. Operationally, we need to take a few steps to reach this target.

Let us, first of all, take the detector output to be a time series, $x(t)$, consisting of a background noise $n(t)$ and a gravitational-wave signal $h(t)$. Letting a tilde denote the Fourier transform of a quantity, as before, we have

---

[3] Given two time series, the correlation quantifies to what extent there are common features. The auto-correlation in (8.8) measures how much the noise has 'in common with itself'.

$$\tilde{x}(f) = \int_{-\infty}^{\infty} x(t)e^{2\pi ift}\, dt. \tag{8.10}$$

Inspired by (8.8) we introduce the correlation

$$c(\tau) = \int_{-\infty}^{\infty} x(t)K(t+\tau)dt, \tag{8.11}$$

where $K(t)$ is a suitable filter. Since $\langle n \rangle = 0$, it then follows immediately that

$$S = \langle c \rangle = \int_{-\infty}^{\infty} h(t)K(t+\tau)dt = \int_{-\infty}^{\infty} \tilde{h}(f)\tilde{K}^*(f)e^{2\pi if\tau}\, df. \tag{8.12}$$

The lag τ now represents the duration by which the filter function lags behind the detector output. The purpose of the correlation integral is to concentrate all the signal energy at one place. It is worth noting that, if the signal arrives at time t_a so that we are matching the filter to $h(t + t_a)$, then we have

$$S = \int_{-\infty}^{\infty} \tilde{h}(f)\tilde{K}^*(f)e^{2\pi if(\tau - t_a)}\, df. \tag{8.13}$$

We see that if we want to maximize the output, we need the lag τ to be equal to the time of arrival of the signal. Hence, we assume this to be the case from now on. Of course, in practice this involves moving any given template shape along the data set to identify the most likely arrival time.

When no signal is present, the detector output is just a realization of the noise, i.e., $x(t) = n(t)$. In order to separate the noise from the signal, we use the variance

$$N^2 = \left\langle (c - \langle c \rangle)^2 \right\rangle. \tag{8.14}$$

We then have

$$N^2 = \int_{-\infty}^{\infty}\int_{-\infty}^{\infty} K(t)K^*(t') \left[\int_{-\infty}^{\infty}\int_{-\infty}^{\infty} \langle \tilde{n}(f)\tilde{n}^*(f') \rangle dfdf' \right] dtdt'$$
$$= \int_{-\infty}^{\infty} \frac{1}{2}S_n|\tilde{K}(f)|^2 df = \int_0^{\infty} S_n|\tilde{K}(f)|^2 df, \tag{8.15}$$

where $\tilde{K}^*(t)$ denotes the complex conjugate of $\tilde{K}(t)$, and we have used

$$\langle \tilde{n}^*(f')\tilde{n}(f) \rangle = \frac{1}{2}\delta(f - f')S_n(f), \tag{8.16}$$

with $\delta(f - f')$ the delta function.

Finally, the signal-to-noise ratio ρ is defined as

$$\rho^2 \equiv S^2/N^2. \tag{8.17}$$

The form of the different integrals suggests that it is natural to define a scalar product of waveforms. Given two functions, $a(t)$ and $b(t)$, we let their scalar product be

$$(a|b) \equiv 4 \, \mathrm{Re} \int_0^\infty \frac{\tilde{a}(f)\tilde{b}^*(f)}{S_n(f)} df. \tag{8.18}$$

We then have

$$S = (h|S_nK/2) = 2 \, \mathrm{Re} \int_0^\infty \tilde{h}(f)\tilde{K}^*(f)df, \tag{8.19}$$

which means that we can express the signal-to-nose ratio as

$$\rho^2 = \frac{(h \, |S_nK)^2}{(S_nK|S_nK)}. \tag{8.20}$$

This is a very important result. We see that the template K that leads to the maximum value of ρ is simply (recalling that we have already used $\tau = t_a$)

$$\tilde{K}(f) = \gamma \frac{\tilde{h}(f)}{S_n(f)}, \tag{8.21}$$

where γ is an arbitrary constant. This represents the optimal (Wiener) filter. It is the best we can hope to do.

We have learned two important things. First, it is the shape of the filter that is important, not the overall amplitude. Second, the optimal filter is not just a copy of the signal; it is weighted by the noise spectral density $S_n(f)$.

Combining the above results, we can work out the optimal signal-to-noise ratio. From (8.20) and (8.21) we get

$$\rho_{\mathrm{opt}}^2 = (h|h) = 4 \int_0^\infty \frac{\left|\tilde{h}(f)\right|^2}{S_n(f)} df. \tag{8.22}$$

It is worth highlighting that this is not just the total energy of the signal (which would be $2 \int_0^\infty |\tilde{h}(f)|^2 \, df$), but rather the integrated signal power weighted by the noise $S_n(f)/2$. This makes intuitive sense. The contribution to the signal-to-noise from a frequency bin where the noise is high is smaller than from a bin where the noise is low. In other words, an optimal filter takes into account the nature of the noise.

The final result (8.22) is powerful but may not be ideal if we want to compare predicted signal strengths to the noise in a given detector. The magnitude $|\tilde{h}(f)|$ provides the raw signal strength as a function of frequency, but (as we will soon see) this instantaneous amplitude can be orders of magnitude smaller than the accumulated signal after some observation time. Hence, a direct comparison with $S_n(f)$ may not be meaningful. The comparison would not make sense, anyway, as the two quantities have different physical dimensions.

In order to make a more meaningful comparison it is common to introduce a *characteristic amplitude* such that

$$|h_c(f)|^2 = 4f^2|\tilde{h}(f)|^2,\tag{8.23}$$

and write the optimal signal-to-noise ratio as

$$\rho_{\mathrm{opt}}^2 = \int_0^\infty \frac{|h_c(f)|^2}{fS_n(f)}d\log f.\tag{8.24}$$

We are now comparing dimensionless quantities, $h_c(f)$ and $\sqrt{fS_n(f)}$, which makes more sense. Moreover, with this convention, if we compare the quantities on a log–log scale, then the area between the source and detector noise curves is simply related to the signal-to-noise ratio. In effect, we can 'integrate by eye' to assess the detectability of any given scenario. Another advantage is that $fS_n(f)$ simply represents the root-mean-square noise in a bandwidth f. Figure 1.4 provides an example of this kind of comparison.

Of course, the characteristic amplitude h_c does not directly represent the amplitude of waves emerging from a given source. As an alternative, we can rearrange (8.24) in such a way that we compare dimensional quantities:

$$\rho_{\mathrm{opt}}^2 = \int_0^\infty \frac{2\left|\sqrt{f}\tilde{h}(f)\right|^2}{S_n(f)}d\log f.\tag{8.25}$$

The advantage of this convention is that the power spectral density S_n, when integrated over all frequencies, gives the mean square amplitude of the detector noise. Of course, the height of the source about the noise curve is no longer trivially related to the signal-to-noise. An example of this kind of comparison, for the black-hole signals detected during the first observation run of Advanced LIGO, is provided in Figure 8.1.

Signals of interest to us involve a number of (a priori unknown) parameters, such as the masses of the two components in a binary and their intrinsic spins, and an optimal filter must agree with both the signal shape and these parameters. If the filter parameters are slightly mismatched with the signal, the signal-to-noise will degrade. For example, even a mismatch of a single cycle in 10^4 during a neutron star binary inspiral can degrade the signal-to-noise ratio by as much as a factor of 2.

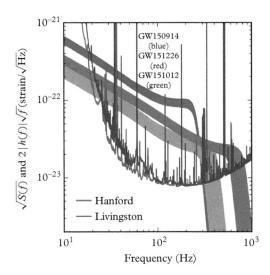

Figure 8.1 *A comparison of the gravitational-wave signals and the detector noise for three black-hole merger events seen during the first LIGO observing run. Note that the GW151012 event was upgraded from a 'candidate' to a detection when the data was reanalysed; see table 9.3. (Reproduced from Abbott et al. (2016a), Creative Commons Attribution 3.0 License.)*

8.3   Applications of matched filtering

As a rule of thumb, matched filtering helps enhance the signal-to-noise ratio as the square root of the number of signal cycles in the detector band, compared to the case in which the signal shape is unknown and all we can do is Fourier transform the detector output and compare the signal energy in a frequency bin to the noise in that bin.

As illustrations of this, let us consider a couple of problems of immediate relevance to gravitational-wave searches: the detection of chirp signals from compact binaries and continuous waves from spinning neutron stars.

8.3.1   Coalescing binaries

In the case of inspiralling binaries, post-Newtonian theory (see Chapter 11) is used to model the dynamics to a high order in v/c, where v is the speed of the binary companions. Given this approximate signal, we can effectively match filter for binaries as long as the system is still 'far' from coalescence. In reality, one takes the waveform to be valid until the innermost stable circular orbit (see Chapter 10). In the case of binaries consisting of two neutron stars, or a neutron star and a black hole, tidal effects might affect the evolution well before reaching this point. We will explore this problem in more detail in Chapter 20, but let us not worry about this issue right now.

As we will see later (Chapters 19–21), numerical relativity simulations provide waveforms for the merger phase of compact binaries, as well. However, the computational cost in matched filtering the merger phase will not be very high, as it only involves a small number of gravitational-wave cycles. Having said that, it is crucial to have the correct waveform to enhance signal visibility and, more importantly, to enable strong-field tests of general relativity.

In the general case of a black-hole binary inspiral the search space is characterized by 17 different parameters. These are the two masses of the bodies, their spins, eccentricity of the orbit, its orientation at some fiducial time, the position of the binary in the sky and its distance from the Earth, the epoch of coalescence and phase of the signal at that epoch, and the polarization angle (quite a list!). However, not all these parameters are important in a matched-filter search. Only those parameters that change the shape of the signal, such as the masses, orbital eccentricity, and spins, or cause a modulation in the signal due to the motion of the detector relative to the source, such as the direction, are to be searched for. Other parameters, such as the epoch of coalescence and the phase at that epoch, are simply reference points in the signal that can be determined without significant burden on the computational power.

For typical neutron star binaries (where the spin of the stars can be ignored, as it is likely to be slow) observed for a short enough period that the detector motion can be neglected, or where the signal lasts only a short time in the detector's sensitivity band, there are essentially two search parameters—the individual component masses.

This problem serves as a useful example of the matched filtering approach. In particular, we can show that the filtering allows us to account for the fact that the binary system spends more time at lower frequencies. If we take ϕ to be the orbital phase, so that $\dot{\phi} = 2\pi f = 2\Omega$, then the number of cycles generated at a given frequency is roughly

$$\mathcal{N} = \frac{f}{2\pi} \frac{d\phi}{df} = \frac{f^2}{\dot{f}}. \tag{8.26}$$

Let us work out the corresponding characteristic amplitude, given the raw amplitude

$$h(t) = h_0 \cos\phi(t), \tag{8.27}$$

with h_0 (approximately) constant. Provided the frequency is slowly evolving, we can work out h_c using the stationary phase approximation. First of all, we have

$$\tilde{h}(f) = \frac{h_0}{2} \int_{-\infty}^{\infty} \left\{ \exp\left[2\pi i \left(ft + \frac{\phi}{2\pi} \right) \right] + \exp\left[2\pi i \left(ft - \frac{\phi}{2\pi} \right) \right] \right\} dt. \tag{8.28}$$

The main contribution to the integrals come from frequencies where the arguments of the exponentials are approximately zero, and we find that (Cutler and Flanagan, 1994)

$$\tilde{h}(f) \approx \frac{h_0}{2} \left(\dot{f} \right)^{-1/2} \exp[2\pi i ft - i\phi(f) - i\pi/4], \tag{8.29}$$

where t is the time at which $\dot\phi = 2\pi f$ (the stationary point) and $\phi(f)$ is shorthand for $\phi[t(f)]$. Given this, it follows from (8.23) that

$$|h_c(f)| \approx \left(\frac{f^2}{\dot f}\right)^{1/2} h_0 = \sqrt{\mathcal N}\, h_0. \tag{8.30}$$

This is the desired result—the characteristic amplitude is enhanced at lower frequencies where the inspiralling system spends more time. From the leading order post-Newtonian result derived in Chapter 5, we know that

$$\mathcal N \sim f^{-5/3}, \tag{8.31}$$

so the enhancement of the signal-to-noise can be significant. We also know that

$$h_0 \sim \frac{\mathcal M^{5/3} f^{2/3}}{d}, \tag{8.32}$$

so

$$h_c \sim \frac{\mathcal M^{5/3} f^{-1/6}}{d}. \tag{8.33}$$

This explains the overall slope of the model inspirals in, for example, Figure 8.1.

8.3.2 Continuous waves

As we discussed in Chapter 6, spinning neutron stars may emit continuous gravitational waves. In this case, the signal shape is simple: a sinusoidal oscillation with small corrections to account for the gradual spin-down of the neutron star. However, the search for such signals is costly as one has to account for the Doppler modulation of the signal caused by the Earth's rotation, the motion of the Earth around the Solar System barycenter, and so on.

Much of the observational literature describes continuous gravitational waves in terms of a intrinsic strain, h_0, associated with the response of a hypothetical detector at either of the Earth's poles to a signal from a spinning star over either pole whose rotation axis is parallel to that of the Earth (Jaranowski *et al.*, 1998). This quantity is simply related to the gravitational-wave luminosity:

$$\dot E = \frac{1}{16\pi}\langle \dot h_+^2 + \dot h_\times^2 \rangle, \tag{8.34}$$

In general, we have (see Chapter 6)

$$h(t) = F_+ \frac{h_0}{2}(1 + \cos^2 i)\cos 2\omega t + F_\times h_0 \cos i \sin \omega t, \tag{8.35}$$

where F_+ and $F_\times$ are the detector antenna patterns from Chapter 7, and ω is the (angular) frequency of the waves. For the assumed configuration we simply have $h_+ = h_\times = h_0$. Next, being less cavalier with numerical factors of order unity than we were in Chapter 3, we average over all possible source orientations. This leads to an additional factor of $2/5$ in the luminosity, so we have

$$\dot{E} = \frac{1}{10}\omega^2 d^2 h_0^2, \qquad (8.36)$$

where d is the distance to the source. Making use of the quadrupole formula result from Chapter 6 we have

$$h_0 = \frac{\epsilon I_0 \omega^2}{d} = \frac{4\pi^2 \epsilon I_0 f^2}{d}, \qquad (8.37)$$

where ϵ is the ellipticity associated with the star's deformation, I_0 is the moment of inertia, and f is the gravitational-wave frequency.

We can now ask how strong the signal would need to be in order for it to be detectable in the ideal case where we know all source parameters to sufficient accuracy that only one template is needed. In this case we can easily integrate for long observation times, T_{obs}. In the case of such a coherent search, with D detectors, the signal-to-noise ratio follows from

$$\rho^2 = \frac{DT_{obs} h_0^2}{S_n}. \qquad (8.38)$$

That is, the effective amplitude of the signal increases as the square root of the observation time. For a one–year search the improvement would be almost a factor of 6,000 over the instantaneous amplitude. This is clearly significant. Note also that the signal-to-noise improves with the number of detectors.

This result allows us, for a given set of (identical) detectors, to work out the minimum h_0 required to achieve a certain signal-to-noise. This minimum amplitude works out to be (Jaranowski *et al.*, 1998)

$$h_0 \approx 11.4 \left(\frac{S_n}{DT_{obs}} \right)^{1/2}. \qquad (8.39)$$

This is a useful measure of the sensitivity of a continuous-wave search. The numerical factor of 11.4 corresponds to a signal-to-noise threshold representing a single trial false alarm rate of 1%, a false dismissal rate of 10%, and a uniform averaging over the possible source orientations and sky positions.

In reality, the continuous-wave problem presents a severe computational challenge. As an illustration, let us estimate the cost of a search for unknown pulsars by working out the number of sky patches one would have to consider (Sathyaprakash and Schutz, 2009).

We want to know how many independent patches we need to observe if we do not want to lose appreciable signal-to-noise. First of all, the baseline of a gravitational-wave detector for continuous-wave sources is essentially $L = 2 \times 1$ AU $\simeq 3 \times 10^{11}$ m. For a source that emits gravitational waves at 100 Hz (near the sensitivity sweet spot of a ground-based detector), the wavelength of the radiation is $\lambda = 3 \times 10^6$ m, and the angular resolution $\Delta\theta$ of the antenna at a signal-to-noise ratio of 1 is $\Delta\theta \simeq \lambda/L = 10^{-5}$, or a solid angle of $\Delta\Omega \simeq (\Delta\theta)^2 = 10^{-10}$. In other words, the number of patches one should search for is $N_{\text{patches}} \sim 4\pi/\Delta\Omega \simeq 10^{11}$. Moreover, for an observation that lasts for about a year ($T_{\text{obs}} \simeq 3 \times 10^7$ s), the frequency resolution is $\Delta f = 1/T_{\text{obs}} \simeq 3 \times 10^{-8}$. If we search a frequency band of (say) 300 Hz we need to consider about 10^{10} frequency bins. We have to search over roughly 10^{11} patches in the sky for each of the 10^{10} frequency bins. This is a formidable task.

In reality, one can only expect to perform a matched-filter search over a short period (days/weeks) of the data or over a restricted region in the sky, or perform targeted searches for known objects. This suggests three possible search strategies. (i) We can target the search on pulsars with known location and spin frequency. (ii) In some cases, like known supernova remnants, we know the location but not necessarily the spin-frequency of the neutron star. In this case the directional information can be used to lower the computational burden. (iii) Finally, and for the reasons given earlier, we have blind all-sky searches, trying to find objects that may not radiate electromagnetically. This kind of search is very costly, but one can still develop hierarchical algorithms that add power incoherently with the minimum possible loss in signal visibility. We will discuss continuous-wave results for all three search strategies in Chapter 14.

8.4 Bursts searches

In many gravitational-wave scenarios a significant amount of energy is emitted as a brief, unpredictable, burst. This is notably the case for core collapse supernovae (see Chapter 20). The very nature of such scenarios means that matched filtering makes little sense. There is no chance of accumulating signal-to-noise by integrating over a large number of cycles, as in the case of binary inspiral. Typically, a search for burst signals cannot be template based.

For burst sources one would typically replace the characteristic amplitude with the root-sum square:

$$h_{\text{rss}}^2 = \int \left(|h_+(t)|^2 + |h_\times(t)|^2 \right) dt].$$ (8.40)

For a linearly polarized wave, with $\tilde{h}(f)$ constant over some bandwidth Δf, it is then approximately the case that

$$h_{\text{rss}}^2 \approx \int |\tilde{h}|^2 df \approx \Delta f |\tilde{h}(f)|^2.$$ (8.41)

Let us take the opportunity to discuss a different kind of averaging, which may be useful when we want to assess the detectability of a given set of sources. In general, the gravitational waves that arrive at a detector will (as long as the detector is small compared to the wavelength; see Chapter 3) be given by

$$h(t) = F_+ h_+ + F_\times h_\times. \tag{8.42}$$

Consider a specific class of burst sources, distributed in space and with random orientation of the polarization axes. In order to work out the detectability of such signals we need to do more than just average over the angles. There may also be a preference for directions and polarizations which leads to a strong signal-to-noise. This is natural because a stronger source can be seen from a larger distance, where the event rate is higher. This can be a significant bias because the event rate increases as d^3 while the gravitational-wave amplitude decreases as $1/d$. The upshot of this is that (see Thorne's review in Hawking and Israel (1989))

$$\left(\frac{S}{N}\right)^2_{\text{strongest}} \approx \left\langle \left(\frac{S}{N}\right)^3 \right\rangle^{2/3} \approx \frac{3}{2}\left(\frac{S}{N}\right)^2. \tag{8.43}$$

Also noting that $\langle F_+^2 \rangle = \langle F_\times^2 \rangle = 1/5$ and $\langle F_+ F_\times \rangle = 0$ we have

$$\left(\frac{S}{N}\right)^2_{\text{strongest}} \approx \frac{3}{10} \int_0^\infty \frac{\langle |\tilde{h}_+|^2 + |\tilde{h}_\times|^2 \rangle}{S_n(f)} df, \tag{8.44}$$

or

$$\left(\frac{S}{N}\right)^2_{\text{strongest}} \approx \frac{3}{10} \frac{\Delta f |\tilde{h}(f)|^2}{S_n(f)}, \tag{8.45}$$

where we have assumed that Δf is sufficiently small that we can take the noise as constant.

Before we move on, it is also worth noting that burst searches often involve time–frequency techniques, where the data from the detector is broken up into segments and then transformed into the frequency domain. After normalizing to the noise spectrum of the detector, a time–frequency plot is produced. Possible signals are 'simply' identified by clusters of pixels that contain excess power. This strategy is useful whenever we are unable to accurately predict the shape of the signal. This can be for a variety of reasons, either technical or arising from our ignorance of the physics. Regardless of the reason, the implication is clear. We cannot use matched filtering to detect such signals. In fact, even in cases where the waveform is known, matched filtering may not be effective if there is a great variety in the shape of the signals. In these cases one is naturally led to suboptimal methods. This means negotiating on the signal-to-noise, but there may be advantages as well. In particular, suboptimal methods are less sensitive to the signal shape and computationally cheaper than matched filtering. The best such methods are sensitive to signal amplitudes a factor of 2 to 3 larger than that required by matched filtering.

The majority of suboptimal techniques build on some form of time-frequency transform, defined as $q(\tau, f)$ for data $x(t)$ using a window $w(t)$, such that

$$q(\tau, f) = \int_{-\infty}^{\infty} w(t - \tau)x(t)e^{2\pi i ft}\, dt. \tag{8.46}$$

The window function $w(t - \tau)$ is centred at $t = \tau$, and one obtains a time–frequency map by moving the window from one end of a data segment to the other. The window is not unique and its efficiency depends on the kind of signal one is looking for. Once the time–frequency map is constructed, one can look for excess power (compared to the average) in different regions, or search for specific patterns.

In practice, the preferred method depends on the signal we are searching for. For unknown signals, the only choice is to look for a departure from the 'average' in different regions of the map. However, if we have some level of knowledge of the spectral/temporal content of the signal, then we can do better. Wavelet-based algorithms, which have been used in searches for unstructured bursts, provide useful examples. One may also develop strategies that improve detection efficiency over a simple search for excess power. For example, chirping signals will leave a characteristic track in the time–frequency plane, with increasing frequency and power as a function of time. An excellent example of this is provided by the data for GW150914, shown in the bottom panels of Figure 1.1. This signal was, in fact, first picked up by a LIGO burst search algorithm.

8.5 Stochastic backgrounds

In terms of search strategies, we also need to consider the possibility of a large number of unresolved signals. This will be important as detectors become sensitive enough to observe a large fraction of the universe. In this case we are dealing with a stochastic gravitational-wave background which can be described in terms of the energy carried by the waves. From the standard flux formula it follows that the energy emitted per unit area is given by

$$E = \frac{c^2}{16\pi G} \int_{-\infty}^{\infty} (2\pi f)^2 |\tilde{h}(f)|^2 df = \int_0^{\infty} \frac{\pi c^2}{4G} f^2 S_n(f) df \tag{8.47}$$

(after integrating over time and using Parseval's theorem). The integrand is often defined in terms of the energy per unit volume of space per unit frequency

$$S_E = \frac{\pi c^2}{4G} f^2 S_n. \tag{8.48}$$

In a cosmological context it is common to work with the dimensionless quantity Ω_{gw}, the energy density per logarithmic frequency interval, normalized to the critical density required to close the universe. This is a straightforward extension of the discussion in Chapter 4. That is, we introduce

$$\Omega_{gw} = \frac{fS_E(f)}{\rho_c c^2} = \frac{f}{\rho_c}\frac{d\rho_{gw}}{df}, \tag{8.49}$$

where ρ_{gw} is the energy density associated with gravitational waves and the critical density is given by (4.72). That is, we have

$$\rho_c = \frac{3H_0^2}{8\pi G}, \tag{8.50}$$

where H_0 is the present value of the Hubble constant.

Before we proceed it is useful to relate the different representations for the gravitational-wave amplitude. We have

$$H_0^2\Omega_{gw}(f) = \frac{8\pi G}{3c^2}fS_E(f) = \frac{2\pi^2}{3}f^2|h_c(f)|^2 = \frac{8\pi^2}{3}\left|\tilde{h}(f)\right|^2. \tag{8.51}$$

We will make frequent use of these relations when we consider cosmological backgrounds in Chapter 22.

In order to detect a stochastic signal, we need to correlate the output from two (or more) detectors. The combined sensitivity can be significantly higher than that obtained for each individual instrument. To see how this goes, let us suppose that each detector output (labelled by I) contains a stochastic signal $s(t)$ along with the usual noise $n(t)$. That is, we have

$$x_I(t) = s_I(t) + n_I(t), \quad I = 1, 2, ... \tag{8.52}$$

Moreover, let us assume that the signal-to-noise ratio is so low that we cannot easily distinguish the signal. It is simply part of the noise. However, all detectors will sense the same gravitational wave so we can use the correlation

$$x = \langle x_1, x_2 \rangle = \int_{-T/2}^{T/2} x_1(t)x_2(t)dt. \tag{8.53}$$

Then we have

$$x \approx \langle s_1, s_2 \rangle + \langle n_2, n_2 \rangle, \tag{8.54}$$

since the cross terms are uncorrelated and should be much smaller than the uncorrelated noise–noise term.

If we run the detectors for a long time T_{obs}, then the signal correlation will grow linearly. Hence we have

$$\langle s_1, s_2 \rangle \sim |s(f)|^2 \Delta f \, T_{\text{obs}}. \tag{8.55}$$

Meanwhile, the noise will execute a random walk, leading to

$$\langle n_1, n_2 \rangle \sim |n(f)|^2 \, (\Delta f \, T_{\text{obs}})^{1/2}. \tag{8.56}$$

This argument tells us that the minimum detectable amplitude will improve as

$$\Omega_{\text{gw}} \sim \frac{|n(f)|^2}{(\Delta f \, T_{\text{obs}})^{1/2}}. \tag{8.57}$$

The main question is whether we can keep observing long enough to make a weak signal detectable. We will return to this problem in Chapter 22.

8.6 Avoiding false alarms

When we are dealing with real detector data we have to accept the presence of glitches and transients that may look like damped sinusoids, or in other ways mimic the signals we are interested in. There will inevitably be false alarms, due to both the instrument and its environment, and we need to figure out efficient ways of dealing with these. As a first step towards ruling out such triggers, let us consider a veto that makes use of the scalar product from matched filtering (Allen, 2005) (see also Lindblom and Cutler (2016)).

Suppose we are looking for binary inspiral signals. For such signals the matched-filter signal-to-noise has contributions from a wide frequency range. However, the result is an integral over the frequency, so it does not distinguish contributions from specific frequency regions. However, imagine dividing the range of integration into a finite number of bins $f_k \le f < f_{k+1}$, $k = 1, \dots, p$ spanning the entire frequency band, in such a way that $f_1 = 0$ and $f_{p+1} = \infty$, and assume that the contribution to the signal-to-noise from each frequency bin is the same. That is

$$4 \int_{f_k}^{f_{k+1}} \frac{|\tilde{h}(f)|^2}{S_n(f)} \, df = \frac{4}{p} \int_0^\infty \frac{|\tilde{h}(f)|^2}{S_n(f)} \, df. \tag{8.58}$$

Then define the contribution from the kth bin as

$$z_k \equiv (K|x)_k \equiv 4\text{Re} \int_{f_k}^{f_{k+1}} \frac{\tilde{K}^*(f)\tilde{x}(f)}{S_n(f)} \, df, \tag{8.59}$$

where $\tilde{x}(f)$ and $\tilde{K}(f)$ are the Fourier transforms of the detector output and the template, respectively. If we sum over all the bins we (obviously) recover the original matched-filter result:

$$z = \sum_{k=1}^{p} z_k. \tag{8.60}$$

Having chosen the bins and the quantities z_k, we can construct a new statistic based on the measured signal-to-noise in each bin compared to the expected value. To be specific, if the background noise is stationary and Gaussian, the quantity

$$\chi^2 = p \sum_{k=1}^{p} \left(z_k - \frac{z}{p} \right)^2 \tag{8.61}$$

obeys the standard χ-square distribution with $p-1$ degrees of freedom. Therefore, the properties of the this statistic are known.[4]

Let us now imagine two triggers with identical signal-to-noise, one caused by a true signal and the other by a glitch with power only in a small frequency range. These triggers will have very different χ^2 values. The first will be much smaller than the second. This idea provides a powerful veto in the search for binaries and has been instrumental in cleaning up the data, in general.

A variant of this kind of analysis is provided in Figure 8.2. The aim of the exercise is to establish the statistical significance of the signal. One can do this by comparing the event (in this case GW150914) to a set of alternative realizations of the data, that should (by construction) not contain an astrophysical signal. The first step involves working out the matched-filter signal-to-noise ratio $\rho(t)$ for a set of inspiral templates in each detector and identifying maxima with respect to the time of arrival of the signal. For each maximum one then calculates a χ^2 statistic to test to what extent the data in different frequency bands are consistent with the matching template. Values of χ^2 near unity indicate that the signal is, indeed, consistent with a merger event. If the result exceeds $\chi^2 = 1$, the signal-to-noise is reweighted as

$$\hat{\rho} = \rho \left[\frac{2}{1 + (\chi^2)^3} \right]^{1/6}. \tag{8.62}$$

The final step involves establishing the likelihood that the event happened by chance. First of all, coincidence between detectors is established by selecting pairs of events from the same template within a 15-ms window—determined by the 10-ms propagation time between the detectors (see Figure 7.2) plus a 5-ms uncertainty in the arrival time of weak signals. Coincident events are simply ranked based on the quadrature sum $\hat{\rho}_c$ of the individual $\hat{\rho}$ from each detector.

To produce background data for the search, the signal-to-noise maxima of one detector are time shifted in such a way that a coincidence with the other detector data cannot have an astrophysical origin. A new set of coincident events is computed on

---

[4] One would typically use a χ^2 statistic to test relationships between variables. The null hypothesis is that the variables are independent.

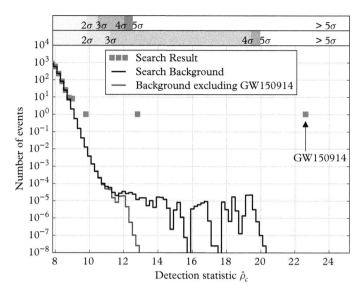

Figure 8.2 *Results from a binary coalescence search on LIGO data including the GW150914 event. The histograms show the number of candidate events (orange markers) and the mean number of background events (jagged black lines) in the search where GW150914 was found as a function of the search detection statistic $\hat{\rho}_c$ (with a bin width of 0.2). The scales at the top of the panel give the significance of an event in Gaussian standard deviations based on the corresponding noise background, showing that the significance of the GW150914 detection was greater than 5σ. The tail in the background is due to random coincidences of GW150914 in one detector with noise in the other detector. The purple curve represents the background excluding those coincidences. This background was used to assess the significance of the second strongest event in the data set. (Reproduced from Abbott et al. (2016b), Creative Commons Attribution 3.0 License.)*

this shifted data set. Repeating the procedure $\sim 10^7$ times produces a noise background equivalent to a search lasting 608,000 years. The background level indicated in Figure 8.2 relates to templates matching the GW150914 event. The inferred value $\hat{\rho}_c = 23.6$ is larger than any background event, which means that one can only infer an upper bound on the false alarm rate. Combining the result with other template sets for the search one arrives at a bound of 1 in 203,000 years. This translates into a false alarm probability of $< 2 \times 10^{-7}$. Expressed in terms of Gaussian standard deviations for the corresponding noise background, the event was detected at the 5σ level.

8.7 Bayesian inference

Up to this point we have laid the foundations for what is often called the 'frequentist' approach to data analysis. This has provided us with useful insights, but if we want to go further it is natural to consider a Bayesian strategy (Sathyaprakash and Schutz, 2009).

In essence, the data analysis problem involves hypothesis testing and it is natural to base the discussion on the construction of probabilities and probability densities. We want to establish strict detection criteria involving quantitative thresholds—any statement of detection must involve a quantified degree of certainty that the event is not an instance of the noise. Similarly, when it comes to parameter extraction, we want to be able to characterize the signal in terms of the most likely parameter values. The Bayesian approach allows us to achieve these goals.

The starting point of all Bayesian analysis, however complex, is Bayes' theorem. It is based on the fundamental principle of calculating joint probabilities of mutually independent events. Two events A and B (say) are said to be mutually independent if the fact that event A happened has no influence on the occurrence (or otherwise) of B. If $P(A)$ and $P(B)$ represent the probabilities for events A and B to happen, then the probability that they both occur, the joint probability of A and B—denoted[5] $P(A,B)$—is the probability of A given that B has occured times the probability of B:

$$P(A,B) = P(A|B)P(B). \tag{8.63}$$

Alternatively, we can write the joint probability as

$$P(A,B) = P(B|A)P(A). \tag{8.64}$$

Equating the two alternative expressions for the joint probability leads to Bayes' theorem

$$P(A|B) = \frac{P(B|A)P(A)}{P(B)}. \tag{8.65}$$

The statistical interpretation of this result has far-reaching consequences. In order to see why this is so, we first of all note that

$$P(B) = P(B|A)P(A) + P(B|\bar{A})P(\bar{A}), \tag{8.66}$$

where the bars denote negation, so $P(\bar{A})$ is the probability that A is not going to happen. Next we identify the first term in the above expression with the numerator on the right-hand side of Bayes' theorem (8.65). This allows us to rewrite the result as

$$P(A|B) = \frac{\Lambda}{\Lambda + P(\bar{A})/P(A)}. \tag{8.67}$$

---

[5] The notation should not be confused with the inner product used for matched filtering.

The new quantity

$$\Lambda = \frac{P(B|A)}{P(B|\bar{A})} \tag{8.68}$$

is known as the likelihood ratio. Rewriting Bayes' theorem in this fashion is powerful because we now see that the outcome $P(A|B)$ only depends on the likelihood ratio Λ and the prior probabilities $P(A)$ and $P(\bar{A})$.

Before moving on it is useful to note that the posterior $P(A|B)$ is always less than or equal to 1, with equality only when $P(B|\bar{A}) = 0$. In all other situations, the result is strictly less than 1.

How do we use these results in practice? Well, let us recast the final statement in terms of a gravitational-wave detection problem. To do this we replace A with h, the expected gravitational-wave signal. B is replaced by x, the data stream, which may contain the signal, $h + n$, or be pure noise, n. We then have

$$P(h|x) = \frac{\Lambda}{\Lambda + P(\bar{h})/P(h)}. \tag{8.69}$$

The conditional probability that a signal of form h is present in the data stream x depends on the likelihood ratio and the a priori probabilities that the signal is present, $P(h)$, and that it is not, $P(\bar{h})$. The likelihood ratio is now given by

$$\Lambda = \frac{P(x|h)}{P(x|\bar{h})}. \tag{8.70}$$

This is the ratio of the probability of detecting the data stream x with the signal present to that of the signal absent. It is notable that the data only enters Λ on the right-hand side of (8.69).

In order to use this result, we need to assess the a priori probabilities and calculate the likelihood ratio. The first step neatly folds in our understanding of the astrophysics and obviously involves a degree of belief. We may not know the probabilities precisely, but we can often make an educated guess. We may, for example, believe that the sources are distributed in some particular way or that a specific event happens at some given rate.

Turning to the calculation of Λ, we first of all note that the probability density for Gaussian noise is given by

$$p_n \propto \exp\left[(n|n)/2\right], \tag{8.71}$$

where we have used the inner product from (8.18). This allows us to quantify $P(x|\bar{h})$. Next we use

$$p_n(x - h) \propto \exp\left[-(x - h|x - h)/2\right], \tag{8.72}$$

to get $P(x|h)$. Putting things together, we have

$$\Lambda = \frac{p_n(x-h)}{p_n} = \exp\left[(x|h) - (h|h)/2\right]. \qquad (8.73)$$

We see that Λ increases monotonically with $(x|h)$. This leads us back to the result for the optimal filter.

The Bayesian analysis correlates the data stream (x) with the signal (h) and allows us to compare the result to some set threshold. Starting from

$$\log \Lambda = (x|h) - \frac{1}{2}(h|h), \qquad (8.74)$$

an often-used criterion is based on the maximum likelihood. The idea is quite simple. Assume that the signal depends on some parameters, collectively labelled θ, so that $h = h(t; \theta)$. Then solve

$$\frac{\partial \log \Lambda}{\partial \theta} = 0, \qquad (8.75)$$

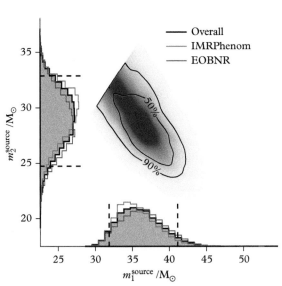

Figure 8.3 *Posterior probability densities (PDF) for the two masses in the GW150914 binary system; m_1^{source} and m_2^{source} (with the convention that $m_2^{source} \leq m_1^{source}$). Results are shown for two specific waveform models, one phenomenological and one based on the effective-one-body framework outlined in Chapter 11, as well as the overall distribution. The dashed vertical lines mark the 90% credible interval for the overall PDF. The two-dimensional plot shows the contours of the 50% and 90% credible regions overlaid on a colour-coded PDF. (Reproduced from Abbott et al. (2016c), Creative Commons Attribution 3.0 License.)*

to find the most likely values. An example is provided in Figure 8.3, again for GW150914. Of course, in reality this is a complex problem. In most cases we are dealing with a large set of parameters—forming an abstract vector θ_i—so calculating the posterior involves a multidimensional integral. As this is computationally expensive, it is often not possible to apply Bayesian techniques to continuously streaming data.

8.8   Geometry in signal analysis

As we are dealing with a challenging multiparameter problem, it is natural that there is a significant literature on the computational requirements (Owen and Sathyaprakash, 1999). In the case of binary inspirals estimates suggest that, for the Advanced LIGO detectors, one will need several thousand templates to search for component masses in the range $[m_{\text{low}}, m_{\text{high}}] = [1, 100] M_{\odot}$. Lowering the mass cut-off leads to an increase in the number of templates that scales as $m_{\text{low}}^{-8/3}$. However, as we are not aware of any formation channels for neutron stars or black holes with lower mass than $m_{\text{low}} \approx 1 M_{\odot}$, it makes sense to use this lower limit.

The inclusion of spins is only important when one or both of the components is rapidly spinning. For neutron stars, one would expect spin effects to be largely unimportant since, in terms of the dimensionless parameter $a/M = \mathcal{J}_{\text{NS}}/M_{\text{NS}}^2 \ll 1$ (see Chapter 17), even the fastest spinning neutron stars are extremely slow. In fact, as we expect the binaries we observe to be old, the neutron stars would have had ample time to slow down and it would be surprising to find a system with significant spin. Regardless, the computational costs, while high, are not formidable for ground-based detectors (even if we include the spins). It should be possible to carry out a real-time search on large computer clusters.

Let us focus on the situation where we have some idea of the character of the signal, so that we can create useful filters. In reality, we will need a bank of filters reflecting the uncertain parameters of the source (like component masses and individual spins in a binary). How many filters do we need in any given situation? The answer depends on the extent to which we are prepared to risk missing out on detections. We would, obviously, like to minimize any losses in this respect, so let us ask how far apart (in parameter space) two filters can be allowed to be without compromising the search. In many ways, this is a geometry problem (Owen, 1996).

8.8.1   Ambiguity function

As a first step towards quantifying how many filters we are likely to need, we introduce the so-called ambiguity function. Well known in the statistical theory of signal detection, this is a powerful tool which helps us assess the number of templates required to span the parameter space of the signal. It can also be used to make estimates of variances and covariances involved in the measurement of various parameters and to compute biases associated with using a family of templates whose shape is not the same as that of the signal.

The ambiguity function is defined as the scalar product of two normalized waveforms maximized over the initial phase of the waveform, in other words, the absolute value of the scalar product. A waveform e is normalized if $(e|e)^{1/2} = 1$, where the inner product is inversely weighted by the power spectral density, as in (8.18). Working with normalized waveforms is useful because it means that it is easy to define the signal strength as h_0 if $h = h_0 e$. It also means that the signal-to-noise ratio is simply given by $(h|h)^{1/2} = h_0$.

Let us suppose that $e(t;\theta)$—where $\theta = \{\theta_i | i = 0, \ldots, p\}$ is the parameter vector made up of $p+1$ parameters—represents a normalized waveform. It is conventional to choose the parameter θ_0 to be the time lag τ, the coordinate time when an event occurs. This is an extrinsic parameter, while the rest of the p parameters are taken to be intrinsic to the source.

Given two normalized waveforms $e(t;\theta)$ and $e(t;\phi)$, whose parameter vectors are not necessarily the same, the ambiguity $\mathcal{A}$ is defined as

$$\mathcal{A}(\theta,\phi) \equiv |(e(t;\theta)|e(t;\phi))|. \tag{8.76}$$

Since the waveforms are normalized, $\mathcal{A}(\theta,\theta) = 1$ and $\mathcal{A}(\theta,\phi) < 1$, if $\theta \neq \phi$. If we want to make contact with the previous discussion, we can think of θ as the parameters of a template while ϕ represents the parameters of a signal. With the template parameters θ fixed, the ambiguity function is a function of the signal parameters ϕ, giving the signal-to-noise obtained by the template for different signals. The region in the signal parameter space for which a template obtains signal-to-noise ratios larger than a chosen value (often called the minimal match) is the span of that template. Template families need to be chosen so that they span the entire signal parameter space of interest with the least overlap of one another's spans. In effect, we may also interpret the ambiguity function as giving the signal-to-noise obtained for a given signal by filters with different parameter values.

The ambiguity function has a local maximum at the 'correct' set of parameters, $\phi = \theta$. Search methods that vary ϕ to find the best fit to the parameter values make use of this property one way or another. But the ambiguity function will usually have secondary maxima as a function of ϕ with fixed θ. If these additional maxima are only slightly smaller than the primary one, noise can lead to confusion. Randomly, a secondary can be elevated to a primary, which would lead to a false measurement of the parameters. Search methods need to be designed to avoid this as much as possible.

As is clear from the definition (8.76), there is no need for the functional forms of template and signal to be the same. The definition works for any signal–template combination. Crucially, the number of template parameters need not be identical (and usually is not) to the number of actual signal parameters. For example, a binary can be characterized by a large number of parameters, while we may use a model waveform involving only the masses. For inspiral waves, $e(t;\phi)$ is the exact waveform emitted by a binary, the form of which we do not know, while the template family can be a post-Newtonian approximation.

8.8.2 Metric on the space of waveforms

When it comes to working out the ideal choice of templates for a given search, it is can be useful to consider the geometric nature of the problem (Owen, 1996). After all, we are interested in figuring out how close neighbouring filters need to be in order for us not to deplete the signal-to-noise of a signal with unknown parameters. In order to make the question quantitative, we may introduce a measure that tells us how close (in parameter space) two filters are. To do this, let x_k, with $k = 1, 2, \ldots, N$, represent the discretely sampled output of a detector. The set of all possible outputs then satisfies the usual axioms of a vector space. This means that we can think of x_k as an N-dimensional vector. In practice, we can assume that this vector is some discrete representation of a continuous problem, with infinite dimensional vectors.

Out of all possible such vectors, we are naturally interested in those that correspond to gravitational waves from a given astronomical source. While every signal can be thought of as a vector in the infinite-dimensional vector space of the detector output, the set of all such signal vectors do not, by themselves, form a vector space. However, the set of all normalized signal vectors (i.e., signal vectors of unit norm) form a manifold, and the parameters of the signal play the role of a coordinate system. The upshot of this is that we have an n-dimensional manifold $\mathcal{S}_n$, where n is the number of independent source parameters, for each kind of source. Following the familiar steps, e.g. from Chapter 2, the manifold $\mathcal{S}_n$ can be endowed with a metric g_{ij} associated with the scalar product from (8.18). In a coordinate system p^i, the components of this metric are defined as

$$g_{ij} \equiv \left(\partial_i \hat{h} | \partial_j \hat{h} \right), \quad \partial_i \hat{h} \equiv \frac{\partial \hat{h}}{\partial p^i}. \tag{8.77}$$

The metric can then be used on the signal manifold as a measure of the proper distance dl between nearby signals with coordinates p^i and $p^i + dp^i$, such that

$$dl^2 = g_{ij} dp^i dp^j. \tag{8.78}$$

Taylor expanding $\hat{h}(p^i + dp^i)$ around p^i, and keeping only terms to second order in dp^i, it is easy to see that the overlap of two infinitesimally close signals can be obtained using the metric:

$$\mathcal{O}(dp^i; p^i) \equiv \left(\hat{h}(p^i) | \hat{h}(p^i + dp^i) \right) = 1 - \tfrac{1}{2} g_{ij} dp^i dp^j. \tag{8.79}$$

This then allows us to determine, in an efficient manner, how close together in parameter space our search templates need to be.

8.8.3 The Fisher matrix

In principle, parameter estimation is not distinct from detection. This is, indeed, the spirit of a Bayesian analysis where one obtains the most likely parameter values by comparing the data against a (usually very large) number of realizations from theoretical models (typically obtain using Markov Chain Monte Carlo simulations). This approach leads to probability distributions for the parameters, as in Figure 8.3. Still, in practice, some methods might be computationally efficient in detecting a signal but not necessarily the best for parameter estimation, while the best parameter estimation methods might not be computationally efficient. Thus, the problem of parameter extraction is often considered separately from detection.

We can never be absolutely certain that a signal is present in a given data set. We can assign a level of confidence, which may be close to 100%, but we cannot know for sure. Moreover, whatever the signal-to-noise ratio may be, we cannot be absolutely certain about the true parameters of the signal. At best we can obtain estimates in a certain range, and this range depends on the level of confidence we require.

For any given measurement, any estimated parameters—however efficient, robust, and accurate—are unlikely to be the actual parameters of the signal, since, at a finite signal-to-noise ratio, noise alters the input signal. In geometric language, the signal vector is being altered by the noise vector and the matched filtering aims at computing the projection of this altered vector onto the signal space.

There is an immediate connection between the geometric approach and the parameter extraction problem. The metric on the signal manifold is nothing but the well-known Fisher information matrix[6] Γ_{ij}, after scaling with the square of the signal-to-noise ratio,

$$g_{ij} = \frac{\Gamma_{ij}}{\rho^2}. \tag{8.80}$$

The information matrix itself is the inverse of the covariance matrix, C_{ij}. These quantities allow us to quantify how accurately we can extract the parameters of a given signal.

For a given waveform model, with unknown parameters to be determined, we can readily construct the Fisher information matrix and determine its inverse. This is the most important quantity from the experimental point of view as its components are directly related to the measurement errors of the parameters. An illustration of this is provided in Chapter 18.

---

[6] The Fisher matrix encodes to what extent different parameters in a given model are related.

Part 2

The dark side of the universe

9

The stellar graveyard

Gravitational waves are generated whenever masses accelerate anywhere in the Universe. As a result, the Earth is bathing in spacetime ripples, stretching and squeezing not just our detectors but all of our everyday reality. If the waves were strong, this would be uncomfortable, but this is obviously not the case. The waves that reach us from the distant Universe are extremely weak. Any effort to detect them must necessarily focus on the most powerful events. We are not interested in subtlety—we seek cosmic drama.

To get an impression of how weak the effect of gravitational-wave emission can be, let us return to the quadrupole formula result (5.33) for the energy loss in a binary system. Recall that

$$\dot{E} = \frac{32}{5} \frac{G}{c^5} \mu^2 a^4 \Omega^6, \tag{9.1}$$

where μ is the reduced mass

$$\mu = \frac{M_1 M_2}{M_1 + M_2}, \tag{9.2}$$

a is the orbital separation, and the (angular) orbital frequency is Ω. Combining this with the energy

$$E = -\frac{G\mu M}{2a}, \tag{9.3}$$

where $M = M_1 + M_2$ is the total mass, we see that the radius of the orbit shrinks at a rate

$$\frac{\dot{a}}{a} = -\frac{\dot{E}}{E} = \frac{64}{5} \left(\frac{\Omega a}{c}\right)^5 \left(\frac{\mu}{M}\right) \Omega. \tag{9.4}$$

So far, this simply repeats the calculation from Chapter 5, although we have written the result in a slightly different way. Now let us consider the implication for the Solar System. Putting in the relevant masses, the distance between the Sun and the Earth and

Gravitational-Wave Astronomy: Exploring the Dark Side of the Universe. Nils Andersson, Oxford University Press (2020).
© Nils Andersson. DOI: 10.1093/oso/9780198568032.001.0001

an orbital period of one year, we find that the rate at which the Earth's orbit decays due to gravitational-wave emission is a tiny fraction of a centimetre in a billion years. This is obviously not impressive (!) and it is easy to see that the associated waves would be very difficult to detect. In fact, there is a practical reason why we would not be able to detect these waves at all. Comparing the wavelength of the waves to the size of the orbit

$$\frac{\lambda_{gw}}{a} \sim \frac{c}{\Omega a} \sim 10^4, \tag{9.5}$$

we see that the Earth is nowhere near in the wave-zone for this problem. If we had access to extreme precision measuring technology, we might be able to monitor the shrinking orbit, but we could never catch the waves because in our neighbourhood they are, in fact, not waves at all.

Looking further afield, it is instructive to ask what kind of astrophysical systems might be better for our purposes. The numerical penalties for the Earth–Sun system are obvious. To do better, we need to have more equal masses (to avoid $\mu/M \ll 1$) and close (ideally relativistic) orbits (to avoid $\Omega a/c \ll 1$). If we consider also the wave amplitude, say from (5.46), we see that it would be good to involve more massive objects. This exercise pretty much tells us that we need to go beyond mainstream stellar astrophysics and focus on compact objects: white dwarfs, neutron stars, and black holes. Representing the possible endpoints of stellar evolution, these objects provide the most promising sources for gravitational-wave astronomy. Given this, we need to understand the physics associated with them better. We need to understand the ways they may radiate gravitational waves, what the amplitude and signature of such events may be, and how frequently they occur within the horizon distance of any given detector.

9.1   White dwarfs

During the first decades of the twentieth century astronomers were puzzled by the nature of the companion of Sirius (imaginatively called Sirius B). The companion's mass could be derived from the binary orbit, and appeared to be similar to that of the Sun, but at the same time it was more than a hundred times dimmer. The obvious explanation that the star was faint because it was cold failed. Stars radiate like black bodies, and cold stars should be unmistakably red. This peculiar star was white. The inevitable conclusion was that the star had to be small.

That the companion of Sirius was, indeed, small was confirmed by redshift measurements in 1925. This remarkable observation demonstrated that extremely dense matter is not only possible, but actually present in the Universe. The new class of stars—the white dwarfs—had to be very compact. We now know that a typical white dwarf has a mass similar to that of the Sun compressed inside a radius of 10^4 km or so. This means that the density inside the star is on the order of 10^6 g/cm^3, making it something like 2,000 times denser than platinum. This may sound exotic, but white dwarfs are actually the second most common stars in the Universe. They represent the endpoint of stellar evolution for virtually all stars. As a star with mass less than about $8M_\odot$ begins to run out of nuclear fuel it swells up to form a red giant (as it continues to burn heavier elements), then it fades and shrinks to form a white dwarf. The remnant gradually fades from view as it radiates away the residual thermal energy over billions of years.

Since the early days, astronomers have found many more white dwarfs. Progress has been particularly swift since the European Space Agency's Gaia telescope started collecting data in 2013.[1] The aim of Gaia is to carefully monitor the position of more than 1 billion stars, also recording spectral information. Results from the second Gaia data release (in 2018) are shown in Figure 9.1. The illustration provides colour–magnitude information—it is a standard Hertzsprung–Russell diagram—bringing out, in addition to the main sequence stars, the white dwarf population. Particularly interesting (for our purposes) is the indication of two distinct populations of white dwarfs, with different masses. The heavier systems could conceivably be the result of white dwarf mergers (Kilic *et al.*, 2018).

The Galaxy contains a large number of low-mass binaries involving white dwarfs. These are, in principle, interesting gravitational-wave sources. However, as the typical orbital timescales ranges from minutes to hours, we cannot expect to detect these systems

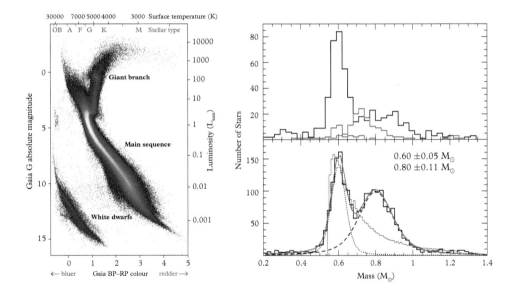

Figure 9.1 *Left: Results from the second Gaia data release as a Hertzsprung–Russell diagram. Right: Evidence of the presence of two populations of white dwarfs, distinguished by the mass. The data includes about 13,000 white dwarfs within 100 pc of the Sun. (Left panel adapted from http://sci.esa.int/gaia/, copyright: ESA/Gaia/DPAC, Creative Commons Attribution-ShareAlike 3.0 IGO License. Right panel reproduced from Kilic et al. (2018).)*

[1] See http://sci.esa.int/gaia/.

from the ground. We need a space-based instrument. In fact, a significant population of double white dwarf systems should be detectable by LISA, to the point where they form a confusion-limited background (see Figure 7.11). In addition, there may be as many as 25,000 individually resolvable systems (Cornish and Robson, 2017). Some of these systems are considered as verification binaries for the LISA mission (Stroeer and Vecchio, 2006). By monitoring this class of binaries, we may improve our understanding of binary evolution leading to the formation of double white dwarf systems. This may, in turn, shed light on the common envelope phase and the stability of the involved mass transfer.

The internal structure of these (no longer very exotic) objects was first understood when Ralph Fowler brought the new quantum statistics of Fermi and Dirac to bear on the problem (Fowler, 1926)—a landmark in the development of our understanding of stellar structure. This drew attention to the fact that the electron gas in matter this dense must be degenerate. When confined to a finite volume even an absolutely cold assembly of electrons retains a spread of momenta (due to Pauli's exclusion principle). This leads to the 'electron degeneracy' pressure that balances gravity in a white dwarf.

9.2   The Fermi gas model

Given its importance, it is worth taking a closer look at the electron degeneracy pressure. We do this by formulating the basic ideas of the Fermi gas model, which plays a key role in solid state physics in regimes where quantum mechanics dominates and it is also essential for dense matter astrophysics.

Due to the exclusion principle, each energy state of a system of fermions can only be occupied by a single particle. In order to determine the ground state of a system of N particles one simply starts filling the energy states from the lowest to higher levels until all particles have been accounted for. Because each state can accommodate two fermions (with distinct spins), the system is degenerate. If we define the Fermi energy E_F as the energy of the highest filled state we have, see Figure 9.2,

$$E_F = \frac{\bar{p}_F^2}{2m},$$

(9.6)

in terms of the Fermi momentum,[2] $\bar{p}_F = \hbar k_F$, where the corresponding wave number is defined as

$$k_F = (3\pi^2 n)^{1/3},$$

(9.7)

and $n = N/V$, where V is the volume, is the number density of particles (each with mass m).[3]

---

[2] We add a bar to the momentum throughout the discussion of Fermi gases in order to make a clear distinction between the momentum $\bar{p}$ and the pressure p.
[3] It is common to work in units such that $\hbar = c = G = 1$, in which case we obviously have $k_F = p_F$. That is, we extend the geometric units to include $\hbar$.

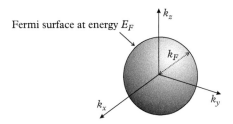

Figure 9.2 *In a system of N fermions the occupied states can be represented by points in a sphere in momentum space, or equivalently, in terms of the wavenumber. In the ground state the particles fill a sphere of radius k_F, where $E_F = \hbar^2 k_F^2/2m$ is the energy of a particle. This is the Fermi energy.*

Now consider a gas made out of these particles. The total energy (density) should simply be $\varepsilon = nE_F$ and we can use the first law of thermodynamics to derive an estimate for the corresponding pressure, p. In order to do this, it is, first of all, natural to introduce the chemical potential μ, which corresponds to the energy required to 'add one more particle to the system'

$$\mu = \frac{d\varepsilon}{dn}. \tag{9.8}$$

Then the pressure follows from the thermodynamical relation

$$p = -\varepsilon + n\mu = \frac{2}{5}nE_F \longrightarrow p \sim n^{5/3}. \tag{9.9}$$

We can use this result together with the equations for fluid dynamics from Chapter 4 to work out the properties of a star supported by electron degeneracy. However, for the moment, a simple back-of-the-envelope estimate will do. If we assume that the density is uniform, then hydrostatic equilibrium immediately leads to

$$\nabla p = -\rho \nabla \Phi \longrightarrow p \approx \frac{GM\rho}{R} \sim \frac{M^2}{R^4}, \tag{9.10}$$

where M is the star's mass and R the radius. Meanwhile, the degeneracy pressure suggests that (using $\rho = mn$)

$$p \sim \rho^{5/3} \sim \frac{M^{5/3}}{R^5}. \tag{9.11}$$

The two results have to balance, so we must have

$$M^{1/3} \sim 1/R. \tag{9.12}$$

We learn that more massive white dwarfs must be smaller. This is contrary to the result for normal stars, and the story is about to become even more peculiar...

9.3   Chandrasekhar's limit

As matter is compressed, the pressure increases. Individual particles move at higher speed and at some point one would expect special relativity to come into play. In order to assess whether we need to worry about this, we can compare the Fermi momentum to the highly relativistic limit. This is equivalent to comparing the interparticle distance d in the gas to the Compton wavelength $\hbar/mc$. Introducing the parameter $x = \bar{p}_F/mc$, such that $x \approx 1$ means that a given particle must be considered as relativistic, we find that the electrons are relativistic above

$$n_e \approx \frac{1}{3\pi^2}\left(\frac{m_e c}{\hbar}\right)^3. \qquad (9.13)$$

For a given white dwarf, with known mass and radius, we also need to contrast the obtained value for n_e (which provides the degeneracy pressure) to the baryon number density, n_b, (which provides the bulk of the mass). Introducing the electron fraction $x_e = n_e/n_b$ (which would be $1/2$ for a pure Helium star), we have

$$\rho \approx m_b n_b \approx \frac{m_b}{x_e} n_e, \qquad (9.14)$$

and we find that relativity becomes important above a density of about 2×10^6 g/cm^3. Given that white dwarfs typically have central densities in the range $10^4 - 10^7$ g/cm^3, we clearly need to pay attention to relativistic effects when we model their interiors.

The question is, how does relativity impact on the Fermi gas model? To answer this, we need to consider the problem in more detail. It is natural to start by considering the density of states $g(\bar{p})$ in momentum space

$$g(\bar{p})d\bar{p}dV = \frac{8\pi}{h^3}\bar{p}^2 d\bar{p}dV, \qquad (9.15)$$

where we have accounted for a factor of 2 for the opposite spins. In general, the electrons obey Fermi–Dirac statistics, which means that the distribution is

$$f(\varepsilon) = \left[\exp\left(\frac{\varepsilon - \mu}{k_B T}\right) + 1\right]^{-1}, \qquad (9.16)$$

where μ is the chemical potential (as before) and $k_B T$ is the thermal energy (k_B is Boltzmann's constant and T is the temperature). At low temperatures the system is completely degenerate, which means that all states up to the surface of the Fermi sphere are filled and all states above it are empty. That is, we have

$$f(\varepsilon) = \begin{cases} 1, & \varepsilon \leq \varepsilon_F, \\ 0, & \varepsilon > \varepsilon_F. \end{cases} \qquad (9.17)$$

Now, on the microscopic scale of a fluid element the pressure, p, follows from the momentum flux through a given surface. If the particles have velocity $v(p)$ this leads to (after integrating out the angular dependence)

$$p = \frac{4\pi}{3h^3} \int_0^{\bar{p}_F} \bar{p}^3 v(\bar{p}) d\bar{p}, \tag{9.18}$$

in the fully degenerate case. For a non-relativistic gas, we have $v = \bar{p}/m$ and if we also make use of the number density (the integrated density of states)

$$n = \frac{8\pi}{h^3} \int_0^{\bar{p}_F} \bar{p}^2 d\bar{p} = \frac{8\pi}{3h^3} \bar{p}_F^3, \tag{9.19}$$

it is easy to obtain Eq. (9.9). However, when the particles move at relativistic speeds, we need account for the Lorentz factor in the (spatial part of the) four-momentum (see Chapter 2). That is, we have

$$\bar{p} = \gamma m v \qquad \text{with} \qquad \gamma = \left(1 - \frac{v^2}{c^2}\right)^{-1/2}, \tag{9.20}$$

which leads to

$$v(\bar{p}) = \frac{\bar{p}}{m}\left[1 + \left(\frac{\bar{p}}{mc}\right)^2\right]^{-1/2}. \tag{9.21}$$

This affects the pressure calculation and we now get

$$p = \frac{\pi m^4 c^5}{6h^3}\left[x(2x^2 - 3)\sqrt{1 + x^2} + 3\sinh^{-1} x\right], \tag{9.22}$$

where $x = \bar{p}_F/mc$, as before. In the highly-relativistic limit, when $x \to \infty$, this reduces to

$$p \approx \frac{1}{16}\left(\frac{3}{\pi}\right)^{1/3} hcn^{4/3}. \tag{9.23}$$

We learn that relativistic effects tend to soften the equation of state (the pressure increases slower as we crank up the density). This has profound implications. Matter finds it harder to withstand its own gravity at higher densities. One can imagine reaching a point where gravity wins and the star has to fall in on itself. This is, indeed, what happens. In order to infer the critical mass, let us again balance the bulk pressure to the, now relativistic, degeneracy pressure. This leads to

$$p \sim \frac{M^2}{R^4} \sim \frac{M^{4/3}}{R^4}. \qquad (9.24)$$

Remarkably, the radius R drops out of the equation and we have a unique result for the mass of a relativistic white dwarf. Whereas we had the freedom to choose the mass and find the corresponding radius in the non-relativistic case, we now appear not to be able to select the mass. A more careful calculation shows that the critical mass is

$$M_{\rm Ch} \approx 5.8 x_e^2 M_{\odot} \approx 1.44 M_{\odot} \text{ for } x_e = 1/2. \qquad (9.25)$$

Real electrons do not exactly obey (9.23), but the approximation should get better as they get more relativistic. In essence, we should consider $M_{\rm Ch}$ as a limit on the mass of all white dwarfs. More massive stars cannot be supported by electron degeneracy pressure. Basically, the maximum mass of a helium white dwarf is just below one and a half solar masses.

   This argument was first worked out by the 19-year-old Subrahmanyan Chandrasekhar in 1929, during the sea voyage from India to Britain, where he was going to study at Cambridge University (Chandrasekhar, 1931). Perhaps understandably, the result was controversial. One of the main opponents was Arthur Eddington, who had led the solar eclipse expedition that lent support to Einstein's curved spacetime theory (see Chapter 10). Eddington famously stated that 'there should be a law of Nature to prevent a star from behaving in this absurd way!' It took time, but eventually the evidence was clear. Chandrasekhar's calculation was correct. Nature does not shy away from absurdity.

9.4   Neutron stars

The remarkable maximum-mass result for white dwarfs threw the door wide open to speculation. What is the fate of more massive stars? Einstein's theory seemed to point in the direction of gravitational collapse, but astronomers were not ready to accept this idea, no matter how persuasive the argument may have been.

   A temporary resolution to the controversy followed James Chadwick's 1932 discovery of the neutron. This led to speculation that entire stars made up of such particles might exist. Scientists argued for the existence of 'neutron stars' for two different reasons. An astrophysically motivated (and visionary) view was taken by Walter Baade and Fritz Zwicky. They suggested that neutron stars might be the remnants of supernova explosions (Baade and Zwicky, 1934). There was, however, no observational support evidence for this suggestion at the time. The association between supernovae and neutron stars would eventually be established, but Zwicky would remain the sole advocate of the idea for many years. Conventional astronomy had no need for anything more exotic than white dwarfs.

   Nuclear physicists had different motivation to discuss neutron stars. In the early 1930s theoretical physics had difficulties to account for the generation of energy in stars in terms of thermonuclear reactions. As an attempt to solve this problem George Gamow and Lev Landau suggested that many, possibly all, stars contained degenerate neutron cores (Landau, 1938; Gamow, 1939). The slow growth of such a core could help fuel the stars' radiation. As matter was neutronized it would reduce in volume and gravitational energy would be released. That this might be a runaway process that would lead to a rapid collapse of the core of the star was not realized until later.

Given what we already know, let us try to estimate the properties of a typical neutron star. For simplicity, we take the Newtonian equation for hydrostatic equilibrium as our starting point

$$\frac{d\rho}{dr} = -G\rho\frac{d\rho}{dp}\frac{m(r)}{r^2}, \tag{9.26}$$

where $m(r)$ is the mass inside radius r and we have assumed that the equation of state is barotropic, $p = p(\rho)$. We can approximate the solution by replacing ρ and $d\rho/dp$ on the right-hand side with their values at the centre of the star (indicated by an index c). Then $m(r) \approx 4\pi\rho_c r^3/3$ and we get

$$\frac{d\rho}{dr} \approx -\frac{4\pi}{3}G\rho_c^2\left(\frac{d\rho}{dp}\right)_c \longrightarrow \rho(r) = \frac{2\pi}{3}G\rho_c^2\left(\frac{d\rho}{dp}\right)_c (R^2 - r^2), \tag{9.27}$$

where we have ensured that the density vanishes at the surface of the star, at radius R. Setting $r = 0$ we find a simple estimate for the star's radius

$$R \approx \left[\frac{2\pi}{3}G\rho\left(\frac{d\rho}{dp}\right)\right]_c^{-1/2}. \tag{9.28}$$

Basically, the radius can be deduced from the central values of the density and the compressibility (the sound speed). A more compressible equation of state leads to a smaller radius. It is common to describe equations of state as either soft or stiff, depending on whether $d\rho/dp$ is large or small. We now see that, for the same central density, a soft equation of state leads to a smaller star than a stiff equation of state. Moreover, the softer the equation of state is, the lower the attainable mass tends to be.

In order to put numbers to these estimates, let us again make use of the Fermi gas model, now assuming that the star is entirely made out of non-relativistic neutrons. Then we find that

$$\frac{1}{\rho}\frac{d\rho}{dp} \approx \frac{(3\pi^2)^{2/3}\hbar^2}{m_b^3}\frac{1}{n^{1/3}}, \tag{9.29}$$

where n is the neutron number density and $m_b \approx 1.67 \times 10^{-27}$ g is the mass of each baryon. Thus, we get

$$R \approx \left[\frac{2\pi}{3} \frac{Gm_b^3 n_c^{1/3}}{(3\pi^2)^{2/3}\hbar^2} \right]^{-1/2}, \tag{9.30}$$

which leads to $R \approx 14$ km for $n_c = 0.6$ fm^{-3}, which would correspond to a central density of about $\rho_c \approx 10^{15}$ g/cm^3. The corresponding mass follows from

$$M = 4\pi \int_0^R \rho r^2 \, dr = \frac{8\pi}{15} \rho_c R^3, \tag{9.31}$$

and we find $M \approx 1.7 M_\odot$ for the suggested value of the central density.

This is a useful exercise. We learn that we should expect a neutron star to have a radius of order of 10–15 km and a mass close to one and a half times that of the Sun. These estimates may be rough, but when compared to more detailed models (see Chapter 12) they turn out to be rather good.

That this kind of object actually exists in the Universe was established (somewhat serendipitously) in 1967 when Jocelyn Bell Burnell found a peculiar periodic signal in the data from a new radio telescope (Hewish *et al.*, 1968). The first signal consisted of a series of equally spaced pulses 3.7 seconds apart. The source was far too rapid to be something like a pulsating star and yet it had to have an astronomical origin because it kept sidereal time. Many possibilities were considered—including that the signal came from a distant civilization—but it was soon suggested that these 'pulsars' were associated with supernova remnants. They were rotating neutron stars (Gold, 1969). We see regular pulses because the star emits radiation (associated with the rotation of the presumably misaligned magnetic field) which is misaligned with the rotation axis. As the star spins, the radiation beam sweeps across the sky, and once every revolution it can be picked up by telescopes on Earth.

The main pulsar observables are the spin period $P = 2\pi/\Omega$ and its derivative $\dot{P}$; see Figure 9.3. If we accept the notion that these systems are rotating neutron stars with a misaligned magnetic field, then we can work out the rate at which they should be losing rotational energy

$$\dot{E} = -\frac{B_p R^6 \Omega^4 \sin^2 \alpha}{6c^3}, \tag{9.32}$$

where α is the misalignment between the rotation axis and the magnetic (dipole) axis and B_p is the strength of the field at the magnetic pole. A more detailed analysis (Spitkovsky, 2006) shows that the star will lose energy even when the magnetic filed is aligned with the rotation axis. However, the estimate from (9.32) will suffice for now. Combining the energy loss with the total rotational energy,

$$E = \frac{1}{2}I\Omega^2, \tag{9.33}$$

where I is the moment of inertia, we can link the magnetic field strength to the observed spin-down rate. Of course, in order to do this we need to assume that the other parameters

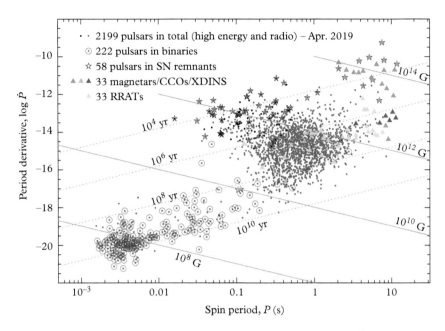

Figure 9.3 *The two main observables, the spin period P and the spindown rate Ṗ, for a collection of observed neutron stars. The rotation-powered radio pulsars make up (by far) the largest population (over 2,500). Fast-spinning (presumably) recycled millisecond pulsars are usually found in binary systems. Magnetars have very strong magnetic fields and tend to be slowly spinning. If we ignore the (likely) evolution of the magnetic field, then a neutron star would evolve from the top left towards the bottom right of the figure, along one of the indicated fixed magnetic field trajectories. This gives us a rough idea of the relative age of the observed systems. (Figure provided by T. Tauris.)*

are known, which they obviously are not. However, we can use 'reasonable' values and assume $M \approx 1.4 M_\odot$ and $R \approx 10$ km (for example). This leads to the moment of energy (for a uniform density sphere) of $I \approx 10^{45}$ g cm^2. In practice, we then obtain a lower limit for the magnetic field ($B > B_p \sin\alpha$) since we do not know the inclination angle[4]

$$B \gtrsim 5 \times 10^{19} \left(\frac{P\dot{P}}{1\text{ s}}\right)^{1/2} \text{ G.} \qquad (9.34)$$

In the case of a small number of young pulsars, observers have also managed to measure the second derivative $\ddot{P}$ (Espinoza *et al.*, 2011*b*). Such observations provide

---

[4] Note that, in the literature it is common to use a slightly more realistic estimate for the moment of inertia than we have done here, leading to the numerical prefactor changing to 3.2×10^{19} (see, for example, Lorimer and Kramer (2012)).

a consistency check of the magnetic spin-down model. It is easy to see that, if we hold the other parameters fixed, then the *braking index* obtained from the power law

$$\dot{\Omega} = -K\Omega^n \tag{9.35}$$

should be $n = 3$. Observed values are close to this, but differ sufficiently from the expectation that the model needs tweaking. Nevertheless, it is clear that one can use the observations to rule out a gravitational-wave-dominated spin-down (leading to $n = 5$ for a rotating deformed star; see Chapter 6), in all cases where $\dot{P}$ has been measured.

Since the early days of radio astronomy more than 2,500 pulsars have been found (see Figure 9.3). The range of observed periods and phenomenology is enormous. Spin periods vary from about 1.4 ms for the fastest known pulsar, PSR J1748-2446ad (Hessels *et al.,* 2006), to several seconds for the slowest pulsars. Some pulsars are remarkably stable, even rivalling the most precise atomic clocks. Others exhibit timing noise (Hobbs *et al.,* 2010*b*) and frequent spin-up glitches (Espinoza *et al.,* 2011*a*). There are fast radio bursts (Lorimer *et al.,* 2007), which release a spike of energy, and systems that seem to switch between different spin-down states (Kramer *et al.,* 2006) (presumably due to processes in the star's magnetosphere). Despite a wealth of observations many key questions remain to be answered. In fact, we do not even know (exactly) what makes the pulsars pulse.

Observations in other wave bands add to the mystery. Some neutron stars have super-strong magnetic fields. These magnetars exhibit X-ray bursts and gamma-ray flares, presumably releasing pent-up magnetic energy (Thompson and Duncan, 1995; Kaspi and Beloborodov, 2017) We see accreting neutron stars emitting X-rays, but we do not (fully) understand the accretion torque and why these stars do not spin faster than they do (Patruno and Watts, 2012). The answer might involve gravitational waves (see Chapter 6). For a small number of isolated neutron stars, we see thermal radiation from the surface, giving us clues to the gradual ageing of the star and perhaps some insight into its composition (Page *et al.,* 2004). With the data from the Fermi satellite we have a growing number of young pulsars that emit gamma-rays (Acero *et al.,* 2015). There is a virtual neutron star zoo out there, and we are doing our best to catalogue its different species.

In order to understand neutron stars we need to consider cutting-edge nuclear, particle, and condensed matter physics, as well as electromagnetism. We need superfluids and superconductors. You only have to list the words to appreciate that this is a challenge for theoretical physics. We also need Einstein's gravity. A neutron star is as close to a black hole as you can get (without collapsing). This is easy to see, since

$$\frac{GM}{Rc^2} \approx 0.2, \tag{9.36}$$

compared to the value of 0.5 for a Schwarzschild black hole; see Chapter 4. Of course, these are all reasons why neutron stars are so exciting. The fact that they are also interesting gravitational-wave sources is icing on the cake. Neutron stars can radiate through a variety of mechanisms and we will explore some of the main ideas later, starting in Chapter 12.

9.5   The rebirth of relativity

When Einstein passed away in 1955 general relativity was not (at least not universally) considered the triumph we perceive it to be today. Of course, it was recognized as a success—the original calculation of the perihelion shift of Mercury and the confirmation of light bending by the Sun's gravity during several eclipse expeditions, had established gravity as a geometric theory—and Einstein spent the last decades of his life as a celebrity, but the general view among astronomers was that his theory had little practical relevance. Besides, the mathematics involved (tensor calculus) was far too complicated for most people to understand.

As we have already seen, our view of the Universe changed dramatically in the decades following Einstein's death. Astronomy underwent a revolution as radar antennae (developed during World War II) were turned towards the heavens and X-ray instruments were developed. The new Universe was very different from the old one. We do not live in a calm Universe where stars serenely glide across the sky, evolving over eons. Our Universe is dramatic and violent. Stars explode in spectacular supernovae as they run out of nuclear fuel. Galaxies collide. Neutron stars and black holes spew out powerful jets. The Universe itself began in a massive explosion. Improved measuring technology enabled these spectacular discoveries. At the same time, the technology reached the level where precision measurements of time and space became possible. This opened the door for more accurate tests of general relativity, e.g. by measuring the slowing down of clocks in a gravitational field and the gravitational redshift (see Chapter 10). Finally, the exciting astronomy stimulated theorists to return to gravity. This led to a deeper understanding of Einstein's theory and established general relativity as one of the two cornerstones of modern physics (alongside quantum mechanics).

Relativistic astrophysics blossomed as a research area in the late 1960s, but several key developments happened much earlier. The suggestion that stars might have neutron cores had prompted Robert Oppenheimer and his students to initiate a series of remarkable investigations already in the late 1930s. The first question they considered was whether there is an upper limit to the possible size of such a neutron core. To answer the question they integrated the relativistic equations for hydrostatic equilibrium numerically (see Chapter 4). Having done this, they found that no equilibrium solutions were possible above $0.7M_\odot$ or so (Oppenheimer and Volkoff, 1939). This showed that there is an upper mass limit, similar to Chandrasekhar's limit for white dwarfs, also for neutron stars. The maximum mass result was substantially below the modern value, because the equation of state they used was unrealistic, but it was an important qualitative conclusion. It brought the issue of gravitational collapse into the limelight.

The follow-up work by Oppenheimer's group remains remarkable. On the 10th of July 1939, Oppenheimer and Snyder submitted a manuscript discussing gravitational collapse to *Physical Review* (Oppenheimer and Snyder, 1939). This paper may be one of the most prophetic ever written in this field of research. More than 80 years later, it needs only modest revision—even the terminology is undated! The scope of the paper (as well as its daring nature) is clear already from the abstract:

> When all the thermonuclear sources of energy are exhausted a sufficiently heavy star will collapse...this contraction will continue indefinitely...the radius of the star approaches asymptotically its gravitational radius; light from the star is progressively reddened, and can escape over a progressively narrower range of angles ... The total time of collapse for an observer co-moving with the stellar matter is finite, and...an external observer sees the star asymptotically shrinking to its gravitational radius.

Oppenheimer and Snyder describe how they integrated Einstein's equations for a collapsing fluid (in spherical symmetry) to obtain a solution that could be joined to Schwarzschild's solution for the exterior vacuum (see Chapter 4). The result established the modern view of gravitational collapse. Unfortunately, the importance of the work would not be appreciated until much later. War had already broken out in Europe when the paper appeared in print (September 1939). Oppenheimer's research group dispersed. By 1941 he had been recruited to the Manhattan Project (and was busy considering other aspects of Einstein's theories). The subject entered a dark age.

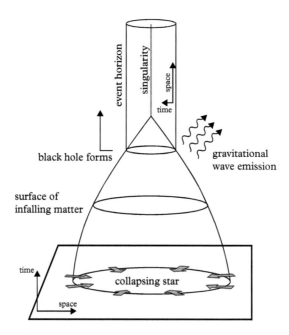

Figure 9.4 *A schematic illustration of gravitational collapse. As a massive star runs out of nuclear fuel it begins to collapse under its own weight. Eventually a black hole forms. If the collapse is asymmetric it generates gravitational waves.*

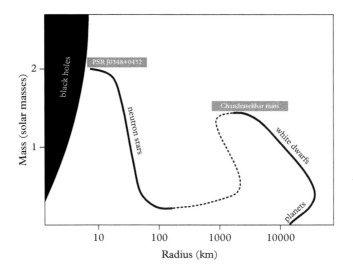

Figure 9.5 *A schematic illustration of the sequence of increasingly compact objects, from white dwarfs to neutron stars and, ultimately, black holes. There are no stable configurations on the dashed part of the curve.*

The rebirth of relativity as a vibrant area of research owed much to a small number of charismatic personalities. One of the main players was John Archibald Wheeler. In the late 1950s Wheeler and his Princeton students became intrigued by the notion of gravitational collapse. In order to understand whether the process could actually happen in nature, they reconsidered the problem using a more realistic equation of state (making use of nuclear physics lessons learned during the war). Also benefiting from advances in computer technology, they managed to graph the stellar mass as a function of central density and found two distinct humps, corresponding to the anticipated white dwarf and neutron star configurations (Harrison *et al.*, 1965). This unified the results into a single picture, see Figure 9.5, which remains valid today.

By the end of the 1960s the landscape had changed dramatically. It was clear that Einstein's theory was essential for an exploration of extreme astrophysics. It had been established that gravitational waves should, indeed, exist. It was understood that massive stars will inevitably undergo gravitational collapse, leading to the formation of black holes (see Figure 9.4). And astronomers were beginning to find the supporting evidence.

9.6 Weighing black holes

Black holes challenge even the liveliest of imaginations. That such objects may exist was, in principle, clear already when Schwarzschild found his solution in 1915. But the notion of an event horizon, acting like a one-way membrane that hides everything that falls through it from the outside Universe, was not taken seriously for most of the following half-century. The simple fact that an infalling clock would appear to freeze as it approached the horizon caused confusion, as it suggested that a collapsing star might actually never quite reach the horizon, either. The resolution to the problem—that the infalling clock itself would measure time as usual and that the conundrum was all a matter of relativity, where it is essential to keep track on who is doing the measurement— eventually came in 1960 when Martin Kruskal suggested a set of spacetime coordinates that describes what actually happens (Kruskal, 1960).

An added reason for the lack of interest in black holes was the lack of evidence for their existence. This has (obviously) changed dramatically in the past decades. Black holes have become mainstream astrophysics. Moreover, black holes are important gravitational-wave sources so we need to understand them better. We need to be able to model how they interact with their surroundings. At first sight this may seem a hopeless task, but these are, in fact, the simplest macroscopic objects we know of. A black hole can be described by three parameters: the mass, the rate of rotation, and the electric charge. However, it is easy to argue that an astrophysical object is unlikely to have significant net charge. If it did it would simply attract particles of the opposite charge from the interstellar medium and neutralize. This leaves us with two parameters. Given values for the mass and the spin, the nature of a black hole is completely determined by the vacuum geometry of spacetime (see Chapters 4 and 17). In contrast, the modelling of other astronomical bodies, like white dwarfs and neutron stars, involves a lot of (often poorly understood) physics.

The question is, how can you expect to pin down an invisible object in the dark cosmic sky? There are two obvious answers. The first involves tracking the motion of a visible object that is influenced by the presence of an unseen companion. The second method becomes relevant when the dark object is close enough to a companion that matter starts flowing towards it (as in the interacting binaries discussed in Chapter 6). The accreted matter generally heats up, and provided that enough gravitational potential energy is released this may lead to detectable X-rays. Both these approaches are relevant for finding black holes.

As it is difficult to distinguish one dark object (say a neutron star) from another (a black hole), one often proceeds by ruling out the neutron star option rather than ruling in the black hole. This can be done whenever the compact object has a binary companion. If the compact object is found to 'weigh' more than the absolute upper limit for neutron stars (usually taken as about $3M_\odot$, maybe adding another 20–30% for a rotating star (Cook et al., 1994)), one has a black-hole candidate.

Working out the mass of a compact, essentially invisible, object is obviously not a trivial matter. However, the main idea behind mass estimates for black-holes in X-ray binaries is clear. We need to infer the binary orbit. Assuming that Newtonian gravity is sufficient to describe the dynamics, let us consider a system with two masses, M_1 and M_2, separated by a distance a. The individual distances from the combined centre of mass then follow from (see Chapter 5)

$$M_1 a_1 - M_2 a_2 = 0. \tag{9.37}$$

Now suppose that we observe variations in the emission from one of the objects, say M_1. Factoring in the inclination angle i between the binary orbit and the line of sight, we can work out the velocity of this body

$$v_1 = \frac{2\pi}{P} a_1 \sin i, \tag{9.38}$$

where P is the orbital period. Combining this with Kepler's law, we arrive at the so-called mass function

$$f(M_1, M_2, i) = \frac{(M_2 \sin i)^3}{(M_1 + M_2)^2} = \frac{P v_1^3}{2\pi}. \tag{9.39}$$

The right-hand side of this expression is a combination of observables, so we can use data to constrain the left-hand side. In systems where both companions are observed, we have two constraints. If we also happen to know the inclination angle, then we can calculate the individual masses. This is only possible in very fortunate cases. Typically, one would observe one partner (say a white dwarf) and involve some model for its properties from stellar structure theory. This would lead to a constraint on the mass of the unseen companion. In many cases the mass function leads to an absolute lower limit on the mass of the compact object. If $f > 3M_\odot$ it would seem unlikely that we are observing a neutron star, and (unless we find some other, probably more exotic, explanation) the X-rays should come from a black hole.

The era of X-ray astronomy truly began with the launch of the Uhuru ('Freedom' in Swahili) satellite off of Kenya's coast in December 1970. By the time of its demise in March 1973, Uhuru had discovered more than 300 discrete sources. The data enabled scientists to make positive identification of X-rays from binary systems and discover two X-ray pulsars, Cen X3 and Her X1 (with periods of 4.84 and 1.24 s, respectively; see Fabian (1975)). These were concluded to be neutron stars since nothing associated with a black hole can lead to a regular pulsing behaviour. It was inferred that most of the galactic X-ray sources ought to be compact objects accreting matter from a binary companion. These conclusions follow from (i) the fact that the sources vary on short timescales, which indicates a small emitting region, (ii) the confirmation that some sources are in binary systems, i.e. there are cases where no optical companion is observed, and (iii) the efficiency with which accretion onto a compact object can convert energy into X-rays.

Since the early 1970s, many other satellites have confirmed the Uhuru data and provided a wealth of additional information. Many other objects—like X-ray burst sources, magnetars, and micro-quasars—have been added to the X-ray zoo. The evidence for solar mass black holes has strengthened and yet the oldest candidate, Cygnus X1, still presents one of the best cases.

Shortly after the detection of X-rays from outside the Solar System in 1962, one of the strongest sources, Cygnus X1, was found to be varying on very short timescales. This led to suggestions that it might be a black hole (Thorne, 1974b). Close observations unveil variability on timescales ranging from months to milliseconds. In order for a source to vary this rapidly it must be compact. Coherent variation is only possible on timescales larger than a light crossing time, and for a millisecond burst we can deduce a size of $R < ct \sim 300$ km. The only (known) astronomical objects that fit this criterion are neutron stars and black holes. Unfortunately, unless one can associate some phenomenon with the presence of a surface (like the explosive events that lead to X-ray bursts from neutron stars) it is not easy to distinguish the signature of a black hole from that of a neutron star.

The upshot of this is that, if the mass function is smaller than $3M_\odot$ (or whatever we take the maximum neutron star mass to be) the dynamical evidence for a black hole is not so convincing. One would have to involve further information, such as the mass of the companion and the inclination angle. The latter can sometimes be constrained by a lack of observed eclipses, and the mass of the companion star may be inferred from its spectral characteristics. But you still have to be lucky. In the case of Cygnus X1, the companion star is a hot blue giant expected to (from its spectral characteristics) have a mass of $24 - 42M_\odot$. From this, and the observed mass function $f = 0.25M_\odot$ we can deduce (using the maximum value $\sin i = 1$) that the mass of the compact object should be larger than $6M_\odot$. Even if we take a devil's advocate point of view, we find that the mass of the compact object in Cygnus X1 remains well above the upper limit for neutron stars. The current mass estimate is, in fact, $14.8 \pm 1.0M_\odot$ (see Table 9.1), a strong argument in favour of the black-hole explanation.

The number of similar black-hole candidates has increased dramatically, see Table 9.1 for a selection. The discovered systems fall in different categories depending on the mass of the companion star. With a companion heavier than 20 $M_\odot$, Cygnus X1 is the prototype for high-mass X-ray binaries (HMXB). In contrast, the companion of a typical low-mass X-ray binary (LMXB) is below one solar mass (usually a white dwarf). Intermediate X-ray binaries (IMXB) obviously lie in between. The classification into high- and low-mass binaries does not specify the nature of the compact accreting object; it can be either a black hole or a neutron star. The prototype LMXB black-hole candidate, A0620-00, was discovered as it flared up to the brightest X-ray source in the sky for two months in 1975 (Elvis *et al.*, 1975). This black hole is estimated to have a mass in the narrow range $6.6 \pm 0.3M_\odot$.

It is also worth highlighting a system that is famous for its similarity with the much larger black holes found in the cores of active galaxies. In July 1994 the bright X-ray source GRO J1655-40 was discovered by the Burst and Transient Source Experiment (BATSE) on board the Compton Gamma-Ray Observatory. Observations of this object led to a precise mass estimate in the range $5.4 \pm 0.3M_\odot$, but the source is interesting for other reasons. A few weeks after the initial X-ray outburst, radio jets were found to emerge from GRO J1655-40 (Tingay *et al.*, 1995). The observed double radio structure resembles the jets seen in many active galactic nuclei. The jet displays superluminal motion which indicates that, just like in the extragalactic counterparts, matter is moving relativistically. This is an interesting illustration of the fact that the flow around black holes can be essentially the same on radically different scales.

Table 9.1 *Black-hole masses in X-ray binaries (classified by the mass of the companion, from Low to Intermediate and High). The objects discussed in the main text are highlighted in bold. (Adapted from Casares et al. (2017), which provides original source references.)*

Object	X-ray binary class	Mass ($M_\odot$)
GRS 1915+105	LMXB/transient	10.6−14.4
V404 Cyg		8.4−9.2
BW Cir		>7.0
GX 339-4		>6.0
XTE J1550-564		7.8−15.6
H1705-250		4.9−7.9
GS 1124-684		9.6−13.1
GS 2000+250		5.5−8.8
A0620-00		6.3−6.9
XTE J1650-500		4.0−7.3
GRS 1009-45		>3.6
XTE J1859+226		> 5.42
GRO J0422+32		>1.6
XTE J1118+480		6.9−8.2
XTE J1819.3-2525	IMXB/transient	5.8−7.0
GRO J1655-40		5.1−5.7
4U 1543-475		2.7−7.5
Cyg X-1	HMXB/persistent	13.8−15.8
LMC X-1		9.5−12.3
LMC X-3		6.4−7.6
M33 X-7		14.2−17.2
MWC 656		3.8−5.6

9.7 The formation of compact binaries

The X-ray data suggests that black holes are common. However, electromagnetic observations provide no handle on black-hole binaries. This is a shame as these are (obviously!) exciting gravitational-wave sources.

In order to estimate the event rates for inspiral signals for different generations of detectors we need to understand how compact binaries form. This question has been attracting attention since the discovery of the binary pulsar PSR1913+16 in 1974 and early estimates played a key part in the arguments for funding the LIGO project and other large interferometers in the first place. As pulsar astronomers continue to identify similar double neutron star systems, see Table 9.2, we get a better idea of the relevant formation rate. Moreover, the detection of GW150914 (Abbott *et al.*, 2016*b*) brought the issue of formation scenarios into sharp focus since the two black holes were found to be more massive than expected. In fact, the black holes (so far) found in gravitational-wave data tend to be more massive than those seen in X-ray binaries (compare the data in Tables 9.1 and 9.3). These systems also appear to be more common than one might have expected, raising questions that need to be resolved by better modelling. This is, however, a difficult problem.

We know that more than 50% of normal stars are found in binaries. Compact binaries should form naturally as a result of stellar evolution of these systems (Tutukov and Yungelson, 1993). They may also form in dense star clusters via dynamical exchanges involving stars and black holes (Sigurdsson and Hernquist, 1993), or in more exotic environments like the disks of active galactic nuclei (see Postnov and Yungelson (2014) for a review of the different scenarios). Population synthesis models allow us to establish how close pairs of compact objects form and whether these systems will merge within a Hubble time. The input parameters for such simulations are: (i) the shape of the galactic gravitational potential, (ii) the initial mass function of massive main sequence stars, (iii) the metallicity of the parent gas cloud, (iv) the fraction of primordial binaries (and triplets), and (v) the distribution of the initial binary separation and eccentricity, which affect the degree of interaction of the two stars over their lifetime.

Stars lose mass through winds, but in binaries they can also exchange mass with the companion (as discussed in Chapter 6). Mass transfer occurs when the most massive star, which first evolves away from the main sequence, fills its Roche lobe. The mass loss to the companion leads to a re-equilibration of the mass ratio and tends to make the less massive star the heaviest in the system. After this period of mass exchange, the faster evolving star becomes a Wolf–Rayet or a helium star (depending on the initial mass) that can evolve towards a supernova. The supernova explosion may unbind the binary due to mass loss and recoil associated with an anisotropic collapse. In fact, we know that the collapse typically involves asymmetries as neutron stars receive birth kicks with mean velocities of several 100 km/s. This may break up the binary. Thus, it is expected that about 90% of potential binaries end up being disrupted after the first supernova explosion. This makes compact binaries rare. Black holes—which form either through fall back of matter (following the supernova) or direct collapse—are likely to receive weaker kicks. The lower level of mass loss that may accompany their formation, and the weaker kicks, may help a heavy binary survive almost intact after the formation of the first compact object. The key implication is that the rate of formation for binary systems with one compact object is not directly set by the initial mass function. This further complicates the population modelling.

Table 9.2 *Observed parameters for known double neutron star systems, including the coalescence time t_m. (Note that the merger time does not simply follow from (5.65) for systems with significant ellipticity, like the Binary Pulsar PSR B1913+16. For these systems one has to combine Eq. (5.62) with the relation for $\dot{a}/a$ from the quadrupole formula in order to determine t_m.) The merger times should be compared to the Hubble time (the age of the Universe) which is about 1.4×10^4 Myr. The two systems below the horizontal line are not yet confirmed to be neutron star binaries. (Based on data from Tauris et al. (2017).)*

Pulsar	P (ms)	$P_{\rm orb}$ (days)	e	$M_{\rm psr}(M_\odot)$	$M_{\rm comp}(M_\odot)$	t_m (Myr)
J0453+1559	45.8	4.072	0.113	1.559	1.174	1.5×10^8
J0737-3039A	22.7	0.102	0.088	1.338	1.249	86
J0737-3039B	2773.5					
J1518+4904	40.9	8.634	0.249			
B1534+12	37.9	0.421	0.274	1.333	1.346	2.7×10^3
J1753-2240	95.1	13.638	0.304			
J1755-2550	315.2	9.696	0.089		>0.40	
J1756-2251	28.5	0.320	0.181	1.341	1.230	1.7×10^3
J1811-1736	104.2	18.776	0.828	<1.64	>0.93	
J1829+2456	41.0	1.176	0.139	< 1.38	>1.22	
J1906+0746	144.1	0.166	0.085	1.291	1.322	310
J1913+1102	27.3	0.206	0.090	<1.84	>1.04	
B1913+16	59.0	0.323	0.617	1.440	1.389	300
J1930-1852	185.5	45.060	0.399	<1.32	>1.30	
J1807-2500B	4.2	9.957	0.747	1.366	1.206	1.0×10^6
B2127+11C	30.5	0.335	0.681	1.358	1.354	220

After the birth of the first compact object, the evolution continues through a common envelope phase (see Figure 9.6). During this phase the second star swells up to a giant and engulfs its partner. The compact object then spirals inwards due to gas dynamical friction, losing orbital angular momentum and energy, which in turn heats the envelope. Next, the remnant either merges with the dense core of the companion—forming what is known as a Thorne–Zytkow object (Thorne and Zytkow, 1975)—or end up in a tight orbit after ejecting the envelope. In this last case, the core of the star evolves into a relic object which may explode in a supernova, form a black hole through the fall-back of matter, or undergo direct collapse, depending on its mass. This route may lead to double neutron star systems or black-hole/mixed binaries. The main uncertainties in the

Table 9.3 *Parameters inferred for the 10 binary black-hole events recorded during the first two Advanced LIGO observing runs, O1–O2. The Virgo detector joined the O2 search at the beginning of August 2017. It is notable that 4 black-hole mergers were observed that month. The energy radiated in each event (in $M_\odot c^2$) is (roughly) given by the difference between the final black hole mass and the total mass of the initial binary system, $M_f - M_1 - M_2$. It is also worth noting that the estimated spin of the final black hole (a_f) is similar in all cases, suggesting that it originates from the orbital angular momentum. (Data from Abbott et al. (2019c) where the estimated errors for each measurement can also be found.)*

Event	M_1 ($M_\odot$)	M_2 ($M_\odot$)	$\mathcal{M}$ ($M_\odot$)	M_f ($M_\odot$)	a_f	d_L (Mpc)
GW150914	35.6	30.6	28.6	63.1	0.69	430
GW151012	23.3	13.6	15.2	35.7	0.67	1060
GW151226	13.7	7.7	8.9	20.5	0.74	440
GW170104	31.0	20.1	21.5	49.1	0.66	960
GW170608	10.9	7.6	7.9	17.8	0.69	320
GW170729	50.6	34.3	35.7	80.3	0.81	2750
GW170809	35.2	23.8	25.0	56.4	0.70	990
GW170814	30.7	25.3	24.2	53.4	0.72	580
GW170818	35.5	26.8	26.7	59.8	0.67	1020
GW170823	39.6	29.4	29.3	65.6	0.71	1850

respective formation rates relate to the common envelope phase of the evolution. The tight binary that forms after common envelope and mass ejection is likely to survive. If the two compact objects remain bound and end up sufficiently close (a few solar radii apart), gravitational waves will drive the binary towards coalescence on timescales that vary between a few million to more than a billion years (see Table 9.2). As the evolutionary scenario is sensitive to complex issues like the kick distribution and the detailed supernova dynamics, predicted binary coalescence rates remain uncertain. In fact, we may need a statistically significant sample of observations to shed light on the uncertain aspects of the process.

Compact binaries may also form as a result of dynamical interaction in dense stellar clusters (Sigurdsson and Hernquist, 1992; Portegies Zwart and McMillan, 2000). The high density of stars in clusters favours the formation via exchange interactions. Three-body interactions can lead to a population of massive binaries. As they are the heaviest objects in the cluster, these binaries sink to the centre on a timescale shorter than the two-body relaxation time. In the process they may encounter other black holes, which can lead to a sequence of mergers and the development of progressively heavier binaries. Scattering off of stars can drive these binaries to coalescence within $\sim 1 - 10$ Gyr. Again, we need observations to establish the relevance of this scenario.

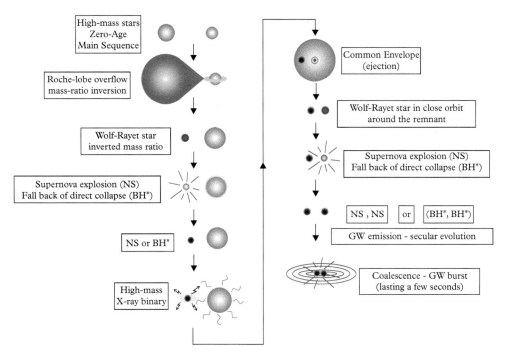

Figure 9.6 *The standard formation path for compact binaries. The initial masses of the stars determine the nature of the binary. To reproduce the data of GW150914, the initial stars must be massive. The evolution is also sensitive to a range of issues, like the metallicity. (Reproduced from Marchant et al. (2016) with permission from Astronomy and Astrophysics, copyright ESO.)*

9.8   Estimating merger rates

In order to understand the chances of detecting gravitational signals from cosmic events we need to have an idea of the signature we should be looking for—to prepare the appropriate data analysis strategy—and we also need to estimate how often these events are likely to happen inside the horizon of our detectors. As we are (mainly) interested in black holes and neutron stars, the latter is a notoriously difficult problem. For isolated objects, we need to understand the supernova mechanism and the relative proportion of events that lead to the direct formation of a black hole. This problem is relatively well constrained; see Figure 9.7. We expect roughly 3–4 core collapse supernova events in our Galaxy each century, so once we consider the local galaxy distribution we see that our detectors need to be sensitive to a signal from a distance of at least 10 Mpc if we want to catch a couple of events each year.

The problem becomes much more difficult when we turn to binaries (Phinney, 1991; Narayan *et al.*, 1991). As we have already seen, different channels may explain the formation of these systems and the details are sensitive to many unknown factors, in particular, the common envelope phase, which is essential in bringing the binary partners into a close enough orbit that gravitational radiation can drive the system to merger inside

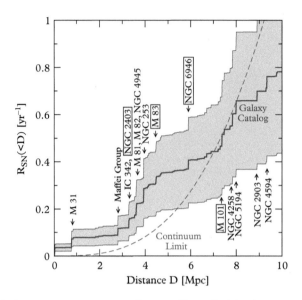

Figure 9.7 *Cumulative supernova rate versus distance. The dashed line is the continuum limit. The stepped line is based on star formation rates for individual galaxies, and the band is the associated uncertainty. Some major galaxies are indicated, and those in boxes have particularly high optical supernova rates (Reproduced from Ando et al. (2005), copyright (2005) by The American Physical Society.)*

the Hubble time. This typically leads to more than an order of magnitude uncertainty in the estimated rates.

For neutron star binaries we can still make progress using known systems. Extrapolations based on the data in Table 9.2 typically lead to about 100 events in the Galaxy every million years. Since the estimate is based on galactic binary data, it is natural to express the extrapolated result in terms of the number of galaxies similar to our own. That is, we anticipate a rate of 100 Myr^{-1} per Milky Way Equivalent Galaxy (MWEG) (Kalogera *et al.*, 2004). Considering the uncertainties the actual merger rate is expected to lie in the range $1 - 1,000$ Myr^{-1} per MWEG (Abadie *et al.*, 2010). If we want to turn this into an estimated detection rate, we need to project the result onto the estimated sensitivity for a given set of detectors. This step is also far from trivial. The detection threshold for a network of interferometers depends on the relative configuration (locations, orientations, and noise power spectral densities of the instruments), the characteristics of the noise, and the employed search algorithms. However, we get a rough idea by expressing the reach of a given gravitational-wave search in terms of the horizon distance: d_h, the distance at which a single detector would detect an optimally oriented source with a signal-to-noise ratio of 8 (as required for a confident detection). We can combine the results from Chapter 8 with a given detector configuration to make this statement quantitative.

In order to estimate rates, we also need to know the number of galaxies inside d_h. This is sometimes expressed in terms of the accessible blue-light luminosity. To convert this luminosity into the number of galaxies within reach, N_g, we need the approximate relation (Abadie *et al.*, 2010)

$$N_g(\text{MWEG}) \approx 1.7 C_L(L_{10}), \tag{9.40}$$

where C_L is the cumulative blue luminosity observable within a given search volume measured in terms of the corresponding luminosity of the Sun (L_{10} is 10^{10} times the blue luminosity of the Sun). This then leads to an estimated number of galaxies

$$N_g \approx \frac{4\pi}{3} \times 0.0116 \times \left(\frac{d_h}{2.26\text{Mpc}}\right)^3 \approx 4 \times 10^{-3} \text{ Mpc}^{-3}, \tag{9.41}$$

where the factor of 2.26 arises from an average over possible sky locations and orientations (see Figure 9.8) and 1.16×10^{-2} Mpc^{-3} is the extrapolated density of MWEGs in space.

Let us see where this gets us. First of all, we find that the expected rate for neutron star inspirals is $10^{-2} - 10$ Mpc^{-3} Myr^{-1}. Taking the initial LIGO horizon distance for neutron star inspirals to be 33 Mpc, we then estimate the expected number of events in a one-year observation to be $2 \times 10^{-4} - 0.2$, and it is not surprising that there were no detections. In fact, the rate estimate illustrates why it is essential to reach the advanced

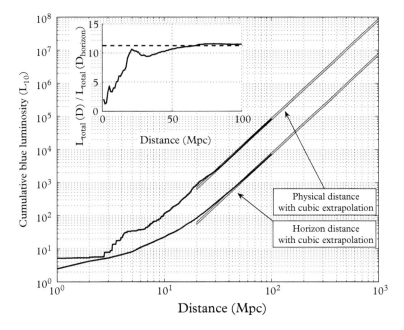

Figure 9.8 *The total blue-light luminosity C_L measured in terms of L_{10} (10^{10} times the blue luminosity of the Sun) within a sphere of a given radius (top curve) and the accessible blue-light luminosity for a given horizon distance d_h, taking location and orientation averaging into account (bottom curve). The inset shows the ratio between the top and bottom curves, which asymptotes to 2.26^3, as used in (9.41). (Reproduced from Kopparapu et al. (2008) by permission of the AAS.)*

LIGO sensitivity, for which $d_h \approx 445$ Mpc, as we then arrive at a binary neutron star detection rate of about 40 events per year (with a range between 0.4 and 400 per year; see Abadie *et al.* (2010)).

Estimates for mixed neutron star–black-hole systems or double black-hole systems are more difficult, as we have no electromagnetic evidence for their existence. As estimated rates tend to be correspondingly uncertain, typical results for the neutron star–black-hole systems merger rate lie in the range $6 \times 10^{-4} - 1$ Mpc^{-1} Myr^{-1}, while the results for black-hole systems is $10^{-4} - 0.3$ Mpc^{-3} Myr^{-1} (Abadie *et al.*, 2010). As these are small numbers and the Advanced LIGO horizon distance for black-hole mergers should reach beyond 1 Gpc, at which point we expect to observe for about a year, it is natural re-express the rate as $0.1 - 300$ Gpc^{-3} yr^{-1}. This estimate predates the breakthrough LIGO observations. However, we can place an improved constraint on the rate already from the data surrounding the GW150914 event. The most conservative assumptions, see Abbott *et al.* (2016e), then suggest a range of $2 - 600$ Gpc^{-3} yr^{-1}, demonstrating that black-hole binaries are more common than we might have expected. The constraint (obviously) improves with further detections. The data from the first two advanced detector observing runs, which include roughly one event for every two weeks of data taking (see the 10 events listed in table 9.3), suggest a merger rate of $9.7 - 101$ Gpc^{-3} yr^{-1} for binary black holes (Abbott *et al.*, 2019c).

We are still waiting for the first observation of a neutron star–black-hole binary, but the LIGO O1 results still provide an upper limit on the merger rate for such systems. For systems with a $1.4M_\odot$ neutron star and black-hole masses of at least $5M_\odot$, the detectors were sensitive to an average distance of at least $d_h \sim 110$ Mpc. This constrains (with 90% confidence) the merger rate of such systems to be less than 3600 Gpc^{-3} yr^{-1} (Abbott *et al.*, 2016e).

We may also weigh in the rate of observed short gamma-ray bursts which are thought to be associated with the merger of systems involving neutron stars (simulations suggest that the presence of a spinning black hole surrounded by an accretion disk is essential for the launch of a jet; see Chapter 21). Observations of short gamma-ray bursts with known redshifts then allow for a better estimate of the merger rate of these systems in the local Universe. Data from the SWIFT satellite lead to a rate in the range $500 - 1,500$ Gpc^{-3} yr^{-1} (Petrillo *et al.*, 2013). One can also compare the observed rates for gamma-ray bursts to the merger rates to constrain the opening angle of the gamma-ray burst jet (Chen and Holz, 2013). This suggests an opening angle of a few degrees (Abbott *et al.*, 2016e).

9.9   Active galaxies

Having focussed on solar-mass compact objects, we sacrificed some of the historical context. The first glimpse of our violent Universe—unveiled in its full glory by the new technology—was Maarten Schmidt's discovery of quasars (Schmidt, 1963). These 'quasi-stellar' radio sources appeared starlike in telescopes but were found at enormous distances so must emit exceptional amounts of energy. These objects are now referred to as Active Galactic Nuclei (AGN) and we know that they can be thousands of times as bright as an ordinary galaxy. They come in different varieties: Seyfert galaxies, broad and narrow line radio galaxies, quasars, BL Lacertae objects, and so on. An outstanding example is M87, a strong radio source that has been identified with an elliptical galaxy about 50 million light years away. Matter is ejected at relativistic speed from the centre of this galaxy, forming two spectacular jets that extend 2,600 light years from the core. The energy associated with these jets is equivalent to that released in something like ten million supernova explosions!

This kind of observation indicates that the nucleus of a galaxy cannot be composed of common stars alone. We need a massive body capable of ejecting a mass many times exceeding that of the Sun. What central engine can possibly release such awesome power? Whatever the central object is, it must be able to radiate continually for a long time. The natural conclusion is that the central object in an active galaxy must be a supermassive black hole feeding from an accretion disk. Within this model, the bewildering variety of AGNs has been unified in a single picture. Different AGNs simply look different because we view them from different angles (Elvis, 2000). It is generally agreed that most galaxy cores harbour supermassive black holes, but it is still legitimate to ask whether the observations are precise enough to rule out other options. So, let us take a brief look at the evidence.

In May 1994 NASA announced that the Hubble Space Telescope (HST) had 'seen' a black hole at the centre of M87 (Kormendy and Richstone, 1995). The gas in the heart of the galaxy was found to whirl rapidly around the centre. Spectrographic measurements showed that the gas moved at velocities of about 500 km/s at a distance of 18 pc. From this one could work out that the centre of M87 must hide an unseen mass of at least $2.4 \times 10^9 M_\odot$ inside a volume of space that contains far too few visible stars to account for this amount of matter. The announcement caused quite a stir, partly because of the remarkable HST images of the central region of the galaxy.

Another impressive case for a supermassive black hole was presented a year later. This involved the use of emission lines from water masers (the microwave equivalent of the laser) to accurately map the gas motion in the spiral galaxy NGC 4258 (Kormendy and Richstone, 1995). Using the Very Long Baseline Array, a resolution more than 100 times sharper than that of the HST was achieved. The radio observations revealed a disk surrounding a compact dark mass, with rotational velocities following an almost exact Keplerian rotation law. The mass of the central object was inferred to be $3.6 \times 10^7 M_\odot$ inside a radius of 0.13 pc. The maser observations make a very strong case for a black hole in NGC 4258. This is, in fact, the most conservative explanation. If the central mass in this galaxy is not a black hole, then it must be something even more exotic.

The case for supermassive black holes in galaxy cores has continued to strengthen in the past couple of decades. The black-hole explanation fits the data and the phenomenology of these systems. It is not easy to come up with a credible alternative. Having said that, we are still far from confirming that these objects are the black holes of Einstein's theory. This would involve observing features directly associated with the presence of an event horizon. This will always be difficult for distant galaxies, but it may well be possible if we look closer to home. The recently announced results from Event Horizon Telescope (a virtual Earth-sized radio telescope; see Psaltis *et al.* (2015)) provide exciting progress in this direction, reconstructing event-horizon-scale images of the supermassive black-hole candidate in M87 and constraining the involved central mass to be $(6.5 \pm 0.7) \times 10^9 M_\odot$ (Akiyama *et al.*, 2019).

Observations of huge black holes at impressive distances tell us that these monsters grow rapidly in the early Universe. How this happens remains a mystery (Rees, 1984). Studies predict black-hole seeds in the mass range $10^3 - 10^5 M_\odot$ at redshifts above $z \approx 10$ (when the Universe was less than a billion years old). These seeds then grow to $10^8 M_\odot$ by accretion and/or repeated mergers, contributing to the formation of cosmic structures in the process. In the standard cold dark matter model (see Chapter 22) galaxies undergo multiple mergers during their lifetime. If one assumes that most galaxies host black holes in their centre (which makes sense given the observational evidence), and that a local galaxy has experienced multiple mergers, then a massive black-hole binary system would be a natural evolutionary stage. After each merger, the central black holes would migrate to the new centre of mass via dynamical friction, followed by (once they get close enough) the emission of gravitational radiation.

Different models take different starting points and make different assumptions about the stages involved. The initial seed black holes may form from the direct collapse of gas (Begelman *et al.*, 2006) (in which case the mass of the black hole may lie in the range $10^4 - 10^6 M_\odot$), gravitational collapse of zero-metallicity population III stars (leading to $\sim 100 M_\odot$ black holes), or mergers of stellar clusters. The evolution may be dominated by gas accretion, black-hole mergers, or the tidal capture of stars. The efficiency of these channels depends on a number of factors (the black-hole spin dictates accretion efficiency, gravitational interactions may lead to slingshot effects and recoils, the galactic environment determines the level of gas supply, and so on). The details are uncertain, but one should be able to constrain the theory with observations. For example, along with any accreted mass the black hole should gain angular momentum, so one might expect massive black holes formed by accretion to spin rapidly (Thorne, 1974*a*). In contrast, the capture of lower mass objects (from randomly oriented orbits) may allow the black hole to remain slowly spinning.

Low-frequency gravitational-wave observations should (eventually) allow us to discriminate between the different models. Any scenario that involves repeated black-hole mergers will (inevitably) lead to gravitational waves in the LISA sensitivity band (see Figure 7.11). Detection of these signals, which should last from hours to months, would help us decode the evolution of supermassive black holes.

9.10   A giant at the centre of the Milky Way

The violent activity of many galaxy cores inevitably draws our attention to the centre of the Milky Way. If most galaxies host a massive black hole, should not our own have one, as well? The centre of our Galaxy is an intriguing part of the sky that displays many astrophysical extremes. It is often studied, but because it is obscured by interstellar dust and gas, detailed observations are difficult. Hardly any of the photons created in the galactic centre are able to pierce the dust along our line of sight. As a consequence, the centre of the Milky Way is the domain of radio and infrared astronomy.

Since the early 1970s we know that an unusual radio source is located at the dynamical centre of the Galaxy. This source, Sgr A*, has long been regarded the prime candidate for our own supermassive black hole. The mere presence of a distinct object close to the centre of the Galaxy is intriguing, but is Sgr A* really a black hole? The source is extremely compact and outshines all other radio sources in the Galaxy by several orders of magnitude. Compared to a typical AGN it is feeble, but the observed radio luminosity is still similar to the nuclear radio sources found in many other nearby galaxies, like Andromeda (M31).

After years of speculation, the object at the centre of the Milky Way emerged as perhaps the strongest massive black-hole candidate of all (Lu *et al.*, 2009; Gillessen *et al.*, 2009). The evidence, which continues to improve, is based on remarkable infrared observations tracking the proper motion of stars near the galactic centre for decades; see Figure 9.9. The observations have much higher spatial resolution than can be achieved with the HST for (say) M31. The detected motion is consistent with a black hole of mass close to $4 \times 10^6 M_\odot$ inside the central 0.015 pc of the Galaxy. The evidence for a black

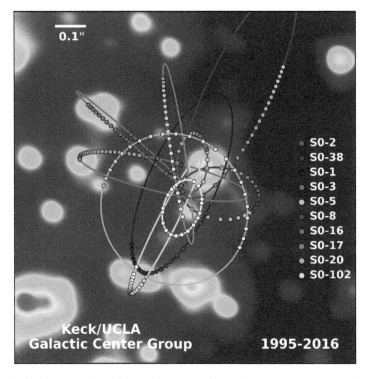

Figure 9.9 *Orbits of stars in the vicinity of the centre of the Milky Way. (Image created by Prof. Andrea Ghez and her research team at UCLA from data sets obtained with the W. M. Keck Telescopes.)*

hole is further strengthened by recent observations of flares in the near infrared (Abuter *et al.*, 2018), indicating a clockwise looped motion on the sky. The flares may originate from a compact hotspot of synchrotron emission from just outside a Schwarzschild black hole.

In essence, observations allow us to establish the presence of a dark mass—most likely a black hole—but they do not allow us to probe the exact nature of this object. To do this, we need to track objects closer to the centre. In principle, this should be possible because the galactic centre is a busy place. In practice, it is difficult because the region is obscured by dust. Still, a single observation of a radio pulsar orbiting close to the central black hole would provide a wealth of information. Finding such an object is one of the exciting targets for the Square Kilometre Array in the next decade (Stappers *et al.*, 2018). If such a pulsar were to venture too close to the black hole, or indeed, end up in a close orbit due to dynamical slingshot interactions, it would be captured. Such gravitational captures should be fairly common in galactic nuclei.

Compact objects (in particular stellar mass black holes and neutron stars, as they are unlikely to tidally disrupt) orbiting a massive black hole may allow us to probe physics that is inaccessible to electromagnetic observations. Usually referred to as *extreme mass-ratio inpirals*, such systems spiral through the strong field region near the event horizon before finally plunging through it. The associated gravitational-wave signal is expected to be very clean, except perhaps in active galaxies containing accreting black holes, where interactions with the accretion disk may impact on the inspiral dynamics. Extreme mass-ratio inspirals trace the geodesics in the black-hole spacetime, and the emitted low-frequency gravitational waves should be observable with an instrument like LISA (see Figure 7.11). Over a typical observation time of months to years, the orbits would highly relativistic, displaying extreme forms of periastron and orbital precession. We will consider the modelling of these systems in Chapter 16.

Observations of extreme-mass-ratio systems will also shed light on the stellar mass black-hole populations in galactic nuclei. We can expect to learn about the mass spectrum of such black holes, which is largely unconstrained—both theoretically and observationally—at present. By they time LISA flies in the mid-2030s, the Large Synoptic Survey Telescope (Tyson and Angel, 2001) should have observed a large number of tidal disruption events, which will also tell us a lot about black holes and stellar populations in galactic centres. However, this information will not have the same precision as that achievable through gravitational-wave observations.

In order to explore this possibility, we need to take a closer look at the theory. It makes sense to start by trying to understand geodesic motion a bit better.

10

Testing relativity

Einstein's theory describes a Universe far from our everyday experience. Not only does gravity move mass, it also bends light and warps time. Clocks run slow in a gravitational field. Gravity makes waves and creates black holes. Ultimately, gravity also helps us explain the cosmos. These are exciting ideas, but how do we know they are right? How do we put the theory to the test?

When Einstein—after a number of false starts—finally formulated general relativity in 1915 the results were presented in a series of short communications to the Prussian Academy. In these papers he suggested several ways that the theory could be tested by experiment.[1]

The first of Einstein's tests involved the perihelion shift of Mercury. It had been known since Le Verrier's work in 1859 that Mercury's orbit marched forwards a tiny bit faster than expected. Once the influence of the other planets was accounted for there was a mismatch of 43 arcseconds per century (about a third of the effect due to the presence of Jupiter). This had led to suggestions of a missing planet and a discussion that lasted until Einstein solved the problem with his new theory.

Einstein's second test involved the bending of light in a gravitational field, essentially a consequence of the equivalence principle. He first calculated this effect already in 1911, well before he completed the theory. He managed to convince astronomers that the prediction could be tested during a solar eclipse. Fortunately, for Einstein, a number of proposed expeditions fell through. This was lucky, because the result was wrong. The 1915 theory corrected the mistake, but by then the world was at war. It was not until 1919 that a British eclipse expedition, led by Arthur Eddington, confirmed the light bending (Dyson et al., 1920). But the measurement was not totally convincing. The observations agreed with Einstein's calculated value with an error of about 30%, a level of uncertainty that would remain for a long time.

The theory also says that gravity makes clocks slow down. Einstein noted that the frequency of light should shift toward the red as it struggles to escape the gravitational pull of the Sun. The first attempt to measure this effect took place in 1917, but the experiment was too difficult for the available technology (the surface of the Sun is 'messy'). The solar redshift was not measured until the 1960s (Brault, 1962).

---

[1] A relatively up-to-date overview of different tests of Einstein's theory can be found in Will (2005).

Gravitational-Wave Astronomy: Exploring the Dark Side of the Universe. Nils Andersson, Oxford University Press (2020).
© Nils Andersson. DOI: 10.1093/oso/9780198568032.001.0001

When Einstein passed away, he left a desk full of unfinished calculations and unanswered questions. Many predictions of the theory remained untested. The effects were simply too small and there was no serious interest in testing the theory, anyway. The renaissance of relativity in the decade that followed Einstein's death was driven by three developments: First, atomic clocks allowed precision measurements of space and time. Second, a new generation of telescopes—building on technology developed during World War II—opened new windows to the Universe, driving a revolution in astronomy (as we have already discussed; see Chapter 9). Third, a generation of talented physicists revisited Einstein's theory, leading to a much better understanding of its implications.

10.1   Geodesics

In order to explore the various tests of Einstein's theory, we need to understand motion in a curved spacetime. We need to consider both light, which moves along null geodesics, and massive bodies (for the moment treated as test particles), which follow timelike trajectories. For clarity, we will focus our attention on the Schwarzschild spacetime. This is natural since the weak field limit of the results can be used to describe the conditions in the Solar System and hence relate directly to the classic tests of the theory. However, the results also apply in the strong-field regime so we gain insight into the physics of black holes along the way.

We already derived the equation that determines geodesics (see Chapter 2), however, when it comes to working things out it is often practical to follow a different route. In Chapter 2 we argued that geodesics trace out the shortest possible paths in spacetime, but we did not back up that statement. It is useful to do so now.

If we want to emphasize that geodesics represent extremal curves, it is natural to use a variational description. Basically, we try to minimize the spacetime interval

$$ds^2 = g_{ab}dx^a dx^b. \tag{10.1}$$

Let us assume that the geodesics are described by some parameter τ, say. Dividing the line element by $d\tau^2$, we then have

$$\left(\frac{ds}{d\tau}\right)^2 = g_{ab}\frac{dx^a}{d\tau}\frac{dx^b}{d\tau} = g_{ab}\dot{x}^a\dot{x}^b \equiv 2\mathcal{L}, \tag{10.2}$$

where dots represent derivatives with respect to τ. This defines the Lagrangian $\mathcal{L}$ for the motion. The spacetime interval between two events, $\mathcal{P}_1$ and $\mathcal{P}_2$, follows from

$$s = \int_{\mathcal{P}_1}^{\mathcal{P}_2} ds = \int_{\mathcal{P}_1}^{\mathcal{P}_2} \frac{ds}{d\tau}d\tau, \tag{10.3}$$

and any extremal path must be such that the variation δs vanishes. Noting that this also means that $\delta s^2 = 0$, we need

$$\delta \int_{P_1}^{P_2} \mathcal{L} d\tau = -\int_{P_1}^{P_2} \left[\frac{d}{d\tau} \left(\frac{\partial \mathcal{L}}{\partial \dot{x}^a} \right) - \frac{\partial \mathcal{L}}{\partial x^a} \right] \delta x^a d\tau = 0, \qquad (10.4)$$

where we have integrated one of the terms by parts. This leads to the Euler-Lagrange equations

$$\frac{d}{d\tau} \left(\frac{\partial \mathcal{L}}{\partial \dot{x}^a} \right) - \frac{\partial \mathcal{L}}{\partial x^a} = 0. \qquad (10.5)$$

At first sight this result seems rather different from (2.72), but it is straightforward to show that the two pictures are identical.

In the case of the Schwarzschild metric (4.10), we have $x^a = [t, r, \theta, \varphi]$ and the Lagrangian can be written

$$2\mathcal{L} = -\left(1 - \frac{2M}{r} \right) \dot{t}^2 + \left(1 - \frac{2M}{r} \right)^{-1} \dot{r}^2 + r^2 \dot{\theta}^2 + r^2 \sin^2 \theta \dot{\varphi}^2. \qquad (10.6)$$

The Euler-Lagrange equations (10.5) for the t, θ, and φ-components become, respectively,

$$\frac{d}{d\tau} \left[\left(1 - \frac{2M}{r} \right) \dot{t} \right] = 0, \qquad (10.7)$$

$$\frac{d}{d\tau} (r^2 \dot{\theta}) - r^2 \sin \theta \cos \theta \dot{\varphi}^2 = 0, \qquad (10.8)$$

and

$$\frac{d}{d\tau} (r^2 \sin^2 \theta \dot{\varphi}) = 0. \qquad (10.9)$$

Finally, we need an equation for the r-component. In principle, we can get this from (10.5), as well, but since the metric elements depend explicitly on r this equation will be a bit messy. Instead, we can make a judicious choice of the parameter τ. In the case of timelike geodesics it is natural to use proper time. Then it immediately follows that $2\mathcal{L} = -1$, which provides us with the equation we need. In the case of light we do not have a meaningful measure of time along the trajectory, but we know that the geodesics should be null so we must have $2\mathcal{L} = 0$, and again we have our final equation.

What do we learn from the Euler-Lagrange equations? Without too much effort, quite a lot. In particular, the equations reflect the symmetry of the spacetime. Equations (10.7) and (10.9) provide us with two conserved quantities (associated with two of the Killing vectors of the spacetime; see Chapter 2). The energy

$$\left(1 - \frac{2M}{r}\right)\dot{t} = \text{constant} = E \tag{10.10}$$

and the angular momentum

$$r^2 \sin^2 \theta \dot{\varphi} = \text{constant} = L \tag{10.11}$$

are both constants of the motion. The quantities are conserved because the spacetime is static and axisymmetric (the metric g_{ab} is independent of t and φ). Meanwhile, the third equation, (10.8), reflects the spherical symmetry of the problem, which implies that motion can be confined to a plane. If we conveniently choose the trajectory to initially be in the equatorial plane, $\theta = \pi/2$, we see that it is possible to find orbits that remain in this plane.

10.2   The gravitational redshift

At the level of geometric optics (discussed in more detail in Chapter 22), light can be described in terms of photons moving along null geodesics. In essence, they are described by a wave vector k^a which satisfies $k^b \nabla_b k^a = 0$. The wave vector is tangent to the geodesics, which means that we can take

$$k^a = \frac{dx^a}{d\lambda} \tag{10.12}$$

(where we are using λ as parameter rather than τ in order to avoid confusion with proper time). The frequency measured by a given observer is then

$$\omega = -g_{ab} u^a_{\text{obs}} k^b, \tag{10.13}$$

where u^a_{obs} is the four-velocity of the measuring device. If we consider a static observer, then the normalization condition immediately leads to

$$u^a_{\text{obs}} = [u^0, 0, 0, 0] \quad \text{with} \quad u^0 = \left(1 - \frac{2M}{r}\right)^{-1/2}. \tag{10.14}$$

Combining the results, and making use of (10.10), we have (reinstating $\hbar$)

$$\hbar\omega = \left(1 - \frac{2M}{r}\right)^{1/2} E. \tag{10.15}$$

However, since the energy E is conserved along a given geodesic, we learn that if a photon is emitted at r_1 and subsequently detected at r_2, then the frequencies are related according to

$$\frac{\omega_2}{\omega_1} = \left(\frac{1 - 2M/r_1}{1 - 2M/r_2}\right)^{1/2}. \tag{10.16}$$

If we, as an example, assume that the photon is emitted from the surface of the Sun (at radius R) and detected far away (say, at infinity), then it follows that

$$\omega_\infty = \left(1 - \frac{2M}{R}\right)^{1/2} \omega_R. \tag{10.17}$$

The observed frequency is lower—it has shifted towards the red.

This is an important result. The gravitational redshift affects any local time-varying phenomenon. For example, we will later consider stellar oscillations. In that case, the observed frequencies will be redshifted. This has important implications for modelling. If we want to make precision predictions, we need to model the seismology dynamics in general relativity.

The first precision test of the gravitational redshift was carried out by Pound and Rebka in 1959 (Pound and Rebka, 1960). They compared the energy shift for photons dropped down a 23-m tower to that of photons launched upwards the same distance (in order to eliminate systematic errors). In order to quantify this effect, we can simplify (10.16) to the situation where the gravitational field is weak and the relative difference between the source and the detector is small. In this case we get

$$\left|\frac{\Delta\omega}{\omega}\right| \approx \frac{gh}{c^2} \approx 2.5 \times 10^{-15}, \tag{10.18}$$

where g is the gravitational acceleration at the surface of the Earth and h is the height of the tower.

10.3   Flying clocks

According to general relativity, clocks run slow in a gravitational field. This suggests an obvious way to test the theory. Keep one accurate (atomic) clock in the laboratory and fly another one at high altitude. If the theory is correct, the flying clock should tick faster than the one left behind. This experiment involves comparing the proper time τ at different altitudes. Assuming that the clocks are at fixed position, the time interval on a local clock is related to a clock at infinity (this is the de facto meaning of the Schwarzschild time coordinate t) according to

$$\Delta \tau = \left(1 - \frac{2M}{r}\right)^{1/2} \Delta t. \tag{10.19}$$

The local clock ticks at a slower rate than the distant time keeper. As an example, the Earth's gravity will cause a clock on the surface to accumulate around 0.0219 fewer seconds each year than a distant clock would.

However, the clock experiment is not as straightforward as it may seem. In reality, the flying clock would be in orbit. To see how this affects the result, let us consider a circular orbit such that $\dot{r} = \dot{\varphi} = 0$ (without any particular loss of generality). In this case, the geodesics are such that (this follows readily from the Euler–Lagrange equation for the radial component)

$$\left(\frac{d\theta}{dt}\right)^2 = \frac{M}{r^3}. \tag{10.20}$$

Combining this result with the Schwarzschild line element, we obtain the rate of proper time associated with the orbit

$$d\tau^2 = -ds^2 = \left(1 - \frac{2M}{r}\right)dt^2 - r^2 d\theta^2 \quad \longrightarrow \quad \Delta\tau = \left(1 - \frac{3M}{r}\right)^{1/2} \Delta t. \tag{10.21}$$

In addition, we need to weigh in the fact that a moving clock slows down. From special relativity, we know that

$$\Delta t_{moving} = \left(1 - \frac{v^2}{c^2}\right)^{-1/2} \Delta t_{rest} \tag{10.22}$$

(this follows from the Lorentz transformation given at the beginning of Chapter 2).

The first successful flying-clock experiment was carried out by Hafele and Keating in 1971 (Hafele and Keating, 1972). They brought one of their atomic clocks on a round-the-world trip—and back again—on commercial airliners. The clock took up two seats, but the scientists had to settle for one each. When they returned and worked out the combined effects of the two relativity theories they found the experiment to be in good agreement with the expectations.

The slowing down of clocks has obviously been tested at much higher precision since those pioneering days, mainly by going into space. The trailblazer for this was Gravity Probe A, a rocket experiment launched in 1976 carrying a hydrogen maser clock.[2] When

---

[2] http://einstein.stanford.edu/content/faqs/gpa1.html.

it splashed down in the Atlantic after a short 2-hour flight, it had tested the theory prediction to better than 0.01%.

The fact that gravity makes your head age faster than your feet may not have much impact on your everyday life, but many of us are regularly using technology that simply would not work if we did not account for relativity. The multibillion dollar Global Positioning System (GPS; see Ashby (2003)) system has 24 satellites orbiting the Earth, each carrying a precise atomic clock, allowing navigation with a precision of 15 m (and local time determination to 50 billionths of a second). The satellites move at 14,000 km per hour in orbits that circle the Earth twice a day, much faster than clocks on the ground. We know that moving clocks tick slower and if we work it out we find that this amounts to 7 millionths of a second per day. However, in the curved spacetime the clocks on the ground move slower. This effect, in turn, makes the orbiting clocks move faster by 45 millionths of a second per day. Combining the results, the moving clocks move faster by 38 millionths of a second every day. The upshot of this is that, if we ignored relativity, navigational errors would accumulate at a rate of more than 10 km every day.

The efforts to reach higher precision are continuing. At the time of writing, the ACES/Pharao clock experiment is flying on the International Space Station (Cacciapuoti *et al.*, 2017). The experiment involves an ensemble of atomic clocks and microwave and optical links to compare the onboard clocks to clocks on the ground. The aim is to achieve as clock signal up to 100 times more stable and accurate than the state-of-the-art technology used in the GPS satellites. This will be the most accurate measurement of time yet.

10.4 Light bending

Let us move on to the problem of light propagation more generally. For practical reasons it makes sense to focus on trajectories in the equatorial plane. Setting $\theta = \pi/2$ and $\dot{\theta} = 0$ and making use of the two conserved quantities from (10.10) and (10.11), we have (with $L = L_z$ for motion in the equatorial plane)

$$\dot{r}^2 = E^2 - \left(1 - \frac{2M}{r}\right)\frac{L_z^2}{r^2}. \tag{10.23}$$

In order to solve this equation, let us rewrite it in terms of a new variable $u = u(\varphi) = 1/r$. Using (as in the corresponding Newtonian problem; see Chapter 5)

$$\dot{r} = -L_z \frac{du}{d\varphi}, \tag{10.24}$$

we arrive at

$$\left(\frac{du}{d\varphi}\right)^2 + (1 - 2Mu)u^2 = \frac{E^2}{L_z^2}. \tag{10.25}$$

This equation is tricky to solve, but if we take a derivative the right-hand side will vanish. We get

$$\frac{d^2u}{d\varphi^2} + u = 3Mu^2. \tag{10.26}$$

In the case of flat space (when $M = 0$) the solution is obviously a straight line

$$u_0 = \frac{1}{D}\sin(\varphi - \varphi_0), \tag{10.27}$$

where D is the distance of closest approach to the origin and φ_0 is the initial infall angle. Suppose we now look for an approximate solution to the curved spacetime equation (taking $\varphi_0 = 0$ without loss of generality). To find such a solution we let

$$u = \frac{1}{D}\sin\varphi + 3Mu_1, \tag{10.28}$$

where the second term is assumed to be sufficiently small that we can use perturbation theory. This then leads to

$$3M\left(\frac{d^2u_1}{d\varphi^2} + u_1\right) = \frac{3M}{D^2}\sin^2\varphi + \underbrace{\frac{6M}{D}\sin\varphi u_1 + 9M^2u_1^2}_{\text{ignore small terms}}, \tag{10.29}$$

or

$$\frac{d^2u_1}{d\varphi^2} + u_1 \approx \frac{1}{2D^2}(1 - \cos 2\varphi), \tag{10.30}$$

and the solution is

$$u_1 = \frac{1}{D^2}\left[A\cos\varphi + B\sin\varphi + \frac{1}{2}\left(1 + \frac{1}{3}\cos 2\varphi\right)\right]. \tag{10.31}$$

Combining this with the leading order result, we have

$$u \approx \frac{1}{D}\sin\varphi + \frac{3M}{D^2}\left[A\cos\varphi + \frac{1}{2}\left(1 + \frac{1}{3}\cos 2\varphi\right)\right]. \tag{10.32}$$

If we are interested in the effect accumulated as light passes near a gravitating body, we only need to consider the asymptotes. Recall that $u \to 0$ as $r \to \infty$ and let the infalling light have $\varphi = 0$. This means that we must have $A = -2/3$. It follows that the light will return to infinity at an angle $\varphi = \pi + \delta$, where

$$\delta = \frac{4M}{D}.$$ (10.33)

If we consider the particular case of light rays grazing the edge of the Sun, we find that light bending amounts to 1.75 arcseconds. This result was confirmed (although not with particularly high precision) by Eddington's eclipse expedition.

Light bending has been tested to high precision since the 1960s. Arrays of radio telescopes have used signals from distant quasars passing close to the Sun to confirm Einstein's prediction to the 0.01% level. ESAs Gaia mission,[3] which aims to track the motion of 1% of the stars in the Galaxy, is expected to test the light bending to the level of one part in a million. This will not leave much wriggle room for alternative models.

Light bending has, in fact, become an important tool for modern astronomy. Gravity can lens distant sources in different ways, forming multiple images, arcs, or, in some cases, complete circles (Schneider *et al.*, 1992). After matching to detailed calculations, the distortion of light from the distant Universe (way back in time!) provides clues of the presence of unseen matter between us and the source. Combined with data from, for example, the Sloan Digital Sky Survey,[4] this gives us one of the best handles we have on dark matter in the cosmos (see Chapter 22).

10.5   Shapiro time delay

In 1964 Irwin Shapiro proposed a new test of Einstein's theory (Shapiro, 1966). Light slows down as it passes through a gravitational field, leading to a signal arriving a little bit later than it would otherwise have done. The Shapiro time-delay has since been measured very precisely. As an example, in 2003 the delay of signals from NASAs Cassini spacecraft were found to agree with the theory to the 0.001% level (Bertotti *et al.*, 2003).

In order to understand the idea behind Shapiro's time delay, let us consider the leading order (straight line) solution to the equation for null geodesics (in the equatorial plane)

$$r \sin \varphi = D.$$ (10.34)

A variation of this leads to

$$dr \sin \varphi + r \cos \varphi d\varphi = 0 \quad \longrightarrow \quad d\varphi = -\frac{D}{\sqrt{r^2 - D^2}} dr.$$ (10.35)

[3] http://sci.esa.int/Gaia
[4] http://www.sdss.org

Combining this with the Schwarzschild line element we have

$$ds^2 = -\left(1 - \frac{2M}{r}\right)dt^2 + \left[\left(1 - \frac{2M}{r}\right)^{-1} + \frac{D^2}{r^2 - D^2}\right]dr^2 = 0. \tag{10.36}$$

After expanding in M/r this leads to

$$dt \approx \pm \frac{r}{\sqrt{r^2 - D^2}}\left[1 + \frac{2M}{r}\left(1 - \frac{D^2}{2r^2}\right)\right]dr. \tag{10.37}$$

The last term in the bracket represents the time delay.

10.6   Light rays and black holes

So far we have analysed the propagation of light in the relatively safe weak-field region. This allowed us to reflect on Solar System tests of the theory. Throwing caution to the wind, let us now venture into the strong-field region.

We take as our starting point the radial equation (10.23), but do not assume that a perturbative solution will suffice. Instead, we define the impact parameter $b = L_z/E$ (and rescale λ accordingly). This allows us to write the equation as

$$\left(\frac{dr}{d\lambda}\right)^2 = \frac{1}{b^2} - \frac{1}{r^2}\left(1 - \frac{2M}{r}\right) = \frac{1}{b^2} - V(r). \tag{10.38}$$

Here we have introduced the effective potential, $V(r)$, that governs the radial motion of a photon outside a Schwarzschild black hole. The potential has a maximum ($= 1/27M^2$) at $r = 3M$, vanishes at the horizon of the black hole, and falls off rapidly towards spatial infinity; see Figure 10.1.

From the requirement that $(dr/d\lambda)^2$ can only be positive or zero, we can deduce that the motion of a photon is restricted to values of r such that $1/b^2 > V(r)$. The upshot of this is that, if a particle with impact parameter $1/b^2 < 1/27M^2$ falls towards the black hole from infinity, there will be a radius $r_0 > 3M$ such that $V(r_0) = 1/b^2$ (case A in Figure 10.1). The point r_0 is a turning point for the radial motion. It is straightforward to analyse what happens at this point. Differentiation of (10.38) leads to

$$\frac{d^2r}{d\lambda^2} = -\frac{1}{2}\frac{dV}{dr} = \frac{r - 3M}{r^4}, \tag{10.39}$$

from which we see that the radial acceleration of the trajectory is always directed outwards for $r_0 > 3M$. Hence, the inwards moving photon turns back to infinity after reaching r_0— it is scattered by the spacetime curvature outside the black hole.

For incoming photons with $1/b^2 > 1/27M^2$ there is no turning point, and they will be swallowed by the black hole (case B in Figure 10.1). The relative smallness of b for such

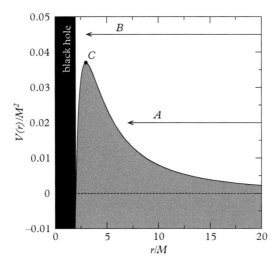

Figure 10.1 *The effective potential for radial motion of light in the Schwarzschild geometry. Three typical trajectories are indicated. In case A, photons are scattered by the black hole, in case B they plunge into the black hole, while case C corresponds to the unstable circular photon orbit at r = 3M.*

plunging trajectories can be interpreted as meaning that the initial orbit must be aimed more or less directly at the black hole.

Finally, circular orbits are only possible at maxima or minima of the potential $V(r)$. In the case of photons, this means that we can only have a circular orbit at $r = 3M$ (case C in Figure 10.1). This orbit is, however, unstable since it is associated with a maximum of the potential. As is clear from Eq. (10.39), a slight deviation from $r = 3M$ will lead to motion away from the maximum.

As we will see later, it is convenient to introduce a new radial coordinate in problems involving wave propagation in a curved spacetime. This so-called *tortoise coordinate*, r_*, encodes the gravitational redshift and its definition follows naturally from the equation for radial geodesics. Taking $L_z = 0$ we see from Eqs. (10.10) and (10.23) that

$$\frac{dr}{dt} = \pm\left(1 - \frac{2M}{r}\right). \tag{10.40}$$

After integration this leads to

$$t = \pm r_* + \text{constant}, \tag{10.41}$$

where we defined the new coordinate r_* by

$$\frac{d}{dr_*} = \left(1 - \frac{2M}{r}\right)\frac{d}{dr}, \tag{10.42}$$

or

$$r_* = r + 2M \log\left(\frac{r}{2M} - 1\right) + \text{constant}. \tag{10.43}$$

The choice of integration constant is irrelevant in most situations.

The tortoise coordinate differs from the ordinary radial variable by a logarithmic term. This has the effect that $r_* \to +\infty$ as $r \to +\infty$ (at spatial infinity), but $r_* \to -\infty$ as $r \to 2M$ (at the event horizon of the black hole). In other words, the event horizon has been 'pushed all the way to $-\infty$'. The tortoise coordinate can be used to explore all the physics that a distant observer will see, but (obviously) not the region inside the event horizon.

10.7 The motion of massive bodies

The trajectories of massive particles in the Schwarzschild geometry share many of the features we have discussed. We readily find that the equation for radial motion can be written

$$\left(\frac{dr}{d\tau}\right)^2 = \tilde{E}^2 - \left(1 - \frac{2M}{r}\right)\left(1 + \frac{\tilde{L}_z^2}{r^2}\right) = \tilde{E}^2 - \tilde{V}(r). \tag{10.44}$$

Here we have used the identification $\lambda = \tau/m$, with τ proper time and m the mass of the particle. We have also rescaled the energy and the angular momentum to a 'per unit mass' basis: $\tilde{E} = E/m$ and $\tilde{L}_z = L_z/m$.

We immediately note a feature that is different from the photon case. For light we could introduce an impact parameter that removed the direct dependence on the angular momentum. As a consequence, the photon results were effectively independent of the value of L_z. For example, the unstable photon orbit will be located at $r = 3M$ regardless of L_z (as long as it does not vanish). Clearly, the massive-particle case does not allow this simplification.

The analysis of massive-particle trajectories splits naturally into two cases. The simplest case corresponds to radial infall, for which φ is constant. Taking $\tilde{L}_z = 0$ we have

$$\frac{dr}{d\tau} = -\left[\tilde{E}^2 - 1 + \frac{2M}{r}\right]^{1/2}. \tag{10.45}$$

The particle plunges into the black hole for all values of $\tilde{E}$, but we can distinguish three cases: (i) When $\tilde{E} < 1$ the particle falls from rest at some finite r, (ii) when $\tilde{E} = 1$ the particle falls from rest at infinity, and (iii) when $\tilde{E} > 1$ the particle falls from infinity with a finite inward velocity. We also see that it takes a finite proper time for the particle to reach $r = 2M$ (or, for that matter, $r = 0$). This is in contrast to the well-known fact that it takes an infinite coordinate time (t) to reach the horizon.

Let us now consider the case of nonvanishing $\tilde{L}_z$. In analogy with the photon problem there will exist both scattered and plunging orbits. These correspond to cases A and B in Figure 10.2, respectively. The condition for circular orbits (the vanishing of the derivative of $\tilde{V}$) leads to

$$r = \frac{\tilde{L}_z^2}{2M} \left[1 \pm \sqrt{1 - \frac{12M^2}{\tilde{L}_z^2}} \right]. \tag{10.46}$$

We see that there are generally two circular orbits for each value of $\tilde{L}_z$. One (the outer one) is stable and the other (the inner one) is unstable. The typical situation is illustrated (for a suitable value of $\tilde{L}_z$) in Figure 10.2. In the figure the unstable circular orbit corresponds to case C, while the stable orbit is case D. Distinct circular orbits only exist if $\tilde{L}_z^2 > 12M^2$. When $\tilde{L}_z^2 = 12M^2$ the two orbits coincide, and when $\tilde{L}_z^2 < 12M^2$ there are no circular trajectories at all. By inserting the minimum value of $\tilde{L}_z = 12M^2$ in Eq. (10.46) we find that the minimum value of r for which circular orbits exist is $r_{\mathrm{isco}} = 6M$. This is the *innermost stable circular orbit* (ISCO for short), When a particle that spirals towards the black hole reaches r_{isco} the character of its motion changes and it plunges through the horizon.

The presence of the ISCO impacts on the inspiral of a binary system and the associated gravitational-wave signal. As a neutron star (say) that approaches a black hole

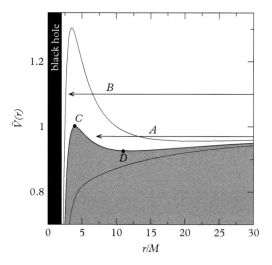

Figure 10.2 *The effective potential for radial motion of a massive particle in the Schwarzschild geometry (the case of orbital angular momentum $\tilde{L}_z = 4$ is shown as a thick solid line, while the cases $\tilde{L}_z = 3$ and 5 are shown as thin solid lines for comparison). Four typical trajectories are indicated. In case A the particle is scattered by the black hole. In case B it plunges into the hole and cases C and D corresponds to the unstable and stable circular orbit, respectively.*

reaches the ISCO the orbital motion will change from a slow adiabatic evolution to a rapid plunge (unless the star is tidally disrupted first; see Chapter 20). The slowly evolving gravitational-wave signal will be cut off roughly at the orbital frequency corresponding to $r_{\rm isco}$. One should be able to probe this feature with observations. Of course, in order to do this we need to first model the ISCO for binaries with comparable masses (Kidder et al., 1993a).

10.8   Perihelion precession

Having considered the general nature of particle orbits, let us briefly return to the Solar System tests of the theory, and repeat one of the calculations Einstein carried out in that frantic month of November 1915. As in the case of photon trajectories, we focus on motion in the equatorial plane, but now we want to solve the two-body problem (in the limit where one body is so much lighter than the other that we can ignore its contribution to the spacetime curvature—it is a test particle). From the variational derivation, we have the two equations

$$r^2\dot\varphi = L_z \tag{10.47}$$

and

$$\dot r^2 + \left(1 - \frac{2M}{r}\right)\left(1 + \frac{L_z^2}{r^2}\right) = E^2. \tag{10.48}$$

Repeating the steps from the light-bending problem (introducing a new variable $u = 1/r$ and taking a derivative of the corresponding equation of motion) we arrive at

$$\frac{d^2u}{d\varphi^2} + u - \frac{M}{L_z^2} = 3Mu^2. \tag{10.49}$$

This is almost, but not quite, the result we obtained in the massless case. The extra term on the left-hand side (which is constant along a given geodesic) has the effect that the general solution to the leading order equation (known as Binet's equation, and which we recognize from the Newtonian problem in Chapter 5) is an ellipse, rather than a straight line. Orienting the coordinate system in such a way that $\varphi = 0$ when the clock starts ticking, we have

$$u_0 = \frac{M}{L_z^2}(1 + e\cos\varphi). \tag{10.50}$$

Considering the right-hand side of (10.49) as small, we use the leading-order solution to get the equation for first-order perturbations (as before). We then see that one of the source terms is a solution to the homogeneous problem. This means that the perturbative solution has a secular term. Specifically, we have

$$u_1 = \frac{M}{L_z^2}\left(1 + \frac{e^2}{2}\right) + \frac{Me}{L_z^2}\underbrace{\varphi \sin\varphi}_{\text{secular}} - \frac{Me^2}{6L_z^2}\cos 2\varphi. \tag{10.51}$$

The second of these terms grows in time. Before long it will dominate the oscillatory terms, which means that we can approximate the overall solution as

$$u \approx \frac{M}{L_z^2}\{1 + e\cos[\varphi(1-\varepsilon)]\} + \mathcal{O}(\varepsilon^2), \tag{10.52}$$

where

$$\varepsilon = \frac{3M^2}{L_z^2} \leq \frac{1}{4}, \tag{10.53}$$

is a small parameter. The period of this solution is clearly no longer 2π. Instead, the smaller body will complete an orbit after

$$\frac{2\pi}{1-\varepsilon} \approx 2\pi\left(1 + \frac{3M^2}{L_z^2}\right). \tag{10.54}$$

In effect, the perihelion of a small planet like Mercury marches forward a little bit each orbit. It is a tiny effect, amounting to 43 arcseconds per century, compared to the overall influence of the other planets, which add up to a perihelion shift of about 5,600 arcseconds per century. Nevertheless, Einstein's calculation resolved a long-standing mystery and gave him confidence in the theory.

10.9 The Double Pulsar

The forwards march of the periastron is much more pronounced in relativistic systems. For example, the effect has been tested in the case of the Binary Pulsar PSR B1913+16 (Hulse and Taylor, 1975). In this system the periastron advances about 4° every year. If we combine this result with the orbital decay due to gravitational-wave emission, then we have two observables in addition to the orbital period P_b and the ellipticity e. This allows us to constrain two of the intrinsic parameters in the system, like the individual masses. This kind of analysis leads to the results listed in Table 9.2.

By extending the analysis of orbital motion to systems of roughly equal masses, one can express the various relativistic effects in terms of so-called post-Keplerian parameters (Damour and Deruelle, 1981; Damour and Taylor, 1991). As indicated by the name, these parameters quantify the deviation from Keplerian motion. However, they can be expressed in terms of the (easily measured) Keplerian parameters: the two masses of the orbiting objects, and the angles that define the direction of the pulsar spin axis. If we are fortunate enough to have more observables than parameters in the model, then we can test the underlying theory.

In the case of the Binary Pulsar one can measure three post-Keplerian parameters: the advance of periastron

$$\dot{\omega} = \frac{3G^{2/3}}{c^2} \left(\frac{2\pi}{P_b} \right)^{5/3} \frac{1}{1-e^2} (M_A + M_B)^2,$$ (10.55)

where the two masses are M_A and M_B; the gravitational redshift

$$\gamma = \frac{G^{2/3}}{c^2} \left(\frac{P_b}{2\pi} \right)^{1/3} e \frac{M_B(M_A + 2M_B)}{(M_A + M_B)^{4/3}};$$ (10.56)

and the change of the orbital period due to gravitational-wave emission

$$\dot{P}_b = -\frac{192\pi}{5} \frac{G^{5/3}}{c^5} \left(\frac{2\pi}{P_b} \right)^{5/3} \frac{1 + \frac{73}{24}e^2 + \frac{37}{96}e^4}{(1-e^2)^{7/2}} \frac{M_A M_B}{(M_A + M_B)^{1/3}}.$$ (10.57)

Since the orbital decay can be determined from the other parameters of the system, while it is also observed directly, we can test the theory and confirm of the existence of gravitational waves. The observational constraints on the masses of the two neutron stars in the system are shown in left panel of Figure 10.3.

A more highly relativistic neutron star system was discovered in 2003 (Lyne *et al.*, 2004). In this case there were pulses from both neutron stars, with periods of 23 ms and 2.8 s, and as a result the system (PSR J0737-3039A/B) was named the Double Pulsar. This system has proved an excellent gravity laboratory. Observers have managed to pin down seven post-Keplerian parameters. In addition to (10.55)–(10.57), we have the range and shape of the Shapiro delay

$$r = \frac{Gm_B}{c^3},$$ (10.58)

and

$$s = \frac{c}{G^{1/3}} \left(\frac{2\pi}{P_b} \right)^{2/3} x_A \frac{(M_A + M_B)^{2/3}}{M_B}$$ (10.59)

(where x_A is the semi-major axis of the pulsar A orbit), which can be precisely measured because of the fortuitous geometry of the system, where pulsar B eclipses the signal

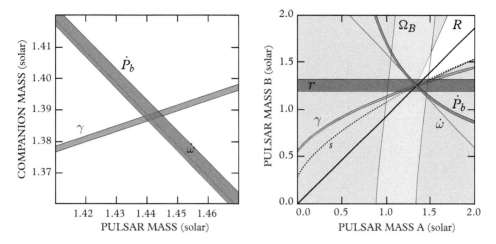

Figure 10.3 *Tests of gravity using double neutron star systems. Left: Constraints on the masses of the two neutron stars in the Binary Pulsar (B1913+16) system, with different curves showing the results for different post-Keplerian parameters (as discussed in the text). In order for the theory to be consistent, the different curves must intersect at a single point. Right: The constraints obtained for the Double Pulsar system (J0737-3039A/B). The white wedge shows the allowed region due to the inclination angle (sin i ≤ 1), and the solid diagonal line comes from the measurement of the mass ratio R. (The left panel is based on the discussed in Will (2014) while the right panel is based on data from Kramer et al. (2006).)*

from pulsar A. The orbital motion determines the mass ratio (see the discussion in Chapter 5)

$$R = \frac{x_B}{x_A} = \frac{M_A}{M_B} + O(v^4/c^4), \tag{10.60}$$

and we also have the geodetic precession of pulsar B (a measure of the impact of the spacetime curvature on the pulsar's spin axis)

$$\Omega_B = \frac{G^{2/3}}{c^2} \left(\frac{2\pi}{P_b} \right)^{5/3} \frac{1}{1-e^2} \frac{M_A(4M_B + 3M_A)}{2(M_A + M_B)^{4/3}}. \tag{10.61}$$

In total, we have seven observables constraining two parameters (the two masses). Five of these relations thus serve as a test of the theory. A beautiful example of how Einstein's theory passes the test (yet again) is provided in the right panel of Figure 10.3.

10.10 Radial infall

With a handle on geodesic motion, we can start to model gravitational-wave signals from bodies moving through a curved spacetime. We are not yet able to account for the backreaction the gravitational-wave emission has on the motion, but we can nevertheless gain useful insight. As an illustration, let us consider a small body falling radially towards a non-rotating black hole. For radial motion we have $L_z = 0$ so the geodesics are determined by

$$\left(\frac{dr}{d\tau}\right)^2 = \tilde{E}^2 - \left(1 - \frac{2M}{r}\right),$$
(10.62)

where τ is proper time. If we assume that the body started out at rest at infinity, then we must have $\tilde{E} = 1$. We are now going to cheat—and replace proper time τ by coordinate time t. In effect, this means that the calculation will not be accurate as we get close to the black hole, but we are only interested in a rough estimate for the moment. For an infalling body, we then need to integrate

$$\frac{dr}{dt} = -\left(\frac{2M}{r}\right)^{1/2}.$$
(10.63)

Let us orient the coordinate system in such a way that the infall is along the z-axis. From the quadrupole formula (3.90) we then see that the only non-vanishing contribution to the gravitational-wave emission comes from the mass multipole

$$M_{zz} = mz^2(t) \quad \longrightarrow \quad \mathcal{I}_{zz} = \frac{2}{3}M_{zz} \quad \longrightarrow \quad \frac{dE}{dt} = \frac{2G}{15c^5}\langle \dddot{M}_{zz}^2 \rangle,$$
(10.64)

where m is the mass of the falling body. The total energy emitted is given by

$$E = \frac{8Gm^2}{15c^5}\int_{-\infty}^{t_{max}} (\dot{z}\dddot{z} + 3\ddot{z}\dddot{z})\, dt.$$
(10.65)

In order to work out the integral, we rewrite (10.63) as

$$\dot{z} = -c\left(\frac{R_s}{z}\right)^{1/2},$$
(10.66)

where $R_s = 2GM/c^2$ is the Schwarzschild radius. Introducing a new variable $u = z/R_s$ and tracking the motion to a final point $r = R$, we have

$$E = \frac{2Gm^2}{R_s}\int_{R/R_s}^{\infty} u^{-9/2}\, du = \frac{4}{105}\frac{Gm^2}{R_s}\left(\frac{R_s}{R}\right)^{7/2}.$$
(10.67)

If we (somewhat simplistically, given the presence of the curvature potential) use this result all the way to the horizon we arrive at

$$E \approx 2 \times 10^{-2} \left(\frac{m}{M}\right) mc^2. \tag{10.68}$$

As a first step towards more similar mass systems, we can also draw on the discussion of binary motion from Chapter 5. That is, we replace m with the reduced mass μ and the central mass M with the total mass $M_1 + M_2$. This way we obtain

$$E \approx 2 \times 10^{-2} \left(\frac{\mu}{M_1 + M_2}\right) \mu c^2. \tag{10.69}$$

In the specific case of an equal mass system, we have $\mu = M_1/2$ so it follows that

$$E \approx 2 \times 10^{-3} M_1 c^2. \tag{10.70}$$

This gives us a first idea of the (not insignificant) amount of energy involved in black-hole mergers. Moreover, as we will see in Chapter 19, this rough estimate is better than one might have expected.

10.11   A bit more celestial mechanics

We could easily adopt the example from the previous section to more general orbits, but we are not going to do this (we will return to the problem in Chapter 16 once we have developed more appropriate tools). Instead, we will take the opportunity to look ahead. Let us ask what happens if we allow the orbiting body to venture closer to the central mass. We have already discussed objects plunging into a black hole. We now consider orbits that remain outside the black hole, yet probe the strong-field regime.

If we want to consider a generic orbit in the Schwarzschild geometry it is natural to begin by noting that the motion is bound as long as

$$\tilde{E} < 1, \quad \text{and} \quad \tilde{L}_z \geq 2\sqrt{3}M. \tag{10.71}$$

The first condition ensures that the body does not escape to infinity, while the second implies the existence of the inner turning point required to prevent plunge into the black hole—the difference between the cases $\tilde{L}_z = 3M$ and $\tilde{L}_z = 4M$ (say) is illustrated in Figure 10.2. When the two conditions from (10.71) are satisfied, the problem has three turning points. Bound motion takes place between the outer two of these, r_p and r_a ($r_p \leq r_a$), which correspond to the periastron and apastron of the orbit, respectively. The third turning point (r_3) is not of great importance as we are not considering plunging orbits.

It is useful to parameterize the orbit in terms of the eccentricity e and the semilatus rectum p. As in Chapter 5, these quantities are defined by

$$e = \frac{r_a - r_p}{r_a + r_p}, \quad \text{and} \quad p = \frac{2 r_a r_p}{r_a + r_p}. \tag{10.72}$$

The first of these parameters describes how circular the orbit is. The second is a measure of the 'size' of the orbit. Together e and p provide a complete description of a bound orbit. In order to solve the equations of motion, e.g. (10.44), we also need

$$\tilde{E}^2 = \frac{(p-2)^2 - 4e^2}{p(p-3-e^2)}, \tag{10.73}$$

and

$$\tilde{L}_z^2 = \frac{p^2 M^2}{p - 3 - e^2}. \tag{10.74}$$

The constraints on $\tilde{E}$ and $\tilde{L}_z$ are now satisfied for all values of p and any $e < 1$. It is also easy to show that a stable circular orbit corresponds to

$$e = 0 \quad \longrightarrow \quad r_{\text{circ}} = pM. \tag{10.75}$$

Meanwhile, the ISCO (corresponding to the merger of r_a and r_3) occurs for

$$p = 6 - 2e \longrightarrow \quad r_{\text{isco}} = \frac{(6+2e)M}{1+e}. \tag{10.76}$$

Note that this orbit does not, in general, have vanishing eccentricity. Since $p \geq 6 - 2e$ for a bound orbit, we have

$$r_p \geq \frac{(6+2e)M}{1+e} > 4M, \tag{10.77}$$

and we see that periastron is always located outside $r = 4M$.

We also learn that bound orbits in the Schwarzschild geometry can be represented by points in the $p - e$ plane which satisfy the conditions

$$0 \leq e < 1 \quad \text{and} \quad p \geq 6 - 2e. \tag{10.78}$$

The latter boundary is often referred to as the separatrix.

The integration of the geodesic equations is complicated by the fact that (10.44) is multivalued, with one branch corresponding to the body moving towards the black hole and the other to it moving away. To avoid this difficulty it is useful to introduce a new parameter χ such that

$$r(\chi) = \frac{pM}{1 + e \cos \chi}. \tag{10.79}$$

With this definition, χ is single-valued and ranges from 0 to 2π as r goes from r_p to r_a and back again. Substituting into (10.44) we find that

$$\frac{d\chi}{d\tau} = \frac{(1 + e \cos \chi)^2}{pM} \left[\frac{p - 6 - 2e \cos \chi}{p(p - 3 - e^2)} \right]^{1/2}. \tag{10.80}$$

A typical example of an elliptic Schwarzschild orbit obtained from this equation is shown in Figure 10.4.

Next we consider the equation for φ, which can be written

$$\varphi(\chi) = p^{1/2} \int_0^\chi \frac{d\chi'}{(p - 6 - 2e \cos \chi')^{1/2}}. \tag{10.81}$$

By working out the angle traversed by the orbit during the passage from periastron to apastron and back, $\Delta\varphi = \varphi(2\pi)$, we find that it is in general not a rational fraction of 2π. We knew this already—typical bound orbits are not closed. It is a minute effect in the case of planets orbiting the Sun. A black-hole spacetime provides much more spectacular

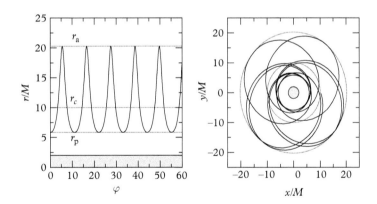

Figure 10.4 *A typical elliptic orbit in the Schwarzschild geometry. This particular case corresponds to parameters close to those what would correspond to a circular orbit at $r_c = 10M$. The angular momentum is the same and the energy has been increased by 1%, leading to an elliptical orbit. The left panel shows r as a function of φ, while the right panel shows the orbit in the equatorial $x - y$ plane. Also shown (as dotted lines) are the periastron r_p and apastron r_a.*

possibilities. In particular, from (10.81) we see that for parameters close to the separatrix, i.e. small values of

$$\varepsilon = p - 6 + 2e, \tag{10.82}$$

the region near periastron ($\chi = 0$ and 2π) will provide the main contribution to the integral and hence to $\Delta\varphi$. By expressing (10.81) as an elliptic integral and expanding for small ε one can show that

$$\Delta\varphi \approx 2 \left(\frac{3+e}{2e}\right)^{1/2} \log \frac{64e}{\varepsilon} \qquad \varepsilon \ll 1. \tag{10.83}$$

We learn that orbits close to the separatrix in the $p - e$ plane, can have changes in φ much larger than 2π. These are known as 'zoom-whirl' orbits (Glampedakis and Kennefick, 2002). The body zooms in from apastron, whirls around the black hole a number of times, and then zooms back out to apastron. This kind of complicated orbital dynamics will be of great interest for an instrument like LISA, which should be able to see small bodies orbiting supermassive black holes in distant galaxy cores (see Chapter 16).

11

Beyond Newton

So far we have only considered linearized models for gravitational-wave sources. Now it is time to get serious. General relativity is fundamentally a nonlinear theory. If we want to achieve the accuracy required for detection and extraction of parameters from astrophysical signals we need to go beyond linear order. This obviously makes the modelling more difficult. Yet, if we (in the first instance) focus our attention on binaries, it is possible to proceed in a systematic manner. The natural way to do this is to build on the approach that led to the quadrupole formula in Chapter 5. We can carry out a low-velocity expansion, order by order in powers of v/c. Basically, we know from Kepler's law that there is an intimate link between the velocity and the gravitational potential

$$v^2 \sim \frac{GM}{a},\qquad(11.1)$$

where M is the total mass of the system. This means that we may equally consider an expansion in the deviation from flat space—essentially M/a where a is the size of the orbit. If we prefer, we can use the source frequency

$$\Omega \sim v/a.\qquad(11.2)$$

At the end of the day, these different parameters are linked. In practice, the post-Newtonian framework[1] we will explore makes use of a small parameter

$$x = \left(\frac{GM\Omega}{c^3}\right)^{2/3} \sim \mathcal{O}\left(\frac{v^2}{c^2}\right) \sim \mathcal{O}\left(\frac{GM}{ac^2}\right).\qquad(11.3)$$

The connection between the different parameters illustrates why the scheme we need to develop is not quite straightforward. It may be easy to outline in words what we want to do, but to carry it out in practice takes some effort. We need to keep track of contributions

---

[1] The principles behind the post-Newtonian approach are clearly laid out by Poisson and Will (2014) while the technical aspects of the binary problem are surveyed by Blanchet (2006). Relevant historical comments are made by Schutz (1996).

Gravitational-Wave Astronomy: Exploring the Dark Side of the Universe. Nils Andersson, Oxford University Press (2020). © Nils Andersson. DOI: 10.1093/oso/9780198568032.001.0001

from the motion as well as the spacetime curvature and, as the underlying problem is nonlinear, the coefficients in the expansion we end up with may be rather complicated.

Higher order post-Newtonian calculations are not for the faint hearted. Yet, it is essential that we take on this challenge. We know from the discussion of matched filtering in Chapter 8 that we need reliable search templates. Effectively, for a binary signal the templates must remain faithful to better than (something like) one wave cycle as the system sweeps through the detector's sensitivity band. In the case of neutron star binaries, this means that we need to accurately represent several thousand wave cycles. This will inevitably involve going to higher orders in the expansion parameter, x. In fact, a key question concerns to what order in the parameter we have to extend the calculation (and, of course, to what extent this is manageable; see Cutler *et al.* (1993)).

11.1   Near and far-zone solutions

The introduction to binary systems in Chapter 5 brought us to a natural transition. We had discussed the nature of gravitational waves and their origin and provided rough estimates of the signal strength and its characteristics. In essence, we covered the problem at the back-of-the-envelope level. It is essential to understand the problem at this level, but we need to dig deeper if we are to consider state-of-the-art aspects. We will now do this—obviously leaving room for further exploration (see Blanchet (2006) for a more detailed discussion).

Let us begin by revisiting the quadrupole formula. At a deeper level, an analysis of a given gravitational-wave source involves subtle issues which may require different kinds of approximations. This is best illustrated by a particular—still fairly general— example. With a slow-motion source in mind, such that $\lambdabar \gg L \gtrsim M$, where $\lambdabar$ is the reduced wavelength, L is the size of the source, and M its mass, let us ask: What is actually involved in calculating the propagation of gravitational waves from the point of origin to the detector?

After a little bit of thought, we see that it makes sense to identify four spacetime regions:

1. *The strong gravity region*: In this region gravity is (obviously) strong. Modelling involves full nonlinear relativity and may require supercomputer simulations, as the equations are too complicated to solve by any other means. We will discuss this approach in Chapters 19–21.

2. *The weak field near zone*: Beyond a distance of $r \approx 10M$ or so it is probably safe to consider gravity as weak. The problem can then be analysed using linearized theory in a, more or less, flat spacetime.

3. *The local wave zone*: It is only in this region (and beyond) that it is meaningful to talk about gravitational waves. The transition to the wave zone is at $r \approx \lambdabar$. Other sources of gravity are assumed to be distant and have no effect on the analysis.

4. *The distant wave zone*: This is the Universe at large. Here we need to worry about additional sources of curvature, interaction with interstellar matter, perhaps the

expansion of the Universe, and so on. As we will discuss in Chapter 22, geometrical optics provides an adequate description of the wave propagation, but other sources of gravity may lead to gravitational lensing and complicate the results.

Why do we need to consider this confusing hierarchy of spacetime regions? The answer lies in the inherent complexity of the Einstein equations. Since we cannot find exact solutions, and since the available computing power is insufficient to allow the construction of a complete numerical solution (not mentioning the technical challenges involved!), we are forced to bring every possible trick and technique to bear on each problem of interest.

In the following we will focus on the transition between regions 2 and 3. This is a key element of the general analysis since it involves going from a region within about a wavelength or so of the source to the region where the concept of gravitational 'waves' makes sense.

As a first step, we return to the integral equation (3.55). In general, the integral on the right-hand side contains more information than we made use of so far; it accounts for all the moments of the stress–energy tensor. We have (first of all) the mass monopole M, the mass dipole M^j, and the momentum dipole P^j,

$$M = \int T^{00} d^3x, \tag{11.4}$$

$$M^j = \int T^{00} x^j d^3x, \tag{11.5}$$

$$P^j = \int T^{0j} d^3x, \tag{11.6}$$

but since mass, linear and angular momentum are all conserved, these multipoles will not contribute to the result. However, all higher multipoles will enter the problem. We can think of this as a formal expansion starting with the mass quadrupole and the leading moment associated with stresses

$$M^{jk} = \int T^{00} x^j x^k d^3x, \tag{11.7}$$

$$S^{jk} = \int T^{jk} d^3x. \tag{11.8}$$

At this level, we also have the momentum contribution

$$P^{jk} = \int T^{0j} x^k d^3x, \tag{11.9}$$

but using the conservation laws (3.59) and (3.60) we can show that

$$\dot{P}^{jk} = S^{jk} \tag{11.10}$$

(where the dot represents a time derivative), so we can replace this with the stress multipole wherever required. Similarly, it is worth noting that

$$\ddot{M}^{jk} = 2S^{jk}. \tag{11.11}$$

Let us now consider the weak-field near zone. Here we are not dealing with radiation. Instead, the field changes quasistatically. This follows since

$$\partial_t \sim \frac{1}{\lambda}, \quad \partial_r \sim \frac{1}{L} \longrightarrow \partial_t \ll \partial_r. \tag{11.12}$$

As long as $\lambda \gg L$ we can neglect time-derivatives compared to spatial ones. The upshot of this is that, in the near zone, we can work with simultaneous expansions in inverse powers of r and the various multipoles.

In order to obtain explicit expressions we need to solve the Lorenz gauge condition (see Chapter 3)

$$\partial_b \bar{h}^{ab} = 0, \tag{11.13}$$

together with

$$\Box \bar{h}^{ab} = \begin{cases} 0 & \text{vacuum,} \\ -16\pi T^{ab} & \text{weak internal gravity.} \end{cases} \tag{11.14}$$

We have already seen that time derivatives are not important in the near zone. Keeping only spatial derivatives we have

$$\partial_j \bar{h}^{0j} = 0, \tag{11.15}$$
$$\partial_k \bar{h}^{jk} = 0, \tag{11.16}$$

from the gauge condition and

$$\nabla^2 \bar{h}^{00} = -16\pi T^{00} = -16\pi\rho, \tag{11.17}$$
$$\nabla^2 \bar{h}^{0j} = -16\pi\rho v^j, \tag{11.18}$$
$$\nabla^2 \bar{h}^{jk} = \mathcal{O}(\rho v^2) \approx 0, \tag{11.19}$$

from the wave equation (since the wave operator $\Box$ reduces to the three-dimensional Laplacian ∇^2 if we neglect the time derivatives). This is now a static problem. Using the standard Green's function for the Poisson equation, we find that

$$\bar{h}^{00}(\boldsymbol{x}) \approx 4 \int \frac{\rho(\boldsymbol{x'})}{|\boldsymbol{x} - \boldsymbol{x'}|} d^3 x' = -4\Phi. \tag{11.20}$$

This was, in fact, clear already from (11.17) since the Newtonian gravitational potential Φ is determined by

$$\nabla^2 \Phi = 4\pi\rho. \qquad (11.21)$$

Next, we turn the integral equation into a multipole expansion by considering $\boldsymbol{x}'$ as a point interior to the source, while $\boldsymbol{x}$ is the exterior field point. Assuming that the centre of mass is close to $\boldsymbol{x}'$, we should have

$$|\boldsymbol{x}| = r \gg r' = |\boldsymbol{x}'|. \qquad (11.22)$$

Under these conditions we have

$$\frac{1}{|\boldsymbol{x}-\boldsymbol{x}'|} = \frac{1}{\sqrt{(\boldsymbol{x}-\boldsymbol{x}')^2}} = \frac{1}{\sqrt{[r^2 - 2\boldsymbol{x}\cdot\boldsymbol{x}' + (r')^2)^2}}$$

$$= \frac{1}{r}\left[1 - \frac{2\boldsymbol{n}\cdot\boldsymbol{x}'}{r} + \left(\frac{r'}{r}\right)^2\right]^{-1/2}$$

$$\approx \frac{1}{r} + \frac{n_j x'^j}{r^2} + \frac{3}{2}\frac{n_j n_k [x'^j x'^k - (r')^2 \delta^{jk}/3]}{r^3}. \qquad (11.23)$$

From this we get

$$\bar{h}^{00}(\boldsymbol{x}) = 4\left\{\frac{1}{r}\int \rho(\boldsymbol{x}')d^3 x' + \frac{n_j}{r^2}\int \rho(\boldsymbol{x}')x'^j d^3 x' \right.$$

$$\left. + \frac{3n_j n_k}{2r^3}\int \rho(\boldsymbol{x}')[x'^j x'^k - (r')^2 \delta^{jk}/3]d^3 x'\right\}. \qquad (11.24)$$

Here we identify the first integral as the mass (monopole moment) M, the second integral is the mass dipole moment M^j, and the third integral is the quadrupole moment M^{jk} (with the trace removed). In other words, we have

$$\bar{h}^{00} = 4\left\{\frac{M}{r} + \frac{M_j n^j}{r^2} + \frac{3}{2}\frac{\mathcal{I}_{jk} n^j n^k}{r^3}\right\}, \qquad (11.25)$$

in the near zone.

As we extend this result into the wave zone, the static parts (the monopole and the dipole) retain their form. But the quadrupole part (which is dynamic) no longer takes the form in (11.25). As r increases we must account for retardation effects and describe the quadrupole component in terms of (outgoing) gravitational waves.

In essence, we want to match (11.25) to an expression that is valid in the wave zone. In this region we can no longer neglect the time derivatives, but gravity is still weak so we can linearize the equations. Thus, we need a solution to

$$\partial_b \bar{h}^{ab} = 0,$$ (11.26)

$$\Box \bar{h}^{ab} = 0,$$ (11.27)

which satisfies the boundary condition

$$\bar{h}^{00} = \frac{6 \mathcal{I}_{jk} n^j n^k}{r^3} \qquad \text{for } r \ll \lambdabar,$$ (11.28)

and which corresponds to purely outgoing waves as $r \to \infty$.

From the simplified analysis that led to (3.67), we already know what form the solution should take. This provides a short cut to the answer. Instead of writing down the general solution, we note that the scalar wave equation $\Box \psi = 0$ admits the outgoing wave solution $\psi = f(t-r)/r$, where f can be any function. In our problem (and in a local inertial frame), each component of $\bar{h}_{ab}$ should satisfy the same wave equation. This inspires us to write the solution as

$$\bar{h}^{00} = 2 \partial_j \partial_k \left[\frac{1}{r} \mathcal{I}^{jk} (t-r) \right].$$ (11.29)

Expanding this expression for small r, using

$$\partial_j \left(\frac{1}{r} \right) = -\frac{n_j}{r^2} = -\frac{x_j}{r^3},$$ (11.30)

$$\partial_j \partial_k \left(\frac{1}{r} \right) = -\frac{3}{r^3} \left[n_j n_k - \frac{\delta_{jk}}{3} \right],$$ (11.31)

and

$$\partial_j [\mathcal{I}^{jk} (t-r)] \sim \frac{\mathcal{I}_{jk}}{\lambdabar} \qquad \text{where} \qquad \lambdabar \gg r,$$ (11.32)

we see that this solution satisfies the boundary condition (11.28) in the near zone.

Finally, given an expression for $\bar{h}^{00}$ we can determine all other components of the perturbed metric. First, we integrate the gauge condition

$$\partial_0 \bar{h}^{00} = -\partial_j \bar{h}^{0j},$$ (11.33)

to get

$$\bar{h}^{0j} = -2 \partial_k \left[\frac{1}{r} \dot{\mathcal{I}}^{jk} (t-r) \right].$$ (11.34)

Then we use

$$\partial_k \bar{h}^{jk} = -\partial_0 \bar{h}^{j0} = -\partial_0 \bar{h}^{0j}, \tag{11.35}$$

to find

$$\bar{h}_{jk} = \frac{2}{r}\ddot{\mathcal{I}}_{jk}(t-r). \tag{11.36}$$

This is the main result. Retaining only the transverse-traceless (TT) part in the wave zone (as discussed in Chapter 3) we have

$$\bar{h}_{jk}^{\mathrm{TT}} = \frac{2}{r}\ddot{\mathcal{I}}_{jk}^{\mathrm{TT}}. \tag{11.37}$$

Not surprisingly, we have arrived at the same expression for the gravitational-wave field as before. Yet, there are subtle differences. In particular, the approach we have used remains valid also when we are dealing with a source with strong internal gravity—we can still deduce the gravitational-wave strength from the quadrupole moment in the weak-field regime. The difference is that in that case we must use the full nonlinear theory to describe the interior dynamics.

Anticipating the need to proceed to higher orders, one may envisage a number of possible strategies. We can add terms to the post-Newtonian v/c expansion, expand in 'powers of G' beyond the flat Minkowski spacetime, decompose in multipoles to deal with the angular behaviour, and use the source 'size' as expansion parameter, expand in powers of $1/r$ to work out the asymptotic behaviour, or focus on a particular class of binaries and treat one body as a small perturbation in the spacetime of the other (this extreme-mass-ratio limit will be discussed in Chapter 16). The different possibilities involve specific steps and choices. If we want to solve the problem we may need to combine several strategies. This becomes apparent as soon as we realize that the post-Newtonian expansion, which would be the natural way to model the (relatively) slow motion of the source, is incompatible with retardation so it cannot be used to describe the outgoing nature of the gravitational waves. This is, of course, a crucial aspect. First of all, these are the waves we are trying to describe and, secondly, we need to be able to balance the emitted energy to work out the impact of radiation reaction on the source. In this sense, the post-Minkowski expansion is more universal (Blanchet, 2006). As long as we are dealing with a weak-gravity source, an expansion in powers of G is valid throughout spacetime. However, the involved mathematics can be confusing.

11.2   A slight aside: symmetric trace-free (STF) tensors

It should be clear, even from our somewhat sketchy derivation, that the notation will become messy as we go to higher order approximations. In order to (at least to some extent) alleviate this problem, it is common to introduce a more 'efficient' notation for the different multipole expressions. In order to introduce the idea, let us consider the problem of tidal interaction in a binary system. Suppose we have two stars, with masses M_A and M_B, respectively, located at x_A and x_B. As long as the stars are far apart, we can expand the two gravitational potentials. Moreover, we can consider the influence of one of the stars (B, say) on the other (A) as due to a point mass. The acceleration of star A due to the presence of star B is then given by

$$U_B(x) = -\sum_{l=0}^{\infty} \frac{1}{l!}(x - x_A)^L \mathcal{E}_L, \qquad (11.38)$$

where we have introduced the tidal moments

$$\mathcal{E}_L = -\left(\partial_L \frac{M_B}{|x - x_B|}\right)_{x=x_A}, \qquad (11.39)$$

and used the short-hand notation $\partial_L = \partial_1 \partial_2 \ldots \partial_l$.

Making contact with our previous derivation, let us use a coordinate system such that $r = |x - x_B|$. Then we have

$$\mathcal{E}_L = -M_B \partial_L \left(\frac{1}{r}\right), \qquad (11.40)$$

and it follows that (as before)

$$\mathcal{E}_j = M_B \frac{n_j}{r^2}, \qquad (11.41)$$

$$\mathcal{E}_{jk} = -\frac{M_B}{r^3}\left(3n_j n_k - \delta_{jk}\right), \qquad (11.42)$$

$$\mathcal{E}_{jkl} = \frac{M_B}{r^4}\left[15n_j n_k n_l - 3\left(n_j \delta_{kl} + n_k \delta_{jl} + n_l \delta_{jk}\right)\right], \qquad (11.43)$$

and so on. It is worth noting that the expressions on the right-hand sides are symmetric and trace-free (STF).

The general pattern from these results suggest the compact expression

$$\mathcal{E}_L = \frac{M_B(-1)^{l+1}(2l-1)!!n^{\langle L \rangle}}{r^{l+1}}. \qquad (11.44)$$

The angular brackets indicate that the object is symmetric and trace-free. It may seems as if we have simply tried to hide the real problem with some clever notation, but this is not the case. There is an efficient (and practical) prescription for working out $n^{\langle L \rangle}$. We need (Thorne, 1980)

$$n^{\langle L \rangle} = n^{\langle j_1 j_2 \cdots j_l \rangle} = \sum_{p=0}^{[l/2]} (-1)^p \frac{(2l - 2p - 1)!!}{(2l - 1)!!} \left[\delta^{j_1 j_2} \cdots \delta^{j_{2p-1} j_{2p}} n^{j_{2p+1}} \cdots n^{j_l} + \text{sym} \right], \quad (11.45)$$

where $[l/2]$ should be taken to mean the largest integer $\leq l/2$. Each of the p terms inside the bracket involves a product of p Kronecker deltas and $l - 2p$ unit vectors.

As we are establishing notation, it is also useful to note that we can translate different expansions into one another. For example, in the case of the external gravitational potential of star A, we have

$$U_A = \sum_{l=0}^{\infty} M_A^L \left(\partial_L \frac{1}{|\boldsymbol{x} - \boldsymbol{x}'|} \right)_{\boldsymbol{x}' = \boldsymbol{x}_A}. \quad (11.46)$$

Making use of our new notation, this can equivalently be written

$$U_A = \sum_{l=0}^{\infty} \frac{(2l - 1)!!}{l!!} M_A^L \frac{n_A^{\langle L \rangle}}{r_A^{l+1}}. \quad (11.47)$$

We can also express the result in spherical harmonics (Thorne, 1980)

$$U_A = \sum_{l=0}^{\infty} \sum_{m=-l}^{l} \frac{4\pi G}{2l + 1} \frac{M_{lm}}{r^{l+1}} Y_{lm}. \quad (11.48)$$

In practice, different representations suit different applications (see Poisson and Will (2014) for more details and a number of examples).

11.3   The relaxed Einstein equations

Formally, the Newtonian limit corresponds to $1/c \to 0$ (as in the case of the fluid equations in Chapter 4). If we take a closer look at the derivation of the quadrupole formula we see that the result was entirely determined by Newtonian dynamics. We used the Newtonian quadrupole moment and worked out the orbital motion in Newtonian gravity. If we want a more accurate description, we need to account for higher multipole contributions. Traditionally, these contributions are labelled in such a way that terms of order $(v/c)^n \sim 1/c^n$ are said to be of $n/2$ post-Newtonian (pN) order. In other words, the gravitational radiation reaction, which occurs at order $1/c^5$ arises at the 2.5 pN order. In order to go beyond this, we need to systematically incorporate higher order terms.

As we set out to devise a scheme that allows us to proceed to high orders, it makes sense to consider the problem from a formal point of view. It is natural to begin by reconsidering the Einstein equations

$$G^{ab} = R^{ab} - \frac{1}{2}g^{ab}R = 8\pi T^{ab}. \tag{11.49}$$

These are, obviously, a set of coupled nonlinear partial differential equations. In order to facilitate an iterative solution, it would be better to have the equations on integral form. This is, however, tricky because we need to somehow invert the differential operator on the left-hand side. However, we can do this by taking a couple of clever steps. We start by introducing a new field (Blanchet, 2006)

$$\mathsf{h}^{ab} \equiv \eta^{ab} - \mathfrak{g}^{ab}, \tag{11.50}$$

such that the 'gothic' metric is given by

$$\mathfrak{g}^{ab} = (-g)^{1/2}g^{ab}. \tag{11.51}$$

The linearized version of the new field leads us back to $\bar{h}^{ab}$

$$\mathsf{h}^{ab} \approx \eta^{ab} - (1+h)^{1/2}\left(\eta^{ab} - h^{ab}\right) \approx h^{ab} - \frac{1}{2}\eta^{ab}h = \bar{h}^{ab}, \tag{11.52}$$

but we are not going to assume that h^{ab} is small. Nevertheless, let us impose the usual harmonic gauge condition

$$\partial_a \mathsf{h}^{ab} = 0, \tag{11.53}$$

where it is important to appreciate that we are using partial derivatives.

With these definitions (and after a fair bit of work) the Einstein equations (11.49) take the form

$$\Box \mathsf{h}^{ab} = -16\pi \tau^{ab}, \tag{11.54}$$

where $\Box$ is the usual flat spacetime wave operator. Meanwhile, the source on the right-hand side is given by

$$\tau^{ab} = (-g)T^{ab} + (16\pi)^{-1}\Lambda^{ab}, \tag{11.55}$$

where Λ^{ab} contains the nonlinear contribution of the gravitational field (it does not depend on the matter). It is explicitly given by

$$\Lambda^{ab} = 16\pi(-g)t_{\mathrm{LL}}^{ab} + (\partial_c \mathsf{h}^{ad}\partial_d \mathsf{h}^{bc} - \mathsf{h}^{cd}\partial_c\partial_d \mathsf{h}^{ab}), \tag{11.56}$$

where t_{LL}^{ab} is the so-called Landau–Lifshitz pseudotensor, defined by

$$16\pi(-g)t_{LL}^{ab} \equiv g_{fc}g^{de}\partial_d h^{af}\partial_e h^{bc}$$
$$+ \frac{1}{2}g_{fc}g^{ab}\partial_e h^{fd}\partial_d h^{ec} - 2g_{cd}g^{f(a}\partial_e h^{b)d}\partial_f h^{ec}$$
$$+ \frac{1}{8}(2g^{af}g^{bc} - g^{ab}g^{fc})(2g_{de}g_{gh} - g_{eg}g_{dh})\partial_f h^{dh}\partial_c h^{eg}. \qquad (11.57)$$

At first sight, it may not seem as if we have achieved much. If anything, the problem looks more complicated than before. We have certainly used more indices than we might be comfortable with. However, this new formulation has a clear advantage. Basically, the differential operator on the left-hand side of (11.54) is well understood so it is (formally) straightforward to invert the problem.

The combination of (11.53) and (11.54) contains all the information from the original Einstein equations, but we now solve the problem in two separate steps. First, we ignore the gauge condition (11.53). We can, for example, assign an arbitrary time dependence to T^{ab}. That is, we can relax the assumption that the matter variables should satisfy the equations of motion. For this reason, Eq. (11.54) is known as the *relaxed Einstein equations*. Of course, in the second step we must impose (11.53) in order to recover the true physics. This logic may seem peculiar, but it makes sense if we note that the condition (11.53) implies that the source term satisfies the conservation law

$$\partial_a \tau^{ab} = 0. \qquad (11.58)$$

Imposing (11.53) is then equivalent to ensuring that the matter behaves according to the usual equations of motion

$$\nabla_a T^{ab} = 0. \qquad (11.59)$$

The derivation of (11.54) involved no approximations. It is a valid alternative description as long as spacetime can be covered by harmonic coordinates. And it is straightforward to write down the formal solution as a functional of source variables (without specifying the motion of the source). Making use of the standard retarded flat space Green's function (with an outgoing-wave boundary condition), we have

$$h^{ab}(t,\boldsymbol{x}) = 4\int \frac{\tau^{ab}(t',\boldsymbol{x}')\delta(t'-t+|\boldsymbol{x}-\boldsymbol{x}'|)}{|\boldsymbol{x}-\boldsymbol{x}'|}d^4x'$$
$$= 4\int \frac{\tau^{ab}(t-|\boldsymbol{x}-\boldsymbol{x}'|,\boldsymbol{x}')}{|\boldsymbol{x}-\boldsymbol{x}'|}d^3x'. \qquad (11.60)$$

Of course, this does not mean that it is easy to find an actual solution. There is no free lunch. We face three main problems. First of all, the 'source' in (11.60) contains terms which depend explicitly on h^{ab}, the quantity we are trying to solve for. We need to know the solution in order to find it. However, this is a common setup for an iterative solution so it should not be much of a concern. The second complication arises from the fact that the fields h^{ab} (and hence τ^{ab}) are likely to have infinite spatial extent. After all, these terms represent the outgoing gravitational waves that we are interested in. As is clear from (11.54), these waves will contribute to the source, generating an additional component of the radiation. We need to account for waves creating waves—we cannot avoid the nonlinearities. Finally, we note that the third term in (11.56) really belongs on the left-hand side of (11.54). Along with the other second derivatives, this term contributes to the principal part of the differential operator. By moving this term to the right-hand side we modify the propagation characteristics of the field from true null cones of curved spacetime to the flat-spacetime characteristics of the d'Alembert operator. We need to keep careful track of this.

11.4   Iterative schemes

Before we consider some of the technical issues, let us set out what we want to achieve. We want to come up with a systematic procedure that extends the orbital motion beyond the Newtonian level. This will involve corrections to the orbital parameters, like the energy E and the angular momentum $\mathcal{J}^i$. Similarly, we need corrections to the gravitational-wave luminosity, for both the rate of energy loss $\mathcal{F}$ and the angular momentum carried by the waves $\mathcal{G}^i$. Once we have these, we can connect them (provided the orbital evolution is slow enough, an assumption that will break down before the bodies merge) by averaging over an orbit. As before, this leads us to the evolution equations

$$\langle \dot{E} \rangle = -\langle \mathcal{F} \rangle, \tag{11.61}$$

and

$$\langle \dot{\mathcal{J}}^i \rangle = -\langle \mathcal{G}^i \rangle. \tag{11.62}$$

From the outset, we have no way of knowing to what order we have to carry out the calculation. We do not know the rate of convergence of the post-Newtonian scheme, or (indeed) if the procedure converges at all. However, we do know that E and $\mathcal{J}^i$ are conserved up to 2pN order. This means that we have to calculate $\mathcal{F}$ and $\mathcal{G}^i$ to 3pN precision, and likely beyond. We need to obtain the equations of motion and work out the wave generation at each order. This requires a mathematical tour-de-force effort involving a number of subtle nonlinear effects.

The general strategy is fairly easy to describe. Because the field h^{ab} appears in the source of the equation, the natural solution is to iterate. Starting from (11.60), we substitute $\mathrm{h}_0^{ab} = 0$ in the integrand and solve for the first-iterated field h_1^{ab}; substitute this into (11.60) and solve for the second-iterated h_2^{ab}; and so on (imposing the gauge condition (11.53) consistently at each order). At each step j, the matter variables are

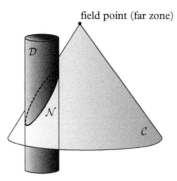

field point (far zone)

Figure 11.1 *The spacetime regions involved in a typical post-Newtonian iteration scheme. The past harmonic null cone $\mathcal{C}$ of an exterior field point intersects the near zone world tube $\mathcal{D}$ at a hypersurface $\mathcal{N}$.*

used to determine T_j^{ab} and Λ_j^{ab}. Equation (11.60) then yields h_j^{ab}. One can extract the motion of the source by substituting h_j^{ab} into the matter stress–energy tensor and working out $_j\nabla_b(T_j^{ab}) = 0$, where $_j\nabla_b$ represents the covariant derivative associated with the jth iterated field.

The approach is logical, but the complexity increases (dramatically) with each step. As an illustration, let us consider a direct integration of (11.60). The equation represents an integration of $\tau^{ab}/|\boldsymbol{x} - \boldsymbol{x}'|$ over the past harmonic null cone $\mathcal{C}$ associated with the field point $(t, \boldsymbol{x})$, as illustrated in Figure 11.1. This past null cone intersects the world tube of the source $\mathcal{D}$, enclosing the near zone at the three-dimensional hypersurface $\mathcal{N}$. In effect, the integral in (11.60) consists of two pieces, an integration over the hypersurface $\mathcal{N}$, and an integration over the rest of the past null cone $\mathcal{C} - \mathcal{N}$. Each requires its own computational strategy.

As an alternative, we could try a strict post-Minkowski expansion with the deviation from flat space increasing at each order. This would be a direct extension of the strategy we used in Chapter 5. But this leads to other technical issues. In particular, treating the bodies in a binary as point masses causes trouble. General relativity abhors singularities and if we try to construct an iterative procedure to solve (11.60) we need to be able to handle powers of singularities. We need to work out integrals that diverge at the location of each point particle. This is a nasty mathematical problem, but luckily it is familiar from quantum electrodynamics. The resolution is the same as in that setting; we need to invoke some kind of regularization procedure (Blanchet, 2006). However, this introduces regularization parameters that remain undetermined by the scheme.

Given the immense importance of the problem for gravitational-wave astronomy, a considerable effort has been invested in high-order post-Newtonian schemes. After decades of work by several groups,[2] the calculations reached the 3.5pN order—the level expected to be sufficient for sensitive gravitational-wave searches.

---

[2] See, for example, Schäfer (1985), Iyer and Will (1995), Jaranowski and Schäfer (1997), Pati and Will (2000), Pati and Will (2002), and Nissanke and Blanchet (2005).

11.5 Inspiralling binaries

Having outlined the post-Newtonian strategy, let us bypass the (sometimes gory) details and translate the results into a form that relates to the 'Newtonian' results from Chapter 5. To do this, we translate the origin of the coordinates to the binary's centre of mass (and enforce the constraint that the dipole moment vanishes). As we expect most binaries to be circular when they enter the sensitivity band of a ground-based detectors we focus on that case.

In order to describe the results (see Blanchet (2006) for a deeper analysis), we first of all introduce the symmetric mass-ratio

$$\eta = \frac{\mu}{M} = \frac{M_1 M_2}{(M_1 + M_2)^2},\tag{11.63}$$

which is constrained to the range $0 \le \eta \le 1/4$. This is always a 'small' quantity and hence we can use it as a formal expansion parameter. We also use the post-Newtonian parameter

$$\gamma = \frac{GM}{ac^2} \sim \mathcal{O}\left(\frac{1}{c^2}\right),\tag{11.64}$$

which quantifies how relativistic the system is.

At 3.5pN order the relative acceleration $\ddot{\boldsymbol{x}} = \ddot{\boldsymbol{x}}_1 - \ddot{\boldsymbol{x}}_2$ of two bodies in circular orbit is then given by

$$\ddot{\boldsymbol{x}} = -\Omega^2 \boldsymbol{x} - \frac{32}{5}\frac{G^3 M^3 \eta}{c^5 a^4}\left[1 - \gamma\left(\frac{743}{336} + \frac{11}{4}\eta\right)\right]\boldsymbol{v} + \mathcal{O}\left(\frac{1}{c^8}\right),\tag{11.65}$$

where $\boldsymbol{x}$ gives the relative separation in harmonic coordinates and Ω is the usual angular frequency. The second term in the square bracket represents the effect of the radiation reaction force at this order. It is worth noting that it acts in the direction opposite to the velocity. This leads a secular decrease of the binary separation a together with an increase of the orbital frequency. We have

$$\dot{a} = -\frac{64}{5}\frac{G^3 M^3 \eta}{c^5 a^4}\left[1 - \gamma\left(\frac{1751}{336} + \frac{7}{4}\eta\right)\right],\tag{11.66}$$

$$\dot{\Omega} = \frac{95}{5}\frac{G^3 M^3 \eta}{c^5 a^4}\left[1 - \gamma\left(\frac{2591}{336} + \frac{11}{12}\eta\right)\right].\tag{11.67}$$

The equations of motion provide a relation between the frequency Ω and the separation a. At 3pN order this leads to a generalized form of Kepler's law

$$\Omega^2 = \frac{GM}{a^3}\left\{1+(-3+\eta)\gamma+\left(6+\frac{41}{4}\eta+\eta^2\right)\gamma^2\right.$$
$$\left.+\left(-10+\left[-\frac{75707}{840}+\frac{41}{64}\pi^2+22\ln\left(\frac{a}{a_0'}\right)\right]\eta+\frac{19}{2}\eta^2+\eta^3\right)\gamma^3\right\}+\mathcal{O}\left(\frac{1}{c^8}\right),$$

$$\tag{11.68}$$

where a_0' is a constant associated with the gauge. Meanwhile, the orbital energy is given by

$$E = -\frac{\mu c^2 \gamma}{2}\left\{1+\left(-\frac{7}{4}+\frac{\eta}{4}\right)\gamma+\left(-\frac{7}{8}+\frac{49}{8}\eta+\frac{\eta^2}{8}\right)\gamma^2\right.$$
$$\left.+\left(-\frac{235}{64}+\left[\frac{46031}{2240}-\frac{123}{64}\pi^2+\frac{22}{3}\ln\left(\frac{a}{a_0'}\right)\right]\eta+\frac{27}{32}\eta^2+\frac{5}{64}\eta^3\right)\gamma^3\right\}\mathcal{O}\left(\frac{1}{c^8}\right).$$

$$\tag{11.69}$$

The energy is an observable quantity and as such it should not depend on the choice of coordinate system. Yet, the gauge constant a_0' appears in the expression. To see that the numerical value of the energy is indeed the same in all coordinate systems, we can replace the post-Newtonian parameter γ, which depends on the separation a in harmonic coordinates, with a parameter related to the frequency (another observable)

$$x = \left(\frac{GM\Omega}{c^3}\right)^{2/3} \sim \mathcal{O}\left(\frac{1}{c^2}\right).$$

$$\tag{11.70}$$

This leads to

$$\gamma = x\left\{1+\left(1-\frac{\eta}{3}\right)x+\left(1-\frac{65}{12}\eta\right)x^2\right.$$
$$\left.+\left[-\frac{675}{64}+\left(\frac{34445}{576}-\frac{205}{96}\pi^2\right)\eta-\frac{155}{96}\eta^2-\frac{35}{5184}\eta^3\right]x^3+\mathcal{O}\left(\frac{1}{c^8}\right)\right\}. \tag{11.71}$$

Substituting this expression into (11.69) we find that the 'unwanted' logarithmic term has disappeared

$$E = -\frac{\mu c^2 \gamma}{2}\left\{1+\left(-\frac{3}{4}-\frac{\eta}{12}\right)x+\left(-\frac{27}{8}+\frac{19}{8}\eta-\frac{\eta^2}{24}\right)x^2\right.$$
$$\left.+\left[-\frac{675}{64}+\left(\frac{34445}{576}-\frac{205}{96}\pi^2\right)\eta-\frac{155}{96}\eta^2-\frac{35}{5184}\eta^3\right]x^3\right\}+\mathcal{O}\left(\frac{1}{c^8}\right). \tag{11.72}$$

Moreover, for circular orbits it turns out that there are no terms of order $x^{7/2}$ so this expression is actually accurate to 3.5pN order. This example highlights the importance of using gauge invariant quantities when we compare different calculations and motivates the use of x as the formal expansion parameter.

The corresponding 3.5pN result for the gravitational-wave luminosity is

$$
\begin{aligned}
\mathcal{F} = \frac{32c^5}{5G}\eta^2 x^5 \Bigg\{ &1 + \left(-\frac{1247}{336} - \frac{35}{12}\eta\right)x + 4\pi x^{3/2} \\
&+ \left(-\frac{44711}{9072} + \frac{9271}{504}\eta + \frac{65}{18}\eta^2\right)x^2 + \left(-\frac{8191}{672} - \frac{583}{24}\eta\right)\pi x^{5/2} \\
&+ \left[\frac{6643739519}{69854400} + \frac{16}{3}\pi^2 - \frac{1712}{105}\gamma_E - \frac{856}{105}\ln(16x)\right. \\
&+ \left(-\frac{134543}{7776} + \frac{41}{48}\pi^2\right)\eta - \frac{94403}{3024}\eta^2 - \frac{775}{324}\eta^3\Bigg]x^3 \\
&+ \left(-\frac{16285}{504} + \frac{214745}{1728}\eta + \frac{193385}{3024}\eta^2\right)\pi x^{7/2} + \mathcal{O}\left(\frac{1}{c^8}\right)\Bigg\}.
\end{aligned} \tag{11.73}
$$

This result provides us with the information we need to work out the all-important phase of the gravitational-wave signal. To do this we 'simply' need to solve the balance Eq. (11.61) for the energy loss. However, as we are dealing with quantities arising from an ordered expansion there are (yet again) different ways of doing this, leading to slightly different results. One option is to introduce a new (dimensionless) time variable

$$
T = \frac{\eta c^3}{5GM}(t_c - t), \tag{11.74}
$$

where t_c represents the (fiducial) time of coalescence, and then turn (11.61) into a differential equation, which can be integrated to give x as a function of the new variable T. However, this is not the result we want. To make contact with the orbital phase, we note that

$$
\frac{d\varphi}{dt} = \Omega \quad \longrightarrow \quad \frac{d\varphi}{dT} = -\frac{5x^{3/2}}{\eta}. \tag{11.75}
$$

This leads to an expression for $\varphi(T)$, which we (finally) recast in terms of x to get

$$
\begin{aligned}
\varphi = -\frac{x^{-5/2}}{32\eta}\Bigg\{ &1 + \left(\frac{3715}{1008} + \frac{55}{12}\eta\right)x - 10\pi x^{3/2} \\
&+ \left(\frac{15293365}{1016064} + \frac{27145}{1008}\eta + \frac{3085}{144}\eta^2\right)x^2 + \left(\frac{38645}{1344} - \frac{65}{16}\eta\right)\pi x^{5/2}\ln\left(\frac{x}{x_0}\right) \\
&+ \left[\frac{12348611926451}{18776862720} - \frac{160}{3}\pi^2 - \frac{1712}{21}\gamma_E - \frac{856}{21}\ln(16x)\right.
\end{aligned}
$$

$$+ \left(-\frac{15737765625}{12192768} + \frac{2255}{48}\pi^2 \right)\eta + \frac{76055}{6912}\eta^2 - \frac{127825}{5184}\eta^3 \Big] x^3$$

$$+ \left(\frac{77096675}{2032128} + \frac{378515}{12096}\eta - \frac{74045}{6048}\eta^2 \right)\pi x^{7/2} + \mathcal{O}\left(\frac{1}{c^8}\right) \Big\}, \qquad (11.76)$$

where x_0 is an integration constant, which can be associated with the initial conditions when the signal enters the detector band. This lengthy expression allows us to work out how many wave cycles each post-Newtonian order contributes as a binary system spirals through the detector bandwidth, adding to the leading order result from Chapter 5.

If we, for example, take the seismic low-frequency cut-off to be 10 Hz and assume that the inspiral lasts until the system reaches the ISCO frequency for the Schwarzschild solution, $f = c^3/(6^{2/3}\pi GM)$, then we arrive at the results in Table 11.1. These estimates show that the 3.5pN results lead to a level of accuracy where less than one accumulated cycle is lost as the signal sweeps through the detector band. This means that there would be no significant loss of signal-to-noise ratio if we were to use the post-Newtonian phase evolution in a matched filter search. We can stop calculating...

In order to describe the waveform itself, we need to account for higher-order post-Newtonian corrections to the amplitude. In addition, there will be harmonics of the orbital frequency. We do not gain much additional insight from the corresponding (rather complicated) expressions. Instead of listing them, let us turn to issues we have not yet accounted for.

We have not considered the spin of either binary partner. Spin is not expected to have a significant effect on the equations of motion for binary neutron stars. These systems

Table 11.1 *Post-Newtonian contributions to the accumulated number of gravitational-wave cycles $\mathcal{N}$ as a binary signal evolves from 10 Hz (representing the seismic cut-off of a fiducial detector) to the ISCO frequency for the Schwarzschild solution, $f = c^3/(6^{2/3}\pi GM)$. Results are shown for a canonical $1.4M_\odot$ neutron-star binary, a mixed binary with a neutron star and a typical $(10M_\odot)$ black hole and an equal-mass black-hole binary. (Data from Blanchet (2006).)*

Order	$1.4M_\odot - 1.4M_\odot$	$1.4M_\odot - 10M_\odot$	$10M_\odot - 10M_\odot$
Newtonian	15952.6	3558.9	598.8
1pN	439.5	212.4	59.1
1.5pN	−210.3	−180.9	−51.2
2pN	9.9	9.8	4.0
2.5pN	−11.7	−20.0	−7.1
3pN	2.6	2.3	2.2
3.5pN	−0.9	−1.8	−0.8

are old, as they take a considerable time to spiral together (see Table 9.2), so when they enter a ground-based detector's sensitivity window they are likely to have spun down. One can imagine different formation channels, e.g. gravitational capture in a globular cluster, where the neutron stars are not given time to spin down, but this may not be very common. The situation is different for black holes. First of all, they can spin much faster than neutron stars, so the maximal spin-effect is larger. Secondly, black holes may not spin down as efficiently as neutron stars.

Spin introduces two new effects (Kidder *et al.*, 1993*b*). The dominant effect is due to the spin-orbit coupling, which is linear in the spin. The spin–spin coupling is quadratic. If we adopt the convention that the spin enters at 0.5pN order, then the leading spin–orbit effect enters at 1.5pN and spin-spin terms are present from 2pN order. These effects modulate the amplitude, phase, and frequency of the gravitational-wave signal. In a binary system with misaligned spins, the orbital plane will precess. In essence, if we want to detect signals from spinning objects, and successfully infer the associated parameters, we need to account for these effects.

For black-hole binaries it is common to characterize the spin in terms of the dimensionless parameter $\chi_i = S_i/M_i^2$ with S_i, each black hole's angular momentum and M_i its mass (and $i = 1 - 2$ labelling the binary partners). In addition, we have the orbital angular momentum L. For a binary system the quantity

$$\chi_{\text{eff}} = \frac{(\chi_1 + q\chi_2) \cdot \hat{L}}{1 + q}, \tag{11.77}$$

where $q = M_2/M_1$ is the mass ratio (assuming $M_1 \geq M_2$), is proportional to the lowest order spin contribution to the gravitational-wave phase. Additionally, it turns out that this quantity is approximately conserved during the inspiral (Racine, 2008).

The actual nature of the bodies in the binary system should also be important. At the Newtonian level, we got away with treating the bodies as point masses. In principle, one might expect this to remain a good approximation also in general relativity. After all, the equivalence principle tells us that the binary partners will fall towards one another in the same fashion regardless of their composition. So why not replace them with point masses? Well, we have to be careful that we do not ignore phenomena associated with the finite size and internal dynamics of the bodies. We know that the gravity of the moon raises tides in the Earth's ocean. Should we be concerned about a similar effect in the case of binary neutron stars? Indeed, we should. The tidal interaction will deform a fluid body, leading to additional quadrupole moments, which may affect the gravitational-wave signal. We will consider that problem in Chapter 21.

The nature of a black hole also introduces new aspects. In particular, black holes have horizons, through which radiation may enter. This leads to energy dissipation from the exterior spacetime, and hence an additional loss of orbital energy compared to the neutron star case. Luckily, this effect is small. For non-rotating black holes the absorption enters at 4pN order, while for Kerr black holes it needs to be accounted for already at 2.5pN order (Poisson and Sasaki, 1995).

11.6   The effective one body approach

When you try to establish the relevance of any given gravitational-wave source, it makes sense to focus on simple approximations. Otherwise the required simulations or data analysis considerations simply become intractable. Of course, given the need for faithful search templates (see Chapter 8), rough estimates eventually need to be turned into precise models. This requirement has been a key driver behind moves towards high-order post-Newtonian models and full nonlinear simulations. At the end of the day, the results of these efforts need to be combined into actual search templates, and both are problematic in this respect. The post-Newtonian approach does not remain valid through the final merger and numerical simulations are too expensive to extend beyond the late stages of inspiral. Moreover, accurate parameter estimation may require millions of waveforms to be compared with the data. In reality, we can never afford to consider the entire parameter space with simulations. Pragmatically, it is natural to turn to some kind of 'hybrid' waveform model drawing on the whatever reliable results we have at hand.

A powerful approach to the problem of 'tuned' gravitational waveforms is the effective-one-body model (EOB) (Buonanno and Damour, 1999; Damour and Nagar, 2016). This is an analytical framework inspired by the classic one-body approach to the binary problem in Newtonian gravity (outlined in Chapter 5). At the heart of the model is an effective one-body Hamiltonian, the form of which is chosen to reproduce known results from the point-particle limit and higher order post-Newtonian approximations. The Hamiltonian has a number of 'calibration' functions that can be adjusted based on, for example, available numerical relativity results.

The effective-one-body model has three key ingredients:

1. a description of the conservative dynamics of two black holes,
2. an expression for the gravitational radiation reaction,
3. a description of the gravitational waves during inspiral.

The main information at each step of the process is provided by the high-order post-Newtonian expansion. However, rather than using the raw results one tends to build the approximation on re-summation (e.g. using Padé approximants; see Damour *et al.* (2001)). Each step in the process can be improved as we gain better understanding of the relevant physics. Recent efforts has paid particular attention to tidal effects in neutron star binaries (see Hinderer *et al.* (2016) and Steinhoff *et al.* (2016), as well as Chapter 21). One can argue that one can extend the resummed model to obtain a sufficiently accurate description of the entire waveform, from inspiral through remnant ringdown, including the nonlinear plunge and merger phases. Such waveforms provide a practical tool for data analysis (Buonanno *et al.*, 2009*a*). However, for obvious reasons, they need to be calibrated against (for example) the output from numerical simulations (Buonanno *et al.*, 2009*b*).

The effective one body approach builds on a one-to-one map between the dynamics of a real binary (in its centre of mass system) and an effective body with mass μ (the reduced mass) moving in some effective metric. Taking this metric to be spherically symmetric, we have (as in the derivation of the Schwarzschild solution in Chapter 4)

$$ds^2 = -A(R)dT^2 + B(R)dR^2 + R^2\left(d\theta^2 + \sin^2\theta\, d\varphi^2\right),\tag{11.78}$$

where the coefficients—which represent a $\hat{\mu} = \mu/M$ deformation of the Schwarzschild metric—are obtained from the (resummed) post-Newtonian results.

In order to describe how radiation reaction is incorporated in the prescription, let us focus on circular binaries. For such systems, it is sufficient to add a radiation reaction force to the p_φ equation of motion (see Chapter 10). To represent the orbital motion, it it natural to use phase space variables R, φ, P_r, and P_φ associated with polar coordinates (in the equatorial plane). In fact, it is practical to replace the radial momentum P_r by the momentum conjugate to the 'tortoise' coordinate (see Chapter 10)

$$R_* = \int (B/A)^{1/2}dR,\tag{11.79}$$

that is

$$P_{r_*} = (A/B)^{1/2}P_r.\tag{11.80}$$

The relevant Hamiltonian is obtained by solving for the energy of the system and re-expressing the result in the preferred variables. This leads to (see Damour and Nagar (2016))

$$\hat{H}_{\text{EOB}}(r, p_{r_*}, \varphi) = \left[1 + 2\hat{\mu}\left(\hat{H}_{\text{eff}} - 1\right)\right]^{1/2},\tag{11.81}$$

with

$$\hat{H}_{\text{eff}} = \left\{ p_{r_*}^2 + A(r)\left[1 + \frac{p_\varphi^2}{r^2} + 2\hat{\mu}\,(4 - 3\hat{\mu})\frac{p_{r_*}^4}{r^2}\right]\right\}^{1/2}.\tag{11.82}$$

In these expressions we have used dimensionless variables $r = R/GM$, $p_{r_*} = P_{R_*}/\mu$, $p_\varphi = P_\varphi/\mu\,GM$, and a rescaled time $t = T/GM$. This then leads to equations of motion of the form

$$\frac{dr}{dt} = \left(\frac{A}{B}\right)^{1/2}\frac{\partial \hat{H}_{\text{EOB}}}{\partial p_{r_*}},$$

$$\frac{dp_{r_*}}{dt} = -\left(\frac{A}{B}\right)^{1/2}\frac{\partial \hat{H}_{\text{EOB}}}{\partial r},$$

$$\Omega \equiv \frac{d\varphi}{dt} = \frac{\partial \hat{H}_{\text{EOB}}}{\partial p_\varphi},$$

$$\frac{dp_\varphi}{dt} = \hat{\mathcal{F}}_\varphi.\tag{11.83}$$

The last of these equations encodes the fact that the system must lose angular momentum as it emits gravitational waves. In order to complete the model, one must construct the resummed version, $\hat{\mathcal{F}}_{\varphi}$, of the known post-Newtonian gravitational-wave flux.

A clear advantage of working with the resummed post-Newtonian results is that the series is likely to converge more rapidly to the true waveform. In fact, effective-one-body waveforms have been shown to accurately reproduce numerical results for both binary black holes and neutron stars (Buonanno *et al.*, 2009*b*; Hinderer *et al.*, 2016). In effect, these waveforms, which are much cheaper to generate, should be accurate enough for both detection and parameter extraction. The procedure necessarily involves some level of parameter fitting to remove the dependence on unknown post-Newtonian terms, but once the model has been calibrated the results should be reliable. In fact, signal templates based on the effective-one-body approach form the basis for several LIGO search strategies.

12

Towards the extreme

We have already considered scenarios that make neutron stars interesting for gravitational-wave astronomy (see Chapter 6). If we want to flesh out these ideas and make our estimates more precise, we need to consider a range of fundamental physics issues. So far we ignored internal composition and dynamics. This led to useful qualitative insight, but if we want quantitative models we need to extend the discussion in two directions. First of all, we need a more realistic matter description. This is tricky—all four fundamental forces of nature (the strong, weak, electromagnetic, and gravitational) impact on the formation, composition, and evolution of a neutron star. We need to account for supranuclear physics, superfluidity/superconductivity, strong magnetic fields, and exotic particle physics. The list of ingredients is rather long. Secondly, if we want to move beyond phenomenological models we need to consider the various astrophysical scenarios in general relativity.

12.1   Matter at supranuclear densities

The problem of understanding the composition and state of matter under the extreme conditions in a neutron star presents a challenge, but we need to make progress on it. In fact, neutron stars may radiate gravitational waves through mechanisms which are, more or less directly, associated with specific physics aspects. Asymmetries in the star's crust will lead to the emission of periodic gravitational waves with frequency twice that of the star's spin (as discussed in Chapter 6). Global non-radial oscillations generate waves that carry a fingerprint which depends directly on, for example, the interior speed of sound (in the case of the pressure p-modes; see Chapter 13) and the star's rotation (in the case of the inertial r-modes; see Chapter 15). The reward associated with a successful detection of these gravitational waves would be significant since it would allow us to test physics that cannot be probed in terrestrial laboratories.

Let us start by taking a cursory peek beneath a neutron star's surface. In many ways, the star can be thought of as a layer cake, with a series of distinct regions from the surface to the deep core see Figure 12.1. As one progresses towards the centre, the density increases monotonically and our understanding of the physics becomes less certain. A cross section of a mature isolated neutron star would unveil an atmosphere, which

Gravitational-Wave Astronomy: Exploring the Dark Side of the Universe. Nils Andersson, Oxford University Press (2020).
© Nils Andersson. DOI: 10.1093/oso/9780198568032.001.0001

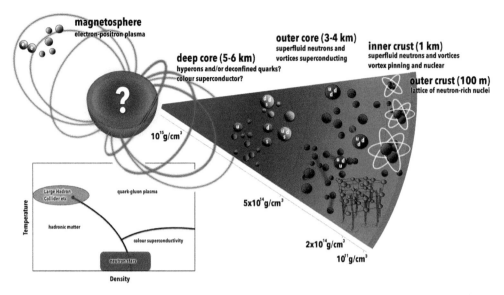

Figure 12.1 *Inset: Sketch of the QCD phase diagram. While colliders like the Large Hadron Collider (LHC) at CERN and the Relativistic Heavy Ion Collider (RHIC) at Brookhaven probe matter at extremely high temperatures but relatively low densities, neutron star physics relies on the complementary low-temperature, high-density regime for highly asymmetric matter; a regime that can never be tested by terrestrial experiments. Main image: Schematic of a neutron star interior. The outer layers consist of an elastic lattice of neutron-rich nuclei. Beyond neutron drip free neutrons form a superfluid that coexists with the lattice. Above about half the nuclear saturation density, the protons form a liquid and should be in a superconducting state. The composition and state of matter in the deep core are not well constrained. The stabilizing effect of gravitational confinement permits (slow) weak interactions (such as electron captures) to reach equilibrium, generating matter that is neutron-rich and may have net strangeness. Hyperons are likely to be present, ultimately giving way to deconfined quarks and a possible colour superconductor. Processes in the pair plasma in the star's magnetosphere give rise to the lighthouse effect observed by radio telescopes. X-rays associated with explosive nuclear burning on—and thermal emission from—the star's surface can be used to infer the properties of the outer layers as well as the internal temperature, which is governed by a range of neutrino processes. Hotspots on the surface of the star can lead to observable pulsations. Large-scale fluid motion in the dense interior may generate detectable gravitational waves. (Reproduced from Graber et al. (2017))*

could be a few meters thick, followed by a 'solid' crust composed of nuclei, neutrons, and electrons. At a density of about 10^{11} g/cm^3 the neutrons begin to drip out of the nuclei to form a superfluid which permeates the crust lattice. The properties of the crust regulate the thermal emission from the star and are also believed to be key to observed pulsar spin glitches (Espinoza *et al.*, 2011*a*). The crust extends to about two-thirds of the nuclear saturation density. If we take the saturation density to be $\rho_0 = 2.8 \times 10^{14}$ g/cm^3 (corresponding to a baryon number density of $n_0 = 0.17$ fm^{-3}), then the crust–core transition takes place at about 1.7×10^{14} g/cm^3. The interior of the star is primarily a neutron fluid, with a small contamination of protons, electrons, and muons. The neutrons are (again) expected to be in a superfluid state, while the protons form a superconductor. At densities beyond a few times nuclear density various exotic states of matter, ranging from a kaon condensate to hyperons and deconfined quarks may be present. In fact, these phases could dominate the deep core as the density of the star may reach about 10^{15} g/cm^3.

12.2   A simple model for npe matter

Inspired by the Fermi gas results from Chapter 9, let us try to build a simple equation of state for a star composed of neutrons with a small fraction of protons and electrons. In doing this, we first of all treat the various species of particles (n, p, and e) as non-interacting Fermi gases, and secondly assume that the baryons are non-relativistic while the electrons are relativistic. This makes sense since the electrons are much lighter than the neutrons and protons. The respective chemical potentials are then given by

$$\mu_x = m_x c^2 + E_{Fx} \qquad x = n, p, \tag{12.1}$$

where m_x are the rest masses, and

$$\mu_e = E_{Fe} = \hbar c (3\pi^2 n_e)^{1/3}. \tag{12.2}$$

We have kept the natural constants in these expressions but in the following we will work in units where $\hbar = c = 1$. As the chemical potentials are given by the partial derivatives of the energy with respect to the individual number densities, it is easy to see that we will have

$$\varepsilon = (m_n n_n + m_p n_p)c^2 + \frac{3}{5}(n_n E_{Fn} + n_p E_{Fp}) + \frac{3}{4} n_e E_{Fe}. \tag{12.3}$$

At each density, the composition is determined by the requirement that the matter is in chemical equilibrium. In the present case we need to consider the Urca reactions

$$n \rightarrow p + e + \nu, \tag{12.4}$$
$$p + e \rightarrow n + \nu. \tag{12.5}$$

In order for these reactions to be in equilibrium—with an equal number of neutrons being destroyed by the first reaction as the number being created through the second—the chemical potentials must satisfy

$$\mu_n = \mu_p + \mu_e, \tag{12.6}$$

where we have assumed that the star is cold enough that the matter is transparent to neutrinos (i.e. that their contribution can be neglected). This should be a valid approximation for all observed neutron stars since the neutrinos only remain trapped for at most a few tens of seconds in a hot newly born star (Burrows and Lattimer, 1986).

In addition, we need to ensure that the matter is charge neutral. If the star were to have a significant net charge it would simply attract particles from its surroundings until charge neutrality was restored. Hence, we impose the (local) condition

$$n_p = n_e. \tag{12.7}$$

Now taking the neutron and proton masses to be equal $m_n = m_p = m_b$ and assuming that $n_p \ll n_n$ (as expected in a neutron star core) one easily arrives at an approximation for the proton fraction

$$x_p = \frac{n_p}{n} \approx 3\pi^2 \left(\frac{\hbar}{2m_b c} \right)^3 n \approx 6 \times 10^{-3} \left(\frac{n}{n_0} \right), \tag{12.8}$$

where $n = n_n + n_p$ is the baryon number density and n_0 is the nuclear saturation density (from before). We see that, in this case, the protons make up about 1% of the baryons at nuclear saturation. More realistic models tend to have slightly larger proton fractions (by a factor of a few) but it is always the case that the neutrons are vastly dominant.

Meanwhile, the pressure and mass density follow from

$$p = -\varepsilon + \sum_x n_x \mu_x = \frac{2}{5} (n_n E_{Fn} + n_p E_{Fp}) + \frac{1}{4} n_e E_{Fe}, \tag{12.9}$$

and

$$\rho \approx m_b n. \tag{12.10}$$

This provides us with the input physics we need in order to build a star. The model may not be particularly realistic, but it achieves two things. It gives us a rough idea of what to expect and also introduces the key building blocks of a more detailed description. In order to improve the model, we must account for particle interactions—the effect that the presence of the protons has on the neutrons, and so on. This is far from straightforward since it requires knowledge of the nuclear N-body interactions at extreme densities. To some extent we can parameterize the effect by introducing 'effective' masses m_x^* for the baryons (replacing the bare masses in the Fermi energies). This would (at least) allow us to account for medium effects in a phenomenological way. In a mixture of Fermi gases, the strong interaction tends to lead to the effective mass being somewhat lower than the bare mass. In addition, the effective mass depends on how many companions of the 'opposite kind' each baryon has. The larger the number, the smaller the effective mass tends to be. As a consequence, the effective proton mass m_p^* will be smaller than the effective neutron mass m_n^*. In a typical neutron star core, one finds that $m_p^* \approx 0.5 - 0.8$

(Baldo *et al.*, 1992). If we, in view of this expectation, reconsider (12.8) we see that medium effects may significantly increase the proton fraction.

12.3   Determining the equation of state

The Fermi gas model highlights the equation of state as the fundamental diagnostic of dense matter interactions. Each theoretical model generates a unique mass-radius relation—through the Tolman–Oppenheimer–Volkoff equations from Chapter 4—which, in turn, predicts a characteristic radius for a range of masses and a maximum mass above which a neutron star collapses to a black hole. It also allows us to work out quantities like the maximum spin rate and the moment of inertia. However, first principle calculations of the interactions for many-body QCD systems are not within reach (due to what is known as the fermion sign problem). Instead one has to resort to phenomenological models, using experiments and observations to test predictions as new models become available (Watts *et al.*, 2016).

Two-body interactions are fairly well constrained by experiment, but the three-body forces still represent the frontier of nuclear physics (Epelbaum *et al.*, 2009; Gandolfi *et al.*, 2012). At low energies, effective field theories based on symmetries from QCD provide a systematic expansion of the nuclear forces, which predict two- and many-nucleon interactions (Hebeler *et al.*, 2010), but it is difficult to extend these calculations to high densities. There are complementary efforts using lattice approaches to the nuclear forces to provide few-body nucleon–nucleon and more generally baryon–baryon interactions, but this approach also remains affected by uncertainties. Predictions of these models can be tested against current nuclear data (like observed particle masses; see Chamel *et al.* (2011)). Where data are not yet available, predictions can be based on the consistency of the approach. As an alternative, one may consider phenomenological models based on as much experimental information as possible (Hebeler *et al.*, 2013).

Exotic neutron-rich nuclei, the target of present and upcoming experiments, provide relevant constraints on effective interactions for many-body systems. Nuclear masses and their charge radii probe symmetric nuclear matter, the neutron skin thickness of lead tests neutron-rich matter (Roca-Maza *et al.*, 2011), while giant dipole resonances and dipole polarizabilities of nuclei also concern largely symmetric matter (Piekarewicz *et al.*, 2012). However, these laboratory experiments probe only matter at densities lower than n_0. Neutron stars can reach densities several times higher. Similarly, heavy-ion collisions probe hot and dense matter, but have uncontrolled extrapolations to the zero-temperature regime relevant for a mature neutron star.

In essence, neutron stars provide a unique environment for testing nuclear physics at high levels of asymmetry, extreme density, and low temperature. Astrophysical constraints on the equation of state can be used to infer crucial aspects, like the nature of the three-nucleon interaction or the presence of free quarks at high densities.

The state of matter adds dimensions to the problem. Mature neutron stars tend to be cold. With a 'typical' core temperature of order 10^8 K they are far below the Fermi temperature of the involved constituents, e.g.

$$T_F = \frac{E_F}{k_B} \approx 10^{12} \left(\frac{n}{n_0}\right)^{2/3} \text{K},$$ (12.11)

in the case of the neutrons. This makes the formation of superfluid/superconducting phases likely throughout the star's core. The involved parameters (e.g. the energy gaps for Cooper pair formation of different constituents (Graber *et al.*, 2017) influence the long-term evolution of the system.

State-of-the-art astrophysical models tend to be based on phenomenological equations of state. Different approaches include nuclear potentials (e.g. the Urbana/Illinois or Argonne forces) that fit two-body scattering data and light nuclei properties (Akmal *et al.*, 1998); phenomenological forces like the Skyrme interaction (Douchin *et al.*, 2000); and microscopic nuclear Hamiltonians that include two- and three-body forces from chiral effective field theory calculations (Hebeler *et al.*, 2013). The challenge for future efforts in this area is to (i) incorporate as much of the predicted microphysics as possible, and (ii) use observations to further constrain the unknown aspects.

Given the uncertainties associated with the nucleon interactions and the difficulties to build an equation of state from first principles, it is helpful to consider the problem in a schematic fashion. As an illustration, consider a Taylor expansion away from symmetric matter (for which $x_p = 1/2$)

$$E(n, x_p) \approx E\left(n, x_p = \frac{1}{2}\right) + \underbrace{\frac{1}{8}\frac{d^2 E}{dx_p^2}}_{=S(n)}(1 - 2x_p)^2 + \ldots .$$ (12.12)

Any realistic model must reproduce the fact that the energy per particle at the saturation density $(n = n_0)$ for symmetric nuclear matter is about -16MeV. Moreover, the coefficient $S(n)$—the 'symmetry energy'—is known to be about 30MeV at nuclear saturation.

As a demonstration of the importance of the symmetry energy (Lattimer, 2014), let us take the total energy (density) to be $\varepsilon(n, x_p)$ so that $E = \varepsilon/n$. Then it is easy to show that

$$\mu_n = \frac{\partial \varepsilon}{\partial n_n}\bigg|_{n_p} = E + n\frac{\partial E}{\partial n} - x_p\frac{\partial E}{\partial x_p},$$ (12.13)

and

$$\mu_p = \frac{\partial \varepsilon}{\partial n_p}\bigg|_{n_n} = E + n\frac{\partial E}{\partial n} + (1 - x_p)\frac{\partial E}{\partial x_p} = \mu_n + \frac{\partial E}{\partial x_p}.$$ (12.14)

From this we see that chemical equilibrium is sensitive to the symmetry energy, since

$$\mu_n - \mu_p = -\frac{\partial E}{\partial x_p} = 4S(1 - 2x_p). \tag{12.15}$$

Adding the lepton energy to (12.12), assuming that the electrons are relativistic (as before), we find that chemical equilibrium requires

$$\mu_e = \hbar c (3\pi^2 n_p)^{1/2} = 4S(1 - 2x_p), \tag{12.16}$$

where we have assumed that the matter is charge neutral. The proton fraction now follows from the solution to the cubic

$$b x_p - (1 - 2x_p)^3 = 0, \tag{12.17}$$

where

$$b = 3\pi^2 n \left(\frac{\hbar c}{4S}\right)^3 \approx 22 \left(\frac{n}{n_0}\right) \left(\frac{30 \text{ MeV}}{S}\right)^3, \tag{12.18}$$

and it follows that $x_p \approx 1/b \approx 4 \times 10^{-2}$ at the nuclear saturation density. The exercise shows that x_p is (essentially) governed by the energy dependence of the symmetry energy.

The symmetry energy has further impact on the composition. For example, when $\mu_e > m_\mu c^2 \approx 105$ MeV (the muon rest mass) it becomes energetically favourable to convert electrons into muons via the reaction

$$e \longrightarrow \mu + \bar{\nu}_\mu + \nu_e. \tag{12.19}$$

From the above results we see that, for small x_p, the threshold corresponds to

$$\mu_e \approx m_\mu c^2 \approx 4S(n) \longrightarrow S(n) \approx \frac{m_\mu c^2}{4}. \tag{12.20}$$

Since the symmetry energy is expected to be about 30 MeV at the saturation density, we learn that muons are likely to appear close to n_0.

The symmetry energy represents the difference between the energies of pure neutron matter and symmetric nuclear matter as a function of density. In (12.12) we truncated the expansion. We can extend the discussion by defining

$$S(n) = E(n, 1/2) - E(n, 0), \tag{12.21}$$

and then expand the result in powers of density near saturation. This way we arrive at (Lattimer, 2014)

$$S(n) \approx S_v(n) + \frac{L}{3}\left(\frac{n - n_0}{n_0}\right) + \frac{K_{\text{sym}}}{18}\left(\frac{n - n_0}{n_0}\right)^2. \tag{12.22}$$

This expression is useful because the two parameters S_v and L can be extracted from a range of nuclear physics experiments (Hebeler *et al.*, 2013). The third parameter, K_{sym}, is much less precisely known.

From rather general arguments we have learned that, at densities below n_0, neutron star matter is dominated by neutrons with a small fraction of protons and electrons (to ensure charge neutrality). Close to nuclear density it becomes favourable for electrons to be converted into muons. We have also seen that the symmetry energy plays a key role in determining the matter composition. As we will now discover, additional degrees of freedom may become relevant at higher densities and as a consequence matter may become even stranger.

Mean-field calculations based on effective Lagrangians (Glendenning, 1996) indicate that hyperons and/or deconfined quarks are likely to appear once the density in a neutron star core reaches a few times the nuclear saturation density. The introduction of these exotic phases of matter considerably enhances the difficulties of determining an accurate representation for the equation of state, as very little is known about the relevant interactions from experiments.

Several possible phases, which may all be present in a neutron star core, endow the matter with net strangeness. We may allow for the presence of hyperons, a Bose condensate of K^- mesons, or deconfined strange quarks (Witten, 1984; Alcock *et al.*, 1986). It is generally thought that such additional states should be present once the density exceeds $\sim 2n_0$. The Λ and Σ^- hyperons may be particularly important. Large hyperon populations are likely to be present once the density approaches that which is typical of heavier neutron stars. The hyperons may, in fact, make up as much as 15–20% of the total baryon population in the star's core.

The presence of hyperons may have significant observable effects. First of all, the Σ^- carries negative charge, which means that the lepton fractions drop following its appearance. In some proposed models there are virtually no electrons left in the core of the star. This affects the different conductivities as the electrons are efficient carriers of heat and electric charge (and impact on the star's gradual magneto-thermal evolution). Secondly, the hyperons can act as an efficient refrigerant. The hyperons may undergo direct Urca reactions essentially as soon as they appear, in contrast to the protons which must exceed a threshold value of $x_p \sim 0.1$ (Lattimer *et al.*, 1991). The upshot of this is that a neutron star with a sizeable hyperon fraction should cool extremely fast. In fact, the star would cool so rapidly that it would be colder than observational data suggests (Page *et al.*, 2004). This could be taken as evidence against a hyperon core, but more likely it suggests that the hyperons are (at least partly) superfluid, in which case the reactions are suppressed (Page *et al.*, 2006).

Beyond some threshold density a neutron star core may contain deconfined quarks. In the quark phase, the difference between the quark masses is significantly smaller than their respective Fermi energies, so the equilibrium composition should involve equal fractions of the three flavours (up, down, and strange), with a strangeness fraction per baryon of almost unity.

In fact, one can argue that if the most stable form of matter at supranuclear densities is a conglomerate of deconfined quarks, then all observed neutron stars ought to be strange stars (Caldwell and Friedman, 1991). If it (more modestly) could be established that strange stars exist in the Universe, it would put important constraints on the QCD parameters, like the so-called bag constant in perturbative QCD. This is a phenomenological parameter, B, in terms of which the equation of state can be written

$$p \approx \frac{1}{3}(\varepsilon - 4B).$$
(12.23)

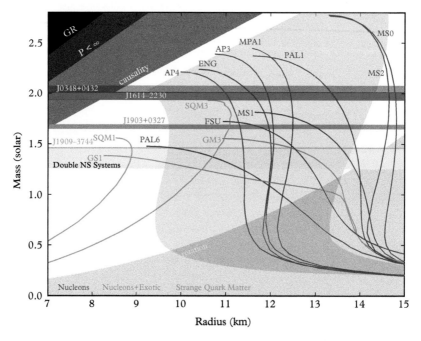

Figure 12.2 *Mass vs radius curves for a sample of realistic neutron star equations of state, compared to constraints from observations. The most severe constraint is set by the most massive known pulsars, PSR J0348+0432 and J1614-2230, which require the equation of state to allow a mass of at least $2M_\odot$. The figure also illustrates the difference between hadronic matter models and strange quark stars (SQM), which is most pronounced for lighter stars. The light red region indicates the expected uncertainty associated with extrapolations to high density from current first principles chiral effective field theory calculations (Gandolfi et al., 2012). Note that some of the included equations of state are ruled out by these constraints. (Adapted from Demorest et al. (2010).)*

The bag constant is usually taken to be in the range 100–250 MeV/fm^{-3}. Note that, in this model, the surface of the star is associated with a finite density. Modern representations of the model include corrections due to, for example, the mass of the strange quark (Alford *et al.*, 2005).

It is obviously important to establish to what extent observations can distinguish a strange star from a neutron star. Unfortunately, this is a delicate problem. Strange stars are held together by both the strong nuclear interaction and gravity, and the corresponding equation of state is quite accurately described by uniform density models. A consequence of this is that, in contrast to the neutron star case, very small strange 'dwarfs' (with mass $M \sim R^3$) can form. However, for the canonical mass of $1.4M_\odot$ gravity dominates the strong interaction, leading to strange stars and neutron stars being similar in size (see Figure 12.2). In other words, it is not clear that one would be able to distinguish between the two cases given observed masses and radii.

12.4   Observational constraints

Turning to the implications of observations for the nuclear physics, one would in the first instance draw on observed masses (which can be obtained if the neutron star has a binary companion; see Chapter 9) and constraints on the radius (a more problematic issue). In terms of mass the current record holders are PSR J0348+0432 at $2.01 \pm 0.04M_\odot$ (Demorest *et al.*, 2010) and (in close second place) PSR J1614-2230 at $1.97 \pm 0.04M_\odot$ (Antoniadis *et al.*, 2013) (although an improved measurement using the NANOGrav nine-year data set leads to a slightly lower preferred mass of around $1.93M_\odot$; see Fonseca *et al.* (2016)). The second case is particularly interesting from the relativity point-of-view as the measurement involves a strong Shapiro delay signature. Knowing that the equation of state must allow a maximum mass above $2M_\odot$ helps us rule out some proposed models (especially ones that include a significant softening phase transition, as expected from the appearance of hyperons) but (as is clear from Figure 12.2) many options remain viable.

If we ignore small effects due to rotation and the star's magnetic field, the neutron star mass–radius relation should be universal. All neutron stars in the Universe are expected to lie on the same curve. From the maximum mass we know how high in the diagram the mass–radius curve has to reach. If we want a tighter constraint on the equation of state, then we need to bracket the radius, as well. This is much more difficult (obviously, because neutron stars are small and distant). Nevertheless, there has been good progress on this problem.

By measuring the X-ray flux and temperature of a system at a known distance, one can work out the radius of the emitting object. The temperature will be redshifted in general relativity (it is an energy, after all...), which means that one would measure the radius as seen at infinity:

$$R_\infty = \left(1 - \frac{2GM}{Rc^2}\right)^{1/2} R. \tag{12.24}$$

In effect, observations lead to a region in the mass–radius plane (Özel *et al.*, 2010; Steiner *et al.*, 2010). The spectrum also depends on the surface gravity in the atmosphere and the redshift.

Recent work on this problem has mainly focussed on quiescent systems in globular clusters (Steiner *et al.*, 2018). One reason for this is that the the distance is fairly well known for globular clusters, and this reduces the uncertainty of the radius constraint. Moreover, it appears that quiescent low-mass X-ray binaries in globular clusters have relatively simple spectra, dominated by thermal surface emission. A number of quiescent systems have been studied in some depth with Chandra and/or XMM-Newton, leading useful constraints on mass and radius. For example, a Bayesian analysis (Steiner *et al.*, 2018) suggests that the neutron star radius (for a $1.4M_\odot$ star) should lie in the range 10–14 km.

12.5 The slow-rotation approximation

If we want to explore real neutron stars, we need to account for rotation. This is important for a number of reasons, ranging from the simple notion that the centrifugal force deforms fluid bodies to the fact that various emission mechanisms rely on the star's spin. If neutron stars did not rotate, we would not see them as radio pulsars—and we know that some pulsars spin at an astonishing rate. The fastest known case is PSR J1748-2446ad with a frequency of 716 Hz (Hessels *et al.*, 2006), leading to the excluded region indicated in Figure 12.2. Rapidly spinning systems could be particularly interesting for gravitational-wave astronomy. After all, we know from our simple solid-body example in Chapter 6 that the gravitational-wave amplitude scales as the square of the spin frequency (everything else being equal). In essence, understanding the impact of rotation is important if we want to understand neutron stars as gravitational-wave sources.

As a first step towards exploring the effect rotation has on a star, let us outline the slow-rotation approximation in Newtonian gravity. The aim is to obtain the equilibrium shape of a rotating configuration. There are different approaches to this problem. The route we will follow is useful because it can be extended to consider magnetically induced deformations.

As long as we are considering a fluid body, the equilibrium equations are the same as before,

$$\frac{1}{\rho}\nabla p = -\nabla \Phi, \tag{12.25}$$

although we now have an effective potential

$$\Phi = \Phi_0 - \frac{1}{2}\Omega^2 r^2 \sin^2\theta, \tag{12.26}$$

where Φ_0 is the gravitational potential and the second term accounts for the centrifugal force.

The first step involves introducing a new radial variable a, such that the level surfaces of the system take the form

$$a(r) = r\left[1 + \sum_l \epsilon_l(r)P_l(\cos\theta)\right], \tag{12.27}$$

where P_l are the Legendre polynomials. Perturbed quantities, like pressure and density, are assumed to be expanded in the same way. As we want to work out how rotation deforms a given star, we insist that the deformation conserves the total mass. Generally, we then require that

$$M_\epsilon = \int_0^R \rho\left(1 + \epsilon_l P_l\right) dV = M. \tag{12.28}$$

Meanwhile, in the case of quadrupole deformations ($l = 2$), the (r and θ) components of the perturbed Euler equations lead to

$$\frac{d\delta p}{dr} + \rho\frac{d\delta\Phi}{dr} + \delta\rho\frac{d\Phi}{dr} = -\frac{2}{3}\rho\Omega^2 r, \tag{12.29}$$

and

$$\delta p + \rho\delta\Phi + \delta\rho\Phi = -\frac{1}{3}\rho\Omega^2 r^2. \tag{12.30}$$

Combining these we have

$$\delta\rho = -\frac{\pi^2}{4\pi GR^2}\left(\delta\Phi + \frac{\Omega^2 r^2}{3}\right). \tag{12.31}$$

Substituting this into Poisson's equations for the gravitational potential, we arrive at

$$\frac{d^2\delta\Phi}{dr^2} + \frac{2}{r}\frac{d\delta\Phi}{dr} - \left(\frac{6}{r^2} + \frac{\pi^2}{R^2}\right)\delta\Phi = -\frac{\pi^2}{R^2}\frac{\Omega^2 r^2}{3}. \tag{12.32}$$

We need to solve this equation, imposing regularity at the centre and matching to the exterior solution

$$\delta\Phi = \frac{A}{r^3}, \tag{12.33}$$

with A some constant. Once we have the solution, we can determine $\delta\rho$ and infer the surface deformation from

$$\epsilon_s = \frac{1}{R}\left[\frac{d\rho}{dr}\right]_R^{-1}\delta\rho(R).$$ (12.34)

For an $n = 1$ polytrope this leads to

$$\epsilon(R) = -\frac{5}{\pi^2}\frac{\Omega^2 R^3}{GM}.$$ (12.35)

As expected, the rotational deformation makes the star oblate.

We can use the same method to obtain the rotational deformation for a uniform density star, but this calculation is a little bit more involved. We need to keep in mind that $\delta\rho$ then has support only between R and the new radius $R(1 + \epsilon_s)$ and we also need to account for the discontinuity in the gravitational potential at the surface. Once this is done, we arrive at (Tassoul, 1978)

$$\epsilon(R) = -\frac{5}{6}\frac{\Omega^2 R^3}{GM}.$$ (12.36)

These may only be rough estimates but they are nevertheless very useful. In particular, we can easily compare the rotational effect to other possible deformations of the star. While doing this in the context of gravitational waves, it is important to stress that the centrifugally induced change in shape does not lead to the kind of asymmetry we need. It is easy to see that the star remains symmetric, no matter which angle we view it from. A rotating star does not emit gravitational waves unless we introduce some additional deformation.

12.6   The virial theorem

Without discussing specific models of rapidly rotating stars—which would require a numerical solution—we can establish a useful relationship between the different energies of the system. The starting point for the analysis is the Euler equation (4.59) and the Poisson equation (4.36). Using the relevant Green's function, we find that the gravitational potential follows from

$$\Phi(\boldsymbol{x}) = -G\int\frac{\rho(\boldsymbol{x}')}{|\boldsymbol{x} - \boldsymbol{x}'|}dV'.$$ (12.37)

Let us now multiply (4.59) by ρx^i and integrate over the entire star. This leads to three distinct contributions. First of all, we have

$$\int \rho x^i D_t v_i dV = \int \rho [D_t(x^i v_i) - v^2] dV = \frac{d}{dt} \int \rho x^i v_i dV - 2T, \qquad (12.38)$$

where V is the volume, we have used the convective derivative

$$D_t = \partial_t + v^j \nabla_j, \qquad (12.39)$$

and identified the kinetic energy

$$T = \frac{1}{2} \int \rho v^2 dV. \qquad (12.40)$$

We have also used the fact that

$$\int (D_t A) dV = \frac{d}{dt} \int A dV, \qquad (12.41)$$

for any scalar function A. The next term leads to

$$\int \rho x^i \left(\frac{1}{\rho} \nabla_i p \right) dV = -3 \int p dV, \qquad (12.42)$$

as long as the pressure vanishes at the surface of the star (as it should).

The gravitational potential energy, W, follows from the remaining term. With a little bit of work, we get

$$
\begin{aligned}
\int \rho x^i \nabla_i \Phi dV &= G \int_V \int_{V'} \rho(\boldsymbol{x}) \rho(\boldsymbol{x}') \frac{\boldsymbol{x} \cdot (\boldsymbol{x} - \boldsymbol{x}')}{|\boldsymbol{x} - \boldsymbol{x}'|^3} dV' dV = \\
&= \frac{G}{2} \int_V \int_{V'} \rho(\boldsymbol{x}) \rho(\boldsymbol{x}') \frac{(\boldsymbol{x} - \boldsymbol{x}') \cdot (\boldsymbol{x} - \boldsymbol{x}')}{|\boldsymbol{x} - \boldsymbol{x}'|^3} dV' dV = \\
&= \frac{1}{2} \int_V \int_{V'} \rho(\boldsymbol{x}) \frac{G\rho(\boldsymbol{x}')}{|\boldsymbol{x} - \boldsymbol{x}'|} dV' dV = -\frac{1}{2} \int \rho \Phi dV = -W. \qquad (12.43)
\end{aligned}
$$

Combining the results we see that the equations of motion imply

$$\frac{d}{dt} \int \rho x^i v_i dV = 2T + W + 3 \int p dV. \qquad (12.44)$$

Now we note that, with $v_i = D_t x_i$, the left-hand side is equal to

$$\frac{1}{2}\frac{d^2}{dt^2}\int \rho|\mathbf{x}|^2 dV = \frac{1}{2}\frac{d^2 I}{dt^2},$$

(12.45)

where I is the moment of inertia.

As a system reaches equilibrium the left-hand side of (12.44) vanishes. Hence, we must have

$$2T + W + 3\int p dV = 0.$$

(12.46)

This is known as the virial theorem. In addition to providing insight into a general equilibrium configuration, the relation serves as a useful accuracy test for numerical solutions.

At this point it is worth introducing a little bit of thermodynamics. Let us assume that the star is described by a barotropic equation of state,[1] such that $\varepsilon = \varepsilon(n)$. Then we have the first law

$$p + \varepsilon = n\frac{d\varepsilon}{dn} = \rho\frac{d\varepsilon}{d\rho},$$

(12.47)

which leads to

$$d\left(\frac{\varepsilon}{\rho}\right) = -pd\left(\frac{1}{\rho}\right).$$

(12.48)

In the case of a polytrope $p = K\rho^\Gamma$, we can integrate the relation to get

$$\frac{\varepsilon}{\rho} = c^2 + \int \frac{p}{\rho^2}d\rho = c^2 + K\frac{\rho^{\Gamma-1}}{\Gamma - 1},$$

(12.49)

where the integration constant has been appropriately chosen. We see that

$$\varepsilon = \rho c^2 + \frac{p}{\Gamma - 1} = \rho c^2 + \varepsilon',$$

(12.50)

and the internal energy of the star readily follows as

$$U = \int \varepsilon' dV = \frac{1}{\Gamma - 1}\int p dV.$$

(12.51)

[1] Realistic neutron star matter can be considered to be barotropic as long as it is charge neutral and in chemical equilibrium.

Given this result, the virial theorem can be written

$$2T + W + 3(\Gamma - 1)U = 0. \tag{12.52}$$

This is an important relation. For example, consider the case of a non-rotating star ($T = 0$). In that case the virial theorem provides a relation between potential and internal energy, which allows us to immediately write down the total energy of the star. We get

$$E_{\text{total}} = U + W = -\frac{3\Gamma - 4}{3(\Gamma - 1)} |W|. \tag{12.53}$$

Since the total energy must be negative in order for the star to remain bound, we learn that we must have $\Gamma > 4/3$. If the adiabatic index is smaller than this, then a small perturbation will lead to disruption.

Before moving on, let us take a closer look at the gravitational potential energy. From the definition (12.43), we have

$$W = \frac{1}{2} \int \rho \Phi dV = -\frac{G}{2} \int \rho(\mathbf{x}) \int \frac{\rho(\mathbf{x}')}{|\mathbf{x} - \mathbf{x}'|} dV' dV. \tag{12.54}$$

In order to evaluate this expression, we first expand the integrand in spherical harmonics. Using $\mu = \cos\theta$ we then get

$$\Phi(\mathbf{x}') = -2\pi G \sum_{l=0}^{\infty} Y_l^m(\mu') \int_{-1}^{1} Y_l^m(\mu) \left[\int_0^{r'} \rho \left(\frac{r}{r'}\right)^{l+1} r dr + \int_{r'}^{R} \rho \left(\frac{r'}{r}\right)^{l} r dr \right] d\mu. \tag{12.55}$$

It is instructive to consider the case of a spherical uniform density star, since we then only need to consider the $l = 0$ term in the sum. With $y = r/R$ we have

$$\Phi = -4\pi G R^2 \rho \left[\frac{1}{y'} \int_0^{y'} y^2 dy + \int_{y'}^{1} y dy \right]. \tag{12.56}$$

Using this in (12.54) to determine the gravitational potential energy we find that the two terms lead to identical contributions, and we have the final result

$$W = -\frac{3}{5} \frac{GM^2}{R}. \tag{12.57}$$

The formalism provided by (12.55) is, of course, much more general. For example, we can show that the centrifugal deformation of a rotating uniform density star affects the gravitational potential energy in such a way that (Tassoul, 1978)

$$\Delta W = \frac{3\epsilon^2 GM^2}{25R} \longrightarrow \frac{\Delta W}{W} = -\frac{\epsilon^2}{5}. \tag{12.58}$$

Given that ϵ is proportional to Ω^2 we learn that the change in potential energy induced by the rotation is proportional to Ω^4. This observation will be useful later, when we discuss deformations of the neutron star's crust (see Chapter 14).

12.7   The Kepler limit

We have already seen that neutron stars can spin rapidly. There is, however, a natural speed limit. If the star rotates so fast that gravity does not overcome the centrifugal force on a particle at the equator, the star will shed mass. This is usually referred to as the Kepler limit. It corresponds to the rotation frequency of the star being equal to the frequency of a particle in orbit around the equator. If the star remained undeformed by the rotation, the limiting frequency would trivially follow from $\Omega^2 = GM/R^3$. Of course, a real star has a centrifugal bulge, so the true Kepler limit corresponds to a lower rotation rate.

It is straightforward to estimate the mass-shedding limit from the rotational deformations we have derived. However, as it is instructive to have different ways of looking at a problem, let us take an alternative route. We will adopt the so-called Roche approximation (Lai et al., 1993), which is known to be accurate for stars governed by a soft equation of state. The idea is that the gravitational potential can be taken as spherical also in the rotating case. Gravity is assumed to be dominated by the stellar core while the relatively low-density envelope is deformed by rotation.

A uniformly rotating model is generally governed by

$$\frac{1}{\rho}\nabla_i p = -\nabla_i \left(\Phi_0 - \frac{1}{2}\Omega^2 r^2 \sin^2\theta \right), \tag{12.59}$$

where we now assume that

$$\Phi_0 \approx -\frac{GM}{r}. \tag{12.60}$$

Introducing the enthalpy

$$\nabla_i h = \frac{1}{\rho}\nabla_i p, \tag{12.61}$$

we have

$$h + \Phi - \frac{1}{2}\Omega^2 r^2 \sin^2\theta = \text{constant} = H. \tag{12.62}$$

The integration constant can be determined by evaluating the expression on the left-hand side at the pole of the star, where $r = R_p$, using the fact that $h = 0$ at the surface. Thus, we get

$$H = -\frac{GM}{R_p}. \tag{12.63}$$

Given that the frequency of a particle in a Keplerian orbit around the equator, with $r = R_e$, follows from

$$\Omega_K^2 = \frac{GM}{R_e^3}, \tag{12.64}$$

we find from (12.62) that we must have

$$-\frac{3}{2}\frac{GM}{R_e} = -\frac{GM}{R_p} \longrightarrow \frac{R_e}{R_p} = \frac{3}{2}. \tag{12.65}$$

If we also assume that the polar radius remains roughly unchanged by the stellar rotation, $R_p \approx R(\Omega = 0) = R_0$ we deduce that the maximum rotation rate of the star is approximated by

$$\Omega_K \approx \left(\frac{2}{3}\right)^{3/2}\sqrt{\frac{GM}{R_p^3}} \approx \left(\frac{2}{3}\right)^{3/2}\sqrt{\frac{GM}{R_0^3}} \approx \frac{2}{3}\sqrt{\pi G\rho_0}, \tag{12.66}$$

where ρ_0 is the average density of the corresponding non-rotating star. That is, we have

$$\Omega_K \approx 2\pi \times 1180 \left(\frac{M}{1.4M_\odot}\right)^{1/2}\left(\frac{R}{10\text{ km}}\right)^{-3/2}\text{s}^{-1}. \tag{12.67}$$

This simple approximation has been shown to be good for rigidly rotating Newtonian bodies. In fact, as we will soon see, it remains reasonably accurate also for relativistic models.

The mass-shedding limit may, however, be significantly different for differentially rotating stars. This is natural since it is the equatorial rate of rotation that plays the key role. Whether the core of the star rotates faster or slower is of hardly any consequence. In fact, the rotation frequency Ω may not be a particularly useful parameter for differentially rotating configurations. Instead, one would often use (in particular in the context of instabilities of a rotating star; see Chapter 13) the ratio β between the kinetic energy, T, and the gravitational potential energy, W. Defining

$$\beta = \frac{T}{|W|}, \tag{12.68}$$

it follows from our previous estimates that, for uniformly rotating constant density stars we have

$$\beta \approx \frac{1}{9}\left(\frac{\Omega}{\Omega_K}\right)^2. \tag{12.69}$$

In other words, the mass-shedding limit corresponds to $\beta \approx 0.11$. This agrees reasonably well with detailed calculations for realistic supranuclear equations of state, which typically lead to a maximum value of β in the range 0.09–0.13 (Stergioulas and Friedman, 1995). However, depending on the adopted rotation law, differentially rotating models may allow much larger values of β. This will be important later.

12.8   Rotating relativistic stars

The different approximations we have outlined provide useful insights into rotating neutron stars, but ultimately we need to move on to relativistic models. This is important for two reasons. First of all, if we want to test our understanding of the microphysics against precise observations, then we need to account for general relativity. Secondly, new effects enter in the relativistic description and these may impact on the phenomenology. We have, in fact, already discussed one such aspect—the maximum mass. Neutron stars do not have a maximum mass in Newtonian gravity.

As in Newtonian theory, rotation can be accounted for in two different ways. Much work has focussed on slowly rotating stars, for which an expansion in the rotation rate may lead to fairly accurate results. This approach has the advantage that one can draw on the result for non-rotating systems and make some progress analytically (Hartle and Thorne, 1968). The dynamics of rapidly rotating stars tends to be much more complicated, and one typically has to resort to full-blown numerical solutions. We will outline both approaches in the following.

In the case of a slowly rotating relativistic star, one may consider the rotation rate Ω as a formal expansion parameter. If we take the rotation to be uniform, then we know from our discussion of the corresponding Newtonian problem that rotational effects associated with the centrifugal force arise at order Ω^2. The relativistic problem is, however, different in that a new effect—the rotational frame-dragging—appears already at linear order in Ω. At the linear level, we have

$$ds^2 = -e^\nu dt^2 + e^\lambda dr^2 - 2\omega r^2 \sin^2\theta \, dt d\varphi + r^2(d\theta^2 + \sin^2\theta \, d\varphi^2), \tag{12.70}$$

where we recognize the non-rotating metric components from the Schwarzchild solution (see Chapter 4) and $\omega = \mathcal{O}(\Omega)$. The fluid four-velocity is given by

$$u^a = [e^{-\nu/2}, 0, 0, \Omega e^{-\nu/2}], \tag{12.71}$$

where $\Omega = d\varphi/dt$ is the angular velocity of the fluid as seen by an observer at rest at infinity. The Einstein equations for a stationary configuration now lead to the familiar Tolman–Oppenheimer–Volkoff equations (at non-rotating order) supplemented by (at first order in rotation) an equation for the frame dragging

$$\varpi' - \left[4\pi (p + \varepsilon) r e^\lambda - \frac{4}{r} \right] \varpi' - 16\pi (p + \varepsilon) e^\lambda \varpi = 0, \tag{12.72}$$

with

$$\varpi = \Omega - \omega. \tag{12.73}$$

In the vacuum exterior, the solution is simply given by

$$\varpi = \Omega - \frac{2\mathcal{J}}{r^3}, \tag{12.74}$$

where $\mathcal{J}$ is the star's angular momentum (as measured by a distant observer). In principle, one can extend the slow-rotation calculation to higher orders. The next order brings in the centrifugal deformation of the star, just as in the Newtonian case (Hartle and Thorne, 1968). Beyond that point the algebra becomes messy and one might as well turn to numerics.

A rapidly rotating star[2] can be described by

$$ds^2 = -e^{2\nu} dt^2 + e^{2\psi} (d\varphi - \omega dt)^2 + e^{2\mu} (dr^2 + r^2 d\theta^2), \tag{12.75}$$

where ν, ψ, ω, and μ are functions of r and θ (not to be confused with the metric functions in (12.70), which were chosen to connect to the Schwarzschild metric in the non-rotating limit.). This form for the line element follows if we assumes that (i) the spacetime has a timelike Killing vector t^a (related to stationarity) and a second axial Killing vector φ^a (associated with axial symmetry), and (ii) the spacetime is asymptotically flat. The fact that the two Killing vectors commute means that one can choose coordinates $x_0 = t$ and

---

[2] The problem is discussed in detail by Friedman and Stergioulas (2013).

$x_3 = \varphi$ and the asymptotic flatness then allows us to represent the three metric potentials ν, ψ, and ω as the invariant combinations

$$g_{tt} = t_a t^a \to -1 \text{ as } r \to +\infty, \tag{12.76}$$

$$g_{\varphi\varphi} = \varphi_a \varphi^a \to +\infty \text{ as } r \to +\infty, \tag{12.77}$$

$$g_{t\varphi} = t_a \varphi^a \to 0 \text{ as } r \to +\infty. \tag{12.78}$$

The final metric potential μ determines the conformal factor associated with the orthogonal 2-surfaces. The metric function ψ also has a geometrical meaning: e^ψ is the proper circumferential radius of a circle around the axis of symmetry. In the nonrotating limit, the metric (12.75) reduces to the metric of a non-rotating star in isotropic coordinates (see Chapter 4).

The four-velocity now takes the form

$$u^a = \frac{e^{-\nu}}{(1 - v^2)^{1/2}} (t^a + \Omega \varphi^a), \tag{12.79}$$

where v is the magnitude of the three-velocity relative to a local zero-angular-momentum observer (see Chapter 17), given by

$$v = (\Omega - \omega) e^{\psi - \nu}. \tag{12.80}$$

Rapidly rotating configurations must be determined numerically. This is typically done via either an implementation of the Hachisu self-consistent field method (Stergioulas and Friedman, 1995) or a spectral approach (Bonazzola et al., 1993; Ansorg et al., 2002). Both methods lead to reliable results, the veracity of which can be established via the relativistic analogue of the virial theorem.

As in the non-rotating case, neutron star models constructed from different equations of state have different bulk properties, associated with uncertainties in the high-density physics. Soft equations of state produce models with low maximum mass, small radius, and potentially large rotation rate. Stiff equations of state lead to models with a high maximum mass, large radius, and low maximum rotation rate. The attainable rotation can differ by as much as a factor of 2 for a reasonable range of equations of state. In general, the effect of rotation is to increase the equatorial radius of the star (as we have already seen) and increase the mass that can be sustained for a given central energy density. In effect, the most massive rotating model tends to be $15 - 20\%$ above the maximum mass non-rotating star. The corresponding increase in radius is $30 - 40\%$, so the difference is significant.

As in the Newtonian case, the maximum rotation rate is reached at the onset of mass-shedding at the equator. In relativity, this corresponds to

$$\Omega_K = \frac{\omega'}{2\psi'} + e^{\nu - \psi} \left[\frac{\nu'}{\psi'} + \left(\frac{\omega'}{2\psi'} e^{\psi - \nu} \right)^2 \right]. \tag{12.81}$$

It is (understandably) not straightforward to come up with an estimate analogous to (12.67) in this case. However, drawing on a collection of numerical results one may infer the empirical relation (Haensel and Zdunik, 1989; Friedman *et al.*, 1989; Haensel *et al.*, 1995)

$$\Omega_K \approx 2\pi \times 1400 \left(\frac{M_{\max}}{1.4M_\odot}\right)^{1/2} \left(\frac{R_{\max}}{10 \text{ km}}\right)^{-3/2} \text{s}^{-1}, \qquad (12.82)$$

where $M_{\max}$ and $R_{\max}$ are the mass and radius of the maximum mass non-rotating model. This estimate typically provides the maximum angular velocity with an accuracy better than 10%. Comparing to the Newtonian estimate (12.67), we see that the results are similar although one should keep in mind that (12.82) involves the parameters for the maximum mass configuration for a given equation of state.

Figure 12.3 provides an illustration of the (dynamically) stable region for a typical sequence of uniformly rotating relativistic stars. The results highlight the different features we have already discussed. In addition, we see that there may exist a class of rotating stars that have no non-rotating counterpart. These 'supramassive' stars must eventually collapse if they are spun down, e.g. by magnetic dipole radiation or gravitational waves.

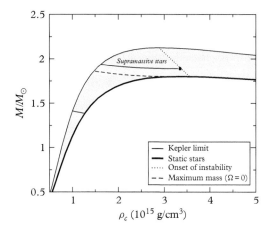

Figure 12.3 *The region of stable relativistic stars for a given equation of state (FPS; see Pandharipande and Ravenhall (1989)). Rotation increases the maximum mass by roughly 20% and also leads to the presence of a family of stars that have no non-rotating counterpart. These configurations are located between the mass-shedding curve and the dashed curve, which represents the most massive rotating star that has a stable non-rotating counterpart. A particular sequence (with constant baryon mass) of such supramassive stars is indicated by a thin solid line. An isolated star that spins down due to magnetic dipole radiation would evolve along this sequence until it reaches the state represented by the filled circle. At this point it will become unstable and undergo gravitational collapse.*

Finally, it is worth recalling that even a small amount of differential rotation can significantly increase the angular velocity required for mass-shedding. This is potentially very important since neutron stars are likely born differentially rotating. Studies of this problem tend to assume a simple analytical rotation law. The relativistic analogue of the so-called j-constant law,

$$A^2(\Omega_c - \Omega) = \frac{(\Omega - \omega)e^{2(\psi-\nu)}}{1 - (\Omega - \omega)e^{2(\psi-\nu)}},$$
(12.83)

where A is a constant, is particularly common. When $A \to \infty$ this leads back to uniform rotation and in the Newtonian limit the rotation is such that the specific angular momentum is constant. This rotation law is computationally convenient, but it should ideally be replaced by a more realistic model deduced from (say) neutron star birth simulations.

12.9   The quasiradial instability

Neutron stars may suffer a number of instabilities, which impact on the star's evolution and may also lead to the generation of gravitational waves. We have already discussed the most familiar case—the instability associated with the maximum mass for a given equation of state, cf. Figure 12.2. Any spherical relativistic star will suffer a dynamical instability before the compactness reaches the Schwarzschild limit (Chandrasekhar, 1964). General relativity does not permit stable stars with $R < 2.25GM/c^2$. In fact, there are no (even remotely reasonable) equations of state which permit stars more compact than $R \approx 3GM/c^2$.

The existence of an upper mass limit is important both for the formation of neutron stars and for mature stars accreting matter from a companion. In the first case, the collapse of a massive core that exceeds the maximum mass should lead to the prompt formation of a black hole. In the latter case, an accreting neutron star which reaches the maximum mass will become unstable and collapse, again leading to the formation of a black hole.

The maximum mass instability is relatively easy to analyse in the case of non-rotating stars. It sets in through the star's radial ($l = m = 0$ in a spherical harmonics expansion) oscillation modes. Assuming that the oscillation is associated with a time-dependence $e^{i\omega t}$, the equations that describe radial oscillations depend only on ω^2. This means that the mode frequencies come in pairs ($\pm\omega$). As one increases the central density of the star (for a given equation of state) the absolute value of the stable (real valued) mode frequencies decreases. It passes through zero exactly at the point at which the mass reaches an extremum (Harrison et al., 1965). Beyond this 'turning point', the mode frequencies become a complex conjugate pair, and one of the modes is unstable. The growth of this unstable mode triggers the collapse of the star.

The 'turning point method' provides a simple way of locating the onset of instability along a sequence of equilibrium models. One simply has to identify the maximum mass model along a sequence with increasing central density. However, it is not immediately obvious that this approach will generalize to rotating stars. The problem is subtle. Nevertheless, in the case of uniform rotation an extremum of the angular momentum J along a sequence with constant baryon number limits the region of stable stars (Sorkin, 1982). This notion is illustrated in Figure 12.3.

If we want to explore the dynamics of a star that reaches the instability point and collapses, we need to resort to nonlinear simulations. We will consider this problem in Chapter 20. Modelling the actual dynamics is particularly important if we want to extract a gravitational-wave signal. We cannot easily estimate the level of emission from back-of-the-envelope arguments.

From an intuitive point of view one might expect gravitational collapse to lead to a very strong gravitational-wave signal. However, the outcome depends entirely on the asymmetry of the collapse. A purely spherical collapse will not radiate gravitationally at all, while the collapse of a strongly deformed body could release a large amount of gravitational waves. The main reason why it is difficult to make 'reliable' estimates for the released energy is that the answer depends entirely on the route the system follows towards the final configuration. This is immediately clear from the post-Newtonian formulas, e.g. (3.90), which show that the gravitational-wave luminosity depends on time-derivatives of the involved multipoles.

As an example of this argument, let us consider the sudden contraction of a neutron star due to a phase transition in the interior. Instead of collapsing completely, the star reaches a new equilibrium at a slightly higher central density.

As a neutron star spins down, e.g. due to magnetic dipole radiation, the central density increases. Various theoretical models indicate that the equation of state may soften significantly once the central density reaches a critical value (likely several times the nuclear saturation density). This could be due to the formation of pion/kaon condensates, the creation of a significant hyperon core or quark deconfinement. Should this happen, it could result in a 'mini-collapse' during which gravitational potential energy may be released as radiation.

Such phase transitions have been suggested as sources for both detectable gravitational waves and gamma-ray bursts (Cheng and Dai, 1998). However, the estimates tend to be overly optimistic. The reason for this is simple. It is typically assumed that the entire change in potential energy incurred during the contraction can be radiated away. However, this is at variance with work that shows that the radiated energy is at best only a few percent of this (Schaeffer *et al.*, 1983). Most of the released potential energy is transferred into internal energy (i.e. heats the star up). That this should be the case is clear from the virial theorem.

Does this mean that the process is irrelevant from the gravitational-wave point of view? Not necessarily. Using the results for a uniform density sphere, we estimate the change in potential energy δW associated with a change in radius δR as

$$\delta W \approx \frac{3}{5} \frac{GM^2}{R^2} \delta R. \tag{12.84}$$

Suppose the contraction associated with a phase-transition in the core of a neutron star leads to $\delta R \approx 10$ m, and that 1% of δW is radiated as gravitational waves (which may not be unreasonable). Combining the result with (3.90), assuming a typical timescale of a millisecond, we find that $h_c \approx 10^{-23}$ (assuming $f = 1$ kHz) for a source at the distance of the Virgo cluster. This is a weak signal, but it may (just) be within reach of advanced instruments.

From this estimate we learn that, even though many statements of the strength of the gravitational waves from neutron star phase transitions are vastly too optimistic, we should not discard the idea. We also need to keep in mind that such events may be more violent than we have assumed. Moreover, a unique event from within our Galaxy could well be detectable. Given that these events are likely to be rare we would obviously be very lucky to catch one, but the information such that an observation would provide about physics at supranuclear densities would be extremely valuable.

12.10   Superfluids and glitches

Mature neutron stars tend to be extremely stable rotators, in some cases rivalling the best atomic clocks. However, in their adolescence they may behave in a less ordered fashion. Many young neutron stars exhibit (more or less) regular glitches, where the spin rate suddenly increases (Espinoza *et al.,* 2011a). These spin-up events tend to be followed by a slow relaxation towards the original spin-down rate. The archetypal glitching neutron star is the Vela pulsar, which has (since the first observed event; see Radhakrishnan and Manchester (1969) and Reichley and Downs (1969)) exhibited a regular sequence of similar size glitches. The most energetic and perhaps also enigmatic system, is the X-ray pulsar PSR J0537-6910 in the Large Magellanic Cloud (associated with the supernova remnant N157B (Marshall *et al.,* 1998)). Spinning at a frequency of 62 Hz this is the fastest spinning non-recycled neutron star and it glitches roughly every 100 days. Data showing the glitch activity of this pulsar, drawing on the complete set of timing observations from the Rossi X-ray Timing Explorer, are shown in Figure 12.4. The results highlight the (almost) predictable regularity of the glitches, the overall glitch-dominated spin-evolution, and the complex inter-glitch behaviour.

When pulsar glitches were first observed in the late 1960s they were thought to be associated with quakes in the star's crust (Ruderman, 1969). In essence, a glitch would be associated with the release of elastic strain built up as the star spun down and the centrifugal force weakened. As repeated large glitches were seen in the Vela pulsar this explanation was no longer viable as there was not enough time to build up the required strain between the events (Baym *et al.,* 1969). Small glitch events may still be explained as starquakes, but larger events (typically associated with a relative change in spin frequency, $\Delta \nu / \nu \sim 10^{-6}$) are now thought to be a manifestation of a superfluid component in the star's interior (Anderson and Itoh, 1975). A glitch is envisaged as a tug-of-war between the tendency of the neutron superfluid to match the spin-down rate of the rest of the star by expelling vortices (by means of which the superfluid rotates) and the impediment experienced by the moving vortices due to 'pinning' to crust nuclei (Alpar *et al.,* 1984a).

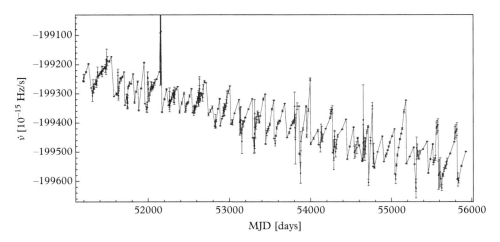

Figure 12.4 *The observed evolution of the spin-down rate of the young X-ray pulsar PSR J0537-6910. The regular glitches (sharp jumps in the spin-frequency and its derivative, $\dot{\nu}$ shown here), roughly every 100 days, are apparent in the data. (Reproduced from Antonopoulou et al. (2018).)*

Strong vortex pinning prevents the neutron superfluid from spinning down, creating a spin lag with respect to the rest of the star (which is spun down electromagnetically). This situation cannot persist (Seveso *et al.*, 2016). The increasing spin lag builds up the Magnus force exerted on the vortices. Above some threshold, pinning can no longer be sustained, the vortices break free, and the excess angular momentum is transferred to the crust, leading to the observed spin-up.

Several decades have passed since the two-component model was first suggested, yet there has been little progress on making it quantitative. Theoretical models are not (yet) at the level where they can be matched to observations like those in Figure 12.4. The outcome is sensitive to issues involving unknown microphysics, like the pinning of superfluid vortices to the nuclear lattice in the star's inner crust (Alpar *et al.*, 1984a) and how mobile the superfluid component is (Chamel, 2012). Moreover, the mechanism responsible for triggering glitches in the first place remains poorly understood and we do not actually know the location of the superfluid reservoir associated with the events (Andersson *et al.*, 2012).

Regardless of the uncertainties, it is clear that we need to understand neutron star superfluidity. Combining the simple fact that neutron stars are very cold—on the nuclear physics temperature scale; see (12.11)—with the standard arguments for laboratory systems, one would expect their outer core, which is dominated by npe matter, to contain a mixed superfluid/superconductor. The relevant critical temperatures inherit an uncertainty from the nuclear interactions that dictate the equation of state, but calculations suggest that the neutrons in the star's crust (pairing in a singlet state) have a critical temperature in the range $T_c = 10^9 - 10^{10}$ K. The critical temperature for the protons to form a (singlet) superconductor is similar, although this phase sets in at higher densities (basically since there are much fewer protons than neutrons at a given density). The neutrons in the star's core may also pair (in a triplet state). The corresponding critical temperature is the most uncertain, with typical estimates in the range $T_c = 10^8 – 10^9$ K. From the schematic illustration of the critical temperatures in Figure 12.5 (see Andersson *et al.* (2005*b*) for a discussion of more detailed models), it is clear that one would expect the bulk of a mature neutron star to contain these macroscopic quantum condensates. However, the relativistic electrons remain in a normal state as their transition temperature lies far below typical neutron star temperatures.

The dissipation channels in a superfluid star are quite different from those of a single fluid system. Basically, the superfluid flows without friction. This will inevitably affect the internal dynamics. Some of the relevant effects are well understood from low-temperature laboratory experiments (Graber *et al.*, 2017). The most familiar low-temperature system is, perhaps, He$_4$, which exhibits superfluidity below a critical temperature near 2K. Experimentally, it has been demonstrated that this system is well described by the Navier–Stokes equations above the critical temperature. Below the critical temperature the behaviour is different, and a 'two-fluid' model is generally required. Superfluid neutron stars are similar. In particular, we know that the second sound in Helium has a set of analogous, more or less distinct, 'superfluid' oscillation modes in a neutron star (Lee, 1995; Andersson and Comer, 2001*a*). The additional modes arise because the different components of a superfluid system are allowed to move 'through' each other.

In a superfluid neutron star, the shear viscosity is dominated by electron–electron scattering (Andersson *et al.*, 2005*b*). The bulk viscosity—which is due to the fluid motion driving the system away from chemical equilibrium and the resultant energy loss due to nuclear reactions—is expected to be (exponentially) suppressed. This has direct

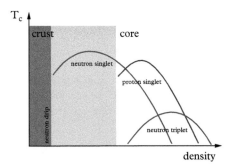

Figure 12.5 *Schematic illustration of the critical temperatures for superfluidity of neutrons in both singlet (red, crust) and triplet (red, core) pairing states and superconducting protons in the singlet state (blue, core). Since neutron stars are well below the Fermi temperatures for the involved baryons, their cores are expected to be dominated by superfluid (or superconducting) components. The density-dependent critical temperatures can be constrained by neutron star cooling data. They also impact on restlessness in the star's spin-down and the engimatic spin-glitches that are seen in (predominantly) young pulsars. The glitches provide information on the mobility of superfluid components (the so-called entrainment effect) and the potential pinning of vortices to nuclei in the star's crust.*

implications for neutron star dynamics. We will return to these aspects later; see for example Chapter 15. It is, however, not the end of the story. A superfluid exhibits an additional dissipation mechanism, usually referred to as 'mutual friction''. The mutual friction is due to the presence of vortices in a rotating superfluid. In a neutron star core, the electrons can scatter dissipatively off of the (local) magnetic field of each vortex (Alpar *et al.*, 1984b; Andersson *et al.*, 2006). This effect may dominate the damping of the interior dynamics of realistic neutron stars.

The basic requirements of a model for superfluid neutron star dynamics is clear. We must account for the additional dynamical degree(s) of freedom, and add in the mutual friction. But this is only the starting point. More realistic models should include finite temperature effects, magnetic fields, and the possible presence of exotic (hyperon and/or quark) cores. The additional physics brings complications, like new fluid degrees of freedom, boundary layers at phase-transition interfaces, and fundamental issues concerning dissipative multifluid systems (Haskell *et al.*, 2012).

As a first step, let us consider the two-fluid model for neutron star cores. This model accounts for two dynamical degrees of freedom, loosely speaking representing the superfluid neutrons (labeled n) and a charge-neutral conglomerate of protons and electrons (labeled p). For simplicity, we assume that the electrons are electromagnetically locked to the protons. Assuming that the individual species are conserved, we have the usual conservation laws for each of the mass densities

$$\partial_t \rho_x + \nabla_i (\rho_x v_x^i) = 0, \tag{12.85}$$

where the constituent index x may be either p or n. Meanwhile, the equations of momentum balance can be written (Andersson and Comer, 2006)

$$\mathcal{E}_i^x = (\partial_t + v_x^j \nabla_j)(v_i^x + \varepsilon_x w_i^{yx}) + \nabla_i (\tilde{\mu}_x + \Phi) + \varepsilon_x w_{yx}^j \nabla_i v_j^x = f_i^x / \rho_x, \quad y \neq x, \tag{12.86}$$

where the velocities are v_x^i, the relative velocity is defined to be $w_{xy}^i = v_x^i - v_y^i$ and $\tilde{\mu}_x = \mu_x / m_x$ represents the chemical potentials (we assume that $m_p = m_n$). Φ is the usual the gravitational potential, and the parameter ε_x encodes the non-dissipative entrainment coupling between the fluids. The force on the right-hand side of (12.86) can be used to represent additional interactions, including dissipation.

We can express the entrainment as a dynamical effective mass (not to be confused with the effective mass associated with the nuclear interactions; see (Prix *et al.*, 2002)). This follows immediately from considering each fluid momentum in the frame moving along with the other fluid (set $v_y^i = 0$). This leads to

$$m_x^\star n_x v_i^x = m_x n_x (1 + \varepsilon_x) v_i^x, \tag{12.87}$$

which gives the effective mass $m_x^\star$. This could be a significant effect. In particular, the effective mass of the superfluid neutrons in the star's inner crust can be much larger than the bare mass (Chamel, 2012). In essence, the neutron superfluid may not be very mobile.

Let us now consider the vortex-mediated mutual friction, which is introduced in the same way as for superfluid helium (Hall and Vinen, 1956; Mendell, 1991; Andersson et al., 2006). This leads to a force of form

$$f_i^n = \rho_n \mathcal{N}_n \kappa \left(\mathcal{B}' \epsilon_{ijk} \hat{\kappa}^j w_{np}^k + \mathcal{B} \epsilon_{ijk} \hat{\kappa}^j \epsilon^{klm} \hat{\kappa}_l w_m^{np} \right) \tag{12.88}$$

acting on the superfluid neutrons (with an equal and opposite force on the charged component). The strength of the friction is intimately linked to the friction on individual vortices. Assuming that the bulk rotation is uniform, we have $\mathcal{N}_n \kappa^i = 2\Omega_n^i$ and each vortex carries a quantum of circulation

$$\kappa = \frac{h}{2m_n} \approx 2 \times 10^{-3} \text{ cm}^2\text{s}^{-1}, \tag{12.89}$$

which means that the (area) density of vortices is

$$\mathcal{N}_n = \frac{2\Omega_n}{\kappa} \approx 6 \times 10^5 \left(\frac{P}{10 \text{ ms}} \right)^{-1} \text{ cm}^{-2}, \tag{12.90}$$

with P the star's rotation period. The mutual friction is often expressed in terms of a dimensionless 'drag' parameter $\mathcal{R}$, such that

$$\mathcal{B}' = \mathcal{R}\mathcal{B} = \frac{\mathcal{R}^2}{1 + \mathcal{R}^2}. \tag{12.91}$$

In the standard picture the mutual friction is due to the scattering of electrons off of the array of neutron vortices (Alpar et al., 1984a). This leads to

$$\mathcal{R} \approx 2 \times 10^{-4} \left(\frac{m_p}{m_p^\star} \right)^{1/2} \left(\frac{\rho}{10^{14} \text{ g/cm}^3} \right)^{1/6} \left(\frac{x_p}{0.05} \right)^{1/6}. \tag{12.92}$$

That is, we have $\mathcal{R} \ll 1$, i.e., $\mathcal{B}' \ll \mathcal{B}$, and hence the first term in the mutual friction force (12.88) can be ignored. It is, however, possible that the problem is in the opposite regime. In particular, if one considers the interaction between the fluxtubes in a type II proton superconductor and the neutron vortices (Ruderman et al., 1998; Link, 2003). In this case one would expect to be in the strong drag regime where $\mathcal{R} \gg 1$, i.e., $\mathcal{B}' \approx 1$ while $\mathcal{B}$ remains small.

In the spirit of the estimates for energy released in a sudden phase transition, it is useful to consider the energy budget for pulsar glitches. We can get some idea of the available energy even though we do not have a detailed understanding of the dynamics. It may not be very sensible to associate these estimates with gravitational-wave emission, but it is interesting to consider the possibility. These are regularly occurring events in galactic systems and the energetics may be suggestive of what one could expect from pulsars in general.

In the particular case of solid-body rotation, with Ω_x^i aligned with the z-axis, energy conservation follows immediately from the two-fluid equations. We have

$$\frac{dE}{dt} = \frac{d}{dt} \left\{ \frac{1}{2} I_n \Omega_n \left[\Omega_n + \varepsilon_n \left(\Omega_p - \Omega_n \right) \right] + \frac{1}{2} I_p \Omega_p \left[\Omega_p + \varepsilon_p \left(\Omega_n - \Omega_p \right) \right] \right\} = 0. \quad (12.93)$$

For simplicity, we have assumed that the entrainment parameters ε_x are constant and defined each constituent moment of inertia as

$$I_{x_i}^j = \int \rho_x \left(\delta_i^j x^2 - x_i x^j \right) dV, \quad (12.94)$$

with $I_x = I_{x_z}^z$. Similarly, we can calculate the angular momentum. To do this we note that

$$\epsilon_{ijk} x^j v_x^l \nabla^k v_l^x = 0 \quad \text{for} \quad i = z, \quad (12.95)$$

as the rotation is aligned with the z-axis. Contracting $\mathcal{E}_i^x$ from (12.86) with $\rho_x \epsilon_{ijk} x^j$ and integrating over the volume V we arrive at

$$\frac{d\mathcal{J}_i^x}{dt} = \int \rho_x \epsilon_{ijk} x^j \mathcal{E}_x^k dV$$
$$= \int \rho_x \epsilon_{ijk} x^j \left[(1 - \varepsilon_x) \frac{\partial v_x^k}{\partial t} + \varepsilon_x \frac{\partial v_y^k}{\partial t} \right] dV = 0 \quad \text{for} \quad i = z. \quad (12.96)$$

This can be rewritten as

$$\frac{d\mathcal{J}_i^x}{dx} = \frac{d}{dt} \left\{ I_{x_i}^j \left[\Omega_j^x + \varepsilon_x \left(\Omega_j^y - \Omega_j^x \right) \right] \right\} = 0 \quad \text{for} \quad i = z. \quad (12.97)$$

From this we see that the total angular momentum is also conserved

$$\frac{d\mathcal{J}_z}{dt} = \frac{d}{dt} \left(I_n \Omega_n + I_p \Omega_p \right) = 0. \quad (12.98)$$

Let us now consider a glitch event, assuming that the key ingredient that brings the two fluids together is the mutual friction. Basically, we take the vortices the be either perfectly pinned or completely unpinned. In such a model, a glitch would proceed as follows. Initially, we take the superfluid vortices to form a uniform, straight, array aligned with the rotation axis. Vortex pinning simply fixes the number of vortices per unit area. This, in turn, dictates the neutron fluid's angular momentum, so the superfluid component rotates at a constant rate. If we assume that the charged fluid is electromagnetically locked to the crust, then the vortices will be rotating with the charged fluid component. As the crust spins down due to an external torque, a velocity difference builds up between the two constituents. This leads to an increasing Magnus force on the vortices. Eventually, when some critical lag, $\Delta\Omega_c$, is reached this force will be strong enough to overcome the nuclear pinning and the vortices are suddenly free to move. At this point the vortex mutual friction transfers angular momentum between the two components. The two components couple, the lag decays, and the crust spins up—leading to the observed glitch. If the system relaxes completely, the end state should be such that the two components rotate at the same rate. The glitch event itself is sudden (Dodson et al., 2002; Palfreyman et al., 2018). On a longer timescale one would expect the vortices to repin. The repinning should determine the long-term relaxation after the event, i.e. the spin evolution on timescales longer than (perhaps) tens of seconds (Alpar et al., 1984a). Eventually, the system will reach a state where the rotational lag increases, and the pulsar may glitch again.

It is relevant to compare the energetics associated with the two main glitch paradigms. In the starquake model, we need to estimate the energy available for radiation based on a single component. The total kinetic energy and angular momentum are (obviously) given by

$$E_{\text{kin}} = \frac{1}{2}I\Omega^2, \quad \text{and} \quad J = I\Omega. \tag{12.99}$$

If we assume that a glitch of size $\Delta\Omega$ results from a change in the moment of inertia ΔI, then—since the total angular momentum is conserved—it is easy to show that the available energy is

$$\Delta E_{\text{quake}} \approx \frac{1}{2}I\Omega\Delta\Omega = \left(\frac{\Delta\nu}{\nu}\right)E_{\text{kin}}. \tag{12.100}$$

Basically, we could release a fraction of $\Delta\nu/\nu \sim 10^{-6}$ (the size of the glitch) of the star's kinetic energy. This level of energy may be interesting for gravitational-wave astronomy (assuming that it is associated with some asymmetry of the system, of course; see Andersson and Comer (2001b)).

However, if we consider the two-component model we get a rather different picture. In this case, for constant I_x, the conservation of angular momentum leads to

$$\Delta\Omega_n = -\frac{I_p}{I_n}\Delta\Omega_p. \tag{12.101}$$

Basically, the superfluid (neutrons) spin down as the crust (protons) spins up. Estimating the available energy, we now find that

$$\Delta E_{\text{sf}} \approx \frac{1}{2}I_p(\Delta\Omega)^2 = \frac{I_p}{I}\left(\frac{\Delta\nu}{\nu}\right)^2 E_{\text{kin}}, \tag{12.102}$$

where we have used $\Delta\Omega = \Delta\Omega_p$ and assumed that $I_p \ll I_n$. For typical parameters, $I_p/I \approx 0.1$ and $\Delta\nu/\nu \sim 10^{-6}$, we see that the available energy is suppressed by a factor of 10^{-7} compared to the starquake case. If this estimate is taken seriously, and glitches really do represent the transfer of angular momentum envisaged in the two-fluid model, then the gravitational-wave signal from a pulsar glitch is unlikely to be detected by any future instrument (Sidery *et al.*, 2010).

Glitches may be more a nuisance than an opportunity (Ashton *et al.*, 2018). Take the case of PSR J0537-6910 as an example. As this is a relatively young neutron star, it is an interesting object for gravitational-wave astronomy (Andersson *et al.*, 2018). However, unless one can figure out how deal with the frequent glitches one will not be able to carry out a coherent search for more than a couple of months (which does not improve the effective gravitational-wave amplitude much). Moreover, a sensitive search requires a reliable timing solution from electromagnetic observations. During the S6 LIGO run this was provided by the Rossi X-ray Timing Explorer (RXTE), leading to an upper limit on the gravitational-wave amplitude (Aasi *et al.*, 2014). However, RXTE was no longer operating during the first advanced detector observing runs so the improvement of instrument sensitivity has not (yet) led to better results in this particular case.

13

From oscillations to instabilities

The oscillations of a compact object, be it a star or a black hole, are of great interest for gravitational-wave astronomy. Plausible scenarios lead to such oscillations being excited to a detectable amplitude. In fact, the late stages of GW150914 showed hints of the ringdown signature of the remnant black hole (see Chapter 16). Similar behaviour is expected from neutron star mergers, like GW170817 (see Chapter 21), with the characteristics of the oscillations depending on whether the merger led to a more massive neutron star or collapse to a black hole. Less dramatic scenarios, like the tidal interaction during the late stages of inspiral (also discussed in Chapter 21), may excite oscillations in the stellar fluid which could leave an observable imprint on the signal. Finally, there are situations where specific oscillation modes become unstable and grow to a large amplitude. Given that typical gravitational-wave signals are weak, this possibility is particularly interesting.

Realistic neutron star models have rich oscillation spectra, with specific modes associated with different aspects of the physics (ranging from the internal composition to the elasticity of the crust, superfluid components, and so on). In the first instance, we will illustrate these ideas in the context of Newtonian gravity. This is natural since the qualitative aspects of the problem mainly depend on the fluid dynamics—the local restoring forces that affect the motion of a given fluid element. This allows us to identify the different classes of oscillation modes and discuss different instabilities that may affect the star. In the process, we arrive at useful estimates for the gravitational-wave emission. However, the ultimate aim is to use gravitational waves from a pulsating neutron star to probe the involved physics. In order to enable actual measurements, we need to carry out fully relativistic mode calculations. We consider that problem in Chapter 18.

13.1 The fundamental f-mode

Stars have complex internal dynamics, often represented in terms of oscillation modes related to waves in the Earth's oceans. Different families of modes can be, more or less directly, associated with specific physics ingredients. This is particularly true for neutron stars, where the composition and state of matter depend on many-body aspects of the nuclear interaction (see Chapter 12). On the one hand, this makes the modelling of such

Gravitational-Wave Astronomy: Exploring the Dark Side of the Universe. Nils Andersson, Oxford University Press (2020).
© Nils Andersson. DOI: 10.1093/oso/9780198568032.001.0001

systems challenging. On the other hand, the effort may bring rich rewards as observations matched to precise theory would provide insight into the star's interior. The ultimate aim is to develop this seismology strategy to a level similar to that which is being used to probe the interiors of the Sun and distant main sequence stars (Aerts *et al.*, 2010).

In order to introduce the key concepts of stellar oscillation theory, let us first consider the case of a constant density star. Neutron stars obviously do not have uniform density, but the results nevertheless help us understand more complicated settings.

The dynamics of a non-rotating incompressible star is governed by the Euler equations (4.59), together with the continuity equation (4.52) and the Possion equation (4.36) for the gravitational potential. However, since the density is constant the continuity equation simplifies and we have

$$\partial_t \rho + \nabla \cdot (\rho \boldsymbol{v}) = 0 \longrightarrow \nabla \cdot \boldsymbol{v} = 0. \tag{13.1}$$

In general, we have the option of considering two different kinds of perturbations. Eulerian perturbations concern changes in the various quantities at a fixed point in space; e.g. for the pressure we have

$$\delta p = p(\boldsymbol{x}, t) - p_0(\boldsymbol{x}, t), \tag{13.2}$$

where p_0 is the unperturbed reference pressure. The Eulerian description is 'macroscopic' in the sense that it does not identify how the fluid elements move. A 'microscopic' description would track the individual fluid elements. We can do this by introducing a Lagrangian displacement $\boldsymbol{\xi}$ which connects the perturbed fluid elements to the corresponding ones in the unperturbed configuration. Lagrangian pressure variations are given by

$$\Delta p = \delta p + \boldsymbol{\xi} \cdot \nabla p. \tag{13.3}$$

For non-rotating stars, the displacement vector is simply related to the perturbed velocity through

$$\partial_t \boldsymbol{\xi} = \Delta \boldsymbol{v} = \delta \boldsymbol{v}. \tag{13.4}$$

Not surprisingly, the case of a rotating star is more complicated. In order to understand that problem it is important to develop the Lagrangian description (for which it is also natural to work in a coordinate basis, making a distinction between co- and contravariant objects). We will consider this framework later. For the moment, as we are taking the first steps, we opt for the Eulerian approach. Thus, we arrive at the perturbed momentum equation

$$\partial_t \delta \boldsymbol{v} + \frac{1}{\rho} \nabla \delta p + \nabla \delta \Phi = 0, \tag{13.5}$$

where $\delta \Phi$ is the variation in the gravitational potential.

We also have Eq. (13.1) and

$$\nabla^2 \delta\Phi = 0. \tag{13.6}$$

For an incompressible fluid the velocity can be determined from a potential χ, such that

$$\delta\boldsymbol{v} = \nabla\chi \tag{13.7}$$

(hence, the motion is often referred to as 'potential flow'). With this definition the perturbed Euler equations lead to

$$\partial_t\chi + \frac{1}{\rho}\delta p + \delta\Phi = D = \text{ constant.} \tag{13.8}$$

Moreover, taking the divergence of (13.5) we see that we must have

$$\nabla^2 \delta p = 0, \tag{13.9}$$

while the continuity equation leads to

$$\nabla^2 \chi = 0. \tag{13.10}$$

We see that the velocity potential χ, the perturbed pressure δp, and the perturbed gravitational potential $\delta\Phi$ all satisfy Laplace's equation. In spherical polar coordinates we then have, for each of these variables $(= X)$,

$$\nabla^2 X = \frac{1}{r}\frac{\partial^2}{\partial r^2}(rX) + \frac{1}{r^2}\nabla_\theta^2 X = 0, \tag{13.11}$$

where we have defined

$$\nabla_\theta^2 X = \frac{1}{\sin\theta}\partial_\theta(\sin\theta\,\partial_\theta X) + \frac{1}{\sin^2\theta}\partial_\varphi^2 X. \tag{13.12}$$

Approaching (13.11) via separation of variables, the angular part can be expressed in terms of the spherical harmonics $Y_l^m(\theta, \varphi)$. Specifically, we get

$$\nabla_\theta^2 Y_l^m = -l(l+1)Y_l^m. \tag{13.13}$$

Then writing the solution to (13.11) as

$$X(r,\theta,\varphi) = \sum_{l,m} X_l(r)Y_l^m, \tag{13.14}$$

we see that the various Y_l^m contributions decouple (since the spherical harmonics form a complete orthogonal set), and we need to solve

$$\frac{1}{r}\frac{d^2}{dr^2}(rX_l) - \frac{l(l+1)}{r^2}X_l = 0. \tag{13.15}$$

The general solution to this equation is

$$X_l = Ar^l + Br^{-l-1}. \tag{13.16}$$

Since all physical quantities must remain regular, we should reject the second term as $r \to 0$. This means that we must take $B = 0$ in the star's interior. Meanwhile, we must have $A = 0$ in the exterior solution for $\delta\Phi$ in order to avoid divergence as $r \to \infty$. Focussing on a solution described by a particular Y_l^m we see that the perturbed quantities we are interested in are given by

$$\chi = a_l r^l Y_l^m, \tag{13.17}$$
$$\delta\Phi = b_l r^l Y_l^m, \tag{13.18}$$
$$\delta p = c_l r^l Y_l^m, \tag{13.19}$$

in the star's interior. If we further assume that the fluid motion has a harmonic dependence on time, i.e. that $\chi \propto e^{i\omega t}$, as one would expect of an oscillation, we get from (13.8)

$$i\omega a_l + \frac{c_l}{\rho} + b_l = 0, \tag{13.20}$$

where we have set $D = 0$ since all functions on the left-hand side of (13.8) vanish as $r \to 0$ (and the constant must have the same value for all r).

To complete our solution we need to consider the boundary conditions at the surface of the star. First of all, the Lagrangian perturbation of the pressure must vanish. This is, in fact, the definition of the surface. We must have

$$\Delta p = \delta p + \xi^r p' = 0 \qquad \text{at } r = R, \tag{13.21}$$

(where the prime indicates a radial derivative). The radial component of the displacement vector then follows from

$$i\omega\xi^r = \delta v^r = \chi' = a_l l r^{l-1} Y_l^m. \tag{13.22}$$

We also know that

$$p' = -\rho\Phi' = -\frac{4\pi G\rho^2 r}{3}. \tag{13.23}$$

Given these relations the boundary condition leads to

$$c_l = \frac{4\pi G\rho^2 l}{3i\omega} a_l. \tag{13.24}$$

The final part of the solution comes from the perturbed gravitational potential. However, since we are considering a uniform density model the condition on the gravitational potential at the surface is somewhat pathological. As a short-cut to the answer, we assume that we can ignore the perturbed gravitational potential. This is called the Cowling approximation (Cowling, 1941). Setting $b_l = 0$ in (13.20) we immediately arrive at

$$\omega^2 = \frac{4\pi G\rho l}{3}. \tag{13.25}$$

For each value of l we have two oscillation modes. These modes, which were first by Lord Kelvin as early as 1863, are known as the fundamental f-modes. The f-mode frequency scales with the average density of the star and it also increases with the multipole l. The first of these scalings provides a hint that it may be possible to use observed data to constrain the physics of the star. Of course, we also learn that the observation of a single mode-frequency would not be enough. It would only constrain the average density, while the real aim is to infer both mass and radius (and perhaps learn about the interior composition, as well).

As we will see later, this qualitative picture remains valid also for more realistic models. In order to demonstrate this, we need to relax the simplifying assumptions. As a step towards this, we may check the accuracy of the Cowling approximation. In order to do this, we need the appropriate matching condition for the perturbed gravitational potential at the surface

$$\delta\Phi' + \frac{l+1}{R}\delta\Phi = -4\pi G\rho\xi^r, \quad \text{at } r = R. \tag{13.26}$$

This leads to

$$b_l = -\frac{4\pi G\rho l}{(2l+1)i\omega} a_l, \tag{13.27}$$

and it follows from (13.20) that

$$\omega^2 = \frac{8\pi G\rho}{3}\frac{l(l-1)}{2l+1}. \tag{13.28}$$

Taking the ratio of the two results for the mode frequency, we see that

$$\left(\frac{\omega_{\text{Cowling}}}{\omega_{\text{non-Cowling}}}\right)^2 \approx \frac{2l+1}{2(l-1)}. \tag{13.29}$$

The error is fairly large for the $l=2$ modes but decreases rapidly as we move to higher multipoles.

For typical neutron star parameters we find that the $l=2$ (quadrupole) f-mode has a frequency of approximately 2 kHz. (We discuss results for realistic equations of state in Chapter 18.) In principle, we can also use the quadrupole formula to estimate how efficiently the mode is damped by gravitational-wave emission (although this exercise a bit messy for a uniform density model; see Detweiler (1975)). This leads to a typical damping timescale

$$\tau \sim \left(\frac{c^2 R}{GM}\right)^{l+1}\frac{R}{c} \sim 0.1 \text{ s}. \tag{13.30}$$

From this we see that the f-modes are rapidly damped by gravitational-wave emission. These results allows us to connect with the flux argument used to estimate the effective gravitational-wave amplitude in Chapter 1.

13.2   General non-rotating stars: p/g-modes

The f-modes are important for several reasons. They are associated with significant density variations and hence lead to efficient gravitational-wave emission. They are expected to be excited in many relevant astrophysical scenarios and may also become unstable in fast-spinning stars. However, this does not mean that we can ignore other classes of oscillation modes. Other modes may be less efficient gravitational-wave emitters and their dynamical role may be more subtle, but they can nevertheless have decisive impact on observations.

More complex stellar models have additional classes of pulsation modes. A useful rule-of-thumb is that each piece of 'physics' added to the model brings (at least) one new family of modes into existence. Thus, there are modes associated with the compressibility of the neutron star fluid. These oscillations correspond to sound waves and are known as the p-modes. If we add internal stratification associated with, for example, temperature or composition gradients, then the so-called gravity g-modes come into play (Reisenegger and Goldreich, 1992). There are also classes of modes directly associated with rotation, the elasticity of the neutron star crust, the magnetic field, the superfluid nature of the star's core, and so on.

Suppose we consider a more general non-rotating perfect fluid star: that is, a body whose dynamics is governed by the Euler equations (4.59), but with the density no longer taken to be constant. Then the perturbation equations become

$$\partial_t^2 \boldsymbol{\xi} = \frac{\delta\rho}{\rho^2}\nabla p - \frac{1}{\rho}\nabla\delta p - \nabla\delta\Phi.$$ (13.31)

As before, $\boldsymbol{\xi}$ is the fluid displacement vector and $\partial_t\boldsymbol{\xi} = \delta\boldsymbol{v}$. We also have the (integrated form of) the continuity equation

$$\delta\rho + \nabla\cdot(\rho\boldsymbol{\xi}) = 0.$$ (13.32)

It is customary to introduce the adiabatic index of the perturbations, Γ_1, as

$$\frac{\Delta p}{p} = \Gamma_1\frac{\Delta\rho}{\rho},$$ (13.33)

or, in terms of the Eulerian variations,

$$\delta p = \frac{p\Gamma_1}{\rho}\delta\rho + \boldsymbol{\xi}\cdot\left[\frac{p\Gamma_1}{\rho}\nabla\rho - \nabla p\right] \equiv \frac{p\Gamma_1}{\rho}\delta\rho + p\Gamma_1(\boldsymbol{\xi}\cdot\boldsymbol{A}),$$ (13.34)

which defines the *Schwarzschild discriminant* $\boldsymbol{A}$. Note also that the sound speed follows as

$$c_s^2 = \frac{\Delta p}{\Delta\rho} = \frac{p\Gamma_1}{\rho}.$$ (13.35)

For spherical stars we can rewrite the perturbed Euler equations as

$$\partial_t^2\boldsymbol{\xi} = -\nabla\left(\frac{\delta p}{\rho}\right) + \frac{p\Gamma_1}{\rho}\boldsymbol{A}(\nabla\cdot\boldsymbol{\xi}) - \nabla\delta\Phi.$$ (13.36)

Once the equation is written in this form we see that the fluid motion is affected by (neglecting $\delta\Phi$) two restoring forces: the pressure variation and the 'buoyancy' associated with $\boldsymbol{A}$. The latter is relevant whenever the star is stratified, by either entropy or composition variations. If we are considering mature neutron stars we can to a good approximation assume that the temperature is zero and neglect internal entropy gradients. Still, we cannot assume that $\boldsymbol{A} = 0$ since any variation of the matter composition will lead to an effective buoyancy force acting on a fluid element. This may be an important effect since we know there will be a varying proton fraction in the neutron star core (see Chapter 12).

In essence, we may consider the equation of state to be of the form $p = p(\rho, x_p)$. This leads to

$$\Delta p = \left(\frac{\partial p}{\partial \rho}\right)_{x_p} \Delta \rho + \left(\frac{\partial p}{\partial x_p}\right)_{\rho} \Delta x_p. \tag{13.37}$$

If the oscillations are fast compared to the nuclear reactions that work to reinstate chemical equilibrium, then individual fluid elements retain their identity. The co-moving proton fraction remains fixed and we have $\Delta x_p = 0$. In this case we identify

$$\Gamma_1 = \frac{\rho}{p}\left(\frac{\partial p}{\partial \rho}\right)_{x_p}. \tag{13.38}$$

The key point is that the proton fraction is held fixed in the thermodynamical derivative, making the result differ from the corresponding derivative in chemical equilibrium.

Returning to the oscillation problem, we want to infer the nature of the various modes of pulsation the star may have from (13.36). In doing this, let us (again) make use of the Cowling approximation, i.e. neglect the variation of the gravitational potential. Since the gravitational acceleration

$$\boldsymbol{g} = -g\hat{\boldsymbol{e}}_r = \frac{1}{\rho}\nabla p, \tag{13.39}$$

is purely radial for a spherical star, the horisontal component of the Euler equation (13.36) leads to

$$\boldsymbol{\xi}_\perp = \frac{1}{\omega^2 \rho}\nabla_\perp \delta p, \tag{13.40}$$

where we have assumed that the perturbation has time-dependence $e^{i\omega t}$ (as before). We have also assumed that the angular dependence of δp can be represented by a single spherical harmonic Y_l^m (which is natural since the pressure variation is a scalar) such that

$$\nabla_\perp^2 \delta p = \frac{1}{r^2}\nabla_\theta^2 \delta p - \frac{l(l+1)}{r^2}\delta p.$$

Making use of

$$\Delta \rho \equiv \delta \rho + \boldsymbol{\xi} \cdot \nabla \rho = -\rho \nabla \cdot \boldsymbol{\xi}, \tag{13.41}$$

we find that

$$\Delta\rho = -\frac{\rho}{r^2}\frac{\partial}{\partial r}(r^2\xi^r) - \rho\nabla_\perp \cdot \boldsymbol{\xi} = -\frac{\rho}{r^2}\frac{\partial}{\partial r}(r^2\xi^r) + \frac{l(l+1)}{\omega^2 r^2}\delta p. \tag{13.42}$$

We also have the radial component of (13.36):

$$-\omega^2\xi^r = -\frac{\partial}{\partial r}\left(\frac{\delta p}{\rho}\right) - \frac{p\Gamma_1 A \Delta\rho}{\rho^2}, \tag{13.43}$$

where $A = |\boldsymbol{A}|$.

We can replace $\Delta\rho$ by noting that

$$\Delta\rho = \frac{\rho}{P\Gamma_1}(\delta p - \rho g \xi^r), \tag{13.44}$$

and we arrive at

$$\frac{1}{r^2}\frac{\partial}{\partial r}(r^2\xi^r) - \frac{\rho g}{p\Gamma_1}\xi^r = \left[\frac{l(l+1)}{\omega^2 r^2} - \frac{\rho}{p\Gamma_1}\right]\frac{\delta p}{\rho}, \tag{13.45}$$

and

$$\frac{1}{\rho}\frac{\partial}{\partial r}\delta p + \frac{g}{p\Gamma_1}\delta p = (\omega^2 + gA)\xi^r = (\omega^2 - N^2)\xi^r, \tag{13.46}$$

where we have introduced the so-called *Brunt–Väisälä frequency*, $N^2 = -gA$.

Let us now introduce new variables

$$\hat\xi^r = \frac{r^2\xi^r}{\phi}, \qquad \delta\hat p = \phi\delta p, \qquad \text{where } \phi = \exp\left[\int g/c_s^2\,dr\right]. \tag{13.47}$$

With these definitions our two equations can be written

$$\frac{\partial\hat\xi^r}{\partial r} = \left[L_l^2 - \omega^2\right]\frac{r^2\delta\hat p}{\rho\omega^2 c_s^2\phi^2}, \tag{13.48}$$

where the *Lamb frequency*, L_l, is given by

$$L_l^2 = \frac{l(l+1)c_s^2}{r^2}, \tag{13.49}$$

and

$$\frac{\partial \delta \hat{p}}{\partial r} = \left[\omega^2 + gA\right]\frac{\rho \hat{\xi}^r \phi^2}{r^2}.$$ (13.50)

It is now easy to reduce the problem to a single differential equation for $\hat{\xi}^r$:

$$\frac{d}{dr}\left\{\frac{\rho \omega^2 c_s^2 \phi^2}{r^2}\left[L_l^2 - \omega^2\right]^{-1}\frac{d\hat{\xi}^r}{dr}\right\} - [\omega^2 - N^2]\frac{\rho \phi^2}{r^2}\hat{\xi}^r = 0.$$ (13.51)

This allows us to draw important conclusions. In particular, we note that the problem reduces to the Sturm-Liouville form both for high and low frequencies. For large ω^2 we get

$$\frac{d}{dr}\left\{\frac{\rho c_s^2 \phi^2}{r^2}\frac{d\hat{\xi}^r}{dr}\right\} + \omega^2\frac{\rho \phi^2}{r^2}\hat{\xi}^r = 0.$$ (13.52)

Standard theory now tells us that there will be an infinite set of modes—which can be labelled by the number of radial nodes (n) of the eigenfunctions—for which $\omega_n \to \infty$ as $n \to \infty$. In the opposite limit, when ω^2 is small, the problem becomes

$$\frac{d}{dr}\left\{\frac{\rho \phi^2}{l(l+1)}\frac{d\hat{\xi}^r}{dr}\right\} + \left(\frac{N^2}{\omega^2}\right)\frac{\rho \phi^2}{r^2}\hat{\xi}_r = 0,$$ (13.53)

and we see that there will be another set of modes, with eigenfrequencies such that $\omega_n \to 0$ as $n \to \infty$.

These are known as the pressure p-modes and gravity g-modes, respectively, and we can estimate their frequencies in the following way: assume that the perturbations have a characteristic wavelength k^{-1}—such that the various functions are proportional to $\exp(ikr)$. Then we can readily deduce the dispersion relation

$$k^2 = \frac{1}{c_s^2 \omega^2}(N^2 - \omega^2)(L_l^2 - \omega^2).$$ (13.54)

Here we must have $\omega^2 > 0$ for stability, and we see that we have oscillation modes ($k^2 > 0$) in two different cases. Either ω^2 must be smaller than both N^2 and L_l^2, or it must be larger than both of them. Solving the quadratic for ω^2, and (for simplicity) assuming that $l \gg kr$, we can estimate that the short-wavelength mode-frequencies are $\omega^2 \approx L_l^2$ for the p-modes and $\omega^2 \approx N^2$ for the g-modes.

As a neutron star cools below a few times 10^9 K the extreme density in the core leads to the formation of various superfluids. The superfluid constituents play a crucial role in determining the dynamical properties of a rotating neutron star. We have already

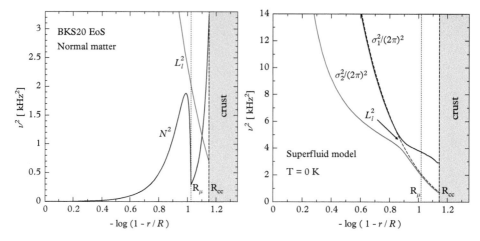

Figure 13.1 *Propagation diagrams for normal (left) and superfluid (right) neutron star cores. The results are obtained for the BSk20 equation of state (Potekhin et al., 2013), and L_l^2 represents the Lamb frequency, while N^2 is the Brunt–Väisälä frequency. In this particular model, muons come into play at radius R_μ and (as is clear from the left panel) their appearance has significant impact on the g-modes. The right panel illustrates the presence of two sets of p-modes in a superfluid star (with frequencies σ_1 and σ_2). It is also worth noting the absence of g-modes in the superfluid case. (Reproduced from Passamonti et al. (2016).)*

considered (see Chapter 12), how the interplay between the lattice nuclei and the superfluid in the inner crust may explain observed pulsar glitches. The presence of a superfluid will also affect the star's oscillations.

Studies of oscillations in superfluid stars have mainly been based on models allowing for two distinct, dynamically coupled, fluids (as outlined in Chapter 12). The two fluids represent the superfluid neutrons and the 'protons', viewed as a conglomerate of all charged components in the star (which are taken to be electromagnetically coupled and hence become co-moving on a very short timescale). A simple counting of degrees of freedom suggests that there ought to exist a new class of modes in a superfluid star. These 'superfluid' modes have the protons moving oppositely to the neutrons, unlike an 'ordinary' fluid mode that has the neutrons and protons moving more or less together (Lee, 1995; Andersson and Comer, 2001a). The right panel of Figure 13.1 demonstrates this feature. At the same time, the varying proton fraction no longer leads to g-modes, because the neutrons and protons have separate dynamical degrees of freedom.

13.3   Calculating stellar oscillation modes

Up to this point, our discussion has been somewhat qualitative. Although we managed to derive the f-modes of an incompressible star, we only provided a rough idea of the frequencies of the p- and g-modes. If we want to determine the rate of gravitational-wave emission we need to do better. However, in general we can no longer solve the problem analytically. We have to resort to numerical integration of the perturbation equations, together with suitable boundary conditions.

When we worked out the f-modes we found that all perturbed scalar quantities could be expanded in spherical harmonics, while the velocity followed from a potential χ, such that $\delta \boldsymbol{v} = \nabla \chi$. If we (quite generally) write[1]

$$\chi = \sum_l \chi_l Y_l^m, \tag{13.55}$$

then we have (for each l)

$$\delta \boldsymbol{v}_l = (\partial_r \chi_l) Y_l^m \hat{\boldsymbol{e}}_r + \frac{\chi_l}{r} \partial_\theta Y_l^m \hat{\boldsymbol{e}}_\theta + \frac{\chi_l}{r \sin \theta} \partial_\varphi Y_l^m \hat{\boldsymbol{e}}_\varphi. \tag{13.56}$$

In the general, compressible, case this class of perturbations is associated with a velocity field of form

$$r \delta \boldsymbol{v}_l = \left(W_l, V_l \partial_\theta, \frac{V_l}{\sin \theta} \partial_\varphi \right) Y_l^m = W_l Y_l^m \hat{\boldsymbol{e}}_r + r V_l \nabla Y_l^m. \tag{13.57}$$

These are called polar (spheroidal) perturbations. Such perturbations tend to be accompanied by significant variations in pressure and density which can, since they are scalar quantities, always be expanded in spherical harmonics. Hence, we have

$$\delta p = \sum_l \delta p_l Y_l^m, \tag{13.58}$$

and

$$\delta \rho = \sum_l \delta \rho_l Y_l^m. \tag{13.59}$$

The nature of a complementary class of perturbations can be deduced from the form of the polar velocity perturbation. Any vector proportional to $\hat{\boldsymbol{e}}_r \times \nabla Y_l^m$ will be orthogonal to the velocity field in (13.57). This leads to the axial (toroidal) perturbations, which are given by

$$r \delta \boldsymbol{v}_l = \left(0, \frac{U_l}{\sin \theta} \partial_\varphi, -U_l \partial_\theta \right) Y_l^m = \frac{1}{\sqrt{l(l+1)}} U_l \boldsymbol{Y}_{lm}^B, \tag{13.60}$$

---

[1] The various m multipoles are never coupled for axisymmetric (linear) systems (like rotating equilibrium stars). Hence, we can separately consider each m component and only need to sum over the permissible l contributions.

where we have introduced the magnetic multipoles (Thorne, 1980)

$$Y^B_{lm} = \frac{1}{\sqrt{l(l+1)}} \hat{e}_r \times \nabla Y_{lm}. \tag{13.61}$$

A non-rotating perfect fluid star has no non-trivial axial modes, but in a model with a solid crust there are distinct axial shear modes (see Chapter 18).

Because of the symmetry of the non-rotating problem, modes corresponding to different l and m decouple (as in the case of the f-modes). Hence, there is no need to sum over the various l and m in the case of perturbed spherical stars. The case of rotating stars is much more complicated. First of all, the symmetry is broken in such a way that the various $-l \le m \le l$ contributions become distinct. Secondly, rotation couples the different l-multipoles. As the rotation rate increases, an increasing number of Y^m_l's are needed to describe a given mode. One must also account for coupling between the polar and axial vectors. These factors make the problem of calculating pulsation modes of rapidly rotating stars a challenge.

An important concept in the study of oscillating rotating stars is the *pattern speed* of a given mode. As each mode is proportional to $e^{i(m\varphi+\omega t)}$ we see that surfaces of constant phase are described by

$$m\varphi + \omega t = \text{ constant.} \tag{13.62}$$

After differentiation this leads to

$$\frac{d\varphi}{dt} = -\frac{\omega}{m} = \sigma_p, \tag{13.63}$$

which defines the pattern speed, σ_p.

Having introduced this quantity, we make two observations concerning the $(l = m)$ f-modes. Let us denote the mode frequency observed in the rotating frame by ω_r, while the inertial frame frequency is ω_i. We then we see from (13.28) that the frequency of the f-modes increases with m roughly as $\omega_r \sim \sqrt{m}$. According to (13.63) this means that the pattern speed of the f-modes decreases as we increase m. As a consequence, one can always find an f-mode with arbitrarily small pattern speed (corresponding to a suitably large value of m) even though the high-order f-modes have increasingly large frequencies. This observation will be important later.

We also see that mode patterns corresponding to opposite signs of m tend to rotate around the star in different directions. Taking the positive direction to be that associated with the rotation of the star we find that the $l = \pm m$ modes are backwards and forwards moving (retro/prograde), respectively, in the limit of vanishing rotation. However, rotation may change the situation for the $l = m$ f-modes. Using (13.28), a very rough estimate of the corresponding mode in a rotating star (observed in the inertial frame) would be

$$\omega_i(\Omega) \approx \omega_r(\Omega = 0) - m\Omega + \text{higher order in } \Omega \text{ corrections,} \tag{13.64}$$

where Ω is the rotation rate. Given this, we estimate that these modes become prograde for rotation rates above

$$\Omega_s \approx \sqrt{\frac{3}{m}} \Omega_K, \qquad (13.65)$$

where Ω_K is the break-up frequency from (12.67). Basically, all but the $l = m = 2$ f-modes are likely to change from backwards to forwards moving (according to an inertial observer) at attainable rates of rotation ($\Omega \lesssim \Omega_K$). This will also be relevant later.

13.4 The r-modes

In general, a slowly rotating star has two classes of low-frequency modes: the g-modes, which arise because of stratification, and a set of inertial modes, which rely on the Coriolis force for their existence (Lockitch and Friedman, 1999). The relation between the two sets is illustrated, for a simple model, in Figure 13.2. In order to understand the distinction, it is useful to translate the Euler equations into the rotating frame. Now

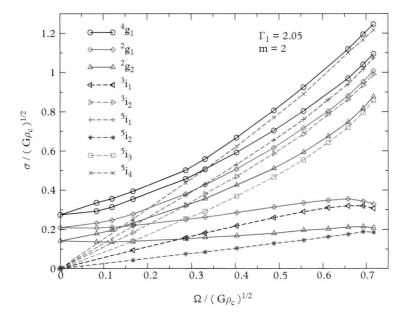

Figure 13.2 *An illustration of the relation between composition g-modes (in a simple model where $\Gamma_1 = 2.05$ while the background is given by a $\Gamma = 2$ polytrope) and inertial modes. The figure shows the frequencies of some selected g-modes (solid lines) and inertial modes (dashed) for the corresponding barotropic model. All frequencies are measured in the rotating frame. The results show that the Coriolis force dominates the buoyancy in sufficiently rapidly rotating stars. In this case, the g-modes are mainly restored by the Coriolis force when $\Omega/\sqrt{G\rho_c} > 0.3$ (where ρ_c is the central density), and thus become similar to the barotropic inertial modes. (Reproduced from Passamonti et al. (2009).)*

working in a coordinate basis and keeping only first-order terms in Ω^i, i.e. ignoring the rotational deformation, we have

$$\partial_t \delta v_i + 2\epsilon_{ijk}\Omega^j \delta v^k + \frac{\delta\rho}{\rho^2}\nabla_i p - \frac{1}{\rho}\nabla_i \delta p + \nabla_i \delta\Phi = 0, \tag{13.66}$$

where the second term is the Coriolis force, together with the continuity equation

$$\partial_t \delta\rho + \nabla_i\left(\rho\delta v^i\right) = 0. \tag{13.67}$$

Let us consider zero-frequency solutions to this problem, i.e. focus on solutions corresponding to neutral convective currents. Such fluid perturbations can be understood as follows. From (13.66) and (13.67) we see that stationary perturbations must be such that

$$\nabla_i(\rho\delta v^i) = 0, \tag{13.68}$$

and

$$\frac{\delta\rho}{\rho^2}\nabla_i p - \frac{1}{\rho}\nabla_i \delta p - \nabla_i \delta\Phi = 0. \tag{13.69}$$

As the two equations decouple we may consider any solution as a superposition of two distinct sets (Lockitch and Friedman, 1999):

$$\delta v^i \neq 0, \quad \delta p = \delta\rho = \delta\Phi = 0,$$
$$\delta v^i = 0, \quad \delta p, \delta\rho, \delta\Phi \text{ nonzero.}$$

In absence of external forcing any static self-gravitating perfect fluid must be spherical. Hence, the second type of solution simply identifies a neighbouring equilibrium. This implies that all stationary nonradial perturbations of an isolated spherical star must be of the first kind.

As we have already discussed, a general mode solution (with time dependence $e^{i\omega_r t}$ in the rotating frame) for an axisymmetric system can be expanded in angular harmonics in such a way that (this is the coordinate basis version of the combination of (13.57) and (13.60))

$$r\delta v^i = \sum_l \left[W_l Y_{lm}\nabla^i r + rV_l \nabla^i Y_{lm} - iU_l \epsilon^{ijk}\left(\nabla_j r\right)\left(\nabla_k Y_{lm}\right)\right] e^{i\omega_r t} \tag{13.70}$$

(the sum is over l only since the m multipoles still separate). As before, the two functions W_l and V_l describe the spheroidal component, while U_l is toroidal. For the stationary modes, one can readily verify that all toroidal displacements satisfy (13.68). In other words, the function U_l is unconstrained. In the case of spheroidal perturbations, one finds that the following relation between W_l and V_l must be satisfied,

$$\frac{d}{dr}(r\rho W_l) - l(l+1)\rho V_l = 0. \tag{13.71}$$

Still, one of the two functions is left unspecified also in this case. This latter class is no longer degenerate for non-barotropic perturbations. In general, the perturbations link the perturbed velocity to δp and $\delta \rho$, which (as we have seen) leads to the presence of g-modes. However, the zero-frequency toroidal modes remain also for non-barotropic stars.

Let us ask what happens when we bring the star into rotation. For spinning stars the Coriolis force provides a weak restoring force that gives the toroidal modes genuine dynamics. This leads to the so-called inertial modes, of which the r-mode is a particular example. A general inertial mode is a mixture of spheroidal and toroidal velocity components to leading order, but the r-modes are special as they remain predominantly toroidal.

The main properties of the r-mode can be understood from a simple exercise. Assume that the mode is purely toroidal to leading order also in the rotating case. Then the motion is essentially horisontal and W_l, V_l, $\delta \rho$, and δp are all of higher order in Ω. To leading order the mode we are interested in is completely determined by U_l.

For simplicity, let us consider a uniformly rotating constant density star. First we take the curl of the Euler equation (13.66) to derive an equation for the vorticity of the fluid. Then we (i) focus on mode-solutions behaving as $\exp[i(\omega_r t + m\varphi)]$, and (ii) make use of the standard recurrence relations for the spherical harmonics:

$$\cos\theta\, Y_l^m = Q_{l+1} Y_{l+1}^m + Q_l Y_{l-1}^m, \tag{13.72}$$
$$\sin\theta\, \partial_\theta\, Y_l^m = l Q_{l+1} Y_{l+1}^m - (l+1) Q_l Y_{l-1}^m, \tag{13.73}$$

with

$$Q_l = \left[\frac{(l+m)(l-m)}{(2l+1)(2l-1)}\right]^{1/2} \tag{13.74}$$

(as well as the ortogonality of the Y_l^m's). This way we end up with two equations:

$$[l(l+1)\omega_r - 2m\Omega]\, U_l = 0 \tag{13.75}$$

and

$$\{[(l-1)\omega_r - 2m\Omega]r\partial_r U_{l-1} + 2m(l-1)\Omega U_{l-1}\}\, Q_l$$
$$- \{[(l+2)\omega_r + 2m\Omega]r\partial_r U_{l+1} + 2m(l+2)\Omega U_{l+1}\}\, Q_{l+1} = 0. \tag{13.76}$$

From Eq. (13.75) we see that the only way to avoid a trivial solution is to have the mode frequency

$$\omega_r = \frac{2m\Omega}{l(l+1)}. \tag{13.77}$$

Moreover, an acceptable solution must be such that only a single U_l is nonzero. Given this, Eq. (13.76) leads to two relations

$$[(l\omega_r - 2m\Omega)r\partial_r U_l + 2ml\Omega U_l] Q_{l+1} = 0, \tag{13.78}$$

$$\{[(l+1)\omega_r + 2m\Omega]r\partial_r U_l + 2m(l+1)\Omega U_l\} Q_l = 0. \tag{13.79}$$

In general, these equations are not compatible—the problem is overdetermined. But in the special case of $l = m$ we have $Q_l = 0$, which means that the second equation is automatically satisfied. In this case an acceptable (non-trivial) mode-solution will exist, and we find that the eigenfunction is

$$U_m = r^{m+1}. \tag{13.80}$$

This solution represents the single r-mode that exists (for $l = m$) in a slowly rotating barotropic star (Lockitch and Friedman, 1999). One can show that this mode is only weakly affected by stratification and remains virtually unchanged in a non-barotropic Newtonian star. However, such stars have many additional r-modes, including various overtones (Saio, 1982; Andersson *et al.*, 1999a).

It is easy to see that the pattern speed for a typical r-mode is

$$\sigma_r = -\frac{2\Omega}{l(l+1)} < 0, \tag{13.81}$$

according to an observer rotating with the star. On the other hand, an inertial observer would find that

$$\sigma_i = \Omega \frac{(l-1)(l+2)}{l(l+1)} > 0. \tag{13.82}$$

The modes appear retrograde in the rotating system but an inertial observer always finds them to be prograde.

13.5   Gravitational-wave emission

If we want to estimate the rate at which gravitational-wave emission damps a given oscillation mode (and we do!), we can use the standard post-Newtonian multipole formulas. If we allow the background star to be rotating and consider both mass- and current multipole radiation, the luminosity associated with a given pulsation mode then follows from (Ipser and Lindblom, 1991; Lindblom *et al.*, 1998)

$$\frac{dE}{dt} = \sum_{l=2}^{\infty} N_l \omega_i^{2l+2} \left(|\delta D_{lm}|^2 + |\delta \mathcal{J}_{lm}|^2 \right),$$ (13.83)

where

$$N_l = \frac{4\pi G}{c^{2l+1}} \frac{(l+1)(l+2)}{l(l-1)[(2l+1)!!]^2}.$$ (13.84)

The first term in the bracket of (13.83) represents radiation due to the mass multipoles (familiar from Chapter 3), which are determined by (with the asterisk representing a complex conjugate)

$$\delta D_{lm} = \int \delta \rho \, r^l \, Y_{lm}^* dV.$$ (13.85)

The second term in the bracket of (13.83) corresponds to the current multipoles, which follow from

$$\delta \mathcal{J}_{lm} = \frac{2}{c} \sqrt{\frac{l}{l+1}} \int r^l (\rho \delta v + \delta \rho \Omega) \cdot \mathbf{Y}_{lm}^{B*} dV.$$ (13.86)

These formulas allow us to draw some general conclusions. First of all, we see that any fluid motion that leads to significant density variations (like the f-mode) will predominantly radiate through the mass multipoles. This follows from the fact that $|\delta \mathcal{J}|^2 \sim |\delta D|^2/c^2$, which means that the current multipole radiation is generally 'one order higher' in the post-Newtonian approximation.[2] However, there are situations where the current multipoles provide the main radiation mechanism. Most notably, this is the case for the r-modes, which will be discussed further in Chapter 15. In fact, the r-modes are unique among expected astrophysical sources of gravitational radiation in radiating primarily by gravito-magnetic effects.

13.6   What do we learn from the ellipsoids?

The general oscillations of a star emit gravitational waves, but under most circumstances the associated signal is too feeble to be detectable. We need scenarios involving

[2] By counting inverse powers of c in each of the terms in (13.83) we see that quadrupole perturbations ($l=2$) will lead to $\dot{E} \sim 1/c^5$ for the mass multipoles, while $\dot{E} \sim 1/c^7$ for the current multipole radiation. Using the standard way of counting orders, see Chapter 11, the mass multipole radiation arises at 2.5pN order while the current multipole radiation is a 3.5pN effect.

large-amplitude oscillations. This could be explosive events, like the boiling cauldron in which a neutron star is born, or the merger of compact stars at the end of binary evolution. We will consider these problems in Chapters 20 and 21, respectively. In addition, neutron stars may suffer a range of instabilities, often associated with specific oscillation modes. As an unstable mode grows it may reach a sufficiently large amplitude that the emerging gravitational waves can be detected. Hence, we need to understand the different classes of instabilities.

The most important instabilities are associated with rotating stars. We have already seen how rotation complicates the stellar oscillation analysis, so we know we are dealing with a messy problem. Fortunately, if we want to understand the nature and origin of different relevant instabilities, we can make progress without specific mode solutions. In fact, the key aspects of the instabilities that may be active in spinning stars can be illustrated by results for rotating ellipsoids—a classic problem in applied mathematics.

The relative mathematical simplicity makes a study of the equilibrium properties and stability of rotating self-gravitating fluid bodies with uniform density analytically tractable. Since the problem provides important insight into the stability properties of rotating stars, we will provide a brief introduction to it. An exhaustive discussion (covering results up to the late 1960s) was provided by Chandrasekhar (1973) and we draw on his results.

Consider an ellipsoid at rest with respect to a Cartesian coordinate system rotating with a constant frequency Ω around the x_3-axis. If the body is homogeneous and incompressible, then it is easy to integrate the equations for hydrostatic equilibrium (cf. the f-mode calculation from the beginning of the Chapter) to get

$$\frac{p}{\rho} + \Phi - \frac{1}{2}\Omega^2(x_1^2 + x_2^2) = \text{constant}. \tag{13.87}$$

To make further progress we need an expression for the gravitational potential. Formally, we know that

$$\Phi = -G \int \frac{\rho(\mathbf{x}')}{|\mathbf{x} - \mathbf{x}'|} dV'. \tag{13.88}$$

This leads to an expression for the interior gravitational potential of an ellipsoid with semi-axes a_1, a_2, and a_3 (for the moment allowing a_2 to be different from a_1, which would make the body triaxial),

$$\Phi = -\pi G \rho \left(I - \sum_{i=1}^{3} A_i x_i^2 \right), \tag{13.89}$$

where

$$I = a_1 a_2 a_3 \int_0^\infty \frac{du}{\Delta}, \tag{13.90}$$

$$A_i = a_1 a_2 a_3 \int \frac{du}{(a_i^2 + u)\Delta}, \tag{13.91}$$

and

$$\Delta^2 = (a_1^2 + u)(a_2^2 + u)(a_3^2 + u). \tag{13.92}$$

Given this result, the isobars correspond to

$$\left(A_1 - \frac{\Omega^2}{2\pi G\rho}\right) x_1^2 + \left(A_2 - \frac{\Omega^2}{2\pi G\rho}\right) x_2^2 + A_3 x_3^2 = \text{constant}. \tag{13.93}$$

Comparing this to a general ellipsoidal surface

$$\frac{x_1^2}{a_1^2} + \frac{x_2^2}{a_2^2} + \frac{x_3^2}{a_3^2} = 1, \tag{13.94}$$

we see that—in order for the two expressions to be compatible—we must have

$$\left(A_1 - \frac{\Omega^2}{2\pi G\rho}\right) a_1^2 = \left(A_2 - \frac{\Omega^2}{2\pi G\rho}\right) a_2^2 = A_3 a_3^2. \tag{13.95}$$

From these equalities we learn that

$$\frac{\Omega^2}{2\pi G\rho} = \frac{a_1^2 A_1 - a_2^2 A_2}{a_1^2 - a_2^2} = \frac{a_1^2 A_1 - a_3^2 A_3}{a_1^2} = \frac{a_2^2 A_2 - a_3^2 A_3}{a_2^2}, \tag{13.96}$$

where the first equality is only relevant if $a_1 \neq a_2$. Removing Ω from (13.95) we can also show that

$$a_1^2 a_3^2 (A_1 - A_2) + a_3^2 (a_1^2 - a_2^2) A_3 = 0. \tag{13.97}$$

Using the definitions for A_i this implies that we must have

$$(a_1^2 - a_2^2) \int_0^\infty \left[\frac{a_1^2 a_2^2}{(a_1^2 + u)(a_2^2 + u)} - \frac{a_3^2}{a_3^2 + u} \right] \frac{du}{\Delta}. \tag{13.98}$$

This equality can be satisfied in two ways. Either we have $a_1 = a_2$, in which case the prefactor vanishes, or $a_1 \neq a_2$ but then we need the integral to vanish identically. The first of these cases leads to the so-called Maclaurin spheroids, which are the most studied (and therefore best understood) figures of rotating, self-gravitating, homogeneous and incompressible fluid bodies in equilibrium. The second case corresponds to triaxial equilibrium configurations known as the Jacobi ellipsoids. These are rigidly rotating about the smallest axis (a_3), and have no vorticity when viewed from a rotating frame in which the figure appears stationary (i.e. the frame in which we performed the calculation). One can show that, for a given angular momentum, mass, and volume a Jacobi ellipsoid has lower energy than the corresponding Maclaurin spheroid.

In addition to these uniformly rotating ellipsoids there exists a number of configurations for which the shape of the surface is supported by internal flow, i.e. differential rotation. For our present purposes, the so-called Dedekind ellipsoids, which have a stationary triaxial shape in the inertial frame, are of particular interest. Their shape is entirely supported by internal motion with uniform vorticity. A Dedekind ellipsoid with the same mass and circulation as the corresponding Maclaurin configuration has lower angular momentum.

Let us return to the Maclaurin spheroids, and recall that it is often useful to introduce the parameter $\beta = T/|W|$; see Chapter 12. Since $a_1 = a_2$ we readily show that

$$I = \frac{2Ma_1^2}{5} \longrightarrow T = \frac{I\Omega^2}{2} = \frac{2}{5}\frac{Ma_1^2\Omega^2}{2}. \tag{13.99}$$

The gravitational potential energy requires more effort. One can show that (Chandrasekhar, 1973)

$$W = -\frac{2\pi G\rho M}{5}(2A_1 a_1^2 + A_3 a_3^2), \tag{13.100}$$

and it follows that

$$\beta = \frac{A_1 - (1 - e^2)A_3}{2A_1 + (1 - e^2)A_3}, \tag{13.101}$$

where we have expressed the ellipticity of the body as

$$e^2 = 1 - \frac{a_3^2}{a_1^2}. \tag{13.102}$$

In terms of this parameter, we have

$$A_1 = \frac{(1 - e^2)^{1/2}}{e^3}\sin^{-1}e - \frac{1 - e^2}{e^2}, \tag{13.103}$$

$$A_3 = \frac{2}{e^2} - \frac{2(1 - e^2)^{1/2}}{e^3}\sin^{-1}e, \tag{13.104}$$

which means that we arrive at the alternative expression

$$\beta = \frac{3}{2e^2}\left[1 - \frac{e(1-e^2)^{1/2}}{\sin^{-1}e}\right] - 1.$$ (13.105)

This parameterization may seem somewhat inconvenient, but it is difficult to invert the last relation to get an expression for β (say) in terms of the rotation rate Ω (which follows from (13.96)). Of course, it is straightforward to solve this problem numerically so we have no difficulties using the above results.

Let us assume that we start with a non-rotating uniform density star and spin it up, keeping the mass constant. Proceeding along this specific sequence of Maclaurin spheroids towards more rapidly rotating configurations, e.g. increasing β, one finds a bifurcation point at $\beta_s \approx 0.14$. At this point the Jacobi and Dedekind ellipsoids both branch off from the Maclaurin sequence; see Figure 13.3. Given the existence of these alternative states (with lower energy/angular momentum) beyond the point of bifurcation

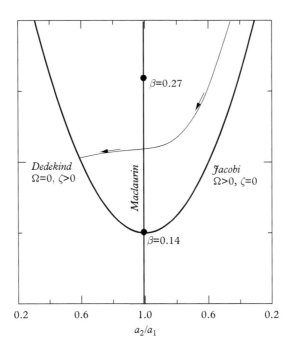

Figure 13.3 *A schematic summary of the instability results for rotating ellipsoids (a_2/a_1 represents the axis ratio, i.e. the ellipticity of the configuration). For values of β greater than 0.14 the Maclaurin spheroids are secularly unstable. Viscosity tends to drive the system towards a triaxial Jacobi ellipsoid, while gravitational radiation leads to an evolution towards a Dedekind configuration. Indicated in the figure is an evolution of this latter kind. Above $\beta \approx 0.27$ the Maclaurin spheroids are dynamically unstable, as there exists a Riemann-S ellipsoid with lower (free) energy. (Reproduced from Andersson (2003).)*

it would be favourable for a perturbed Maclaurin spheroid to move towards either the Jacobi or the Dedekind sequence. However, this is not possible as long as the system conserves circulation and angular momentum. As a result, the Maclaurin spheroid remains stable unless we add dissipation to the dynamical equations. In other words, the bifurcation point at β_s indicates the onset of *secular instabilities*.

Viscosity dissipates energy while preserving angular momentum. The Maclaurin spheroids are therefore susceptible to a viscosity-driven instability once $\beta > \beta_s$, and the instability drives the system towards the Jacobi sequence. Gravitational waves, on the other hand, radiate angular momentum while conserving the internal circulation. Thus, the Maclaurin spheroids also suffer a gravitational-wave-driven instability when $\beta > \beta_s$ (Chandrasekhar, 1970). The gravitational-wave instability tends to drive the system towards the Dedekind sequence (the members of which do not radiate gravitationally).

These secular instabilities actually set in through the quadrupole f-modes of the ellipsoids. Just like the equilibrium configurations, oscillations of rotating ellipsoids have been studied in great detail (Lai and Shapiro, 1995). Figure 13.4 shows the frequencies of the $l = |m| = 2$ Maclaurin spheroid f-modes. These modes are often referred to as the 'bar-modes', due to the way that they deform the star. The figure illustrates several general features of the pulsation problem for rotating stars. In particular we note (i) the rotational splitting of modes that are degenerate in the non-rotating limit, i.e. the $m = \pm 2$

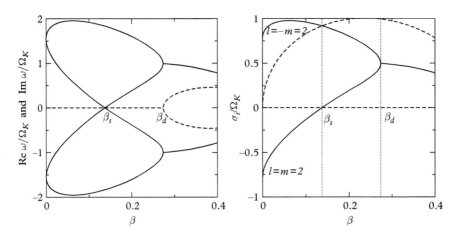

Figure 13.4 *Results for the $l = |m| = 2$ f-modes of a Maclaurin spheroid. The left frame shows the oscillation frequencies (solid lines) and imaginary parts (dashed lines) of the modes, while the right frame shows the mode pattern speed σ_i in the inertial frame for the two modes that have positive frequency in the non-rotating limit (the pattern speeds for the modes which have negative frequency in the non-rotating limit are obtained by reversing the sign of m). The dashed curves in the right frame represent a vanishing pattern speed (i) in the inertial frame (the horizontal line), and (ii) in the rotating frame (the circular arc, which shows Ω/Ω_K as a function of β). The points where the Maclaurin ellipsoid becomes secularly (β_s) and dynamically (β_d) unstable are indicated by vertical dotted lines. (Reproduced from Andersson (2003).)*

modes are distinct in the rotating case, and (ii) the symmetry with respect to $\omega = 0$, which reflects the fact that the equations are invariant under the change $[\omega, m] \to [-\omega, -m]$. In Figure 13.4 we also show the pattern speed for the two modes that have positive frequency in the non-rotating limit, cf. (13.28). We see that the $l = -m = 2$ mode, which is always prograde in the inertial frame, has zero pattern speed in the rotating frame at β_s ($\sigma_p = \Omega$). At this point the mode becomes unstable to the viscosity-driven instability. Meanwhile, the gravitational-wave instability sets in through the originally retrograde $l = m = 2$ mode. At β_s this mode has vanishing pattern speed in the inertial frame ($\sigma_p = 0$).

The evolution of the secular instabilities depends on the relative strength of the dissipation mechanisms. This tug-of-war is typical of these kinds of problems and tends to make them complex (see Chapter 15). Since the gravitational-wave-driven mode involves differential rotation it is damped by viscosity, and since the viscosity-driven mode is triaxial it is damped by gravitational-wave emission. A detailed understanding of the dissipation mechanisms is crucial for any study of secular instabilities of spinning stars.

In a non-dissipative model, the Maclaurin spheroids remain stable up to $\beta_d \approx 0.27$. At this point there exists a bifurcation to yet another family of ellipsoids that have lower 'free energy' than the corresponding Maclaurin spheroid for the same angular momentum and circulation. A dynamical transition to a lower energy state may take place without violating any conservation laws. In other words, at β_d the Maclaurin spheroids become dynamically unstable to $m = 2$ perturbations. This instability is usually refered to as the *dynamical bar-mode instability* (Toman et al., 1998; New et al., 2000)

In terms of the oscillation modes, the dynamical instability sets in at a point where two real-frequency modes merge, cf. Figure 13.4. At the bifurcation point β_d the two modes have identical frequencies and their angular momenta will vanish. Given this, one of the degenerate modes can grow without violating the conservation of angular momentum. The physical conditions required for the dynamical instability are easily understood. The instability occurs when the originally backwards moving f-mode (which has $\delta \mathcal{J} < 0$ for $\beta < \beta_d$) has been dragged forwards by rotation so much that it has 'caught up' with the originally forwards moving mode (which has $\delta \mathcal{J} > 0$ for $\beta < \beta_d$). In order for the modes to merge and become degenerate the perturbation must have vanishing angular momentum at β_d ($\delta \mathcal{J} = 0$). We will discuss the bar-mode instability in more detail when we turn to numerical simulations in Chapter 20.

13.7 Lagrangian perturbation theory for rotating stars

If we want to study rotational instabilities in detail we need to analyse the problem within the Lagrangian perturbation formalism developed by John Friedman and Bernard Schutz in the 1970s (Friedman and Schutz, 1978a). This approach is more powerful that the 'naive' approach we have used so far. In particular, it enables us to derive key conserved quantities for rotating stars.

The Lagrangian variation (Δ) of a quantity is related to the Eulerian variation (δ) by

$$\Delta = \delta + \mathcal{L}_\xi,$$

(13.106)

where the Lie derivative $\mathcal{L}_\xi$ has the meaning

$$\mathcal{L}_\xi p = \xi^i \nabla_i p \qquad (13.107)$$

for scalars,

$$\mathcal{L}_\xi v^i = \xi^j \nabla_j v^i - v^j \nabla_j \xi^i \qquad (13.108)$$

for contravariant vectors, and

$$\mathcal{L}_\xi v_i = \xi^j \nabla_j v_i + v_j \nabla_i \xi^j \qquad (13.109)$$

for covariant objects. The idea is illustrated in Figure 13.5.
   The Lagrangian change in the fluid velocity follows from

$$\Delta v^i = \partial_t \xi^i. \qquad (13.110)$$

Given this, and (note the similarity to the infinitesimal gauge transformation discussed in Chapter 3)

$$\Delta g_{ij} = \nabla_i \xi_j + \nabla_j \xi_i, \qquad (13.111)$$

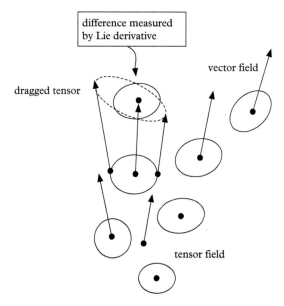

Figure 13.5 *An illustration of the Lie derivative. The derivative is taken with respect to a smooth vector field. The Lie derivative measures how a quantity (like a tensor field) actually changes, compared to the quantity dragged along by the vector field.*

where g_{ij} is the flat three-dimensional metric, we have

$$\Delta v_i = \Delta(g_{ij}v^j) = \partial_t \xi_i + v^j \nabla_i \xi_j + v^j \nabla_j \xi_i. \tag{13.112}$$

It is also useful to note that

$$\delta v^i = \partial_t \xi^i + v^j \nabla_j \xi^i - \xi^j \nabla_j v^i = g^{ij} \delta v_j. \tag{13.113}$$

Given the Lagrangian formalism, one can prove that

$$\Delta(\partial_t + \mathcal{L}_v)v_i = (\partial_t + \mathcal{L}_v)\Delta v_i, \tag{13.114}$$

a result that will be useful when deriving the perturbed equations of motion.

Let us consider a perfect fluid star, described by the familiar Euler equations from Chapter 4. The perturbed equation for mass conservation leads to

$$\Delta \rho = -\rho \nabla_i \xi^i \longrightarrow \delta\rho = -\nabla_i(\rho \xi^i), \tag{13.115}$$

while the perturbed gravitational potential follows from[3]

$$\nabla^2 \delta\Phi = 4\pi G \delta\rho = -4\pi G \nabla_i(\rho \xi^i). \tag{13.116}$$

In order to perturb the momentum equations we first rewrite (4.59) as

$$(\partial_t + \mathcal{L}_v)v_i + \frac{1}{\rho}\nabla_i p + \nabla_i \Phi - \frac{1}{2}\nabla_i v^2 = 0. \tag{13.117}$$

With the equation in this form we can make use of (13.114). Perturbing (13.117) we then have—after a bit of algebra—the perturbed Euler equations,

$$\rho \partial_t^2 \xi_i + 2\rho v^j \nabla_j \partial_t \xi_i + \rho (v^j \nabla_j)^2 \xi_i - \nabla_i[p\Gamma_1 \nabla_j \xi^j]$$
$$+ (\nabla_j \xi^j)\nabla_i p - (\nabla_i \xi^j)\nabla_j p + \rho \nabla_i \delta\Phi + \rho \xi^j \nabla_i \nabla_j \Phi = 0, \tag{13.118}$$

where Γ_1 is defined as before. It is useful to write this as (suppressing indices since there should be little risk of confusion)

$$A\partial_t^2 \xi + B\partial_t \xi + C\xi = 0. \tag{13.119}$$

---

[3] For future reference it is worth noting that it is natural to work with Eulerian variations of the gravitational field.

Defining the inner product

$$\langle \eta^i, \xi_i \rangle = \int (\eta^i)^* \xi_i dV, \tag{13.120}$$

where the asterisk denotes complex conjugation, one can show that

$$\langle \eta, A\xi \rangle = \langle \xi, A\eta \rangle^*, \tag{13.121}$$
$$\langle \eta, B\xi \rangle = -\langle \xi, B\eta \rangle^*. \tag{13.122}$$

The proof of the latter requires the background relation

$$\nabla_i(\rho v^i) = 0, \tag{13.123}$$

and holds as long as $\rho \to 0$ at the surface of the star. A slightly more tedious calculation leads to

$$\langle \eta, C\xi \rangle = \langle \xi, C\eta \rangle^*. \tag{13.124}$$

Assuming that η and ξ both solve the perturbed Euler equation (13.118), it is easy to show that the quantity

$$W(\eta, \xi) = \left\langle \eta, A\partial_t \xi + \frac{1}{2}B\xi \right\rangle - \left\langle A\partial_t \eta + \frac{1}{2}B\eta, \xi \right\rangle, \tag{13.125}$$

is conserved. That is, we have

$$\partial_t W = 0. \tag{13.126}$$

The fact that W is conserved motivates the definition of the *canonical energy* as

$$E_c = \frac{1}{2}W(\partial_t \xi, \xi) = \frac{1}{2}[\langle \partial_t \xi, A\partial_t \xi \rangle + \langle \xi, C\xi \rangle]. \tag{13.127}$$

This leads to

$$E_c = \frac{1}{2}\int \left[\rho|\partial_t \xi|^2 - \rho|v \cdot \nabla \xi|^2 + \Gamma p|\nabla \cdot \xi|^2 + (\xi^* \cdot \nabla p)(\nabla \cdot \xi) \right.$$
$$\left. + (\xi \cdot \nabla p)(\nabla \cdot \xi^*) + \xi^{i*}\xi^j(\nabla_i \nabla_j p + \rho \nabla_i \nabla_j \Phi) - \frac{1}{4\pi G}|\nabla \delta \Phi|^2 \right] dV. \tag{13.128}$$

For axisymmetric systems (like rotating stars) one can also show that the *canonical angular momentum*

$$J_c = \frac{1}{2} W(\partial_\varphi \xi, \xi) \tag{13.129}$$

is conserved. The proof of this relies on the fact that (i) $W(\eta, \xi)$ is conserved for any two solutions to the perturbed Euler equations, and (ii) ∂_φ commutes with $\rho v^j \nabla_j$ in axisymmetry, which means that if ξ solves the Euler equations, then so does $\partial_\varphi \xi$. Explicitly, the canonical angular momentum can be written

$$
\begin{aligned}
J_c &= \frac{1}{2} \left\{ \left\langle \partial_\varphi \xi, A \partial_t \xi + \frac{1}{2} B \xi \right\rangle - \left\langle A \partial_{t\varphi}^2 \xi + \frac{1}{2} B \partial_\varphi \xi, \xi \right\rangle \right\} \\
&= - \operatorname{Re} \left\langle \partial_\varphi \xi, A \partial_t \xi + \frac{1}{2} B \xi \right\rangle.
\end{aligned}
\tag{13.130}
$$

Before we move on, we need to mention a factor that complicates a stability analysis. The Lagrangian perturbation formalism permits the presence of so-called 'trivial' displacements (Friedman and Schutz, 1978*a*). In a sense, the trivials can be thought of as 'integration constants' representing a relabeling of the physical fluid elements. They correspond to displacements which leave the physical quantities unchanged, i.e. that are such that $\delta \rho = \delta v^i = 0$. The trivials cause trouble because they affect the canonical energy. Before one can use the canonical energy to assess the stability of a rotating configuration one must deal with this 'gauge problem'. The way to do this is to ensure that the displacement vector ξ is orthogonal to all trivials. Having said this, as long as we are considering modes of oscillation we are safe, because one can prove that all mode solutions are orthogonal to the trivials.

13.8   The CFS instability

The importance of the canonical energy stems from the fact that it can be used to test the stability of the system. In particular, we note that:

(i) If the system is coupled to radiation (e.g. gravitational waves) which carries away positive energy (which should be taken to mean that $\partial_t E_c < 0$), then any initial data for which $E_c < 0$ will lead to an instability.

(ii) Dynamical instabilities are only possible for motions such that $E_c = 0$. This makes intuitive sense since the amplitude of a mode for which E_c vanishes can grow without bounds and still obey the conservation laws.

Consider a complex normal-mode solution to the perturbation equations. That is, a solution of form

$$\xi^j \to \tilde{\xi}^j e^{i\omega t}, \tag{13.131}$$

with ω possibly complex. Then the associated canonical energy becomes

$$E_c = \omega \left[\operatorname{Re} \omega \langle \tilde{\xi}, A\tilde{\xi} \rangle - \frac{i}{2} \langle \tilde{\xi}, B\tilde{\xi} \rangle \right], \qquad (13.132)$$

where the expression in the bracket is easily shown to be real valued.

For the canonical angular momentum we get, in a similar way,

$$\mathcal{J}_c = -m \left[\operatorname{Re} \omega \langle \tilde{\xi}, A\tilde{\xi} \rangle - \frac{i}{2} \langle \tilde{\xi}, B\tilde{\xi} \rangle \right]. \qquad (13.133)$$

Combining these two relations we see that, for real frequency modes we have

$$E_c = -\frac{\omega}{m}\mathcal{J}_c = \sigma_p \mathcal{J}_c, \qquad (13.134)$$

where σ_p is the pattern speed of the mode.

Moreover, for real frequency normal modes, Eq. (13.133) can be rewritten as

$$\frac{\mathcal{J}_c}{\langle \tilde{\xi}, \rho\tilde{\xi} \rangle} = -m\omega + m\frac{\langle \tilde{\xi}, i\rho v \cdot \nabla\tilde{\xi} \rangle}{\langle \tilde{\xi}, \rho\tilde{\xi} \rangle}. \qquad (13.135)$$

Using cylindrical coordinates, and $v^j = \Omega\varphi^j$, one can show that

$$-i\rho\tilde{\xi}_i^* v^j \nabla_j \tilde{\xi}^i = \rho\Omega[m|\tilde{\xi}|^2 + i(\tilde{\xi}^* \times \tilde{\xi})_z]. \qquad (13.136)$$

However,

$$|(\tilde{\xi}^* \times \tilde{\xi})_z| \le |\tilde{\xi}|^2, \qquad (13.137)$$

so we must have (for uniform rotation)

$$\sigma_p - \Omega\left(1 + \frac{1}{m}\right) \le \frac{\mathcal{J}_c/m^2}{\langle \tilde{\xi}, \rho\tilde{\xi} \rangle} \le \sigma_p - \Omega\left(1 - \frac{1}{m}\right). \qquad (13.138)$$

This result forms an integral part of the proof that rotating perfect fluid stars are generically unstable in the presence of radiation (Friedman and Schutz, 1978b). The argument proceeds as follows: consider modes with finite frequency in the $\Omega \to 0$ limit. Then (13.138) implies that co-rotating modes (with $\sigma_p > 0$) must have $\mathcal{J}_c > 0$, while counter-rotating ones (for which $\sigma_p < 0$) will have $\mathcal{J}_c < 0$. In both cases $E_c > 0$ (from (13.134)), which means that the two classes of modes are stable. Now consider a small region near a point where $\sigma_p = 0$ (at a finite rotation rate). Typically, this corresponds to a point where the initially counter-rotating mode becomes co-rotating. In this region $\mathcal{J}_c < 0$. However, because of (13.134), E_c will change sign at the point where σ_p (or, equivalently, the frequency ω) vanishes. Since the mode was stable in the non-rotating limit the change of sign indicates the onset of instability.

The mechanism for gravitational-wave-driven instability can be understood in the following way. First consider a non-rotating star. The mode-problem leads to eigenvalues for ω^2, which, in turn, gives equal values $\pm|\omega|$ for the forwards and backwards propagating modes (corresponding to $m = \pm|m|$). These two branches of modes are affected by rotation in different ways, as illustrated in Figure 13.4. A backwards moving mode will be dragged forwards by the rotation, and if the star spins sufficiently fast the mode will move forwards with respect to the inertial frame. But the mode is still moving backwards in the rotating frame. The gravitational waves from such a mode carry positive angular momentum away from the star but, since the perturbed fluid actually rotates slower than it would in absence of the perturbation, the angular momentum of the mode is negative. The emission of gravitational waves makes the angular momentum increasingly negative and leads to the instability.

As we have already argued, one can always find an f-mode with a low pattern speed, even though the high-order (large l) modes have arbitrarily large frequencies. This means that, when the star is rotating there must be an unstable f-mode no matter how slow the rotation rate is (Friedman and Schutz, 1978b). Hence, the instability is generic in rotating stars. Of course, the high-order short-wavelength modes that become unstable first are not efficient gravitational-wave emitters. If we want to explore realistic situations where the gravitational-waves may dominate, we need to consider the physics of the neutron star interior beyond equilibrium. In order to facilitate a move in this direction, we need to explore more of the relevant physics.

14

Building mountains

If we want to model scenarios relevant to gravitational-wave physics, we (often) need to go beyond the global properties like mass and radius of a neutron star (recall Figure 12.2). In fact, this was clear from the very beginning—a stationary axisymmetric rotating star does not radiate gravitational waves. The fact that the star's shape is deformed by the rotation is irrelevant. In order to radiate gravitationally, the star must have some additional asymmetry. Such asymmetries may develop in a number of ways. An obvious possibility involves deformations of the elastic outer region of the star—the neutron star crust. In order to understand this problem we need to explore the properties of the crust, figure out how deformations may form, and (try to) estimate the likely size of 'mountains' on a realistic neutron star. This requires us to go beyond the fluid dynamics we used to describe stellar oscillations in the previous chapter and account for imposed stresses. The discussion also leads us to consider the internal magnetic field—which may deform the star, perhaps setting a natural lower limit for neutron star asymmetries.

14.1 The crust

At densities below about 2×10^6 g/cm^3 the lowest energy state of matter is in the form of iron (^{56}Fe). This is the matter of the neutron star 'skin'. As the density increases, it becomes energetically favourable for nuclei to capture electrons and undergo inverse beta decay

$$(A,Z) + e \rightarrow (A,Z-1) + \nu_e, \tag{14.1}$$

leading to the production of increasingly neutron-rich nuclei deeper in the crust; see Figure 14.1. In order to find the composition of the crust one has to work out the ground state of matter at various densities (Chamel and Haensel, 2008). This is complicated, as it involves accounting for the presence of nuclei which would be unstable in the laboratory but which become stable in the star because the electron Fermi sea has no available states for electrons produced by beta decay. Detailed calculations show that the mean nuclear weight A increases dramatically, while the total charge Z remains roughly constant, as

Gravitational-Wave Astronomy: Exploring the Dark Side of the Universe. Nils Andersson, Oxford University Press (2020).
© Nils Andersson. DOI: 10.1093/oso/9780198568032.001.0001

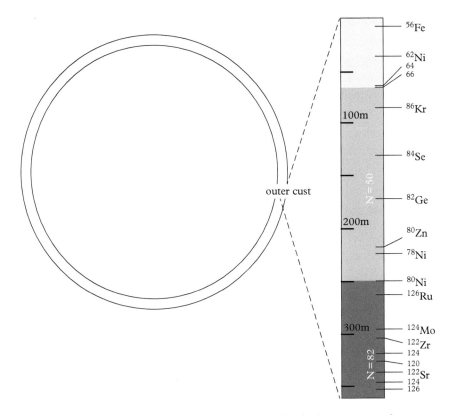

Figure 14.1 *A schematic illustration of the changing composition in the outer crust of a neutron star. Layers of increasingly neutron-rich nuclei—stabilized by the extreme pressure—are present at higher densities.*

the base of the crust is approached. The neutron number $N = A - Z$ tends to be close to the magic numbers $N = 50$ and 82 throughout the crust.

Once the crust composition has been determined, one can move on to the melting temperature and the shear modulus. The former is important because it tells us how soon after the star's birth the crust forms. The latter determines to what extent the crust lattice can withstand shear stresses.

As a starting point, it is typically assumed that the nuclei in the neutron star crust form a homogeneous body-centred cubic (bcc) lattice. This Coulomb lattice has interaction energy

$$E_{\text{Coul}} \approx \frac{Z^2 e^2}{a},\tag{14.2}$$

where e is the unit charge and a the mean spacing between nuclei—the lattice constant. A bcc lattice has two nuclei per unit cube (one in the centre plus 1/8 per corner times 8 corners), so given the number density of nuclei, n_N, we have

$$n_N a^3 = 2. \tag{14.3}$$

The crust's melting temperature is obtained by balancing the Coulomb energy to the thermal energy. The crust freezes when E_{Coul} exceeds $E_{th} = k_B T$ by a critical factor Γ. If we take the melting temperature to be T_m, we have

$$k_B T_m \approx \frac{1}{\Gamma} E_{Coul}, \tag{14.4}$$

where the value of Γ is empirically found to be about 175 (Strohmayer *et al.*, 1991). In order to allow for the fact that neutrons drip out of nuclei above a density of $4 \times 10^{11} \text{g/cm}^3$ (neutron drip), we may use

$$n_N = \frac{1 - x_f}{A} n, \tag{14.5}$$

where x_f is the fraction of 'free' neutrons and n is the baryon number density. This way we arrive at

$$T_m \approx 6.4 \times 10^9 \left(\frac{180}{\Gamma}\right) (1 - x_f)^{1/3} \left(\frac{Z}{20}\right)^2 \left(\frac{100}{A}\right)^{1/3} n^{1/3} \text{ K}. \tag{14.6}$$

This estimate tells us that the crust will begin to form once the star cools to a few times 10^9 K—within minutes after the neutron star is born; see Figure 18.7.

The rigidity of the crust is an issue of obvious importance for gravitational-wave estimates. Together with the breaking strain, the rigidity determines how large an asymmetry the nuclear lattice is able to sustain. This information is encoded in the shear modulus. Unfortunately, the neutron star crust turns out to be more like jelly than a solid. It can support shear stresses, but they are unlikely to be large.

In order to quantify this statement, we draw on the shear modulus of a bcc lattice. For a solid with cubic symmetry one can determine three independent elastic constants. One of these is directly related to the compressibility of the material, while the remaining two correspond to volume-preserving distortions of a lattice cell. Monte-Carlo simulations of the deformation of the neutron star crust (Ogata and Ichimaru, 1990; Horowitz and Hughto, 2008) show that we need to work with an effective shear modulus based on averaging over the degrees of freedom. This leads to

$$\mu = 0.1194 \left(\frac{4\pi}{3}\right)^{1/3} (Ze)^2 \left(\frac{1 - x_f}{A}\right)^{4/3} n^{4/3}. \tag{14.7}$$

The scaling with density makes sense if we consider the results for the Fermi gas model from Chapter 12.

Since the ratio

$$\tilde{\mu} = \frac{\mu}{\rho} \approx 10^{16} \text{ cm}^2/\text{s}^2, \tag{14.8}$$

only varies by a factor of a few across the relevant range of densities, the speed of shear waves is nearly constant throughout the crust. In fact, as a first approximation, it is quite reasonable to treat $\tilde{\mu}$ as a constant. It is useful to keep this in mind.

In some instances, e.g. for very young neutron stars, we may need to account for temperature dependence. Again, Monte-Carlo calculations show that one should then replace (14.7) by (Strohmayer *et al.*, 1991)

$$\mu_{\text{eff}} \approx \frac{0.1194}{1 + 1.781(100/\Gamma)^2} \left(\frac{4\pi}{3}\right)^{1/3} (Ze)^2 \left(\frac{1 - x_f}{A}\right)^{4/3} n^{4/3}. \tag{14.9}$$

It is instructive to ask how important the presence of the crust is likely to be in various dynamical scenarios. The answer depends on the extent to which the elasticity can overcome other forces at play. Let us, for example, quantify the crust's rigidity in terms of the ratio Λ (say) between the speed of shear waves and a characteristic rotational velocity Ω_0, i.e.

$$\Lambda \equiv \frac{\tilde{\mu}}{R^2 \Omega^2}. \tag{14.10}$$

The crust behaves almost like a fluid when $\Lambda \ll 1$ and is essentially rigid in the opposite limit. Using

$$\Omega_0^2 = \frac{GM}{R^3} \rightarrow \Omega_0^2 R^2 \approx 10^{20} \text{ cm}^2/\text{s}^2, \tag{14.11}$$

(such that the Kepler limit is about two-thirds of Ω_0; see Chapter 12) we have

$$\Lambda \approx 10^{-4} \left(\frac{\Omega_0}{\Omega}\right)^2. \tag{14.12}$$

This shows that the $\Lambda \ll 1$ limit is appropriate as long as we do not let $\Omega \rightarrow 0$. This conclusion is important in the context of stellar oscillations as it shows that the crust is likely to partake in global oscillations of the star without significantly altering the fluid motion. Of course, since the crust is elastic it can support shear waves so there will exist a distinct family of crustal modes of oscillation; see Chapter 18.

The deep crust region is expected to contain nuclear structures with different geometries (see Figure 14.2). These arise as the matter becomes frustrated and undergoes a series of transitions. Since these geometries resemble spaghetti and lasagne, this region has become known as the 'nuclear pasta' (Iida *et al.*, 2001). The pasta phases are likely to be disordered, which has implications for a range of phenomena. For example, electrons

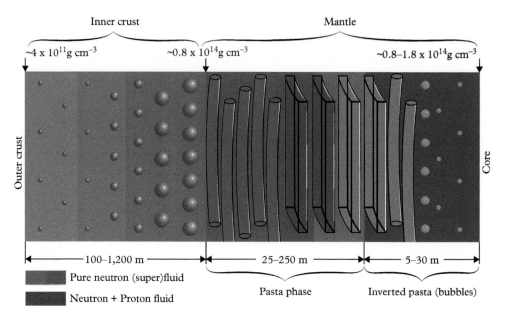

Figure 14.2 *Zooming in on the structure of the inner crust of a neutron star. At low densities, a lattice of neutron-rich nuclei is immersed in superfluid neutrons and a relativistic electron gas. At high densities the nuclei might deform and connect along specific directions. This leads to the formation of extended tubes, sheets, and bubbles of nuclear matter. These so-called 'nuclear pasta' phases are expected to form a layer at the base of the neutron star crust. (Reproduced from Newton (2013).)*

scattering off impurities in the lattice provide the dominant contribution to the electrical resistivity in the crust. The disordered nature of the pasta region may lead to a high electrical resistivity and faster than (otherwise) expected magnetic field dissipation (Pons *et al.*, 2013). The presence of nuclear pasta will also have implications for the crustal breaking strain, although the astrophysical importance of this is not yet clear.

14.2 Energetics

We have seen that the crust forms very early in a neutron star's life, at a time when the star may still spin rapidly. At this point, the star is more oblate than it will be once it spins down due to electromagnetic braking. As the star slows down, the centrifugal force decreases, leading to gravity trying to pull the crust into a less oblate shape. The rigidity of the crust will resist the change, making it remain less spherical than it would like to be. Once stresses build up to a critical level, the crust yields to relieve excess oblateness. As a result, the moment of inertia is suddenly reduced, and the star may be seen to spin up due to the conservation of angular momentum.

This scenario provided the first explanation for the enigmatic glitches observed in the Vela and Crab pulsars already in 1969 (Radhakrishnan and Manchester, 1969; Reichley and Downs, 1969). Shortly after the observations, it was proposed that the entire excess strain would be released in a quake (Ruderman, 1969). In order for this to happen, the crust would have to break into pieces much smaller than the radius of the star. After such an event, the crust presumably refreezes and new stress builds along with the star's continued spin-down. In this picture one would expect glitches to occur at roughly uniform time intervals (for any given pulsar), representing the time it takes to build up the stress from zero to the breaking point. However, this model has been ruled out as explanation for the large glitches seen in, for example, the Vela pulsar, as it cannot explain regularly occurring events of the observed magnitude (Baym and Pines, 1971). As discussed in Chapter 12, large glitches are now thought to be associated with the star's superfluid interior. Nevertheless, crust quakes remain a possible explanation for smaller glitches in, for example, the Crab pulsar. The effect may also be relevant for other pulsar timing irregularities (Hobbs *et al.*, 2010*b*).

The crust quake scenario provides a useful framework for discussing likely asymmetries of a rotating star. Of course, we are interested in gravitational waves, so we should focus on quadrupole deformations. Luckily, this happens to be the simplest case as we can parameterize the crust deformation in terms of a single parameter ϵ, the ellipticity introduced in Chapter 6. For the present discussion, it is convenient to let this parameter represent the departure of the moment of inertia of the crust I_c from the value it would have in the non-rotating case I_{c0}. That is, we define

$$\epsilon = \frac{I_c - I_{c0}}{I_{c0}}. \tag{14.13}$$

Let us now imagine that the crust solidified in the distant past, leaving it with a zero-strain oblateness ϵ_0 (the reference shape). This parameter will change only through crust failure or gradual plastic creep, both of which we ignore for the time being. If the actual oblateness ϵ differs from ϵ_0, the crust will store a strain energy (Pines and Shaham, 1972)

$$E_{\text{strain}} = B(\epsilon - \epsilon_0)^2, \tag{14.14}$$

where B is a constant related to the shear modulus, which must to be determined from the equation of state.

Similarly, we can write down an expression for the total energy of the star

$$E = E_0 + \frac{\mathcal{J}^2}{2I} + A\epsilon^2 + B(\epsilon - \epsilon_0)^2. \tag{14.15}$$

Here E_0 represents the energy of the spherical star, the second term is the kinetic energy ($\mathcal{J}$ is the angular momentum and I the total moment of inertia, as usual), the third term represents the increase in gravitational potential energy due to the star's shape no longer being spherical, and the fourth term gives the elastic strain energy. We can make these quantities precise by building on the virial argument from Chapter 12, but for now this (simpler) description will suffice. It is, however, useful to relate the energy expression to slowly rotating stellar models (also discussed in Chapter 12). First of all, we then see that, if I_0 denotes the moment of inertia of a spherical star and we define the oblateness ϵ in accordance with (14.13), the moment of inertia about the rotation axis is $I_0(1+\epsilon)$. Comparing this to the expression for the moment of inertia of a slowly spinning uniform density star, e.g. (12.36), we see that

$$\epsilon = \frac{5\Omega^2}{8\pi\rho}. \tag{14.16}$$

Moving on, the constant A depends on the equation of state but will generally be the same order of magnitude as the gravitational potential energy of the star. In fact, from (12.58) we see that for an incompressible solid star A is $-1/5$ of the gravitational potential energy of the non-rotating star.

The equilibrium configuration of the system can be found by minimizing the energy (14.15) at fixed angular momentum:

$$\left.\frac{\partial E}{\partial \epsilon}\right|_{\mathcal{J}} = 0, \tag{14.17}$$

leading to

$$\epsilon = \frac{I_0\Omega^2}{4(A+B)} + \frac{B}{A+B}\epsilon_0 \equiv e_\Omega + b\epsilon_0. \tag{14.18}$$

Here we have introduced the 'rigidity parameter' (Haensel, 1997)

$$b = \frac{B}{A+B}, \tag{14.19}$$

which vanishes for a fluid star (when $B=0$) and is equal to 1 for a perfectly rigid body (when $B/A \to \infty$). Clearly, the oblateness is made up of two parts. The first, e_Ω, scales as Ω^2 and we identify it as the centrifugal bulge due to rotation. The second term, $b\epsilon_0$, is due to the stresses of the nuclear lattice. The first contribution does not generate gravitational waves, as the deformation is symmetric with respect to the rotation axis. The second one can, but we need the deformation to be asymmetric.

As expected from (14.12), realistic equation of state calculations lead to $B \ll A$, with b a steeply varying function of the mass. Any stresses in the crust will only slightly change the shape away from that of the corresponding fluid body. The relative smallness of the Coulomb deformation is due to the fact that the involved forces are much weaker than the gravitational and centrifugal ones (on length scales typical to the star).

The rigidity parameter b is key to estimating the gravitational-wave emission from a deformed rotating star, so we need to understand how to work it out. Let us, first of all, consider the simple case of a uniform density star. We then have (again, see the discussion of the virial theorem in Chapter 12)

$$A = \frac{3}{25} \frac{GM^2}{R}. \tag{14.20}$$

Real neutron stars are, of course, compressible. The enhanced central concentration of mass leads to a value of A somewhat larger than this rough estimate. The parameter B is more difficult to determine. From the definition, we see that (for a thin uniform crust with volume V_c) it may be reasonable to use

$$B \approx \frac{\mu V_c}{2} \approx 2\pi \mu R^2 \Delta R, \tag{14.21}$$

where $\Delta R \ll R$ is the thickness of the crust. This leads to

$$b = \frac{B}{B+A} \approx \frac{B}{A} \approx \frac{25}{2} \frac{\mu}{\rho} \frac{\Delta R}{GM} \approx \frac{25}{2} \left(\frac{c^2 R}{GM} \right) \left(\frac{\Delta R}{R} \right) \left(\frac{\mu}{\rho c^2} \right). \tag{14.22}$$

In this expression the first factor is the inverse compactness of the star, typically close to 5, while the second term is the relative thickness of the crust, about $1/10$. The magnitude of the final factor can be gleaned from (14.8), which leads to the required factor being about 10^{-5}. Combining these factors, we arrive at a rough estimate of $b \approx 7 \times 10^{-5}$ for a typical neutron star.

More detailed calculations (Cutler *et al.*, 2003) show that we have overestimated the value of b by about a factor of 40. First of all, the rigidity parameter is reduced by about a factor of 5 due to the star's compressibility. An extra factor of 8 or so is due to cancellations in the integral over the stress tensor components. Combined, these reductions lead to $b \approx 2 \times 10^{-6}$ for quadrupole deformations of the crust.

Building on these estimates, we need to figure out how large a deformation a neutron star may sustain. As b is small we expect to have $\epsilon \approx \epsilon_\Omega$. One would not expect the shape of a rotating elastic star to deviate much from the shape of the corresponding fluid body. This is an important insight, but it does not tell us anything about the expected level of gravitational-wave emission. In order to emit gravitational waves we need the deformation to be asymmetric with respect to the rotation axis. In effect, we can ignore the effect of rotation and assume that that star's relaxed shape is spherical. As ϵ and ϵ_0 can differ at most by the breaking strain σ_{br} of the crust we can obtain the estimate we want from (14.18). However, the breaking strain is difficult to estimate. For terrestrial materials the breaking strain lies in the range $10^{-4} \leq \sigma_{br} \leq 10^{-2}$. In comparison to this, the nuclear lattice appears to be super strong. Molecular dynamics simulations for high-pressure Coulomb crystals (Horowitz and Kadau, 2009) suggest that

$$\sigma_{br} \approx 0.1. \tag{14.23}$$

Given this we arrive at the estimate

$$\epsilon \leq 2 \times 10^{-7} \left(\frac{\sigma_{br}}{0.1} \right). \tag{14.24}$$

Our simple energy estimates suggest that neutron stars may only sustain 'mountains' a fraction of a centimetre high. Not very impressive, but if we compare to the detectability arguments from Chapter 6, we see that these deformations may be large enough that the gravitational waves can be detected. We will consider the status of actual searches shortly.

14.3   Modelling elastic deformations

If we want to make more precise statements—based on the actual nuclear physics of the crust—we need to improve on the simple energetics argument. In order to do this, we have to solve the equations of elasticity in the solid phase, allowing for gravitational and pressure forces and including the degeneracy pressure, as well. This is not straightforward, but it should not be beyond our means. However, it is difficult to make the problem 'calculable' because it involves poorly understood evolutionary aspects. We need to know why the star is deformed in the first place. Plausible scenarios tend to involve additional (typically less controlled) physics. We will discuss some of the possibilities later.

As a first step towards more realistic models of a strained neutron star crust, let us focus on a question that does not rely on a specific formation scenario. What is the maximal allowed crust deformation? This may not be exactly the problem we would like to solve, but it nevertheless provides useful insight. In particular, it gives us a handle on upper limits on the gravitational-wave signal. This could, in turn, help us fine-tune our search strategy.

In this setup, we assume that the star has a (quadrupole) deformation imposed by some unknown agent. The crust is stressed away from its preferred reference shape, leading to the build-up of strain. At some point, we reach the breaking strain σ_{br} and the crust fails. The corresponding configuration tells us what the largest permissible mountain will be. In order to quantify the strain, we need to keep track of the difference between the unstrained shape and the actual shape of the crust. It is natural to represent this difference in terms of a Lagrangian displacement vector ξ^i (see Chapter 13). If we assume that the crust responds elastically, and work in Newtonian theory, it is useful to define the stress tensor of the solid as (Ushomirsky *et al.*, 2000)

$$\tau_{ij} = -pg_{ij} + t_{ij}, \tag{14.25}$$

where g_{ij} is the flat metric and we have isolated the isotropic pressure from the trace-free tensor

$$t_{ij} = \mu \left(\nabla_i \xi_j + \nabla_j \xi_i - \frac{2}{3} g_{ij} \nabla^k \xi_k \right), \tag{14.26}$$

which describes the shear stresses. As the expression is given in a coordinate basis, the covariant derivative ∇_i is the one associated with g_{ij}. In addition, we have the Poisson equation for the gravitational potential and the continuity equation. We need to solve the equations subject to the condition that the traction vanishes at the top and base of the crust. This is essentially a perturbation problem (as we expect the maximum deformation to the small). If we, for simplicity, assume that the relaxed configuration is spherical, then we can treat the deformation (and hence t_{ij}) as a first-order quantity. This leads to

$$\delta \tau_{ij} = -\delta p g_{ij} + t_{ij}. \tag{14.27}$$

Once we have solved the (now linearized) equations for a given deformation, we can calculate the corresponding quadrupole moment from

$$Q_{22} = \int \delta \rho r^4 dr. \tag{14.28}$$

Finally, we can relate the answer to our previous estimates by identifying[1]

$$\epsilon I_0 = \left(\frac{8\pi}{15} \right)^{1/2} Q_{22}. \tag{14.29}$$

In practice, we can pretty much ignore the numerical factor as the error it introduces is smaller than many of the other uncertainties. For example, taking the neutron star radius to be in the range suggested by X-ray observations, $R = 10 - 14$ km, introduces an uncertainty of a factor of a few in the moment of inertia. The actual density distribution also affects the result (recall that we simply assumed a uniform density sphere). For actual pulsars we also need the distance. This is typically obtained from a dispersion measure, which may also be wrong by a factor of a few.

If we opt to work with the quadrupole moment—rewriting the various gravitational-wave formulas accordingly—then we do not need to introduce the moment of inertia I_0. But this does not give us the intuitive connection with the magnitude of the deformation of the star encoded in ϵ.

---

[1] This identification depends on the convention for the spherical harmonics, whether we work with the complex form or take the real part from the outset. Here we use the convention from Ushomirsky *et al.* (2000), which is different from that in Thorne (1980). The relation we use has been adopted in LIGO observational papers, following Owen (2010).

This formulation of the problem allows us to make quantitative statements for any given deformation, equation of state, and crust model. However, if our main interest is in the largest possible mountain, then there is an elegant alternative (Ushomirsky *et al.*, 2000). This argument has the additional advantage that it leads us to introduce some of the tools we will need later. In essence, we need to expand the tensor t_{ij} in a suitable set of harmonics. The basic idea is that different parts of t_{ij} transform differently under rotations on the two-sphere. Schematically, we have

$$t_{ij} \sim \begin{pmatrix} \boxed{S} & \boxed{V} \\ \boxed{V} & \boxed{T} \end{pmatrix}.$$

The part marked S transforms as a scalar under rotation, V transform as a two-vector, and T transforms as a two-tensor. We have seen the first two contributions already. Quantities which transform as scalars can be expanded in spherical harmonics and the vectors we need are the same as in the case of stellar oscillations (see Chapter 13). Basically, we have the two vectors (momentarily suppressing the indices on the spherical harmonics, $Y_l^m \to Y$, to avoid confusion)

$$V_i^1 = \tilde{\nabla}_i Y = \partial_i Y, \tag{14.30}$$

and

$$V_i^2 = \epsilon_{ij} \gamma^{jk} \tilde{\nabla}_k Y, \tag{14.31}$$

where we have used the flat metric on the sphere (which defines the covariant derivative $\tilde{\nabla}_i$ in these expressions)

$$\gamma_{ij} = \begin{pmatrix} 1 & 0 \\ 0 & \sin^2\theta \end{pmatrix}, \tag{14.32}$$

and the corresponding anti-symmetric tensor

$$\epsilon_{ij} = \begin{pmatrix} 0 & -\sin\theta \\ \sin\theta & 0 \end{pmatrix}. \tag{14.33}$$

For the tensor part (on the sphere) we have three distinct (symmetric) contributions

$$T_{ij}^1 = \tilde{\nabla}_i \tilde{\nabla}_j Y, \tag{14.34}$$

$$T_{ij}^2 = \gamma_{ij} Y, \tag{14.35}$$

and

$$T_{ij}^3 = \frac{1}{2}\left(\epsilon_i{}^l T_{lj}^1 + \epsilon_j{}^l T_{li}^1\right).$$

(14.36)

Making use of these contributions we have, from V_i^2 and T_{ij}^3, respectively,

$$e_{ij}^1 = \begin{pmatrix} 0 & -\frac{1}{\sin\theta}\partial_\varphi Y_{lm} & \sin\theta\,\partial_\theta Y_{lm} \\ \text{sym} & 0 & 0 \\ \text{sym} & 0 & 0 \end{pmatrix},$$

(14.37)

$$e_{ij}^2 = \begin{pmatrix} 0 & 0 & 0 \\ 0 & \frac{1}{\sin\theta}X_{lm} & -\frac{\sin\theta}{2}W_{lm} \\ 0 & \text{sym} & -\sin\theta\,X_{lm} \end{pmatrix},$$

(14.38)

where

$$X_{lm} = (\partial_\theta\partial_\varphi - \cot\theta\,\partial_\varphi)Y_{lm},$$

(14.39)

and

$$W_{lm} = \left(\partial_\theta^2 - \cot\theta\,\partial_\theta - \frac{1}{\sin^2\theta}\partial_\varphi^2\right)Y_{lm}.$$

(14.40)

Both e_{ij}^1 and e_{ij}^2 are manifestly trace-free and they transform as $(-1)^{l+1}$ under the spatial inversion $(\theta,\varphi) \to (\pi-\theta, \pi+\varphi)$. In the literature the corresponding solutions are commonly referred to as *odd parity* (as they change sign under inversion for $l=2$), *axial* (as they can be associated with rotation), or *toroidal*.

Similarly, we have from the remaining building blocks

$$f_{ij}^1 = \begin{pmatrix} Y_{lm} & 0 & 0 \\ 0 & 0 & 0 \\ 0 & 0 & 0 \end{pmatrix},$$

(14.41)

$$f_{ij}^2 = \begin{pmatrix} 0 & \partial_\theta Y_{lm} & \partial_\varphi Y_{lm} \\ \text{sym} & 0 & 0 \\ \text{sym} & 0 & 0 \end{pmatrix},$$

(14.42)

$$f_{ij}^3 = \begin{pmatrix} 0 & 0 & 0 \\ 0 & Y_{lm} & 0 \\ 0 & 0 & \sin^2\theta\,Y_{lm} \end{pmatrix}',$$

(14.43)

and

$$f_{ij}^4 = \begin{pmatrix} 0 & 0 & 0 \\ 0 & \partial_\theta^2 Y_{lm} & X_{lm} \\ 0 & \text{sym} & Z_{lm} \end{pmatrix}.$$ (14.44)

where

$$Z_{lm} = (\partial_\varphi^2 + \sin\theta \cos\theta \, \partial_\theta) Y_{lm}.$$ (14.45)

These solutions are often called *even parity*, *polar*, or *spheroidal*. They all transform as $(-1)^l$ under space inversion. It is worth noting that f_{ij}^4 is not trace-free, which is inconvenient as we prefer to work in Transverse-Traceless (TT) gauge. However, a second suitable tensor (alongside e_{ij}^2) is easily obtained from

$$\tilde{f}_{ij}^4 = f_{ij}^4 + \frac{l(l+1)}{2} f_{ij}^3.$$ (14.46)

Similarly, we have the trace-free combination

$$\tilde{f}_{ij}^1 = f_{ij}^1 - \frac{1}{2} f_{ij}^3.$$ (14.47)

After this fairly lengthy detour, let us return to the mountain problem. For the kind of deformations we are interested in, we need the second set of tensor harmonics. Moreover, as we want the trace of t_{ij} to vanish we need to use a combination of $\tilde{f}_{ij}^1$, f_{ij}^2 and $\tilde{f}_{ij}^4$. Following Ushomirsky *et al.* (2000) we use[2]

$$e_{ij} = g_{ij} - r_i r_j = r^2 f_{ij}^3 \quad \longrightarrow \quad \left(r_i r_j - \frac{1}{2} e_{ij} \right) Y_{lm} = r^2 \tilde{f}_{ij}^1,$$ (14.48)

$$f_{ij} = \frac{r^2}{\beta} f_{ij}^2, $$ (14.49)

where $\beta = \sqrt{l(l+1)}$, and

$$\Lambda_{ij} = \frac{r^2}{\beta^2} f_{ij}^4 + \frac{1}{\beta} f_{ij} \quad \longrightarrow \quad \Lambda_{ij} + \frac{1}{2} e_{ij} Y_{lm} = \beta^{-2} r^2 \left(\tilde{f}_{ij}^4 + f_{ij}^2 \right).$$ (14.50)

---

[2] This is a bit of a sideways step, but it is useful as it allows a direct comparison to the original work.

In terms of this set of harmonics, we have

$$t_{ij} = t_{rr}\left(r_i r_j - \frac{1}{2}e_{ij}\right)Y_{lm} + t_{r\perp}f_{ij} + t_\Lambda\left(\Lambda_{ij} + \frac{1}{2}e_{ij}Y_{lm}\right). \tag{14.51}$$

The equations we need to solve for a deformed equilibrium are:

$$\nabla^i \delta t_{ij} = \delta\rho g(r)r^i + \rho\nabla^i \delta\Phi, \tag{14.52}$$

where ρ represents the background density and $g(r) = Gm(r)/r^2$ (with $m(r)$ the mass inside radius r) is the gravitational acceleration. Rearranging this expression, we can integrate the radial component to get the quadrupole moment. Thus, we have (making the Cowling approximation, $\delta\Phi = 0$)

$$Q_{22} = \int\left[\frac{r^4}{g(r)}\left(\frac{3}{2}\frac{dt_{rr}}{dr} - \frac{4}{\beta}\frac{dt_{r\perp}}{dr} - \frac{r}{\beta}\frac{d^2 t_{r\perp}}{dr^2} + \frac{1}{3}\frac{dt_\Lambda}{dr} + \frac{3}{r}t_{rr} - \frac{\beta}{r}t_{r\perp}\right)\right]dr. \tag{14.53}$$

If we also integrate by parts (assuming that the shear modulus μ vanishes at the edges of the crust, which may not be true but is 'convenient' as it allows us to ignore surface terms) we arrive at

$$Q_{22} = -\int_{rb}^R \frac{r^3}{g}\left[\frac{3}{2}(4-U)t_{rr} + \frac{1}{3}(6-U)t_\Lambda\right.$$
$$\left. + \sqrt{\frac{3}{2}}\left(8 - 3U - \frac{1}{3}U^2 - \frac{r}{3}\frac{dU}{dr}\right)t_{r\perp}\right]dr, \tag{14.54}$$

where

$$U = \frac{d\ln g}{d\ln r} + 2. \tag{14.55}$$

So far we have estimated the elastic response of the crust. However, real solids behave elastically only up to the maximum strain, σ_{br}, beyond which they either fail or deform plastically. In order to estimate the highest neutron star mountain we assume that the crust fails upon reaching a certain yield strain. We then need to consider the strain tensor rather than the stresses, as this allows us to impose a simple criterion to establish when the crust breaks. The strain tensor is simply defined as:

$$\sigma_{ij} = t_{ij}/\mu. \tag{14.56}$$

If we define the strain scalar as $\sigma^2 = \frac{1}{2}\sigma_{ij}\sigma^{ij}$, the standard Von Mises criterion states that the crust yields when

$$\sigma \geq \sigma_{\text{br}}. \tag{14.57}$$

Our expression for the stresses (14.51) leads to

$$\sigma_{ij}\sigma^{ij} = \frac{3}{2}\sigma_{rr}^2[\text{Re}(Y_{lm})]^2 + \sigma_{r\perp}^2[\text{Re}(f_{ij})]^2 + \sigma_{\Lambda}^2[\text{Re}(\Lambda_{ij} + \frac{1}{2}Y_{lm}e_{ij})]^2, \tag{14.58}$$

where σ_{rr}, $\sigma_{r\perp}$, and σ_{Λ} are the components of the strain tensor associated with the tensor spherical harmonics from (14.51). By assuming that the quadrupole reaches its largest value when the equality of Eq. (14.57) is satisfied, we find that the maximum is obtained when (Ushomirsky et al., 2000)

$$\sqrt{3}\sigma_{rr} = \sqrt{6}\sigma_{r\perp} = \sigma_{\Lambda} = \sqrt{\frac{96\pi}{5}}\sigma_{\text{br}}. \tag{14.59}$$

The Von Mises criterion, together with the expression for the maximum quadrupole in (14.114), leads to a maximum quadrupole deformation

$$Q_{22}^{\text{max}} \approx 10^{39}\left(\frac{\sigma_{\text{br}}}{0.1}\right)\text{ g cm}^2, \tag{14.60}$$

no matter how the strain arises, under the assumption that all the strain is in the $l = m = 2$ harmonic. Strain in other harmonics would push the crust closer to the yield point without contributing to the quadrupole. The estimate is an upper limit for the quadrupole deformation the crust can sustain, and hence on the energy the star may emit in gravitational waves. Taking the (canonical) moment of inertia to be $I_0 \approx 10^{45}$ g cm^2 we have

$$\epsilon_{\text{max}} \approx 10^{-6}\left(\frac{\sigma_{\text{br}}}{0.1}\right). \tag{14.61}$$

The implications of this estimate are clear from Figure 14.5. The spin-down limit (obtained from observed P and $\dot{P}$ for known pulsars; see Chapter 6) suggests an overly optimistic level of gravitational-wave emission at frequencies below 100 Hz or so. However, the more detailed analysis leads to a result that only differs from the basic energetics argument by a factor of a few. This could be an indication that we are close to the real answer. Of course, we have made a number of simplifications and it is legitimate to ask how they affect the outcome. For example, we need to quantify the impact of the Cowling approximation. It would also be desirable to explore the effects of general relativity—especially if we want to make estimates using realistic equation of state. While it may be reasonable to treat the crust elasticity at the Newtonian level, this is clearly not the case for the star's core (which will also be deformed at some level). More detailed

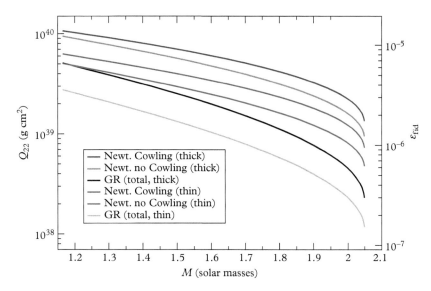

Figure 14.3 *Comparing results for the maximum quadrupole moment Q_{22} (and fiducial ellipticity) for a deformed neutron star in the Newtonian Cowling approximation, including the perturbed gravitational potential (no Cowling), and the full relativistic calculation (including stress contributions, GR) due to crustal stresses vs mass for two choices of crustal thickness. The models are obtained for the SLy equation of state and assumes a breaking strain of 0.1. (Reproduced from Johnson-McDaniel and Owen (2013), copyright (2013) by the American Physical Society.)*

modelling (Haskell *et al.*, 2006; Johnson-McDaniel and Owen, 2013) shows that the typical uncertainty is at the level of a factor of a few; see Figure 14.3.

14.4   Searches for known pulsars

As we make sluggish progress on the modelling, we can seek guidance from observations. The signal from a spinning neutron star is unavoidably weak, but we have seen (in Chapter 8) that the effective amplitude (after matched filtering) improves (roughly) as the square-root of the observation time. Given the expected maximum deformation, we can easily work out that—for a neutron star emitting gravitational waves at 100 Hz at a distance of 1 kpc—we need observations lasting at least one year. This tells us that we need to analyse long stretches of data, which makes the problem computationally demanding. On the other hand, we know the location and spin rate of many radio pulsars, so at least we have some idea of what we are looking for.

With a known position and spin evolution, one can use long stretches of data in a coherent way. One can dig deeper into the noise. Taking the introduction to the data analysis problem from Chapter 8 as our starting point, we know that we have four unknown signal parameters: the gravitational-wave amplitude h_0, the unknown phase

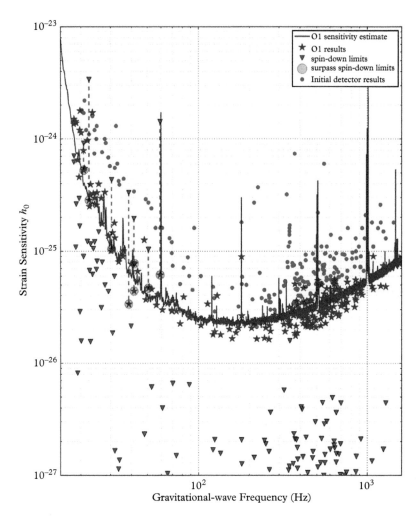

Figure 14.4 *A summary of upper limits from targeted pulsar searches in O1 data. Stars show 95% credible upper limits on the gravitational-wave amplitude, h_0, for 200 pulsars. Upside-down triangles give the spin-down limits for all pulsars (based on distance values taken from the ATNF pulsar catalog (apart from in a few cases; see the original paper for details) and assuming the canonical moment of inertia. The upper limits shown within shaded circles are those for which the spin-down limits (linked via the dashed vertical lines) are surpassed with the observations. The grey curve provides an estimate of the detector strain sensitivity, combining representative amplitude spectral density measurements for the two LIGO instruments. Results from the initial detector era (Aasi et al., 2014) are shown as red circles. (Reproduced from Abbott et al. (2017e) by permission of the AAS.)*

of the wave ϕ_0, the inclination of the spin axis i, and the polarization angle ψ. One of the advantages of dealing with known systems is that we may have prior information about some of the parameters. This alleviates the computational challenge. For example, in the case of the Crab pulsar one can infer the orientation of the polarization angle from X-ray observations of the pulsar wind nebula. This reduces the computational cost of a search. Of course, one has to use any such information judiciously. After all, it may be that the star is not aligned with the observed structures.

An observational milestone was reached when LIGO used data from the first 9 months of the S5 science run (2005–7) to beat the Crab pulsar spin-down limit (Abbott *et al.*, 2008a). It may have been obvious from the beginning that there was no real possibility that 100% of the observed Crab pulsar spin-down was gravitational-wave powered, as this would conflict with the measured braking index (see Chapter 6). Nevertheless, the demonstration that less than 6% of the available power is radiated as gravitational waves was the first step into unknown territory. The summary of the results from the entire S5 run (about 500 days of data; see Abbott *et al.* (2010)), provided upper limits on the gravitational-wave emission for 116 pulsars. The Crab result was tightened to a maximum of 2% of the available power and we learned that some pulsars have an astonishing degree of symmetry. In the case of PSR J2124-3358 (spinning at 200 Hz) the maximum allowed ellipticity was found to be $\epsilon = 7 \times 10^{-8}$. Even though this is above the spin-down limit for this system, it is an impressive result. The equator of this pulsar is one of the roundest things in the Universe.

At the end of the initial interferometer era, the spin-down limit had also been beaten for the Vela pulsar (crucially involving data from the Virgo instrument, with better sensitivity at lower frequencies; see Abadie *et al.* (2011a)). The first observing run of the advanced detectors provided upper limits that improved on the spin-down estimates for another 6 pulsars (based on about 70 days of data; Abbott *et al.* (2017e)). The gravitational-wave contribution to the Crab pulsar spin down was limited to less than 0.2% and the total catalogue of upper limits included 200 pulsars. The O1 results are illustrated, and compared to the detector sensitivity, in Figure 14.4. The corresponding limits on the ellipticities (ϵ) and mass quadrupole moments (Q_{22}) are shown in Figure 14.5. It is instructive to compare these results to the estimated maximum allowed deformations from Figure 14.3.

14.5 All-sky searches

Targeted pulsar searches (obviously) make use of information gleaned from electromagnetic observations—sky location, spin-down rate, and so on. Having this information is an advantage, as it allows us to dig deeper into the data. However, even though this allows us to set precise limits on the gravitational-wave strain, one can argue that we already 'know' that some of these systems are unlikely to be exciting gravitational-wave sources. Given the (admittedly small sample of) braking index measurements for young systems, one would expect the spin-down to be dominated by the electromagnetic torque. Let us (for the moment) suppose that this is the case. Where does it leave us? We would

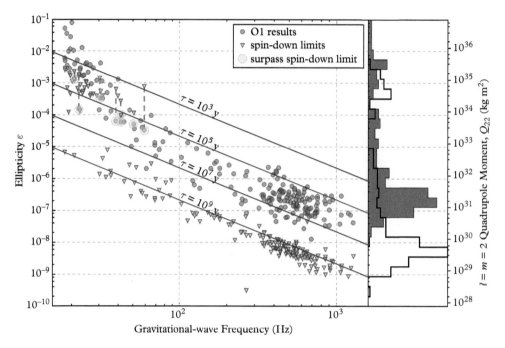

Figure 14.5 *Limits on fiducial ellipticities (ε) and mass quadrupole moments (Q₂₂) for pulsars targeted in the O1 LIGO run. Upside-down triangles represent the spin-down limits for these systems. Pulsars for which the spin-down limit is beaten are highlighted within larger shaded circles and linked to their spin-down limit values with dashed vertical lines. The diagonal lines show the constant characteristic age, τ, for hypothetical systems that spin down entirely due to gravitational waves (with braking indices of n = 5). (Reproduced from Abbott et al. (2017e) by permission of the AAS.)*

immediately know that we have to beat the inferred spin-down limit in order to have a chance of detecting a continuous gravitational-wave signal, but without a more detailed understanding of the evolutionary scenario(s) that lead to the formation of asymmetries in the system (e.g. in the star's crust) we cannot know how sensitive our searches have to be in order to be successful. As we will see later, the internal magnetic field sets a lower limit on quadrupole deformation, but such estimates are also associated with significant uncertainties. This seems rather pessimistic.

Taking a more upbeat view, we might assume that electromagnetic observations do not tell us the whole story. Could it perhaps be that there are systems that spin down predominantly due to gravitational waves, without an observable electromagnetic counterpart signal? In principle, there is no reason why this could not be the case (although one would again have to think carefully about the evolution leading to the formation of such an object). Suppose we want to search for electromagnetically silent, yet spinning down objects, how do we go about it? The answer is clear. As we are literally looking in the dark, we have to search across the entire sky.

In general, a blind search over the plausible parameter space is severely limited by computing cost. One has to cover as large a portion of the plausible parameter space as possible, including the signal frequency and its time derivative. In order to facilitate a sensitive search, observers need to constrain the parameters. One way to do this is to set an upper limit on the spin-down rate. For systems where we do not see pulses this is tricky, but we can make use of a simple energetic argument. If we assume that all the star's kinetic energy is released as gravitational waves due to a quadrupole deformation, we have (see Chapter 6)

$$\frac{d}{dt}\left(\frac{1}{2}I_0\Omega^2\right) = -\frac{32G}{5c^5}\Omega^6(I_0\epsilon)^2, \tag{14.62}$$

where ϵ is the star's ellipticity and I_0 is the moment of inertia. If we assume that the star has spun down significantly since its birth (a natural assumption if the system is emitting continuous gravitational waves) we see that the characteristic spin-down timescale is

$$\tau = \frac{5c^5}{2^7 G}\frac{1}{I_0\epsilon^2\Omega_0^4}, \tag{14.63}$$

where Ω_0 is the initial spin rate. Combining this energy loss with the usual formula for the gravitational-wave strain, we have

$$h_0 \approx \frac{4G}{c^4}\frac{I_0\epsilon}{d}\left(1+\frac{t}{\tau}\right)^{-1/2}\Omega_0^2. \tag{14.64}$$

Again taking the $t/\tau \gg 1$ limit and using (14.63) we arrive at

$$h_0 \approx \left(\frac{5G}{8c^3}\right)^{1/2}\left(\frac{I_0}{t}\right)^{1/2}\frac{1}{d}. \tag{14.65}$$

Basically, we have an upper limit on the gravitational-wave strain from a neutron star of known age and at a given distance. We do not need either the spin frequency or the ellipticity.

An amplitude estimate like (14.65) is helpful, as it provides a plausible constraint on the parameters for an all-sky search, but the problem is still challenging. In order to alleviate the computational severity, all-sky searches tend to be hierarchical (coherent search strategies are prohibitively expensive). One would typically start from a fast Fourier transform over a fixed length data set, to identify 'interesting' peaks in the corresponding time–frequency maps, and then drill deeper using some combination of coherent and incoherent steps to reduce the computational burden (Wette *et al.*, 2008). This inevitably leads to a loss of sensitivity. As a rough guide, one would expect an all-sky search to be up to an order of magnitude less sensitive than a targeted pulsar search. The most sensitive search to date, based on LIGO O1 data, sets upper limits on continuous gravitational waves in the frequency range 475–2,000 Hz (Abbott *et al.*, 2018c). At the highest frequencies the search was sensitive to a neutron star with ellipticity 1.8×10^{-7} at a distance of 1 kpc. With improved computing power, future searches will involve longer coherence times and even better sensitivities.

The most 'ambitious' all-sky search effort is Einstein@Home (Abbott *et al.*, 2009*a*, 2017*d*), the goal of which is to carry out coherent searches for signals using wasted CPUs on idle computers at homes, offices, and university departments around the world. The project has been successful in attracting a large number of subscriptions and provides powerful computational infrastructure for the search for continuous-wave signals. The Einstein@Home framework has also been adapted for the search for pulsars in gamma-ray data from the Fermi satellite, leading to the discovery of a number of new systems (Pletsch *et al.*, 2013; Lazarus *et al.*, 2016; Clark *et al.*, 2018). An example of the achievable gravitational-wave sensitivity (in the frequency range 20–100 Hz) is provided in Figure 14.6. In this case, the search rules out neutron stars with an ellipticity larger than 10^{-5} or so within 100 pc of the Earth (Abbott *et al.*, 2017*d*). This figure is also

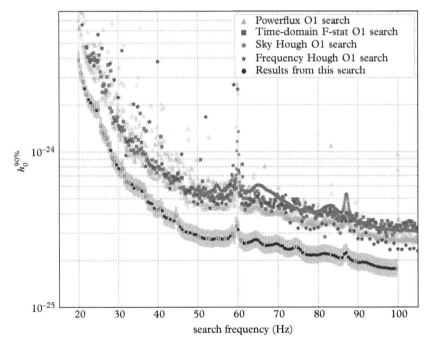

Figure 14.6 *The 90% confidence upper limits on the amplitude of continuous gravitational-wave signals with frequency in 0.5-Hz bands and with spin-down values in the range [−2.65 × 10⁻⁹, 2.64 × 10⁻¹⁰] Hz/s. The lowest set of points (black circles) are the Einstein@Home distributed computing results. The most recent upper limit results in this frequency range from O1 data obtained with various search pipelines are also shown for comparison. These searches covered a broader frequency and spin-down range. All inferred upper limits are averaged over the full sky and source polarization. (Reproduced from (Abbott et al., 2017d), Creative Commons Attribution 4.0 licence.)*

useful as it compares different search strategies used for the LIGO data (although one should keep in mind that the frequency range considered for the Einstein@Home search is narrower).

14.6 The magnetic field

From the gravitational-wave point of view, it is important to have an idea of what the highest expected neutron star mountain may be. It is encouraging that one can think of 'reasonable' scenarios for such deformations being generated and the upper limits from targeted pulsar searches (Figure 14.4) are obviously interesting. However, the discussion comes with a major caveat—there is (at least, not yet) an obvious reason why the crust should be stretched to the limit. It could, for example, be that plastic flow relaxes the system (Chugunov and Horowitz, 2010). Or perhaps a sequence of crust failures leads to deformations at smaller and smaller scales, making the system a less efficient gravitational-wave emitter. These issues lead us to the obvious question: Is there a minimum deformation we should expect in a realistic neutron star? Interestingly, the answer is (at least, in principle) yes, and as so often is the case in astrophysics the explanation is the magnetic field.

The (interior) magnetic field tends to deform a star and this should lead to some level of gravitational-wave emission, It is easy to see why this has to be the case. A predominately dipolar field, as required to explain pulsar spin down, is axisymmetric, but if the magnetic axis is misaligned with the spin-axis the associated deformation will not be aligned with the centrifugal bulge and hence will radiate gravitational waves.

Unfortunately, the magnetic deformation tends to be small for typical pulsar field strengths. A simple energy argument—based on comparing the magnetic energy to the gravitational potential energy—leads to (Haskell *et al.*, 2008)

$$\epsilon \sim \frac{\int B^2 dV}{GM^2/R} \approx 10^{-12} \left(\frac{B}{10^{12}\,\mathrm{G}} \right)^2. \tag{14.66}$$

One would expect any magnetic deformation to be small as the neutron star has a tremendous self-gravity. However, it is the *internal*, rather than the external magnetic field that counts. This means that we have little guidance from the dipole field inferred from pulsar spin down. We need the internal configuration and this is a tricky issue. For example, the above estimate assumes a normal fluid core while real neutron stars are expected to harbour a proton superconductor (see Chapter 12). This complicates the picture, but it could be good news as superconductivity may lead to larger asymmetries. Basically, the stresses in a type II superconductor are different (as the field is carried by quantized fluxtubes; see Glampedakis *et al.* (2011)). Simplistically, this leads to (Cutler, 2002)

$$\epsilon \approx 10^{-9} \left(\frac{B}{10^{12}\,\mathrm{G}} \right) \left(\frac{H_{\mathrm{c}}}{10^{15}\,\mathrm{G}} \right), \tag{14.67}$$

where $H_c \approx 10^{15}$ G is the so-called critical field. The main problem is that we really do not know what the internal magnetic field configuration is.

In order to illustrate the problem, we take as our starting point the equation for hydrostatic equilibrium and add the Lorentz force associated with the magnetic field. This leads to

$$\nabla p + \rho \nabla \Phi = f_L, \tag{14.68}$$

where, if we assume ideal magnetohydrodynamics (ignore the displacement current in Ampère's law),

$$f_L = j \times B = \frac{1}{4\pi} B \times (\nabla \times B), \tag{14.69}$$

with j the charge current and B the magnetic field. We also have the usual Poisson equation for the gravitational potential. For a barotropic model, we can introduce the enthalpy

$$\nabla h = \frac{1}{\rho} \nabla p, \tag{14.70}$$

which means that the left-hand side of (14.68) can be written as a gradient, and we must have

$$\nabla \times \left(\frac{f_L}{\rho}\right) = \nabla \times \left[\frac{B \times (\nabla \times B)}{\rho}\right] = 0. \tag{14.71}$$

Meanwhile, the Maxwell equation

$$\nabla \cdot B = 0, \tag{14.72}$$

(the statement that there are no magnetic monopoles) implies that the magnetic field only has two degrees of freedom. Working in cylindrical coordinates $\{\varpi, \varphi, z\}$, we can introduce a pair of scalar stream functions $u(\varpi, z)$ and $f(\varpi, z)$ and write an axisymmetric field in a form that is automatically divergence-free:

$$B = \frac{1}{\varpi} \left[\nabla u \times \hat{\varphi} + f \hat{\varphi}\right]. \tag{14.73}$$

The assumed axisymmetry requires $f_L^\varphi = 0$, which in turn leads to the functional dependence $f = f(u)$. Then it is easy to show that

$$f_L = \rho \nabla M, \tag{14.74}$$

where $M = M(u)$ is another scalar function (which makes (14.69) an identity).

Using (14.73) in (14.74) to calculate f_L leads to the so-called Grad–Shafranov equation, which governs the hydromagnetic equilibrium in the star's interior (see, for example, Lander and Jones (2009)):

$$\frac{\partial^2 u}{\partial \varpi^2} - \frac{1}{\varpi}\frac{\partial u}{\partial \varpi} + \frac{\partial^2 u}{\partial z^2} = -4\pi\rho\varpi^2\frac{dM}{du} - f\frac{df}{du}. \tag{14.75}$$

In this equation, the two functions $M(u)$ and $f(u)$ may be freely specified (up to requirements of regularity and symmetry). Through specific choices one may introduce restrictions on the equilibrium solutions.

If we want to understand the meaning of particular choices, it is instructive to express Ampère's law in terms of the stream functions:

$$\nabla \times \boldsymbol{B} = \frac{4\pi}{c}\boldsymbol{j} = \frac{df}{du}\boldsymbol{B} + 4\pi\rho\varpi\frac{dM}{du}\hat{\varphi}. \tag{14.76}$$

The first term on the right-hand side describes the force-free part of the current while the second term represents a purely azimuthal plasma flow.

In order to determine a magnetic equilibrium configuration, it is common to use the vector identity for axisymmetric systems

$$\frac{\varpi}{\sin\varphi}\nabla^2\left(\frac{u\sin\varphi}{\varpi}\right) = \left(\frac{\partial^2}{\partial\varpi^2} - \frac{1}{\varpi}\frac{\partial}{\partial\varpi} + \frac{\partial^2}{\partial z^2}\right)u, \tag{14.77}$$

and rewrite the Grad–Shafranov equation as a 'magnetic Poisson equation' involving the Laplace-type operator:

$$\nabla^2\left(\frac{u\sin\varphi}{\varpi}\right) = -\left(\frac{f}{\varpi}\frac{df}{du} + 4\pi\varpi\rho\frac{dM}{du}\right)\sin\varphi. \tag{14.78}$$

We need to solve this equation together with the usual equations for a barotropic fluid equilibrium. This is typically done using the iterative method that was originally designed to determined rotating equilibria (see, for example, Stergioulas and Friedman (1995)). Crucially, the calculation must account for the back reaction of the magnetic field on the fluid. It is not sufficient to solve for the field holding the fluid configuration fixed.

The nature of the magnetic equilibrium depends heavily on how we specify the toroidal function $f(u)$. In calculations where the exterior is assumed to be vacuum— which would seem a natural starting assumption—one can fit $f(u)$ inside the *last closed* poloidal field line (ensuring the absence of exterior currents) by using (Tomimura and Eriguchi, 2005)

$$f(u) = \begin{cases} a(u - u_{\text{int}})^\zeta & u > u_{\text{int}}, \\ 0 & u \le u_{\text{int}}, \end{cases} \tag{14.79}$$

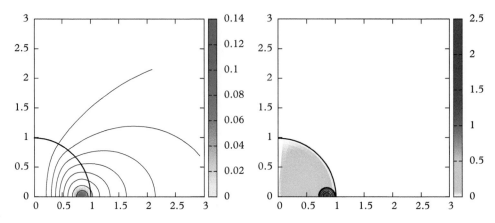

Figure 14.7 *An illustration of a typical twisted torus magnetic field equilibrium. The direction of poloidal field/current is illustrated with lines and toroidal-field/current magnitude are shown as colour scales. The numerical domain is expressed in units of the stellar radius R. Left: The magnetic field configuration, with $\zeta = 0.1$ (see (14.79)) and with a vacuum exterior ($\nabla \times \boldsymbol{B} = 0$); no exterior current or toroidal field. The magnetic energy contained in the toroidal field component is about 3% of the total magnetic energy. Right: The electric current distribution, $\boldsymbol{j}/c = \nabla \times \boldsymbol{B}/4\pi$, for the magnetic equilibrium shown in the left-hand panel. (Reproduced from Glampedakis et al. (2014).)*

where a and ζ are constant and u_{int} is the value of the stream function associated with the last closed poloidal line. An example of this 'twisted torus' equilibrium is shown in Figure 14.7 (left panel) for the specific choice $\zeta = 0.1$ (chosen to give the strongest possible toroidal field; the value of the amplitude a sets the overall scale and is of less importance). The corresponding current distribution is shown in the right panel of Figure 14.7. It is easy to see that the currents are confined to the star. Meanwhile, the poloidal field extends from (most of) the interior to the exterior. The poloidal component is strongest in the centre and only vanishes in a small region at the edge of the star (seen as the pair of semicircular contours on the equator at $r/R \approx 0.8$). The toroidal field is contained within the small region where the poloidal field vanishes.

Within this scheme we can work out the deformation associated with a given magnetic field configuration. The results tend to confirm rough estimates like (14.66). As we are not dealing with a uniform field, it is natural to quantify the result in terms of an average. For example, we may use the volume average

$$\bar{B}^2 = \frac{1}{V} \int B^2 \, dV, \tag{14.80}$$

(where the integral includes the external field). For a purely poloidal field, we then have (Lander and Jones, 2009)

$$\epsilon \approx 5 \times 10^{-12} \left(\frac{\bar{B}}{10^{12} \, \text{G}} \right)^2, \tag{14.81}$$

while a purely toroidal field leads to

$$\epsilon \approx -3 \times 10^{-12} \left(\frac{\bar{B}}{10^{12} \text{ G}} \right)^2 . \tag{14.82}$$

The sign indicates that the toroidal configuration is prolate rather then oblate. This will be of interest to us later.

What do these estimates mean for the observational effort? We immediately see that we need the average magnetic field to be very strong ($\sim 10^{15}$ G) in order for the magnetic deformation to be competitive with the maximum allowed elastic one. But a neutron star with such a strong external magnetic field would rapidly spin down and exit the sensitivity band of ground-based interferometers (Stella *et al.*, 2005). If we assume a typical radio pulsar magnetic field, then the magnetic deformation will be too small to ever be detected through gravitational waves. We need nature to be less conservative.

Of course, the magnetic field results come with a range of caveats. Most importantly, numerical work (Lander and Jones, 2012) suggests that most (perhaps all) configurations we can build are unstable. This is a problem. The issue could be related to the toroidal contribution to the field. It is generally expected (based on stability analyses) that one would need a significant toroidal field component to stabilize the configuration. But in models obtained from the Grad–Shafranov approach, the toroidal component appears to be bounded (typically to less than 10% of the overall field). Of course, we have made a number of assumptions. Perhaps it is not appropriate to use ideal magnetohydrodynamics. Indeed, it cannot be if the star's interior is superconducting (Glampedakis *et al.*, 2011). Perhaps we should not assume that the equation of state is a barotrope. Indeed, if we relax this assumption, then one can imagine stratification providing an additional force to balance the magnetic contribution (Mastrano *et al.*, 2011; Glampedakis and Lasky, 2016). Perhaps we should not assume that the star is in equilibrium. Strictly speaking, it will not be. A neutron star evolves (cools) due to the emission of neutrinos. Perhaps we have simply not been sufficiently imaginative in our choice of the magnetic stream functions.

The discussion of magnetic 'mountains' also leads us to an observational question. How large can we expect a neutron star magnetic field to be?

14.7   The birth of a magnetar

Magnetars are young and strongly magnetized neutron stars with intriguing phenomenology (Kaspi and Beloborodov, 2017). They burst and flare across the spectrum, from X-rays to gamma rays. While many of the details remain to be understood, their activity can be explained by the evolution of a super-strong magnetic field, that stresses and fractures the crust as it untwists (Thompson and Duncan, 1995). The typical activity ranges from millisecond X-ray bursts to month-long outbursts and occasional, incredibly powerful, gamma-ray flares. Since the first discovery of such a system—SGR0526-66, which exhibited an enormous flare on 5 March 1979 (Barat *et al.*, 1983)—the population of observed magnetars has grown to about 30 (Olausen and Kaspi, 2014). Recent evidence provides a link to highly magnetized radio pulsars and there are, in fact, objects that bridge the two classes (Rea *et al.*, 2012).

With an inferred (exterior) magnetic field of order $10^{14} - 10^{15}$G, a magnetar rapidly spins down to a period of several seconds (in 10^4 years or so). In effect, even though they may have interestingly large asymmetries from the gravitational-wave point-of-view, they are likely to be outside the sensitivity band of any current (or future) ground-based detector. However, these systems may make up as much as 10% of the young neutron star population in the Galaxy. In essence, if we assume that they are born in supernovae—an assumption supported by the association between several magnetars and observed supernova remnants—then the magnetar birth rate could be as large as a tenth of the supernova rate. If this is the case, then one would expect a magnetar to be born in our Galaxy every few hundred years. It would not be a regular event.

In a slightly different context, the birth of a magnetar has been used to explain observed light curves of superluminous supernovae and some gamma-ray bursts (Kasen and Bildsten, 2010). The idea is simple. An additional source of energy input—over an extended period of time—is needed to explain the observed evolution of the brightness and/or the light curve. Sustained injection of energy from a hypothesized central engine would explain the observed features. As the formation of a strong magnetic field requires some kind of dynamo action, it is generally agreed that magnetars should be born rapidly spinning. Magnetic dipole radiation would drain the star of rotational energy, injecting energy into the surroundings. While the details of the scenario remains to be worked out, the idea seems plausible, and if the star is born rapidly spinning, gravitational-wave emission may also influence the evolution.

If we apply our estimates for a rotating deformed neutron star to this scenario, we can estimate how large the gravitational-wave component is allowed to be before the model is no longer viable. Assuming the magnetar model, we can connect the required extra energy and timescale to the initial magnetic field, B, and the initial spin frequency, ν_0. These quantities then provide initial conditions for the evolution. This leads to a constraint on the allowed ellipticity (Ho, 2016). We must have

$$\epsilon < 10^{-4} \left(\frac{B}{10^{14}\, \text{G}} \right) \left(\frac{1\, \text{kHz}}{\nu_0} \right). \tag{14.83}$$

If the deformation is larger than this, then the light curve would be strongly affected, leading to a significant decrease in peak luminosity and the time it takes to reach this peak. Constraints for different observed systems are shown in Figure 14.8. These constraints are (obviously) not as tight as those obtained from direct pulsar searches, but they are nevertheless interesting.

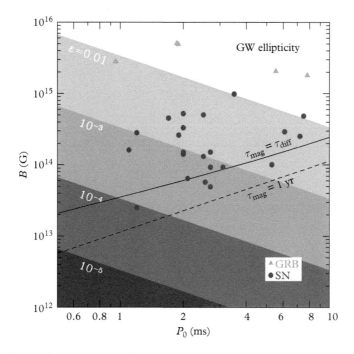

Figure 14.8 *Constraints on the allowed quadrupole deformation ε as a function of the initial neutron star spin period P_0 and the magnetic field B. The dashed line indicates when the magnetic spin-down timescale is equal to 1 yr. Circles represent P_0 and B for observed superluminous supernovae while triangles are data for GRBs. (Reproduced from Ho (2016).)*

14.8 Modelling accretion

A neutron star's magnetic field dictates the general spin evolution. It is also key to how the neutron star interacts with its environment. In particular, the magnetic field may determine the flow of matter accreted onto the star. This problem is of great importance. First of all, the accretion of angular momentum from a binary partner may spin a neutron star up. We need to understand the accretion torque if we want to explain the origin of the fastest observed pulsars (Alpar *et al.*, 1982; Radhakrishnan and Srinivasan, 1982). Secondly, the accreted matter may be channeled onto the star's magnetic poles and if the magnetic field is misaligned with the spin-axis, then this may provide a mechanism for deforming the star. Hence, accreting fast-spinning neutron stars may be interesting gravitational-wave sources (Vigelius and Melatos, 2010). We have already argued (see Chapter 6) that gravitational-wave emission may balance the accretion torque. We are now going to complicate the story by showing that the magnetic field may put a brake on the spin-up, as well.

In order to outline the role of the magnetic field in the accretion problem, we consider the interaction between a geometrically thin disk and the neutron-star magnetosphere. The basic picture is that of a rotating magnetized neutron star surrounded by a

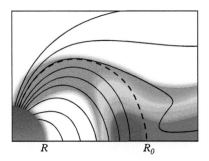

R R_0

Figure 14.9 *A schematic illustration of the accretion problem for a magnetically threaded disk. In the phenomenological model discussed in the main text, we take R_0 to be equal to R_M from (14.86). More detailed calculations suggest that this is reasonable (up to a factor of order unity).*

magnetically threaded accretion disk (Ghosh and Lamb, 1978); see Figure 14.9 for a schematic illustration. In the magnetosphere, accreting matter follows the magnetic field lines and gives up angular momentum on reaching the surface, exerting a spin-up torque. The material torque at the inner edge of the disk (taken to be the point where the magnetic pressure begins to dominate the fluid pressure) is usually approximated by (Ghosh and Lamb, 1978, 1979)

$$\dot{j} = \dot{M}\sqrt{GMR_M},\tag{14.84}$$

where the magnetosphere radius, R_M, is given by

$$R_M = \left(\frac{\mu^4}{2GM\dot{M}^2}\right)^{1/7},\tag{14.85}$$

and $\mu \sim BR^3$ is the magnetic dipole moment (with B the strength of the star's dipole field). This leads to

$$R_M \approx 7.8 \left(\frac{B}{10^8 \text{ G}}\right)^{4/7} \left(\frac{R}{10 \text{ km}}\right)^{12/7} \left(\frac{M}{1.4M_\odot}\right)^{-1/7} \left(\frac{\dot{M}}{\dot{M}_{\text{Edd}}}\right)^{-2/7} \text{ km}.\tag{14.86}$$

Clearly, this torque can be significantly stronger than the rough estimate from Eq. (6.43) if $R_M \gg R$. In order to see whether this is likely to be the case, we need to estimate the accretion rate. An approximation of the maximum accretion rate follows from balancing the pressure due to spherically infalling gas to that of the emerging radiation. This leads to the Eddington limit

$$\dot{M}_{\text{Edd}} \approx 1.5 \times 10^{-8} \left(\frac{R}{10 \text{ km}}\right) M_\odot \text{ yr}^{-1},\tag{14.87}$$

which would correspond to an X-ray luminosity

$$L_x = \eta \frac{GM\dot{M}}{R} \approx 1.8 \times 10^{38} \eta \left(\frac{M}{1.4M_\odot}\right)\left(\frac{\dot{M}}{\dot{M}_{\mathrm{Edd}}}\right) \text{ erg/s} \qquad (14.88)$$

(where the efficiency factor η is usually taken to be 1, in absence of a more detailed understanding). From these estimates we see that, for accretion at a fraction ϵ_{eff} of the Eddington rate, e.g. $\dot{M} = \epsilon_{\mathrm{eff}}\dot{M}_{\mathrm{Edd}}$, the magnetic field must be accounted for (in the sense that $R_M > R$) as long as it is stronger than

$$B \geq 1.6 \times 10^8 \epsilon_{\mathrm{eff}}^{1/2} \text{G}. \qquad (14.89)$$

Since observations indicate that rapidly rotating neutron stars have magnetic fields of order 10^8 G, and many transient low-mass X-ray binaries accrete with $\epsilon_{\mathrm{eff}} \sim 0.01$, the magnetic field may well play a role in these systems.

The magnetosphere radius is not the only important scale in the problem. Outside the co-rotation radius,

$$R_c \approx 17 \left(\frac{P}{1 \text{ ms}}\right)^{2/3}\left(\frac{M}{1.4M_\odot}\right)^{1/3} \text{ km}, \qquad (14.90)$$

the field lines rotate faster than the local Keplerian speed of the disk matter, resulting in a negative torque. In fact, if $R_M > R_c$ the accretion flow will be centrifugally inhibited and matter may be ejected from the system. This will happen if the spin period becomes very short, or the rate of flux of material onto the magnetosphere drops. It is known as the propeller regime. In this phase, accreting matter is flung away from the star, leading to a spin-down torque. In order to account for this effect we alter the material torque (Ho *et al.*, 2014)

$$\dot{j} = \dot{M}R_M^2[\Omega_K(R_M) - \Omega] = \dot{M}R_M^2\Omega_K(1 - \omega_s)$$
$$= \dot{M}\sqrt{GMR_M}\left[1 - \left(\frac{R_M}{R_c}\right)^{3/2}\right], \qquad (14.91)$$

where Ω is the angular frequency of the star, Ω_K is the angular velocity of a particle in a Keplerian orbit

$$\Omega_K(r) = \left(\frac{GM}{r^3}\right)^{1/2}, \qquad (14.92)$$

and the so-called fastness parameter is given by $\omega_s = \Omega/\Omega_K$. Even though this expression only accounts for the propeller regime in a phenomenological way, it agrees with the expectation that accretion will not spin the star up beyond the point $R_M = R_c$ (White and Zhang, 1997). The model is now able able to explain why an accreting system would attain equilibrium before the star reaches the breakup limit. Simply setting $R_M = R_c$, we find the equilibrium period

$$P_{eq} \approx 0.30 \left(\frac{B}{10^8 \, G}\right)^{6/7} \left(\frac{R}{10 \, km}\right)^{18/7} \left(\frac{M}{1.4 M_\odot}\right)^{-5/7} \left(\frac{\dot{M}}{\dot{M}_{Edd}}\right)^{-3/7} \quad ms, \qquad (14.93)$$

or the limiting spin frequency

$$\nu_{eq} \approx 530 \left(\frac{B}{10^8 \, G}\right)^{-6/7} \left(\frac{R}{10 \, km}\right)^{-18/7} \left(\frac{M}{1.4 M_\odot}\right)^{5/7} \left(\frac{\dot{M}}{\dot{M}_{Edd}}\right)^{3/7} \quad Hz. \qquad (14.94)$$

It is clear that, in order to reach a spin rate of several 100 Hz, we need the star to have a weak magnetic field and accrete at a high rate.

Conversely, given an observed spin period and mass accretion rate we can (assuming that the system has reached equilibrium) deduce the neutron star's magnetic field. In order to carry out this exercise, we need both the spin and an estimate of the accretion rate. The spin is, quite naturally, the easiest to determine. For X-ray pulsars, we can use the measured pulse frequency. Many of the observed low-mass X-ray binaries are not pulsars, but we nevertheless have a handle on the spin rate through observed burst oscillations (assumed to be associated with explosive burning in the neutron star ocean; see Patruno and Watts (2012)). Finally there are systems which exhibit neither pulsations nor burst oscillations, but where high-frequency (kHz) quasiperiodic oscillations are observed. There has been suggestions that the separation of observed pairs of frequencies is related to the star's spin, but this identification is not particularly reliable (Watts *et al.*, 2008).

The accretion rate can be estimated from X-ray observations. However, the luminosity can be highly variable. It is clear that the estimated equilibrium period is shortest when the accretion rate is highest (alternatively, for a given spin rate, when the inferred magnetic field is maximal). Hence, we may perhaps assume that the observed spin rate is the equilibrium period associated with the maximum accretion rate for a given source, even for sources that are transient or highly variable. After all, the main contribution to the spin-up torque ought to be associated with the phase where the star accretes at the fastest rate.

If we combine the observed data with the assumed mass and radius for a canonical neutron star ($1.4 M_\odot$ and 10 km) we infer magnetic fields similar to those of millisecond radio pulsars for low-mass X-ray binaries accreting at the level of $10^{-2} \dot{M}_{Edd}$ and below (Andersson *et al.*, 2005a). The model does not do quite so well for systems accreting with $\dot{M} \approx \dot{M}_{Edd}$ for an extended period. In these cases the estimated magnetic fields appear to be too large. This may hint towards an additional spin-down torque in systems that accrete near the Eddington rate, possibly the presence of a gravitational-wave torque.

Are observations consistent with the idea that these systems are in magnetic spin equilibrium? The answer is not clear. Overall, the model works well, but it does not completely explain observations. In the case of two systems, XTE J1751$-$305 (which spins at 435 Hz) and IGR J00291+5934 (which spins at 599 Hz), we have sufficient data to carry out a 'consistency' check. In these systems, a strong spin-up during an outburst of accretion has been observed to be followed by a slow drop-off in the spin. This behaviour would be naturally explained in terms of a dramatic rise in the accretion torque during the initial outburst and standard magnetic dipole spin-down in between outbursts. Within this scenario we can constrain each star's magnetic field—both from the accretion model and the dipole braking. If the results are consistent, there would be no need to invoke an additional gravitational-wave torque.

This argument leads to the results in Figure 14.10. The vertical bands show the inferred magnetic field from the observed spin-down rate along with the level of theoretical uncertainty. If the spin up of the neutron star is the result of accretion, then the magnetic field can be inferred from the associated torque. The magnetic field inferred from (14.84) is shown as dotted lines in Figure 14.10. Of course, the simple accretion torque does not account for the magnetic field threading the accretion disk. The more realistic torque estimate (involving the fastness parameter ω_s) from (14.91) is shown as solid lines and cross-hatched regions. We see that the accretion torque produces a spin-up rate significantly below the observed one for both XTE J1751−305 and IGR J00291+5934. This suggests that our understanding of the problem remains incomplete. Of course, an additional gravitational-wave spin-down torque would not help solve the problem.

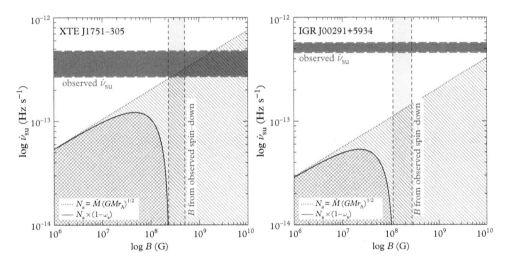

Figure 14.10 *The inferred magnetic field of XTE J1751−305 (left) and IGR J00291+5934 (right) as determined by the observed spin-down rate, spin-up rate, and fastness parameter ω_s. The observed spin-down rate (with 90% uncertainty) following an outburst produces the vertical band. The observed spin-up rate (with 90% uncertainty) during outburst is represented by the horizontal band. The dotted and solid lines show the maximum calculated spin-up rates due to accretion torques (14.84) and (14.91), respectively. The magnetic field determined by $\omega_s = 1$ is inferred from the point where the solid line crosses the bottom axis. (Reproduced from Andersson et al. (2014).)*

As we consider these issues, we need to keep in mind that the model is phenomeno-logical. In order to do better, we have to resolve a number of issues. First of all, the description of the accretion problem is inconsistent since our various estimates, e.g., of the size of the magnetosphere, are based on spherical infall of matter. The model can be improved, albeit at the cost of introducing several new (largely unknown) parameters. For example, we need a description of the viscosity in the disk. Viscosity is the main agent that dissipates energy and angular momentum, enabling matter to flow inwards. Since the microphysical viscosity (likely due to the magneto-rotational instability in some form; see Chapter 20) is difficult to determine, it is common to use the so-called α-viscosity from Shakura and Sunyaev (1973). This introduces an unknown parameter. Moreover, in the case of a magnetically threaded disk, we can improve the description of the interaction between the disk flow and the magnetic field. Figure 14.9 provides a schematic illustration of the problem. To develop a more precise model of this complicated physics problem is not a simple task. Basically, one would expect the magnetic field to influence the infalling matter over a range of radii, as in Figure 14.9. This means that we should replace R_M with some appropriate average value R_0 (say). In order to determine this average, we need to estimate the radius at which magnetic stresses balance the actual material stresses in the disk. This is tricky, but most calculations suggest that the phenomenological model fares rather well. One typically finds that $R_0 \approx \xi R_M$, with ξ of order unity (Wang, 1995). The upshot of this is that the estimated spin equilibrium will involve unknown factors, but as long as these factors are of order unity our estimates may not change very much.

14.9   The low-mass X-ray binaries

Fast-spinning neutron stars are found either as radio millisecond pulsars or in their (presumed) progenitor systems, accreting neutron stars in low-mass X-ray binaries. As we have outlined, accretion theory predicts the spin-up of these systems (Alpar *et al.*, 1982; Radhakrishnan and Srinivasan, 1982). It also suggest that, due to magnetic field channeling, the matter flows onto hotspots on the star, and the majority of accreting neutron stars should pulsate and be seen as accreting X-ray pulsars. This expectation is not brought out by observations. The total number of accreting millisecond pulsars is only 19 out of about 200 non-pulsating low-mass X-ray binaries (Patruno and Watts, 2012). In addition, about half of these systems are nuclear-powered pulsars, exhibiting short-lived 'burst oscillations' associated with runaway thermonuclear explosions on the neutron star surface (see Table 14.1). A key problem concerns to what extent the spin distribution of these accreting neutron stars can be reconciled with detailed accretion torque models (White and Zhang, 1997; Andersson *et al.*, 2005a).

The spin-distribution of known fast-spinning accreting neutron is shown in the right-hand panel of Figure 6.3. The data hints at a clustering at the fastest observed spins (Patruno *et al.*, 2017). This may be suggestive of an additional braking torque operating in these systems. Whatever the mechanism is that causes the clustering, it must set in sharply when the stars reach a given spin rate. Gravitational-wave emission, which scales with a high power of the spin frequency (ν^5 for a deformed star) may lead to exactly

Table 14.1 *Accretion and nuclear-powered millisecond pulsars spinning faster than 200 Hz. Systems above the horizontal line are seen as millisecond X-ray pulsars. Those below the line are nuclear-powered systems exhibiting X-ray bursts. The estimated outburst duration Δt is given in days and the average accretion rate $\langle \dot{M} \rangle$ (in units of $10^{-10} M_{\odot} yr^{-1}$). We also list the distance d (in kpc) of the system and the spin frequency v_s (in Hz). The mass accretion rates is linked to the X-ray flux through (14.115). Given the uncertainties associated with this relation, the estimates should be seen as indicative. (Based on data from Haskell et al. (2015), where original references can be found.)*

Source	v_s	d	$\langle \dot{M} \rangle$	Δt
IGR J00291+5934	599	5	6	14
Aql X-1	550	5	10	30
Swift J1749.4-2807	518	6.7	2	20
XTE J1751–305	435	7.5	10	10
SAX J1808.4–3658	401	3.5	4	30
IGR J17498–2921	400.9	7.6	6	40
HETE J1900.1–2455	377	5	8	3000
XTE J1814–338	314	8	2	60
IGR J17511–3057	244.9	6.9	6	24
NGC 6440 X-2	204.8	8.5	1	4
4U 1608–52	620	3.6	20	700
SAX J1750.8–2900	601	6.8	4	100
4U 1636–536	581	5	30	Persistent
EXO 0748–676	552	5.9	3	8760
KS 1731–260	526	7	11	4563
4U 0614+091	415	3.2	6	Persistent
4U 1728–34	363	5	5	Persistent
4U 1702–429	329	5.5	23	Persistent

this behaviour (Bildsten, 1998). However, this is a phenomenological argument. It is entirely possible that the answer has nothing whatsoever to do with gravitational waves. The accretion torque may simply become much less efficient as soon as the star reaches above 500 Hz, perhaps due to wind emission.

Additional observational evidence suggests that some of these systems might, in fact, lack a magnetosphere strong enough to affect the dynamics of the plasma in the accretion disk. Recent searches for pulsations in several low-mass X-ray binaries have found no evidence for accretion-powered pulsations (Messenger and Patruno, 2015). It is difficult to reconcile this observation with the presence of a magnetosphere strong enough to channel the matter flow (and hence impact on the accretion torque). This would argue against a pure accretion explanation for the observed spins. But this might be problematic also for the gravitational-wave argument. If no magnetosphere (or a very weak one) is present, then we have to use the torque (14.84). The change in spin scales linearly with the amount of mass transferred and the neutron star keeps spinning up (not even spinning down during quiescence, since the pulsar spin-down mechanism from magnetic dipole radiation is also suppressed—in contrast with observations). Moreover, if there is no magnetosphere, then the accretion flow does not lead to asymmetries on the star's surface and therefore one may not expect the system to develop the deformation required for gravitational-wave emission (although oscillation mode instabilities, see Chapter 15, can still play a role and frozen-in compositional asymmetries may lead to the neutron star being deformed).

With these caveats in mind, let us turn to one of the possible mechanisms for forming a significant quadrupole moment in an accreting neutron star. The idea is simple (even though the details are not): during an accretion phase, matter—originally composed of light elements—is compressed to higher densities where it undergoes a series of nuclear reactions (Haensel and Zdunik, 2003). An asymmetric accretion flow may lead to inherited asymmetries in the internal composition, which in turn deform the star. An approximate expression for the quadrupole due to asymmetric crustal heating from nuclear reactions in the crust is given by (Ushomirsky *et al.*, 2000):

$$Q_{22} \approx 1.3 \times 10^{35} \left(\frac{R}{10 \text{ km}} \right)^4 \left(\frac{\delta T_q}{10^5 \text{ K}} \right) \left(\frac{Q}{30 \text{ MeV}} \right)^3 \text{ g cm}^2, \tag{14.95}$$

where δT_q is the *quadrupole* component of the temperature variation due to nuclear reactions and Q is the reaction threshold energy. Higher threshold energies correspond to reactions at higher densities. In general, the reactions will heat the region by an amount (Brown and Bildsten, 1998)

$$\delta T \approx 10^6 C_k^{-1} p_d^{-1} Q_n \Delta M_{22} \text{ K}, \tag{14.96}$$

where C_k is the heat capacity per baryon (in units of the Boltzmann constant k_B), p_d is the pressure (in units of 10^{30} erg cm^{-3}) at which the reaction occurs, Q_n is the heat released per baryon (in MeV) deposited by the reactions, and ΔM_{22} is the deposited mass (in units of 10^{22} g). Note that (14.96) gives the total increase in temperature. Only a small fraction of this is likely to be asymmetric and associated with the quadrupole moment. The estimates of Ushomirsky *et al.* (2000) suggest that $\delta T_q/\delta T \leq 0.1$, but the true ratio is not known.

As the system returns to quiescence after an accretion outburst, the deformations decay on the crust's thermal timescale (Bildsten, 1998)

$$t_{th} \approx 0.2 \, p_d^{3/4} \text{ yr.} \tag{14.97}$$

If the system is in quiescence longer than this, the quadrupole deformation is erased before the next outburst. Meanwhile, a shorter recurrence time could lead to an additional accumulation of material. It is also entirely plausible that composition asymmetries are frozen into the crust and not erased on a thermal timescale. Such a scenario would predict the formation of large quadrupoles in all transient systems (with $10^{38} \lesssim Q \lesssim 10^{40}$ g cm^2). However, this level of gravitational-wave induced spin-down is already excluded by measurements between outbursts in four transient systems (SAX J1808.4–3658, XTE J1751–305, IGR J00291+5934, and SWIFT J1756.9–2508; see Patruno and Watts (2012)). The detectability of typical transient systems is indicated in Figure 14.11. The results suggest that thermal mountains on neutron stars in transients will be challenging to detect (even with third-generation detectors, like the Einstein Telescope). This conclusion accords with the detailed analysis of Watts *et al.* (2008).

The gravitational waves from transient accreting neutron stars may be difficult to detect, but the associated energy loss could still be significant. Recent observations of the pulsar J1203+0038 (Haskell and Patruno, 2017), which spins at 592 Hz and has been observed to transition between a radio state (during which it is visible as a millisecond radio pulsar) and a low-mass X-ray binary state (during which X-ray pulsations are visible) are particularly interesting in this respect. Timing during the two phases shows that the neutron star is spinning down at a rate 27% faster during the accreting phase than during the radio phase. This is at odds with the standard accretion model. The increased

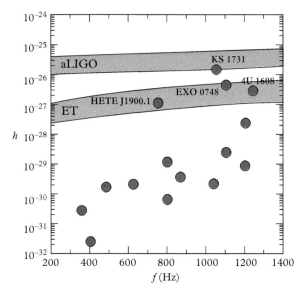

Figure 14.11 *Estimated gravitational-wave strain as function of frequency for deformed accreting neutron stars in low-mass X-ray binaries. The data represents transient sources, for which the mountain is the largest that can be created during an outburst, in the case of a shallow capture layer. Deep capture layers lead to slightly weaker signals. The upper and lower ends of each detector band represent a one-month and a two-year integration, respectively. (Based on data from Haskell et al. (2015).)*

spin-down rate would be compatible with gravitational-wave emission, perhaps due to the creation of a 'mountain' during the accretion phase. This would require a quadrupole moment (or ellipticity) (Haskell and Patruno, 2017)

$$Q_{22} \approx 4.4 \times 10^{35} \text{ g cm}^2 \quad \longrightarrow \quad \epsilon \approx 5 \times 10^{-10}, \tag{14.98}$$

for canonical neutron star parameters. Thermal deformations at this level would be consistent with the accretion history of the system. The corresponding gravitational-wave strain $h_0 \approx 6 \times 10^{-28}$ (at a distance of 1.4 kpc) is, however, too weak to be detected by current instruments (see Figure 14.11).

14.10   Magnetic field burial and confinement

A possible explanation for the absence of pulsations in many accreting systems is that the magnetic field is buried by the matter flow. The problem of magnetic field burial has been considered in detail for young neutron stars (Geppert *et al.*, 1999; Ho, 2015) following supernova fall-back accretion and in some simplified form also for accreting neutron stars.

To get an idea to what extent this idea is viable for the much lower accretion rates relevant for low-mass X-ray binaries, we can adapt the usual argument. We first estimate the depth at which the field would be buried by balancing the timescale associate with the inflowing matter to that of Ohmic dissipation (Geppert *et al.*, 1999). We then have

$$t_{\text{flow}} = \frac{L}{v_r}, \tag{14.99}$$

where L is a typical length scale of the problem, and in the case of accretion we have

$$v_r = \frac{\dot{M}}{4\pi r^2 \rho}. \tag{14.100}$$

Secondly, we need

$$t_{\text{Ohm}} = \frac{4\pi \sigma L^2}{c^2}, \tag{14.101}$$

where σ is the conductivity. If $t_{\text{flow}} < t_{\text{Ohm}}$, the magnetic field is dragged along by the inward flowing matter. The matter piles up faster than the field can diffuse out and hence we have burial. If the accretion stops, the field emerges on the timescale (t_{Ohm}) associated with the burial depth.

As we do not expect the field to be buried deep, we consider the neutron star envelope where the ions are liquid (the electrons are degenerate and relativistic). Then we have (Geppert *et al.*, 1999)

$$\sigma \approx 9 \times 10^{21} \left(\frac{\rho_6}{AZ^2}\right)^{1/3} \text{s}^{-1}, \tag{14.102}$$

with $\rho_6 = \rho/(10^6 \text{ g/cm}^3)$. This leads to

$$\frac{t_{\text{Ohm}}}{t_{\text{flow}}} \approx 6 \times 10^5 \frac{L_5 \dot{M}/\dot{M}_{\text{Edd}}}{r_6^2 \rho_6^{2/3} (AZ^2)^{1/3}}, \tag{14.103}$$

where $\dot{M}_{\text{Edd}} \approx 10^{-8} M_\odot/\text{yr}$, A is the mass number of the nuclei and Z is the proton number. As we are interested in the outer region it makes sense to consider Fe^{56}, with $A = 56$ and $Z = 26$. We can also set $r_6 = r/10^6$ cm ≈ 1, as we are near the star's surface. Then we have

$$\frac{t_{\text{Ohm}}}{t_{\text{flow}}} \approx 2 \times 10^4 \frac{L_5}{\rho_6^{2/3}} \frac{\dot{M}}{\dot{M}_{\text{Edd}}}, \tag{14.104}$$

where $L_5 = L/10^5$ cm.

We also need to know how the density increases with depth. As we only want a rough estimate, we use the pressure scale height

$$H = \frac{p}{\rho g}, \tag{14.105}$$

where the gravitational acceleration

$$g = \frac{GM}{R^2}, \tag{14.106}$$

can be taken as constant. From Brown and Bildsten (1998) we take

$$H \approx 265 \left(\frac{2Z}{A}\right)^{4/3} \rho_6^{1/3} \text{ cm} \longrightarrow \rho_6^{2/3} \approx 2 \times 10^5 H_5^2, \tag{14.107}$$

with $H_5 = H/10^5$ cm. Finally, it makes sense to let $L \approx H$, so we are left with

$$\frac{t_{\text{Ohm}}}{t_{\text{flow}}} \approx 0.1 H_5^{-1} \frac{\dot{M}}{\dot{M}_{\text{Edd}}}, \tag{14.108}$$

and we learn that the magnetic field is buried up to a density

$$\rho_{\text{burial}} \approx 7 \times 10^{10} \left(\frac{\dot{M}}{\dot{M}_{\text{Edd}}} \right)^3 \text{ g/cm}^3. \tag{14.109}$$

This estimate agrees well with (extrapolations of) more detailed models (Geppert *et al.*, 1999).

We also have

$$L_{\text{burial}} \approx H_{\text{burial}} \approx 8 \times 10^3 \left(\frac{\dot{M}}{\dot{M}_{\text{Edd}}} \right) \text{ cm}, \tag{14.110}$$

which means that, once accretion stops, the field will re-emerge after

$$t_{\text{Ohm}} \approx 10^{10} \left(\frac{\dot{M}}{\dot{M}_{\text{Edd}}} \right)^2 \text{ s}. \tag{14.111}$$

That is, the field will emerge a few hundred years after a star accreting at the Eddington rate goes into quiescence. This may be an indication of the lifetime of an accretion-induced asymmetry.

We also need to know how long it takes to bury the field in the first place. Somewhat simplistically, this follows from the accreted mass corresponding to the estimated burial depth. This is roughly given by

$$\Delta M \approx 4\pi \rho_{\text{burial}} R^2 H \approx 4 \times 10^{-6} \left(\frac{\dot{M}}{\dot{M}_{\text{Edd}}} \right)^{4/3} M_\odot, \tag{14.112}$$

and we see that it would also take a few hundred years to bury the field at the Eddington accretion rate. Lower accretion rates, such as those of many of the observed systems (see Table 14.1), would lead to a more shallow burial, and the field being buried and re-emerging on much shorter timescales.

These estimates obviously come with several caveats. A number of complicating factors may come into play, like possible plasma instabilities and the tension of the internal magnetic field, which can lead to sharp gradients and reduce the typical length scale L, thus reducing the amount of mass and the timescale needed for burial. However, the general scenario should still apply. A modest level of accretion may lead to a shallow field burial, with the magnetic field re-emerging shortly after a system goes into quiescence.

An interesting— potentially highly relevant—question concerns what happens when the external magnetic field is squeezed inside the crust. Intuitively, one might expect accretion to deform the magnetic field. As matter is accreted and spreads towards the equator it drags the field with it, compressing it. This could lead to a *locally* strong field that may sustain a 'magnetic' quadrupole deformation (Vigelius and Melatos, 2010). The quenching of an external dipolar field, B_{ext}, follows from (Haskell *et al.*, 2015)

$$B_{ext} = B_* \left(1 + \frac{M_a}{M_c}\right)^{-1}, \tag{14.113}$$

where M_a is the accreted mass. This leads to a mass quadrupole

$$Q_{22} \approx 10^{45} A \left(\frac{M_a}{M_\odot}\right) \left(1 + \frac{M_a}{M_c}\right)^{-1} \text{g cm}^2, \tag{14.114}$$

where $A \approx 1$ is a geometric factor that depends on the equation of state and the accretion geometry, while M_c is the critical amount of accreted matter at which the mechanism saturates. The size of the generated mountain is strongly dependent on the strength of the magnetic field when accretion begins, B_*. Observations of both low-mass X-ray binaries and millisecond radio pulsars suggest exterior fields of order $B_{ext} \approx 10^8$ G. Estimates suggest that one would need an initial field $B_* \approx 10^{12}$ G in order for the emitted gravitational waves to be detectable with Advanced LIGO (or even the Einstein Telescope). The presence of such strong internal magnetic fields in the low-mass X-ray binaries seems unlikely.

14.11 Persistent sources

We have argued that transient accreting systems are unlikely to develop detectable asymmetries. The situation may be different for persistently accreting sources. In such systems, ongoing accretion may lead to the build-up of mountains close to the breaking strain of the crust. However, that level of deformation would comfortably exceed the torque balance limit (see Chapter 6). As a result, one should perhaps not expect these systems to be rapidly spinning. The problem is that we do not know! The spin frequency of many of the persistent accretors is not known. When it comes to gravitational-wave searches, this make the problem computationally expensive.

We can estimate the gravitational-wave strain using the observed (bolometric) X-ray flux F, which follows from

$$\dot{M} = \frac{4\pi R d^2 F}{GM}, \tag{14.115}$$

where d is the source distance. Combining this with the gravitational-wave luminosity from the quadrupole formula (as in Chapter 6), we have

$$h_0 = 3 \times 10^{-27} \left(\frac{F}{10^{-8} \text{ erg cm}^{-2} \text{ s}^{-1}}\right)^{1/2} \left(\frac{\nu_s}{300 \text{ Hz}}\right)^{-1/2} \tag{14.116}$$

(for canonical neutron star parameters). Basically, if we accept the assumptions involved, then the gravitational-wave signal strength depends only on two observables; the X-ray flux and the neutron star spin rate. In the case of Sco X1, one of the brightest X-ray point sources in the sky, we have $F = 3.9 \times 10^{-7}$ erg cm^{-2} s^{-1}. Assuming that gravitational waves are emitted at twice the spin frequency (at a frequency $f = 2\nu_s$), we then have

$$h_0 \approx 3.4 \times 10^{-26} \left(\frac{f}{600 \text{ Hz}} \right)^{-1/2}. \tag{14.117}$$

This estimate is compared to the current best sensitivity LIGO searches in Figure 14.12.

In essence, we learn that only a few of the persistently bright neutron stars, accreting at rates near the Eddington limit, are likely to be within reach of Advanced LIGO. Moreover, we need them to emit gravitational waves at a rate matching that of the accretion torque. Given that the latter is poorly understood, it should (by now) be clear that this is a challenging problem that requires further work. We need to advance the theory and at the same time design clever search strategies.

The most sensitive results in Figure 14.12 are based on a semi-coherent search method using details of a parameterized continuous model signal to combine data separated by less than a specified coherence time (Abbott *et al.*, 2017*l*). This allows the analyst to dig into the noise below the level of sensitivity of an individual data segment. The current upper limit is a factor of a few above the predicted torque balance. Future observing runs should improve on this limit. The most important enhancement will come with better detector sensitivity. Longer observation times will help, but for this search method the

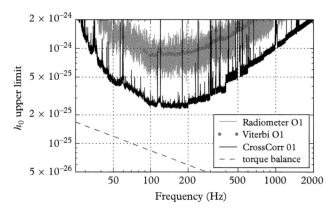

Figure 14.12 *Upper limits from directed searches for gravitational waves from Sco X1 in O1 LIGO data. The data show upper limits on h_0, after marginalizing over the neutron star spin inclination. The dashed line shows the nominal expected level assuming torque balance (14.116) as a function of the (unknown) frequency. For comparison, the most sensitive cross-correlation results (CrossCorr), results from the Viterbi analysis of Abbott et al. (2017k) (dark green dots), and the radiometer analysis of Abbott et al. (2017b) (broad light magenta curve) are also shown. (Reproduced from Abbott et al. (2017l), Creative Commons Attribution 3.0 licence.)*

effective amplitude scales as $T_{\text{obs}}^{1/4}$. Given this weak scaling, there is an obvious trade-off between the assumed coherence time and computational cost.

14.12   Free precession

So far we have discussed gravitational waves from rotating deformed neutron stars, radiating at twice the spin frequency. In many ways, this is an idealized situation. The gravitational-wave signal from an asymmetric spinning star may be more complicated. In fact, one would generically expect (from basic classical mechanics) a rotating rigid body to undergo free precession. It would not be too surprising if neutron stars were found to be precessing, too. And if we find that they are not, then this could provide insight into their interior structure (Jones and Andersson, 2001). Hence, it is useful to understand (the basics of) the free precession problem.

We know from the discussion in Chapter 6 that we can write the moment of inertia tensor of any asymmetric rigid body as a spherical piece and a part representing the deformation. In the case of an axisymmetric body, this leads to

$$I^{ij} = I_0 \delta^{ij} + \Delta(n_{\text{d}}^i n_{\text{d}}^j - \delta^{ij}/3), \qquad (14.118)$$

where the unit vector n_{d}^i points along the body's symmetry axis (here taken to be the $\hat{e}_3$-axis, as in Figure 14.13). The principal moments follow from $I_1 = I_2 = I_0 - \Delta/3$ and $I_3 = I_0 + 2\Delta/3$, where $\Delta = I_3 - I_1$ is negative for a prolate body and positive for an oblate one. It follows that the angular momentum is related to the angular velocity according to

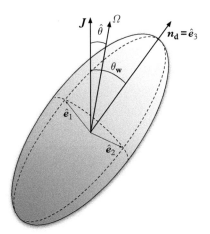

Figure 14.13 *A schematic illustration of a body undergoing free precession. The 'reference plane' contains the deformation axis n_{d}^i ($\hat{e}_3$ in this case), the angular velocity vector Ω^i, and the fixed angular momentum $\mathcal{J}^i$. The angle θ_{w} is known as the wobble angle.*

$$\mathcal{J}^i = (I_0 - \Delta/3)\Omega^i + \Delta n^i_{\mathrm{d}}(n^j_{\mathrm{d}}\Omega_j) = I_1\Omega^i + \Delta n^i_{\mathrm{d}}(n^j_{\mathrm{d}}\Omega_j), \tag{14.119}$$

and we see that the three vectors $\mathcal{J}^i$, Ω^i and n^i_{d} must be coplanar, as in Figure 14.13. Given that the angular momentum is fixed, this reference plane must revolve around $\mathcal{J}^i$. As a consequence, free precession can be parameterized by the angle θ_{w} between n^i_{d} and $\mathcal{J}^i$. This is known as the wobble angle. For a nearly spherical body the angle $\hat{\theta}$ between Ω^i and $\mathcal{J}^i$ is much smaller than the wobble angle. If $\Delta \ll I_0$ we have (Jones and Andersson, 2001)

$$\hat{\theta} \approx \frac{\Delta}{I_1}\sin\theta_{\mathrm{w}}\cos\theta_{\mathrm{w}}. \tag{14.120}$$

Note that, for a prolate body, $\hat{\theta} + \theta_{\mathrm{w}} < \theta_{\mathrm{w}}$, as in Figure 14.13.

Let us now denote the unit vector along the angular momentum by n^i_{j}. Then we can decompose the angular velocity in such a way that

$$\Omega^i = \dot{\phi}n^i_{\mathrm{j}} + \dot{\psi}n^i_{\mathrm{d}}. \tag{14.121}$$

Since the angular momentum is constant $(= \mathcal{J}n^i_{\mathrm{j}})$ we find from (14.119) that

$$\mathcal{J} = I_1\dot{\phi}, \tag{14.122}$$

and

$$\dot{\psi} = -\frac{\Delta}{I_3}\dot{\phi}\cos\theta_{\mathrm{w}}. \tag{14.123}$$

That is, the symmetry axis n^i_{d} rotates about $\mathcal{J}^i$ along a cone of half-angle θ_{w} with angular frequency $\dot{\phi}$. This is the inertial precession frequency and the period $P = 2\pi/\dot{\phi}$ is the spin period. At the same time, the body spins about the symmetry axis n^i_{d} at the angular velocity $\dot{\psi}$. This is usually referred to as the body-frame precession frequency. The corresponding period,

$$P_{\mathrm{fp}} = \frac{2\pi}{\dot{\psi}}, \tag{14.124}$$

is the free precession period. For a nearly spherical body, Eq. (14.123) shows that $\dot{\psi} \ll \dot{\phi}$, or equivalently $P \ll P_{\mathrm{fp}}$. Slightly deformed stars should undergo long-period precession.

Since astrophysical neutron stars tend to rotate significantly below the break-up limit, and crustal deformations are expected to be very small, it is natural to simplify the analysis by making the approximations of a small wobble angle and a nearly spherical star. Then Eqs. (14.120) and (14.123) become

$$\hat{\theta} \approx \frac{\Delta}{I_0}\theta_w, \qquad (14.125)$$

and

$$\dot{\psi} \approx -\frac{\Delta}{I_0}\dot{\phi}. \qquad (14.126)$$

Moving on to the gravitational waves emitted from a precessing body, we need to generalize the quadrupole formula results from Chapter 6. As in the simpler case, it is useful to take advantage of the fact that the moment of inertia tensor I_{ij} is constant in the body frame (Zimmermann and Szedenits, 1979). After cranking through the algebra of the transformation (from the body frame to the inertial frame), we find that the energy carried away from the wobbling neutron star is given by

$$\frac{dE}{dt} = \frac{2G}{5c^5}\Delta^2\dot{\phi}^6\sin^2\theta_w(\cos^2\theta_w + 16\sin^2\theta_w). \qquad (14.127)$$

In the case of a small wobble angle we have

$$\frac{dE}{dt} \approx \frac{2G}{5c^5}\Delta^2\dot{\phi}^6\theta_w^2. \qquad (14.128)$$

We see that, in the limit of a vanishing wobble angle ($\theta_w \to 0$) there will be no gravitational-wave emission. This makes sense since the body is then spinning around the $\hat{e}_3$-axis, which makes the situation axisymmetric. If, on the other hand, we take the limit $\theta_w \to \pi/2$, then the body spins around the $\hat{e}_1$-axis and we retain the result (6.24) from before. We simply need to identify $\epsilon^2 I_0^2 = \Delta^2$ and $\Omega = \dot{\phi}$.

When it comes to working out the actual gravitational-wave signal, the problem turns out to be both simpler and more difficult than the evaluation of the radiated energy. The problem is easier because we need only two time derivatives of the moment of inertia tensor and the wave amplitude is linear in the components rather than quadratic. At the same time, the problem is more involved since we require information about the location of the observer relative to the source. This means that the rotational transformation, e.g. in terms of the Euler angles θ, ϕ, and ψ from Figure 14.14, appears explicitly in the calculation. We also need to introduce a new parameter, the observer's inclination angle i.

Again working out the algebra, we arrive at (Zimmermann and Szedenits, 1979)

$$h_+ = \frac{2G\Delta\dot{\phi}^2}{rc^4}\sin\theta_w\left[(1+\cos^2 i)\sin\theta_w\cos 2\phi + \sin i\cos i\cos\theta_w\sin\phi\right], \qquad (14.129)$$

and

$$h_\times = \frac{2G\Delta\dot{\phi}^2}{rc^4}\sin\theta_w\left[2\cos i\sin\theta_w\sin 2\phi + \sin i\cos\theta_w\sin\phi\right]. \qquad (14.130)$$

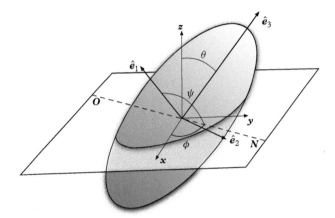

Figure 14.14 *The two coordinate systems used in the description of a freely precessing body. The inertial coordinate system has basis vectors* $\hat{\boldsymbol{e}}_x, \hat{\boldsymbol{e}}_y,$ *and* $\hat{\boldsymbol{e}}_z,$ *while the body frame is represented by basis vectors* $\hat{\boldsymbol{e}}_1, \hat{\boldsymbol{e}}_2,$ *and* $\hat{\boldsymbol{e}}_3$. *The Euler angles* $\theta, \phi,$ *and* ψ *which relate the two coordinate systems are defined as in the illustration. The line ON represents the line of nodes formed by the intersection of the* $\hat{\boldsymbol{e}}_x - \hat{\boldsymbol{e}}_y$*-plane and the* $\hat{\boldsymbol{e}}_1 - \hat{\boldsymbol{e}}_2$*-plane. Note that the angle* θ *is equal to the wobble angle* θ_w *in the free precession problem.*

We see that, for small wobble angles, the radiation is dominated by the contribution at the spin frequency. This is an important lesson. Observers need to be aware of the possibility that the strongest signal may not be found at twice the spin-frequency (Jones and Andersson, 2002; Jones, 2010). However, we also see that, if the body flips over to become an orthogonal rotator ($\theta_w \to \pi/2$), then this contribution vanishes and we retain our previous results. Finally, as the wobble angle is likely to be small, we can use the order-of-magnitude estimate

$$h \approx \frac{2G\Delta\dot{\phi}^2}{rc^4}\theta_w, \tag{14.131}$$

to asses the relevance of a precession signal (Jones and Andersson, 2002).

The question is if there is any evidence that real neutron stars are precessing. Sure, on theoretical grounds one would expect free precession to be generic in spinning bodies, but nature may not care too much about our theorizing. Indeed, it turns out that the expectation is not brought out by observations. Radio observations tell us that pulsars do not tend to undergo long period free precession. Nevertheless, there are a few cases where the precession interpretation would seem to fit the data. The best evidence is provided by PSR B1828-11, which spins with a frequency of 2.5 Hz and exhibits timing variability that may be explained as slow precession (Stairs *et al.*, 2000). Since the discovery, several regular cycles have been observed; see Figure 14.15. The data show a strong periodicity at about 500 and 1,000 days. A possible explanation for the two periodicities is that the magnetic dipole is very nearly orthogonal to the star's deformation axis (Jones and Andersson, 2001; Link and Epstein, 2001). In this case, both the phase- and the

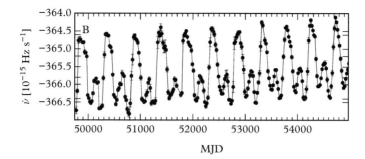

Figure 14.15 *Observed timing variability of PSR B1828-11 through 14 years of observations. The regular behaviour may be interpreted as evidence for free precession. (Adapted from Ashton et al. (2016), based on data from Lyne et al. (2010).)*

amplitude modulations have significant components at $2\dot{\psi}$. This scenario leads to an inferred wobble angle in good agreement with the value estimated from the amplitude modulation. Of course, for this scenario to apply, we need near-perfect orthogonality between the deformation axis and the magnetic dipole axis.

While the PSR B1828-11 data may suggest a precessing pulsar, there are other explanations. A likely alternative involves state switching in the pulsar magnetosphere, leading to different spin-down rates (Lyne *et al.*, 2010). This explanation is attractive as it also fits the phenomenology of other systems. Moreover, if we want to interpret this kind of data within the precession model, we need to appreciate that reality is more complicated than we have so far pretended. Real neutron stars are not rigid bodies. In order to make the model more realistic, we need to account for the fact that that mature neutron stars have an elastic crust shielding a (compressible) fluid core (Jones and Andersson, 2001). The core fluid is made up of (at least in the outer region) a viscous electron–proton plasma coexisting with a neutron superfluid. The crust itself may contain a (potentially pinned) neutron superfluid. The star's magnetic field will thread the core, in a way which depends on the properties of the superfluid phase. In principle, the core will couple to the crust through friction. The free precession of this kind of system is much more complicated than that of a rigid body. A key issue concerns the moment of inertia associated with superfluid vortices, which may change the effective oblateness of the star by an enormous factor (Shaham, 1977; Jones and Andersson, 2001). If a significant fraction of the crust superfluid is pinned, corresponding to a few percent of the star's total moment of inertia, then the effective oblateness parameter will be of order unity, and we will have $P_{\text{fp}} \sim P$. That is, the precession period would be similar to the spin period. There is currently no evidence for such behaviour from observations, but the simple argument should provide motivation for further investigation of the possibility.

14.13 Evolution of the wobble angle

Having derived an estimate of the energy loss due to gravitational-wave emission, it is natural to ask what the back-reaction on the precessing motion may be. We already know from the case of eccentric binaries (see Chapter 5) that this question is likely to be non-trivial. In the case of a body spinning around the principal axis associated with the largest moment of inertia it was clear that the energy lost to gravitational radiation had to spin the object down. In the more general case, when the body is precessing, the emitted waves will also affect the wobble angle. As we will now show, the wobble angle generally decays much faster than the star spins down.

Let us begin by writing down the spin-down timescale, i.e. the timescale on which $\dot{\phi}$ changes. Since we have $\mathcal{J} = I_1\dot{\phi}$, and $\mathcal{J}$ is conserved, it follows that

$$\ddot{\phi} = \frac{\dot{\mathcal{J}}}{I_1} = \frac{\dot{E}}{I_1\dot{\phi}} = -\frac{2G}{5c^5}\frac{\Delta^2}{I_1}\dot{\phi}^5 \sin^2\theta_w(\cos^2\theta_w + 16\sin^2\theta_w), \tag{14.132}$$

where we have noted that the gravitational waves drain energy from the motion and we have also used the general relation (6.8), i.e. $\dot{E} = \dot{\phi}\dot{\mathcal{J}}$. Moreover, we have assumed that the energy carried away by the radiation—given by (14.127) or (14.128)—correponds exactly to that given up by the local motion. This should be true in our simple case but one still has to be careful with this 'flux-balance' argument (Cutler and Jones, 2001).

It is straightforward to see that the timescale on which the star spins down is

$$t_{sd} = \left|\frac{\dot{\phi}}{\ddot{\phi}}\right| \approx \frac{5c^5}{2G}\frac{I_0}{\Delta^2}\frac{1}{\dot{\phi}^4\theta_w^2}, \tag{14.133}$$

in the limit of a small wobble angle.

In order to estimate the timescale on which the wobble angle evolves, we note that the precessing system is completely described by $\mathcal{J}$ and θ_w. Assuming that $E = E(\mathcal{J}, \theta_w)$ we then have

$$\dot{E} = \left(\frac{\partial E}{\partial \mathcal{J}}\right)_{\theta_w}\dot{\mathcal{J}} + \left(\frac{\partial E}{\partial \theta_w}\right)_{\mathcal{J}}\dot{\theta}_w. \tag{14.134}$$

Hence, we have

$$\dot{\theta}_w = \dot{E}\left[1 - \frac{1}{\dot{\phi}}\left(\frac{\partial E}{\partial \mathcal{J}}\right)_{\theta_w}\right]\left(\frac{\partial E}{\partial \theta_w}\right)_{\mathcal{J}}^{-1}. \tag{14.135}$$

Considering that the kinetic energy of the system is given by (6.6), we have

$$E = \frac{\mathcal{J}^2}{2I_1}\left(1 + \frac{\Delta}{I_1}\cos^2\theta_w\right), \tag{14.136}$$

from which we can work out the required partial derivatives. This leads to (since the gravitational waves cary energy away from the system)

$$\dot{\theta}_{\mathrm{w}} = \frac{\dot{E}}{I_0 \dot{\phi}^2} \frac{\cos\theta_{\mathrm{w}}}{\sin\theta_{\mathrm{w}}} \approx -\frac{2G}{5c^5} \frac{\Delta^2}{I_0} \dot{\phi}^4 \theta_{\mathrm{w}}. \tag{14.137}$$

The result shows that gravitational-wave emission always leads to a decreasing wobble angle, regardless of whether the deformation is prolate or oblate. Moreover, the timescale for alignment, $t_{\mathrm{w}} = \theta_{\mathrm{w}}/\dot{\theta}_{\mathrm{w}}$, is much faster than the spin-down

$$t_{\mathrm{w}} \approx \theta_{\mathrm{w}}^2 t_{\mathrm{sd}}. \tag{14.138}$$

Parameterizing the alignment timescale we find that

$$t_{\mathrm{w}} = 1.8 \times 10^6 \left(\frac{10^{45}\,\mathrm{g\,cm}^2}{I_0}\right)\left(\frac{10^{-7}}{\epsilon}\right)^2\left(\frac{P}{1\mathrm{ms}}\right)^4 \mathrm{yr}. \tag{14.139}$$

For example, for a star with $\epsilon \approx 10^{-7}$, a canonical moment of inertia $I_0 = 10^{45}\,\mathrm{g\,cm}^2$, and fast spin $P = 1$ ms, this corresponds to damping in 6×10^9 free precession periods.

If we ignore the gravitational waves the star would, once excited, precess forever. The motion of a real star will, of course, be different. Comparing our estimates to the damping rate due to various friction mechanisms (which we discuss in Chapter 15), it is easy to see see that the gravitational radiation reaction is unlikely to be the dominant driver of alignment in any neutron star of physical interest. A number of dissipation mechanisms will act to damp the precession. Understanding this internal damping is important because it impacts on observations. It is also an interesting question in principle, and it turns out that the outcome may be surprising. This becomes apparent as soon as we consider the general timescale associated with internal dissipation. Key to the argument is the simple fact that viscosity drains energy from the system—it involves the conversion of mechanical energy into heat or radiation—but conserves angular momentum. This means that we can assume that the angular momentum is nearly constant. That is, we have

$$\dot{\theta}_{\mathrm{w}} = \dot{E}\left(\frac{\partial E}{\partial \theta_{\mathrm{w}}}\right)_{\mathcal{J}}^{-1}, \tag{14.140}$$

where E now denotes the total energy of the deformed crust plus a possibly pinned superfluid (say). Combining this with the precession model, we have

$$\dot{\theta}_{\mathrm{w}} = \frac{\dot{E}}{\dot{\phi}^2 \theta_{\mathrm{w}} I_0 \epsilon_{\mathrm{eff}}}. \tag{14.141}$$

This result has an interesting implication. If the amount of superfluid pinning is negligible, then we have $I_0 \epsilon_{\mathrm{eff}} \approx \Delta$. For oblate deformations, we know that $\Delta > 0$ and we see that θ_w decreases. This is as expected. However, in the case of a prolate deformation we have $\Delta < 0$, and hence the wobble angle will *increase* as energy is dissipated (Jones, 1976). As discussed by Cutler (2002), this would eventually lead to the body becoming an orthogonal rotator. The question is whether there are situations where the deformation of a real neutron star is prolate.

Making contact with the discussion of the magnetic field, a strong toroidal field could provide the prolate deformation we need for the spin flip. One can estimate that the toroidal field must be (Cutler, 2002)

$$B_t \geq 3.4 \times 10^{12} \left(\frac{f}{300 \text{ Hz}} \right)^2 \text{ G}, \tag{14.142}$$

in order for the (prolate) magnetic deformation to overcome the (oblate) one due to crustal stresses. This is likely too strong for the mechanism to act in fast-spinning accreting stars, but it may be relevant for a young magnetar (Dall'Osso *et al.*, 2018). The question then is whether the viscous timescale can be short enough, compared to the fast spin-down, that the star may actually become an orthogonal rotator (Lasky and Glampedakis, 2016).

The problem of determining the rate at which energy is dissipated from the precession motion shares many aspects with the problem of viscous damping of neutron star oscillations. This makes sense since free precession can be viewed as a mode of oscillation of the system, and the damping is directly associated with the induced fluid flow (Sedrakian *et al.*, 1999). Given this connection we need to consider dissipative aspects of neutron star fluid dynamics in more detail, and we will do this in the context of the gravitational-wave driven r-mode instability.

15

The r-mode instability

The dynamics of rotating stars are of obvious relevance to gravitational-wave astronomy. One can think of many different astrophysical scenarios in which the oscillations of a star may be excited to a significant level. These range across the life of a neutron star—from its violent birth in a core-collapse supernova to the formation of a hot, rapidly spinning, remnant at the end of binary inspiral. They may also involve different instabilities. Of particular interest may be the gravitational-wave driven Chandrasekhar–Friedman–Schutz (CFS) instability (see Chapter 13). The mere notion that the gravitational waves that drive an oscillation mode unstable in the first place may be detectable is intriguing.

We have already seen that the $(l > 2)$ fundamental f-modes would become unstable in (Newtonian) neutron stars spinning close to the break-up rate. This instability has been studied in considerable detail—and we will summarize the state of the art in Chapter 18—but we already know that it is unlikely to provide an explanation for the apparent speed limit for astrophysical neutron stars. The f-mode instability sets in at too high a rotation rate. However, this does not mean that the CFS scenario does not play a role. In fact, the inertial r-modes are likely to become unstable at lower spin rates (Andersson, 1998; Friedman and Morsink, 1998; Lindblom et al., 1998; Andersson et al., 1999a). As a result the instability of these modes may dominate, even though they are (in principle) less efficient gravitational-wave emitters.

In retrospect, the instability of the r-modes may seem obvious. We have already shown that the mode frequency is negative in the rotating frame and positive in the inertial frame (see Chapter 13). Hence, the modes satisfy the CFS instability criterion for all rotation rates. Nevertheless, the implications of this were overlooked for a long time, possibly due to an 'unconscious bias' that mass multipoles dominate the gravitational-wave emission. Be that as it may, we now know that the r-modes are an interesting source of gravitational waves and different aspects of the problem continue to be explored (Andersson and Kokkotas, 2001; Ho et al., 2011a).

There are several reasons why it is worth discussing the r-mode problem in detail. The intricate balance between gravitational-wave emission—which drives the instability—and various damping agents—which serve to suppress it—provides insight into the physics of the star's interior. This is important, as the outcome depends on equation of state issues beyond the assumed pressure-density relation (for chemical equilibrium). We need

Gravitational-Wave Astronomy: Exploring the Dark Side of the Universe. Nils Andersson, Oxford University Press (2020).
© Nils Andersson. DOI: 10.1093/oso/9780198568032.001.0001

to consider a range of transport properties for supranuclear matter. In addition, it is interesting to ask how astrophysical observations, e.g. in X-rays and radio, constrain the instability scenario and how this influences our understanding of the theory.

15.1   The instability window

Once you identify the presence of an instability, it makes sense to ask whether it has any actual physical relevance. We have learned that the r-modes are unstable at all rates of rotation in a perfect fluid star. In principle, this implies that—unless nature intervenes—such stars should not be able to rotate. The unstable modes would grow, radiate more gravitational waves, and spin the star down. However, this is not reality. The fact that an instability is present in principle does not mean that it is important in practice.

First of all, we need to establish that the instability grows fast enough to be relevant. Intuitively, one may expect the r-mode instability to be weak since the associated fluid motion does not involve large density variations and therefore should not be associated with strong gravitational waves.

In order to estimate the relevant instability timescales we assume that the mode-solution is well represented by the non-dissipative perturbation equations (Ipser and Lindblom, 1991). Then we use these solutions to evaluate the effect of various dissipation mechanisms, adding their respective contributions to the rate of change of the mode energy, dE/dt. Finally, we verify that the first assumption is justified by checking that the estimated growth/damping times are considerably longer than the oscillation period of the mode.

Assume that the r-mode eigenfunctions are proportional to $\exp(-t/t_d)$, with t_d the timescale associated with a specific dissipation mechanism. Then recall that the mode-energy follows from the square of the perturbation, which means that

$$\frac{dE}{dt} = -\frac{2E}{t_d},$$
(15.1)

where

$$E \approx \frac{1}{2}\int \rho|\delta v|^2 dV \approx \frac{l(l+1)}{2}\omega_r^2\int_0^R \rho|U_l|^2 dr,$$
(15.2)

with U_l the r-mode eigenfunction from Chapter 13, is the energy of an r-mode measured in the rotating frame and ω_r is the corresponding frequency (recall that, for a barotropic model, there will be a single r-mode for each $l = m$; see Chapter 13). Next, introduce a suitable normalization—in terms of a dimensionless amplitude α—such that

$$\delta v \approx \alpha \Omega R \left(\frac{r}{R}\right)^l \mathbf{Y}_{ll}^B e^{i\omega_r t},$$
(15.3)

where $\mathbf{Y}_{ll}^B$ is the magnetic multipole from (13.61).

This way we can determine the r-mode energy for any given stellar model. If we focus our attention on the simple case of an $n = 1$ polytrope, which is useful as we can obtain simple scaling relations in terms of the stellar parameters (Lindblom *et al.*, 1998; Andersson *et al.*, 1999a), we have

$$E \approx 10^{51} \alpha^2 \left(\frac{M}{1.4 M_\odot} \right) \left(\frac{R}{10 \text{ km}} \right)^2 \left(\frac{v_s}{1 \text{ kHz}} \right)^2 \text{ erg,} \qquad (15.4)$$

where M and R are the mass and radius of the star, respectively, and v_s is the star's spin frequency. For simplicity, we only keep the leading order terms in (15.4) and the dissipation integrals below. We also assume the Cowling approximation, as this simplifies the analysis (although one has to be careful, as density variations play a key role in estimates of the bulk viscosity).

As a first step towards establishing the relevance of the r-mode instability, let us show that the modes grow on an astrophysically interesting timescale. This is, essentially, an exercise in applying the post-Newtonian multipole formulas from Chapter 13. The energy change from the mode (in the rotating frame) follows from

$$\frac{dE}{dt} = -\frac{\omega_r}{\omega_i} \frac{dE}{dt} \bigg|_{\text{gw}}. \qquad (15.5)$$

We have already argued that the leading contribution to the r-mode emission comes from the current multipoles. After inserting the leading order eigenfunction in the relevant current multipole term, we find that

$$\frac{dE}{dt} \bigg|_{\text{gw}} \approx -4l^2 N_l \omega_r^3 \omega_i^{2l+1} \left| \int_0^R \rho r^{l+1} U_l dr \right|^2, \qquad (15.6)$$

with N_l given by (13.84). This leads to an estimated growth timescale for the instability of the $l = m = 2$ r-mode (Andersson *et al.*, 1999a)

$$t_{\text{gw}} \approx -47 \left(\frac{M}{1.4 M_\odot} \right)^{-1} \left(\frac{R}{10 \text{ km}} \right)^{-4} \left(\frac{v_s}{1 \text{ kHz}} \right)^{-6} \text{ s,} \qquad (15.7)$$

where the sign indicates that the mode is unstable. The timescale increases by roughly one order of magnitude with each l, so higher order multipoles lead to significantly weaker instabilities.

The main lesson from (15.7) is that the unstable r-modes grow fast enough that the instability could be relevant for astrophysics (a few tens of seconds is considerably shorter than the typical evolutionary timescale for a mature neutron star). This is encouraging, but we still have work to do. In a real star, a number of mechanisms compete with the instability and prevent it from growing. Unfortunately, many of the relevant mechanisms involve physics that is poorly understood and difficult to model in a realistic fashion.

Conversely, the instability may provide a probe of these unknown aspects. This would be exciting as it may allow us to constrain the physics.

Let us (first of all) explore the possibility that the r-mode instability is active in a newly born neutron star. In order for the r-modes to be relevant they must then grow fast enough that they are not completely damped out by shear- and bulk viscosity. These are due to rather different physical mechanisms and hence we consider them in turn.

At relatively low temperatures (below a few times 10^9 K) the main viscous dissipation in a fluid star arises from momentum transport due to particle scattering. This leads to friction that can be modelled in terms of a macroscopic shear viscosity. The effect of shear viscosity on the r-modes can be estimated from (Lindblom *et al.*, 1998; Andersson *et al.*, 1999a)

$$\left.\frac{dE}{dt}\right|_{\rm sv} = -2\int \eta\delta\sigma^{ij}\delta\sigma_{ij}^* dV, \tag{15.8}$$

where η is the viscosity coefficient and the shear, $\delta\sigma_{ij}$, associated with the perturbed fluid motion is given by

$$\delta\sigma_{ij} = \frac{i\omega_r}{2}\left(\nabla_i\xi_j + \nabla_j\xi_i - 2g_{ij}\nabla_k\xi^k\right), \tag{15.9}$$

with ξ^i the fluid displacement. After working out the angular integrals we are left with

$$\left.\frac{dE}{dt}\right|_{\rm sv} = -\omega_r^2 l(l+1)\left\{\int_0^R \eta|\partial_r U_l|^2 dr + (l-1)(l+2)\int_0^R \frac{\eta}{r^2}|U_l|^2 dr\right\}. \tag{15.10}$$

In order to make use of this result, we have to know the density dependence of the viscosity coefficient.

Above the transition temperature at which the neutron star becomes superfluid (several times 10^9 K; see Chapter 12), the appropriate viscosity coefficient is due to neutron–neutron scattering (Flowers and Itoh, 1976, 1979; Cutler and Lindblom, 1987; Andersson *et al.*, 2005b). We then have

$$\eta_{\rm n} = 2\times 10^{18}\left(\frac{\rho}{10^{15}\ {\rm g/cm^3}}\right)^{9/4}\left(\frac{T}{10^9\ {\rm K}}\right)^{-2}\ {\rm g/cm s}, \tag{15.11}$$

which leads to an estimated shear-viscosity damping timescale

$$t_{\rm sv}\approx 6.7\times 10^7\left(\frac{M}{1.4M_\odot}\right)^{-5/4}\left(\frac{R}{10\ {\rm km}}\right)^{23/4}\left(\frac{T}{10^9\ {\rm K}}\right)^2\ {\rm s} \tag{15.12}$$

(assuming that the star is isothermal). This estimate is expected to be relevant for the first months of the life of a hot young neutron star. As soon as the core temperature drops sufficiently that the neutrons in the star's core become superfluid, the neutron–neutron scattering is suppressed. The above viscosity coefficient must then be replaced by

$$\eta_e = 6 \times 10^{18} \left(\frac{\rho}{10^{15} \text{ g/cm}^3} \right)^2 \left(\frac{T}{10^9 \text{ K}} \right)^{-2} \text{ g/cms}, \tag{15.13}$$

which follows from an analysis of electron–electron scattering. This leads to

$$t_{sv} \approx 2.2 \times 10^7 \left(\frac{M}{1.4M_\odot} \right)^{-1} \left(\frac{R}{10 \text{ km}} \right)^5 \left(\frac{T}{10^9 \text{ K}} \right)^2 \text{ s}. \tag{15.14}$$

We see that superfluidity changes the shear viscosity timescale by a factor of a few, but it is clear that—for sufficiently rapidly spinning stars and moderate (to high) temperatures—gravitational-wave emission drives the r-mode growth faster than shear viscosity can damp it.

At high temperatures (above a few times 10^9 K) bulk viscosity is the dominant dissipation mechanism. Bulk viscosity arises as the oscillation drives the fluid away from beta equilibrium. It corresponds to an estimate of the extent to which energy is dissipated from the fluid motion as weak interactions try to re-establish equilibrium. The energy lost through bulk viscosity is carried away by neutrinos. Bulk viscosity is a resonant mechanism, which becomes very strong when the typical timescale associated with the mode is similar to that of the reactions. In effect, this leads to rapid damping in a specific temperature range which sensitively depends on the involved reactions (and the matter composition).

We have seen that we can estimate the shear viscosity timescale from the leading order contribution to the fluid motion. This is not the case for the bulk viscosity. In order to assess its relevance we need the Lagrangian density perturbation and for the r-mode this arises at a higher order in Ω. Explicitly, we need

$$\left. \frac{dE}{dt} \right|_{bv} = - \int \zeta |\delta\sigma|^2 dV, \tag{15.15}$$

where $\delta\sigma$ is the expansion associated with the mode. In the standard case, where β-equilibrium is regulated by the modified Urca reactions, the relevant bulk viscosity coefficient is (Sawyer, 1989)

$$\zeta = 6 \times 10^{25} \left(\frac{\rho}{10^{15} \text{ g/cm}^3} \right)^2 \left(\frac{T}{10^9 \text{ K}} \right)^6 \left(\frac{\omega_r}{1 \text{ s}^{-1}} \right)^{-2} \text{ g/cms}. \tag{15.16}$$

In order to evaluate the integral in (15.15), we need

$$\delta\sigma = -i\omega_r \frac{\Delta\rho}{\rho}, \tag{15.17}$$

where $\Delta\rho$ is the Lagrangian density variation associated with the mode. However, this means that we have to work out the spheroidal corrections to the (predominantly toroidal) r-mode. Since these arise at the same order in the slow-rotation expansion as the centrifugal change in shape of the star, the calculation of the bulk viscosity is quite messy (Lindblom *et al.*, 1999). Accounting for the various effects (although still in the Cowling approximation) we arrive at the estimate (Andersson *et al.*, 1999a)

$$t_{bv} \approx 2.7 \times 10^{11} \left(\frac{M}{1.4M_\odot}\right)\left(\frac{R}{10\text{ km}}\right)^{-1}\left(\frac{v_s}{1\text{ kHz}}\right)^2\left(\frac{T}{10^9\text{ K}}\right)^{-6}\text{ s.} \qquad (15.18)$$

At this point it is worth making two comments. First, the proton fraction in the star's core may be large enough to make direct Urca reactions possible. In this case, the bulk viscosity coefficient becomes significantly larger than (15.16). However, this is unlikely to happen in the outer regions where the r-modes are mainly located (recall that $U_l \sim r^{l+1}$). Due to the density factors in the different integrands the main r-mode damping originates in the star's outer core. As a result, one would not expect the direct Urca reactions to be particularly important for the r-mode instability. The second point concerns very hot stars. At high temperatures, the star is no longer transparent to neutrinos. This basically shuts off the bulk viscosity. However, a newly born neutron star is likely to cool too fast for this low-viscosity region to be relevant for the development of an instability.

Equipped with these estimates we can address the main issue of interest: Should we expect the r-mode instability to be relevant for astrophysical neutron stars? Well, it is easy to see that the modes will only be unstable in a certain temperature range. To have an instability we need t_{gw} to be smaller in magnitude than both t_{sv} and t_{bv}. This leads to the notion of an instability window, often illustrated by the critical rotation period above which the mode is unstable as a function of temperature. We simply find the relevant critical rotation rate by solving for the roots of

$$\frac{1}{2E}\frac{dE}{dt} = \frac{1}{t_{gw}} + \sum\frac{1}{t_{diss}} = 0, \qquad (15.19)$$

throughout the relevant temperature range. This typically leads to the result illustrated in Figure 15.1.

From our estimates we find that shear viscosity will completely suppress the r-mode instability at core temperatures below 10^5 K. Similarly, bulk viscosity will prevent the mode from growing in a star hotter than a few times 10^9 K. In the intermediate region there is a temperature window where the growth time due to gravitational radiation is short enough to overcome the viscous damping and drive the mode unstable.

If we take our estimated timescales at face value, then a nascent neutron star would be unstable at rotation periods shorter than 25 ms. This is interesting, because the critical rotation rate is rather low. Moreover, it is reminds us of the rotation period we would infer for the Crab Pulsar at birth, $P \approx 19$ ms (if we extrapolate the current spindown and measured braking index back to 1054). However, the predicted instability

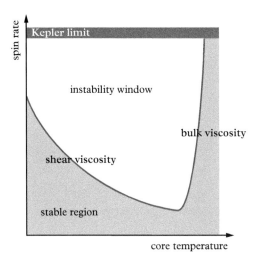

Figure 15.1 *Schematic illustration of the r-mode instability window. At low temperatures dissipation due to shear viscosity counteracts the instability. At temperatures above a few times 10^9 K bulk viscosity suppresses the instability. At very high temperatures the nuclear reactions that lead to the bulk viscosity are suppressed and an unstable mode can, in principle, grow. This region is not shown since it may only be relevant for the first few tens of seconds following the birth of a neutron star. The main instability window is expected at core temperatures near $T_c \approx 10^9$ K. Provided that gravitational radiation drives the unstable mode rapidly enough, the instability may govern the spin-evolution of a hot young neutron star. (Reproducing the instability window as discussed in the first work on the problem; Lindblom et al. (1998) and Andersson et al. (1999a).)*

window is problematic. If we focus on young neutron stars, then we also need to consider the X-ray pulsar J0537-6910 which currently spins with a period of 16 ms (Marshall *et al.*, 1998). In this case we infer a birth spin in the range 6–9 ms, which would place this object firmly inside the instability region. Moreover, when we consider accreting neutron stars in low-mass X-ray binaries we find that many systems would have to be unstable if the results illustrated in Figure 15.1 were the whole story (Ho *et al.*, 2011*a*). In order to resolve this conundrum, we need to consider the physics in more detail.

15.2 Complicating factors

The ability of the r-mode instability to act in a realistic neutron star depends on the actual state and composition of matter. As we add features to the model, the problem becomes increasingly complex. Different aspects may depend on one another and—as much of the involved physics is poorly known—one has to tread carefully. In order to illustrate this point, we consider the impact of four specific improvements to the simple fluid model. First of all, let us explore the role of the star's elastic crust.

15.2.1   The Ekman layer

The neutron star crust, which forms shortly after the neutron star is born, may have significant effect on the oscillations of the star. In particular, if the crust is assumed to be rigid, the viscous fluid motion must drop to 0 at the base of the crust. The same effect is observed in any fluid in a rotating container (or after stirring a cup of tea!), so it is not mysterious. Viscosity leads to the formation of a boundary layer close to the fluid–solid boundary. This is the so-called Ekman layer (Bildsten and Ushomirsky, 2000). The thickness of the boundary layer, δ, can be deduced by balancing the Coriolis force and the shear viscosity:

$$\delta \sim \left(\frac{\eta}{\rho\Omega}\right)^{1/2}, \tag{15.20}$$

where η is the shear viscosity coefficient from (15.11). Putting numbers into this relation we see that δ will typically be a few centimetres for a rapidly rotating neutron star (taking the crust–core transition to be at about half the nuclear saturation density). In essence the viscous boundary layer enhances the shear viscosity damping by a potentially decisive factor. For an r-mode the dissipation timescale due to the presence of such an Ekman layer would be very short (Bildsten and Ushomirsky, 2000):

$$t_{\mathrm{Ek}} \approx 830 \left(\frac{T}{10^9\mathrm{K}}\right)\left(\frac{\nu_s}{1\mathrm{kHz}}\right)^{-1/2} \mathrm{s}. \tag{15.21}$$

This effect overwhelms the standard shear viscosity and leads to all neutron stars with a rigid crust being stable at rotation periods longer than roughly 5 ms. This would go a long way towards reconciling the theory with observed accreting systems; see Figure 15.8. However, as is often the case, the devil is in the detail. The rough estimate of the boundary-layer damping does not hold up to closer scrutiny. We know from the discussion in Chapter 12 that the neutron star crust is not rigid. In reality, the Coriolis force will be strong enough that the crust takes part in the r-mode motion (Levin and Ushomirsky, 2001). As a result, the viscous damping is weaker (depending on the extent to which the core fluid 'slips' relative to the crust). The true dissipation may be (at least) a factor of about a hundred weaker than (15.21).

The effect that the slippage has on the Ekman layer damping is indicated by the rough estimate (Levin and Ushomirsky, 2001):

$$t_{\mathrm{Ek}} \approx 3 \times 10^5 \left(\frac{T}{10^9\mathrm{K}}\right)\left(\frac{\nu_s}{1\mathrm{kHz}}\right)^{-1/2} \mathrm{s}. \tag{15.22}$$

We arrive at this estimate by taking the rigid Ekman layer estimate and assuming a 'slippage' factor $S_c = 0.05$ (see Figure 1 in Levin and Ushomirsky (2001)). The impact of this slippage factor is shown in Figure 15.8. However, there are additional complications. As the spin of the star increases, the r-modes will undergo a series of so-called avoided crossings with shear modes in the crust (see Chapter 18). Close to such resonances, the character of the r-mode changes and it may be more rapidly damped. This can lead to sharp resonances and narrow horizontal regions of stability near the resonant frequencies. The main lesson is that the shape of the instability window may be more complicated than Figure 15.1 suggests.

15.2.2   Superfluid mutual friction

As the star cools, the formation of superfluids in the core brings additional degrees of freedom (see Chapter 12) and new dissipation mechanisms. Of particular importance in this respect is the so-called mutual friction (Alpar *et al.*, 1984*b*). In the standard picture, this mutual friction is linked to the fact that a superfluid rotates by forming a dense array of quantized vortices. Combining the vortices with the superfluid entrainment, due to which a momentum induced in one of the constituents causes some of the mass of the other component to be carried along—see, for example, Eq. (12.86)—the flow of superfluid neutrons around the quantized vortices induces a flow in a fraction of the protons. This, in turn, leads to magnetic fields forming around the vortices. Mutual friction encodes the dissipative scattering of electrons off of the magnetized vortices (Mendell, 1991; Andersson *et al.*, 2006).

Superfluid mutual friction has been shown to suppress the instability of the f-mode in a rotating (Newtonian) star (Lindblom and Mendell, 1995) and it was originally thought to be the dominant damping agent for the r-modes, as well. However, detailed calculations show that the r-mode instability window is essentially unaffected by the mutual friction (Lindblom and Mendell, 2000; Haskell *et al.*, 2009). Intuitively, it is easy to see why this should be so. To leading order in the slow-rotation approximation, the r-mode remains close to its normal-fluid counterpart. The two fluid components move in sync. This means that the mutual friction, which relies on the relative flow, requires an analysis to second order in the slow-rotation approximation (including the centrifugal deformation of the star). As the effect enters at higher orders, it makes sense that it is weaker.

A careful analysis of the superfluid r-mode problem to the required order shows that the mutual friction is unlikely to suppress the instability (Haskell *et al.*, 2009). In order to have significant effect, the dimensionless parameter, $\mathcal{R}$, which describes the efficiency of the dissipation, see (12.92), must be much larger than expected. This is clear from the representative results in Figure 15.2. If our current understanding of the relevant parameters is correct, then mutual friction has no real impact on the r-mode instability. However, if for some reason, $\mathcal{R}$ is enhanced by a factor of 100 or so, the effect would be significant. One possible reason for a stronger mutual friction could be the fact that the neutron superfluid coexists with a proton superconductor (in which case friction could arise through interactions between vortices and fluxtubes; see Epstein and Baym (1992) and Link (2003), but this effect is difficult to quantify (Haskell *et al.*, 2014)).

As in the case of the elastic crust, the superfluid problem has a level of fineprint that remains to be understood. For example, along with the r-mode, a neutron star core has a large set of superfluid inertial modes. These may exhibit resonances with the r-modes as

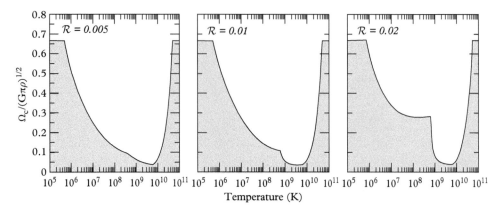

Figure 15.2 *The r-mode instability window for a superfluid neutron star with entrainment $\varepsilon_p = 0.6$ (see (12.86)) calculated as a function of the core temperature T and for a range of drag parameters $\mathcal{R}$. The results are for a canonical neutron star and relatively weak superfluidity (represented by a small triplet pairing gap for the neutrons). For the expected value of the drag parameter, $\mathcal{R} \approx 2 \times 10^{-4}$, mutual friction has no impact on the r-mode instability (left). However, if (for some reason) $\mathcal{R}$ is enhanced by a factor of 100 or so the effect would be significant (right). (Based on the results of Haskell et al. (2009).)*

the star cools. This could lead to more efficient damping (as the r-mode adopts features of the mode it is in resonance with), leading to sharp vertical peaks of stability (Gusakov et al., 2014).

15.2.3   Exotica: hyperons and quarks

The presence of exotic particles in the core of a neutron star may lead to significantly stronger viscous damping than assumed in the standard instability analysis. Of particular relevance may be the presence of hyperons or deconfined quarks, because in each case one would expect fast nuclear reactions to operate. In fact, it has been suggested that hyperon bulk viscosity would completely wipe out the r-mode instability (Lindblom and Owen, 2002). However, the problem is multifaceted. The reactions that lead to the claimed result would also efficiently cool the star, making neutron stars significantly colder than observations suggest. The discrepancy can be avoided if the hyperons are superfluid and the relevant nuclear reactions are quenched. But this would also suppress the bulk viscosity. The current understanding is that the presence of hyperons would not have a strong impact on the r-mode instability (Haskell and Andersson, 2010).

Interestingly, strong viscous damping could work in favour of the r-modes as a gravitational-wave source. This is illustrated by the case of strange stars. The observational evidence for the existence of strange stars is tenuous, but they may exist if strange matter is the most stable form of matter at high densities (Witten, 1984; Alcock et al., 1986). The r-mode instability window is very different for deconfined quark matter (Madsen, 1998). In particular, as in the case of the hyperons, the bulk viscosity of strange matter is many orders of magnitude stronger than its neutron star counterpart (and the associated resonance happens at lower temperatures).

The shear viscosity damping is now given by (Madsen, 1998)

$$t_{sv} \approx 7.4 \times 10^7 \left(\frac{\alpha_s}{0.1}\right)^{5/3} \left(\frac{M}{1.4M_\odot}\right)^{-5/9} \left(\frac{R}{10\text{km}}\right)^{11/3} \left(\frac{T}{10^9\text{K}}\right)^{5/3} \text{s}, \qquad (15.23)$$

where α_s is the fine-structure constant for the strong interaction. Meanwhile, the situation is more complicated for the bulk viscosity (which now is a result of the change in concentration of down and strange quarks in response to the mode oscillation). For low temperatures we have (Madsen, 1998)

$$t_{bv}^{low} \approx 7.9 \left(\frac{M}{1.4M_\odot}\right)^2 \left(\frac{R}{10\text{km}}\right)^{-4} \left(\frac{v_s}{1\text{kHz}}\right)^{-2} \left(\frac{T}{10^9\text{K}}\right)^{-2} \left(\frac{m_s}{100\text{MeV}}\right)^{-4} \text{s}, \quad (15.24)$$

where m_s represents the mass of the strange quark. These results lead to the main instability window being shifted towards lower temperatures. The suppression of the bulk viscosity also opens up an additional unstable region at high temperatures (estimating this region requires a more detailed calculation).

We learn that the r-mode instability would not—in contrast to the case for neutron stars—be active in strange stars with a core temperature of 10^9 K. In a strange star the r-modes are unstable at lower temperatures (between 10^5 and 5×10^8 K) and also at temperatures above a few times 10^9 K. In principle, the instability would be active for a brief period after a strange star is born. Then the mode would become stable until thousands of years later when the star has cooled sufficiently to enter the low-temperature instability window. An interesting consequence of shifting the instability window to lower temperatures is that an accreting neutron star may heat up sufficiently that—while still inside the instability region—it evolves to the part of the instability curve that increases with temperature. This may lead to the formation of a persistent gravitational-wave source (Andersson *et al.*, 2002). The main lesson is that the details of the interior physics may impact on the astrophysics phenomenology.

15.2.4 The magnetic field

Intuitively, one might expect the star's (interior) magnetic field to significantly affect the r-mode instability. However, given that neither the field configuration nor the state of matter is particularly well understood, it is not so easy to quantify this expectation. In order to address the problem we need to make progress on the modelling of the field structure in the first place. Nevertheless, there has been a fair bit of work on r-modes in magnetic stars. The efforts essentially fall into three categories.

First, we can ask what effect the magnetic field may have on the r-mode itself. However, the problem of oscillating magnetic neutron stars is computationally complex. Due to the combination of a spherical star with the cylindrical symmetry of the magnetic field there is no natural set of basis functions one can use to decompose the various perturbations—one has to deal with large sets of coupled harmonics. Moreover, one can argue that (at least in ideal magnetohydrodynamics; see Chapter 20) the problem may have a continuous spectrum (Levin, 2007). If this is the case, then the usual stellar oscillation strategy will fail.

We can estimate how strong the magnetic field has to be to affect r-mode motion. If we treat the influence of the magnetic field as a small perturbation of the usual fluid mode we find that (in the simple case of a uniform magnetic field aligned with the spin axis; see Morsink and Rezania (2002))

$$\omega_r = -\frac{2\Omega}{l+1}\left\{1 - \frac{\mathcal{M}}{T}\frac{(l+1)(2l+3)}{10}\right\}, \tag{15.25}$$

where the ratio of magnetic to rotational energy is

$$\frac{\mathcal{M}}{T} \approx \frac{5}{6}\frac{B^2 R}{M\Omega^2} \approx 8 \times 10^{-12}\left(\frac{B}{10^{12}\,\mathrm{G}}\right)^2\left(\frac{\nu_s}{1\,\mathrm{kHz}}\right)^{-2}, \tag{15.26}$$

for canonical neutron star parameters. This estimate suggests that it may be safe to ignore the impact of the magnetic field in most cases. Perhaps not surprisingly, we need magnetar level field strengths for the correction to the mode frequency to be significant. For example, for a field of 10^{14} G and a star spinning at 1 Hz we have $\mathcal{M}/T \approx 0.1$.

Second, the presence of a magnetic field penetrating the star's crust into the fluid core may change the dynamics of any viscous boundary layer. In essence, the magnetic field should prevent slippage of the core fluid relative to the crust (at least in ideal magnetohydrodynamics). This simple observation has not yet been incorporated in actual models, but estimates show that the magnetic field may have decisive impact (Mendell, 2001).

Third, the interaction between an unstable mode and the magnetic field may have an interesting consequence. The r-mode fluid motion induces differential rotation which leads to a winding up of the interior magnetic field (Rezzolla et al., 2000, 2001). In some situations, where the initial field is strong enough, this may lead to the magnetic field preventing the r-mode from growing further. The problem relies on the so-called Stokes drift (due to which fluid elements undergo a secular drift when a wave is present in the system), the magnitude of which depends on the latitude of the fluid element and the r-mode amplitude. Key issues concern how the differential drift affects the magnetic field of the star, and what the back reaction on the mode may be (Friedman et al., 2017).

15.3   A simple spin-evolution model

Having established that the r-modes may be unstable in real neutron stars, let us turn to the problem of the evolution of the instability. Provided it can grow to a significant amplitude, one would expect the instability to influence the spin-evolution of the star. At the same time, the induced shear viscosity will heat the fluid. In order to get an idea of how this may work, we need a model that evolves the unstable mode along with the star's rotation and temperature. Such a model will necessarily be somewhat qualitative, but it may still provide useful insights.

While the early growth phase of an unstable mode can be described by perturbation theory, an understanding of many effects (such as the coupling between different modes) that may eventually dominate the dynamics requires nonlinear calculations. As such calculations are beyond reach (given the timescales involved; see Lindblom *et al.* (2001), Lindblom *et al.* (2002), and Lin and Suen (2006)) we will try to capture the essential features of the problem phenomenologically. Intuitively, one would expect the growth of an unstable mode to be halted at some amplitude. As the mode saturates it seems plausible that the excess angular momentum will be radiated away and the star will spin down. This motivates a simple three-parameter description of a spinning star governed by the r-mode instability—using the spin rate Ω (assumed uniform, for simplicity), the mode-amplitude α, and the star's core temperature T as the key quantities (Owen *et al.*, 1998; Ho and Lai, 2000).

We first of all consider the (canonical) angular momentum of the unstable mode (see Chapter 13)

$$J_c = -\frac{3\Omega\alpha^2 \tilde{J}MR^2}{2}, \tag{15.27}$$

where the dimensionless quantity $\tilde{J}$ is given by

$$\tilde{J} = \frac{1}{MR^4}\int_0^R \rho r^6 \, dr = 1.635 \times 10^{-2}, \tag{15.28}$$

for an $n=1$ polytrope (which we take as our base model throughout this discussion). The angular momentum evolves as gravitational-wave emission drives the instability (on a timescale given by t_{gw}) while viscosity tries to damp the motion. In order to describe the damping timescale we introduce

$$\frac{1}{t_{diss}} = \frac{1}{t_{sv}} + \frac{1}{t_{bv}} + \text{other dissipation terms.} \tag{15.29}$$

The evolution is then given by

$$\frac{dJ_c}{dt} = -2J_c\left(\frac{1}{t_{gw}} + \frac{1}{t_{diss}}\right) \tag{15.30}$$

(recall that the we use a negative timescale to indicate growth).

A second evolution equation follows from the total angular momentum of the system

$$J = I\Omega + J_c = \left(1 - Q\alpha^2\right)\tilde{I}\Omega MR^2, \tag{15.31}$$

where

$$\tilde{I} = I/MR^2 \approx 0.261 \tag{15.32}$$

(again for an $n = 1$ polytrope) and $Q = 3\tilde{\mathcal{J}}/2\tilde{I}$. The first term in (15.31) represents the bulk rotation of the star and the second is due to the canonical angular momentum of the r-mode. This representation makes sense since $\mathcal{J}_c$ corresponds to the change in angular momentum due to the presence of the mode (in absence of viscosity or radiation).

Taking a time derivative of (15.31) we have

$$\frac{d\mathcal{J}}{dt} = I\frac{d\Omega}{dt} + \frac{d\mathcal{J}_c}{dt} + N, \tag{15.33}$$

where the last term allows us to include additional torques on the system, e.g. due to electromagnetic dipole emission or accretion.

The two equations, (15.30) and (15.33), lead to the coupled system

$$\frac{d\alpha}{dt} = -\frac{\alpha}{t_{gw}} - \alpha\frac{1 - Q\alpha^2}{t_{diss}} - \frac{N}{2I\Omega}, \tag{15.34}$$

and

$$\frac{d\Omega}{dt} = -\frac{2Q\Omega\alpha^2}{t_{diss}} + \frac{N}{I}. \tag{15.35}$$

These equations govern the mode evolution in the phase where the amplitude of the r-mode grows, as well as in the late phase where the temperature has decreased sufficiently to make the mode stable.

We also need to account for the fact that the mode may heat the star as it stirs the core fluid (noting that the viscosities are highly sensitive to temperature changes). This heating must be combined with the gradual cooling expected as the star ages. Thus we need an equation for the thermal energy. Combining the main mechanisms, we have

$$\frac{dE_{th}}{dt} = C_v\frac{dT}{dt} = \left|\frac{2E_c}{t_{gw}}\right| - L_v - L_\gamma + H, \tag{15.36}$$

where C_v is the heat capacity (averaged over the star's core), and

$$E_c = \frac{1}{2}\alpha^2\Omega^2 MR^2\tilde{\mathcal{J}} \tag{15.37}$$

is the canonical energy of the r-mode (again; see Chapter 13). The two terms, L_v and L_γ, represent cooling due to neutrino emission and photons radiated from the star's surface, respectively, while H accounts for additional heating, e.g. due to nuclear burning of accreted material.

Since they should be adequately described by perturbation theory, the early and late parts of the evolution of an unstable mode are reasonably well understood. However, we are mainly interested in the nonlinear phase as this is where the main spin-down would occur. To make progress, we need to model this nonlinear regime. As a (somewhat pragmatic) starting point we take the view that the mode-amplitude saturates at a critical value $\alpha_s \leq 1$, noting that $\alpha_s = 1$ would correspond to the r-mode carrying a large fraction of the angular momentum of the system (Owen *et al.*, 1998). This is easy to see from

$$\left| \frac{\mathcal{J}_c}{I\Omega} \right| \approx \frac{3\alpha^2 \tilde{\mathcal{J}}}{2\tilde{I}} \approx 0.1\alpha^2. \tag{15.38}$$

If we assume that α stays constant throughout the saturated phase (which may not be at all realistic), then we see from (15.35) that the star will spin down on the viscous timescale (as long as we ignore external torques). Furthermore, ignoring the $Q\alpha^2$ term in (15.34), as it is likely to be small, we see that we must have $|t_{\mathrm{gw}}| = t_{\mathrm{diss}}$ throughout the saturated phase. As the mode heating acts on the same timescale, the system may reach thermal balance, in which case we have from (15.36)

$$\left| \frac{2E_c}{t_{\mathrm{gw}}} \right| = \left| \frac{2E_c}{t_{\mathrm{diss}}} \right| = L_v + L_\gamma \tag{15.39}$$

(ignoring additional heating). For a given saturation amplitude α_s, this provides a 'heating equals cooling' curve in the $\Omega - T$ plane (Bondarescu *et al.*, 2007)

$$\alpha_s = \left(\frac{t_{\mathrm{gw}}}{\Omega^2 M R^2 \tilde{\mathcal{J}}} \right)^{1/2} (L_v + L_\gamma)^{1/2}. \tag{15.40}$$

If we want to use this result, we need a better idea of the luminosities. First of all, we need the modified Urca neutrino emission process (which is likely to dominate the cooling unless the star is massive), for which the luminosity is (Ho *et al.*, 2011a)

$$L_v \approx 7.4 \times 10^{31} \left(\frac{T}{10^8 \text{ K}} \right)^8 \text{ erg s}^{-1} \tag{15.41}$$

(for canonical neutron star parameters). Meanwhile, the photon luminosity follows from the effective surface temperature. In general, this leads to

$$L_\gamma = 4\pi R^2 \sigma T_{\mathrm{eff}}^4, \tag{15.42}$$

where σ is the Stefan–Boltzmann constant and R is the star's radius. The left-hand side can (at least sometimes) be inferred from X-ray observations. However, we need to relate this effective temperature to the core temperature used in (15.36). To do this, we make assumptions about the outer regions of star. The composition of the star's envelope is important as this region acts as a heat blanket that shields the (generally hotter) interior from the surface. The composition depends on the amount of accreted material. For an isolated star (with an iron heat blanket), we have (Gudmundsson *et al.*, 1983)

$$T \approx 1.288 \times 10^8 \left[\left(\frac{T_{\text{eff}}}{10^6 \text{ K}} \right)^4 \left(\frac{g}{10^{14} \text{ cm s}^{-2}} \right)^{-1} \right]^{0.455} \text{ K}, \tag{15.43}$$

where

$$g = \frac{GM}{R^2} \left(1 - \frac{2GM}{Rc^2} \right)^{-1/2} \approx 2.4 \times 10^{14} \text{ cm s}^{-2} \tag{15.44}$$

is the (redshifted) surface gravity (and the numbers relate to a canonical neutron star). The result depends on the chemical composition as the thermal conductivity becomes lower with increasing Z. Accreted envelopes composed of light elements are more transparent (i.e. have higher T_{eff} for a given T). The difference is, however, not expected to be larger than about a factor of 2. Combining the results, we have

$$L_\gamma \approx 9.8 \times 10^{32} \left(\frac{T}{10^8 \text{ K}} \right)^4 \text{ erg s}^{-1}, \tag{15.45}$$

and we see that the neutrino emission dominates the surface emission for stars with core temperature above 2×10^8 K or so.

If we focus on the regime where neutrino emission dominates, and use the fact that

$$L_{\text{gw}} \approx 4 \times 10^{45} \left(\frac{\alpha_s}{10^{-2}} \right)^2 \left(\frac{\nu_s}{1 \text{ kHz}} \right)^8 \text{ erg s}^{-1}, \tag{15.46}$$

we find that the heating balances the cooling when

$$\nu_s \approx 56 \left(\frac{\alpha_s}{10^{-2}} \right)^{-1/4} \left(\frac{T}{10^8 \text{ K}} \right)^{1/2} \text{ Hz.} \tag{15.47}$$

The trajectory is not very sensitive to the saturation amplitude. This weak scaling is good news given that the actual level of saturation is not well known.

Next, we can work out the typical spin-down timescale. In the saturated regime, we have

$$\frac{d\nu_s}{dt} = -\frac{2Q\alpha_s^2 \nu_s}{t_{\text{gw}}} \approx -4\alpha_s^2 \left(\frac{\nu_s}{1 \text{ kHz}} \right)^7 \text{ s}^{-2}. \tag{15.48}$$

If we assume that the initial spin was much faster than the final one, ν_{final}, the spin down takes place on a timescale

$$t_{\text{sd}} \approx 42 \left(\frac{\alpha_s}{10^{-2}}\right)^{-2} \left(\frac{\nu_{\text{final}}}{1\text{ kHz}}\right)^{-6}\text{ s.} \tag{15.49}$$

For example, if we take $\nu_{\text{final}} = 50$ Hz—as may be appropriate for the Crab Pulsar—and assume $\alpha_s = 10^{-2}$, then the spin-down would take about 10 years. Of course, if the saturation amplitude is smaller then the spin-down timescale can be much longer. Each order of magnitude decrease in α_s increases the timescale by a factor of 100. This is important, as it means that the spin-down may take place on a long timescale compared to any gravitational-wave observation. The r-mode instability may effectively become a continuous wave source.

We also see that (15.48) leads to a typical braking index for an r-mode evolution of 7, rather than the 3 that would be expected for pure magnetic dipole radiation or the 5 that would follow for a deformed star spinning down due to gravitational waves (see Chapter 6). However, this result will change if the saturation amplitude is spin dependent.

Finally, by comparing the gravitational-wave torque in (15.35) to the standard magnetic dipole result, i.e. taking

$$N \approx -\frac{B^2 R^6 \Omega^3}{6c^3}, \tag{15.50}$$

we find that magnetic braking dominates the gravitational-wave emission for

$$B > 3.5 \times 10^{13} \left(\frac{\alpha_s}{10^{-2}}\right) \left(\frac{\nu_s}{1\text{ kHz}}\right)^2\text{ G.} \tag{15.51}$$

That is, the r-mode instability is unlikely to have a major impact on the spin-down of young magnetars (Watts and Andersson, 2002).

We can also work out an upper limit on the allowed amplitude in the magnetar scenario for superluminous supernovae and gamma-ray bursts discussed in Chapter 14. If we use the gravitational-wave luminosity for r-modes we find that we must have (Ho, 2016)

$$\alpha < 0.01 \left(\frac{B}{10^{14}\text{ G}}\right) \left(\frac{\nu_0}{1\text{ kHz}}\right)^{-2}, \tag{15.52}$$

where ν_0 is the initial spin frequency. If the r-mode is allowed to grow larger than this, then the observed light curves would be affected, leading to a decrease in peak luminosity and also the time it takes to reach this peak.

15.4   Nonlinear saturation

As we bring the phenomenological spin-evolution model to bear on the r-mode problem we see that, not surprisingly, the key parameter is the saturation amplitude. Provided that α_s is sufficiently large, the r-modes will rapidly spin down a young neutron star. If we assume that the neutron star is born with core temperature above 10^{10} K and that it initially spins at the Kepler limit, with ν_s of order 1 kHz, the r-mode instability comes into play within a few seconds as the star cools and enters the instability window. The mode then grows to the saturation level in a few minutes. Once the mode has saturated, the star spins down. After some time, when the star has cooled (or spun down) sufficiently that the r-mode is again stable, the mode amplitude decays and the star enters a phase where magnetic braking takes over and dominates the spin-evolution. We already have some idea of this evolution from the estimates in the previous section. Let us now try to develop a more detailed picture.

The first studies of the r-mode instability assumed a saturation amplitude in the range $\alpha_s = 0.01 - 1$ (Owen *et al.*, 1998). As we have seen, this would lead to a newly born neutron star spinning down on a timescale of a few weeks to months and the associated gravitational-wave signal would potentially be detectable by Advanced LIGO for sources in the Virgo cluster (at a distance of 15–20 Mpc). The question is if the unstable mode can grow to such large amplitudes. The answer requires us to figure out how nonlinear fluid dynamics enters the problem. One way to do this would be to carry out fully nonlinear fluid simulations (Lindblom *et al.*, 2001, 2002; Lin and Suen, 2006). However, this approach is limited in two important ways. First of all, the instability timescale ($\gg$ minutes) is much longer than the typical dynamical timescale of the fluid ($\sim$ milliseconds). Secondly, multidimensional simulations are expensive so we may not afford to resolve the physics we are interested in. In order to overcome the first problem, we can artificially accelerate the gravitational-wave emission (by scaling G to a large value). Such simulations have shown that an r-mode with amplitude $\alpha \approx 1$ would create shocks on the stellar surface, and these shocks would sap energy from the mode (Lindblom *et al.*, 2001, 2002). However, if one introduces a more gentle driving of the mode, then the effect becomes weaker (Arras *et al.*, 2003) and the mode is expected to saturate well before it reaches the amplitude where shocks form.

As the early phase of r-mode evolution should be well modelled by perturbation theory, it makes sense to consider what happens at second order in the perturbation amplitudes. In effect, we can consider the problem as a system of nonlinearly coupled oscillators, where the r-mode is driven by radiation reaction but other modes may be efficiently damped by viscosity. The mode-coupling problem can then be formulated as a set of coupled ordinary differential equations for the amplitudes of the various modes in the system.

The nonlinear mode-coupling problem is subtle. The unstable r-mode may couple to a large number of other modes and the nature of these modes can be sensitive to the physics of the neutron star core (in contrast to the r-mode itself, which is rather robust in this respect). The most complete analysis (by Brink *et al.* (2005)) considers the coupling between the r-mode and a very large network of inertial modes for an incompressible star (including almost 5,000 inertial modes with nearly 150,000 direct couplings to the r-mode).

The discussion is often framed in terms of the various mode amplitudes, parameterized in such a way that

$$E_a = MR^2 \Omega |c_a|^2, \tag{15.53}$$

where the subscript a labels the different modes (not to be confused with a spacetime index). As long as the problem remains weakly nonlinear, it can be formulated as a set of coupled oscillators (Schenk *et al.*, 2002; Arras *et al.*, 2003),

$$\dot{c}_a(t) - i\omega_a c_a + \gamma_a c_a = -i\frac{\omega_a}{\epsilon_a} \sum_{bc} \kappa_{abc} c_b c_c, \tag{15.54}$$

where κ_{abc} are coupling coefficients, γ_a encodes the driving/damping of each mode, ϵ_a is the scaled mode energy, and ω_a is the frequency. The strength of the coupling, and the efficiency with which the r-mode drives a set of daughter modes, essentially depends on (i) a set of selection rules following from symmetry, (ii) proximity to resonance, and (iii) overlap integrals involving the eigenfunctions of the modes.

In the coupled system there exists, for each 'parent' mode, a threshold amplitude below which there are no oscillations in the amplitude. Above this threshold, a parametric instability excites two 'daughter' modes, as illustrated in Figure 15.3.. These grow until they begin to influence the amplitude of the parent. The interplay between the modes leads to a long-term evolution which is difficult to track. However, we can establish the amplitude at which the three-mode coupling sets in (Arras *et al.*, 2003),

$$|c_a|^2 = \frac{\gamma_b \gamma_c}{4\omega_b \omega_c \kappa^2} \left[1 + \left(\frac{\delta\omega}{\gamma_b + \gamma_c} \right)^2 \right], \tag{15.55}$$

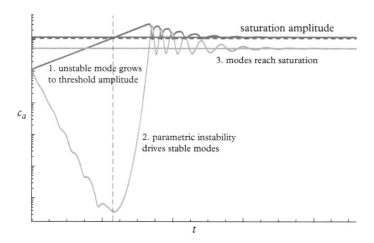

Figure 15.3 *Schematic illustration of mode saturation due to nonlinear coupling. (Figure provided by P. Pnigouras.)*

where $\delta\omega = \omega_a - \omega_b - \omega_c$ is the detuning and κ^2 encodes the coupling coefficients. It is notable that the threshold amplitude does not depend on the growth rate of the parent mode.

The fact that the daughter modes tend to have short wavelengths explains why the mechanism is difficult to observe in nonlinear simulations. Given the coarse resolution of a typical numerical grid the short wavelength oscillations may simply not be resolved.

The mode-coupling results allow us to predict the point at which nonlinear effects become relevant, but they do not by themselves provide the saturation amplitude for the r-mode instability. This is more complicated, as it is possible (perhaps even likely) that the system is driven beyond the first three-mode threshold. If this happens, other parametric instabilities come into play and the r-mode amplitude may end up exhibiting large and unpredictable variations. This is a concern as it could make the signal difficult to detect using a search based on theoretical templates.

Interestingly, the problem could end up being similar to that of the Fermi–Ulam pasta (Arras *et al.*, 2003), extensively studied since the 1950s. In that case, the numerical study of a large set of coupled nonlinear oscillators led to surprises. It was originally expected that the system would relax to equipartition, but instead it was found that only a small number of oscillators were excited. Moreover, after a sufficiently long time the system returned to its initial state with a single driven mode. The large number of remaining modes never grew above the noise. These conclusions serve as important caution for the r-mode analysis. Even though we may expect the r-mode to be driven strongly enough that it triggers many parametric instabilities, it is not yet clear that this will be the outcome. In fact, numerical results (Brink *et al.*, 2005) suggest that the r-mode amplitude does, indeed, return to the first parametric threshold after a long time of evolution, rather than approaching a state of equipartition. In effect, the analysis suggests that the r-mode saturates when (Bondarescu *et al.*, 2007, 2009)

$$E_{\text{r-mode}} \approx 1.5 \times 10^{-5} MR^2\Omega^2, \tag{15.56}$$

which translates into a (notably frequency-independent) saturation amplitude

$$\alpha_s \approx 1.6 \times 10^{-4}. \tag{15.57}$$

However, given that we do not yet fully understand the problem we need to consider what might happen if the coupling to a small number of modes fails to saturate the instability. If a large number of parametric resonances come into play, a significant set of daughter modes may be excited until the system becomes turbulent. In this case, estimates suggest that the saturation amplitude is given by (Arras *et al.*, 2003)

$$\alpha_s \approx 8 \times 10^{-3} \left(\frac{\alpha_e}{0.1}\right)^{1/2} \left(\frac{\nu_s}{1\text{kHz}}\right)^{5/2}, \tag{15.58}$$

where the parameter α_e may be as small as 4×10^{-4}. This would lead to a lower saturation amplitude than the three-mode coupling for systems with $\nu_s \leq 200$ Hz or so.

15.5   Are the gravitational waves detectable?

Having suggested that the r-mode instability may spin a newly born neutron star down to a fraction of its initial spin rate in a few months to years we (obviously) want to know whether the gravitational waves that carry away the star's angular momentum are detectable.

We can assess the detectability of the emerging waves in the standard way. First of all, we note that the gravitational-wave frequency is (for the main $l = m = 2$ r-mode)

$$f_{gw} = \frac{4\nu_s}{3} = \frac{2\Omega}{3\pi}. \tag{15.59}$$

We can combine this with the gravitational-wave flux formula, making use of the idealized source-detector configuration we used for deformed spinning stars in Chapter 8, i.e.

$$h_0^2 = \frac{10G}{c^3}\left(\frac{1}{2\pi f_{gw}d}\right)^2 \dot{E}, \tag{15.60}$$

where d is the distance to the source. Combining this with the gravitational-wave luminosity for the r-modes, we arrive at (Owen *et al.*, 1998)

$$h_0 \approx \frac{3\alpha}{4d}\left(\frac{10GMR^2\tilde{J}}{c^3 t_{gw}}\right)^{1/2}. \tag{15.61}$$

Scaling to suitable parameter values, we have

$$h_0 \approx 2.5 \times 10^{-24}\alpha\left(\frac{M}{1.4M_\odot}\right)\left(\frac{R}{10\text{ km}}\right)^3\left(\frac{\nu_s}{1\text{ kHz}}\right)^3\left(\frac{15\text{ Mpc}}{d}\right). \tag{15.62}$$

In order to assess the detectability of the signal we also need to average over all possible sky locations. As discussed in Chapter 8, this lowers our estimated strain by a factor of about $\sqrt{3/10}$.

We have scaled the gravitational-wave strain in (15.62) to the distance to the Virgo cluster. At this distance one would expect to see several neutron stars being born each year. However, the amplitude is too weak to be observed without a detailed data analysis strategy. To assess to what extent one may be able to make progress in this direction, let us first consider the standard matched-filtering approach from Chapter 8. This will give us an idea of the improvement that a tailored data analysis approach may bring, but in reality matched filtering is unlikely to be possible for signals as complex and unpredictable as that of the r-modes.

In order to compare the predicted strain to the sensitivity of different detectors, let us first consider the Fourier transform of the signal, making use of the stationary-phase result (see Chapter 8)

$$\tilde{h}_0^2(f) = \left|\frac{dt}{df}\right| h_0^2(t). \tag{15.63}$$

Given that

$$\frac{df}{dt} = \frac{3f\alpha^2\tilde{\mathcal{J}}}{\tilde{I}t_{gw}}, \tag{15.64}$$

we find that (including the averaging over sky location)

$$\tilde{h}_0 \approx \frac{3}{4d}\left(\frac{GI}{fc^3}\right)^{1/2}, \tag{15.65}$$

or

$$\tilde{h}_0 \approx 6.9 \times 10^{-25}\left(\frac{f_{gw}}{1\text{ kHz}}\right)^{-1/2}\left(\frac{15\text{ Mpc}}{d}\right). \tag{15.66}$$

As discussed in Chapter 8, we need to compare $f\tilde{h}_0$ to the dimensionless noise in the detector, $h_{rms} = \sqrt{fS_n(f)}$, as in the left panel of Figure 15.4. The result can be interpreted as saying that the detectability of an almost periodic signal improves as the square root of the number of cycles radiated in the time it takes the frequency to change by f—a natural extension of the intuition we developed in Chapter 8.

The result indicates that the r-mode signal from a young, fast-spinning, neutron star would be detectable at this distance (Owen *et al.*, 1998). However, we have to be careful with this conclusion. Basically, the analysis assumes that the entire spin-down takes place on a timescale shorter than the observation time. This would be true for large r-mode saturation amplitudes ($\alpha_s \approx 0.01 - 1$) and observation times of the order of one year, cf. (15.49). The situation would be quite different for smaller saturation amplitudes. A frequency-dependent α_s, as in the case of the turbulent cascade leading to (15.58), would also impact on the result. In fact, in this case, the star may be spinning down very slowly. For example, if we take (15.58) at face value, then the spin-down timescale changes to

$$t_{sd} \approx 0.36\left(\frac{\alpha_e}{0.1}\right)^{-1/2}\left(\frac{\nu_{final}}{1\text{ kHz}}\right)^{-11}. \tag{15.67}$$

In this case, a spin-down from the break-up velocity to 50 Hz would take over 2 million years. In effect, for most of the spin-down phase, the system would not evolve much during a one-year observation. Hence, we may assess the detectability by considering the system as (effectively) a continuous wave source at fixed frequency. In this case, it would be appropriate to compare $\sqrt{t_{obs}}h(t)$, using (15.62), to $11.4\sqrt{S_n}$ (see Chapter 8 for discussion). This comparison is shown in the right panel of Figure 15.4. The results

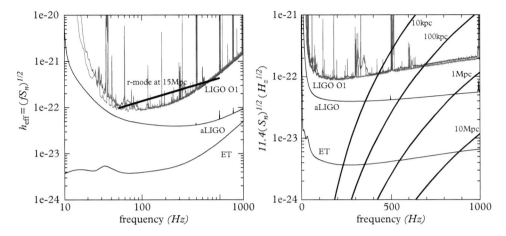

Figure 15.4 *Left: Comparing the effective r-mode amplitude $f\tilde{h}$ obtained from (15.66) to representative detector sensitivities, given by $\sqrt{fS_n(f)}$. The source is assumed to be at the distance to the Virgo cluster (at 15 Mpc). The result indicates that the r-mode signal from a young, fast-spinning, neutron star would be detectable at this distance. However, this conclusion comes with significant caveats. Right: Comparing the effective r-mode amplitude for a finite observation time $\sqrt{t_{obs}}h(t)$ for a one-year observation (obtained from 15.62) to representative detector sensitivities, given by $11.4\sqrt{S_n(f)}$. These results suggest that the r-mode signal from a young, fast-spinning, neutron star is unlikely to be detected from sources far beyond the Galaxy.*

suggest that the r-mode signal from a young star is unlikely to be detectable from large distances (Bondarescu *et al.*, 2007, 2009).

15.6 Astrophysical constraints for young neutron stars

Early discussions of the r-mode instability (Lindblom *et al.*, 1999; Owen *et al.*, 1998; Andersson *et al.*, 1999a) noted that a young neutron star would enter the instability window shortly after birth and exit at a frequency close to the suggested 'birth' spin rate of the Crab Pulsar, $\nu_{\text{final}} \approx 50$ Hz. This idea would be consistent as long as the r-mode saturates at a relatively large amplitude, $\alpha_s \geq 0.01$. If this were the case, gravitational-wave emission would significantly spin down the star in the first year. However, for low—more realistic—saturation amplitudes, the spin down would be much less dramatic, in conflict with the notion that the Crab Pulsar has spun down from an initial spin near the break-up velocity. According to (15.67) the gravitational-wave spin-down would take far too long, seeing as the Crab Pulsar is less than 1,000 years old. Moreover, we know from the observed braking index (see Chapter 6) that the current spin evolution of the Crab pulsar is not dominated by gravitational-wave emission. In the case of an r-mode driven spin down, it follows from (15.48) that we should have a braking index of $n = 7$. As there is no evidence for this we must conclude that the Crab Pulsar does not currently reside inside the instability window. This may not be a very restrictive statement, but it is the first hint that we may be able to use observational data to constrain the uncertain r-mode theory.

We may also consider the young 16-ms X-ray pulsar J0537-6910 in the supernova remnant N157B (Marshall *et al.*, 1998). It seems reasonable to assume that this system had an initial spin period of a few milliseconds. As we have already discussed (see Chapter 12), this is a very interesting system because the neutron star exhibits regular glitches, roughly every 100 days. The overall spin-down evolution is, in fact, dominated by the glitches and their recovery (see Figure 12.4). If one tries to extract a braking index for this system, then the answer appears to be negative (Antonopoulou *et al.*, 2018). However, one can focus on the inter-glitch periods. This leads to much larger braking indices—typically too large to be explained by any proposed spin-down mechanism. The data following the first observed (and largest) glitch seem to lead to an asymptotic value of $n \approx 7$ (Andersson *et al.*, 2018). This may not be a very robust observation, but it is nevertheless natural to speculate that the r-modes may play a role in this system. A closer look at this possibility suggests that an r-mode scenario for this pulsar may be (borderline) consistent, provided one negotiates on the saturation amplitude. In order to explain the observed spin-down rate one would need $\alpha_s \approx 0.1$, significantly larger than the predicted value.

In absence of a detailed theoretical understanding, it makes sense to search for gravitational waves from young neutron stars. A particularly interesting class of systems to consider are the central compact objects associated with supernova remnants. A celebrated example of this class of sources is the remnant in Cassiopeia A. Likely associated with a supernova observed by the British astronomer John Flamsteed in 1680, this is the youngest known neutron star in the Galaxy. It is a very interesting case because X-ray data hints at accelerated cooling, which may be associated with the onset of superfluidity in the star's core (Page *et al.*, 2011; Shternin *et al.*, 2011). However, even though we know the remnant is a neutron star, we do not know its current spin rate. No evidence for pulsations have been found in the data. This may be due to the magnetic field having been buried by fallback accretion following the supernova (Geppert *et al.*, 1999; Ho, 2015), in which case one would expect the magnetic field to gradually emerge. The absence of a known spin period makes a gravitational-wave search challenging (see Chapter 8).

Earlier (in Chapter 14) we suggested an energy argument that could be used to put an upper limit on the gravitational-wave emission in absence of a known spin-down rate. It is straightforward to adapt the logic to the r-mode case (Owen, 2010). That is, we can use the estimate (14.65) as a guide for directed gravitational-wave searches. Directed searches represent a middle ground between all-sky searches and targeted searches for known pulsars. They are more sensitive than blind searches, as one can make use of directional information and age constraints, but less precise than a targeted search where one has reliable timing data from electromagnetic observations. At the back-of-the-envelope level a blind search tends to be about one order of magnitude less sensitive than a targeted search. Using directional information one may regain about a factor of 2 of the lost sensitivity. This may not seem like much, but every bit of improvement helps.

In the case of young neutron stars associated with supernova remnants, we can make use of the sky location (and thus the detector-frame Doppler modulation) but we may not have the spin frequency and other parameters. This is the case for the Cas A remnant. Searches for this object must cover a range of gravitational-wave frequencies. The first constraints on gravitational waves from this system were obtained using 12 days of data from the fifth LIGO science run (S5). The search covered frequencies from 100 to 300 Hz and a wide range of first and second frequency derivatives. The search was specifically designed to beat the indirect upper limit set by (14.65). For Cas A, which is at a distance of 3.4 kpc, we can turn (14.65) into an upper limit on the r-mode amplitude (Abadie *et al.*, 2010),

$$\alpha \lesssim 3.9 \times 10^{-4} \left(\frac{I_0}{10^{45} \text{ g cm}^2} \right)^{-1/2} \left(\frac{t}{300 \text{ yr}} \right)^{-1/2} \left(\frac{f}{100 \text{ Hz}} \right)^{-2}, \qquad (15.68)$$

where f is the gravitational-wave frequency. The S5 search was limited to 12 days for computational reasons. The cost of a coherent search scales with the 7th power of the observation time (Wette *et al.*, 2008), while the sensitivity improves only as the square root (see Chapter 8). In essence, increasing the search time beyond 12 days would dramatically increase the computational cost for a negligible gain in sensitivity.

The S5 search provided the first upper limit on the amplitude of an unstable r-mode, constraining the amplitude to $\alpha \leq 0.005 - 0.14$ in the considered frequency range. These upper limits improve on the indirect limit from (14.65) across the frequency range. A similar search was carried out using data from the sixth LIGO science run (Aasi *et al.*, 2015*b*). However, the results did not lead to a significant improvement of the upper limit for an r-mode in the Cas A neutron star (basically because of the restricted search time), but similar results were obtained for 8 other supernova remnants (and for a wider frequency range). The tightest upper limit on the r-mode amplitude was $\alpha \approx 4 \times 10^{-5}$. This was further improved using data from the first observing run of Advanced LIGO (Abbott *et al.*, 2019*b*); see Figures 15.5 and 15.6. These results demonstrate that observations are beginning to constrain the theory.

In addition to the indirect limit (14.65), electromagnetic observations may constrain the allowed r-mode signal. The argument is quite simple—and serves as a useful reminder that we need to base our search strategies on as much information as possible. From the X-ray luminosity of the Cas A remnant we can infer the maximum allowed r-mode amplitude by matching the observations against (15.46). This leads to

$$\alpha \leq 5 \times 10^{-8} \left(\frac{\nu_s}{1 \text{ kHz}} \right)^{-4} \left(\frac{L_\gamma}{10^{35} \text{ ergs}^{-1}} \right)^{1/2}. \qquad (15.69)$$

Systems like the remnant in Cassiopeia A typically have X-ray luminosities in the range $10^{32} - 10^{34}$ erg s^{-1}. This leads to the constraints shown in Figure 15.7. Basically, we see that the detection of gravitational waves from these systems could be a real challenge.

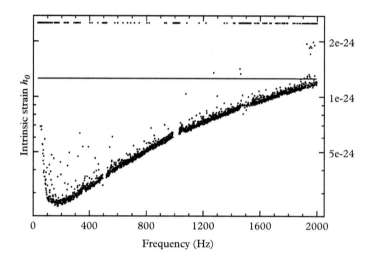

Figure 15.5 *Direct observational (95% confidence) upper limits on the intrinsic strain for the Cas A neutron star, based on data from the first Advanced LIGO observing run. The horizontal line indicates the indirect limit from energy conservation; see (14.65). Scattered points on a higher line indicate 1-Hz bands where no upper limit was set due to data quality issues. (Reproduced from Abbott et al. (2019b).)*

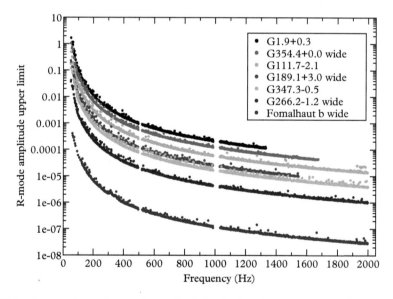

Figure 15.6 *Upper limits on the r-mode amplitude for the Cas A neutron star (in the supernova remnant G111.7-2.1) and five other supernova remnants as well as Fomalhaut b, an extrasolar planet candidate which might possibly be a nearby old neutron star. The results are based on data from the first observing run of Advanced LIGO. (Reproduced from Abbott et al. (2019b).)*

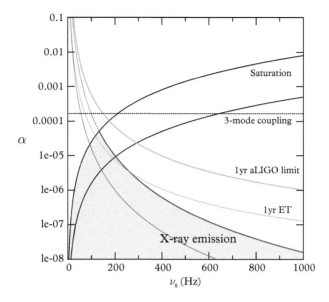

Figure 15.7 *Constraints on the r-mode amplitude from X-ray observations of compact central objects, like the remnant in Cas A. The r-modes are limited by the predicted (frequency-dependent) saturation amplitude, here taken to be given by (15.58)—although we also indicate the level from (15.57)—and the limit on the X-ray luminosity, which constrains the mode through (15.46). Also shown are the sensitivities obtainable with a one-year integration over data at the Advanced LIGO design sensitivity and with a third-generation instrument like the Einstein Telescope.*

15.7   r-modes in accreting systems

The r-mode instability may also, quite naturally, play a role for accreting neutron stars (Bildsten, 1998; Andersson *et al.*, 1999b). The problem can be analysed following the steps from Chapter 14. As the basic idea is the same—the accretion torque is balanced by gravitational-wave emission—we will focus on aspects that are unique to the r-mode problem.

Let us, first of all, note that the existence of the r-mode instability calls into question a possible formation route for millisecond pulsars. A millisecond pulsar may, in principle, form by accretion-induced collapse of a white dwarf, but this scenario is inconsistent with the r-mode picture. The collapse would form a star hot enough to spin down because of the instability (Andersson *et al.*, 1999b). Of course, the idea may still work, provided that the r-mode saturation amplitude is small.

Turning to accreting neutron stars in low-mass X-ray binaries, it is easy to argue that (given what we think we know) the r-mode instability should be relevant. Balancing the r-mode gravitational-wave emission with the simple accretion torque from (14.84), we see that we need

$$\alpha \approx 10^{-6} \left(\frac{\nu_s}{1\ \mathrm{kHz}} \right)^{-7/2} \left(\frac{\dot{M}}{10^{-8} M_\odot/\mathrm{yr}} \right)^{1/2}, \qquad (15.70)$$

in order for the unstable r-mode to halt accretion-driven spin-up. This is much smaller than the expected saturation amplitude, e.g. (15.58), for all reasonable accretion rates. On the one hand, this is good news because there is no fundamental reason why the r-modes cannot grow large enough to balance the accretion torque. On the other hand, it is problematic. There is no reason why the r-modes should not grow to a larger amplitude, in which case they would rapidly spin these stars down. In essence, it is difficult to reconcile the notion of torque balance with our understanding of the r-mode instability, unless, of course, the observed spin limit turns out to coincide with the edge of the instability window.

The problem has additional twists. In addition to generating gravitational waves that dissipate angular momentum from the system, the unstable r-modes heat the star up. Since the impact of viscosity gets weaker as the temperature increases, the mode heating triggers a thermal runaway (Levin, 1999; Andersson *et al.*, 2000). As the star heats up, it becomes increasingly unstable. This process continues until the mode saturates and the

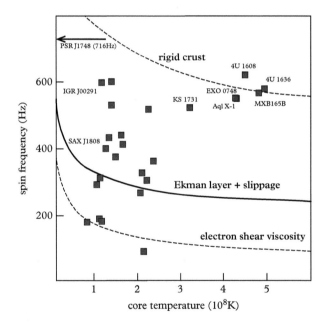

Figure 15.8 *The r-mode instability window (spin frequency vs core temperature) for accreting neutron stars. We show results for the simplest (fluid) model (assuming that shear viscosity due to electron–electron scattering dominates, thin dashed curve), as well as for a star with a crust (assuming a standard Ekman layer with a slippage correction, thick solid line) and a (not realistic) rigid crust (upper dashed curve). The squares indicate specific low-mass X-ray binaries, for which the temperature is derived from the observed X-ray luminosity. (Adapted from Ho et al. (2011a).)*

star spins down. This could mean that, even if the r-modes are relevant for these systems, we may not catch them in the act. If the saturation amplitude is large, these systems may only radiate for a tiny fraction of the their lifetime. With a smaller amplitude the duty cycle looks more promising (Heyl, 2002), but the gravitational-wave amplitude is (obviously) weaker. It is a delicate balance. We also need to consider the fact that many of the observed low-mass binaries are transients (Patruno and Watts, 2012). This means that we need to consider episodes of accretion and quiescence. This is relevant because one can use X-ray observations to place upper limits on the temperature of the neutron stars in these systems. Typical results are shown in Figure 15.8. The estimates show that a significant number of systems should be unstable unless some additional dissipation mechanism impacts on the instability window. This problem remains to be solved. The r-mode instability may play a role in the evolution of a low-mass X-ray binary (see Figure 15.9) but we need to make progress on a range of issues—from the accretion torque to the dissipation mechanisms that determine the instability window and the saturation amplitude.

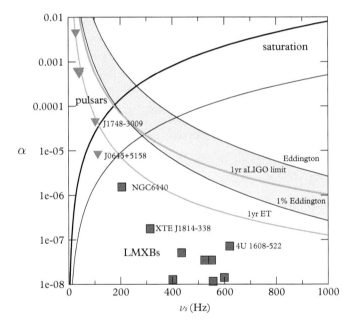

Figure 15.9 *Constraints on the r-mode amplitude from X-ray observations of low-mass X-ray binaries. As in Figure 15.7, the r-modes are assumed to be limited by the theoretical saturation amplitude, again taken to be given by (15.58), while the limit on the X-ray luminosity constrains the mode through (15.46). Also shown are the sensitivities obtainable with a one-year integration over data at the Advanced LIGO design sensitivity and with a third generation instrument like the Einstein Telescope. If these constraints represent reality, then gravitational waves from most of these systems will not be detectable.*

Do the different arguments (somehow) suggest that the r-modes do not play a role in accreting systems? Perhaps. However, there is intriguing evidence that a non-radial oscillation mode may be present in the X-ray data associated with the 2002 discovery outburst of XTE J1751-305 (Strohmayer and Mahmoodifar, 2014a). The observed frequency ($0.5727597 \times \nu_s$) is quite close to the frequency of the star's quadrupole r-mode ($2\nu_s/3$ in Newtonian theory). At first sight, this interpretation may seem unlikely, because the observed frequency is close to the mode frequency in a frame co-rotating with the star, whereas one might expect a distant observer to see the inertial frame frequency (which is $4\nu_s/3$ to leading order). However, it has been argued that a non-radial oscillation mode can indeed lead to modulations of an X-ray hotspot being observed at the rotating frame frequency (Lee, 2014). Adding in relativistic effects (e.g. the gravitational redshift and the rotational frame dragging) and the rotational deformation of the star (Andersson *et al.*, 2014), one finds that the observed frequency is consistent with an r-mode, without adjustments to our understanding of the physics. Moreover, one would infer sensible values for the star's mass and radius.

However, the association becomes problematic when we consider the spin-evolution of the system (Andersson *et al.*, 2014). Two features of the observations are relevant: (i) The mode amplitude α has a typical value of $\sim 10^{-3}$ over the duration of the outburst, and (ii) the star spins up during this phase. An r-mode with the suggested amplitude ought to lead to strong spin down, which would overwhelm any reasonable accretion torque. The (seemingly) unavoidable conclusion is that, within the phenomenological model for r-modes (be they stable or unstable), it is not possible to reconcile the presence of a mode excited to the suggested amplitude with the observed spin evolution. Adding to the 'confusion', a similar quasi-periodic feature in the 582-Hz system 4U1636-536 (Strohmayer and Mahmoodifar, 2014b), would seem to be consistent with the inertial frame r-mode frequency. This calls the r-mode interpretation into question. Of course, it also means that we are left with an interesting problem to solve.

16

Black-hole dynamics

Black holes are mysterious objects hidden from view. Yet, they are remarkably simple. An astrophysical black hole is fully described by two parameters, the mass M and the angular momentum $\mathcal{J}$ (often considered in terms of the dimensionless quantity $a/M = c\mathcal{J}/GM^2 < 1$). An important step towards understanding these objects, first addressed in the 1950s—long before there was any observational evidence to worry about—involved establishing that they were stable (Regge and Wheeler, 1957). In order to be astrophysically relevant, black holes would have to be stable to external perturbations. If they simply 'exploded if an ant sneezed in the vicinity' (Vishveshwara, 1970b), gravitational collapse would have to lead to some different end state.

Many astrophysical scenarios involve dynamical deviations away from a given background equilibrium (or, at least, slowly evolving) state. We have already seen how perturbation theory plays a key role in gravitational-wave physics—in fact, the very 'definition' of the waves is based on the notion of a linear deviation away from a background geometry (see Chapter 3). We are now going to develop our understanding of perturbation theory in general relativity a little bit further. We do this by considering the problem for curved backgrounds, like the Schwarzschild solution. This leads to the introduction of several new concepts, like the quasinormal modes of a compact object (Kokkotas and Schmidt, 1999; Nollert, 1999; Berti *et al.*, 2009). These differ from the normal modes of, say, an oscillating Newtonian star (from Chapter 13) due to the fact that we now have to account for the emission of gravitational waves. In essence, the modes are damped at a rate that reflects how efficiently they radiate.

16.1 Issues of stability

Let us begin by outlining a simple stability argument. To address the issue one typically considers linear perturbations away from a known background metric, such that

$$g_{ab} = g_{ab}^{\mathrm{B}} + h_{ab}, \qquad \text{where } |h_{ab}| \ll 1. \tag{16.1}$$

This is the same procedure that we used to define gravitational waves in Chapter 3. The only difference is that we now take the background solution to be given by the

Gravitational-Wave Astronomy: Exploring the Dark Side of the Universe. Nils Andersson, Oxford University Press (2020).
© Nils Andersson. DOI: 10.1093/oso/9780198568032.001.0001

Schwarzschild metric (for a non-rotating black hole) from Chapter 4. As we will see, this leads to a wave equation with an effective potential V (associated with the spacetime curvature)

$$\frac{\partial^2 u}{\partial r_*^2} - \frac{\partial^2 u}{\partial t^2} - V(r)u = 0, \tag{16.2}$$

where u encodes the perturbation h_{ab} and the tortoise coordinate (from Chapter 10);

$$\frac{d}{dr_*} = \left(1 - \frac{2M}{r}\right)\frac{d}{dr}, \tag{16.3}$$

translates the part of spacetime accessible to a causal observer into the range $-\infty \le r_* \le \infty$. Basically, the event horizon has been pushed all the way to $r_* = -\infty$.

Given this differential equation we can ask whether a solution remains stable or if it becomes unbounded as it evolves. All indications point towards black holes being stable. The first hints of this involved the construction of suitable energy integrals (Regge and Wheeler, 1957; Vishveshwara, 1970b). If we multiply (16.2) with its complex conjugate and integrate, we arrive at

$$\int_{r_*=-\infty}^{r_*=+\infty} \left[\left|\frac{\partial u}{\partial t}\right|^2 + \left|\frac{\partial u}{\partial r_*}\right|^2 + V|u|^2\right] dr_* = \text{constant}. \tag{16.4}$$

As long as V is positive definite, the integral bounds $\partial u/\partial t$, which excludes exponentially growing solutions. This suggests that there are no unstable 'modes' of a non-rotating black hole. However, the stability problem is a bit more intricate. The energy argument leaves loopholes through which an instability might sneak in. For example, perturbations that grow linearly (or slower) with time are not ruled out. Also, we have only provided a bound for integrals of u. The perturbation may still blow up in an ever-narrowing spatial region. We can obtain a stronger stability argument, but it would require a more sophisticated analysis than we may be comfortable with at this point (Kay and Wald, 1987; Beyer, 2001).

16.2   Scalar field dynamics

The problem we are interested in—that of perturbed black holes—obviously has no analogue in Newtonian gravity. This is in contrast to the dynamics of relativistic stars that remain qualitatively close to the Newtonian problem (see Chapter 13). The frequencies of fluid oscillations may be altered, and should become complex-valued to incorporate damping due to gravitational-wave emission, but one would not expect dramatic changes. The black-hole problem is different, as we are exploring the dynamics of the gravitational field itself.

Luckily, the problem has features familiar from classic scattering situations and it turns out that the key concepts can be understood from a study of massless scalar waves. This may seem somewhat odd given that our main interest is in gravitational waves which have a tensorial character, but it turns out that the linear equations that govern different physical fields can be reduced to wave equations of the form (16.2).

If we want to study the dynamics of a massless scalar field Φ in the geometry of a non-rotating black hole we need to solve the wave equation

$$\Box \Phi = 0, \tag{16.5}$$

where $\Box = g^{ab}\nabla_a\nabla_b$ is the wave operator associated with the Schwarzschild metric (in geometric units $c = G = 1$, as usual)

$$ds^2 = -\frac{\Delta}{r^2}dt^2 + \frac{r^2}{\Delta}dr^2 + r^2(d\theta^2 + \sin^2\theta\,d\varphi^2), \tag{16.6}$$

and we have used $\Delta = r(r - 2M)$, M being the mass of the black hole (see Chapter 4). Since the metric is spherically symmetric it is natural to decompose the field in spherical harmonics, Y_{lm}. It also turns out that the problem is degenerate in the azimuthal harmonics (again because of the symmetry), so it is sufficient to consider the $m = 0$ case. Thus, we have

$$\Phi = \sum_{l=0}^{\infty}\Phi_l = \sum_{l=0}^{\infty}\frac{u_l(r,t)}{r}P_l(\cos\theta), \tag{16.7}$$

where P_l are the Legendre polynomials. With this decomposition the function $u_l(r,t)$ is governed by the one-dimensional wave equation

$$\frac{\partial^2 u_l}{\partial r_*^2} - \frac{\partial^2 u_l}{\partial t^2} - V_l(r)u_l = 0, \tag{16.8}$$

with the effective potential V_l given by

$$V_l(r) = \frac{\Delta}{r^2}\left[\frac{l(l+1)}{r^2} + \frac{2M}{r^3}\right]. \tag{16.9}$$

It should be noted that we have used the 'tortoise' coordinate r_* to simplify the wave operator, but the potential is still a function of the radial coordinate r.

In order to understand the solutions to (16.8), let us assume a harmonic time dependence $u_l(r,t) = \hat{u}_l(r,\omega)e^{-i\omega t}$ (strictly speaking, we take the Fourier transform). We then arrive at the ordinary differential equation

$$\frac{d^2 \hat{u}_l}{dr_*^2} + \left[\omega^2 - V_l(r)\right]\hat{u}_l = 0. \tag{16.10}$$

This problem looks simple. In fact, the effective potential $V_l(r)$ is similar to the effective potential for photon geodesics from Chapter 10. It represents a potential barrier with maximum roughly located at the position of the unstable circular photon orbit, $r = 3M$. Moreover, it is easy to understand the asymptotics of the solution. Since the potential V_l vanishes at both spatial infinity and the event horizon, the two linearly independent solutions to (16.10) behave as

$$\hat{u}_l \sim e^{\pm i\omega r_*} \qquad \text{as } r_* \to \pm\infty. \tag{16.11}$$

Before moving on it is worth noting that all astrophysical black holes were (obviously) formed by gravitational collapse at some point in the past. This means that there must be a limit to how far back we can assume that the spacetime is appropriately described by (say) the Schwarzschild metric. The description is only appropriate in a region outside matter that undergoes perfectly spherical collapse. This is, of course, an idealized case with little (or no) physical relevance. An astrophysical collapse is expected to be non-spherical, and thus the Schwarschild metric cannot be an adequate description of the spacetime until the dust from the collapse event has settled. However, when one studies the physics in this 'late time' region it is often convenient to ignore the distant past, including the messy details of the collapse. We will adopt this attitude and consider 'eternal' black holes throughout this Chapter.

We can construct a set of solutions to the scalar field equation (16.10) by imposing suitable boundary conditions at the horizon and spatial infinity. In order to illustrate the solutions we are interested in, it is useful to use conformal (Carter–Penrose) diagrams. These are two-dimensional diagrams that capture the causal relationship between different spacetime points. The vertical direction represents time, while the horizontal direction represents space, and lines at an angle of 45° correspond to light rays. Locally, the metric is conformal to the actual metric, with the conformal factor (see Chapter 4) chosen in such a way that the entire infinite spacetime is transformed into a finite-size diagram. For spherically symmetric spacetimes, like the Schwarzschild geometry we are considering here, every point in the diagram corresponds to a two-sphere.

The advantage of using conformal diagrams is that it is easy to illustrate the asymptotics of a given problem. In our case, one would start by imposing conditions at past infinity, $\mathcal{J}^-$, and the past horizon, $\mathcal{H}^-$; see Figure 16.1. Once such conditions are introduced we can deduce the corresponding behaviour in the late time region, at $\mathcal{J}^+$ and $\mathcal{H}^+$, respectively.

Given the asymptotic behaviour (16.11) it is easy to define solutions to (16.10) that satisfy the relevant boundary conditions. One such solution satisfies the natural condition that no waves should emerge from the horizon of the black hole. We refer to this as the IN-mode (Chrzanowski, 1975), and define it by

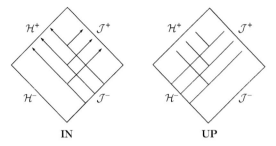

Figure 16.1 *An illustration of the IN- (left) and UP-modes (right). By combining the radial solutions to (16.10) with the assumed time-dependence* $\exp(-i\omega t)$ *we see that the UP solution satisfies the boundary condition of purely ingoing waves at* $\mathcal{H}^+$*. Accompanying this radiation there are scattered waves reaching* $\mathcal{J}^+$*. To achieve this solution, exactly the right amount of radiation should emerge from* $\mathcal{H}^-$*. The amplitudes of the various waves are such that the UP-mode is an acceptable solution to (16.10). The situation is analogous for the IN-mode.*

$$\hat{u}_l^{\text{in}}(r_*,\omega) \sim \begin{cases} e^{-i\omega r_*} & r_* \to -\infty, \\ A_{\text{out}}(\omega)e^{i\omega r_*} + A_{\text{in}}(\omega)e^{-i\omega r_*} & r_* \to +\infty. \end{cases} \tag{16.12}$$

The complex conjugate of this solution, the OUT-mode, corresponds to purely outgoing waves at $\mathcal{H}^-$. A second pair of basic solutions (the UP- and DOWN-modes) can be defined in a similar way. The UP-mode corresponds to purely outgoing waves at spatial infinity, and is specified by

$$\hat{u}_l^{\text{up}}(r_*,\omega) \sim \begin{cases} B_{\text{out}}(\omega)e^{i\omega r_*} + B_{\text{in}}(\omega)e^{-i\omega r_*} & r_* \to -\infty, \\ e^{+i\omega r_*} & r_* \to +\infty, \end{cases} \tag{16.13}$$

while the DOWN-mode is the complex conjugate of this solution. Figure 16.1 illustrates the IN- and UP-modes.

Since two linearly independent solutions to (16.10) can be used to represent any other solution, the coefficients in (16.13) cannot be independent of those in (16.12). Using the fact that the Wronskian of two linearly independent solutions to (16.10) must be a constant, it is easy to relate the coefficients. For example, we have

$$W \equiv \hat{u}_l^{\text{in}} \frac{d\hat{u}_l^{\text{up}}}{dr_*} - \hat{u}_l^{\text{up}} \frac{d\hat{u}_l^{\text{in}}}{dr_*} = 2i\omega A_{\text{in}}(\omega) = 2i\omega B_{\text{out}}(\omega). \tag{16.14}$$

Thus, we find that (assuming for the moment that ω is real)

$$B_{\text{out}}(\omega) = A_{\text{in}}(\omega) \tag{16.15}$$
$$B_{\text{in}}(\omega) = -\bar{A}_{\text{out}}(\omega) = -A_{\text{out}}(-\omega), \tag{16.16}$$

where a bar denotes complex conjugation. We see that the solution to (16.10) can always be expressed in terms of a combination of any two of the IN-, UP-, OUT-, and DOWN-modes. Most studies use the combination IN–UP, since they incorporate the causal boundary conditions. The IN-mode describes how a wave coming in from infinity is scattered by the effective potential (16.9), while the UP-mode corresponds to waves emerging from the horizon (or its vicinity). The outcome is either a combination of waves falling across the future horizon or reaching future null infinity.

When a black hole is perturbed by an external agent, or when it settles down after formation following supernova core collapse or binary merger, it oscillates. These oscillations are associated with the 'quasinormal modes' of the black hole, first seen in a simple scattering problem involving scalar waves (Vishveshwara (1970a); see also Press (1971)), as illustrated in Figure 16.2. For a given perturbing field, the frequencies of the quasinormal modes depend only on the parameters of the black hole (Kokkotas and Schmidt, 1999; Nollert, 1999; Berti *et al.*, 2009). This is important as it may allow us

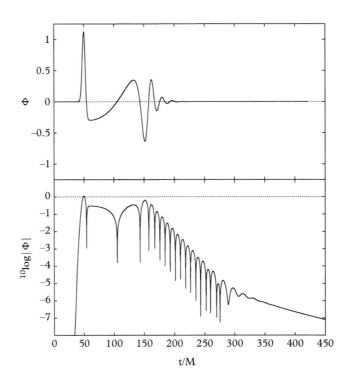

Figure 16.2 *A reproduction of the scalar-wave scattering experiment that provided the first illustration of black-hole quasinormal modes (Vishveshwara, 1970a). The illustration shows how the presence of a Schwarzschild black hole affects a Gaussian wavepacket. At roughly t = 50M the initial Gaussian passes the observer on its way towards the black hole. Quasinormal-mode ringing dominates the signal after t ≈ 150M. At very late times (after t ≈ 300M) the signal is dominated by a power-law fall-off with time. (Reproduced from Andersson (1995), copyright (1995) by the American Physical Society.)*

to extract the mass M and rotation rate from observational data, thus identifying the black hole. As we will discuss later, nonlinear simulations confirm that quasinormal-mode ringing dominates the radiation from most processes that involve a black hole. No matter how you kick a black hole, the response is dominated by quasinormal modes. This is good news since the mode problem can be understood at the level of linear perturbations.

For astrophysical (presumably stable) black holes, one would expect the solutions to (16.8) to be damped in time. Hence, all acceptable mode solutions should have complex frequencies with $\text{Im}\,\omega < 0$. This simple fact leads to one of the practical difficulties associated with the determination of quasinormal modes.

Intuitively, a black-hole oscillation mode should not depend on the character of the waves falling in from infinity, or emerging from the horizon. This means that one would expect a quasinormal mode solution to (16.10) to be given by (say) the IN-mode with $A_{\text{in}} = 0$. Combining this with the required damping in time at each spatial location, r_*, we immediately see that the desired solution to the radial equation grows exponentially as we approach $r_* = \pm\infty$. Although this conclusion may, at first, seem peculiar (and indeed, undesirable), it is easy to see why this happens. Properly defined, each mode (with frequency ω_n) corresponds to purely outgoing waves reaching $\mathcal{H}^+$ and $\mathcal{J}^+$. For example, at $\mathcal{J}^+$ we expect to have $u_l(r_*,t) \sim \exp[-i\omega_n(t-r_*)]$. From this it is clear that, if we assume a harmonic time dependence we must require solutions that behave as $\hat{u}_l \sim \exp(i\omega_n r_*)$. For $\text{Im}\,\omega_n < 0$ this solution diverges as $r_* \to +\infty$. In other words, the problem is one of our own making. It is simply due to the way we decoupled the time dependence of the solution. Because a mode solution is expected to be damped with time at any fixed value of r_* it must diverge as $r_* \to +\infty$ at any fixed value of t (i.e. on a spacelike hypersurface). In the physical solution, the apparent divergence is balanced by the fact that it takes an infinite time for the signal to reach $\mathcal{J}^+$.

Building on this intuition, we turn to the associated initial-value problem. Suppose we are given a specific scalar field at some time, say $t = 0$, and we want to deduce the evolution of this field. We then require a scheme for calculating (for each l multipole) $u_l(r_*,t)$ once we are given $u_l(r_*,0)$ and $\partial_t u_l(r_*,0)$. This problem can be solved in terms of a Green's function $G(r_*,y,t)$. The time-evolution of $u_l(r_*,t)$ follows from (Ching et al., 1995; Andersson, 1997)

$$u_l(r_*,t) = \int G\partial_t u_l(y,0)dy + \int \partial_t G u_l(y,0)dy, \tag{16.17}$$

for $t > 0$, where the (retarded) Green's function is defined by

$$\left[\frac{\partial^2}{\partial r_*^2} - \frac{\partial^2}{\partial t^2} - V_l(r)\right]G = \delta(t)\delta(r_* - y), \tag{16.18}$$

together with the (causality) condition $G = 0$ for $t \leq 0$. Finding the Green's function is now the main task. Once we know it, we can study the evolution of any initial field by evaluating the integrals in (16.17).

The first step involves reducing (16.18) to an ordinary differential equation. To do this we use the Laplace transform[1]

$$\hat{G}(r_*, y, \omega) = \int_{0^-}^{+\infty} G(r_*, y, t)e^{i\omega t} dt.$$ (16.19)

The transform is well defined as long as $\text{Im } \omega \geq 0$, and can be inverted using

$$G(r_*, y, t) = \frac{1}{2\pi} \int_{-\infty+ic}^{+\infty+ic} \hat{G}(r_*, y, \omega)e^{-i\omega t} d\omega,$$ (16.20)

where c is some positive number.

The frequency-domain Green's function can now be expressed in terms of two linearly independent solutions to the homogeneous equation (16.10). In terms of the IN- and UP-modes from (16.12) and (16.13) we have

$$\hat{G}(r_*, y, \omega) = -\frac{1}{2i\omega A_{\text{in}}(\omega)} \begin{cases} \hat{u}_l^{\text{in}}(r_*, \omega)\hat{u}_l^{\text{up}}(y, \omega), & r_* < y, \\ \hat{u}_l^{\text{in}}(y, \omega)\hat{u}_l^{\text{up}}(r_*, \omega), & r_* > y, \end{cases}$$ (16.21)

where we have used the Wronskian from (16.14).

The problem can, in principle, be solved by integration of (16.10) for (almost) real values of ω and subsequent inversion of (16.20). This should lead to a good representation of the evolution, as long as some care is taken at each step. However, as we are mainly interested in explaining the features of the emerging waves, we will try to isolate the behaviour of the Green's function in different time intervals. A useful trick that helps us achieve this aim is to use analytic continuation and 'bend' the integration contour in (16.20) into the lower half of the complex ω-plane; see Figure 16.3.

What do we learn by extending the Green's function into the complex frequency plane? First of all, we find that $\hat{G}(r_*, y, \omega)$ has an infinite number of distinct singularities in the lower half of the ω-plane. These correspond to the quasinormal modes and are associated with the zeros of the Wronskian $W(\omega)$ (Leaver, 1985; Andersson, 1992; Nollert and Schmidt, 1992). For a quasinormal mode, the two solutions $\hat{u}_l^{\text{in}}$ and $\hat{u}_l^{\text{up}}$ are linearly dependent. The mode frequencies do not, however, contain all the information required to evaluate the Green's function. While it is formally straightforward to use the residue theorem to determine the mode contribution it is, in practice, not quite so easy to evaluate the various excitation coefficients (Leaver, 1986; Andersson, 1995).

We must also account for a branch-cut in the Green's function. This branch-cut emerges from the origin and is usually taken to be along the negative imaginary ω-axis. The branch-cut leads to a power-law tail that dominates the field at late times (Price, 1972); see Figure 16.2. Physically, the late-time tail arises because of backscattering off

---

[1] In order to make contact with the previous discussion, we have chosen to write the transform as a one-sided Fourier transform. The usual textbook Laplace transform follows if we replace $i\omega \rightarrow -s$.

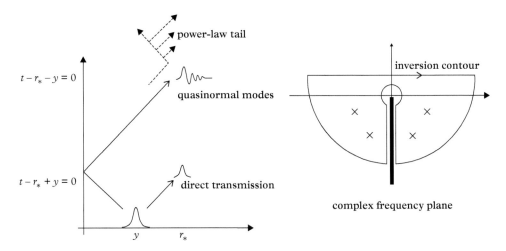

Figure 16.3 *Left: A schematic description of a black hole's response to initial data with compact support. The directly transmitted wave (from a source point y) arrives at a distant observer (at r_*) roughly at $t - r_* + y = 0$. The black hole's response, which is dominated by quasinormal-mode ringing, reaches the observer at roughly $t - r_* - y = 0$. At very late times the signal falls off as an inverse power of time. This power-law tail arises because of backscattering off the spacetime curvature. Right: Integration contours in the complex frequency plane. The original inversion contour for the Green's function lies above the real frequency axis. When analytically continued in the complex plane this contour can be replaced by the sum of (i) the quasinormal modes (the singularities of $\hat{G}(r_*,y,\omega)$; the first few are represented by crosses), (ii) an integral along the branch cut (a thick line along the negative imaginary ω axis in the figure), that leads to the power law tail, and (iii) high-frequency arcs (which lead to roughly 'flat space propagators' at early times). (Reproduced from Andersson (1997), copyright (1997) by the American Physical Society.)*

of the slightly curved spacetime in the region far away from the black hole.[2]. This means that the tail will not depend on the exact nature of the central object (Gundlach *et al.*, 1994) A neutron star of a certain mass will give rise to the same late-time tail as a black hole of the same mass.

The behaviour at late times can be obtained from a low-frequency approximation of the integral along the branch cut (Leaver, 1986; Andersson, 1995; Barack and Ori, 1999). The leading order contribution is

$$G^C(r_*,y,t) = (-1)^{l+1} \frac{(2l+2)!}{[(2l+1)!!]^2} \frac{4M(r_*y)^{l+1}}{t^{2l+3}}. \qquad (16.22)$$

From the example in Figure 16.2 it may seem unlikely that the power-law tail will have much interest for gravitational-wave astronomy. This is probably true if we expect a direct detection, but it turns out to be false when we consider the bigger picture. The

---

[2] This effect also explains the presence of logarithmic terms in the higher order post-Newtonian expansion; see Chapter 11.

low-frequency contribution to the Green's function plays a decisive role in determining the gravitational self-force, a problem we will consider shortly.

16.3   Gravitational perturbations

The scalar-field problem illustrates many of the concepts we need in order to understand the gravitational-wave signal from a slightly perturbed black hole. In fact, the equations that govern gravitational perturbations can be cast in the form (16.10). At first, this may seem surprising—gravitational perturbations ought to be fundamentally different from massless scalar waves. After all, the latter correspond to fields evolving in a fixed background geometry, while gravitational perturbations correspond to changes in the metric itself. Of course, we have already seen (see Chapter 3) that (weak) gravitational waves can also be thought of as living in the fixed background spacetime, so the two problems really are similar.

We want to solve the linearized Einstein equations in vacuum. This means that the perturbed Ricci tensor must vanish, so we have

$$\delta R_{ab} = 0. \tag{16.23}$$

In order to work out the perturbed Ricci tensor, we recall the discussion of the nature of gravitational waves from Chapter 3, and consider the metric as a sum of the unperturbed background metric g_{ab}^{B} and a perturbation h_{ab}, as before. At this point, we only require that the background metric is a solution of the Einstein field equations. As in Chapter 3, we use the background metric to raise and lower indices on first-order quantities.

Since we already know that the background metric satisfies the Einstein equations, we can compute the Ricci tensor to linear order (keeping only quantities that are linear in h_{ab}). From the definition (2.86) it follows that we need

$$\delta R_{ab} = -\partial_c \delta \Gamma_{ab}^c + \partial_b \delta \Gamma_{ac}^c - \delta \Gamma_{ab}^d \Gamma_{dc}^c - \Gamma_{ab}^d \delta \Gamma_{dc}^c + \delta \Gamma_{ac}^d \Gamma_{db}^c + \Gamma_{ac}^d \delta \Gamma_{db}^c. \tag{16.24}$$

The problem seems a bit messy. However, we can make use of a clever 'trick'. Let us consider a local inertial frame, in which the Christoffel symbols all vanish. Then the equation reduces to

$$\delta R_{ab} = -\partial_c \delta \Gamma_{ab}^c + \partial_b \delta \Gamma_{ac}^c. \tag{16.25}$$

One can show that $\delta \Gamma_{bc}^d$ is a tensor—even though Γ_{bc}^d is not! – so (16.25) remains valid in all coordinate systems once the partial derivatives are replaced by covariant ones. That is, the variation of the Ricci tensor is governed by

$$\delta R_{ab} = -\nabla_c \delta \Gamma_{ab}^c + \nabla_b \delta \Gamma_{ac}^c. \tag{16.26}$$

After some algebra, we find that

$$\delta\Gamma^c_{ab} = \frac{1}{2}g^{dc}\left(\nabla_a h_{db} + \nabla_b h_{da} - \nabla_d h_{ab}\right), \tag{16.27}$$

and we have all the elements we need to construct an explicit form for (16.23) to linear order in h_{ab}.

However, even though we expect there to be two gravitational-wave degrees of freedom, we are still dealing with six coupled partial differential equations. In order to make progress we would like to separate the variables. Naively, one might think that we could simply expand all components of h_{ab} in spherical harmonics. However, this does not work. Because of the underlying spherical symmetry of the problem, it should not change under a rotation around the origin. But a spherical harmonics decomposition does change under such rotations. In fact, it transforms in a way that explicitly depends on the coordinates. The upshot of this is that the expansion coefficients still depend on the angular variables, which prevents separation and defeats the purpose of the exercise.

Luckily, we have already dealt with this kind of problem. To make progress we simply need to recall the discussion of the fluid velocity and the stresses in Chapter 14. In a similar way, we need to decompose h_{ab} in functions which have the correct behaviour under rotation. As a first step, we note that different parts of h_{ab} transform differently. Schematically, we have (Nollert, 1999)

$$h \sim \begin{pmatrix} \boxed{\text{S}} & \boxed{\text{S}} & \boxed{\text{V}} \\ \boxed{\text{S}} & \boxed{\text{S}} & \boxed{\text{V}} \\ \boxed{\text{V}} & \boxed{\text{V}} & \boxed{\text{T}} \end{pmatrix}$$

The parts marked S transform as scalars under rotation, V transform as two-vectors and T as a two-tensor. The scalars can be expanded in spherical harmonics, as usual. For the remaining parts we need a basis of 'tensor spherical harmonics'—precisely the one we developed for the elastic problem in Chapter 14. The required tensor basis can be written in different ways. Looking ahead to the discussion of numerical relativity in Chapter 19 we choose to decompose the metric into a lapse function $\alpha = g_{tt}$, a shift vector $\beta_i = g_{ti}$, and the spatial metric $\gamma_{ij} = g_{ij}$.

Let us first consider the perturbations $\delta\alpha$ and $\delta\beta_i$. We know that the first transforms as a scalar and hence can be expanded in spherical harmonics. To get a corresponding expression for $\delta\beta_i$ we use the same decomposition as for the fluid velocity in Chapter 14. Again, there will be two classes of perturbations with distinct parity. Axial perturbations correspond to

$$\delta\alpha = 0, \tag{16.28}$$

$$\delta\beta_i = \left(0, -\frac{h_0}{\sin\theta}\partial_\varphi Y_{lm}, h_0 \sin\theta \, \partial_\theta Y_{lm}\right), \tag{16.29}$$

where the coefficient h_0 is a function of t and r. For polar perturbations, the corresponding expressions are

$$\delta\alpha = \frac{\Delta}{r^2}H_0\,Y_{lm},\tag{16.30}$$

$$\delta\beta_i = \left(H_1\,Y_{lm}, c_0\partial_\theta\,Y_{lm}, c_0\partial_\varphi\,Y_{lm}\right).\tag{16.31}$$

As in the fluid problem, terms belonging to different l and different parity will not mix (as long as the background is spherically symmetric). Hence, we expect the perturbations of a non-rotating black hole to be described by two sets of equations that decouple and can be studied separately.

To complete the perturbed metric, we need expressions for the three-metric, h_{ij}. These are (simply) given by

$$h_{ij}^{\text{axial}} = h_1 e_{ij}^1 + h_2 e_{ij}^2\tag{16.32}$$

and

$$h_{ij}^{\text{polar}} = \frac{r^2}{\Delta}H_2 f_{ij}^1 + c_1 f_{ij}^2 + r^2 K f_{ij}^3 + r^2 G f_{ij}^4,\tag{16.33}$$

with e_{ij}^I and f_{ij}^I defined as in Chapter 14.

Now that we have a general expression for the perturbed spacetime metric, we can consider the separation of variables and derive the equations for the perturbations. As a first step it is useful to decide what gauge to work in. The freedom to choose coordinates can be turned into a suitable set of conditions on the perturbed metric. The traditional choice is to remove the highest angular derivatives from the problem by setting $h_2 = 0$ for axial perturbations, while taking $c_0 = c_1 = G = 0$ in the polar case. This is known as the Regge–Wheeler gauge (Regge and Wheeler, 1957).

Since the background metric is spherically symmetric, the various spherical harmonics decouple. In fact, as in the scalar case, it is sufficient to consider the $m = 0$ case. In effect, all φ-derivatives can be taken to vanish and each spherical harmonic Y_{lm} can be replaced by the corresponding Legendre function $P_l(\cos\theta)$. For axial perturbations we then get

$$h_{ab}^{\text{axial}} = \begin{pmatrix} 0 & 0 & 0 & h_0 \\ 0 & 0 & 0 & h_1 \\ 0 & 0 & 0 & 0 \\ \text{sym} & \text{sym} & 0 & 0 \end{pmatrix} \times \sin\theta\,\partial_\theta P_l(\cos\theta).\tag{16.34}$$

Substituting this expression into (16.23) we find three nontrivial equations for the variables h_0 and h_1

$$\delta R_{t\varphi} = 0 \longrightarrow h_0' - \dot{h}_1' - \frac{2}{r}\dot{h}_1 - \frac{l(l+1)r - 4M}{r\Delta}h_0 = 0,\tag{16.35}$$

$$\delta R_{\theta\varphi} = 0 \longrightarrow \dot{h}_0 - \frac{\Delta^2}{r^4}h_1' - \frac{2M\Delta}{r^4}h_1 = 0,\tag{16.36}$$

and

$$\delta R_{r\varphi} = 0 \longrightarrow \dot{h}'_0 - \ddot{h}_1 - \frac{2}{r}\dot{h}_0 - \frac{n\Delta}{r^4}h_1 = 0, \tag{16.37}$$

where a dot represents a time-derivative, a prime is a derivative with respect to r, and we have defined

$$n = (l-1)(l+2). \tag{16.38}$$

It is easy to show that the first equation is a consequence of the other two, so we have two equations for two unknown variables. These equations can be cast in the form of a wave equation with an effective potential. Eliminating h_0 we find

$$\ddot{h}_1 - r^2 \frac{\partial}{\partial r}\left[\frac{\Delta}{r^4}\frac{\partial}{\partial r}\left(\frac{\Delta}{r^2}h_1\right)\right] + \frac{n\Delta}{r^4}h_1 = 0. \tag{16.39}$$

This equation simplifies further if we use the tortoise coordinate r_* (as in the scalar-field problem). If we also introduce a new dependent variable

$$u = \frac{\Delta}{r^3}h_1, \tag{16.40}$$

we arrive at a wave equation with no first derivatives with respect to r. This is known as the Regge–Wheeler equation and it can be written

$$\frac{\partial^2 u}{\partial r_*^2} - \frac{\partial^2 u}{\partial t^2} - V(r)u = 0, \tag{16.41}$$

with

$$V(r) = \frac{\Delta}{r^2}\left[\frac{l(l+1)}{r^2} - \frac{6M}{r^3}\right]. \tag{16.42}$$

The Regge–Wheeler equation is remarkably similar to the wave equation for a massless scalar field (16.10). In fact, the 'centrifugal' (l-dependent) part, which dominates at large distances, is exactly the same. The two problems share qualitative features, even though they are physically rather different.

However, we need to exercise some caution. Our equation for gravitational perturbations was derived using a specific gauge and it is legitimate to ask how this affects the outcome. The choice we made was convenient because it simplifed the calculation, but we need to establish to what extent the final equation (16.41) describes the true physics—which should not depend on our choice of gauge/coordinates.

The issue is emphasized by the fact that the Regge–Wheeler gauge has pathologies that make it unsuitable for discussing (say) gravitational waves far away from the black

hole (which is ultimately what we want to do). That we have a problem is easily seen
from (16.40). We have

$$h_1 = \frac{r^2}{r - 2M} u,$$ (16.43)

and since $u \sim e^{\pm i\omega r_*}$ as $r_* \to \infty$ the metric perturbation h_1 diverges towards spatial
infinity, e.g. $h_{r\theta} \sim r$ as $r \to \infty$. This is in obvious contrast with the expectation that the
gravitational-wave amplitude should fall off as $1/r$ away from the source.

To fix this problem, we need to analyse the behaviour of the perturbations under a
gauge transformation. As discussed in Chapter 3, the infinitesimal transformation $x^a \to$
$x^a + \xi^a$ affects the metric perturbations in such a way that

$$h_{ab} \to h_{ab} + \nabla_b \xi_a + \nabla_a \xi_b = h_{ab} + \delta h_{ab}.$$ (16.44)

We combine this result with the generator of an arbitrary axial gauge transformation,
which can be written (cf. the expression for $\delta \beta_i$)

$$\xi_a = C(t,r)\left(0, 0, -\frac{1}{\sin\theta}\partial_\varphi Y_{lm}, \sin\theta\, \partial_\theta Y_{lm}\right).$$ (16.45)

The fact that there is one unspecified function C is an indication that we should expect
to be able to construct a single gauge-invariant function for the problem. From (16.44)
it follows that

$$\delta h_0 = \dot{C},$$ (16.46)

$$\delta h_1 = C' - \frac{2C}{r},$$ (16.47)

$$\delta h_2 = -2C.$$ (16.48)

To construct a gauge-invariant perturbation quantity, let us call it a_1, we need to find a
combination of h_0, h_1 and h_2 such that $\delta a_1 = 0$. This can be done in several ways. One
option is to combine the last two equations and use

$$a_1 = h_1 + \frac{1}{2}h_2' - \frac{h_2}{r}.$$ (16.49)

It then follows that

$$u = \frac{\Delta}{r^3}a_1$$ (16.50)

satisfies the Regge–Wheeler equation.

In essence, this exercise shows that one can derive gauge-invariant perturbation variables for the Schwarzschild problem (Moncrief, 1974; Gerlach and Sengupta, 1979; Gundlach and Martín-García, 2000) Our derivation of the Regge–Wheeler equation was carried out in a specified gauge, but the information it contains is actually gauge invariant. Hence, the information we extract represents the true physics.

However, the problem of the pathological behaviour as $r \to \infty$ remains. The issue can be resolved by a coordinate transformation into a so-called radiation gauge (Chrzanowski, 1975). Pragmatically—assuming that we are mainly interested in the waves far from the source—we may simply transform the result into the Lorenz gauge used to discuss the weak-field dynamics in Chapter 3.

The analysis of polar perturbations is not quite as straightforward, mainly since we are starting out with a larger number of equations (Zerilli, 1970a; Vishveshwara, 1970b). Nevertheless, the introduction of a new variable,

$$\frac{(nr+6M)}{r^3}u = -\frac{\partial K}{\partial r_*} + \frac{nr+6M}{r^2}K + \left(1 - \frac{2M}{r}\right)\frac{H_0}{r} \tag{16.51}$$

(and a bit of algebra), leads to (16.41) with the effective potential

$$V(r) = \frac{\Delta}{r^2}\frac{n^2(n+2)r^3 + 3n^2Mr^2 + 18nM^2r + 36M^3}{r^3(nr+6M)^2}. \tag{16.52}$$

This is known as the Zerilli equation (Zerilli, 1970a). As in the axial case one can construct gauge-invariant polar quantities which satisfy this wave equation (Moncrief, 1974). One can also show that the axial and polar problems are closely related (Chandrasekhar, 1992).

16.4   Quasinormal modes

Let us now return to the quasinormal modes of the black hole. Since they tend to dominate the signal from a perturbed black hole we need to understand these modes. However, as we have 'diverging boundary conditions' we are not dealing with a standard eigenvalue problem, and the identification of quasinormal modes is somewhat involved. In order to find a mode one must ensure that no contamination of ingoing waves remains at infinity—and that no waves are coming out of the horizon—but these unwanted contributions are exponentially small. There are different ways to overcome this technical challenge (Leaver, 1985; Andersson, 1992; Nollert and Schmidt, 1992), but we will not go into the details here. Instead, we focus on a simple approximation.

The quasinormal-mode problem is essentially one of wave scattering off of a potential barrier. The modes themselves are analogous to scattering resonances associated with 'energies' close to the top of the potential barrier. The analogy with the quantum problem means that we can use the celebrated Bohr–Sommerfeld quantization formula from WKB theory (Schutz and Will, 1985)

$$\int_{t_1}^{t_2} Q dr_* = \left(n + \frac{1}{2}\right)\pi, \tag{16.53}$$

where n is the 'quantum number' of the mode (not to be confused with the variable from (16.38)) and

$$Q^2 = \omega^2 - V(r). \tag{16.54}$$

The limits of integration, t_1 and t_2, are zeros of the function Q^2. Let us assume that the quasinormal modes are, indeed, associated with the peak of the potential. Then we can Taylor expand the integrand around this point, r_*^0, to get

$$Q \approx \sqrt{\omega^2 - V_0} - \frac{V_0''}{4\sqrt{\omega^2 - V_0}}(r_* - r_*^0)^2, \tag{16.55}$$

where $V_0 = V(r_*^0)$ and

$$V_0'' = \left(\frac{d^2 V}{dr_*^2}\right)_{r_*^0}. \tag{16.56}$$

We can easily solve for the zeros of this function and work out the integral in (16.53). This way we find that the mode frequencies can be approximated by

$$\omega_n^2 \approx V_0 + \frac{3}{8}(V_0'')^{1/2}\left(n + \frac{1}{2}\right)\pi. \tag{16.57}$$

Neglecting the 'field-dependent term' in the different effective potentials, we use

$$V(r) \approx \frac{\Delta}{r^2}\frac{l(l+1)}{r^2}. \tag{16.58}$$

Then $V_0'' < 0$ and assuming that $\mathrm{Re}\,\omega \gg \mathrm{Im}\,\omega$ we arrive at the approximate result

$$\mathrm{Re}\,\omega_n \approx \pm\sqrt{V_0} \approx \left(\frac{l}{3\sqrt{3}}\right)M \tag{16.59}$$

$$\mathrm{Im}\,\omega_n \approx \frac{3}{16}\left(-\frac{V_0''}{V_0}\right)^{1/2}\left(n + \frac{1}{2}\right)\pi. \tag{16.60}$$

From these expressions we learn that the real part of the quasinormal mode frequency (the oscillation frequency) increases linearly with l. For $l = 2$, presumably the most important multipole for gravitational-wave detection, the frequency is $\mathrm{Re}\,\omega_n M \approx 0.38$. Meanwhile, the imaginary part of the mode frequency (the damping rate) depends on the curvature of the effective potential. For the fundamental ($n = 0$) mode we find

Table 16.1 *The first five quasinormal-mode frequencies, $\omega_n M$, for the three lowest radiating multipoles ($l \geq 2$) of a gravitationally perturbed Schwarzschild black hole. The frequencies are given in units of $[GM/c^3]^{-1} = (32{,}312/2\pi \; Hz) \times (M/M_\odot)^{-1}$. (Data from Andersson (1992).)*

n	$l = 2$	$l = 3$	$l = 4$
0	$0.373672 - 0.088962i$	$0.599443 - 0.092703i$	$0.809178 - 0.094164i$
1	$0.346711 - 0.273915i$	$0.582643 - 0.281298i$	$0.796631 - 0.284334i$
2	$0.301052 - 0.478276i$	$0.551683 - 0.479093i$	$0.772710 - 0.479904i$
3	$0.251504 - 0.705148i$	$0.511956 - 0.690338i$	$0.739835 - 0.683916i$
4	$0.207514 - 0.946845i$	$0.470173 - 0.915660i$	$0.701524 - 0.898240i$

that Im $\omega_n M \approx 0.08$, independently of the value of l. The damping rate of the higher overtones then increases linearly with n. These approximations may not be very precise but they explain the features brought out by more detailed calculations; see Table 16.1.

In physical units, the fundamental quadrupole gravitational-wave quasinormal mode of a Schwarzschild black hole has frequency

$$f \approx 12 \left(\frac{M_\odot}{M} \right) \text{kHz}, \tag{16.61}$$

while the associated e-folding time is

$$\tau \approx 0.05 \left(\frac{M}{M_\odot} \right) \text{ms}, \tag{16.62}$$

and the various overtones damp much faster. The quasinormal modes of a black hole are extremely short lived—spacetime is a poor oscillator compared to other systems in nature. If we define the quality factor

$$q \approx \frac{1}{2} \left| \frac{\text{Re } \omega_n}{\text{Im } \omega_n} \right|, \tag{16.63}$$

then quasinormal modes have $q \sim l \sim 2$. In contrast, the fundamental fluid pulsation mode of a neutron star (see Chapter 13) has $q \sim 1000$, and the typical value for an atom is $q \sim 10^6$.

16.5   Test particle motion

Back in the days when numerical relativity was still in its infancy (throughout the 1970s and into the 1980s) a significant body of work explored gravitational-wave emission from a small body moving along geodesics in a black-hole spacetime (Zerilli, 1970*b*; Davis *et al.*, 1971). This problem provides intuition that helps us understand more complex nonlinear work. It is also relevant for gravitational captures by supermassive black holes in galaxy cores. For the present discussion, the key point is that we can (as long as we ignore the impact of the mass of the moving body on the background curvature) use perturbation theory to attack the problem.

In order to work out the gravitational waves emitted by a moving body, we need to add the relevant stress–energy tensor to the perturbation problem. As we have already seen in Chapter 3, this is straightforward. In the case of a point mass we have

$$T^{ab} = m \int_{-\infty}^{\infty} \frac{\delta^4(x - z(\tau))}{\sqrt{-g}} \frac{dz^a}{d\tau} \frac{dz^b}{d\tau} \, d\tau, \tag{16.64}$$

where the δ-function gives the location of the particle, τ is proper time along the geodesic, and m is the rest mass. As a sanity check, we may confirm that the divergence

$$\nabla_a T^{ab} = 0, \tag{16.65}$$

leads to the background geodesics.

In (16.64), the vector x represents a field point with coordinates $[t, r, \theta, \varphi]$. The space-time coordinates of the moving mass are $z^a(\tau)$. The components of z^a are $[t', r', \theta', \varphi']$. The determinant of the background metric tensor is g, so that, in the Schwarzschild case, we have $\sqrt{-g} = r^2 \sin\theta$.

It is useful to simplify the stress–energy tensor. To do this, we first change the variable of integration from τ to t' using (Zerilli, 1970*b*)

$$\int_{-\infty}^{\infty} d\tau \to \int_{-\infty}^{\infty} dt' \frac{d\tau}{dt'} = \int_{-\infty}^{\infty} \frac{dt'}{\gamma}, \tag{16.66}$$

where

$$\gamma = \frac{dt'}{d\tau}. \tag{16.67}$$

Using the chain rule to rewrite the velocities as

$$\frac{dz^a}{d\tau} = \gamma \frac{dz^a}{dt'}, \tag{16.68}$$

we have

$$T^{ab} = m \int_{-\infty}^{\infty} \frac{\delta^4(x - z(t'))}{\sqrt{-g}} \frac{dz^a}{dt'} \frac{dz^b}{dt'} \gamma \, dt'. \tag{16.69}$$

Integrating out the delta function $\delta(t - t')$, we get

$$T^{ab} = m\gamma \frac{\delta(r - r'(t))\delta^2(\Omega - \Omega'(t))}{r^2} \dot{z}^a \dot{z}^b, \tag{16.70}$$

where

$$\delta^2(\Omega - \Omega') = \frac{\delta(\theta - \theta')\delta(\varphi - \varphi')}{\sin\theta}, \tag{16.71}$$

and we have defined

$$\dot{z}^a = \frac{dz^a}{dt}. \tag{16.72}$$

We have already introduced the tensor harmonics required to decompose the perturbations of the spacetime metric. In principle, we can use the same basis for the stress–energy tensor. This would require us to use the orthogonality of the tensor harmonics to evaluate integrals of inner products. An alternative would be to expand the angular delta function, $\delta^2(\Omega - \Omega')$, (which determines the θ and φ dependence of the stress–energy tensor), in terms of spin-weighted spherical harmonics.

The notion of spin-weighted spherical harmonics arises naturally in the tetrad-based Newman–Penrose formalism, which we will discuss in Chapter 17. We introduce $_sY_{lm}(\theta,\varphi)$, where s is the spin weight, as a generalization of the usual spherical harmonics, which have spin weight 0. That is, $Y_{lm}(\theta,\varphi) = {}_0Y_{lm}(\theta,\varphi)$. In principle, we may allow half-integer spins but we will only need integer values here. Harmonics of different spin weight are related by raising and lowering operators. The raising operator, $\eth$, is defined by

$$\eth\, {}_sY_{lm}(\theta,\varphi) = -(\sin\theta)^s \left[\frac{\partial}{\partial\theta} + i\csc\theta \frac{\partial}{\partial\varphi} \right] (\sin\theta)^{-s}\, {}_sY_{lm}(\theta,\varphi). \tag{16.73}$$

This increases the spin weight by 1, so that

$$\eth\, {}_sY_{lm}(\theta,\varphi) = \sqrt{(l-s)(l+s+1)}\, {}_{s+1}Y_{lm}(\theta,\varphi). \tag{16.74}$$

Meanwhile, the lowering operator, $\bar\eth$, which follows from the complex conjugate of (16.73), lowers the spin weight by 1. Using $\eth$ and $\bar\eth$, we can construct spin-weighted spherical harmonics of non-zero s from the spherical harmonics $Y_{lm}(\theta,\varphi)$. Note that, (16.73) and its conjugate imply that

$$_sY_{lm}(\theta,\varphi) = 0, \text{ for } |s| > l. \tag{16.75}$$

The spin-weighted spherical harmonics satisfy a second-order differential equation

$$\bar{\eth}\eth\,_sY_{lm} = \left[\frac{\partial^2}{\partial\theta^2} + \cot\theta\frac{\partial}{\partial\theta} - \frac{m^2}{\sin^2\theta} - \frac{2ms\cos\theta}{\sin^2\theta} - s^2\cot^2\theta + s\right]\,_sY_{lm}. \tag{16.76}$$

These harmonics form a complete set for angular functions of spin weight s on the unit sphere. The completeness relation is

$$\delta^2(\Omega - \Omega') = \sum_{l\geq|s|}^{\infty}\sum_{m=-l}^{l}\,_s\overline{Y}_{lm}(\theta',\varphi')\,_sY_{lm}(\theta,\varphi), \tag{16.77}$$

where the bar indicates complex conjugation. Harmonics of the same spin weight are orthonormal in the sense that

$$\int\,_s\overline{Y}_{l'm'}(\theta,\varphi)\,_sY_{lm}(\theta,\varphi)\,d\Omega = \delta_{ll'}\,\delta_{mm'}, \tag{16.78}$$

where the integration is over the sphere.

16.6   Taking the plunge

The radiation emitted as a small body falls radially into a Schwarzschild black hole was calculated as one of the first astrophysical applications of the perturbation equations in the early 1970s (Davis *et al.*, 1971). The energy spectrum (averaged over all directions) as measured by a distant observer is shown in Figure 16.4 (for the case where the falling body starts off from rest at infinity). The spectrum peaks at the slowest damped quadrupole quasinormal mode (at $\omega M \approx 0.37$) and the total energy emitted is $E \approx 0.01m^2/M$. Roughly 90% of this energy is radiated through the quadrupole. This result shows that our simple estimate (10.70) was much better than we had any reason to expect. Of course, now we are also accounting for the spacetime curvature. The calculation shows that as much as 97% of the total energy goes into quasinormal-mode oscillations.

Qualitatively, the result hardly changes at all if the particle is given some initial angular momentum or an initial velocity (Ruffini, 1973; Kojima and Nakamura, 1983). The energy spectrum in each multipole still peaks at the relevant quasinormal-mode frequency. One generally finds that the contribution from the $l = m$ multipole dominates the radiation. As $\tilde{L}_z$ (the angular momentum per unit mass; see Chapter 10) increases, the importance of the higher multipoles is enhanced. As a result, the radiated energy may increase by as much as a factor of 50.

In a similar way, one can work out the radiation from bodies that are 'scattered' by the black hole (Kojima and Nakamura, 1984). When the particle (again, initially at rest at infinity) has initial angular momentum $\tilde{L}_z > 2$ it is not captured by the black hole—it

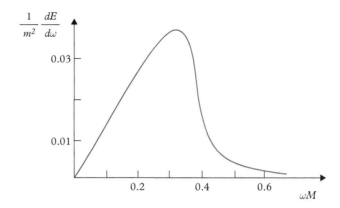

Figure 16.4 *The spectrum of quadrupole (l = 2) gravitational radiation emitted when a particle falls radially into a Schwarzschild black hole (starting out at rest at infinity). The spectrum (averaged over all directions) peaks at the slowest damped quasinormal-mode frequency ($\omega M \approx 0.37$). (Based on the results of Davis et al. (1971).)*

escapes to infinity. The larger the initial value of $\tilde{L}_z$ is, the further away from the black hole the particle reaches periastron. Hence, one would expect less radiation to be emitted as $\tilde{L}_z$ increases. This is, indeed, brought out by detailed calculations—when $\tilde{L}_z$ becomes very large, the particle never gets close enough to the black hole to induce relativistic effects.

These exercises tell us that the bulk of the energy radiated when a particle falls into a black hole emerges through the quasinormal modes. We also learn that the modes are hardly excited at all when a particle is scattered off to infinity. This is true even when periastron is quite close to the black hole. Instead, the peak frequency depends on the initial angular momentum $\tilde{L}_z$ of the particle (Kojima and Nakamura, 1984). Basically, one can show that the spectrum depends on the angular velocity Ω_p at periastron. The position of all peaks in the spectrum are explained in the same way. The source term contains a factor $\cos(\omega t - m\varphi)$. Close to periastron it is reasonable to approximate $\varphi \approx \Omega_p t$. It then follows that the spectrum will peak at $\omega \approx m\Omega_p$, or since the $l = m$ term dominates, $\omega_{\max} \approx l\Omega_p$.

This argument also explains why the quasinormal modes are not excited. For a particle that falls into the black hole the quasinormal modes are excited as the particle passes the peak of the curvature potential barrier. For a scattered particle, the modes can only be excited as the gravitational radiation emitted by the particle motion reaches the black hole. But the frequency of these waves is typically such that they get reflected off the black-hole potential barrier before getting close. For example, for $l = 2$ one gets $\omega_{\max} M <$ 0.2 and such waves will not really excite the quasinormal modes. For this reason, the quadrupole formula—which should be valid for particles moving at $v \ll c$ in flat space (and obviously ignores quasinormal modes)—provides a reasonable approximation of the radiated energy in many situations.

16.7   The self-force problem

While an understanding of geodesic motion provides us with a way of estimating the gravitational-wave emission from a moving body, an important question remains to be answered. How does the energy/angular momentum loss affect the motion? We know what we expect to happen—the energy loss will draw the orbit closer towards the eventual merger—and we have a good idea of the relevant timescale from post-Newtonian calculations (see Chapter 11). When the evolution is slow, we can use the averaged emission to estimate the (adiabatic) evolution (Cutler *et al.*, 1993), but this approximation breaks down as the two bodies get close. Eventually, we need to consider the relativistic two-body problem in full. However, this problem is complicated. General relativity is a non-linear theory, and we typically need to resort to numerical simulations (see Chapter 19). However, when the mass ratio is small, the widely separated scales make the problem difficult—at the very least, extremely expensive—to resolve numerically. Instead, we can make clever use of perturbation theory.

The extreme mass-ratio regime is naturally explored perturbatively (see Poisson *et al.* (2011), Barack (2009), and Barack and Pound (2019) for detailed reviews), as we have a prescribed small parameter. The leading order configuration is that of a test particle, with mass m, moving along a geodesic of the fixed spacetime of the larger object, with mass M. This then serves as a basis for a perturbation scheme, where corrections due to the finite mass of the smaller body are included order by order in the (supposedly small) mass ratio $\eta = m/M \ll 1$. At first order in η the gravitational field of the small object is a linear perturbation of the background geometry. The back-reaction from this perturbation gives rise to an effective gravitational self-force that diverts the small body from its geodesic motion. It is this self-force that is responsible for the radiative decay of the orbit, as illustrated in Figure 16.5.

Interest in the extreme mass-ratio problem has grown since the mid-1980s, when the space-based detector LISA was first proposed (Faller *et al.*, 1985). LISA is expected to observe signals from the inspiral of compact objects into massive black holes in galaxy cores (Amaro-Seoane *et al.*, 2017). The inspiralling objects have to be compact

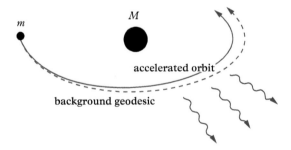

Figure 16.5 *A schematic illustration of the gravitational self-force problem. The emission of gravitational waves drives the orbit of a small moving body away from the geodesics of the spacetime of the more massive companion.*

to avoid tidal disruption before they reach a relevant signal strength. In fact, LISA is expected to detect hundreds of extreme-mass-ratio events, out to cosmological distances, $z \sim 1$; see Figure 22.7. As an example, a $10M_\odot/10^6 M_\odot$ system will spend the last few years of inspiral in a very tight orbit emitting $10^5 - 10^6$ gravitational-wave cycles in the LISA band. These inspiral trajectories exhibit extreme versions of periastron precession, Lense-Thirring precession of the orbital plane, and other strong-field effects (Barack and Pound, 2019). This complex dynamics is encoded in the gravitational waves, see Figure 16.6 for an illustration, which carry a detailed map of the spacetime geometry around the massive black hole. It has been argued that LISA will be able to measure

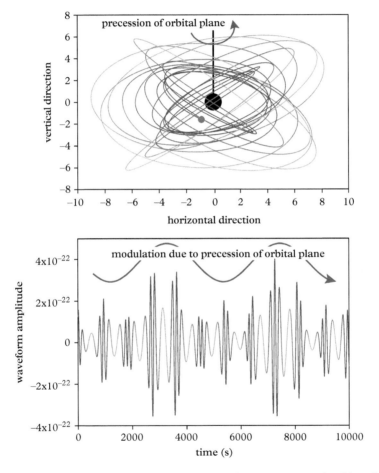

Figure 16.6 *An illustration of the intricate detail of a typical extreme mass-ratio orbit and the corresponding gravitational-wave signal. The top panel shows the geometrical shape of the relativistic orbit. The lower panel provides the gravitational-wave amplitude as a function of time. (Reproduced with permission from Amaro-Seoane et al. (2013).)*

fractional deviations as small as one part in a thousand in the quadrupole moment of the black-hole spacetime, setting tight bounds on the parameters of alternative theories of gravity. A precise understanding of the gravitational self-force is essential for this problem. In a typical extreme mass-ratio inspiral, the self-force drives the orbital decay over a timescale of months, but dephases the orbit already after a few hours (Barack and Pound, 2019).

Given the importance of the problem, let us take a closer look at how we can model the gravitational self-force. At first 'post-geodesic' order, one would expect the equation of motion to have the form

$$mu^b \nabla_b u^a = f^a_{\text{self}}, \tag{16.79}$$

where u^a is the small body's four-velocity in the background spacetime, ∇_b is the background covariant derivative, and f^a_{self} is the force we want to determine. From fundamental principles, one can argue that the self-force can be expressed in terms of the 'tail' part of the physical metric perturbation (Barack et al., 2002), arising from the part of the Green's function that is supported inside (rather than on) the past light-cone of the source. This makes intuitive sense, but it is far from straightforward to turn the words into a practical computational framework.

First of all, the self-force is a gauge-dependent notion, as is the accelerated trajectory in the background geometry. A gauge transformation in the perturbed geometry leads to a distinct accelerated trajectory. This means that meaningful information about the motion requires the combination of the self-force and the metric perturbation (in a specific gauge). In fact, the intuition leading to (16.79) is based on using the harmonic (Lorenz) gauge (see Chapter 3). Secondly, the definition (16.79) only holds locally, close to a given point along the trajectory. The formulation of a faithful scheme for the long-term evolution of the orbit involves subtle issues, like the fact that the Lorenz-gauge condition cannot be consistently imposed when the source's world-line is accelerating.

Despite the challenges, there has been significant progress on the self-force problem since the 1990s, much of it stemming from a formal derivation of the required equations of motion (Mino et al., 1997; Quinn and Wald, 1997) and a reinterpretation of the problem in terms of geodesic motion in a smooth perturbed spacetime (Detweiler and Whiting, 2003). As we have seen, it is natural to treat the orbiting body (with mass m) as a point mass. This causes the perturbations to diverge at the location of the particle, which is exactly where the force must be evaluated. Additionally, general relativity does not 'allow' point masses—a black hole must form if we arbitrarily shrink a body with a fixed mass. However, we can split the perturbation into two pieces, a direct part and a tail part (Detweiler and Whiting, 2003):

$$h_{ab} = h^{\text{dir}}_{ab} + h^{\text{tail}}_{ab}. \tag{16.80}$$

The direct part is divergent—it is the relativistic analogue of the singularity in the Newtonian potential. The tail part is an integral over the past history of the orbiting body. As illustrated in Figure 16.3, a wave propagating in a curved spacetime will scatter off the background curvature, rather than propagating as a sharp pulse. It develops a 'tail' which may interact with the mass that generated the radiation in the first place. The interaction between the mass m, its radiation field, and the background spacetime gives rise to the gravitational self-force.

Since the tail term is not a homogeneous solution of the field equations (Detweiler and Whiting, 2003), it makes sense to divide the perturbation in a different way

$$h_{ab} = h_{ab}^S + h_{ab}^R, \tag{16.81}$$

where the first term is singular (behaving like $\sim m/r$ close to the particle, with r the distance to the worldline). The second part is regular and given by a homogeneous solution to the field equations, including backscattered waves inside the light cone of the background geometry. This is the piece that leads to the self-force. It also involves the prior history of the orbiting mass. In this formulation, the particle can be thought of as moving along a geodesic of $g_{ab}^B + h_{ab}^R$, where g_{ab}^B is the background metric. The equations of motion are then simply reformulated as

$$m\tilde{u}^b\tilde{\nabla}_b\tilde{u}^a = 0, \tag{16.82}$$

where the tildes indicate the four-velocity and covariant derivative associated with the complete first-order metric. However, the perturbation h_{ab}^R is not the physical metric perturbation induced by the particle (for instance, it is not causal)—it is a mathematical construction that serves as an effective potential for the motion.

The formulation of a suitable framework for the problem is in progress, but we need to turn it into actual calculations. This poses additional challenges. For example, since the gravitational self-force is gauge dependent, one has to take care in interpreting the results. In order to make a meaningful comparison of different calculations, it is necessary to identify gauge-invariant quantities (representing physical observables). Such observables include the orbital frequency and the rate of periastron advance (Barack and Sago, 2009; Le Tiec *et al.*, 2011). There is also a gauge-invariant relation between energy and angular momentum. The final step of the analysis involves incorporating the corrections to the orbital motion into the gravitational waveforms. Detailed studies shows that, in order to provide waveforms with the precision required for LISA data analysis, the problem has to be solved to second order in the mass ratio. However, higher order relativistic perturbation theory is messy, so this requirement adds to the complexity of the calculation (especially as one should proceed in a gauge-invariant manner; see Pound (2017)). Ultimately, one would like to directly integrate the (second-order) equation of motion in a self-consistent manner. However, this is also tricky.

While the appropriate computational tools are being developed, we can make progress using further approximations. As an example, orbital evolutions have been carried out using the method of 'osculating geodesics', where the motion is obtained as a smooth sequence of geodesics, each a tangent to the 'true' orbit at a particular time. This leads to results like those in figure 16.7—an important step towards determining reliable waveforms to inform data analysis.

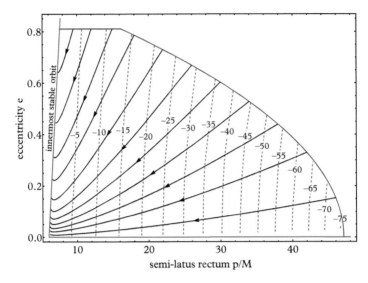

Figure 16.7 *Evolutionary tracks for inspiral orbits around a Schwarzschild black hole for a mass ratio $m/M = 10^{-5}$. Each solid black curve tracks the evolution of a particular extreme-mass-ratio orbit in the e–p plane (eccentricity vs. semilatus rectum), from the point at which it enters the LISA band (blue curve, assuming a black-hole mass of $M = 10^6 M_\odot$), until it reaches the innermost stable orbit (red line on the left). During the evolution, the conservative piece of the self-force acts to decrease the rate of periastron advance. The dashed contour lines (with associated numerical values) indicate the total amount of periastron phase (in radians) accumulated due to this effect, from a given moment until plunge (Reproduced from Barack and Pound (2019), based on data from Osburn et al. (2016), copyright (2016) by the American Physical Society.)*

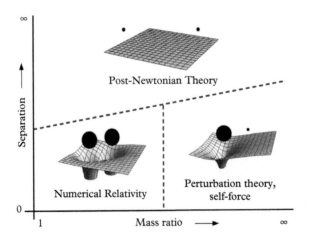

Figure 16.8 *Schematic representation of the different domains of the binary black-hole problem. Numerical solutions of the full Einstein equations are required for close binaries of comparable masses. Widely separated binaries are most efficiently treated within post-Newtonian theory, while strongly gravitating binaries with large mass ratios are the realm of perturbative self-force theory. (Illustration provided by L. Barack.)*

In addition to its immediate relevance for low-frequency gravitational-wave astronomy, the self-force programme informs efforts in complementary parts of parameter space. In particular, one can use the self-force results to (i) fix unknown regularization parameters in high-order post-Newtonian calculations (Blanchet *et al.*, 2010), and (ii) calibrate effective-one-body waveforms (see Chapter 11). The shift in the frequency of the innermost stable circular orbit due to the self force (Barack and Sago, 2009), has already been used to discriminate between post-Newtonian calculations (Favata, 2011). The perturbative results also complement full nonlinear simulations, which become challenging for small mass ratios. When combined, results from the different approaches provide a 'complete' understanding of the relativistic two-body problem, as sketched in figure 16.8.

17

Spinning black holes

Stars rotate and since black holes form through gravitational collapse one would expect them to rotate, as well. The spin of a black hole is associated with a range of phenomena. In particular, it is thought to play a key role in the formation of the powerful jets seen emerging from active galactic nuclei. A spinning black hole accreting matter from a torus of fallback material is also thought to be the central engine that drives observed gamma-ray bursts (see Chapter 21). In essence, if we want to understand these problems we need to consider spinning black holes.

Fortunately, much of the rotating black hole phenomenology is similar to that for Schwarzschild black holes. The rotating case tends to be a bit more complex, but the underlying behaviour is often similar. In view of this, we will focus on the main differences and features that enter when we add a twist to the black-hole problem.

17.1 The Kerr solution

One of the early breakthroughs of general relativistic astrophysics was Roy Kerr's discovery of a solution to the Einstein field equations for spinning black holes (Kerr, 1963). The line element of the Kerr solution can be written

$$ds^2 = -\frac{\Delta - a^2 \sin^2\theta}{\Sigma} dt^2 + \frac{\Sigma}{\Delta} dr^2 - \frac{4aMr\sin^2\theta}{\Sigma} dt d\varphi$$

$$+ \Sigma d\theta^2 + \frac{(r^2 + a^2)^2 - a^2\Delta\sin^2\theta}{\Sigma} \sin^2\theta d\varphi^2, \qquad (17.1)$$

where

$$\Delta = r^2 - 2Mr + a^2, \qquad (17.2)$$

$$\Sigma = r^2 + a^2 \cos^2\theta. \qquad (17.3)$$

The rotation is encoded in the parameter $a = \mathcal{J}/M$, which lies in the range $0 \leq a \leq M$. If we take $a = 0$, the black hole is not spinning and we recover the Schwarzschild solution

Gravitational-Wave Astronomy: Exploring the Dark Side of the Universe. Nils Andersson, Oxford University Press (2020).
© Nils Andersson. DOI: 10.1093/oso/9780198568032.001.0001

from Chapter 4. Maximal spin corresponds to $a = M$. The horizon of the black hole follows from $\Delta = 0$, and we see that there are two solutions

$$r_{\pm} = M \pm \sqrt{M^2 - a^2}. \tag{17.4}$$

Adding to the familiar event horizon (r_+) we also have an inner Cauchy horizon (r_-). However, the latter is not of any consequence unless we venture inside the black hole.

The coordinates $[t, r, \theta, \varphi]$ used in (17.1) are called Boyer–Lindquist coordinates (Boyer and Lindquist, 1967): φ is the angle around the symmetry axis, t is the time coordinate in which everything is stationary, and (even though they are not immediately associated with a geometrical definition) r and θ are related to their counterparts in the spherically symmetric case (discussed in Chapter 4). These coordinates are the natural generalization of the Schwarzschild ones, and may be suitable for many purposes. For example, they are ideal for studying the asymptotic behaviour of various fields, which is what we tend to base our physical interpretations on. But the Boyer–Lindquist coordinates are somewhat 'unphysical' close to the black hole, which means that they should be replaced by some other coordinates if one is interested in the region near the horizon.

17.2 Inertial framedragging

Rotation brings about an important qualitative difference between Newtonian physics and general relativity. A spinning object drags spacetime along with it. This influences the motion of objects in the vicinity of a rotating black hole. The effect is associated with the presence of off-diagonal elements in the metric. In the case of the Kerr solution (17.1), we have

$$g_{t\varphi} = -\frac{2aMr\sin^2\theta}{\Sigma}. \tag{17.5}$$

In order to introduce the notion of frame dragging, it is useful to consider particle motion in the Kerr spacetime. This problem is a little bit more involved than the Schwarzschild case from Chapter 10, but the starting point is the same. For a particle with (rest) mass m, the momentum is given by $p^a = mu^a$, where u^a is the four-velocity. However, the symmetries of the Kerr spacetime admits two Killing vectors. These can be taken to be

$$t^a = (1,0,0,0) \quad \text{and} \quad \varphi^a = (0,0,0,1), \tag{17.6}$$

associated with the stationarity and axisymmetry of the spacetime, respectively. As discussed in Chapter 2, we can use these Killing vectors to construct conserved quantities. In the present case, we have

$$\tilde{E} = -t_a u^a = -u_t \tag{17.7}$$

and

$$\tilde{L}_z = \varphi_a u^a = u_\varphi, \tag{17.8}$$

representing the energy and angular momentum (per unit mass), as in Chapter 10. It is easy to show that these quantities are conserved by combining the geodesic equation with Killing's equation (2.73), but we can demonstrate this is a different way. Recall that geodesic motion can be obtained through a variational argument starting from a Lagrangian

$$\mathcal{L} = \frac{1}{2} g_{ab} \dot{x}^a \dot{x}^b, \tag{17.9}$$

where dots are derivatives with respect to a suitable parameter λ (say) along the trajectory. For massive particles it is natural to take this parameter to be proper time, but we will leave it unspecified for the moment as this allows us to consider the massless case, as well. The momentum that is conjugate to the coordinates x^a follows from

$$p_a = \frac{\partial \mathcal{L}}{\partial \dot{x}^a} \quad \longrightarrow \quad p_a = g_{ab} \dot{x}^b, \tag{17.10}$$

and we retain the usual definition for the momentum of a particle if we let $\lambda = \tau/m$.

The Euler–Lagrange equations lead to (see Chapter 10)

$$\frac{\partial p_a}{\partial \lambda} = \frac{\partial \mathcal{L}}{\partial x^a} = \frac{1}{2} \partial_a g_{bc} \dot{x}^b \dot{x}^c. \tag{17.11}$$

Since the Kerr metric is independent of t and φ it follows that $\tilde{E}$ and $\tilde{L}_z$ are conserved along geodesics.

Let us now consider a particle with vanishing angular momentum that falls towards the black hole. That is, we take $\tilde{L}_z = 0$. The angular (coordinate) velocity of the particle then follows from

$$\Omega = \frac{\partial \varphi}{\partial t}, \tag{17.12}$$

and since

$$p^\varphi = g^{\varphi\varphi} p_\varphi + g^{\varphi t} p_t, \tag{17.13}$$

$$p^t = g^{tt} p_t + g^{t\varphi} p_\varphi, \tag{17.14}$$

we see that

$$\Omega = \frac{p^\varphi}{p^t} = \frac{g^{\varphi t}}{g^{tt}} \equiv \omega(r). \tag{17.15}$$

That is, in a rotating background a particle with zero angular momentum can have non-vanishing angular velocity! Inverting the Kerr metric we find that

$$g^{rr} = \frac{\Delta}{\Sigma} \qquad g^{\theta\theta} = \frac{1}{\Sigma},$$

$$g^{tt} = -\frac{(r^2 + a^2)^2 - a^2 \Delta \sin^2\theta}{\Sigma\Delta},$$

$$g^{t\varphi} = -\frac{2aMr}{\Sigma\Delta} \qquad g^{\varphi\varphi} = \frac{\Delta - a^2\sin^2\theta}{\Sigma\Delta\sin^2\theta}, \tag{17.16}$$

which means that

$$\omega(r) = \frac{2aMr}{(r^2 + a^2)^2 - a^2\Delta\sin^2\theta} \sim \frac{2aM}{r^3} \text{ as } r \to \infty. \tag{17.17}$$

This is the expression for the inertial frame dragging. It shows that, even if a particle is dropped 'straight in' towards the rotating black hole it will be dragged along with the rotation. The denominator is positive for all values of r, so the frame dragging has the same sign as the spin parameter a; a particle is always dragged along with the black hole. We also see that ω falls off rapidly with distance. In principle, the frame dragging provides a means for measuring the rotation rate of any object, but the effect is small in most cases. Nevertheless, the Gravity Probe B experiment confirmed the frame-dragging induced by the spin of the Earth to within 15% of the prediction of the theory (Everitt *et al.*, 2011).

17.3 Kerr geodesics

The general trajectory of a particle moving in the Kerr geometry is more complicated than in the Schwarzschild case, but because of the axial symmetry we still expect to have $p_\theta = 0$ for motion confined to the equatorial plane. A simple symmetry argument suggests that, if a particle is initially moving in the equatorial plane, it should stay there. Such equatorial trajectories are obviously special cases, but they are nevertheless a useful starting point for an exploration of particle motion around a rotating black hole.

As we have seen, the fact that the spacetime is stationary and axisymmetric means that we have two constants of motion $p_t = -E \; (= -m\tilde{E})$ and $p_\varphi = L_z \; (= m\tilde{L}_z)$, the 'energy measured at infinity' and the component of the angular momentum along to the symmetry axis of the spacetime. Given this we can readily deduce two of the required equations of motion:

$$p^t = \frac{dt}{d\lambda} = \frac{(r^2+a^2)^2 - a^2\Delta\sin^2\theta}{\Sigma\Delta}E - \frac{2aMr}{\Sigma\Delta}L_z, \tag{17.18}$$

$$p^\varphi = \frac{d\varphi}{d\lambda} = \frac{2aMr}{\Sigma\Delta}E + \frac{\Delta - a^2\sin^2\theta}{\Sigma\Delta\sin^2\theta}L_z. \tag{17.19}$$

17.3.1   Light propagation

Let us, first of all, consider photons moving in the equatorial plane. The above results—together with $p_a p^a = 0$ and $p_\theta = 0$—lead to an equation for the radial motion that can be factorized as (Schutz, 2009)

$$\left(\frac{dr}{d\lambda}\right)^2 = \frac{(r^2+a^2)^2 - a^2\Delta}{r^4}(E - V_+)(E - V_-), \tag{17.20}$$

where

$$V_\pm(r) = \frac{2aMr \pm r^2\Delta^{1/2}}{(r^2+a^2)^2 - a^2\Delta}L_z. \tag{17.21}$$

These potentials fall off as $1/r$ as $r \to \infty$, and we already know that the rotational effects enter at higher orders. In essence, the rotation of the black hole has little effect on a distant photon. But as the photon approaches the black hole the potentials have a much stronger influence and we can distinguish two cases.

The way the rotation of the black hole affects an incoming photon depends on the direction of L_z relative to the sense of rotation of the black hole. When $aL_z > 0$ the photon moves around the black hole in a prograde orbit, and we have the situation illustrated in Figure 17.1. It follows from (17.21) that

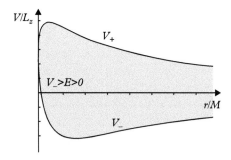

Figure 17.1 *The effective potentials for a photon moving in the equatorial plane of a rotating black hole. The figure illustrates the case where the photon has angular momentum directed in the same sense as the rotation of the black hole (and a = 0.5M). The result for a retrograde photon (with aL_z < 0) is obtained by turning the figure upside down.*

$$V_- = 0 \text{ at } r = 2M, \tag{17.22}$$

$$V_+ = V_- = \frac{aL_z}{2Mr_+} = \Omega_H L_z \text{ at } r = r_+, \tag{17.23}$$

where we have defined the angular velocity of the horizon, Ω_H.

Since the left-hand side of (17.20) must be positive (or zero) we infer that the photon must move either in the region $E > V_+$ or in $E < V_-$. In the first case, the rotation of the black hole leads to nothing new. An incoming photon from infinity can either be scattered by, or plunge into, the black hole (cf. Figure 10.1). But what about the region $V_- > E > 0$, which would also seem to be accessible? An analysis of this possibility requires a bit of care. It turns out that it is not sufficient to require that $E > 0$, as one might (initially) think. The reason is easy to understand. E is the energy measured at infinity, and as we get closer to the black hole this becomes a less useful measure. We need an observer located close to the horizon to do the measurements for us.

A convenient choice of local observer is one that has zero angular momentum and resides at a fixed distance from the black hole (at constant r). Such zero angular momentum observers (ZAMOs; Bardeen *et al.* (1972)) do not follow geodesics, and consequently must maintain their position by means of either a rocket or the hand of a supreme being. The character of a ZAMO means that it must have four-velocity:

$$u^t = A, \qquad u^\varphi = \omega A, \qquad u^r = u^\theta = 0, \tag{17.24}$$

with the unknown coefficient A specified by the usual normalization

$$u_a u^a = -1, \tag{17.25}$$

In the Kerr case, we find that

$$A^2 = \frac{g_{\varphi\varphi}}{(g_{\varphi t})^2 - g_{tt}g_{\varphi\varphi}}. \tag{17.26}$$

We are now better equipped to address the question of photons in the region $V_- > E > 0$ in Figure 17.1. A ZAMO will measure the energy of a photon as

$$E_{\text{zamo}} = -p_a u^a = -(p_t u^t + p_\varphi u^\varphi) = A(E - \omega L_z). \tag{17.27}$$

This 'locally measured' energy must be positive, which means that we must have $E > V_+$ in Figure 17.1. In other words, the $V_- > E > 0$ region is not physically acceptable and we conclude that the case $aL_z > 0$ only involves the kind of photon trajectories we found in the Schwarzschild case.

This is not, however, true for the case $aL_z < 0$, when the photon is inserted in a retrograde orbit around the black hole. (The potentials for this case are obtained by turning those in Figure 17.1 upside down.) We then find from (17.21) that

$$V_+ = 0 \text{ at } r = 2M, \qquad (17.28)$$

and it is clear that some forward-moving photons (that must lie above V_+ according to our previous analysis) can have $E < 0$. That is, negative energy (as measured at infinity) photons may exist close to the black hole! As should be clear from Figure 17.1 these negative energy photons cannot escape to infinity, but the fact that they may exist in the first place has an interesting consequence.

Let us suppose that a pair of photons, the total energy of which is zero, are created in the region $r_+ < r < 2M$. The positive energy photon can escape to infinity, while the negative energy one must be swallowed by the black hole. The net effect would be that rotational energy is carried away from the black hole, and it slows down. This energy extraction process, first suggested by Roger Penrose (Penrose, 1969; Penrose and Floyd, 1971), can be extended to other objects. One can simply assume that a body breaks up into two or more pieces. If one of them is injected into a negative energy orbit the total energy of the remaining pieces must be greater than the energy of the original body, since E is a conserved quantity. The extra energy is mined from the rotation of the black hole.

17.3.2   The ergosphere

As we have just seen, energy arguments may become weird in a region close to a rotating black hole (in the region $r < 2M$ in the equatorial plane). This is the so-called ergosphere, and it is clear that there are interesting effects associated with it.

Consider a photon emitted at some point r in the equatorial plane ($\theta = \pi/2$) of a Kerr black hole. Assume that the photon is initially moving in the $\pm\varphi$ direction. That is, it is inserted in an orbit that is tangent to a circle of constant r. In this situation only dt and $d\varphi$ will be nonzero, and we readily find from $ds^2 = 0$ that

$$\frac{d\varphi}{dt} = -\frac{g_{t\varphi}}{g_{\varphi\varphi}} \pm \sqrt{\left(\frac{g_{t\varphi}}{g_{\varphi\varphi}}\right)^2 - \frac{g_{tt}}{g_{\varphi\varphi}}}. \qquad (17.29)$$

Something interesting happens if g_{tt} changes sign. When $g_{tt} = 0$ we have the two solutions

$$\frac{d\varphi}{dt} = -2\frac{g_{t\varphi}}{g_{\varphi\varphi}}, \qquad \text{and} \qquad \frac{d\varphi}{dt} = 0. \qquad (17.30)$$

The first case corresponds to a photon moving in the direction of the rotation of the black hole. The second solution, however, indicates that a photon sent 'backwards' does not (initially) move at all! The frame dragging has become so strong that the photon cannot move in the direction opposite to the rotation. As light sets the speed limit in relativity, this means that all bodies must rotate along with the black hole. No observers can remain at rest (at constant r, θ, φ) in the ergosphere.

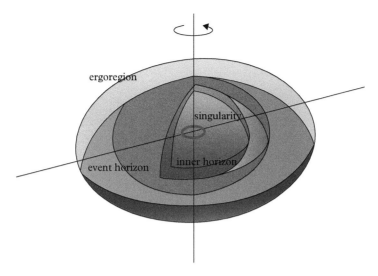

Figure 17.2 *An illustration of the ergosphere that surrounds a rotating black hole.*

As the example suggests, the boundary of the ergosphere follows from $g_{tt} = 0$. In the Kerr case this corresponds to

$$\Delta - a^2 \sin^2 \theta = 0, \tag{17.31}$$

or

$$r_{\text{ergo}} = M \pm \sqrt{M^2 - a^2 \cos^2 \theta}. \tag{17.32}$$

It follows that the ergosphere always lies outside the event horizon (even though it touches the horizon at the poles); see Figure 17.2.

17.3.3   More general orbits: Carter's constant

So far we have only considered photons moving in the equatorial plane of a rotating black hole. This is clearly a very special case, due to the axial symmetry of the Kerr spacetime. What can we say about more general orbits? First of all, we should recall that one must specify four quantities to fully describe the motion of a particle. In the cases covered so far (the orbits around a non-rotating black hole and equatorial trajectories in the Kerr geometry) the four constants of motion were: the rest mass m, the energy E, the angular momentum L_z, and (conveniently) $p_\theta = 0$. The first three remain useful for a general trajectory around a Kerr black hole, but we need a fourth constant of motion. That such a fourth constant exists was first shown by Brandon Carter (1968). Specifically, he proved that the quantity

$$Q = p_\theta^2 + \cos^2\theta \left[a^2(m^2 - E^2) + \frac{L_z^2}{\sin^2\theta} \right] \qquad (17.33)$$

is conserved. This is now known as the *Carter constant*.

Having found the required fourth constant of motion we can write down the equations of motion for a particle in a general orbit around a rotating black hole. First, we note that the equations for $dt/d\lambda$ and $d\varphi/d\lambda$ remain unchanged, so we can still use (17.18) and (17.19). As a second step we solve (17.33) for p_θ, to find

$$\Sigma p^\theta = \Sigma\frac{d\theta}{d\lambda} = \pm\left\{ Q - \cos^2\theta \left[a^2(m^2 - E^2) + \frac{L_z^2}{\sin^2\theta} \right] \right\}^{1/2}. \qquad (17.34)$$

Finally, it follows from the normalization $p_a p^a = -m^2$ that

$$\Sigma p^r = \Sigma\frac{dr}{d\lambda} = \pm V_r^{1/2}, \qquad (17.35)$$

where

$$V_r(r) = [(r^2 + a^2)^2 - a^2\Delta]E^2 - 4aMrEL_z + (a^2 - \Delta)L_z^2 - \Delta(Q + m^2 r^2). \qquad (17.36)$$

Recall that we can rescale the energy and the angular momentum to $\tilde{E} = E/m$ and $\tilde{L}_z = L_z/m$ and identify the affine parameter as $\lambda = \tau/m$, with τ the proper time. The relation (17.35) then becomes an equation for $dr/d\tau$.

It is difficult to assign a specific physical meaning to Carter's fourth constant of motion, Q, in general, but in the non-rotating limit $a \to 0$ we find that $Q + p_\varphi^2$ corresponds to the square of the total angular momentum. We can also deduce some details about particle trajectories for different values of Q. From (17.34) we see that motion is restricted in such a way that

$$Q \geq \cos^2\theta \left[a^2(m^2 - E^2) + \frac{L_z^2}{\sin^2\theta} \right]. \qquad (17.37)$$

This suggests three possibilities:

(i) The simplest case corresponds to $Q = 0$. Motion in the equatorial plane ($\cos^2\theta = 0$) belongs to this case. Furthermore, we can see that if θ is to vary, it must do so in such a way that the square bracket in (17.37) vanishes identically. That is, we must have

$$\sin\theta = \pm\frac{L_z}{a(E^2 - m^2)^{1/2}}. \qquad (17.38)$$

These are trajectories that lie either above or below the equatorial plane. They can touch the $\theta = \pi/2$ plane, but never cross it.

(ii) For $Q > 0$ we can also distinguish two possibilities. Since $\cos\theta = 0$ at the equator it is clear that there will always be solutions that cross the equator. Also, since $\sin\theta$ diverges on the symmetry axis, we see that these orbits cannot reach the axis of symmetry, unless $L_z = 0$. The second possibility also allows motion along the symmetry axis, but this is only possible if $L_z = 0$ and $Q \geq a^2(m^2 - E^2)$.

(iii) Finally, we have the case $Q < 0$. Then there can be no solutions unless $a^2(E^2 - m^2) > L_z^2$. We must also have (Carter, 1968)

$$Q \geq -\left\{\left[a^2(m^2 - E^2)\right]^{1/2} + |L_z|\right\}^2. \tag{17.39}$$

In the case of a strict inequality, θ can vary in a range between the symmetry axis and the equatorial plane (but not touch it!), unless $L_z = 0$, in which case the motion is along the symmetry axis.

To conclude the discussion of particle trajectories in the Kerr geometry let us consider possible circular orbits for massive particles. For simplicity, we again restrict ourselves to the equatorial plane. A circular orbit then corresponds to

$$V_r(r) = 0, \quad \text{and} \quad V_r'(r) = 0. \tag{17.40}$$

Solving these two equations for $\tilde{E}$ and $\tilde{L}_z$ we find (Bardeen *et al.*, 1972)

$$\tilde{E} = \frac{r^{3/2} - 2Mr^{1/2} \pm aM^{1/2}}{r^{3/4}(r^{3/2} - 3Mr^{1/2} \pm 2aM^{1/2})^{1/2}}, \tag{17.41}$$

$$\tilde{L}_z = \pm\frac{M^{1/2}(r^2 \mp 2aM^{1/2}r^{1/2} + a^2)}{r^{3/4}(r^{3/2} - 3Mr^{1/2} \pm 2aM^{1/2})^{1/2}}, \tag{17.42}$$

where the upper signs correspond to prograde orbits ($\tilde{L}_z > 0$), and the lower sign is for retrograde orbits ($\tilde{L}_z < 0$).

What do we learn from this? First, we note that circular orbits cannot exist if the argument of the square root in the denominator is negative. We must have

$$r^{3/2} - 3Mr^{1/2} \pm 2aM^{1/2} \geq 0. \tag{17.43}$$

When this is an equality we have a photon orbit, since $\tilde{E} = E/m \to \infty$. This orbit corresponds to

$$r_{\text{ph}} = 2M\left\{1 + \cos\left[\frac{2}{3}\cos^{-1}(\mp a/M)\right]\right\}. \tag{17.44}$$

In the limit $a = 0$ we recover the familiar photon orbit at $r = 3M$, and for an extreme Kerr black hole $(a = M)$ we find that $r_{\text{ph}} = M$ for a prograde photon but $r_{\text{ph}} = 4M$ for a photon in a retrograde orbit.

For massive particles we know that an orbit is unbound if $\tilde{E} > 1$. Given a small outward perturbation, such a particle will escape. Hence, we can deduce that marginally bound circular orbits correspond to $\tilde{E} = 1$ or

$$r_{\text{mb}} = 2M \mp a = 2M^{1/2}(M \mp a)^{1/2}. \tag{17.45}$$

Any particle that penetrates to $r < r_{\text{mb}}$ must plunge into the black hole. For $a = 0$ we find, $r_{\text{mb}} = 4M$, while for $a = M$ and a prograde orbit we have $r_{\text{mb}} = M$. A retrograde orbit for an extreme Kerr black hole leads to $r_{\text{mb}} = 5.83M$.

Finally, a circular orbit is not stable unless $V_r'' < 0$. This leads to the condition

$$r^2 - 6Mr \pm 8aM^{1/2}r^{1/2} - 3a^2 \geq 0. \tag{17.46}$$

For the extreme case $a = M$ this leads to an innermost stable circular orbit at

$$r_{\text{isco}} = \begin{cases} M, & \text{prograde orbit,} \\ 9M, & \text{retrograde orbit.} \end{cases} \tag{17.47}$$

From this, as well as the results for r_{ph} and r_{mb}, we draw an important conclusion about Kerr black holes: Particles in pro- and retrograde orbits will differ in their view of the 'size' of the black hole. A rotating black hole appears larger to a particle in a retrograde orbit.

Finally, it should be pointed out that the apparent equality between the various circular orbits (and the event horizon) for an extreme $(a = M)$ Kerr black hole, $r_{\text{ph}} = r_{\text{mb}} = r_{\text{isco}} = r_+$, is an artefact of the Boyer–Lindquist coordinates. In reality, the orbits are distinct, and we have $r_{\text{isco}} > r_{\text{mb}} > r_{\text{ph}} > r_+$ (Bardeen et al., 1972).

17.4   The Newman–Penrose formalism

The problem of perturbed spinning black holes is more involved than the corresponding Schwarzschild problem. If we try to expand in tensor spherical harmonics we find that the rotation couples the different multipoles and the equations become messy. To make progress it is useful to consider an alternative strategy. Instead of working with a coordinate-based approach, we introduce a tetrad description where all tensors are projected onto a complete vector basis at each spacetime point. The basis can be chosen to reflect the symmetries of spacetime, which may simplify the mathematics of the observables. The particular description we use is called the Newman–Penrose formalism (Newman and Penrose (1962); Newman and Penrose (2009)).[1] It makes use of a tetrad with four null vectors (two real ones and a complex-conjugate pair). The two real vectors point (asymptotically) radially inwards and outwards, respectively. This makes the description ideal for a study of wave propagation and it is natural to use this approach for black-hole perturbations, as well. In fact, in the Newman–Penrose description the Kerr problem turns out not to be much harder than the Schwarzschild one. This is basically because—from the Newman–Penrose point of view—the Kerr and Schwarzschild geometries are similar (they are both classified as Petrov type D vacuum metrics; see Chandrasekhar (1992)).

In a systematic way, the Newman–Penrose approach starts by separating the Riemann tensor into a traceless part, called the Weyl tensor, and a 'Ricci' part. The Weyl tensor is obtained as

$$C_{abcd} = R_{abcd} - \frac{1}{2}(g_{ac}R_{bd} + g_{bd}R_{ac} - g_{bc}R_{ad} - g_{ad}R_{bc})$$

$$+ \frac{1}{6}(g_{ac}g_{bd} - g_{bc}g_{ad})\,R. \tag{17.48}$$

In four spacetime dimensions, the Riemann tensor has 20 independent components. These are now split between the Ricci and Weyl tensors, which have 10 components each.

Next we introduce a null tetrad $\boldsymbol{l}, \boldsymbol{n}, \boldsymbol{m}, \bar{\boldsymbol{m}}$ (where the bar denotes complex conjugation) which is orthogonal

$$\boldsymbol{l} \cdot \boldsymbol{m} = \boldsymbol{l} \cdot \bar{\boldsymbol{m}} = \boldsymbol{n} \cdot \boldsymbol{m} = \boldsymbol{n} \cdot \bar{\boldsymbol{m}} = 0, \tag{17.49}$$

and null

$$\boldsymbol{l} \cdot \boldsymbol{l} = \boldsymbol{n} \cdot \boldsymbol{n} = \boldsymbol{m} \cdot \boldsymbol{m} = \bar{\boldsymbol{m}} \cdot \bar{\boldsymbol{m}} = 0, \tag{17.50}$$

together with the normalization conditions (which are not required, but convenient)

$$\boldsymbol{l} \cdot \boldsymbol{n} = 1, \quad \text{and} \quad \boldsymbol{m} \cdot \bar{\boldsymbol{m}} = -1. \tag{17.51}$$

Given the tetrad, we can replace the Einstein equations with a set of first-order equations for projected tensor components. This involves introducing the so-called Weyl scalars

---

[1] In the Newman–Penrose approach it is conventional to use the opposite sign convention to that assumed throughout the rest of the book. That is, the metric is taken to have signature $+---$ rather than $-+++$. In order to avoid confusing readers that may consult the original papers we will use this other signature in the discussion of the Newman–Penrose approach. However, as this only affects a fairly self-contained part of the book, this should not be a major issue.

$$\Psi_0 = -C_{abcd}l^a m^b l^c m^d, \tag{17.52}$$

$$\Psi_1 = -C_{abcd}l^a n^b l^c m^d, \tag{17.53}$$

$$\Psi_2 = -C_{abcd}l^a m^b \bar{m}^c n^d, \tag{17.54}$$

$$\Psi_3 = -C_{abcd}l^a n^b \bar{m}^c n^d, \tag{17.55}$$

$$\Psi_4 = -C_{abcd}n^a \bar{m}^b n^c \bar{m}^d. \tag{17.56}$$

These five complex scalars completely specify the ten degrees of freedom of the Weyl tensor. In particular, they encode the radiative aspects of the problem. Their physical interpretation (Szekeres, 1965) is such that Ψ_2 is a 'Coulomb-like' term that represents the gravitational field far from the source. Meanwhile, Ψ_1 and Ψ_3 are in- and outgoing 'longitudinal' radiation and Ψ_0 and Ψ_4 represent in- and outgoing 'transverse' waves. Based on this interpretation, we expect all Weyl scalars apart from Ψ_2 to vanish for an unperturbed black hole. Moreover, as gravitational waves are transverse in Einstein's theory, we are mainly interested in Ψ_0 and Ψ_4. Making contact with the discussion of gravitational waves in Chapter 3, we have (asymptotically)

$$\Psi_4 = -\ddot{h}_+ - i\ddot{h}_\times. \tag{17.57}$$

In order to complete the description, we need to introduce derivatives. This involves using the tetrad as a basis, and thinking of the basis vectors as directional derivatives. Thus, we introduce the derivatives

$$D = e_1 = e^2 = n, \tag{17.58}$$

$$\Delta = e_2 = e^1 = l, \tag{17.59}$$

$$\delta = e_3 = -e^4 = m, \tag{17.60}$$

$$\bar{\delta} = e_4 = -e^3 = \bar{m}. \tag{17.61}$$

Finally, we need the so-called Ricci rotation coefficients (which are related to the covariant derivative; see Chapter 2). At each spacetime point the basis vectors have components $e_{\hat{a}}^a$, where the tetrad indices are indicated by hats. These are, naturally, such that

$$e_{\hat{a}}^a e_a^{\hat{b}} = \delta_{\hat{a}}^{\hat{b}}, \quad \text{and} \quad e_{\hat{a}}^a e_b^{\hat{a}} = \delta_b^a. \tag{17.62}$$

This allows us to project any tensor, A^a say, onto the tetrad in such a way that

$$A_{\hat{a}} = e_{\hat{a}}^a A_a, \tag{17.63}$$

which means that the directional derivatives are simply given by

$$e_{\hat{a}} = e_{\hat{a}}^a \partial_a. \tag{17.64}$$

For scalar fields we get

$$\partial_{\hat{a}}\Phi = e^a_{\hat{a}}\partial_a\Phi, \tag{17.65}$$

while for vectors we need

$$\partial_{\hat{b}}A_{\hat{a}} = e^a_{\hat{a}}e^b_{\hat{b}}\nabla_a A_b + \left[e^a_{\hat{a}}e^c_{\hat{c}}\nabla_a e_{\hat{a}c}\right]A^{\hat{c}} = e^b_{\hat{b}}e^a_{\hat{a}}\nabla_a A_b + \gamma_{\hat{c}\hat{a}\hat{b}}A^{\hat{c}}, \tag{17.66}$$

where the $\gamma_{\hat{c}\hat{a}\hat{b}}$ are the Ricci rotation coefficients (or simply spin coefficients). In the spirit of the formalism, these are designated by symbols

$$\kappa = \gamma_{311}, \quad \rho = \gamma_{314}, \quad \epsilon = \frac{1}{2}(\gamma_{211} + \gamma_{341}), \tag{17.67}$$

$$\sigma = \gamma_{313}, \quad \mu = \gamma_{243}, \quad \gamma = \frac{1}{2}(\gamma_{212} + \gamma_{342}), \tag{17.68}$$

$$\lambda = \gamma_{244}, \quad \tau = \gamma_{312}, \quad \alpha = \frac{1}{2}(\gamma_{214} + \gamma_{344}), \tag{17.69}$$

$$\nu = \gamma_{242}, \quad \pi = \gamma_{241}, \quad \beta = \frac{1}{2}(\gamma_{213} + \gamma_{343}). \tag{17.70}$$

So far, this may seem like an exercise in the Greek alphabet, but we are about to discover the power of the approach.

For the Kerr spacetime, it is customary to use the Kinnersley tetrad (Kinnersley, 1969). This makes use of two of the principal null directions[2] of (the equatorial plane of) the geometry. In particular,

$$\frac{dt}{dr} = \pm\frac{r^2 + a^2}{\Delta}, \tag{17.71}$$

$$\frac{d\varphi}{dr} = \pm\frac{a}{\Delta}. \tag{17.72}$$

We choose the vectors l^a and n^a in these directions and complete the tetrad in such a way that the spin-coefficient ϵ vanishes. Imposing the orthogonality conditions and the normalization, as discussed earlier, we then arrive at

---

[2] These are the eigenvectors of the Weyl tensor, which can be used to classify the nature of a given spacetime (Chandrasekhar, 1992).

$$l^a = \left[\frac{r^2+a^2}{\Delta},1,0,\frac{a}{\Delta}\right],$$

(17.73)

$$n^a = \frac{1}{2\Sigma}\left[r^2+a^2,-\Delta,0,a\right],$$

(17.74)

$$m^a = \frac{1}{\sqrt{2}(r+ia\cos\theta)}\left[ia\sin\theta,0,1,\frac{i}{\sin\theta}\right],$$

(17.75)

and we can write the Kerr metric as

$$g^{ab} = l^an^b + n^al^b - m^a\bar{m}^b - \bar{m}^am^b.$$

(17.76)

Making use of the various definitions, we have the spin coefficients

$$\kappa = \sigma = \lambda = \nu = \epsilon = 0,$$

(17.77)

and

$$\rho = -\frac{1}{r-ia\cos\theta},$$

(17.78)

$$\beta = -\frac{1}{2\sqrt{2}}\cot\theta\,\bar{\rho},$$

(17.79)

$$\pi = \frac{ia}{\sqrt{2}}\sin\theta\rho^2,$$

(17.80)

$$\tau = -\frac{ia}{\sqrt{2}\Sigma}\sin\theta,$$

(17.81)

$$\mu = \frac{\Delta}{2\Sigma}\rho,$$

(17.82)

$$\gamma = \mu + \frac{r-M}{2\Sigma},$$

(17.83)

$$\alpha = \pi - \bar{\beta}.$$

(17.84)

We also find that the only non-vanishing Weyl scalar for a stationary black hole is

$$\Psi_2 = M\rho^3.$$

(17.85)

However, when the black hole is perturbed, quantities that vanished in the background become non-zero. Furthermore, Ψ_2 and the remaining eight spin-coefficients all pick up small increments. For a perturbed Kerr black hole we expect Ψ_0 and Ψ_4 to be small. They represent the gravitational waves that we are interested in.

In order to determine an equation for Ψ_0 we use three of the original Newman–Penrose equations (see, for example, Chandrasekhar (1992)):

$$(\bar{\delta} - 4\alpha + \pi)\Psi_0 - (D - 2\epsilon - 4\rho)\Psi_1 = 3\kappa\Psi_2, \qquad (17.86)$$

$$(\Delta - 4\gamma + \mu)\Psi_0 - (\delta - 4\tau - 2\beta)\Psi_1 = 3\sigma\Psi_2, \qquad (17.87)$$

$$(D - \rho - \bar{\rho} - 3\epsilon + \bar{\epsilon})\sigma - (\delta - \tau + \bar{\pi} - \bar{\alpha} - 3\beta)\kappa = \Psi_0. \qquad (17.88)$$

Somewhat fortuitously, these equations are already 'linearized'. When we use the Kinnersley tetrad the scalars $\Psi_0, \Psi_1, \kappa,$ and σ are all first-order quantities. Hence, we only need to replace all other quantities in the above equations with their background values to arrive at the relevant perturbation equations. We will assume that this is done, and not label first and zeroth order quantities explicitly (as there should be little risk of confusion).

To proceed, we note that the background Ψ_2 satisfies (by virtue of the Bianchi identities)

$$D\Psi_2 = 3\rho\Psi_2, \qquad (17.89)$$

$$\delta\Psi_2 = 3\tau\Psi_2, \qquad (17.90)$$

which means that we can rewrite (17.88) as

$$(D - 4\rho - \bar{\rho} - 3\epsilon + \bar{\epsilon})\sigma\Psi_2 - (\delta - 4\tau + \bar{\pi} - \bar{\alpha} - 3\beta)\kappa\Psi_2 = \Psi_0\Psi_2. \qquad (17.91)$$

The second step towards an equation for Ψ_0 is to eliminate Ψ_1 from (17.86) and (17.87). This can be done (Teukolsky, 1973) by making use of the commutation relation[3]

$$[D - (p+1)\epsilon + \bar{\epsilon} + q\rho - \bar{\rho}](\delta - p\beta + q\tau)$$

$$- [\delta - (p+1)\beta - \bar{\alpha} + \bar{\pi} + q\tau](D - p\epsilon + q\rho) = 0, \qquad (17.92)$$

where p and q are any two constants. If we operate with $(\delta + \bar{\pi} - \bar{\alpha} - 3\beta - 4\tau)$ on (17.86) and $(D - 3\epsilon + \bar{\epsilon} - 4\rho - \bar{\rho})$ on (17.87), and then subtract the results, all terms in Ψ_1 can be made to vanish by means of (17.92) with $p = 2$ and $q = -4$. This way we arrive at the final equation (after using (17.91) to remove κ and σ)

$$[(D - 3\epsilon + \bar{\epsilon} - 4\rho - \bar{\rho})(\Delta - 4\gamma + \mu)$$

$$- (\delta + \bar{\pi} - \bar{\alpha} - 3\beta - 4\tau)(\bar{\delta} + \pi - 4\alpha) - 3\Psi_2]\Psi_0 = 0. \qquad (17.93)$$

This is a single decoupled (second-order) equation for the Weyl scalar Ψ_0.

An analogous equation for Ψ_4 can be derived by using the fact that the Newman–Penrose equations are invariant under the change $l \rightarrow n$ and $m \rightarrow \bar{m}$. With this interchange we also get

---

[3] One can show that this relation holds for any type D metric.

$$D \to \Delta \quad , \quad \delta \to \bar{\delta},$$

(17.94)

and

$$\epsilon \to -\gamma \quad , \quad \pi \to -\tau \quad , \quad \rho \to -\mu \quad , \quad \alpha \to -\beta,$$

(17.95)

and vice versa. Hence, we have

$$[(\Delta + 3\gamma - \bar{\gamma} + 4\mu + \bar{\mu})(D + 4\epsilon - \rho)$$

$$- (\bar{\delta} - \bar{\tau} + \bar{\beta} + 3\alpha + 4\pi)(\delta - \tau + 4\beta) - 3\Psi_2]\Psi_4 = 0 .$$

The two perturbation equations (17.93) and (17.96) for Ψ_0 and Ψ_4 are (it turns out) independent of both the choice of tetrad and the specific coordinates. They provide gauge-invariant measures of the gravitational-wave content of a perturbed Kerr spacetime.

17.5 The Teukolsky equation

In the case of non-spinning black holes we showed that the perturbations could be described in terms of (relatively simple) decoupled wave equations (see Chapter 16). Given the the added complexities, one might not expect this to be possible also for Kerr black holes. That it can (nevertheless) be done was first demonstrated by Saul Teukolsky (1973).

We took the first steps of the calculation in the previous section. Completing the argument (essentially by making the various quantities explicit for the Kerr metric), and allowing for the presence of matter, we arrive at the so-called Teukolsky equation, which describes the evolution of scalar, electromagnetic, and gravitational perturbations of a Kerr black hole. This equation takes the form (Teukolsky, 1973)

$$\left[\frac{(r^2 + a^2)^2}{\Delta} - a^2 \sin^2\theta\right]\frac{\partial^2\psi}{\partial t^2} + \frac{4Mar}{\Delta}\frac{\partial^2\psi}{\partial t\partial\varphi} + \left[\frac{a^2}{\Delta} - \frac{1}{\sin^2\theta}\right]\frac{\partial^2\psi}{\partial\varphi^2}$$

$$- \Delta^{-s}\frac{\partial}{\partial r}\left(\Delta^{s+1}\frac{\partial\psi}{\partial r}\right) - \frac{1}{\sin\theta}\frac{\partial}{\partial\theta}\left(\sin\theta\frac{\partial\psi}{\partial\theta}\right) - 2s\left[\frac{a(r - M)}{\Delta} + \frac{i\cos\theta}{\sin^2\theta}\right]\frac{\partial\psi}{\partial\varphi}$$

$$- 2s\left[\frac{M(r^2 - a^2)}{\Delta} - r - ia\cos\theta\right]\frac{\partial\psi}{\partial t} + (s^2\cot^2\theta - s)\psi = 4\pi\Sigma T.$$

(17.96)

Here $a \leq M$ represents (as usual) the black hole's angular momentum and s is the spin weight of the perturbing field ($s = 0, \pm 1, \pm 2$). In the case of gravitational perturbations (spin $s = +2$), we have

$$\psi = \Psi_0,$$

(17.97)

$$T = 2(\delta + \bar{\pi} - \bar{\alpha} - 3\beta - 4\tau)[(D - 2\epsilon - 2\bar{\rho})T_{lm} - (\delta + \bar{\pi} - 2\bar{\alpha} - 2\beta)T_{ll}]$$
$$+ 2(D - 3\epsilon + \bar{\epsilon} - 4\rho - \bar{\rho})[(\delta + 2\bar{\pi} - 2\beta)T_{lm} - (D - 2\epsilon + 2\bar{\epsilon} - \bar{\rho})T_{mm}]. \quad (17.98)$$

We have projected the matter stress–energy tensor onto the tetrad in such a way that $T_{ll} = T_{ab}l^a l^b$ and so on. Similarly,

$$\psi = \frac{1}{\rho^4}\Psi_4, \quad (17.99)$$

$$\rho^4 T = 2(\Delta + 3\gamma - \bar{\gamma} + 4\mu + \bar{\mu})[(\bar{\delta} - 2\bar{\tau} - 2\alpha)T_{nm} - (\Delta + 2\gamma - 2\bar{\gamma} + \bar{\mu})T_{\bar{m}\bar{m}}]$$
$$+ 2(\bar{\delta} - \bar{\tau} + \bar{\beta} + 3\alpha + 4\pi)[(\Delta + 2\gamma - 2\bar{\mu})T_{n\bar{m}} - (\bar{\delta} - \bar{\tau} + 2\bar{\beta} + 2\alpha)T_{nn}], \quad (17.100)$$

for $s = -2$.

If the existence of the decoupled wave equation (17.96) was a surprise, the fact that it is possible to separate the variables, just as one can do for Schwarzschild black holes, was totally unexpected.

Let us consider the source-free case $(T = 0)$ and use an Ansatz inspired by the Fourier transform

$$_s\psi_{lm} = {}_sR_{lm}(r, \omega)\, {}_sS_{lm}(\theta, \omega)e^{-i\omega t + im\varphi}. \quad (17.101)$$

Then the equations that govern the radial function $_sR_{lm}$ and the angular function $_sS_{lm}$ are (suppressing all indices for clarity)

$$\Delta^{-s}\frac{d}{dr}\left(\Delta^{s+1}\frac{dR}{dr}\right) - VR = 0, \quad (17.102)$$

with

$$V(r) = \frac{K^2 - 2is(r - M)K}{\Delta} + 4is\omega r - \lambda, \quad (17.103)$$

and

$$\frac{1}{\sin\theta}\frac{d}{d\theta}\left(\sin\theta\frac{dS}{d\theta}\right) + \left[a^2\omega^2\cos^2\theta - \frac{m^2}{\sin^2\theta} - 2a\omega s\cos\theta\right.$$
$$\left. - \frac{2ms\cos\theta}{\sin^2\theta} - s^2\cot^2\theta + E - s^2\right]S = 0. \quad (17.104)$$

We have also introduced

$$K = (r^2 + a^2)\omega - am, \quad (17.105)$$

and

$$\lambda = E - s(s+1) + a^2\omega^2 - 2am\omega. \tag{17.106}$$

The angular equation (17.104) reduces to that for spin-weighted spherical harmonics discussed in Chapter 16 when $a = 0$. The main difference is that the eigenvalues are now frequency dependent. Given the connection, the solutions to (17.104) are referred to as 'spin-weighted spheroidal harmonics'.

The Kerr problem thus reduces to an analysis of the radial equation (17.102). One complication is immediately obvious. The 'effective' potential explicitly depends on the frequency ω. It is also clear that rotation breaks the angular degeneracy that was present for Schwarzschild black holes. Hence, we must consider all $2l+1$ values of m for each integer $l \geq |s|$.

Reassuringly, as in the Schwarzschild case, it is useful to introduce a tortoise coordinate. We now have

$$\frac{d}{dr_*} = \frac{\Delta}{r^2 + a^2}\frac{d}{dr}. \tag{17.107}$$

With this definition, and introducing a new dependent variable

$$u = (r^2 + a^2)^{1/2}\Delta^{s/2}R, \tag{17.108}$$

the radial equation (17.102) transforms into

$$\frac{d^2u}{dr_*^2} + \left[\frac{K^2 - 2is(r-M)K + \Delta(4ir\omega s - \lambda)}{(r^2+a^2)^2} - G^2 - \frac{dG}{dr_*}\right]u = 0, \tag{17.109}$$

where the function G is

$$G = \frac{s(r-M)}{r^2+a^2} + \frac{r\Delta}{(r^2+a^2)^2}. \tag{17.110}$$

The asymptotic behaviour of (17.109) implies that the solutions behave as

$$u \sim r^{\pm s}e^{\mp i\omega r_*}, \tag{17.111}$$

or

$$_sR_{lm} \sim \begin{cases} e^{+i\omega r_*}/r^{2s+1}, \\ e^{-i\omega r_*}/r, \end{cases} \quad \text{as } r \to \infty. \tag{17.112}$$

The power-law behaviour of the out- and ingoing-wave solutions is in accordance with the so-called 'peeling theorem' (Newman and Penrose, 1962).

Close to the event horizon, r_+ (the outer solution to $\Delta = 0$), the behaviour is

$$u \sim \Delta^{\pm s/2} e^{\pm i \varpi r_*}, \tag{17.113}$$

which means that

$$_s R_{lm} \sim \begin{cases} e^{+i\varpi r_*}, \\ \Delta^{-s} e^{-i\varpi r_*}, \end{cases} \quad \text{as } r \to r_+. \tag{17.114}$$

Here ϖ is defined as $\varpi = \omega - m\Omega_H$.

For rotating black holes the formulation of the physical boundary condition at the horizon requires some care. A wave must be ingoing 'in the frames of all physical observers' (which will be dragged along by the rotation). One can show that this means that $_s R_{lm} \sim \Delta^{-s} e^{-i\varpi r_*}$ as $r \to r_+$ (Teukolsky, 1973).

The asymptotic behaviour from (17.112) demonstrates that the Penrose process for extracting rotational energy from a spinning black hole has a simple wave analogue. Imagine setting up a scattering experiment. As a wave with a given frequency approaches the black holes, some part of it will be scattered back to infinity and some of it will disappear through the horizon. We can quantify this using (17.112) (and its complex conjugate). This leads to

$$\left(1 - \frac{m\Omega_H}{\omega}\right) T = 1 - R, \tag{17.115}$$

where $T = 1/|A_{\text{in}}|^2$ and $R = |A_{\text{out}}/A_{\text{in}}|^2$ are the transmission and reflection coefficients, respectively (see the discussion in Chapter 16). We see that we can have $R > 1$, i.e. more waves coming out of the black hole than we threw into it, if

$$0 < \omega < m\Omega_H = \frac{ma}{2Mr_+}. \tag{17.116}$$

This effect is known as superradiance (Press and Teukolsky, 1972). The energy gain in the scattered waves corresponds to extraction of rotational energy from the black hole.

The amplification tends to be small, but if we fine-tune the setup the gravitational-wave amplitude may be almost doubled (although this requires a specific frequency and a near extreme Kerr black hole). It is interesting to ask whether superradiance may have astrophysical relevance. This may seem somewhat far fetched, but it is conceivable. In particular, it is known that perturbations due to a massive scalar field become unstable (essentially due to feedback coupling to the superradiance). This may have interesting implications for ultralight bosons, should they exist (Brito *et al.*, 2017).

Table 17.1 Quasinormal-mode frequencies ($\omega_n M$) for gravitational perturbations of Kerr black holes with various spin rates. The results are for $l = 2$, and illustrates that the co-rotating modes ($m > 0$) become long(er) lived as the black hole approaches extreme rotation. In contrast, the counter-rotating modes ($m < 0$) are much less affected by the rotation. (Data from Glampedakis and Andersson (2003).).

a/M	$m = 2$	$m = 1$	$m = 0$	$m = -1$	$m = -2$
0	$0.37367 - 0.08896i$	$0.37367 - 0.08896i$	$0.37367 - 0.08896i$	$0.37367 - 0.08896i$	$0.37367 - 0.08896i$
0.10	$0.38702 - 0.08871i$	$0.38043 - 0.08880i$	$0.37403 - 0.08890i$	$0.36781 - 0.08900i$	$0.36177 - 0.08911i$
0.20	$0.40215 - 0.08831i$	$0.38825 - 0.08849i$	$0.37512 - 0.08870i$	$0.36274 - 0.08894i$	$0.35105 - 0.08918i$
0.30	$0.41953 - 0.08773i$	$0.39733 - 0.08800i$	$0.37699 - 0.08835i$	$0.35837 - 0.08876i$	$0.34133 - 0.08918i$
0.40	$0.43984 - 0.08688i$	$0.40798 - 0.08726i$	$0.37968 - 0.08783i$	$0.35463 - 0.08848i$	$0.33246 - 0.08913i$
0.50	$0.46412 - 0.08564i$	$0.42063 - 0.08617i$	$0.38332 - 0.08707i$	$0.35149 - 0.08809i$	$0.32431 - 0.08903i$
0.60	$0.49405 - 0.08377i$	$0.43597 - 0.08456i$	$0.38805 - 0.08600i$	$0.34891 - 0.08757i$	$0.31678 - 0.08889i$
0.70	$0.53260 - 0.08079i$	$0.45512 - 0.08209i$	$0.39413 - 0.08445i$	$0.34687 - 0.08688i$	$0.30981 - 0.08872i$
0.80	$0.58602 - 0.07563i$	$0.48023 - 0.07796i$	$0.40192 - 0.08216i$	$0.34536 - 0.08600i$	$0.30331 - 0.08851i$
0.90	$0.67163 - 0.06486i$	$0.51629 - 0.06980i$	$0.41200 - 0.07848i$	$0.34436 - 0.08487i$	$0.29724 - 0.08828i$
0.99	$0.87086 - 0.02951i$	$0.57272 - 0.04621i$	$0.42369 - 0.07270i$	$0.34389 - 0.08355i$	$0.29211 - 0.08805i$

17.6 Kerr quasinormal modes

Just like a non-rotating black hole, a perturbed Kerr black hole settles down in a way that is dominated by its quasinormal modes. The problem is more involved than the Schwarzschild one, but we can nevertheless determine the quasinormal modes. The main difference is that, when the black hole rotates, the azimuthal degeneracy is broken. For a given multipole l there are $2l + 1$ distinct modes for each of the Schwarzschild modes. These modes correspond to specific values of m, where $-l \leq m \leq l$.

It is relevant to consider the behaviour of the Kerr quasinormal modes for increasing spin. The sample of numerical results provided in Table 17.1 show that rotation has a pronounced effect on modes with $m = l > 0$. These can be thought of as co-rotating with the black hole (in analogy with the discussion of waves in rotating stars in Chapter 13). In the limit $a \to M$ the co-rotating modes become long-lived (Im $\omega_n M \to 0$) and accumulate at the frequency $m\Omega_H$ In contrast, the counter-rotating modes (with $m < 0$) are much less affected by rotation. They remain relatively close to the Schwarzschild quasinormal modes for all values of a. The general behaviour of the slowest damped $l = 2$ modes is illustrated in Figure 17.3.

In order to compare the theory to observations—and facilitate parameter extraction— it is useful to have a simple approximation of the numerical results. The leading co-rotating quadrupole Kerr mode is well approximated by a frequency (Echeverria, 1989; Finn, 1992)

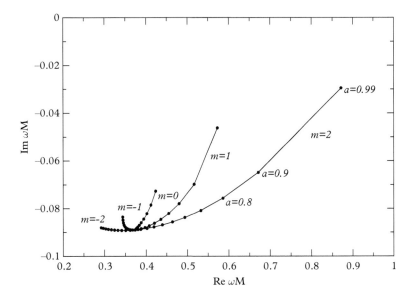

Figure 17.3 *The trajectories (Im ωM as a function of Re ωM) for the slowest damped $l = 2$ gravitational quansinormal mode of a Kerr black hole. Results are shown for a range of rotation rates, $0 \leq a/M \leq 0.99$. The position of each mode at spin $a/M = 0,0.1,0.2...0.9,0.99$ is represented by dots. The data is taken from Table 17.1.*

$$f \approx 32 \left[1 - \frac{63}{100}(1 - a/M)^{3/10} \right] \left(\frac{M_\odot}{M} \right) \text{ kHz} , \qquad (17.117)$$

with damping time

$$\tau = \frac{1.9 \times 10^{-2}}{(1 - a/M)^{9/10}} \left[1 - \frac{63}{100}(1 - a/M)^{3/10} \right]^{-1} \left(\frac{M}{M_\odot} \right) \text{ ms.} \qquad (17.118)$$

In principle, the fact that the modes of a black hole rotating close to the extreme limit may be very long lived could—provided that these modes are actually excited in a realistic process (which is not at all clear)—improve the chances of detection. We have already seen that the effective amplitude of a periodic signal improves as the square root of the number of observed cycles. In this sense, the presence of long-lived modes might seem promising. However, one can argue that these modes are difficult to excite to a large amplitude (given the mismatch between the timescale of the typical dynamics and the damping time; see Sasaki and Nakamura (1990) and Glampedakis and Andersson (2001)).

17.7   GW150914: a faint fingerprint

An observational identification of quasinormal ringdown would provide clear evidence for the presence of a black hole and a direct way of inferring its parameters (Finn, 1992; Baibhav *et al.*, 2018). Excitingly, the breakthrough detection event GW150914 provided the first opportunity to test this idea. However, the analysis of the problem is complicated by the spin effects.

The spin of the binary companions in a merger impacts on both the amplitude and the phase of the gravitational-wave strain. If the two spins are misaligned with the orbital angular momentum, L, they cause the orbital plane to precess around the (almost constant) direction of the total angular momentum, $J = L + S_1 + S_2$ (the sum of the orbital angular momentum and the contributions from the spin of the two binary partners). In the case of GW150914 the limited signal-to-noise (~ 24) made it difficult to untangle the individual black-hole spins. The results were, in fact, consistent with a non-precessing fit. In essence, the value of the final black-hole spin may have been a simple consequence of the conservation of angular momentum (with the orbital angular momentum converted into the spin of the final black hole). As a result, the final spin is more precisely determined than the spins of the individual black holes.

In the case of GW150914, the two black holes merged to form a remnant with mass $62M_\odot$ and spin $a/M \approx 0.67$; see Figure 17.4. The inferred black-hole spin was a first in astronomy[4]—a tight constraint on the rotation rate of a black hole.

From the posterior distributions of the mass and spin of the final black hole one can predict the frequency and damping time of the leading quasinormal mode (most likely

---

[4] The spin of accreting black holes has been inferred via observations of broadened iron lines (Miniutti *et al.*, 2004), but this method has uncertain systematics.

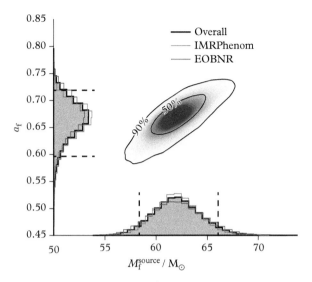

Figure 17.4 *Probability density functions for the source-frame mass and the dimensionless spin of the black hole formed in the GW150914 merger event. Results are shown for three different waveform models (see the original paper for details). The dashed vertical lines mark the 90% credible interval for the probability density. The two-dimensional plot shows the contours of the 50 and 90% credible regions. (Reproduced from Abbott et al. (2016c), Creative Commons Attribution 3.0 licence.)*

the fundamental $l = m = 2$ mode, cf., Table 17.1). The GW150914 observations suggest that $f = 251 \pm 8$ Hz and $\tau = 4.0 \pm 0.3$ ms at 90% confidence. Figure 17.5 shows the 90% credible contours in the frequency-damping rate plane as a functions of a possible time offset to the start of the ringdown phase, alongside the contour for the slowest-damped quasinormal mode from (17.117) and (17.118) for the estimated mass and spin parameter of the remnant. In essence, the contours begin to overlap with the theory prediction about 3 ms after the merger. This does not in itself prove the observation of this particular oscillation mode, but the signal is at least consistent with numerical simulations.

The GW150914 results demonstrate the potential of future observations. Given that a strong constraint on both the frequency and the damping time allows an independent extraction of the final black hole's mass and spin, the identification of additional mode features provide a test of the theory. The basic idea is analogous to the argument for the Double Pulsar; see Figure 10.3. In principle, a multi-mode observation may allow us to test of the no-hair theorem in general relativity (Berti *et al.*, 2016; Dhanpal *et al.*, 2018). However, one should be aware that this is a non-trivial exercise. In particular, the quasinormal-mode contribution to the signal is not easily separated from (say) the merger part. Higher order modes are more rapidly damped and therefore play a more prominent role at early times, precisely when it may not be clear to what extent the signal is accurately represented by a sinusoidal ringdown. A detection of comparably

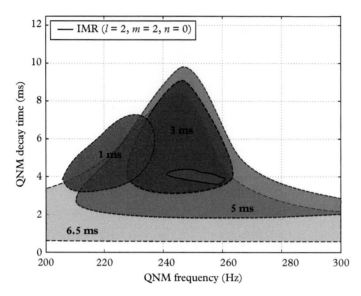

Figure 17.5 *Regions of 90% credibility (in the joint posterior distributions) for the frequency (f) and damping time (τ) of the main quasinormal l = m = 2 mode of the black hole formed in the GW150914 event. Results are shown assuming that the ringdown starts at various times after the merger (as indicated). The red filled-in region shows the 90% credible region for the frequency and decay time of the slowest damped mode as derived from the posterior distributions of the remnant mass and spin parameters (shown in Figure 17.4). (Based on the results of Abbott et al. (2016d).)*

weak overtone quasinormal modes may require third-generation detectors or a space-based instrument, like LISA.

Given the fundamental importance of this kind of observation, it is interesting to consider other possible scenarios. Suppose we establish that we are not seeing a black hole as described by Einstein's theory, then what? It is well known that you cannot squeeze a material object much inside $R \approx 3M$ (ignoring rotation) without invoking 'non-standard' physics. Hence, alternative scenarios tend to be speculative, introducing deviations from the classical black-hole solution on the horizon scale. This can be done by introducing a surface (e.g. supported by a negative pressure (Mazur and Mottola, 2015)) or by invoking quantum effects (Saravani *et al.,* 2014), and typically leads to a different quasinormal-mode response, e.g. revealing itself through the emergence of secondary pulses in the late-time waveform (Cardoso *et al.* (2016); Abedi *et al.* (2017)). These so-called *echoes* are likely to be faint and difficult to observe (see Abedi *et al.* (2017) and Westerweck *et al.* (2018) for recent contributions to the discussion).

18

Relativistic asteroseismology

We have seen that a detailed understanding of neutron star physics necessarily involves all four fundamental forces of nature. This makes realistic modelling complicated. At the same time it offers a unique opportunity to explore extreme physics. By developing more precise neutron star models and matching them to observations we can probe physics in a regime that may never be within the reach of terrestrial laboratories. A key part of this enterprise involves extracting neutron-star parameters from observed data to constrain the underlying microphysics, e.g. the equation of state. In this respect, gravitational-wave asteroseismology offers a promising strategy (Andersson and Kokkotas, 1998; Benhar et al., 2004; Doneva et al., 2013).

Neutron stars have rich oscillation spectra, with more or less distinct families of modes depending, often rather sensitively, on the detailed physics (see Chapter 13). However, if we want to develop an accurate strategy for inverting observed data to constrain the physics, we need reliable models. Given our relative ignorance of issues like the state and composition of matter at extreme densities this is a severe challenge, but one thing is absolutely clear: the models have to be relativistic. The Newtonian models we considered in, for example, Chapter 13 provide qualitative insight and help us decide where we should focus our efforts, but it makes little sense to couple a Newtonian study to realistic microphysics. Since relativistic effects have crucial impact on issues like the neutron star radius for a given central density, it would make no sense to base a quantitative strategy for parameter extraction on Newtonian results.

18.1 Relativistic fluid perturbations

In order to develop realistic models for neutron star dynamics we need to account for the different physics aspects—outlined in Chapter 12—in general relativity. As a first step towards this goal, it is natural to consider the problem for a perfect fluid. In this case, the unperturbed equilibrium is provided by a solution to the Tolman–Oppenheimer–Volkoff equations from Chapter 4. The perturbed fluid motion is governed by the linearized Einstein equations

$$\delta G_a{}^b = 8\pi \delta T_a{}^b, \tag{18.1}$$

Gravitational-Wave Astronomy: Exploring the Dark Side of the Universe. Nils Andersson, Oxford University Press (2020).
© Nils Andersson. DOI: 10.1093/oso/9780198568032.001.0001

in terms of the Eulerian perturbation (δ) of the Einstein tensor $G_a{}^b$ and the stress–energy tensor $T_a{}^b$. We may also want to use the perturbed fluid equations of motion

$$\nabla_b \delta T_a{}^b = 0, \tag{18.2}$$

where it is worth noting that the (local) perturbation (δ) commutes with the covariant derivative.

In the first instance, we will consider non-rotating perfect fluid stars. This means that the perturbations can be expressed in terms of deviations from a metric of the Schwarzschild form (see Chapter 4), with metric coefficients ν and λ (functions of the radial coordinate r only), and involves working with the covariant derivative with respect to this spacetime, as well (as outlined in Chapter 3).

In the case of a perfect fluid, we have

$$T_a{}^b = (p + \varepsilon) u_a u^b + p \delta_a^b, \tag{18.3}$$

where p is the (isotropic) pressure and ε is the energy density. That is,

$$\delta T_a{}^b = (\delta\varepsilon + \delta p) u_a u^b + \delta p \delta_a^b + (p + \varepsilon)\left(u^b \delta u_a + u_a \delta u^b\right). \tag{18.4}$$

In principle, the information in (18.2) is already contained in the perturbed Einstein equations (18.1), but the relation between the results is not obvious. This means that, in practice, it may be useful to work with different combinations of the equations. To some extent it is a matter of choice. However, we have to tread carefully if we want to make simplifications, like the Cowling approximation. In the Newtonian case this involved omitting the perturbed gravitational potential (with the argument that the fluid motion was predominantly horizontal). In the relativistic case it may seem natural to take this to mean that we can ignore the metric perturbations, i.e. take $\delta g_{ab} = h_{ab} = 0$. But this could be too drastic. After all, the left-hand side of (18.1) then vanishes identically which would seem to suggest that the right-hand vanishes, as well. This (obviously) makes the 'approximation' meaningless. However, we would have drawn similar conclusions in the Newtonian case if we had taken the Poisson equation for the gravitational potential as our starting point, so there is no reason to panic. We can simply choose to ignore the Einstein equations and work with (18.2) instead. In this case, the problem remains 'consistent' even if we leave out the metric variations, and it closely resembles its Newtonian counterpart. However, we still have to be careful. The stress–energy tensor involves components associated with the fluid momentum, and these may be directly associated with changes in the spacetime. As a result one can argue that the natural generalization of the Cowling approximation would be to retain the 'momentum part' of the perturbed metric (Finn, 1988). This would involve only setting $\delta g_{ij} = 0$. We will see an example where this distinction becomes important later.

In the case of a non-rotating (spherically symmetric) star the perturbation problem splits into axial and polar sectors, just like in the black-hole case. In the polar case (and Regge–Wheeler gauge), we have (as in Chapter 16)

$$h_{ab} = \begin{pmatrix} e^\nu H_0 & H_1 & 0 & 0 \\ \text{sym} & e^\lambda H_2 & 0 & 0 \\ 0 & 0 & r^2 K & 0 \\ 0 & 0 & 0 & r^2 \sin^2\theta K \end{pmatrix} Y_l^m, \tag{18.5}$$

where the perturbations are assumed to be functions of t and r. Due to the spherical symmetry of the problem, the different multipoles do not couple. In fact, we can, without loss of generality, assume that the perturbations are axisymmetric $(m = 0)$.

We also need the perturbed fluid velocity. The polar components we need are (see Chapter 13, although we are now using a coordinate basis)[1]

$$\delta u^r = W, \tag{18.6}$$

$$\delta u^\theta = \frac{1}{r^2} V \partial_\theta Y_l^m, \tag{18.7}$$

and

$$\delta u^\varphi = \frac{1}{r^2 \sin^2\theta} V \partial_\varphi Y_l^m. \tag{18.8}$$

After cranking through the algebra, we find that (18.1) leads to a set of coupled differential equations (see, for example, Thorne and Campolattaro (1967), Lindblom and Detweiler (1983), or Kojima (1992))

$$\delta G_t^{\ t} = 8\pi \delta T_t^{\ t} \longrightarrow$$

$$e^{-\lambda} r^2 K'' + e^{-\lambda} \left(3 - \frac{r\lambda'}{2} \right) r K' - \frac{n}{2} K - e^{-\lambda} r H_2'$$

$$- \left[\frac{1}{2} l(l+1) + e^{-\lambda} (1 - r\lambda') \right] H_2 = -8\pi r^2 \delta\varepsilon, \tag{18.9}$$

where primes are radial derivatives, dots are time derivatives, and we have used $n = (l-1)(l+2)$ (recall Eq. 16.38),

$$\delta G_r^{\ r} = 8\pi \delta T_r^{\ r} \longrightarrow$$

$$r^2 e^{-\nu} \ddot{K} - 2e^{-(\nu+\lambda)} r \dot{H}_1 - e^{-\lambda} \left(1 + \frac{r\nu'}{2} \right) r K' + \frac{n}{2} K$$

$$+ e^{-\lambda} r H_0' - \frac{1}{2} l(l+1) H_0 + e^{-\lambda} (1 + r\nu') H_2 = -8\pi r^2 \delta p, \tag{18.10}$$

---

[1] In principle, all perturbation variables should have indices l and m to indicate which multipole they refer to, but as we know the multipoles decouple in spherical symmetry we suppress these indices here.

$$\delta G_t^{\,r} = 8\pi \,\delta T_t^{\,r} \longrightarrow$$

$$\dot{K}' + \frac{1}{r}(2 - rv')\dot{K} - \frac{1}{r}\dot{H}_2 - \frac{l(l+1)}{2r^2}H_1 = 8\pi(p+\varepsilon)e^{\lambda+v/2}W. \qquad (18.11)$$

We also have

$$\delta G_\theta^{\,\theta} + \delta G_\varphi^{\,\varphi} = 8\pi\left(\delta T_\theta^{\,\theta} + \delta T_\varphi^{\,\varphi}\right) \longrightarrow$$

$$r^2 e^{-\lambda}\left(H_0' - K'\right) + r^2 e^{-v}\left(\ddot{K} + \ddot{H}_2\right) - 2r^2 e^{-\lambda-v}\dot{H}_1' - re^{-\lambda}\left[\frac{1}{2}r(v'-\lambda')+2\right]K'$$

$$+ e^{-\lambda}\left(1 + rv' - \frac{1}{2}r\lambda'\right)rH_0' + re^{-\lambda}\left(1 + \frac{1}{2}rv'\right)H_2'$$

$$+ \frac{1}{2}l(l+1)(H_2 - H_0) + 2e^{-\lambda}\left(e^\lambda - 1 - rv'\right)H_2 = -16\pi r^2 \delta p, \qquad (18.12)$$

$$\delta G_\theta^{\,\theta} - \delta G_\varphi^{\,\varphi} = 8\pi\left(\delta T_\theta^{\,\theta} - \delta T_\varphi^{\,\varphi}\right) \longrightarrow$$

$$H_2 - H_0 = 0, \qquad (18.13)$$

$$\delta G_t^{\,\theta} = 8\pi\,\delta T_t^{\,\theta} \longrightarrow$$

$$-e^{-\lambda}H_1' + \dot{H}_2 + \dot{K} + \frac{1}{2}(\lambda' - v')e^{-\lambda}H_1 = 16\pi(p+\varepsilon)e^{v/2}V, \quad (18.14)$$

and finally

$$\delta G_r^{\,\theta} = 8\pi\,\delta T_r^{\,\theta} \longrightarrow$$

$$-e^{-v}\dot{H}_1 - K' + e^{-v}\left[e^v H_0\right]' + \frac{1}{r}\left(1 + \frac{r}{2}v'\right)(H_2 - H_0) = 0. \quad (18.15)$$

Two of these equations, (18.11) and (18.14), provide the perturbed velocity components (W and V) in terms of the metric variables. The remaining equations can be expressed as coupled wave equations for H_0 and K. The equations are, however, rather messy and we do not learn very much from them (see, for example, Kojima (1992) or Allen *et al.* (1998) for explicit expressions). It is easy to count the degrees of freedom to confirm that we have a well-posed problem, but if we choose to work with the Einstein equations and not involve the fluid equations of motion, then the fluid dynamics becomes a bit convoluted. This is not surprising as we expect the problem to involve coupled sound waves and gravitational waves. The first set is not naturally expressed in terms of the spacetime metric. Despite these caveats, the set of equations we have written down provides a useful system for studying the oscillations of a relativistic star (Detweiler and Lindblom, 1985; Andersson *et al.*, 1995; Krüger *et al.*, 2015).

18.2   f- and p-modes in relativity

The problem of relativistic neutron star seismology was first explored in the late 1960s (Thorne and Campolattaro, 1967). As in the Newtonian case, the typical strategy involves assuming a harmonic time-dependence $e^{i\omega t}$ for all perturbed quantities and searching for mode solutions to the perturbation equations. In the relativistic case, these solutions are no longer normal modes. Instead, they are quasinormal modes, corresponding to purely outgoing (gravitational) waves at spatial infinity. The analysis of the exterior problem is completely analogous to that for black holes (see Chapter 16). In essence, one can extract both an oscillation frequency and a characteristic damping time for each stellar oscillation mode.

Much of the literature has focussed on perfect fluid stars, often without consideration of the interior composition and actual state of matter. In fact, many studies are based on phenomenological models (typically polytropes) which only capture the rough properties of a realistic equation of state. Nevertheless, this has led to an understanding of the nature of the stellar spectrum, like the fundamental f-mode and the acoustic p-modes. The gravity g-modes (arising because of composition gradients in a mature neutron star) have also been considered (Finn, 1988; Krüger *et al.*, 2015), as has the role of the crust elasticity (Schumaker and Thorne, 1983; Samuelsson and Andersson, 2007), superfluidity (Comer *et al.*, 1999; Lin *et al.*, 2008), and the star's magnetic field (Gabler *et al.*, 2012; Colaiuda and Kokkotas, 2012). We will not discuss all these aspects. Rather, we will focus on the general strategy for solving the perturbation equations.

When the problem is approached in the frequency domain, it is common to recast it as a set of coupled first-order differential equations (Lindblom and Detweiler, 1983; Detweiler and Lindblom, 1985). For polar perturbations, we end up with a fourth-order system so we need four independent variables. From the set of equations, (18.9)–(18.35), it is easy to see that a natural choice may be to work with $[K, H_0, H_1, \delta p]$. One would typically integrate this system from the centre of the star, implementing regularity conditions through a Taylor expansion for small values of r (see Lindblom and Detweiler (1983)), repeating the procedure with linearly independent initial vectors in order to provide a basis that can be used to express the general solution. In order to avoid numerical difficulties, it is also common to initiate an integration from the surface of the star inwards. Finally, the obtained solutions are matched at an interior point. This requires the solution of a linear system for the 'amplitudes' of the different solutions. The boundary conditions at the centre and the surface of the star determine a unique (up to amplitude) solution throughout the star's interior. At the surface, this solution is matched to an exterior solution obtained from the equations from black-hole perturbation theory (e.g. the Zerilli equation from Chapter 16). The procedure is iterated until one finds a complex frequency such that the amplitude of the ingoing wave amplitude at infinity vanishes. The computational strategy is described in detail by, for example, Lindblom and Detweiler (1983) and Krüger *et al.* (2015).

In order to facilitate gravitational-wave astronomy, it is relevant to explore the 'inverse problem' for gravitational waves from oscillating stars. If we observed these waves, could we infer the properties of the star from which they originated? To answer this question we need to consider the frequencies and damping times of oscillation modes that may realistically be excited to a detectable level for a range of plausible equations of state.

As an illustration, let us consider the results from the first detailed consideration of the seismology strategy for neutron stars (Andersson and Kokkotas, 1998). The starting point is a family of stellar models for some set of equations of state, as shown in Figure 18.1. It should be immediately clear that this particular set of equations of state is not up to date. In particular, most of the models do not allow for neutron stars heavier than $2M_\odot$, which we now know is an observational requirement (see Chapter 12). The range of stiffness is also larger than expected. Conservatively, one may argue that X-ray observations constrain the neutron star radius to lie in the range 10–14 km (the grey vertical band in Figure 18.1, Steiner *et al.* (2018)). It is important to keep these issues in mind, but they do not have much impact on the general argument.

Ignoring the caveats, we calculate the various oscillation modes we are interested in for each of these stellar models. This leads to the results in Figure 18.2, showing the

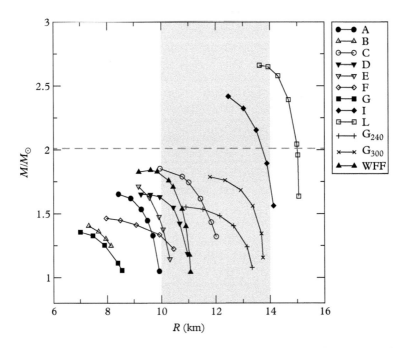

Figure 18.1 *The mass–radius relation for the set of neutron star models used in the discussion of empirical relations for neutron star oscillations. Note that, most of the equations of state used here are outdated as they do not allow for neutron stars heavier than $2M_\odot$ (the dashed line gives the mass of the heaviest known neutron star). The range of stiffness is also larger than expected—X-ray observations constrain the neutron star radius to lie in the range 10–14 km (the grey vertical band (Steiner et al., 2018). The collection is nevertheless useful as it represents a wide range in stiffness. (Based on data from Andersson and Kokkotas (1998).)*

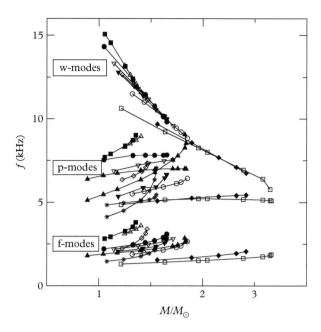

Figure 18.2 *The frequencies of the fundamental f-mode, the lowest order pressure p-mode, and the first w-mode determined for the stellar models from Figure 18.1. (Based on data from Andersson and Kokkotas (1998))*

fundamental f-mode, the lowest order pressure p-mode, and the first w-mode (more about this later). So far so good. Now let us ask what we would conclude from an observation of a specific mode frequency, say at 2.5 kHz. This would allow us to put a horizontal line through Figure 18.2. We would safely be able to say that we are dealing with an f-mode and may be able to rule out some equations of state, but we would clearly be left with several options.

We can do better by folding in some understanding of the physics. For example, in the case of the f-modes we expect the frequency to depend (roughly) on the average density of the star, cf. Chapter 13. That this scaling remains robust also for realistic equations of state (and in relativity) is demonstrated by the results in the left panel of Figure 18.3, where we show the frequency as a function of the average density of the star. Similarly, we may assume that the quadrupole formula gives a reasonable indication of the gravitational-wave emission. This suggests a scaling for the damping time of the mode

$$t_f \sim \frac{\text{oscillation energy}}{\text{power emitted in GWs}} \sim R\left(\frac{R}{M}\right)^3 = t_{\text{gw}}, \tag{18.16}$$

which also turns out to be fairly reliable—see the results in the right panel of Figure 18.3.

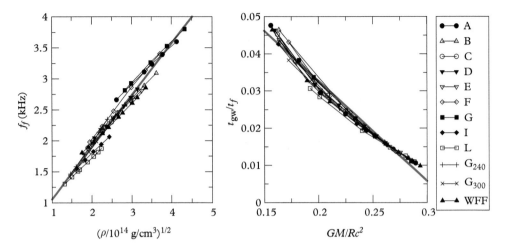

Figure 18.3 *Left: Numerically obtained f-mode frequencies as functions of the average stellar density. Right: The scaled f-mode damping times as functions of the stellar compactness. The straight (red) lines represent the fits to the data given in (18.17) and (18.18). (Based on data from Andersson and Kokkotas (1998).)*

Based on these results, we can infer empirical relations which may form the basis for a solution to the inverse problem. A linear fit to the calculated f-mode frequencies[2] leads to (Andersson and Kokkotas, 1998)

$$\frac{f_f}{1\ \text{kHz}} \simeq 0.22 + 2.16 \left(\frac{M_{1.4}}{R_{10}^3}\right)^{1/2},\qquad (18.17)$$

where $M_{1.4} = M/1.4M_\odot$ and $R_{10} = R/10$ km. We see that the typical f-mode frequency is around 2.4 kHz. Meanwhile, we have for the damping time

$$\frac{1\text{s}}{t_f} \simeq \left(\frac{M_{1.4}^3}{R_{10}^4}\right)\left[22.85 - 14.65\left(\frac{M_{1.4}}{R_{10}}\right)\right].\qquad (18.18)$$

The small deviation of the numerical data from this fit is apparent in Figure 18.3, and one can easily see that a typical value for the damping time of the f-mode is a tenth of a second. Given the Newtonian estimates from Chapter 13, these results should not be surprising.

---

[2] This is an improved fit of the original data. Similar empirical relations have been discussed by (amongst others) Tsui and Leung (2005) and Lau *et al.* (2010), making use of different scalings with the stellar parameters. However, as our main interest here is to illustrate the principle we base the discussion of the original results from Andersson and Kokkotas (1998).

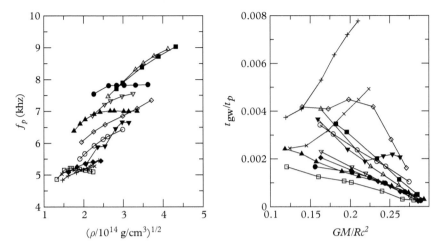

Figure 18.4 *Left: Numerically obtained p-mode frequencies as functions of the average density of the star. Right: The corresponding damping times as functions of the stellar compactness. (Based on data from Andersson and Kokkotas (1998).)*

The fact that these empirical relations are robust reflects the expectation that the fundamental mode depends on the star's bulk properties. It is relatively insensitive to the interior physics. In contrast, the frequency of the pressure p-modes depends on the local sound speed (see Chapter 13) and the damping of these modes is sensitive to the detailed perturbation amplitudes inside the star (Andersson *et al.*, 1995). As a result, different equations of state lead to rather different p-mode results, cf. Figure 18.4. It is difficult to construct useful empirical relations from these results. However, in the overall scheme of things, this may not be bad news. The distinct nature of the p-modes could be an advantage. For example, if the mass and the radius of the star are already obtained by other means, then an observed p-mode can be used to identity the equation of state. Of course, we have to keep in mind that the p-modes reside at higher frequencies where gravitational-wave detectors tend to be less sensitive.

18.3   The inverse problem

Making contact with the idea of determining a black hole's mass and spin from observed quasinormal modes (see Chapter 17), one may envisage combining an observed neutron star f-mode frequency and damping time with the empirical relations (18.17) and (18.18) to invert the problem and constrain the mass and the radius. The precision of the result obviously depends on statistical uncertainties which we have not yet considered.

As an example of this kind of parameter extraction—illustrating the main ideas—let us suppose that we want to detect a signal from an oscillating star (or, indeed, a black hole; see Kokkotas *et al.* (2001)). The signal would then be rather simple (at least in principle). For each individual mode we have a damped sinusoid,

$$
h(t) = \begin{cases} 0 & \text{for } t < t_a, \\ A e^{-(t-t_a)/\tau} \sin[2\pi f(t - t_a)] & \text{for } t > t_a, \end{cases} \tag{18.19}
$$

where A is the initial amplitude of the signal, t_a is its arrival time (the earliest time at which the damped oscillation is present in the signal), and f and τ are the frequency and damping time of the oscillation, respectively. From the standard flux formula from Chapter 3, we have

$$
A \approx 2.4 \times 10^{-20} \left(\frac{E}{10^{-6} M_\odot c^2} \right) \left(\frac{10 \text{ kpc}}{d} \right) \left(\frac{1 \text{ kHz}}{f} \right) \left(\frac{1 \text{ ms}}{\tau} \right)^{1/2}. \tag{18.20}
$$

That is, we can relate the amplitude of the oscillation to the energy carried by gravitational waves.

In a realistic situation, e.g. after neutron star merger, the assumed signal may only be relevant at late stages when the remnant is settling down and its pulsations can be accurately described as a superposition of the various modes that have been excited. At earlier times ($t < t_a$) the waves are likely to have a more complex character that is completely uncorrelated with the intrinsic noise of the detector. This partly justifies the simplification of setting the waveform equal to 0 for $t < t_a$.

Following the matched-filter strategy from Chapter 8, we use templates of the same form as the expected signal. This leads to the signal-to-noise ratio

$$
\left(\frac{S}{N} \right)^2 = \rho^2 \equiv (h|h) = \frac{4Q^2}{1 + 4Q^2} \frac{A^2 \tau}{2S_n}, \tag{18.21}
$$

where

$$
Q \equiv \pi f \tau, \tag{18.22}
$$

is the quality factor of the oscillation, S_n is the spectral density of the detector noise (assumed to be constant over the bandwidth of the signal in this example), and we have adapted the strategy we outlined for burst signals in Chapter 8. That is, we have used

$$
\begin{aligned}
(h_1(t)|h_2(t)) &= 4 \text{Re} \int_0^\infty \frac{\tilde{h}_1 \tilde{h}_2}{S_n} df \approx \frac{4}{S_n(f)} \text{Re} \int_0^\infty \tilde{h}_1 \tilde{h}_2^* df \\
&= \frac{4}{S_n(f)} \text{Re} \int_0^\infty h_1(t) h_2^*(t) dt.
\end{aligned} \tag{18.23}
$$

In order to work out the accuracy with which the parameters of the signal can be determined, it is useful to first introduce dimensionless variables such that

$$f_o\epsilon \equiv f - f_o, \quad \tau_o\eta \equiv \tau - \tau_o, \quad t_o\zeta \equiv t_a - t_o \quad \mathcal{A}_o\xi \equiv \mathcal{A} - \mathcal{A}_o, \qquad (18.24)$$

where f_o, τ_o, t_o, and $\mathcal{A}_o$ are the true values of the four quantities f, τ, t_a, and $\mathcal{A}$. The new parameters are simply relative deviations of the measured quantities from their true values. Once we have decided to work with these parameters we can construct the Fisher information matrix Γ_{ij} (see Chapter 8) and by inverting it, obtain all possible information about the measurement accuracy of each parameter of the signal, and the correlations between the associated errors.

First of all, for a system with Gaussian noise the components of the (symmetric) Fisher matrix follow from

$$\Gamma_{ij} \equiv \left(\frac{\partial h}{\partial \theta_i} \middle| \frac{\partial h}{\partial \theta_j} \right), \qquad (18.25)$$

where $\theta_i = (\epsilon, \eta, \zeta, \xi)$ are the signal parameters. Given our simple model signal it is straightforward to work out the components of this matrix, The inverse of the Fisher matrix, $C_{ij} \equiv \Gamma_{ij}^{-1}$, gives the covariance matrix. This is the most important quantity from the experimental point of view, since its components are directly related to the measurement errors. In the present case, if we want to know how accurately we can determine the mode frequency f and the damping time τ, we need $C_{\epsilon\epsilon}$ and $C_{\eta\eta}$. Working out the algebra (Kokkotas *et al.*, 2001), we find that these are given by

$$C_{\epsilon\epsilon} = \frac{1 - 2Q^2 + 8Q^4}{2Q^4(1 + 4Q^2)} \frac{1}{\rho^2}, \qquad (18.26)$$

and

$$C_{\eta\eta} = \frac{4(5 + 4Q^2)}{(1 + 4Q^2)} \frac{1}{\rho^2}. \qquad (18.27)$$

These results represent the squares of the relative measurement errors of f and τ, respectively.

As an illustration, let us consider a typical f-mode with frequency $f = 2.4$ kHz and damping time $\tau = 0.1$ s. Then we have $Q \approx 750$ and if we consider a galactic source at a distance of $d = 10$ kpc and a fiducial detector with $S_n^{1/2} = 10^{-23}$ Hz$^{-1/2}$ at the mode frequency (roughly the level of Advanced LIGO), then we expect a signal-to-noise ratio

$$\rho \approx 30 \left(\frac{E}{10^{-6} M_\odot c^2} \right)^{1/2}. \qquad (18.28)$$

That is, even if the system radiates as little as $4 \times 10^{-7} M_\odot c^2$ through this mode we would detect the signal with $\rho \approx 10$. From (18.26) and (18.27) it follows that

$$\delta f / f \approx 1.3 \times 10^{-3} / \rho, \qquad (18.29)$$

and

$$\delta \tau / \tau \approx 2 / \rho. \qquad (18.30)$$

For a signal leading to $\rho \approx 10$ we would accurately infer the mode frequency but the damping rate would only be known at the 20% level. If we want to extract both the frequency and the damping to (say) the 1% level, then we need a signal-to-noise of at least 200. This would require the release of an energy of order $4 \times 10^{-5} M_\odot c^2$, which may be unrealistic. The main lesson is that, while we may be able to extract the f-mode frequency, we are much less likely to constrain the damping rate. In essence, a realistic asteroseismology proposal needs to involve a different combination of parameters.

18.4   The w-modes

Up to this point we have assumed that the oscillations of relativistic stars are similar to the well-established Newtonian results. Relativistic effects impact on the oscillation frequencies through the gravitational redshift and the modes are damped by gravitational-wave emission, but the results remain qualitatively similar to the Newtonian case. There is, however, more to this story.

Relativistic stars have additional oscillation modes, with no relation to the fluid modes from Newtonian theory (Kokkotas and Schutz, 1992). Rather, the new class of modes is due to the dynamical spacetime. In the case of these modes the fluid hardly pulsates at all (Andersson *et al.*, 1996*b*). Like the quasinormal modes of a black hole, these so-called (gravitational-wave) w-modes reflect the nature of the curved spacetime.

In order to explain the w-modes, let us turn our attention to the axial perturbations. In this case, we have (see Chapter 14)

$$h_{ab} = \begin{pmatrix} 0 & 0 & -h_0 \partial_\varphi Y_l^m / \sin\theta & h_0 \sin\theta \, \partial_\theta Y_l^m \\ 0 & 0 & -h_1 \partial_\varphi Y_l^m \sin\theta & h_1 \sin\theta \, \partial_\theta Y_l^m \\ \mathrm{sym} & \mathrm{sym} & 0 & 0 \\ \mathrm{sym} & \mathrm{sym} & 0 & 0 \end{pmatrix}, \qquad (18.31)$$

while the axial velocity field is given by

$$\delta u^\theta = -\frac{1}{r^2 \sin\theta} U \partial_\varphi Y_l^m \qquad (18.32)$$

and

$$\delta u^\varphi = \frac{1}{r^2 \sin\theta} U \partial_\theta Y_l^m.$$ (18.33)

Once we account for the assumed axisymmetry (setting $m = 0$), we arrive at three equations for the metric perturbations:

$$\delta G_t^\varphi = 8\pi \delta T_t^\varphi \longrightarrow$$

$$h_0' - \dot{h}_1' - \frac{1}{2}(\nu' + \lambda')(h_0' - \dot{h}_1) - \frac{2}{r}\dot{h}_1 - \left[\frac{n}{r^2}e^\lambda + \frac{1}{r}(\nu' + \lambda') + \frac{2}{r^2}\right]h_0$$

$$= 16\pi (p + \varepsilon)e^{\nu/2+\lambda} U,$$ (18.34)

$$\delta G_r^\varphi = 8\pi \delta T_r^\varphi \longrightarrow \dot{h}_0' - \ddot{h}_1 - \frac{2}{r}\dot{h}_0 - \frac{n}{r^2}e^\nu h_1 = 0,$$ (18.35)

and

$$\delta G_\theta^\varphi = 8\pi \delta T_\theta^\varphi \longrightarrow e^{-\nu}\dot{h}_0 - e^{-\lambda}h_1' - \frac{1}{2}(\nu' - \lambda')e^{-\lambda}h_1 = 0.$$ (18.36)

It is easy to see that the last of these equations leads to

$$\dot{h}_0 = e^{(\nu-\lambda)/2}\left[e^{(\nu-\lambda)/2}h_1\right]',$$ (18.37)

and it follows from (18.35) that

$$\ddot{h}_1 - r^2 \frac{\partial}{\partial r}\left\{\frac{e^{(\nu-\lambda)/2}}{r^2}\frac{\partial}{\partial r}\left[e^{(\nu-\lambda)/2}h_1\right]\right\} - \frac{n}{r^2}e^\nu h_1 = 0.$$ (18.38)

Meanwhile, (18.34) leads to

$$\dot{U} = -e^{-\nu/2}\dot{h}_0,$$ (18.39)

where we have used the background relation (see Chapter 4)

$$\nu' + \lambda' = 8\pi (p + \varepsilon)re^\lambda.$$ (18.40)

We learn that the fluid motion is slaved to the metric variations. This is as expected. Fluids cannot support shear stresses so one would not expect the axial problem to have interesting dynamics. We have already seen this in the case of the r-modes (see Chapter 13), which require rotation to become distinct. However, the relativistic problem brings something new. We have a wave equation for the perturbed metric. This should not come as a surprise—we are discussing gravitational waves, after all—but it is an interesting hint that there are features of the relativistic problem which do not have a Newtonian counterpart. In fact, it is easy to show that, if we make the star sufficiently compact (e.g. by assuming a uniform density), then the corresponding effective potential develops a well inside the star. Gravitational waves may be temporarily trapped in this well, leading to a set of slowly damped oscillation modes (Chandrasekhar and Ferrari, 1991; Andersson *et al.*, 1996a).

The w-modes come in both the axial and the polar variety and for less compact stars they are rapidly damped (Kokkotas and Schutz, 1992). The curvature of the stars is not able to efficiently trap gravitational waves. Typical results for the lowest polar w-mode are compared to the f- and p-modes in Figure 18.2. The corresponding frequencies and damping times are shown in Figure 18.5.

Since they are mainly due to the spacetime curvature and they do not excite significant fluid motion, one would expect the w-modes to be relatively independent of the matter composition. This should lead to robust empirical relations between the mode frequencies and stellar parameters. Indeed, numerical results (Andersson and Kokkotas, 1998) suggest that the frequency of the (main) w-mode is inversely proportional to the size of the star; see Figure 18.5. Meanwhile, the damping time is related to the compactness of the star; i.e. the more relativistic the star is, the longer the w-mode oscillation lasts. This is also shown in Figure 18.5.

From numerical results we obtain the following relation for the frequency of the first w-mode, (Andersson and Kokkotas, 1998)

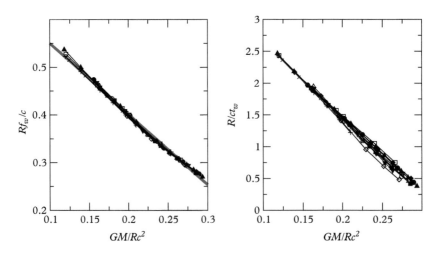

Figure 18.5 *Left: Numerically obtained w-mode frequencies. We show Rf_w as functions of the compactness of the star. Right: The corresponding damping times, in terms of R/ct_w as functions of the compactness of the star. (Based on data from Andersson and Kokkotas (1998).)*

$$\frac{f_w}{1 \text{ kHz}} \simeq \frac{1}{R_{10}} \left[20.95 - 9.17 \left(\frac{M_{1.4}}{R_{10}} \right) \right], \tag{18.41}$$

while the damping rate of the mode is well described by

$$\frac{M_{1.4}}{t_w \text{ (ms)}} \simeq 3.90 + 104.06 \left(\frac{M_{1.4}}{R_{10}} \right) - (67.28 \pm 3.84) \left(\frac{M_{1.4}}{R_{10}} \right)^2. \tag{18.42}$$

We see that a typical value for the w-mode frequency is 12 kHz, but since the frequency depends strongly on the radius of the star it varies greatly for different equations of state. For example, for a very stiff equation of state the w-mode frequency may be as low as 6 kHz while for the softest equation of state in our set the typical frequency is around 14 kHz. Perhaps not surprisingly, the w-mode damping time is comparable to that of an oscillating black hole with the same mass, i.e. typically less than a tenth of a millisecond. Given the high frequencies involved the w-modes may prove difficult to detect, but they have the potential to provide a distinct signature of a neutron star spacetime.

18.5   The evolving spectrum of adolescent neutron stars

Neutron stars evolve as they age. Born in the furnace of a supernova core collapse, they start out extremely hot but cool rapidly (over 10s of seconds) through neutrino emission (Burrows and Lattimer, 1986; Pons *et al.*, 1999). As they cool, internal entropy gradients are smoothed out and thermal effects become less prominent. At the same time, other physics aspects come into play. The crust of the star forms and the core fluids become superfluid/superconducting. During the early life of a neutron star, these changes are rapid but they slow down to become much more gradual as the star matures (Ho *et al.*, 2012).

   As much of the relevant physics is poorly understood, it is difficult to build models that faithfully track this evolution. Nevertheless, many of the key features are clear and one can make (some) progress on understanding how the evolution of the star impacts on its oscillation spectrum. We will demonstrate this by considering two related problems. First, we discuss the early evolution of a proto-neutron star, where entropy gradients support a family of thermal g-modes (Ferrari *et al.*, 2003). Second, we ask how the g-mode spectrum changes once the thermal effect become weak and the internal composition variation provides the main source of buoyancy (Reisenegger and Goldreich, 1992). In essence, we track the evolution of the low-frequency features of a neutron star through the first millennium of its existence.

   Let us start by considering the first few tens of seconds. During this early phase neutrino diffusion cools the neutron star, and as a result the neutrino mean-free path increases. After less than a minute, the mean-free path becomes comparable to the stellar radius. This means that the star is effectively transparent, so the neutrinos can escape. In this initial phase, a neutron star contracts and its gravitational mass decreases slightly. As a result—given that their frequency scales as $(M/R^3)^{1/2}$—the f-modes evolve.

In addition, the internal entropy gradient supports a set of thermal g-modes, which also evolve as the star cools. The composition of a hot proto-neutron star is also different—the internal stratification is more prominent, leading to higher frequency g-modes (Ferrari *et al.*, 2003). The impact of these effects on the leading modes is illustrated in Figure 18.6 (ignoring the, potentially significant, effect of rotation).

The results show that the frequencies of the f-, p-, and w-modes are much lower than those of a cold neutron star. Initially, they cluster in a narrow range of 900 – 1500 Hz. The behaviour of the f-mode frequency is particularly interesting, because it does not show the usual scaling at these early times. During the first second the mass of the star remains approximately constant while the radius rapidly decreases (Pons *et al.*, 1999). One would expect the f-mode frequency to decrease, but the opposite appears to happen. However, the behaviour can be understood if we also consider the leading g-mode. After about 0.5 s this mode exhibits a so-called avoided crossing with the f-mode. At this point, the modes exchange properties. In this particular example, the expected increase in the f-mode frequency is probably obscured by this mode-crossing. The fact that the frequencies evolve over the first few seconds is important. We need to keep this in mind, especially if the mode features are expected to persist on this kind of timescale. That the p- and w-mode frequencies are much lower than usual could be a bonus for observers—it means that they (briefly) radiate in a regime where the detectors are more sensitive.

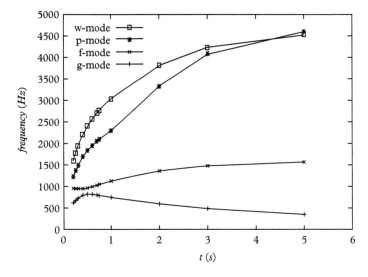

Figure 18.6 *Frequencies of the lowest order oscillation modes of a proto-neutron star shown as functions of the time elapsed from the gravitational collapse. The frequencies initially cluster in a narrow range but move apart after the first second of evolution. After about 5 seconds the spectrum of a mature neutron star is established and the evolution becomes more gradual. It is worth noting that, at least in this model, the f-mode and the g-mode exhibit an avoided crossing after about 0.5 s. (Reproduced from Ferrari et al. (2003).)*

We have seen that a neutron star's oscillation spectrum evolves during the proto-neutron star stage, when the emission of neutrinos leads to a loss of thermal support. However, once the star cools to about 10^{10} K the thermal support becomes insignificant. The star has radiated much of its binding energy. It has settled down to its final mass and radius—a neutron star has been born. This does not mean that the star's spectrum ceases to evolve. The star continues to mature for several hundred years. We can explore this phase by tracking the cooling of a given star from a minute or so after birth through hundreds of years, paying particular attention to the changes in the thermal pressure and the formation of that star's elastic crust (Ho *et al.*, 2012). State-of-the-art cooling simulations lead to the results shown in Figure 18.7. We see that the star is (close to) isothermal after about 100 years. We also need to keep track of the internal state of matter, in particular, the formation of the star's crust. Cooling calculations typically start above the melting temperature of the crust, but it begins to crystallize after approximately one day. Typically, it takes at least 100 years for the bulk of the crust to form. The plot thickens further if we consider the formation of superfluid/superconducting components in the star's core. These also form gradually. In particular, the core superfluid may develop over 1,000 years (Ho *et al.*, 2012). The upshot of this is that some observed young neutron stars may still be in their formative period. After all, the remnant in Cas A is only about 300 years old.

We can use the cooling data as input for a detailed seismology analysis (Krüger *et al.*, 2015), accounting for density discontinuities associated with distinct phase-transitions, interior composition gradients, thermal pressure, and the elastic crust which grows in thickness as the neutron star cools.

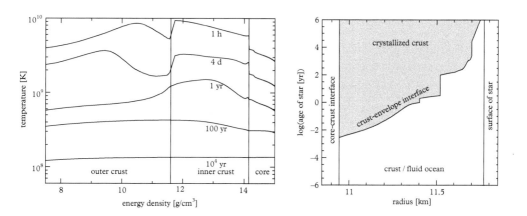

Figure 18.7 *Left: The thermal evolution of a neutron star model with $M = 1.45\,M_\odot$. As is apparent from this graph, the neutron star is nearly isothermal after 100 years (without additional heat sources). Right: The gradual formation of the solid crust over time (showing only the outer layers of a particular model star). The area where the crust is crystallized (shaded in grey) is obtained using a sharp threshold of $\Gamma > 173$ (see Chapter 14). (Reproduced from Krüger et al. (2015), copyright (2015) by the American Physical Society.)*

At different points in the thermal evolution we output the temperature profile and feed it into the mode calculation (Krüger *et al.*, 2015). The results of this exercise shed light on the influence of thermal effects on the various oscillation modes of the star. As expected, the fundamental f-mode and the various p-modes are only weakly affected by the gradual change in temperature. Meanwhile, the gravity g-mode spectrum changes completely as the thermal pressure weakens. We also find a set of interface (i-) modes (arising from density 'discontinuities' associated with, for example, the edges of the crust (McDermott *et al.*, 1988)). The presence of the elastic crust enriches the spectrum by shear modes, which also evolve as the star cools and the crust region grows. The results provide a sequence of snapshots of how the star's oscillation spectrum evolves as the star ages.

Typical results for (a large set of) low-frequency modes are shown in Figure 18.8. These modes are all rather different from those of a nearly perfect fluid star. This is not surprising—we have changed the physics. It is notable that all modes exhibit avoided crossings (easily visible in the high-frequency part of the graph). Early on in the evolution, we can identify a set of thermal g-modes. Their frequency decreases as the star cools and after about 100 years almost all of them have frequencies below about 20 Hz. At this point the temperature has decreased so much that the thermal pressure is negligible. We are left with g-modes which owe the internal composition for their existence. There is also a set of interface modes (distinguished by having more localized eigenfunctions inside the star). As the star continues to cool, the frequencies of these interface modes also change slightly, finally leaving us with the spectrum of a cold star.

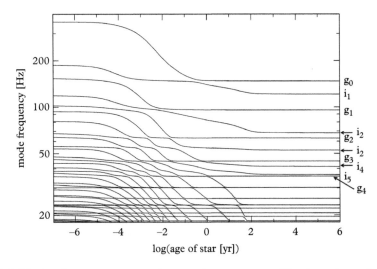

Figure 18.8 *The low-frequency spectrum of a maturing neutron star. The thermal g-modes rapidly decrease in frequency and fall below about 20 Hz after 100 years or so. The four identified interface modes are also affected by the thermal pressure (here labelled $i_1, i_2, i_4,$ and i_5). After about 100 years of evolution, the remaining g-modes are all due to the internal composition. All modes exhibit avoided crossings during the evolution. (Reproduced from Krüger et al. (2015), copyright (2015) by the American Physical Society.)*

18.6   Magnetar seismology

We have argued that asteroseismology provides a strategy for probing stellar physics
through observed oscillation modes. The promise of the idea is clear, but for neutron
stars it may not be so easy to execute it in practice. We need precise observations along
with detailed models to test the observations against. In the case of neutron stars we have
until very recently had neither reliable observations nor sufficiently detailed models.

We have discussed how progress is made as the various theory issues are addressed.
When it comes to observations, the situation changed with the observations of quasiperiodic oscillations following giant flares in three soft gamma-ray repeaters (magnetars).
Analysis of the X-ray data unveiled a set of periodicities (Duncan, 1998; Israel *et al.*,
2005; Strohmayer and Watts, 2005; Strohmayer and Watts, 2006) with frequencies that
agree reasonably well with the expected torsional shear modes of the neutron star crust.
This was exciting because it allowed us to test—and improve—our models for neutron
star seismology. However, it also raised warning flags about the 'missing physics'. As the
observed flares are thought to originate from the crust in a magnetar yielding due to
stresses that build up as the field evolves (Thompson and Duncan, 1995), one would
expect the strong magnetic field to play a role. This means that we (most likely) need to
account for the (likely superconducting) nature of the star's interior magnetic field. This
is far from easy.

As the problem is complex, let us focus on the elastic aspects in order to illustrate
the involved principles. We need to include the crust elasticity in a relativistic calculation
of axial oscillations. As the crust region has low density it seems reasonable to assume
that we can neglect the dynamical nature of the spacetime. This simplifies the problem.
The main technology required for this calculation is the same as that used to quantify
the point at which the crust yields during binary inspiral (see Chapter 21) although we
must now consider the time-dependent situation. To make progress, we need the axial
perturbation equations for an elastic solid in the Cowling approximation (Samuelsson
and Andersson, 2007). These can be written

$$F' + A'F' + BF = 0, \tag{18.43}$$

where a prime denotes a derivate with respect to the radial (Schwarzschild) coordinate
r and

$$e^A = r^4 e^{\nu-\lambda}(\varepsilon + p)v^2 , \tag{18.44}$$

$$B = \frac{e^{2\lambda}}{v^2}\left[e^{-2\nu}\omega^2 - \frac{v^2(l-1)(l+2)}{r^2} \right]. \tag{18.45}$$

The amplitude of the oscillation is F, ε is the energy density, p is the pressure, and v is the
shear speed (for simplicity we assume that the crust lattice is isotropic; see Chapter 14).
The integer l is the usual multipole that enters when we expand in spherical harmonics
and ω is the angular frequency.

In the case we are considering, the boundary conditions require the traction to vanish at the top and the bottom of the crust. This leads to

$$e^A F' = 0 \quad \text{at } r = R_c \text{ and } r = R. \tag{18.46}$$

We can use these equations to calculate axial crust modes for stellar models with given core mass and radius (as long as we ignore the magnetic field and viscosity, the core fluid does not couple to the crust motion). This allows us to consider a variety of supranuclear equations of state in the core (Samuelsson and Andersson, 2007). We can then combine the results to outline a workable strategy to (i) identify the key parameters that govern the various modes, and (ii) try to represent the results in such a way that a parameter 'inversion' becomes possible given actual observations. As in the case of the f-modes, which we have already considered, it is useful to work with approximate relations based on the numerical data. In order to justify such relations, it would be helpful to find an approximate solution. To do this, we consider Eq. (18.43) and introduce a new (tortoise-type) coordinate x through

$$\frac{dx}{dr} = e^{-A}. \tag{18.47}$$

The perturbation equation then becomes

$$\frac{d^2 F}{dx^2} + e^{2A} BF = 0. \tag{18.48}$$

It is now written on a form that lends itself to a WKB-type approximation. Thus, we assume that the solution can be written

$$F = C_1 e^{iw(x)} + C_2 e^{-iw(x)}, \quad w(x) = \int_{R_c}^{x} e^A B^{1/2} dx = \int_{R_c}^{r} B^{1/2} dr. \tag{18.49}$$

At the base of the crust ($r = R_c$) we need to ensure the vanishing of the traction. Hence, we impose the boundary condition

$$F' = iB^{1/2}(C_1 - C_2) = 0 \quad \Rightarrow \quad C_1 = C_2. \tag{18.50}$$

The analogous condition at the surface ($r = R$) implies that

$$F' = iC_1 B^{1/2} \left[e^{iw(R)} - e^{-iw(R)} \right] = 0 \quad \Rightarrow \quad w(R) = \int_{R_c}^{R} B^{1/2} dr = n\pi. \tag{18.51}$$

To make further progress, we make the approximation that the shear speed is constant (an approximation which is good throughout much of the crust) and that the metric coefficients ν and λ are constant (which is a reasonable assumption since the crust mass is negligible compared to that of the core). Then, assuming that

$$\omega^2 \gg e^{2\nu} v^2 \frac{(l-1)(l+2)}{r^2}, \tag{18.52}$$

we may Taylor-expand $B^{1/2}$ to get

$$B^{1/2} \approx e^{\lambda-\nu} \frac{\omega}{v} \left[1 - \frac{e^{2\nu} v^2 (l-1)(l+2)}{2\omega^2 r^2} \right], \tag{18.53}$$

which may be integrated to yield (after using the boundary condition at the surface)

$$\omega^2 - e^{\nu-\lambda} \frac{n\pi v}{\Delta} \omega - e^{2\nu} \frac{v^2 (l-1)(l+2)}{2RR_c} \approx 0, \tag{18.54}$$

where $\Delta = R - R_c$. This provides a useful first approximation to the frequencies of the axial crust modes.

Let us first consider the fundamental crust mode, which corresponds to $n = 0$. For this case we immediately find that

$$\omega^2 \approx \frac{e^{2\nu} v^2 (l-1)(l+2)}{2RR_c}. \tag{18.55}$$

This provides useful insight into the scaling with various parameters. In particular, it gives the dependence on l required to match the observed data with specific crust modes; see Table 18.1.

Turning to the overtones, $n \neq 0$ for any given l, we need to solve the quadratic equation (18.54). Expanding the resulting square root in the small parameter Δ/R and ignoring the negative root we arrive at

Table 18.1 *Observed quasiperiodic oscillations frequencies for the giant flares in SGR 1806−20 and SGR 1900+14 and the suggested corresponding elastic crust modes. The identified modes are denoted as $_n t_l$ with l the multipole and n the order of the mode. (Data from Samuelsson and Andersson (2007).)*

SGR 1806−20		SGR 1900+14	
f (Hz)	Mode	f (Hz)	Mode
29	$_0 t_2$	28 ± 0.5	$_0 t_2$
92.7 ± 0.1	$_0 t_6$	53.5 ± 0.5	$_0 t_4$
150.3	$_0 t_{10}$	84	$_0 t_6$
626.46 ± 0.02	$_1 t_1$	155.1 ± 0.2	$_0 t_{11}$

$$\omega \approx e^{\nu - \lambda} \frac{n\pi v}{\Delta} \left[1 + e^{2\lambda} \frac{(l-1)(l+2)}{2\pi^2} \frac{\Delta^2}{RR_c} \frac{1}{n^2} \right],$$ (18.56)

where the second term in the bracket is negligible for moderate l. In order to estimate the overtone frequencies for any given M and R (say) we need the crust thickness Δ which then leads to R_c. One can show that that the crust thickness is reasonably well approximated by

$$\frac{\Delta}{R} \approx \left(1 + \frac{M}{\alpha R} e^{2\lambda} \right)^{-1},$$ (18.57)

where α is a parameter that depends on the equation of state (essentially measuring the average compressibility of the crust). For the (realistic) crust model considered by Samuelsson and Andersson (2007) the relevant value is $\alpha = 0.02326$.

It is straightforward to use the analytic approximations to analyse the observed magnetar oscillation features. The requirement that a single model must allow for the presence of all observed frequencies then leads to constraints on the neutron star parameters. Let us first assume that we observe the fundamental quadrupole mode together with the first overtone ($n = 1$). In the data for SGR 1806−20; see Table 18.1, it seems reasonable to assume that the first is represented by the 29-Hz oscillation, while the latter corresponds to the 626-Hz mode. From our approximate formulae we see that the ratio of these modes provides an expression that depends only on the compactness. In effect, we arrive at a curve in the $M - R$ plane on which the true stellar model should lie; see Figure 18.9. Solving this constraint for the compactness we find that $M/R \approx 0.12$, so $R \approx 8.1M$. From Eq. (18.57) we also see that the relative crust thickness is $\Delta/R \approx 0.17$. We can insert this value for β in the expression for (say) the fundamental mode. This allows us to solve for the radius, and we find that $R \approx 11.4$ km, which means that $M \approx 1.41$ km $\approx 0.96M_\odot$ and $\Delta \approx 1.9$ km. These values do not seem unreasonable, although the inferred mass is a bit too low and the crust would seem surprisingly thick. Of course, we have used simple approximations and we also ignored the impact of the star's magnetic field. Be that as it may, the example demonstrates the main steps of this kind of analysis.

Next, consider the fundamental modes for different values of l, noting that the scaling with l allows us to immediately work out the ratio of the different mode frequencies. If we assign the $\approx$30-Hz feature to the fundamental $l = 2$ mode we can infer the multipoles of the various higher frequency modes in the observed data. This leads to the identifications suggested in Table 18.1. Once we have determined the various multipoles, our approximate formula for the fundamental modes can (again) be used to constrain the stellar parameters. Let us take the quadrupole mode as an example. If the frequency is known, then the approximate formula gives the stellar radius as a function of the compactness β. Since $M = \beta R$ we again have a constraint curve in the $M - R$ plane. In order to be consistent with the results for the $n = 1$ overtone, the two curves must intersect. The point of intersection immediately provides us with the mass and radius of

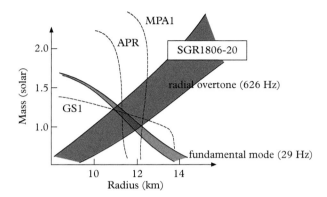

Figure 18.9 *Seismology analysis based on numerical axial crust-mode results. For the flare in SGR 1806−20 we associate the lower frequency oscillations (approximately 29, 93, and 150 Hz) with fundamental (n = 0) modes with l = 2, 6, 9, or 10, respectively. The inclusion of the 150-Hz mode does not substantially change the picture. The higher frequency oscillation (626 Hz) is assumed to be associated with an n = 1 mode of arbitrary l. Models allowed by the set of low frequencies form a rather broad region orthogonal to the line corresponding to models that have an n = 1 overtone of the right magnitude. The true stellar parameters should lie in the region of overlap. (Adapted from the results of Samuelsson and Andersson (2007).)*

the star. This is illustrated in Figure 18.9. The results for higher l-modes provide further constraints that can be used to verify the consistency of the result. Of course, our model is an idealization (i.e. a spherical star with an isotropic crust). One would expect the use of real data to lead to a spread of the various curves in the $M - R$ plane. This uncertainty can (to some extent) be used to assess the faithfulness of the parameter extraction.

Let us now take a leap of faith and assume that the mechanism that generates the magnetar flare also leads to gravitational-wave emission. On the one hand, this would be expected as the process must involve asymmetries. On the other hand, if the event mainly involves the crust, then the density is low and one would not expect significant gravitational dynamics. It could perhaps be that the observed oscillations are associated with shear waves in the crust, but are triggered by a more global event. For example, a large-scale adjustment of the magnetic field configuration (tapping into the gravitational potential energy) could lead to a change in the moment of inertia and the emission of gravitational waves (Ioka, 2001), perhaps releasing as much as $10^{48} - 10^{49}$ ergs of energy (Corsi and Owen, 2011). If this energy were to be channeled through the star's f-mode, the signal could be within reach of current detectors. Unfortunately, at the time of the giant flare in SGR 1806−20 (27 December 2004), only one of the two LIGO instruments were operational. The data from the Hanford detector sets an upper limit of 7.7×10^{46} erg on the energy release in gravitational waves (Abbott *et al.*, 2007*a*). An upper limit was also set, albeit in a more narrow frequency range, by the AURIGA bar detector (Baggio *et al.*, 2005). As we have not seen a magnetar giant flare in the Advanced LIGO era, other searches have focused on smaller (but more regular) burst events (Abbott *et al.*, 2008*b*; Abbott *et al.*, 2009*b*; Abadie *et al.*, 2011*b*). Future searches would benefit from theory developments, ideally providing robust estimates of the character and strength of the expected signal.

18.7   The relativistic r-modes

Continuing the theme of axial modes, let us briefly consider how the r-modes are affected by relativistic effects. If we assume that the modes can be obtained by a leading order slow-rotation calculation (as in Chapter 13), then we need to consider two effects. The frequency of the modes should exhibit the gravitational redshift and the rotational effects have to allow for the frame-dragging (see Chapter 17). We will demonstrate the impact of both effects at the first-order post-Newtonian level. We will also show that there are no pure r-modes (modes whose limit for a spherical star is purely axial) in a barotropic star (Lockitch *et al.*, 2001). This is in contrast with the Newtonian case for which we found a set of such modes (one for each for $l = m$); see Chapter 13. The Newtonian r-modes with $l = m \geq 2$ pick up relativistic corrections (with both axial and polar contributions), making their relation to general inertial modes of a rotating star (which have this 'mixed' nature) more apparent (Lockitch and Friedman, 1999).

In order to understand the nature of the relativistic r-modes, it is helpful to start by noting that stationary non-radial ($l > 0$) perturbations of a spherical star must have[3]

$$H_0 = H_2 = K = \delta\varepsilon = \delta p = 0, \tag{18.58}$$

and satisfy

$$0 = H_1 + \frac{16\pi(\varepsilon + p)}{l(l+1)}e^{\lambda}rW, \tag{18.59}$$

$$0 = e^{-(\nu-\lambda)/2}\left[e^{(\nu-\lambda)/2}H_1\right]' + 16\pi(\varepsilon + p)e^{\lambda}V, \tag{18.60}$$

$$h_0'' - \frac{1}{2}(\nu' + \lambda')h_0' + \left[\frac{(2 - l^2 - l)}{r^2}e^{\lambda} - \frac{1}{r}(\nu' + \lambda') - \frac{2}{r^2}\right]h_0 = \frac{2}{r}(\nu' + \lambda')U, \tag{18.61}$$

where a prime denotes a derivative with respect to r, as before. If we use (18.59) to eliminate H_1 from (18.60) we obtain

$$V = \frac{e^{-(\nu+\lambda)/2}}{l(l+1)(\varepsilon + p)}\left[(\varepsilon + p)e^{(\nu+\lambda)/2}rW\right]'. \tag{18.62}$$

---

[3] These results follow immediately from the time-independent version of the perturbation equations for polar and axial perturbations.

This result generalizes the conservation of mass equation from Newtonian theory. The other two equations relate the dynamical degrees of freedom of the spacetime metric to the perturbed fluid velocity. They vanish in the Newtonian limit.

The perturbations must be regular everywhere and satisfy the boundary condition that the Lagrangian change in the pressure vanishes at the surface of the star, $r = R$. In this case, this leads to $W(R) = 0$. If W and U are specified, then the functions H_1, h_0, and V follow from the above equations. The solutions for the metric variables are also subject to matching conditions to the exterior spacetime, which must be regular at infinity.

We want to understand how the problem changes when the star is slowly rotating. The relevant slow-rotation metric was already discussed in Chapter 12. We know that slow rotation should be taken to mean that Ω is small compared to the Kepler velocity, $\Omega_K \sim \sqrt{M/R^3}$, the angular velocity at which the star is dynamically unstable to mass shedding at the equator. Neglecting quantities of order Ω^2 and higher, the star remains spherical. This means that the Tolman–Oppenheimer–Volkoff equations from Chapter 4 remain valid. In addition, we need to work out the frame dragging, represented by $\varpi(r) = \Omega - \omega(r)$, by solving (12.72).

Turning to the perturbations, rotation couples the different multipoles, leading to the equations being more complicated than their non-rotating counterparts (Kojima, 1992). In general, an l mode that is axial to leading order picks up polar rotational corrections corresponding to the $l \pm 1$ multipoles. In order to illustrate the results, let us consider a uniform density star with

$$\varepsilon = \frac{3M}{4\pi R^3}. \tag{18.63}$$

In this case, we can write down an analytic solution for the background configuration (Lockitch *et al.*, 2001). We also find that if we truncate the solution at first order in a $2M/R$ expansion (the first post-Newtonian order), we have

$$\frac{\varpi(r)}{\Omega} = 1 - \left(1 - \frac{3r^2}{5R^2}\right)\left(\frac{2M}{R}\right) + \mathcal{O}\left(\frac{2M}{R}\right)^2. \tag{18.64}$$

Working to this order of approximation, we find that the post-Newtonian corrections to the Newtonian r-modes from Chapter 13 are such that (Lockitch *et al.*, 2001)

$$\kappa = \frac{2}{(m+1)}\left[1 - \frac{4(m-1)(2m+11)}{5(2m+1)(2m+5)}\left(\frac{2M}{R}\right) + \mathcal{O}\left(\frac{2M}{R}\right)^2\right], \tag{18.65}$$

provides the rotating frame frequency $\kappa\Omega \equiv \sigma + m\Omega$ (here we use σ for the mode frequency in order to avoid confusion with the frame dragging).

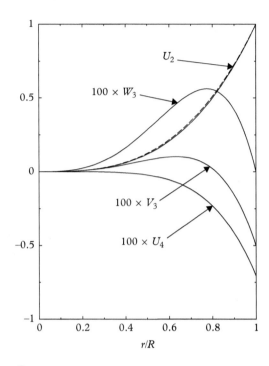

Figure 18.10 *The $(r/R)^3$ dependence of the Newtonian $l = m = 2$ r-mode eigenfunction (dashed curve) compared to the post-Newtonian corrections to this mode for a uniform density star of compactness $2M/R = 0.2$ (the coefficients $U_l(r)$, $W_l(r)$, and $V_l(r)$ with $l \leq 4$, solid curves). The vertical scale is set by normalizing $U_2(r)$ to unity at the surface of the star, and the other coefficients have been scaled by a factor of 100. The results show that, while the relativistic corrections to the equilibrium structure of the star are of order 20%, the relativistic corrections to the r-mode are at the 1% level. (Reproduced from Lockitch et al. (2001), copyright (2001) by the American Physical Society.)*

The corresponding eigenfunctions demonstrate the expected mixing of axial and polar terms; see Figure 18.10. In addition, we see from (18.65) that the r-mode frequency decreases with increasing neutron star compactness. It is natural that general relativity has this effect. First of all, the gravitational redshift will tend to decrease fluid oscillation frequencies measured by a distant inertial observer. Also, because these modes are rotationally restored they will be affected by the frame dragging. The local Coriolis force is determined not by the angular velocity Ω of the fluid relative to a distant observer but by its angular velocity relative to the local inertial frame, $\varpi(r)$. Thus, the Coriolis force decreases, and the modes are seen to oscillate less rapidly as the frame dragging becomes more pronounced.

For $m = 2$ and canonical neutron star parameters we have $\kappa \approx 0.6$ so the frequency shift away from the Newtonian result is significant. In fact, calculations for a range of proposed equations of state (Idrisy et al., 2015) show that the expected range of

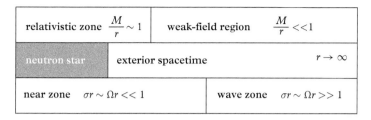

relativistic zone $\dfrac{M}{r} \sim 1$	weak-field region $\dfrac{M}{r} <<1$
neutron star　exterior spacetime	$r \to \infty$
near zone　$\sigma r \sim \Omega r << 1$	wave zone　$\sigma r \sim \Omega r >> 1$

Figure 18.11 *The spatial regions relevant to the relativistic r-mode problem.*

frequencies for the relativistic r-mode does not include the Newtonian value. It is important to keep this in mind when designing search strategies.

The overall conclusions are supported by detailed numerical calculations. In addition, one can work out a consistent relativistic estimate for the growth rate of the unstable r-modes (Lockitch *et al.*, 2003). This part of the analysis is conceptually interesting because it draws on the principles we used in our discussion of post-Newtonian theory. Basically, the r-mode perturbation problem is somewhat different in different regions of spacetime (see Figure 18.11). In the 'near zone', where $\sigma r \ll 1$, we can ignore second time derivatives, whereas we cannot do this in the 'wave zone', the region where $\sigma r \gg 1$. However, the inertial modes we are interested in are restored by the Coriolis force, so their frequencies scale with the angular velocity of the star, $\sigma \sim \Omega$. For slow rotation, this implies that the near zone extends far away from the star into the non-relativistic region ($M/r \ll 1$) and that the wave zone will be located entirely within this weak-field region. These distinctions are crucial for a calculations of the energy radiated as gravitational waves and the timescale on which gravitational radiation reaction drives an unstable modes. The near-zone equations are sufficient to determine the modes, but we need to extend these solution to the wave zone if we want to quantify the gravitational-wave emission (Lockitch *et al.*, 2003).

In order to illustrate the results for the r-mode growth timescale we keep the baryon mass, M_B, fixed at $1.4 M_\odot$ and set $\Omega^2 = \pi G \varepsilon$. A log–log plot then reveals that the timescale depends on the star's compactness as $(M/R)^{-(l+3)}$ for low M/R. This can be seen from the results in Figure 18.12, which compares the Newtonian and relativistic growth timescales of the modes whose Newtonian analogues are the first five $l = m$ r-modes (Lockitch *et al.*, 2003).

The results suggest that for very compact stars, the relativistic calculation tends to give a slightly *longer* growth timescale than that of a Newtonian star with the same equation of state, baryon mass, and compactness. General relativity tends to make the r-mode instability (slightly) weaker than expected. That this should be the case is natural. Inertial modes have relatively low frequencies, since $\sigma \sim \Omega$. One would generally expect low-frequency modes to radiate less efficiently as the star becomes more compact, since the gravitational waves will suffer backscattering by the spacetime curvature as they escape to infinity.

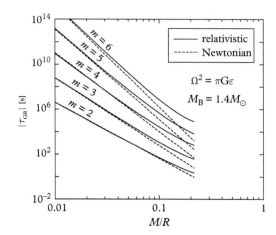

Figure 18.12 *Gravitational radiation reaction timescales for the fastest growing $l = m$ Newtonian r-modes (dashed lines) and their relativistic counterparts (solid curves). The timescales are shown as a function of compactness for $n = 1$ polytropes with fixed baryon mass, $M_B = 1.4M_\odot$. (Reproduced from Lockitch et al. (2003), copyright (2003) by the American Physical Society.)*

18.8 The unstable f-modes

If we neglect viscosity, the gravitational-wave-driven CFS instability is generic in rotating stars (see Chapter 13). It may act through both polar and axial modes. We have already discussed the axial case—in the form of the r-modes—in Chapter 15. Let us now consider the polar case. Since polar modes, like the f-mode and the p-modes, are present already in a non-spinning star, the instability sets in above some critical angular velocity, when the frequency of the mode goes through 0 in the inertial frame (according to the instability criterion from Chapter 13). The critical angular velocity is smaller for increasing mode multipole l. Thus, there will always be a large enough value of l for which a slowly rotating star will be unstable (Friedman and Schutz, 1978b).

Of course, just as in the case of the r-modes, shear and bulk viscosity suppress the growth of the CFS instability outside a certain temperature window (Ipser and Lindblom, 1991). The effect of shear viscosity, in particular, increases for higher order multipoles, as we are dealing with smaller scales. Combined with the fact that the gravitational-wave emission is much less efficient for large values of l, one would expect only the lowest multipoles to exhibit significant instabilities. Newtonian studies show that the f-mode may become unstable near the mass-shedding limit, with the $l = 4$ mode leading to the largest instability window (Ipser and Lindblom, 1991). The quadrupole f-mode is never unstable in a uniformly rotating Newtonian star.

The situation is slightly different in a relativistic star. Relativity enhances the instability, allowing it to occur in stars with (somewhat) lower rotation rates, so the quadrupole f-mode may also come into play. In order to establish the point at which the instability sets in, we may try to find neutral modes along a rotating star sequence. The instability sets in when a model has zero frequency in the inertial frame, so the problem is effectively time independent. This simplifies the calculation. In full general relativity, neutral modes have been determined for polytropic equations of state using a scheme which involves finding an appropriate gauge in which the time-independent perturbation equations can be solved numerically for δg_{ab}. The results confirm that relativity strengthens the instability (Stergioulas and Friedman, 1998; Yoshida and Eriguchi, 1999).

For $n = 1$ polytropes, the critical angular velocity ratio Ω_c/Ω_K, where Ω_K is the mass-shedding limit at the same central energy density, can be reduced by as much as 15%. The empirical formula

$$\left(\frac{T}{W}\right)_{l=2} = 0.115 - 0.048\frac{M}{M_{\max}}, \tag{18.66}$$

where $M_{\max}$ is the maximum mass for a spherical star allowed by a given equation of state, gives the critical value of T/W for the $l = 2$ f-mode instability with an accuracy of about 5%, for a wide range of realistic equations of state (Koranda *et al.*, 1997).

The results suggest that the f-modes—which efficiently emit gravitational waves—may be unstable in rapidly rotating neutron stars. However, exploring this instability is more challenging than studying, for example, the r-modes. We can no longer rely on slow-rotation results and we obviously need a fully relativistic analysis. Until quite recently, the effects of rapid rotation were not tested properly in linear perturbation theory. Almost all formulations of the perturbation equations were prone to numerical instabilities either at the surface or along the rotation axis of the neutron star. The first results for the oscillations of rapidly rotating stars were, in fact, obtained from evolutions of the full nonlinear equations (Font *et al.*, 2000), albeit in axial symmetry ($m = 0$) where the effects of rotation are less pronounced and where there are no gravitational-wave-induced instabilities.

Eventually a tour-de-force effort led to progress in the modelling of non-axisymmetric perturbations of rapidly rotating neutron stars. The basic strategy was to focus on time-evolutions of the perturbation equations in 2+1 dimensions (Gaertig and Kokkotas, 2008; Gaertig and Kokkotas, 2011). The azimuthal angle can always be decoupled (into the usual $e^{im\varphi}$ Fourier modes) given that the unperturbed rotating star is axisymmetric, but the polar angle generally leads to a complex system of equations coupled by the rotation. So, rather than decomposing the problem in spherical harmonics one formulates an initial-value problem in two spatial dimensions. With this computational technology, the oscillation spectra of fast-rotating relativistic stars have been calculated. The effect of rotation on f- and r-modes has been demonstrated and the critical point for the onset of the f-mode instability has been determined (Gaertig *et al.*, 2011). In parallel, there has been progress on g-modes (Gaertig and Kokkotas, 2009) and differentially rotating stars (Krüger *et al.*, 2010). The current state-of-the-art assumes the Cowling approximation, where the spacetime is assumed to be frozen. This approximation is expected to be good for r- and g-modes, as well as higher order p-modes, but the frequency of the quadrupole f-mode may be altered by as much as 30%. We need to keep this in mind as we move on.

A key advantage of working in the frequency domain, as one would usually opt to do for non-rotating stars, is that mode frequencies and damping times are directly obtained from the real and imaginary parts of each complex eigenfrequency. In contrast, in a time-dependent evolution formulation one has to extract the modes through post-processing. The frequency of a specific oscillation mode is obtained by Fourier transforming the time series at different points inside the star into the frequency domain (so that the time dependence of a mode is $e^{i\sigma t}$, as in the previous section) and identifying the corresponding peaks in the power spectrum. A typical result of this procedure is shown in Figure 18.13. The results are extracted in a coordinate frame co-rotating with the

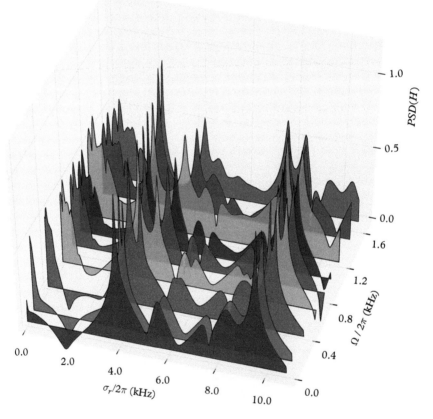

Figure 18.13 *An illustration of the splitting of the (power spectral density for) non-axisymmetric $|m| = 2$ modes obtained from time evolutions of the perturbation equations of a sequence of rapidly spinning neutron stars. The results are shown in a rotating reference frame with corresponding mode frequency σ_r. The mass-shedding limit for this particular sequence corresponds to $\Omega/2\pi = 2.18$ kHz. (Reproduced from Gaertig and Kokkotas (2011), copyright (2011) by the American Physical Society.)*

star. As before, the frequency σ_r is related to the inertial frame mode-frequency, σ_i, through

$$\sigma_i = \sigma_r - m\Omega, \tag{18.67}$$

where Ω is the rotation rate of the star according to a distant observer.

The results in Figure 18.13 show that, while the modes are degenerate in the non-rotating limit (all values of m lead to the same frequency), the frequencies of modes with the same multipole l but opposite azimuthal index $m = \pm|m|$ (co- and counter-rotating modes) move apart as the spin of the star increases. In the non-rotating limit, one can identify various peaks in the power spectrum. In the illustrated example, the sharpest ones are located at $\sigma_1/2\pi = 3.837$ kHz and at $\sigma_2/2\pi = 9.432$ kHz. Inspection of the corresponding eigenfunctions shows that the peak at σ_1 belongs to the quadrupole f-mode while σ_2 matches the first p-mode. The rotational splitting can be clearly seen for these two modes. Other modes are obviously also split, but as the corresponding peaks are less pronounced this effect is more difficult to track.

One key motivation for working in the co-moving frame is that one can construct a model-independent relation between the mode frequency and the rotation rate. This is demonstrated in Figure 18.14, which shows how the f-mode frequency is affected by

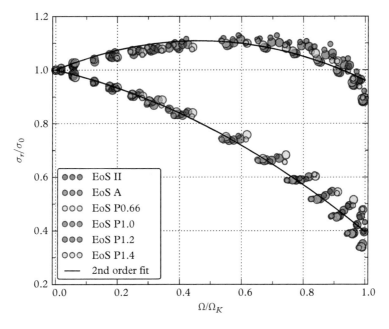

Figure 18.14 *Normalized mode frequencies and fitting curves in the co-moving frame. Larger circles represent more compact stellar models while the small circles are for less compact ones. σ_0 is the frequency of the f-mode in the non-rotating limit and Ω_K represents the Kepler limit. (Reproduced from Gaertig and Kokkotas (2011), copyright (2011) by The American Physical Society.)*

rotation for a set of equations of state. Quadratic fits to the results lead to (Gaertig and Kokkotas, 2011) (similar relations for realistic equations of state can be found in Doneva *et al.* (2013))

$$\frac{\sigma_r}{\sigma_0} \approx \begin{cases} 1.0 - 0.27\left(\frac{\Omega}{\Omega_K}\right) - 0.34\left(\frac{\Omega}{\Omega_K}\right)^2, & m = -2, \\ 1.0 + 0.47\left(\frac{\Omega}{\Omega_K}\right) - 0.51\left(\frac{\Omega}{\Omega_K}\right)^2, & m = 2. \end{cases} \tag{18.68}$$

We can use these relations to establish the onset of instability. Working out the mode pattern speed in the inertial frame, we find that the co-rotating $m = -2$ mode is always stable (as expected) while the counter-rotating $m = 2$ mode may become unstable above a critical rotation rate. In order to estimate the critical rotation rate, we need the non-rotating mode frequency, which is given by

$$\frac{1}{2\pi}\sigma_0 \text{ (kHz)} \approx 0.498 + 2.418 \left(\frac{M}{1.4M_\odot}\right)^{1/2} \left(\frac{R}{10 \text{ km}}\right)^{-3/2}, \tag{18.69}$$

as well as an empirical relation for the Kepler frequency[4]

$$\frac{\Omega_K}{2\pi} \text{ (kHz)} \approx 1.015 \left[1 + 0.334\left(\frac{M}{1.4M_\odot}\right)\left(\frac{R}{10 \text{ km}}\right)^{-1}\right] \left(\frac{M}{1.4M_\odot}\right)^{1/2}\left(\frac{R}{10 \text{ km}}\right)^{-3/2}. \tag{18.70}$$

This exercise tells us that more compact stars become unstable earlier than less centrally condensed ones. As an example of the onset of the instability, consider a $2M_\odot$ star described by the (often-used) APR equation of state (Akmal *et al.*, 1998) with radius 10.88 km. From the above results it follows that this star becomes unstable above 97% of the Kepler limit. This result is illustrated in Figure 18.15, alongside the results for the $l = 3$ and 4 modes.

As we have already outlined for non-rotating stars, empirical relations like (18.68) can be employed in an asteroseismology analysis, determining (say) the mass, radius, and rotation rate of a neutron star from observed frequencies and/or damping times. Since we now need to constrain three parameters we need (at least) three observables. However, not all combinations of frequencies and damping times are suitable for solving the inverse problem. For example, in the simplest case one might try to use three frequencies of different modes. But we see from the empirical relations that this would only allow us to infer the rotation rate Ω and the average density M/R^3— not mass and radius independently. The effect of the compactness M/R on the Kepler frequency is too weak to have significant impact on the analysis. In order to break the degeneracy, we need some additional observable (like one of the damping times).

---

[4] Note that the parameters in this relation are different from those used in (12.82). However, the two results are close for a given neutron star model.

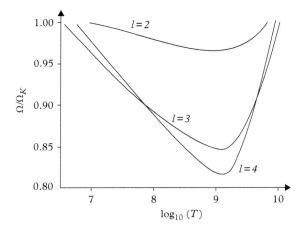

Figure 18.15 *Instability window for f-modes of a rotating star. The results are for a star described by the APR equation of state with gravitational mass, in the non-rotating limit, of $M = 2.0\,M_\odot$. (Adapted from Doneva et al. (2013).)*

As we are working within the Cowling approximation, we need to use the quadrupole formula to estimate the damping times of the modes. The analysis is analogous to that for the r-modes (see Chapter 15) although now the main contribution comes from the mass multipoles. For rotating stars, the damping time τ_{gr} depends crucially on how the mode frequencies change with the rotation rate. Estimated damping times corresponding to the modes from Figure 18.14 are provided in Figure 18.16. It is worth noting that the behaviour is rather different for the two branches. The point at which the counter-rotating $m = 2$ modes become unstable corresponds to the point where the damping time crosses the dashed line in the left panel.

In order to explore the f-mode instability window, we need to also consider the effect of shear- and bulk viscosity. This analysis follows the same steps as in the case of the r-mode instability in Chapter 15. An example of the instability regions for the $l = 2, 3$, and 4 f-modes is provided in Figure 18.15. We see that, as per the estimate following equation (18.70), the quadrupole modes are only marginally unstable—the instability window only reaches down to about 97% of the Kepler limit. The $l = 3$ and 4 f-modes lead to larger instability regions, reaching down to $80 - 85\%$ of the Kepler limit. In these cases, a newly born and rapidly rotating neutron star may remain in the instability window long enough to make the gravitational-wave signal observable.

It makes sense to conclude the discussion of the f-mode instability with an astro-physically motivated example. Consider the gradual evolution of a binary neutron star remnant, on a timescale longer than can be tracked with fully nonlinear evolutions (see Chapter 20). The remnant formed by merger is likely to be supramassive—it will collapse once it has lost enough angular momentum—but it may survive long enough to leave an observational signature. We can track the evolution of an unstable f-mode in such a supramassive star with a set of equations analogous to those used for the r-modes in

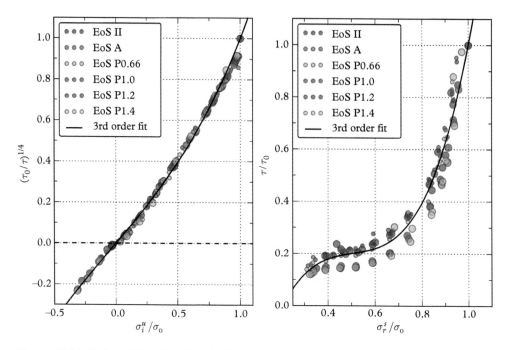

Figure 18.16 *Estimated damping times for the f-modes of rotating neutron stars. Larger circles represent the compact models while small circles are for the less compact configurations of each equation of state. Left: Damping times (τ) for the counter-rotating m = 2 branch (scaled in terms of the damping time of the mode in a non-rotating star, τ₀), which becomes unstable above the horizontal dashed line. Right: Damping times for the co-rotating m = −2 branch, which is always stable. (Reproduced from Gaertig and Kokkotas (2011), copyright (2013) by the American Physical Society.)*

Chapter 15. Moreover, as the evolution time is likely much shorter than the observation time, we can make use of the usual argument leading to the effective gravitational-wave amplitude. If we quantify the f-mode saturation amplitude in terms of an energy $E_{sat}^f = 10^{-6} Mc^2$ we find that the signal may possibly be detectable at the distance to the Virgo cluster with advanced detectors (Doneva *et al.*, 2015). Of course, the formation of a strong magnetic field and the associated dipole torque will impact on this signal (cf. the discussion of constraints from the magnetar model for superluminous supernovae in Chapter 14). The presence of an unstable r-mode will also sap angular momentum from the system, potentially weakening the f-mode signal.

The first of these effects is quantified in the left panel of Figure 18.17, which shows the gravitational-wave signal-to-noise ratio (associated with the quadrupole f-mode) as a function of the dipole component of the magnetic field. The distance to the source is taken to be $d = 20$ Mpc and the assumed sensitivity is that of Advanced LIGO (the result for a third-generation instrument like the Einstein Telecope would be about an order of magnitude larger). The most important conclusion is that, as long as the dipole

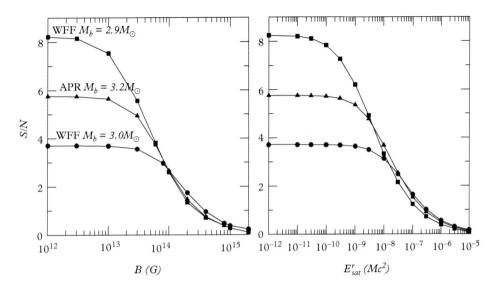

Figure 18.17 *Left: The estimated f-mode afterglow signal-to-noise ratio for Advanced LIGO as a function of the dipole component of the magnetic field on the surface, B. Results are shown for the $l = m = 2$ f-modes of three neutron star models: (i) WFF2 with baryon mass $M_b = 2.9 M_\odot$ (squares), (ii) WFF2 with $M_b = 3.0 M_\odot$ (circles), and (iii) APR with $M_b = 3.2 M_\odot$ (triangles). The saturation amplitude is taken to be $E_{sat}^f = 10^{-6}$ and the assumed distance to the source is $d = 20$ Mpc. Right: The analogous signal-to-noise ratios for the case when the r-mode instability is active. The assumed r-mode saturation amplitude is E_{sat}^r from (18.71). (Based on data provided by D. Doneva, taken from Doneva et al. (2015).)*

component of the magnetic field is below 10^{14} G, i.e. we are not dealing with a magnetar, the gravitational-wave signal may reach a detectable level (typically requiring a signal-to-noise ratio of about of 8) for Advanced LIGO. For stronger magnetic fields, the dipole radiation efficiently drains rotational energy from the system (as discussed in Chapter 14). Hence, less energy is emitted through gravitational waves and the signal becomes more difficult to detect.

The impact of the r-mode instability is illustrated in the right panel of Figure 18.17. In this case, the unstable r-modes may drain a substantial amount of rotational energy from the star (depending on the saturation amplitude), which again shortens the evolution time and makes the f-mode gravitational-wave signal less detectable. However, we see from Figure 18.17 that, as long as the r-mode saturation amplitude, E_{sat}^r, remains below roughly 10^{-8} the signal-to-noise ratio is interesting for Advanced LIGO and significant for the Einstein Telescope. In general, we need the r-mode saturation amplitude to be roughly two orders of magnitude smaller than E_{sat}^f in order for the r-modes to not significantly affect the f-mode evolution. Comparing the amplitudes used here (in terms of energy) to the r-mode discussion in Chapter 15 we have

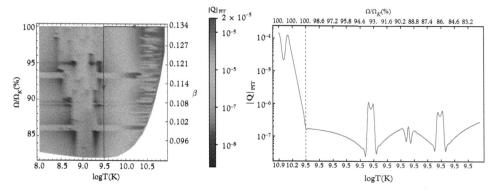

Figure 18.18 *Left: The lowest parametric instability threshold for f-mode coupling and hypothetical evolution of a toy model for a supramassive neutron star formed following binary merger (in this example, $M = 2.5\,M_\odot$). The stellar model is a polytrope with $\Gamma = 3$ and adiabatic exponent $\Gamma_1 = 3.1$. The star enters the instability window during its cooling phase, rotating at its maximum angular velocity, until thermal equilibrium is established (indicated by the vertical dashed line), at which point it descends through the window at $T \approx 3 \times 10^9\,K$. Right: The corresponding evolution of the f-mode amplitude. (Reproduced from Pnigouras and Kokkotas (2016), copyright (2016) by the American Physical Society.)*

$$E^r_{\text{sat}} = 2.4 \times 10^{-5}\alpha_s^2 \left(\frac{R}{10\ \text{km}}\right)^2 \left(\frac{v_s}{1\ \text{kHz}}\right)^2. \tag{18.71}$$

Recalling that we expect α_s to be smaller than 10^{-3} we see that the r-mode instability is unlikely to have an adverse effect on any f-mode afterglow from supramassive merger remnants (Doneva *et al.*, 2015). This should be good news for gravitational-wave searches.

Of course, we also need to establish at what level the unstable f-mode reaches saturation. The theoretical framework for answering this question is pretty much the same as in the case of the r-modes. We need to consider the nonlinear coupling between the f-mode and other modes in the system (Pnigouras and Kokkotas, 2015). As the f-mode is only unstable at high rates of rotation we have to account for rotational corrections to mode frequencies and eigenfunctions. Moreover, as in the case of the r-modes, the specific couplings that lead to the most efficient saturation will evolve as the system evolves. Hence, the saturation problem is also key if we want to understand the evolutionary path of stars that exhibit the f-mode instability. Figure 18.18 provides a typical result from the only detailed study of this problem (Pnigouras and Kokkotas, 2016). The colour-coded results for the lowest parametric instability threshold exhibit a complex brush-stroke pattern which arises from the fact that the coupling sensitively depends on the parameters of the problem. This leads to a, potentially rapidly, varying saturation amplitude along any evolutionary path. As in the case of the r-mode instability, this leads to a complex gravitational-wave signal, for which it may be difficult to develop reliable search templates.

19

Colliding black holes

We have arrived at a point where we understand many aspects of gravitational-wave astronomy. However, this understanding is incomplete. We used approximations to explore the nature of gravitational waves and a range of relevant astrophysical sources. This led to important insights and useful intuition, but, as Einstein's theory is ultimately nonlinear, the scope of this kind of modelling is limited. There will always be a concern that we might be missing crucial aspects—lost in the linearization/simplification. In fact, the most promising gravitational-wave sources involve nonlinear dynamics—explosive hydrodynamics and/or nonlinear aspects of strong gravity.

If we want to explore scenarios like compact binary mergers or gravitational collapse to form a neutron star or a black hole, we must resort to numerical simulations. As we turn to this problem, we face a new set of challenges. We need to figure out how we best 'put Einstein's equations on the computer' and how we implement realistic physics in large-scale simulations. The issues are both conceptual and practical. Yet, decades of effort have led to (fairly) reliable simulations, either with (supernovae or neutron star mergers; see Fryer and New (2011), Ott (2009), and Baiotti and Rezzolla (2017)) or without (black-hole dynamics; see Sperhake (2015)) a realistic matter description or the inclusion of magnetic fields (gamma-ray bursts; see Gehrels and Mészáros (2012)). In order to understand the developments behind state-of-the-art simulations—which often involve runs lasting months on the largest available supercomputers—it is natural to break the discussion into two parts. The first part introduces the main strategy and leads us through to successful simulations of the inspiral and merger black holes—a pure vacuum problem. The second step adds flesh to these bones by providing a description for matter degrees of freedom and a connection to the underlying physics. This chapter focuses on the first part of the problem.

Before going into detail, it is worth asking why the problem is so difficult. In order to appreciate the answer, let us strip away any complex matter physics. By focussing on vacuum problems, we do not have to worry about hydrodynamics, equations of state, matter shocks, thermodynamics, and so on. We are left with the vacuum Einstein equations

$$R_{ab} = 0.$$

Gravitational-Wave Astronomy: Exploring the Dark Side of the Universe. Nils Andersson, Oxford University Press (2020).
© Nils Andersson. DOI: 10.1093/oso/9780198568032.001.0001

We know that this is a set of coupled nonlinear partial differential equations, second order in space and time, representing 'wave equations' for the metric variables. Naively, one might think this problem should be straightforward. However, this is not the case. It is clear from the outset that any numerical simulations will be both technically challenging and computationally expensive. The difficulties are easy to demonstrate. Consider a set of grid points in spacetime, each associated with a worldline of a fiducial observer (as in Figure 19.1). The (time) step of a simulation is fundamentally limited by the speed of light. Each numerical grid point is associated with a light cone. When the future worldline of one of the observers (B, say) intersects the light cone of another (A) we must update the information in the simulation to account for the history (and physics) of the second observer. This constrains the time steps of a simulation. Ultimately, it also explains the local nature of the problem and why making maximal use of parallel computing is tricky.

Simulations involving black holes introduce additional (practical) issues (see, for example, Alcubierre (2008)). The evolution variables may develop large gradients, degrading the numerical accuracy. The presence of singularities, e.g. at the 'centre' of a black hole, will make a code unstable. Even seemingly 'simple' problems require serious thinking. For example, how do we move a black hole across a numerical grid? As soon as we realize that this would involve grid points emerging from inside the event horizon, we see that the question involves non-causal behaviour (at some level) which may (obviously) not be trivial.

As a first step towards exploring these issues, we will outline the basic strategy for solving the Einstein equations numerically, without introducing approximations or

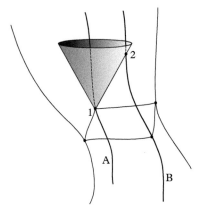

Figure 19.1 *An illustration of the fundamental limitation of the time step in a relativistic spacetime simulation. As soon as the worldline of a fiducial observer associated with one of the numerical grid points (B) intersects the (future) light cone of the observer associated with another grid point (A), we must update the simulation to account for the history (and physics) associated with the different observers. The speed of light ultimately limits all simulations by rendering the problem relatively local.*

imposing particular symmetries. The basic idea is first to cast the Einstein equations as an 'initial value problem' and then develop an algorithm that allows the computer to march from one time level to the next. In Newtonian physics this is straightforward—we have a universal concept of time. In general relativity, the problem is not so easy as the notion of time depends on the observer. We also have to deal with the coordinate freedom of the problem. However, to some extent this is good news as we can turn the choice of 'coordinates' to our advantage.

In order to get started we have to reinstate the concept of time. Intuitively, Einstein's equations represent the 'evolution' of a given spacetime, and in order to carry out a numerical evolution (tell the computer how to take a step towards the 'future') it is necessary to introduce a slicing of spacetime. The standard approach to this problem is based on our everyday intuition (Arnowitt *et al.*, 2008). We introduce a time coordinate and foliate spacetime into an ordered sequence of (spacelike) hypersurfaces, as in the left panel of Figure 19.2. This leads to a fairly standard initial value problem with boundary conditions to be imposed at some finite distance. Still, as we are dealing with nonlinear problems, these boundary conditions may cause trouble. If we want to model gravitational-wave signals, then we need to figure out a way to extract the waves at a finite distance from the source. We will consider this issue in more detail in the following. We will also outline an alternative formulation based on null coordinates, illustrated in the right panel of Figure 19.2. This approach has the advantage of allowing us to include spatial infinity on the computational grid, which enables a more precise measure of the emitted waves.

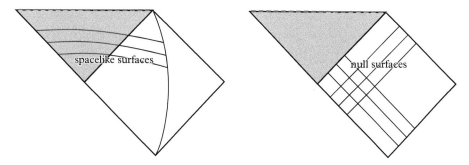

Figure 19.2 *Two different slicings (thin lines) of a compactified black-hole spacetime. The grey area represents the inside of the black hole. The left panel shows the standard 3+1 decomposition, based on introducing a timelike coordinate and solving the evolution problem on a sequence of spacelike hypersurfaces. The right panel illustrates a null slicing, where the problem is solved on null hypersurfaces. This simplifies the analysis of the asymptotic behaviour and the extraction of emerging gravitational waves.*

19.1   The 3+1 decomposition

The majority of simulations in numerical relativity have been based on the approach introduced by Arnowitt, Deser, and Misner in the early 1960s. (hereafter ADM[1]) The first step involves foliating spacetime into a family of spacelike hypersurfaces, Σ_t, which arise as level surfaces of a new scalar time, t. Given the normal to each surface[2]

$$N_a = -\alpha \nabla_a t, \tag{19.1}$$

we have

$$N_a = (-\alpha, 0, 0, 0), \tag{19.2}$$

and the normalization $N_a N^a = -1$ immediately leads to $\alpha^2 = -1/g^{tt}$. The function α is called the lapse. The dual to $\nabla_a t$ leads to a time vector

$$t^a = \alpha N^a + \beta^a, \tag{19.3}$$

which introduces the so-called shift vector β^a. This vector is spatial, so, we have $N_a \beta^a = 0$, and it follows that

$$N^a = \alpha^{-1}\left(1, -\beta^i\right). \tag{19.4}$$

The spacetime can now be written in the standard ADM form

$$ds^2 = -\alpha^2 dt^2 + \gamma_{ij}\left(dx^i + \beta^i dt\right)\left(dx^j + \beta^j dt\right), \tag{19.5}$$

where the (induced) metric on each spatial hypersurface is

$$\gamma_{ab} = g_{ab} + N_a N_b. \tag{19.6}$$

For future reference, it is worth noting that γ_b^a provides a projection orthogonal to N_a and (as a result) γ_{ab} and its inverse can be used to raise and lower indices of purely spatial tensors. For example, we have

$$\beta_i = \gamma_{ij}\beta^j. \tag{19.7}$$

---

[1] See Arnowitt *et al.* (2008) for a reprinted version of the original paper and York (1979) for a seminal contribution. Modern approaches are described in Alcubierre (2008), Baumgarte and Shapiro (2010), and Shibata (2016).
[2] The sign is chosen in such a way that time flows into the future.

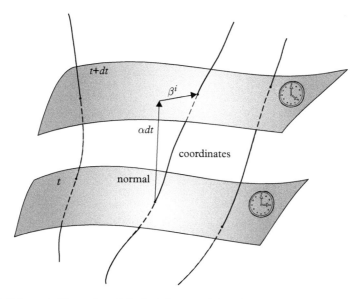

Figure 19.3 *Schematic illustration of the 3+1 decomposition of spacetime, indicating the meaning of the lapse α and the shift β^i.*

It is also worth stressing the geometrical interpretation of the new variables. The lapse determines how proper time advances from one time slice to the next along the normal vector, while the shift vector determines how the coordinates are shifted on the next slice; see Figure 19.3. Together, these two functions encode the coordinate freedom of general relativity. They do not influence the physics.

Reading off the metric from the line element (19.5), we have

$$g_{ab} = \begin{pmatrix} -\alpha^2 + \beta^2 & \beta_j \\ \beta_i & \gamma_{ij} \end{pmatrix},$$

$$(19.8)$$

with inverse

$$g^{ab} = \begin{pmatrix} -1/\alpha^2 & \beta^j/\alpha^2 \\ \beta^i/\alpha^2 & \gamma^{ij} - \beta^i\beta^j/\alpha^2 \end{pmatrix}.$$

$$(19.9)$$

We want to write down the equations of general relativity in this new framework. In doing this it is natural to work with derivatives within each hypersurface. Hence, we introduce the (totally) projected derivative

$$D_a = \gamma_a^b \nabla_b,$$

$$(19.10)$$

where (in general) all free indices should be projected into the surface. This new derivative is 'compatible' with the spatial metric;

$$D_a\gamma_{bc} = \gamma_a^d\gamma_b^e\gamma_c^f\nabla_d\gamma_{ef} = \gamma_a^d\gamma_b^e\gamma_c^f\nabla_d(g_{ef} + N_eN_f)$$
$$= \gamma_a^d\gamma_b^e\gamma_c^f\left(N_f\nabla_dN_e + N_e\nabla_dN_f\right) = 0, \tag{19.11}$$

which follows since $\gamma_a^bN_b = 0$. This means that we have a covariant derivative in the surface orthogonal to N^a and we can follow the derivation from the four-dimensional case and construct a tensor algebra for the three-dimensional spatial slices. For example, the Christoffel symbols associated with D_a are simply obtained by repeating the steps from Chapter 2.

19.2   Evolving the spacetime

Having laid out the required tools, let us consider the left-hand side of the Einstein equations. First of all, we can introduce a (projected) three-dimensional Riemann tensor which satisfies the condition

$$N_aR^a{}_{bcd} = 0. \tag{19.12}$$

As in the full spacetime problem, this curvature tensor represents the failure of second covariant derivatives to commute.

Have we now introduced all the quantities we need to represent Einstein's equations? Not yet, because we have not explained how each hypersurface is embedded in spacetime. The missing piece of information is encoded in the extrinsic curvature, K_{ab}. This is a spatial tensor, such that

$$N^aK_{ab} = 0, \tag{19.13}$$

which measures (roughly speaking) how the Σ_t surfaces curve relative to spacetime. In practice, we measure how the normal N_a changes as it is parallel transported along the hypersurface. Hence, we define

$$K_{ab} = -D_aN_b = -\gamma_a^c\gamma_b^d\nabla_cN_d = -\nabla_aN_b - N_a(N^c\nabla_cN_b), \tag{19.14}$$

where the second term is an analogue of the four acceleration. It is worth noting that K_{ab} is symmetric.

We also have the trace

$$K = K_a^a = g^{ab}K_{ab} = \gamma^{ab}K_{ab} = -D_aN^a = -\nabla_aN^a. \tag{19.15}$$

Alternatively, we can use the properties of the Lie derivative[3] to show that

$$K_{ab} = -2\mathcal{L}_N \gamma_{ab}. \tag{19.16}$$

Since

$$\mathcal{L}_N = \frac{1}{\alpha}(\mathcal{L}_t - \mathcal{L}_\beta) = \frac{1}{\alpha}(\partial_t - \mathcal{L}_\beta), \tag{19.17}$$

this relation is ideal for our present purposes. It links the extrinsic curvature to the way the spatial metric γ_{ab} changes along the vector field N^a. As we are thinking of N^a as representing time, this is exactly what one would want to have for an evolution problem.

Noting that it is natural to use t^a as the time coordinate (rather than N^a), we have

$$\partial_t \gamma_{ij} = -2\alpha K_{ij} + \mathcal{L}_\beta \gamma_{ij}. \tag{19.18}$$

This relation tells us how γ_{ij} evolves and links this evolution to K_{ij}. The fact that the two quantities act like conjugate variables, in turn, suggests that we should try to identify an evolution for the extrinsic curvature.

It is also worth noting that the trace of (19.18) leads to

$$\alpha K = -\partial_t \ln \gamma^{1/2} + D_i \beta^i, \tag{19.19}$$

where $\gamma = g^{ab}\gamma_{ab}$ and

$$\gamma^{ab}\partial_t \gamma_{ab} = \partial_t \ln \gamma. \tag{19.20}$$

We now have the ingredients we need to write down a 3+1 decomposition of the Einstein equations. In principle, the building blocks of the construction are obtained by acting on the indices of the Einstein tensor, either by contraction with N^a or by projecting into the spatial hypersurface. If we project both indices into Σ_t we arrive at the so-called Gauss equation. We have

$$\mathcal{H} = 2N^a N^b G_{ab} = R + K^2 - K_{ij}K^{ij} = 16\pi N^a N^b T_{ab} = 16\pi\rho, \tag{19.21}$$

where we have defined ρ as the energy density of any matter source (as measured by an observer moving along with N^a; see Chapter 20).

Similarly, the so-called Codazzi equation results from one projection and one contraction

$$\mathcal{M}_i = D_j K^j_{\ i} - D_i K = -\gamma^a_{\ i} N^b T_{ab} = 8\pi S_i, \tag{19.22}$$

---

[3] The definition of the Lie derivative from Chapter 13 is easily extended to the present case. We only have to replace spatial indices with spacetime ones.

where S_i is the momentum density of matter moving relative to an observer with N^a.

To complete the description of the problem, we need evolution equations. One such equation, (19.18), is already in hand. In addition, we have what is known as the Ricci equation, which follows from two contractions and two projections on the indices of the Riemann tensor. Combining the result with the Einstein equations we have an evolution equation for the extrinsic curvature

$$(\partial_t - \mathcal{L}_\beta)K_{ij} = -D_i D_j \alpha + \alpha(R_{ij} + KK_{ij} - 2K_{ik}K^k_j)$$

$$-8\pi\alpha \left[S_{ij} - \frac{1}{2}\gamma_{ij}(S - \rho) \right], \tag{19.23}$$

where we have defined the spatial stress tensor as

$$S_{ij} = \gamma_{ia}\gamma_{jb}T^{ab}, \tag{19.24}$$

and $S = \gamma^{ij}S_{ij}$.

We now have a set of equations, (19.18) and (19.23), that describe the evolution from one hypersurface to the next, assuming the matter contribution is known. In addition, we have the Hamiltonian constraint (19.21) and the momentum constraint (19.22). As these constraints do not involve time derivatives they must hold on each spacelike hypersurface without reference to neighbouring time levels. In fact, one can show that (essentially by virtue of the Bianchi identities) if the constraints are satisfied on the initial slice, then they will be preserved during the evolution. Of course, this is not guaranteed in a numerical evolution—the constraints are never exactly satisfied. Numerical errors introduce violations which may grow as we proceed. This suggests that, while the 3+1 formulation allows a free evolution (with no account of the constraints other than for the initial data), it is important to monitor constraint violations as the evolution marches on. In fact, in some cases, it may be advantageous to carry out a constrained evolution, which actively enforces (19.21) and (19.22).

Before we move on, it is useful to sanity check the formulation by counting equations and degrees of freedom. Focussing on the spacetime, the fundamental variables are the three-metric γ_{ij} and the extrinsic curvature K_{ij}. After a bit of counting, we see we have 12 evolution equations, 4 constraints, and the 4 degrees of gauge freedom (the lapse and shift) that we are used to. Moreover, given that the evolution equations are first order, one can deduce that the number of unconstrained dynamical variables is consistent with the presence to two gravitational-wave degrees of freedom (second order in time).

19.3   Initial data

The construction of relevant initial data for numerical simulations is a tricky problem. The reason for this is obvious. Any astrophysical system has evolved to our initial time

($t = 0$, say) and the (possibly nonlinear) evolution up to this point will influence the nature of spacetime (e.g. the gravitational-wave content) at some level.

Without access to the evolution that precedes the set initial time, the specification of initial data becomes a non-trivial 'guess'. There is, however, a well-developed framework for obtaining suitable information (York, 1971; Cook, 2000). In order to explain the procedure, recall that we need a solution to the Hamiltonian and momentum constraints, (19.21) and (19.22). The data consist of the spatial metric γ_{ij} and the extrinsic curvature K_{ij} alongside possible matter fields. These initial data clearly contain too much freedom. Only 4 out of the 12 components of (γ_{ij}, K_{ij}) are fixed by the constraints. The remaining parts—including matter variables—are freely specifiable and can be adjusted to the situation under consideration. Furthermore, it is not obvious which 4 components of (γ_{ij}, K_{ij}) should be obtained by solving the constraints. We have a lot of choice.

The standard procedure separates the freely specifiable data from that fixed by the constraints (Cook, 2000). This method has the advantage that it provides a natural framework for 'superposing' solutions. For instance, given data (γ_{ij}, K_{ij}) for individual black holes, it is possible to 'add' the solutions and solve the constraints to obtain self-consistent, fully nonlinear, initial data for a black-hole binary system.

In order to illustrate the method, let us focus on the vacuum problem. First of all, the spatial metric γ_{ij} is conformally transformed to 'factor out' a scalar component ψ,

$$\gamma_{ij} = \psi^4 \tilde{\gamma}_{ij}, \tag{19.25}$$

where the tilde denotes a conformal quantity. Next, we split the extrinsic curvature in such a way that

$$K_{ij} = A_{ij} + \frac{1}{3}\gamma_{ij}K, \tag{19.26}$$

noting that we have $K = 0$ in vacuum, and

$$A^{ij} = \psi^{-10}\tilde{A}^{ij} \quad \leftrightarrow \quad A_{ij} = \psi^{-2}\tilde{A}_{ij}. \tag{19.27}$$

The new quantity A_{ij} is traceless and we now find that the two constraints decouple. If we further split $\tilde{A}^{ij}$ into a longitudinal and a transverse-traceless part, then the momentum constraint simplifies. Specifically, we can take the conformal metric $\tilde{\gamma}_{ij}$ and $\tilde{A}^{ij}$ as free data. Assuming conformal and asymptotic flatness, i.e. $\tilde{\gamma}_{ij} = \eta_{ij}$ (where η_{ij} is the flat metric in arbitrary coordinates) and $\psi|_\infty = 1$, one can arrive at an analytic solution to the momentum constraint (Bowen and York, 1980)

$$\begin{aligned}\tilde{A}_{ij} = {} & \frac{3}{2r^2}\left[P_i n_j + P_j n_i - 2(\eta_{ij} - n_i n_j)P^k n_k\right] \\ & + \frac{3}{r^3}(\epsilon_{kil}S^l n^k n_j + \epsilon_{kjl}S^l n^k n_i),\end{aligned} \tag{19.28}$$

where $n^i = x^i/r$ with r the radial coordinate. A closer inspection of this solution shows that P^i and S^i represent the linear and angular momentum of the spacetime, respectively. Finally, we need to solve the Hamiltonian constraint, which now takes the form

$$\tilde{D}^2\psi - \frac{1}{8}\psi^{-7}\tilde{A}_{ij}\tilde{A}^{ij} = 0, \tag{19.29}$$

where $\tilde{D}_i$ is the covariant derivative associated with $\tilde{\gamma}_{ij}$. This is an elliptical problem, which is typically solved using a standard relaxation scheme or a spectral approach (Bonazzola *et al.*, 1997).

Since the momentum constraint (19.22) is linear, we can superpose solutions. The associated momenta simply add up. This way we can construct initial data for multiple black holes from single black-hole solutions. A number of different multi-black-hole solutions have been proposed and used in evolutions. We will consider three possibilities. First of all, we can build relevant solutions by joining separate 'universes' through a 'wormhole'. This idea is natural given that the Schwarzschild solution has exactly this kind of wormhole structure. The first such solution, proposed by Misner (1960), represents two isometric, time-symmetric, conformally flat sheets connected by the black holes. The solution is essentially given as an analytic series expansion, representing two momentarily stationary 'throats' with proper separation L in a spacetime with ADM mass[4] M (so each black hole has mass $M/2$). It is also natural to introduce the dimensionless separation $\mu_0 = L/M$. Initial data with larger values of μ_0 then represent infall from a larger separation.

An often used alternative is the initial data proposed by Brill and Lindquist (1963), inspired by the form of the Schwarzschild solution in isotropic coordinates (see Chapter 4):

$$ds^2 = -\frac{M-2r}{M+2r}dt^2 + \left(1+\frac{M}{2r}\right)\left[dr^2 + r^2(d\theta^2 + \sin^2\theta\, d\varphi^2)\right]. \tag{19.30}$$

The spatial part of this metric suggests the conformally flat, time-symmetric solution

$$\psi = 1 + \sum_{i=1}^{N}\frac{M_i}{2r_i}, \tag{19.31}$$

where M_i is the mass of each black hole and r_i is its coordinate distance.

---

[4] The definition of mass is delicate in general relativity, especially if we are interested in a local measurement. In numerical simulations, one would typically consider the so-called ADM mass (see Alcubierre (2008)), which is obtained from the integral over a sphere (S) at a large distance (='infinity'),

$$M = \frac{1}{16\pi}\int_S (\partial_j g_{ij} - \partial_i g_{jj})dV,$$

where g_{ij} is the metric in geodesic coordinates.

A final initial data set—of particular importance for modern evolutions—is based on the idea of punctures (Brandt and Brügmann, 1997). In this scheme, one does not imagine the presence of a wormhole. Instead, one introduces the central singularity into the problem. The assumed conformal factor is

$$\psi = \frac{1}{\chi} + u, \qquad \text{with} \qquad \frac{1}{\chi} = \sum_{i=1}^{N} \frac{M_i}{2r_i}, \qquad (19.32)$$

where the function u (assumed to be smooth) follows as a solution to the Hamiltonian constraint

$$\tilde{\nabla}^2 u - \frac{1}{8} \chi^7 \tilde{A}_{ij} \tilde{A}^{ij} (1 + \chi u)^{-7} = 0. \qquad (19.33)$$

19.4 Slicing conditions

Having formulated the evolution problem and considered the initial data, we have everything we need to carry out an evolution (apart from boundary conditions, which we will discuss later). However, we have not yet discussed the freedom associated with the lapse α and the shift vector β^i. From a formal point of view we know that these variables have no physical meaning, but this does not mean that we can make arbitrary choices. In fact, the gauge freedom has to be handled with care. This is perhaps the most important and yet least understood aspect of numerical relativity (Alcubierre, 2008). We want the coordinates to be chosen to our advantage; for example, to avoid singularities which would cause a code to crash. It is also preferable to avoid stretching of the numerical grid, as this would decrease the accuracy. The problem is complicated since choices that aim to avoid singularities often lead to grid stretching, turning a desired property into an undesired one.

A seemingly natural starting point would be to work with a set of freely falling observers, such that the normal to each hypersurface is a geodesic. This *geodesic slicing* simply involves setting $\alpha = 1$. It is easy to see that this leads to

$$N^b \nabla_b N^a = D^a \ln \alpha = 0. \qquad (19.34)$$

Moreover, if we set $\beta^i = 0$ (exercising our right to ignore the shift vector), then t^a also satisfies the geodesic equation. However, this is not an ideal choice, because geodesics can focus to create coordinate singularities. In order to avoid this problem, we may consider the so-called *maximal slicing*, which specifically aims to prevent focussing. This involves imposing $K = 0$, which ensures that volume elements remain constant. It has the advantage that the slicing 'avoids' strong-field regions and, hence, is a common choice for spacetime evolutions. However, the desirable properties come at a cost—we now have to solve an elliptic equation for the lapse. Still keeping $\beta^i = 0$, we have (see (19.23))

$$D_i D^i \alpha = \alpha K_{ij} K^{ij}. \tag{19.35}$$

Implementing this may be computationally expensive, as we have to do the calculation at each time step. We also have to introduce suitable boundary conditions, and this may not be straightforward on a finite size grid. A third option would be to provide the lapse, or its time derivative, as an algebraic function of the geometric variables.

In order to understand the different options for the lapse, it is instructive to consider the gauge degree of freedom for harmonic coordinates in linearized theory (see Chapter 3). We know that, after introducing the Lorenz gauge conditions on the linearized metric, there is still some freedom associated with the infinitesimal gauge vector ξ^a. Specifically, any solution that satisfies $\Box \xi^a = 0$ will do. Inspired by this we note that, combining the ADM decomposition with

$$\nabla_a \nabla^a x^b = 0, \tag{19.36}$$

we have

$$(\partial_t + \mathcal{L}_\beta)\alpha = -\alpha^2 K. \tag{19.37}$$

This leads to the so-called *harmonic slicing*. The key step is adding the lapse to the set of evolved variables. This is advantageous as it avoids the computational issues associated with solving an elliptic equation at each time step.

In fact, we can generalize the idea by introducing a function $f(\alpha)$ such that

$$(\partial_t + \mathcal{L}_\beta)\alpha = -\alpha^2 f(\alpha) K. \tag{19.38}$$

Modern evolutions tend to use slicing conditions based on this prescription, with a particularly popular choice being $f(\alpha) = 2/\alpha$. This is known as *1+log slicing* (Alcubierre, 2003). It is an attractive option because it is singularity avoiding and also prevents the formation of (gauge) shocks.

Let us turn to the shift vector, which allows us to adjust the coordinate movement on subsequent time slices. It is immediately clear that simply setting $\beta^i = 0$ is not a clever strategy. The shift vector controls the coordinate motion of specific physical features and we have already met situations where such aspects play a key role. Consider, for example, the frame dragging associated with a spinning black hole or a rotating star. The shift vector can also be used to delay unwanted grid stretching. A classic choice is the so-called *minimal distortion shift*. The idea is to minimize changes in the conformal metric throughout the evolution. This leads to a condition that can be written

$$D^j D_j \beta^i + \frac{1}{3} D^i D_j \beta^j + R^i_{\ j} \beta^j - 2\left(K^{ij} - \frac{1}{3}\gamma^{ij} K\right) D_j \alpha = 0. \tag{19.39}$$

However, yet again, we have to solve an elliptic equation so this choice may not be ideal. As in the case of the lapse function, we may draw inspiration from the harmonic gauge and consider dynamical choices for β^i. A fairly natural way to approach this problem is to, first of all, identify the property one would like to preserve (e.g. minimal distortion) and then make the problem hyperbolic by introducing (phenomenologically) a suitable time derivative, e.g. $(\partial_t - \mathcal{L}_\beta)\beta^i$. As this is a phenomenological construction—aimed at enabling a successful evolution—there is still a lot of freedom, and many different options have been suggested. We will describe a particularly successful choice, the so-called gamma-driver shift (Alcubierre *et al.*, 2003), later.

19.5   Wave extraction

In the 3+1 formulation we inevitably have to deal with the presence of boundaries at the edges of the computational domain. Inaccuracies introduced near any boundary travel inwards and may (rather quickly) degrade an evolution. In simple one-dimensional problems, the boundary issue can often be ignored. One can make the computational grid large enough that the simulation of the relevant physics is unaffected on the timescale of interest. For full three-dimensional simulations this is typically not affordable. Moreover, we have to ensure that the constraint equations are satisfied. Hence, we need a detailed analysis of the behaviour far away from a given system. We would like to understand the problem to the point where we can ensure that the behaviour near the boundary is appropriate. This is a complex issue. In practice, most present day simulations are carried out either assuming the boundary is 'far enough away' or imposing (approximate) outgoing-wave conditions for all fields (possibly including some kind of damping to prevent reflection into the computational domain).

The asymptotic behaviour is also important if we want to work out the gravitational-wave signal for a given scenario. As we have already discussed, it is only meaningful to discuss the gravitational waves in the wave-zone, relatively far away from the source. Intuitively, we can carry out this analysis sufficiently far away that we may consider the problem at the linear perturbation level. This may not be practical for numerical simulations, which are likely to be resource limited, but we can still try to adopt the perturbation strategy. The problem is that we do not necessarily have an idea of what the 'unperturbed' background may be.

There are two distinct, and commonly used, wave-extraction procedures (see Bishop and Rezzolla (2016) for a recent review). The first strategy is based on black-hole perturbation theory. We know that vacuum perturbations are governed by wave equations like the Zerilli equation from Chapter 16. Of course, we do not necessarily know the mass of the 'central object'. Nevertheless, assuming that we are in the weak-field region (so that we can neglect second-order perturbations) we can assume that the numerically generated spacetime represents a perturbation of the Schwarzschild solution. After projecting on the tensor spherical harmonics, we can integrate over a sphere to extract the corresponding multipole amplitudes at some finite extraction radius. In general, this procedure provides us with axial and polar amplitudes, u_{lm}^- and u_{lm}^+, respectively, for each value of l and m. These amplitudes then provide the gravitational-wave strain in terms of spin-weighted spherical harmonics (see Chapter 16)

$$h_+ - ih_\times = \frac{1}{\sqrt{2}r} \sum_{l=0}^{\infty} \sum_{m=-l}^{l} \left(u_{lm}^+ - i \int u_{lm}^\times dt \right) {}_{-2}Y_l^m. \tag{19.40}$$

In order to test the reliability of the result, we can compare the inferred amplitude for a sequence of increasing radii (see Figure 19.4). If we truly are in the linear regime, then we should arrive at the same amplitudes regardless of the radius of the extraction sphere. If the numerical results converge in this sense, we have a reliable result.

An alternative approach builds on the Newman–Penrose formalism. From the discussion in Chapter 17, we know that the outgoing-wave content is encoded in the Weyl scalar Ψ_4. Asymptotically, we have

$$\ddot{h}_+ - i\ddot{h}_\times = \Psi_4. \tag{19.41}$$

In order to use this result, we need to introduce a suitable null tetrad. However, we can usually assume that the spacetime is close to the Kerr geometry in the wave zone. As long as this is the case, we can use the Kinnersley tetrad from Chapter 17 to determine Ψ_4 from the numerical data. The veracity of the result can be confirmed by using a set of extraction radii, just like in the Zerilli equation approach.

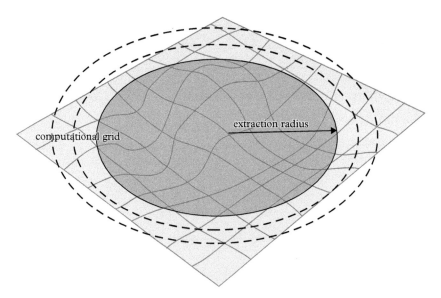

Figure 19.4 *An illustration of the idea of wave extraction. Gravitational waves are extracted (essentially by projecting the numerical data onto the basis used in a perturbation calculation; see Chapter 16) at several radii and the results are considered as reliable when they converge.*

Once the gravitational waveform has been extracted, we have access to all related quantities, like radiated energy, momentum, and angular momentum, from the standard linear theory relations.

The presence of spacetime singularities also requires a careful treatment. Singularities typically leads to diverging (or vanishing) metric components—recall the form of g_{rr} in Schwarzschild coordinates, see Chapter 4—leading to numerical errors which rapidly render a simulation useless. An intuitive way to handle this problem is to remove the divergence from the computational domain. Since one would expect (due to Penrose's cosmic censorship hypothesis) any singularity to be cloaked by an event horizon, the exterior spacetime should be causally disconnected from anything that goes on inside the horizon. In particular, the exterior region should not be affected by the removal of a finite region around the singularity from a numerical evolution. In practice, this involves excising a region inside the apparent horizon, the location of which can be obtained locally.[5] The practical implementation of this kind of scheme is (still) non-trivial, but it has been demonstrated to work in large-scale simulations (Alcubierre *et al.*, 2001).

Improved technology has also allowed detailed studies of the geometrical and dynamical properties of the black-hole horizon (Ashtekar and Krishnan, 2004). The event horizon itself remains tricky as it (strictly speaking) requires an infinite evolution for a precise identification, but the appearance of an apparent horizon provides a useful diagnostic. Precise studies of the horizon dynamics allow the determination of both the mass and angular momentum of the black hole, which help set upper limits on the energy and angular momentum radiated during (for example) gravitational collapse.

19.6 2 + 2 and the Bondi news

An alternative formulation of the evolution problem is based on the use of null coordinates (Bondi, 1960; Bondi *et al.*, 1962; Sachs, 1962). The idea is inspired by the fact that the speed of light is fundamental in Einstein's theory and gravitational waves are expected to travel along null hypersurfaces. The light-cone structure provides a fundamental building block for the theory.

Instead of working with a traditional timelike coordinate, as in the 3+1 formulation, we choose to work with outgoing null rays (analogues of the flat space coordinate $u = t - r$; see Figure 19.5). Such a formulation has a number of advantages. In particular, we can compactify spacetime, as indicated in Figure 19.2, which means that the issue of boundary conditions becomes easier to deal with. It also turns out that the Einstein equations lead to a simple hierarchy of equations with a minimal number of variables

---

[5] In order to determine the location of the event horizon we need to know the complete future evolution of the spacetime. The upshot of this is that it cannot be located from local geometry. In contrast, the apparent horizon can be determined locally. Given a surface in a spacetime, a marginally trapped surface is one whose future-pointing outgoing null geodesics have zero expansion. The apparent horizon is simply the outermost boundary of these trapped surfaces. The presence of an apparent horizon always implies the existence of an event horizon outside of it.

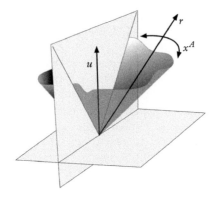

Figure 19.5 *An illustration of the light-cone-inspired spacetime decomposition used in the 2+2 approach.*

(Winicour, 2012; Mädler and Winicour, 2016). For our present purposes, perhaps the most important feature is that we have a well-defined energy at future infinity, meaning that we can read off the emerging gravitational waves without a specific extraction procedure.

Let us start by assuming that $x^0 = u$ is a null coordinate. With $k_a = -\partial_a u$, the normal to $u =$ constant surfaces, we then have

$$g^{ab} k_a k_b = 0,\qquad(19.42)$$

which implies that $g^{00} = g_{11} = 0$. Next, let the angular coordinates $x^A = (\theta^2, \theta^3)$ be constant along these null rays. That is,

$$k^a \nabla_a x^A = 0,\qquad(19.43)$$

which implies that $g^{0A} = g_{1A} = 0$, as well. These angular coordinates label the null surfaces. Finally, let $x^1 = r$ be an areal coordinate, in the sense that

$$\frac{\det(g_{AB})}{r^4} = f(x^A),\qquad(19.44)$$

from which it follows that $g^{01} \neq -1$ and $g_{AB} = r^2 h_{AB}$.

With these assumptions, the line element takes the form

$$ds^2 = -\left(\frac{V}{r} e^{2\beta} - r^2 h_{AB} U^A U^B\right) du^2$$
$$-2e^{2\beta} du\,dr - 2r^2 h_{AB} U^A dx^B + r^2 h_{AB} dx^A dx^B.\qquad(19.45)$$

The spacetime is expressed in terms of the six functions β, V, U^A, and h_{AB}, with a clear physical interpretation. V is an analogue of the Newtonian gravitational potential. The scalar β measures the expansion of the light cone between the asymptotic frame and the world tube, U^A are angular shift components, and h_{AB} encodes the two gravitational-wave polarizations.

In order to develop the relevant evolution equations, let us focus on vacuum problems.[6] We then have the Einstein tensor, $G_{ab} = 0$, which we know satisfies the Bianchi identities

$$\nabla_a G^a{}_b = \frac{1}{\sqrt{-g}} \partial_a \left(\sqrt{-g} \, G^a{}_b \right) + \frac{1}{2} \left(\partial_b g^{ad} \right) G_{ad} = 0. \tag{19.46}$$

Inspired by the corresponding problem in electromagnetism (Tamburino and Winicour, 1966), one next introduces the so-called main equations

$$G^u{}_a = 0, \tag{19.47}$$

and

$$G_{AB} - \frac{1}{2} g_{AB} (g^{CD} G_{CD}) = 0. \tag{19.48}$$

If these equations are satisfied, it follows that, for the metric (19.45), we have

$$G^a{}_r = -e^{2\beta} G^{ua} = 0, \tag{19.49}$$

and therefore the $a = r$ component of (19.46) reduces to

$$\left(\partial_r g^{AB} \right) G_{AB} = -\frac{2}{r} g^{AB} G_{AB} = 0. \tag{19.50}$$

Next, the retarded time and angular components of (19.46) lead to

$$\partial_r \left(r^2 e^{2\beta} G^r{}_X \right) = 0, \qquad \text{where} \qquad X = u, A. \tag{19.51}$$

If they are satisfied at (say) future infinity, $\mathcal{J}^+$, then the $G^r{}_X$ equations are satisfied everywhere. This means that they play a secondary role in the evolution scheme and, hence, they are referred to as the supplementary equations. Physically, they are related to the conservation of energy and angular momentum.

The problem now reduces to a hierarchy, which can be solved in two steps. First we have the hypersurface equations (19.47), which are solved by radial integration along

---

[6] Problems involving matter are not naturally described in this formalism as fluids (say) have their own characteristics associated with the speed of sound.

each null surface. This provides us with β, U^A, and V on the surface. After a bit of massaging, Eq. (19.48) can be cast into evolution equations (involving the retarded time derivative) for the two gravitational-wave degrees of freedom in h_{AB}. These equations are used to update the system to the next slice.

A conceptual advantage of the characteristic formulation is that we can carry out measurements at infinity ($\mathcal{J}^+$). Asymptotically, the Bondi variables approach 0 with known fall-off rates. Expressing the two degrees of freedom from h_{AB} as a complex scalar $\mathcal{J}$, we have (anticipating the fact that we are dealing with radiation)

$$\mathcal{J} = \frac{\mathcal{J}_1}{r} + \mathcal{O}(1/r^2). \tag{19.52}$$

Based on this behaviour, we define the Bondi news function as

$$\mathcal{N} = \frac{1}{2}\partial_u \mathcal{J}_1 = -\lim_{r\to\infty} \frac{1}{2}r^2 \partial_r \partial_u \mathcal{J}, \tag{19.53}$$

and the mass loss associated with the emerging radiation is determined by

$$\frac{dm}{du} = \int_{\mathcal{J}^+} |\mathcal{N}|^2. \tag{19.54}$$

From the fact that the integrand is positive, we see that the Bondi mass of the system (m) must decrease if gravitational waves are emitted (if there is news). If there is no news, $\mathcal{N} = 0$, then the Bondi mass is constant. One can also show that the news function is related to the outgoing wave Weyl scalar through

$$\partial_u \mathcal{N} = -\frac{1}{2}\bar{\Psi}_4, \tag{19.55}$$

with the bar representing complex conjugation. We see that the integral of the Bondi news, $\int \mathcal{N} du$, provides a direct measure of the gravitational-wave strain. Of course, we still need to work out this integral and this may be tricky for noisy numerical data.

19.7   Milestones and breakthroughs

Numerical evolutions based on the original 3+1 formulation turn out to be problematic. With the advantage of hindsight, there is no particular reason why it should not be so. After all, we did not consider the underlying character of the problem. However, knowing that the ADM formulation fails, we are forced to consider alternatives and/or fixes. How do we achieve stable long-term evolutions?

As a starting point, it is useful to better understand the origin of the problem. After all, this may provide useful hints at the solution. From a formal point of view the issue is fairly clear. We need to ensure that we are dealing with a well-posed hyperbolic problem

(Gundlach and Martín-García, 2006). In essence, we should be able to (schematically) write the equations as

$$\partial_t \boldsymbol{u} + \boldsymbol{A}\partial_i \boldsymbol{u} = \boldsymbol{s}(\boldsymbol{u}), \tag{19.56}$$

where $\boldsymbol{u}$ is a vector containing the evolution variables. The characteristics of the problem follow from the eigenvalues of the matrix $\boldsymbol{A}$. An ideal formulation would be 'strongly' hyperbolic, in which case all eigenvalues are real. This ensures the existence of unique solutions that vary smoothly with smooth changes of the initial data. The ADM equations are, however, only weakly hyperbolic and this causes problems.

19.7.1   Head-on collisions

The issues with the ADM formulation became increasingly pressing as simulation technology developed. Nevertheless, there was progress and one can identify a number of milestone results.

The first attempts to collide black holes on the computer were carried out by Smarr and Eppley already in the late 1970s (Smarr, 1979). The results were not particularly precise, due to a low resolution, but they nevertheless brought out the expectation that the signal would be dominated by quasinormal modes (see Chapter 16).

It would take 15 years and a grand alliance of dedicated numerical relativity groups to make decisive progress. The first major milestone involved the head-on collision of non-rotating black holes (Anninos *et al.*, 1993; Anninos *et al.*, 1995; Anninos and Brandt, 1998). The problem was simplified by appealing to symmetries—for equal mass black holes described by Misner-type initial data, the head-on problem is both time- and axi-symmetric. The simulations allowed the first reliable extraction of gravitational waveforms and emitted energy. The results confirmed the dominant role of the black-hole quasinormal modes. Waves extracted at different radii agreed to 10–20%, indicating the level of accuracy of large-scale simulations from the mid-1990s. Good enough for qualitative insights, but limited in precision and (more critically) restricted to short evolution times.

For two black holes, each with mass M, the total energy emitted was found to be of order $\Delta E \approx 2 \times 10^{-3} M$. It is useful to contrast this with the expectations from perturbation theory. Recall the result for a point particle falling radially into a black hole, discussed in Chapter 16, for which the radiated energy was found to be

$$\Delta E \approx 0.01 \frac{m^2}{M} \quad \text{for} \quad m \ll M. \tag{19.57}$$

From the quadrupole formula (see Chapter 3) we know that, if the two masses are similar it would be natural to replace m with the reduced mass μ. In the equal mass case, this leads to the predicted energy being a factor of $1/4$ lower, in fairly good agreement with the numerical simulations (Anninos *et al.*, 1995).

The obvious way to probe the nonlinear aspects of binary inspiral and merger is to perform direct simulations of Einstein's equations. Of course, it an expensive endeavour. The physics of a binary merger typically involves a range of physical scales. The wavelength of gravitational waves from near merger is about 100 times the size of the black holes, so the simulated region should be at least an order of magnitude larger than this. For fluid systems the problem is even more acute, as we then need to keep track of local matter properties. In order to facilitate multidimensional simulations one may have to draw on an aggressive use of adaptive mesh refinement. In addition, we need a handle on the reliability of the results.

However, we have seen that we can get useful insight from perturbation theory. Taking another step in this direction, we may assume that the two black holes start off so close together that they are surrounded by a common horizon. Then one can consider the situation as corresponding to a single perturbed black hole (Price and Pullin, 1994). That the 'perturbation' is sufficiently 'small' to make the approach sensible is not obvious, but the idea leads to surprisingly accurate results—and the strategy is relatively straightforward. After expressing the initial data as 'Schwarzschild background + something else' one ends up with an initial-value problem for the Zerilli equation from Chapter 16.

In the case of Misner data—characterized by the parameter $\mu_0 = L/M$ where L is the initial separation of the two black holes and M is the total mass of the spacetime, as before—one finds that the black holes are surrounded by a common horizon for $\mu_0 < 1.36$. However, it turns out that the parameter μ_0 enters the calculation through a multiplicative factor. As a result, it affects only the amplitude of the gravitational waves and a single perturbation calculation can be used to describe all initial separations. This leads to significant savings compared to nonlinear simulations.

The perturbative evolutions do not provide additional qualitative insight—the waveforms are dominated by quasinormal-mode ringing and at late times they follow the expected power-law tail (discussed in Chapter 16). However, the agreement with numerical simulation of the full nonlinear equations is surprising. Moreover, the agreement extends well beyond the small μ_0 regime. The perturbation calculation remains reasonably accurate even when the two black holes are not initially surrounded by a single horizon. In essence, the close-limit approach provides an efficient way of getting a handle on the gravitational waves from a merger event.

The seemingly unreasonable accuracy of the perturbative approach has a simple explanation. The spacetime is strongly distorted only in the region close to the horizon. Because of the existence of the potential barrier (with peak at roughly $r = 3M$) in the Zerilli equation most of this initial perturbation never escapes to infinity. Much of the 'error' associated with the approximation leads to radiation that is swallowed by the black hole—the predicted waveform in the outer region is quite accurate.

The close-limit approach can be extended to allow the black holes to have initial linear and angular momentum (Baker et al., 1997; Brandt et al., 2000). This way it was found that the radiation efficiency ($\Delta E/E$) saturates at about 2%. This suggests that, no matter how large the initial momentum is, it is not possible to achieve high efficiency in black-hole collisions.

19.7.2   Inspiral and merger

The head-on collision results showed promise, but the problem was still far from solved. Progress stalled (again), largely due to the presence of instabilities preventing longer simulations. In order to avoid the problems, different groups explored different directions and options ranging from minor fixes to complete reformulations of the evolution equations (Baumgarte and Shapiro, 2010; Shibata, 2016). In essence, there are many approaches to the problem. This is easy to see as we have the freedom to add any multiple of the constraints to the evolution equations without affecting the physics. The new equations will have the same physical solutions, but they may be better behaved, both mathematically and from a stability point of view.

Current state-of-the-art simulations adopt one of two competing strategies. The first strategy aims to cure the problems associated with the original ADM approach. The standard approach is inspired by electromagnetism and involves the introduction of additional evolution variables (Shibata and Nakamura, 1995; Baumgarte and Shapiro, 1999). The second approach is quite different. It takes us back to the discussion of linearized theory, which provides the underlying wave nature of dynamical solutions to Einsteins' equations. One can try to preserve this aspect by generalizing the harmonic coordinates (Pretorius, 2005). Both strategies have merits (and perhaps also drawbacks).

Let us first outline how we can fix the issues associated with the ADM formulation. An important step in this direction was taken by in the late 1990s. Baumgarte and Shapiro (1999) achieved dramatically improved stability by adapting an idea that had been pioneered for neutron star simulations by Shibata and Nakamura (1995). The key idea behind this (so-called BSSN) formulation is to conformally decompose the three-metric γ_{ij} in such a way that

$$\tilde{\gamma}_{ij} = e^{-4\phi}\gamma_{ij}, \tag{19.58}$$

and use the conformal factor as a new evolution variable, along with the constraint det $\tilde{\gamma}_{ij} = 1$. The trace of the extrinsic curvature $K = \gamma^{ij}K_{ij}$ is also used as a separate variable. Introducing

$$\phi = \frac{1}{4}\log\psi, \tag{19.59}$$

we have

$$\tilde{A}_{ij} = e^{-4\phi}\left(K_{ij} - \frac{1}{3}\gamma_{ij}K\right). \tag{19.60}$$

Finally, the conformal connections

$$\tilde{\Gamma}^{i} = \tilde{\gamma}^{jk}\tilde{\Gamma}^{i}_{jk} = -\partial_{j}\tilde{\gamma}^{ij} \tag{19.61}$$

are also evolved.

This leads to a system of equations that is larger than the ADM set—we now have as many as 17 evolution equations. However, the formulation fixes many of the problems with the original approach. In particular, the stability is much improved. As a result, about 10 years after the head-on collisions, the first simulations of orbiting black holes were carried out (Brügmann *et al.*, 2004). This involved adding another key ingredient—a clever use of co-moving coordinates. The simulations combined the 1+log slicing along with a version of the so-called *gamma-driver* condition for the shift (Alcubierre *et al.*, 2003)

$$\partial_t \beta^i = \frac{3}{4} \alpha^p \psi^{-n} B^i, \tag{19.62}$$

with

$$\partial_t B^i = \partial_t \tilde{\Gamma}^i - \eta B^i. \tag{19.63}$$

Here ψ is the time-independent conformal factor of the Brill–Lindquist initial data used in the simulations. The main idea is that, the coefficient η introduces damping that relaxes the solution towards the desired behaviour.

With a black hole moving across the computational grid, the stage was set for a breakthrough.

The first successful simulation of inspiralling and merging black holes came from an unexpected direction (Pretorius, 2005). Instead of following the mainstream, Frans Pretorius developed a numerical code based on a generalization of harmonic coordinates. This has obvious conceptual advantages. For example, one can use a compactified coordinate system where the outer boundaries of the grid are at spatial infinity, so the physically correct boundary conditions can be used. In his first simulations, Pretorius was able to extract the gravitational waves from the final orbit and merger of two equal mass, non-spinning black holes; see Figure 19.6. The simulations showed the formation of a Kerr black hole with angular momentum parameter $a \approx 0.7$ and the emitted energy was estimated to be about 5% of the initial rest mass of the system. As it turns out, these results are in very good agreement with the level of energy associated with the discovery event GW150914 (see Chapter 1).

When the breakthrough came, the standard 3+1 simulations were close, but they still required a minor tweak. Successful simulations followed the introduction of a clever 'trick'—the use of moving punctures (Baker *et al.*, 2006*b*; Campanelli *et al.*, 2006). Recall the discussion of punctures for black-hole initial data. In that problem, the metric on the initial slice was given by

$$\gamma_{ab} = (\psi + u)^4 \delta_{ab}, \tag{19.64}$$

with ψ given by (19.31). The key difference between the puncture data and (say) the description of a black hole in isotropic coordinates is the additional control asso-

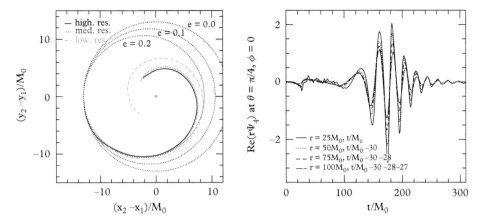

Figure 19.6 *Left: The orbit of the first successful inspiral+merger simulation. The figure shows the coordinate position of the center of one black-hole apparent horizon relative to the other, in the orbital plane. Units have been scaled to the mass (M_0) of a single black hole, and curves are shown from simulations with three different resolutions. Overlaid on the figure are reference ellipses of eccentricity 0, 0.1, and 0.2, suggesting that if one were to attribute an initial eccentricity to the orbit it would be in the range $0 - 0.2$. Right: A sample of the gravitational waves emitted during the merger, as estimated by the Newman–Penrose scalar Ψ_4. The real part of Ψ_4 multiplied by the coordinate distance r from the centre of the grid is shown at a fixed angular location, for several distances r. The waveform has been shifted in time (by amounts shown in the plot) so that the oscillations overlap. (Reproduced from Pretorius (2005), copyright (2005) by the American Physical Society.)*

ciated with the smooth (and finite) function u. If the puncture positions are fixed through an evolution, then the singular behaviour in the metric can be treated analytically. However, this leads to coordinate distortions and simulations tend to crash before a common horizon forms. The new approach 'simply' allowed the punctures to move. In practice, this was still far from straightforward, but the key step was conceptual—many of the established building blocks, like lapse and shift conditions, could be immediately brought to bear on the simulations. Before long, several groups had adapted their codes, allowing them to carry out increasingly precise simulations of inspiral and merger. The black-hole problem was, for all practical purposes, solved.

Numerical relativity waveforms are crucial to the development of data analysis strategies and they also provide important validation for any claimed detection. The results of a modern simulation (from the group at Georgia Tech; see Jani *et al.* (2016)), aimed at reproducing the GW150914 event, are shown in Figure 19.7. This kind of simulation remains expensive and it may be that we will never be able to use numerical data across the entire parameter space of interest. Instead, as a more efficient alternative, sets of high-resolution simulations are used to guide the development of phenomenological schemes, like the effective-one-body approach discussed in Chapter 11.

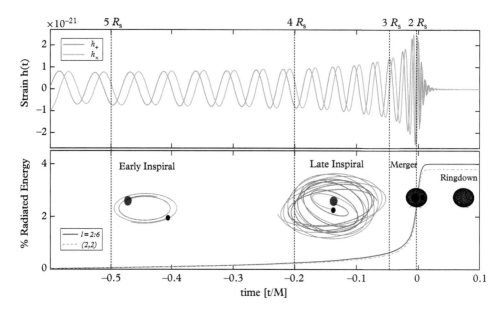

Figure 19.7 *The different stages of binary black-hole coalescence. The upper panel shows the emitted gravitational-wave strain for the two polarizations, h_+ and $h_\times$. The waveform has been scaled to the total mass and distance of the discovery event GW150914. The bottom panel shows the energy emitted at each stage alongside the corresponding trajectories and apparent horizon shapes of the individual black holes. The energy contained in the radiated modes ($l = 2 - 6$, $m = -l - l$, red line) is compared to the dominant mode of radiation ($l = 2, m = \pm2$, grey dashed line). (Figure provided by K. Jani, based on data from Jani et al. (2016).)*

19.8   Recoil and kicks

When we discussed how gravitational waves carry energy and angular momentum away from a system (back in Chapter 3), we did not consider the linear momentum. Yet, it seems perfectly reasonable to ask whether there are situations where gravitational waves carry linear momentum and, as a result, impart a 'kick' on the emitting system. Indeed, this should be the generic case for a binary inspiral as long as there is some asymmetry in the system.

Linear momentum is generated due to a beating of the mass octupole and current quadrupole moments. The mechanism is quite intuitive. Consider a system with one body lighter than the other. If the two objects are in a circular orbit, the lighter mass will be moving faster and will be more effective in beaming gravitational radiation in the forward direction. This leads to net momentum ejection in the direction of motion of the lighter mass, causing a recoil of the system in the opposite direction; see Figure 19.8. The rate of momentum emission is (as usual) obtained by integrating the flux over a sphere far away from the system. We have

$$\frac{dP^i}{dt} = \int \frac{dE}{d\Omega dt} n^i d\Omega, \tag{19.65}$$

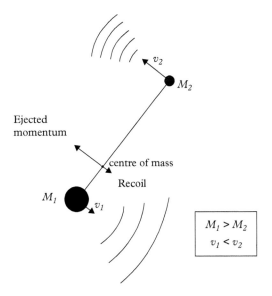

Figure 19.8 *Asymmetric momentum ejection leading to recoil of a binary system. (Adapted from Wiseman (1992).)*

where n^i is the (unit) normal vector. We also have (in the TT-gauge)

$$\frac{dE}{d\Omega dt} = \frac{r^2}{32\pi} \left(\dot{h}_{ij}^{TT} \dot{h}_{TT}^{ij} \right).$$ (19.66)

From the (schematic) form of the waveform

$$h_{TT}^{ij} \approx \frac{2\mu}{r} \left[Q^{ij} + \left(\frac{\delta M}{M} \right) P_{1/2}^{ij} + P_1^{ij} + \left(\frac{\delta M}{M} \right) P_{3/2}^{ij} + \dots \right],$$ (19.67)

where μ is the reduced mass (as in Chapter 11), Q^{ij} is the leading quadrupole moment, P_n^{ij} are the post-Newtonian corrections, and the mass difference is $\delta M = M_1 - M_2$, we see that the leading order momentum emission arises from the cross term of Q^{ij} and $P_{1/2}^{ij}$. The effect is linear in δM. A slow-motion weak-field analysis, essentially leads to (Wiseman, 1992)

$$\dot{P}_n \sim \eta^2 \left(\frac{M}{r} \right)^5 \frac{\delta M}{M} \left(\frac{M}{r} \right)^{1/2},$$ (19.68)

where $M = M_1 + M_2$ is the total mass and $\eta = M_1 M_2 / M^2$ is the symmetric mass ratio.

Recalling that the energy flux scales as (see Chapter 3)

$$\dot{E} \sim \eta^2 \left(\frac{M}{r}\right)^5, \tag{19.69}$$

we have the rough estimate

$$\frac{dP}{dE} \sim \left(\frac{\delta M}{M}\right)\left(\frac{M}{r}\right)^{1/2}, \tag{19.70}$$

which leads to (Wiseman, 1992)

$$v_{\text{kick}} \sim c\left(\frac{dP}{dE}\right)\left(\frac{\Delta E}{M}\right) \sim 180 \text{ km/s} \left(\frac{dP/dE}{0.02}\right)\left(\frac{\Delta E/M}{0.03}\right). \tag{19.71}$$

We see that a significant velocity may be imparted on a merger remnant.

As the result may have important astrophysical consequences, the kick problem has been explored with numerical simulations. Typical results are shown in Figure 19.9. The linear momentum can be inferred via the usual Weyl scalar Ψ_4. In general, we have

$$\Psi_4 = \kappa F(t)_{-2}Y_2^2 + \lambda \bar{F}(t)_{-2}Y_{-2}^2, \tag{19.72}$$

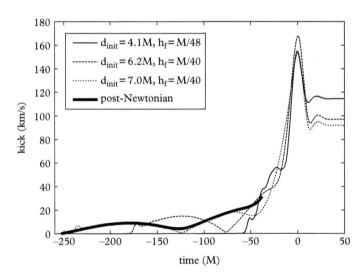

Figure 19.9 *Magnitude of the radiated momentum, as a function of time, extracted from three different black-hole merger simulations, with different initial coordinate separations leading to slightly different kick velocities. For comparison, the second-order post-Newtonian radiated momentum is also shown. The anti-kick during the late stages (after the peak at time ≈ 0) is significant in all simulations. (Reproduced from Baker et al. (2006a) by permission of the AAS.)*

in terms of spin-weighted spherical harmonics. This leads to

$$\frac{dE}{dt} = \frac{r^2}{16\pi}\left(\kappa^2 + \lambda^2\right)\left|\int_{-\infty}^{t} F(t')dt'\right|^2 \tag{19.73}$$

and

$$\frac{dP}{dt} = \frac{2}{3}\frac{r^2}{16\pi}\left(\kappa^2 - \lambda^2\right)\left|\int_{-\infty}^{t} F(t')dt'\right|^2. \tag{19.74}$$

It follows that

$$v_{\text{kick}} \approx c\frac{\kappa^2 - \lambda^2}{\kappa^2 + \lambda^2}\left(\frac{\Delta E}{M}\right). \tag{19.75}$$

Numerical results broadly bring out the expectations, see Figure 19.9, but there are surprises. In particular, simulations show that the final merger event leads to an 'anti-kick' which reverses some of the linear momentum induced during the inspiral. This effect can be understood from the horizon dynamics (Rezzolla *et al.*, 2010). In essence, an anisotropic curvature distribution on the horizon correlates with the direction and intensity of the recoil.

A fit to numerical results for non-spinning black holes suggests that (Sopuerta *et al.*, 2006; González *et al.*, 2007)

$$v \approx 750 \text{ km/s } (4\eta)^2\sqrt{1 - 4\eta}(1 - 0.93\eta). \tag{19.76}$$

This indicates that the maximal kick velocity will be $v \approx 175$ km/s for $\eta \approx 0.195$. We can also use the fit to estimate the kick associated with particular gravitational-wave events. For example, in the case of GW150914 the two masses, $M_1 = 36M_\odot$ and $M_2 = 29M_\odot$, lead to $\eta \approx 0.247$ and a modest kick of $v \approx 61$ km/s. However, this may be a poor estimate as the result can be significantly different for spinning black holes.

The black-hole spin enters the problem is several distinct ways: (i) the spin terms contribute to the orbital decay, and therefore contribute to the accumulated phase of the gravitational waveform, (ii) the spins cause the orbital plane to precess, which changes the orientation of the orbital plane with respect to an observer, thus causing modulation of the shape of the waveform, and (iii) the spins contribute directly to the gravitational-wave amplitude of the waveform. The effects can be quite complicated, but the main impact on the kick velocity can be estimated from (Kidder *et al.*, 1993*b*; Kidder, 1995)

$$\dot{P}_{SO} \sim \eta^2 \left(\frac{M}{r}\right)^5 \left(|\hat{r} \times \Delta| + |\hat{v} \cdot \Delta|\right), \tag{19.77}$$

where

$$\mathbf{\Delta} = \frac{1}{M}(M_2\mathbf{a}_2 - M_1\mathbf{a}_1), \tag{19.78}$$

with $\mathbf{a}_1$ and $\mathbf{a}_2$ the two spin vectors.

Spin effects introduce a second surprise. By fine-tuning the two spins one can demonstrate the existence of super-kicks, with velocities well above 1,000 km/s (González et al., 2007). In order for this to happen, the spins must be directed away from each other in the orbital plane, which may seem an unlikely configuration. One would expect the spins to be aligned with orbital angular momentum if the inspiral is driven by torques from a circumbinary disk. Still, different configurations may arise if the accretion is chaotic. One may also get a random spin orientation in binaries formed through captures, e.g. in clusters. Noting that a super-kick would likely lead to the final black hole breaking free from the host galaxy, see Figure 19.10, it is interesting to ask whether there is any astrophysical evidence for such roaming black holes.

Even relatively modest kicks on the order of 100 km/s may have implications for the growth of massive black holes in the early Universe (Merritt et al., 2004). In essence, dark

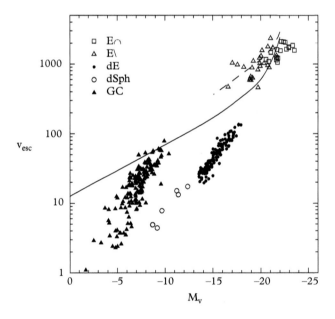

Figure 19.10 *Central escape velocities (in units of km/s) for four types of stellar system that could harbour merging black holes: giant elliptical (E), dwarf elliptical (dE), and dwarf spheroidal (dSph) galaxies and globular clusters (GCs). The solid line provides the mean escape velocity from the dark matter halos associated with the luminous matter. It is easy to see that a kick in excess of 100 km/s may lead to a central black hole becoming unbound. (Reproduced from Merritt et al. (2004) by permission of the AAS.)*

matter halos are expected to merge as part of a hierarchical structure formation, and their central black holes are also assumed to come together. If the resulting kick exceeds the escape velocity of the merged halo, the black hole is likely to depart. This may impact on the number of black-hole mergers in the early Universe, as issue that will be probed by LISA (see Figure 22.7).

20

Cosmic fireworks

Many relevant astrophysical phenomena involve violent matter dynamics. The modelling of such scenarios requires fully nonlinear multidimensional simulations taking into account both the live spacetime of general relativity and the matter degrees of freedom. We have already considered the main issues associated with the spacetime dynamics. Let us now see what happens when we add substance to the discussion.

In recent years there has been considerable progress in developing the tools required for the modelling of archetypal gravitational-wave sources like a supernova core collapse (Ott, 2009) or a neutron star merger (Baiotti and Rezzolla, 2017). The computational technology has reached the level where more detailed matter issues are being considered. In the case of supernova modelling, it is well known that neutrinos play an important role in triggering the explosion (Janka, 2012) and the role of magnetic fields may also be significant (Burrows *et al.*, 2007). For neutron star mergers, finite temperature effects are crucial as shock heating ramps up the temperature to levels beyond that expected even during core collapse. In addition, dynamical magnetic fields are likely to have decisive impact on the post-merger dynamics and may leave an observational signature, e.g. in terms of a short gamma-ray burst (Rezzolla *et al.*, 2011). Similarly, viscous effects may impact on the violent dynamics of a neutron star merger (Alford *et al.*, 2018; Fujibayashi *et al.*, 2018).

20.1 Simulating fluids

The astrophysical problems we want to model involve (often rather complex) matter components. Hence, it is important to understand how matter is incorporated in the 3+1 formulation. To some extent, we have seen this already. The evolution equations for the spacetime (from Chapter 19) introduce a decomposition of the stress–energy tensor

$$T^{ab} = \rho N^a N^b + 2N^{(a}S^{b)} + S^{ab}. \tag{20.1}$$

This expression is general enough that it can account for (pretty much) anything we want. However, the description is adapted to the Eulerian observers associated with N^a

Gravitational-Wave Astronomy: Exploring the Dark Side of the Universe. Nils Andersson, Oxford University Press (2020).
© Nils Andersson. DOI: 10.1093/oso/9780198568032.001.0001

and involves quantities—the energy ρ, the momentum flux S^a and the stresses S^{ab}—measured accordingly. These variables are unlikely to represent the local physics of a given fluid element.

To make progress we need to relate the ingredients from the 3+1 foliation to a specific matter model. As an example, we consider a perfect fluid represented by a barotropic equation of state $\varepsilon = \varepsilon(n)$ where ε is the energy density and n is (for example) the baryon number density (see Chapter 12)—including the various many-body interactions prescribed by the microphysics. One of the main steps involves relating these thermodynamic variables, which are naturally associated with an observer that moves along with the flow, to the quantities used by the (Eulerian) observer associated with N^a.

In order to work out the required 'translation', we need the four-velocity of the fluid, u^a, which can be decomposed as

$$u^a = W(N^a + v^a),\tag{20.2}$$

where $N_a v^a = 0$ and the Lorentz factor is given by

$$W = -N_a u^a = \alpha u^t = (1 - v^2)^{-1/2},\tag{20.3}$$

(the last equality follows from $u^a u_a = -1$). From this we see that

$$v^t = 0, \qquad v^i = \frac{u^i}{W} - N^i = \frac{1}{\alpha}\left(\frac{u^i}{u^t} + \beta^i\right),\tag{20.4}$$

and it follows that

$$v_t = g_{ti} v^i = \beta_i v^i, \qquad v_i = \gamma_{ij} v^j = \frac{\gamma_{ij}}{\alpha}\left(\frac{u^j}{u^t} + \beta^j\right).\tag{20.5}$$

Next we want to translate the equations of fluid dynamics into the 3+1 form. Let us start with the simple case of baryon number conservation. We have

$$\nabla_a(nu^a) = \nabla_a[Wn(N^a + v^a)] = 0,\tag{20.6}$$

and, noting that the particle number density measured by the Eulerian observer is

$$\hat{n} = -N_a(nu^a) = nW,\tag{20.7}$$

we have

$$N^a\nabla_a\hat{n} + \nabla_i(\hat{n}v^i) = -\hat{n}\nabla_a N^a = \hat{n}K,\tag{20.8}$$

since v^a is spatial and we recall that K is the trace of the extrinsic curvature K_{ab}. Making use of the Lie derivative from (19.17), we have

$$N^a \nabla_a \hat{n} = \mathcal{L}_N \hat{n} = \frac{1}{\alpha}(\partial_t - \mathcal{L}_\beta)\hat{n} = -\nabla_i(\hat{n}v^i) + \hat{n}K ,$$
(20.9)

or

$$\partial_t \hat{n} + (\alpha v^i - \beta^i)\nabla_i \hat{n} + \alpha \hat{n} \nabla_i v^i = \alpha \hat{n} K.$$
(20.10)

Finally, since v^a and β^a are already spatial, we have

$$\partial_t \hat{n} + (\alpha v^i - \beta^i)D_i \hat{n} + \alpha \hat{n} D_i v^i = \alpha \hat{n} K = -\hat{n}\partial_t \ln \gamma^{1/2} + \hat{n}D_i \beta^i ,$$
(20.11)

where D_i is the spatially projected derivative (as in Chapter 19). Thus, we have

$$\partial_t \left(\gamma^{1/2}\hat{n}\right) + D_i \left[\gamma^{1/2}\hat{n}(\alpha v^i - \beta^i)\right] = 0,$$
(20.12)

where we have used the fact that

$$(-g)^{1/2} = \alpha \gamma^{1/2},$$
(20.13)

so

$$\nabla_a(-g)^{1/2} = \nabla_a(\alpha \gamma^{1/2}) = 0.$$
(20.14)

Now that we have seen what is involved, the other fluid equations readily follow. In the case of a perfect fluid, we have the stress–energy tensor

$$T^{ab} = (p + \varepsilon)u^a u^b + pg^{ab},$$
(20.15)

where p is the pressure. After the relevant projections, cf. (20.1), we have

$$\rho = N_a N_b T^{ab} = \varepsilon W^2 - p\left(1 - W^2\right),$$
(20.16)

$$S^i = -\gamma^i_a N_b T^{ab} = (p + \varepsilon)\, v^i,$$
(20.17)

and

$$S^{ij} = \gamma^i_a \gamma^j_b T^{ab} = p\gamma^{ij} + (p + \varepsilon)\, W^2 v^i v^j,$$
(20.18)

With these relations in hand, it is practical to work out the equations of motion from (20.1). Starting from

$$\nabla_a T^{ab} = 0, \qquad (20.19)$$

we first of all project along N_b. This leads to an equation for the energy

$$N^a \nabla_a \rho + \nabla_a S^a = \rho K - S^b N^a \nabla_a N_b - S^{ab} \nabla_a N_b, \qquad (20.20)$$

which becomes

$$\frac{1}{\alpha} \left(\partial_t - \mathcal{L}_\beta \right) \rho + \nabla_a S^a = \rho K - S^i D_i \ln \alpha + S^{ij} K_{ij}, \qquad (20.21)$$

and finally

$$\partial_t \left(\gamma^{1/2} \rho \right) + D_i \left[\gamma^{1/2} \left(\alpha S^i - \rho \beta^i \right) \right] = \gamma^{1/2} \left(\alpha S^{ij} K_{ij} - S^i D_i \alpha \right). \qquad (20.22)$$

If we instead make an orthogonal projection (using γ^a_b), we arrive at an equation for the momentum. We get

$$\rho N^a \nabla_a N^c + \gamma^c_b N^a \nabla_a S^b + S^c \nabla_a N^a + S^a \nabla_a N^c + \gamma^c_b \nabla_a S^{ab} = 0, \qquad (20.23)$$

which leads to

$$\left(\partial_t - \mathcal{L}_\beta \right) S_i - S^j \left(\partial_t - \mathcal{L}_\beta \right) \gamma_{ij} - \alpha K S_i + \rho D_i \alpha + \alpha \gamma_{ij} D_k S^{ik} = 0, \qquad (20.24)$$

and the final result

$$\partial_t (\gamma^{1/2} S_i) + D_j \left[\gamma^{1/2} \left(\alpha S^j_i - S_i \beta^j \right) \right] = \gamma^{1/2} \left(S_j D_i \beta^j - \rho D_i \alpha \right). \qquad (20.25)$$

As in the case of spacetime simulations from Chapter 19, the evolution equations may require massage to get into suitable shape. In the fluid problem, sharp gradients may lead to the formation of shocks and a numerical simulation must be able to handle such features. In order to deal with this issue, the relativistic hydrodynamics equations are usually written in flux-conservative form (Font, 2008). In effect, the evolved (conservative) variables $\boldsymbol{U}$ (say) satisfy equations of the form

$$\partial_t \boldsymbol{U} + \nabla \cdot \boldsymbol{F}(\boldsymbol{U}) = \boldsymbol{S}(\boldsymbol{U}), \qquad (20.26)$$

where the fluxes $\boldsymbol{F}$ and sources $\boldsymbol{S}$ are functions of $\boldsymbol{U}$ but not of their derivatives. A conservative shock-capturing scheme is then built around a reconstruction method, where fluid quantities on either side of a computational cell are fed into a solution to the classic Riemann problem that provides the required fluxes across each cell face.

Once we have the equations we need to evolve the fluid from one hypersurface to the next, we can turn our attention to the thermodynamics. We need to connect the evolved variables, $\hat{n}$ and S_i, to the physics encoded in the equation of state. This problem can be messy. In our simple model setting, the equation of state leads to the chemical potential

$$\mu = \frac{d\varepsilon}{dn},$$

(20.27)

and the pressure p is defined by

$$p = n\mu - \varepsilon.$$

(20.28)

In order to connect with the evolution equations, we need to work out the number density. The evolution system (20.12) and (20.25) provides (assuming that $\gamma^{1/2}$ is known from the evolution of the Einstein equations)

$$\hat{n} = nW = n(1 - v^2)^{-1/2},$$

(20.29)

and

$$S^i = \gamma^{1/2}(p + \varepsilon)v^i = \gamma^{1/2}n\mu v^i.$$

(20.30)

We need to invert these two relations to get n and v^i. This will enable us to work out the source terms in the evolution equations, as required to take another step. Of course, we need to know n in order to work out μ from the equation of state, so this inversion is not going to be a simple algebraic operation. In principle, we have to solve

$$n^2 = \hat{n}^2 + \frac{1}{\gamma}\left(\frac{S}{\mu}\right)^2.$$

(20.31)

This is a one-dimensional root-finding problem. Starting from a guessed value for n we calculate μ for the right-hand side and then carry out an iteration to identify the correct value for the number density.

This procedure can be computationally expensive, and we have only outlined the very simplest case. As we make the model more realistic – the more complex the matter model is—the more involved the conversion from conservative (evolved) to primitive variables becomes. In particular, the introduction of magnetic fields significantly complicates the analysis (Font, 2008; Dionysopoulou *et al.*, 2013).

20.2   The bar-mode instability

An archetypal fluid problem—that has attracted considerable attention—is that of the bar-mode instability of a rotating star. We outlined the theory behind this instability in Chapter 13, establishing that a spinning ellipsoid is dynamically unstable above a critical value $\beta_d \approx 0.27$, where the parameter β represents the ratio of kinetic to potential energy ($= T/|W|$; see Chapter 12). In addition to the obvious astrophysical interest, there is a very practical reason why the bar-mode scenario is attractive for simulations. The instability is dynamical. It sets in by its own accord and grows rapidly (on the dynamical timescale). Through simulations, one would hope to explore how the instability evolves once it reaches a nonlinear amplitude and what the associated gravitational-wave signal may be (New *et al.*, 2000; Fryer and New, 2011).

The first step of a bar-mode simulation involves building a rotating model, along the lines discussed in Chapter 12. Doing this, we find that realistic neutron star equations of state do not allow values of β much larger than 0.1 in uniformly rotating neutron stars (see Chapter 12 and Stergioulas and Friedman (1995)). Different scenarios may nevertheless lead to a compact star becoming dynamically unstable to the bar-mode. For example, since $\beta \sim 1/R$ one might expect a collapsing star to suffer the instability at some point during its evolution. Moreover, the maximum attainable β changes dramatically if the star is differentially rotating. Differential rotation may lead to an increase of the mass-shedding limit simply by allowing the equator to rotate slower than the central parts of the star.

Newtonian simulations have established that the critical value at which the bar-mode becomes dynamically unstable remains close to the result for Maclaurin spheroids, $\beta_d \approx 0.27$, for models with varying compressibility. Moreover, it turns out that the onset of instability is only weakly dependent on the chosen differential rotation law. Having said this, there are extreme angular momentum distributions for which β_d becomes very small (Centrella *et al.*, 2001; Shibata *et al.*, 2002). This is interesting, as it indicates that dynamical instabilities could play a role also for relatively slowly rotating stars. This possibility has been demonstrated by numerical simulations of differentially rotating polytropes which indicate the presence of a dynamical instability for $\beta_d \approx 0.14$ (close to the point where secular instabilities are expected to set in; see Chapter 13). These simulations show that an $m = 1$ mode plays a dominant role in determining the evolution of the system (Ou and Tohline, 2006). Closer studies of this result have established a new class of dynamical instabilities, thought to be associated with the nature of the shearing flow (Watts *et al.*, 2005). The nature of these low-T/W instabilities, and whether they are likely to play a role in neutron star astrophysics, continues to be investigated.

Let us return to the classic bar-mode instability. Simulations show that the nature of the bar-mode instability depends on the magnitude of β compared to the critical value (New *et al.*, 2000; Baiotti *et al.*, 2007). For large values, $\beta \gg \beta_d$, the initial exponential growth of the unstable mode (on the dynamical timescale) is followed by the formation of spiral arms. Gravitational torques on the spiral arms lead to the shedding of mass and angular momentum. The unstable mode saturates and the star reaches a dynamically stable state. In this scenario gravitational waves are emitted in a relatively short burst. For some time it was thought that this was the typical behaviour, but more recent work indicates that, when β is only slightly larger than β_d, a long-lived ellipsoidal structure may be formed. If this is the case, the bar-mode could decay rather slowly (on the viscosity/gravitational-wave timescale) until the star reaches the point where it is secularly stable. This may, in turn, lead to a relatively long-lasting gravitational-wave signal. Snapshots from a typical bar-mode evolution are shown in Figure 20.1. However, throwing a slight spanner in the works, other simulations indicate that a long-lived instability requires extreme fine-tuning (Baiotti *et al.*, 2007). The generic bar-mode configuration, quenched by the nonlinear coupling to other modes in the system, may not last long (see Figure 20.2).

It is fairly straightforward to estimate the strength of the gravitational waves emitted by a sizeable bar-mode. Let us assume that the mode saturates at an amplitude η represented by the axis ratio of the ellipsoidal structure. Typical values may lie in the

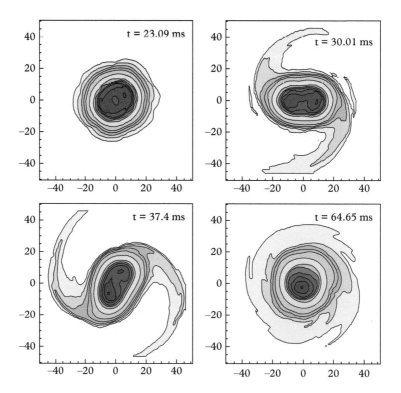

Figure 20.1 *Snapshots from a nonlinear simulation showing the development of the bar-mode instability. The different panels show sets of isodensity contours from which the development of a (temporary) bar structure is evident. (Reproduced from Baiotti et al. (2007), copyright (2007) by the American Physical Society.)*

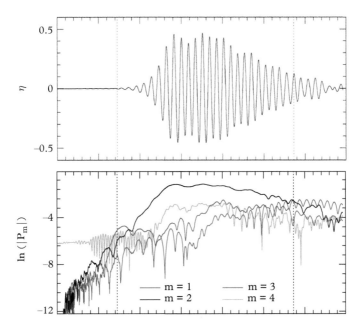

Figure 20.2 *Time evolution of the bar-mode instability. The top panel shows the behaviour of the quadrupole distortion parameter η while the bottom panel provides the power in the m = 1 − 4 modes. The results suggest that the growth of higher order modes saturates the instability and prevents a long-lasting gravitational-wave signal. (Reproduced from Baiotti et al. (2007), copyright (2007) by the American Physical Society.)*

range $\eta \approx 0.2 - 0.4$ (Shibata *et al.*, 2000). From the standard results for a rotating solid body with a given ellipticity (see Chapter 6) we then have

$$\frac{dE}{dt} \approx -\eta^2 \frac{GM^2 R^4 \Omega^6}{c^5}, \tag{20.32}$$

which leads to

$$h \approx 4 \times 10^{-23} \left(\frac{\eta}{0.2}\right) \left(\frac{f}{2\,\text{kHz}}\right)^2 \left(\frac{M}{1.4 M_\odot}\right) \left(\frac{R}{10\,\text{km}}\right)^2 \left(\frac{15\,\text{Mpc}}{d}\right), \tag{20.33}$$

where we have used the fact that the gravitational-wave frequency f is twice the rotation frequency. This estimate compares reasonably well with results from simulations. A signal with this strength could be detectable from sources in local galaxy group. Of course, the detectability of the signal would be significantly improved if the instability leads to the formation of a persistent bar-like structure. Should a long-lived bar form and last for hundreds of rotation periods, one can easily gain a factor of 10 in the signal-to-noise ratio. Such factors could be crucial, so it is important that the long-term evolution of the bar-mode instability, and the nonlinear saturation, are understood in detail.

Given the breakthroughs in numerical relativity it is feasible to carry out bar-mode simulations in full general relativity, with a live spacetime. Such work—for varying degrees of differential rotation—tends to show good agreement with the Newtonian results (Shibata *et al.*, 2000). Relativistic effects enhance the dynamical instability only very slightly and hardly change the critical value β_d at all. Overall, the results of the simulations also seem to be in agreement with the Newtonian estimates of the strength of the gravitational-wave signal.

20.3 Tidal disruption

As a further step along the road to relativistic matter simulations with a dynamical spacetime, let us consider tidal disruption events. This is a problem where the main features may, to some extent, be understood by considering matter flowing in a fixed spacetime (representing the gravitational interaction with a binary partner).

To get a first impression, we extend the discussion of Roche lobe overflow from Chapter 6, adding a twist to the discussion of binary inspiral. When we considered compact binaries in Chapter 5 we assumed that the two stars (or black holes) would gradually approach each other until they touched, or reached the innermost stable orbit, at which point they would crash together. However, the evolution of the system may be affected by mass transfer at an earlier stage. As this could leave an imprint on the gravitational-wave signal—and there may be an electromagnetic counterpart signal, as well (Kobayashi *et al.*, 2004; Shcherbakov *et al.*, 2013)—it is relevant to explore the range of possibilities.

The estimates in Chapter 6 showed that, if the mass ratio exceeds $q = M_2/M_1 = 5/6$ then the mass transfer will be dramatic (for a brief period). This may be relevant, as it means that an inspiralling system could be tidally disrupted before merger. This would lead to a cut-off in the gravitational-wave chirp, providing a measure of the size of the compact star (essentially a handle on the equation of state; see Kyutoku *et al.* (2011), Pannarale *et al.* (2015)). In this context, mixed binaries—with a neutron star spiralling into a black hole—are of particular interest. However, in order to properly understand such systems we need dynamical simulations. Before we consider that problem, let us illustrate how the matter equation of state makes a difference. This also provides us with an opportunity to comment on issues that impact on any attempt to make back-of-the-envelope estimates for this kind of problem.

To be specific, we focus on the case of stable mass transfer (see Chapter 6), assuming that the lighter star expands sufficiently rapidly that it continues to fill its Roche lobe. The orbital separation then increases, which means that the gravitational-wave amplitude and frequency should both decrease. The signature of stable mass transfer would therefore be rather different for gradual tidal disruption and a direct plunge.

As an illustration of how the equation of state enters the problem, let us contrast a system where the donor star is a neutron star to one with a self-bound quark star (Prakash and Lattimer, 2004). As usual, we assume that the orbit decays by the emission of gravitational radiation, according to the quadrupole formula (to leading order; see Chapter 3). In terms of the total mass $M = M_1 + M_2$ and the reduced mass $\mu = M_1 M_2 / M$, the rate of change of orbital angular momentum $\mathcal{J}$ is then (see Chapter 5)

$$\dot{\mathcal{J}}_{\rm gw} = -\frac{32}{5}\frac{G^{7/2}}{c^5}\frac{\mu^2 M^{5/2}}{a^{7/2}} = -\frac{32}{5}\frac{G^{7/2}}{c^5}\frac{q^2 M^{9/2}}{(1+q)^4 a^{7/2}}, \tag{20.34}$$

where a is the orbital separation, and the angular momentum of the system is

$$\mathcal{J}^2 = GM\mu^2 a = \frac{GM^3 a q^2}{(1+q)^4}. \tag{20.35}$$

We focus on circular orbits since we know that the timescale for decay of orbital eccentricity is much shorter than the inspiral timescale.

The binary orbit gradually shrinks until tidal disruption or the system reaches the innermost stable circular orbit (ISCO, see Chapter 10), whichever happens first. Mass transfer sets in when the compact star fills its Roche lobe, at which point matter begins to flow onto the companion. Given that the Roche radius is approximated by (see Chapter 6 and Paczyński (1971))

$$R_L \approx \frac{2a}{3^{4/3}}\left(\frac{M_2}{M}\right)^{1/3} = \frac{2a}{3^{4/3}}\left(\frac{q}{1+q}\right)^{1/3}, \tag{20.36}$$

mass overflow begins when $R_L \approx R_2$, the radius of the secondary star. Mass transfer continues in a stable fashion (for fixed M and $\mathcal{J}$) if the star's radius, after a period of mass transfer, is less than R_L, so that continued gravitational-wave emission maintains the process. For this to happen, we need

$$\frac{d\ln R}{d\ln M_2} \equiv \alpha \le \frac{d\ln R_L}{d\ln M_2} = \frac{d\ln a}{d\ln M_2} + (1+q)\frac{d\ln(R_L/a)}{d\ln q}, \tag{20.37}$$

where we have introduced the parameter α, which is a function of the equation of state and M_2, the companion mass. This parameter determines both the onset of stable mass transfer and the subsequent evolution.

If the mass transfer conserves angular momentum the evolution of the system is determined by

$$\frac{d\ln\mathcal{J}}{dt} = \frac{1}{2}\frac{d\ln a}{dt} + \frac{1-q}{1+q}\frac{d\ln q}{dt} = \frac{\dot{\mathcal{J}}_{\rm gw}}{\mathcal{J}} = -\frac{32}{5}\frac{G^3}{c^5}\frac{qM^3}{(1+q)^2 a^4}. \tag{20.38}$$

Combining this with Eq. (20.37), we arrive at

$$\frac{\dot{\mathcal{J}}_{gw}}{\mathcal{J}} \geq \left(\frac{d\ln q}{dt}\right)\frac{1}{2(1+q)}\left[\alpha + 2(1-q) - (1+q)\frac{d\ln(R_L/a)}{d\ln q}\right], \qquad (20.39)$$

and, since both $\dot{\mathcal{J}}_{gw}$ and $\dot{q}$ are negative, the condition for stable mass transfer becomes

$$\alpha \geq 2q - \frac{5}{3} \qquad (20.40)$$

(which follows from making sure that the square bracket in (20.39) is not negative).

Let us now consider specific 'equations of state'. For hadronic matter, the equation of state is (typically) such that the radius is relatively constant for a range of masses around $1.4 M_\odot$ (recall Figure 12.2). In effect, we have $\alpha \approx 0$, which means that $q \lesssim 5/6$ is the condition for stable mass transfer (just as in the conservative case from Chapter 6). Strange quark stars, on the other hand, are nearly incompressible (away from the maximum mass) so $\alpha \approx 1/3$, which leads to the condition $q < 1$. These simple estimates suggest that binaries containing quark stars would be more likely to have epochs of stable mass transfer. During such an episode, $\dot{a} > 0$ and, as a result the binary companions spiral apart, leaving a (potentially) clean imprint on the gravitational-wave signal (Prakash and Lattimer, 2004).

The question is to what extent these Newtonian estimates can be trusted for neutron star binaries. Our approximations gloss over a few well known facts. First of all, we need the mass transfer to set in before the merger. As a rough estimate of the likelihood of this, we can compare the tidal disruption radius to the ISCO. From (20.36) we have

$$\frac{a}{R_2} = \frac{a}{R_L} \lesssim \frac{3^{4/3}}{2}\left(\frac{M}{M_2}\right)^{1/3}. \qquad (20.41)$$

In the case of a neutron star binary, the masses will be similar so mass transfer will set in roughly when

$$a \lesssim 2.7 R_2 \approx 14 M_2, \qquad (20.42)$$

where we have used $R_2/M_2 \approx 5$ (in geometric units), a typical value for a neutron star. If the primary is also a neutron star, and we use the test particle result for the ISCO, $a_{isco} \approx 6 M_1$, then we see that the mass transfer should start before the end of the inspiral. Of course, it may well be that the late stage inspiral proceeds so rapidly that the actual effect will be small.

Turning to the case of a mixed binary, with a neutron star falling towards a black hole, it is natural to assume that $M_1 \gg M_2$ in which case mass transfer kicks in when

$$\frac{a}{a_{\text{isco}}} \lesssim 1.8 \left(\frac{M_2}{M_1}\right)^{2/3}. \tag{20.43}$$

Basically, only low-mass black holes can disrupt neutron stars outside the ISCO. Of course, as we are discussing an effect that would only be relevant at the late stage of inspiral, we have to use these estimates with caution. To do better, we need simulations.

20.4   Black hole–neutron star mergers

We have seen that, for massive black holes, the tidal forces are typically too weak to disrupt a neutron star before it reaches the ISCO. If we want to make more precise statements, we need to include relativistic effects (noting, for example, that the location of the ISCO shifts when we account for the mass of the secondary). The spin of the black hole may also affect the result.

The development of simulations of mixed binaries, with a live spacetime, have tended to lag behind pure vacuum black-hole binary work. The reason for this is easy to understand as we have to combine the difficulties associated with black-hole 'singularities' and the event horizon with the subtleties of hydrodynamics. In order to study the tidal disruption problem, we need to track several binary orbits. Moreover, the initial orbital separation must be large enough that the tidal disruption takes place during the evolution. These requirements make simulations costly. Adding to this, simulations of systems with large mass ratios are more expensive than their equal mass cousins. Before disruption the numerical time step is limited by the minimum spacing of the grid, which scales as the size of the smaller object. Large mass ratio simulations require many more time steps per orbit in order to reproduce the same accuracy. The merger phase adds to the challenge. The rapid accretion of matter onto the black hole can only be resolved with a fine numerical grid, again requiring small time steps. The prohibitive cost has led to a focus on low-mass black holes. Large mass ratio binaries have almost exclusively been studied using approximations (like a fixed black-hole spacetime).

Interestingly, the 'moving puncture' breakthrough for vacuum simulations (see Chapter 19) also led to progress on the mixed binary problem (Shibata and Uryu, 2007; Etienne *et al.*, 2008). Geometrical arguments suggest that the standard 1+log slicing condition leads to dynamical simulations approaching finite area surfaces around black-hole singularities. That is, they never reach the singularity itself, which means that moving puncture simulations may be successful also for problems involving the accretion of matter. Since the simulations only cover 'regular' regions of the spacetime, the matter flow never encounters the black-hole singularity. This approach may seem somewhat pragmatic but it is appealing as it does not require explicit excision of the black-hole interior.

Largely as expected, numerical simulations demonstrate that mergers of low eccentricity black hole–neutron star binaries fall into two categories (Kyutoku *et al.*, 2011). In the first, the neutron star is tidally disrupted before the system reaches the ISCO. Material from the star is then either accreted onto the black hole or ejected in a tidal tail.

As matter in the tail falls towards the black hole, it forms an accretion disk. These disks tend to be thick, with temperatures of a few MeV (significantly above 10^{10} K), and may contain a significant fraction of the initial mass of the neutron star (up to a few tenths of a solar mass). Moreover, the evolution may maintain a baryon-free region along the black hole's rotation axis—basically setting the system up in such a way that the interaction with the star's magnetic field may lead to the launch of a relativistic jet (Rezzolla *et al.*, 2011). An example of this kind of simulation is provided in Figure 20.3. In the second category, the star simply reaches the ISCO before it fills its Roche lobe. No tidal tail is formed, and there is a prompt merger.

To change the behaviour one must either consider eccentric orbits (for which partial disruption is possible) or ramp up the black-hole spin. It is easy to explain why disk formation tends to be favoured for large spins. We know from the geodesics of the Kerr spacetime (see Chapter 17) that the ISCO radius is smaller for a prograde orbit around a rotating black hole—it approximately halves when the spin increases from $a = 0$ to $0.75M$. Meanwhile, the orbital separation at the onset of mass transfer depends only weakly on the black-hole spin. The decrease of the ISCO radius thus enhances the possibility of disruption before merger compared to the non-spinning case. A retrograde spin plays the opposite role. The ISCO radius increases slightly so tidal disruption is less likely.

The range of behaviour is encoded in the emerging gravitational-wave signal from the late stages of inspiral; see Figure 20.4. Low-spin simulations have spectra similar

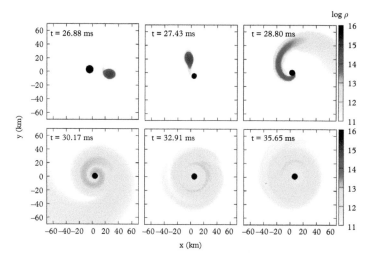

Figure 20.3 *Evolution of the rest-mass density profile (in units of g/cm³) and the location of the apparent horizon (filled circle) of the black hole (in the equatorial plane) for a typical tidal disruption event involving a neutron star falling towards a more massive black hole. In this case the mass ratio is $Q = M_{bh}/M_{ns} = 3$, with $M_{ns} = 1.35M_\odot$ and black-hole spin $a = 0.75M$. (Adapted from Kyutoku et al. (2011), copyright (2011) by the American Physical Society.)*

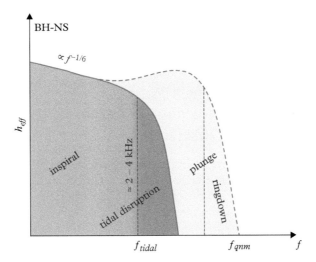

Figure 20.4 *Schematic spectrum of effective amplitude h_{eff} for mixed neutron star-black hole mergers. During the inspiral phase, h_{eff} scales as $f^{-1/6}$. If the neutron star is tidally disrupted before reaching the ISCO, the gravitational-wave emission will cut off at $f_{tidal} \sim 2 - 4\,kHz$. If the neutron star plunges into the black hole without being tidally disrupted, the plunge cuts off the emission and excites the quasinormal modes of the black hole, which ring down emitting gravitational waves at frequency f_{qnm}. (Reproduced from Bartos et al. (2013).)*

to that of a black-hole binary. The power slowly decreases with increasing frequency during inspiral and then peaks at the time of merger (at approximately 1 kHz). It then drops exponentially as the remnant black hole rings down. In contrast, when the spin is significant the neutron star disrupts and there is no longer a peak in the gravitational-wave spectrum. Instead, the high-frequency signal depends on the details of the tidal disruption. Lower spins lead to less disruption, with most of the mass rapidly falling into the black hole, while higher spins lead to more significant disruption and a lower frequency cut-off in the spectrum. Information of the tidal disruption is reflected in a clear relation between the compactness of the neutron star (M_{ns}/R_{ns}) and an appropriately defined 'cut-off frequency' in the gravitational-wave spectrum, above which the spectrum falls off exponentially.

These merger simulations are of interest beyond gravitational-wave astronomy. The formation of a black hole surrounded by a massive, hot accretion disk from the remains of a disrupted neutron star provides the ideal conditions to power short gamma-ray bursts. Such systems could radiate a significant amount of energy (10^{48} erg) via neutrino emission or the Blandford–Znajek mechanism (Blandford and Znajek, 1977) in a timescale of a couple of seconds. This could launch the jet associated with observed gamma-ray bursts. The remnant disk, which may be as massive as $0.1M_\odot$ and involve velocities of $0.2 - 0.3c$, may also produce an electromagnetic counterpart signal in the form of a kilonova (Metzger and Berger, 2012). This emission is powered by decay of unstable r-process elements and non-thermal radiation from electrons accelerated at blast waves between the merger ejecta and the interstellar medium. The decay of unbound neutron-rich material powers an infrared transient, observable days after the merger.

These are exciting ideas, but they make complete simulations even more challenging. In fact, the different objectives demand very different kinds of simulations. To model the gravitational-wave signal, we need accurate simulations of the final tens of orbits before merger. During this phase, the main physical effects can be recovered using relatively simple models for the neutron-star matter. Meanwhile, to assess to what extent a given binary can power detectable electromagnetic signals and to predict nucleosynthesis yields, we need shorter inspiral simulations with a more detailed description of the physics: magnetic fields, neutrino emission, nuclear reactions, and the composition and temperature dependence of the properties of neutron-rich, high-density material. Ejected material must also be tracked far from the merger site, requiring reliable evolutions in a much larger region than during the inspiral.

20.5   Magnetohydrodynamics

As we turn to possible counterpart signals, it is clear that we need to make our simulations more sophisticated. This raises the level of technical difficulty and there is a legitimate concern that the physics we require may not be that well understood. Nevertheless, the first step towards increased realism is clear—we have to account for electromagnetism. We have already seen how a neutron star's magnetic field dictates much of the observed phenomenology, and we know that electromagnetism is of central importance for many astrophysical scenarios. However, as soon as we start thinking about the problem, we realize that we may not want to work directly with the electric and magnetic fields, E^a and B^a. The concepts are intuitive, which helps understanding and interpretation of results, but they are observer dependent—a moving electric field generates a magnetic field and view versa, and this can be confusing.

Electromagnetic dynamics is fully specified in terms of the vector potential, A^a, through the anti-symmetric Faraday tensor

$$F_{ab} = 2\nabla_{[a}A_{b]}, \tag{20.44}$$

and in the 3+1 decomposition, where the observer is associated with N^a, we have

$$F_{ab} = 2N_{[a}E_{b]} + \epsilon_{abcd}N^c B^d, \tag{20.45}$$

That is, the electric and magnetic fields (measured in the Eulerian frame) are

$$E_a = -N^b F_{ba}, \tag{20.46}$$

and

$$B_a = -N^b \left(\frac{1}{2} \epsilon_{abcd} F^{cd} \right) \equiv \frac{1}{2} \epsilon_{acd} F^{cd}, \tag{20.47}$$

where we have introduced the short-hand notation ϵ_{abc} for later convenience. Both fields are manifestly orthogonal to N^a, so each has three components—just as in non-relativistic physics. The electromagnetic contribution to the stress–energy tensor is

$$T_{ab}^{\text{EM}} = \frac{1}{\mu_0} \left[g^{cd} F_{ac} F_{bd} - \frac{1}{4} g_{ab} (F_{cd} F^{cd}) \right], \tag{20.48}$$

where the permeability μ_0 is typically taken to be constant.[1] This leads to

$$\nabla_a T_{\text{EM}}^{ab} = j_a F^{ab} \equiv -f_{\text{L}}^b, \tag{20.49}$$

which defines the Lorentz force, f_{L}^a.

In addition to the fluid equations of motion, which follow from the divergence of the total stress–energy tensor (as in Chapter 4 and which now also account for the Lorentz force), we need Maxwell's equations. First of all,

$$\nabla_a F^{ba} = \mu_0 j^b, \tag{20.50}$$

with

$$j^a = \sigma N^a + \mathcal{J}^a, \qquad \text{and} \qquad \mathcal{J}^a N_a = 0, \tag{20.51}$$

leads to

$$\gamma^{ab} \nabla_b E_a = \mu_0 \sigma + \epsilon^{abc} (\nabla_a N_b) B_c, \tag{20.52}$$

where σ is the charge density. In effect, we have

$$\gamma_a^b \nabla_b E^a - \mu_0 \sigma = -\epsilon^{abc} K_{ab} B_c = 0, \tag{20.53}$$

---

[1] In general, Maxwell's equations involve not just E^a and B^a, but $D^a = \varepsilon_0 E^a$ and $H^a = B^a/\mu_0$, as well. The permittivity ε_0 and the permeability μ_0 combine to give the speed of light (in vacuum)

$$c^2 = 1/(\mu_0 \varepsilon_0) \, .$$

In astrophysics it is common to work in so-called gauss units where $\varepsilon_0 = 1/(4\pi c)$ and (it follows that) $\mu_0 = 4\pi/c$. In geometric units, which are natural for spacetime, we then only have to keep track of factors of 4π. Connecting with physical measurements, it is natural to work with cgs units in which case the magnetic field is obtained in gauss (G).

since K_{ab} is symmetric, and it follows that

$$D_i E^i = \mu_0 \sigma. \tag{20.54}$$

We also get

$$\gamma_{ab} N^c \nabla_c E^b - \epsilon_{abc} \nabla^b B^c + \mu_0 \mathcal{J}_a = -E^b K_{ba} + E_a K + \epsilon_{abc}(N^d \nabla_d N^b) B^c, \tag{20.55}$$

where $\mathcal{J}^a$ is the spatial charge current. This leads to

$$\left(\partial_t - \mathcal{L}_\beta \right) E^i - \epsilon^{ijk} D_j(\alpha B_k) + \alpha \mu_0 \mathcal{J}^i = \alpha K E^i. \tag{20.56}$$

The second pair of Maxwell equations follow from

$$\nabla_{[a} F_{bc]} = 0. \tag{20.57}$$

First of all, we have

$$\gamma_a^b \nabla_b B^a = \epsilon^{abc} E_a K_{bc} = 0, \tag{20.58}$$

or

$$D_i B^i = 0. \tag{20.59}$$

Finally,

$$\gamma_{ab} N^c \nabla_c B^b + \epsilon_{abc} \nabla^b E^c = -\epsilon_{abc}(N^d \nabla_d N^b) E^c - B^b K_{ba} + B_a K, \tag{20.60}$$

leads to

$$\left(\partial_t - \mathcal{L}_\beta \right) B^i + \epsilon^{ijk} D_j(\alpha B_k) = \alpha K B^i. \tag{20.61}$$

Large-scale simulations tend to make use of the magnetohydrodynamics approximation. This involves simplifying the dynamics by (essentially) ignoring the inertia of the electromagnetic charge current. In practice, this involves introducing a closure condition for the equations by assuming that the current follows from an Ohm's law of form (depending on the assumptions involved; see Andersson *et al.* (2017))

$$E^i + \epsilon^{ijk} v_j B_k = \mathcal{R} \mathcal{J}^i, \tag{20.62}$$

where $\mathcal{R}$ represents the resistivity. In the ideal magnetohydrodynamics limit we have $\mathcal{R} = 0$, so the electric field vanishes according to an observer moving along with the fluid. In this case, the Lorentz force simplifies to

$$f_i^{\text{L}} = \epsilon_{ilm} \mathcal{J}_l B^m,$$ (20.63)

which leads to the usual expression

$$\boldsymbol{f}_L = \boldsymbol{J} \times \boldsymbol{B}.$$ (20.64)

Similarly, the weak-field version of Maxwell's equations follow by setting $\alpha = 1$ and $\beta = K = 0$ in the various relations. We then have

$$\nabla \cdot \boldsymbol{E} = \mu_0 \sigma,$$ (20.65)
$$\partial_t \boldsymbol{E} - \nabla \times \boldsymbol{B} = -\mu_0 \boldsymbol{J},$$ (20.66)
$$\nabla \cdot \boldsymbol{B},$$ (20.67)

and

$$\partial_t \boldsymbol{B} + \nabla \times \boldsymbol{E} = 0.$$ (20.68)

At this level, the assumptions of magnetohydrodynamics involve charge neutrality (effectively ignoring small-scale variations in the charge density by setting $\sigma = 0$) and neglecting the displacement current (leaving out $\partial_t \boldsymbol{E}$ in (20.66)). The current is inferred from the curl of the magnetic field and one can write down a closed system of equations, without involving the electric field.

20.6 The magnetorotational instability

The presence of a magnetic field impacts on many of the phenomena we have considered. It also adds new features. Of particular importance is the so-called magnetorotational instability (MRI), which generates turbulence and may lead to a dramatic amplification of the magnetic field in a dynamical setting (Balbus and Hawley, 1991). The MRI is thought to be the main mechanism for angular momentum transport in accretion disks. It can also play a role in core-collapse supernovae—powering the explosion in the first place, or generating magnetically driven outflows. This is particularly important in the context of gamma-ray bursts.

In order to establish the presence of the instability, let us carry out a plane-wave analysis of an axisymmetric magnetized system with a shearing flow. That is, we assume a fluid rotation profile $\Omega(r)$, with r the cylindrical radius, and consider small perturbations (away from a given equilibrium) such that $\exp(-i\omega t + ikz)$, where z represents the symmetry axis. Filtering out the sound waves (e.g. by means of a low-Mach approximation where the waves are assumed to be slow compared to the speed of sound) we arrive at a dispersion relation

$$\omega^4 - \left(\kappa^2 + 2k^2 v_{\text{A}}^2 \right) \omega^2 + k^2 v_{\text{A}}^2 \left(\kappa^2 - 4\Omega^2 + k^2 v_{\text{A}}^2 \right) = 0,$$ (20.69)

where we have introduced the Alfvén wave velocity

$$v_{\mathrm{A}}^2 = \frac{B^2}{4\pi\rho}, \tag{20.70}$$

with ρ the mass density, and the epicyclic frequency

$$\kappa^2 = 4\Omega^2 + 2r\Omega\frac{d\Omega}{dr}. \tag{20.71}$$

Let us first consider uniform rotation; i.e. take Ω to be constant. In that case, we have the two roots

$$\omega^2 = 2\Omega^2 + k^2 v_{\mathrm{A}}^2 \pm \left[\left(2\Omega^2 + k^2 v_{\mathrm{A}}^2\right)^2 - k^4 v_{\mathrm{A}}^4\right]^{1/2}, \tag{20.72}$$

and it is easy to see that, if we ignore the rotation, we are left with the Alfvén waves

$$\omega \approx \pm k v_{\mathrm{A}}. \tag{20.73}$$

Meanwhile, for a rotating star with a weak magnetic field we basically retain a set of modified inertial modes (see Chapter 13).

Switching on the shearing flow, it is easy to identify the onset of instability. In order for a mode to become unstable, the frequency must pass through the origin (an originally positive ω^2 must become negative). Thus, the instability comes into play when

$$\kappa^2 - 4\Omega^2 + k^2 v_{\mathrm{A}}^2 \le 0, \tag{20.74}$$

(since the coefficient of ω^2 in (20.72) is always positive). We see that we need the rotational velocity to decrease with increasing distance. In this case, wavelengths such that (Balbus and Hawley, 1991)

$$k^2 \le \frac{1}{v_{\mathrm{A}}^2}\left|\frac{d\Omega^2}{d\ln r}\right|, \tag{20.75}$$

are unstable and the fastest growing mode grows on a timescale

$$\frac{1}{\tau_{\mathrm{MRI}}} \approx \frac{1}{2}\left|\frac{d\Omega}{d\ln r}\right|. \tag{20.76}$$

The astrophysical relevance of the MRI is obvious since the condition required for its presence should be satisfied in a Keplerian accretion disk (Balbus and Hawley, 1991). The instability may also be relevant for compact binary mergers. If the merger produces some kind of disk the MRI may induce turbulence, leading to angular momentum transport and dissipation that drives accretion. The MRI may also enhance the magnetic field in the merger remnant (Kiuchi *et al.*, 2015).

However, resolving the length scales required to confirm these expectations—and tracking the saturation of the instability—is challenging with current simulations. The wavelength of the fastest growing mode is proportional to the magnetic field strength, and is typically much smaller than the size of system under consideration. As a result, the instability has mostly been explored using local simulations—generally demonstrating agreement with the expectations (Riquelme *et al.*, 2012)—but so far there have been few multidimensional studies. Nevertheless, there is convincing evidence (Siegel *et al.*, 2013;

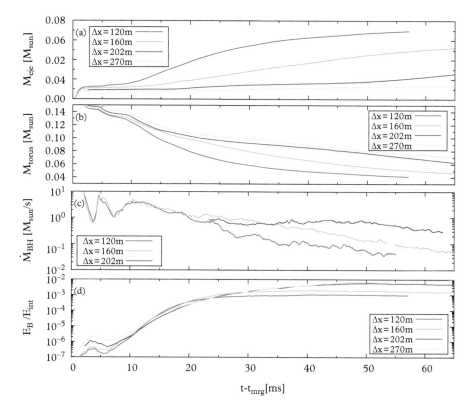

Figure 20.5 *Numerical evolutions demonstrating the presence of the magnetorotational instability in the case of a neutron star which is tidally disrupted by a black hole to form a thick accretion torus. The four panels show, from top to bottom; the ejecta mass, torus mass, mass accretion rate onto the black hole, and ratio of the magnetic field energy, E_B, to internal energy, E_{int}, as functions of time. Crucially, the bottom panel shows that (irrespective of grid resolution) the magnetic field energy grows exponentially and eventually saturates, at E_B about 0.1% of E_{int}. The growth rate observed between 10 ms $\lesssim t - t_{mrg} \lesssim$ 20 ms corresponds to a conversion of 7–8% of the orbital angular velocity. This accords reasonably well with the expectations from a perturbative analysis. (Reproduced from Kiuchi et al. (2015), copyright (2015) by the American Physical Society.)*

Kiuchi *et al.*, 2015; Kiuchi, 2018) that the instability acts in the hot remnant following a neutron star merger. Rapidly growing periodic structures are seen to form, reproducing features associated with the MRI in local simulations. The growth time and wavelength of the fastest growing mode can be extracted and compared to analytical predictions. An example is provided in Figure 20.5, which shows results from simulations of black hole–neutron star mergers.

20.7   Gravitational collapse

Gravitational collapse of rotating stars to form black holes is a central problem for any theory of gravity, and it is (naturally) a long-standing issue in general relativity (May and White, 1966). We touched upon some of the early work in Chapter 12 and we have now advanced to the point where we can consider numerical simulations. The problem is of obvious relevance for gravitational-wave astronomy, it is central to the gamma-ray burst phenomenon and it is of conceptual importance for the theory itself. It is only through numerical simulations that we may hope to improve our understanding of issues like cosmic censorship and the black-hole no-hair theorems.

If we want to understand the collapse process in absence of symmetries and for a realistic matter description, we have to turn to numerical relativity. This is not a simple task, but there has been huge progress towards understanding collapse dynamics since the landmark simulations of Stark and Piran (1985). They used an axisymmetric code to calculate the gravitational radiation produced when a rotating star collapsed to a black hole, triggering the collapse by (arbitrarily) reducing the star's pressure by at least a factor of 60% (up to 99% for the fastest spinning models). The results provide a picture that remains relevant today. The main features are simple—the gravitational waveforms show the familiar 'precursor-burst-ringdown' signature of a particle falling into a black hole (see Chapter 16).

One of the challenges of the collapse problem involves dealing with matter falling through the black-hole horizon. Once inside the black hole, this matter should no longer influence the exterior dynamics—it would require superluminal motion for information to exit the horizon. However, the physical speed limit may not be respected by a numerical scheme. One has to treat the horizon with care. The typical solution involves removing the black-hole interior (along with the central singularity, which would also cause grief) together with a suitable boundary condition that prevents an unphysical flow of information (see Chapter 19). However, despite significant improvements of such excision of the black-hole interior along with the use of an adaptive mesh (which automatically refines the resolution as required) the qualitative picture remains unchanged. The main difference is that full 3D simulations tend to be less 'optimistic' when it comes to the amplitude of the emerging gravitational waves. This is likely due to the use of less severe pressure reductions to trigger the collapse in the first place. For a uniformly rotating neutron star near the mass-shedding limit collapsing at a distance of 10 kpc (in our Galaxy) the signal-to-noise ratio is expected to be at the level of a few for

Advanced LIGO (Fryer and New, 2011). Given the expected event rate, one would be very lucky to observe the gravitational waves from such an event.

The generic collapse dynamics is summarized by spacetime diagrams like that in Figure 20.6—essentially an accurate version of the sketch from Figure 9.4. Initially, the matter contracts in an almost homologous way. In the illustrated case it maintains its axisymmetric distribution until $t \sim 175 M_\odot$. At high densities ($r \lesssim 2 M_\odot$) the contour lines slightly expand before collapsing. An apparent horizon is first found at $t \sim 188 M_\odot$. Shortly after horizon formation, all the matter has fallen into the black hole.

While the gravitational-wave signal from a collapsing compact star will be difficult to detect, the situation may be different for massive stars collapsing in the early Universe. This process is related to the formation of the supermassive black holes found in the centre of galaxies—a long-standing problem in astrophysics (Rees, 1984). A possible scenario involves the direct collapse of a supermassive star of mass $\sim 10^5 M_\odot$ to form a heavy seed black hole. Recent star-formation models suggest that, if a high mass-accretion rate of $\sim 0.1 M_\odot$/yr continues through the period of nuclear burning ($\sim 2 \times 10^6$ yr), a supermassive star with mass $\sim 2 \times 10^5 M_\odot$ may, indeed, form (Hosokawa *et al.*, 2013). The core of this star would eventually collapse directly to a black hole.

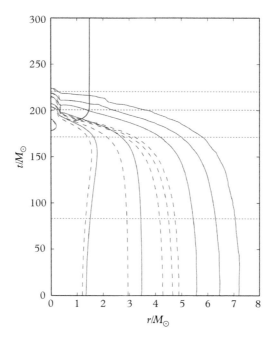

Figure 20.6 *Spacetime diagram illustrating typical collapse dynamics—an actual simulation version of the schematic illustration from Figure 9.4. A set of fixed density contours are traced in both the equatorial plane (black solid) and the perpendicular plane (black dashed). The apparent horizon forms at 188M_\odot (blue line). (Reproduced from Dietrich and Bernuzzi (2015), copyright (2015) by the American Physical Society.)*

Simulations of rotating supermassive stellar cores collapsing to a black hole have been carried out (Shibata *et al.*, 2016). The results show that the peak gravitational-wave amplitude is $h \approx 5 \times 10^{-21}$ at a frequency of $f \approx 5$ mHz for an event at a cosmological redshift, $z = 3$, if the collapsing core is in the hydrogen-burning phase. Such gravitational waves could be detectable by an instrument like LISA (with a signal-to-noise ratio ≈ 10; see Figure 20.7), indicating that future observations may be able to test the direct collapse scenario for the formation of massive seed black holes.

As in the case of the collapse of a star with a more modest mass, the gravitational-wave signal is characterized by a quasinormal-mode ringdown with frequency (cf. the results from Chapter 16)

$$f \approx 20 \left(\frac{M}{6.3 \times 10^5 M_\odot} \right)^{-1} (1+z)^{-1} \text{ mHz}, \tag{20.77}$$

and strain amplitude

$$h \approx 5 \times 10^{-21} \left(\frac{M}{6.3 \times 10^5 M_\odot} \right) \left(\frac{d}{25\,\text{Gpc}} \right)^{-1}, \tag{20.78}$$

where M is the mass of the original stellar core, z is the cosmological redshift, and d is the luminosity distance (about 26 Gpc for $z = 3$ in the standard ΛCDM model; see Chapter 22). The total energy emitted as gravitational waves is

$$\Delta E \approx 1.1 \times 10^{-6} M, \tag{20.79}$$

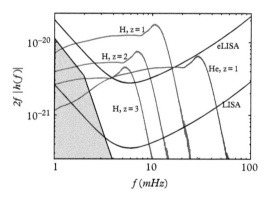

Figure 20.7 *The gravitational-wave spectrum for collapsing supermassive stars. The results represent a hydrogen-burning model (H) at $z = 1, 2$ and 3, and a Helium-burning model (He) at $z = 1$. The gravitational-wave strain, represented by $2f|h(f)|$, is compared to the expected noise level for two space-based detector configurations; eLISA (upper) and LISA (lower). The unresolved contribution of gravitational waves emitted by galactic binaries is also shown (as a shaded region in the lower left part of the figure). (Reproduced from Shibata et al. (2016), copyright (2016) by the American Physical Society.)*

far below the level expected for black-hole mergers, for which as much as $\sim 0.1M$ may be radiated (see Chapter 19).

20.8 Supernova core collapse

Towards the end of its life—as it runs out of light elements to fuel the nuclear furnace—a massive star (above $8 - 10 M_\odot$) develops an iron core that eventually becomes unstable and collapses from 1000 km or so to 30 km in a few 100 ms (Janka *et al.*, 2016). For relatively low-mass progenitors with highly degenerate cores, the collapse is triggered by a reduction in electron degeneracy pressure following electron captures. For more massive stars, radiation pressure and photo-disintegration of heavy nuclei also contribute to the gravitational instability. As the core collapses and matter is compressed, electron captures lower the lepton fraction, release neutrinos, and make the matter increasingly neutron rich. The core shrinks homologously (as in Figure 20.6) until it reaches supranuclear densities (enclosing a mass of about half that of the Sun). At this point, the equation of state stiffens due to the strong nuclear force and the infalling matter rebounds ('bounces'). The collapse abruptly halts, sending a shock wave through the (supersonically) infalling material. However, this prompt shock is not strong enough to push through the entire star. It loses energy and stalls. What happens next is key to the supernova story. We need to understand how the shock is revived and a successful explosion is launched.

The post-bounce dynamics depends on the interplay of a number of mechanisms. We have to combine aspects of stellar structure and evolution, nuclear and neutrino physics, hydrodynamics, kinetic theory, and strong gravity. There are (at least) two possible routes to explosion (Janka *et al.*, 2016). The shock may revive, on a timescale of hundreds of milliseconds, because of neutrino heating or (in potentially rare cases of rapidly rotating progenitors) magnetohydrodynamic effects. Hours later, the shock reaches the stellar surface and the supernova becomes visible across the electromagnetic spectrum.

The complex processes involved in the supernova scenario may produce gravitational waves through different channels (Ott, 2009). Some are associated with the dynamics of the proto-neutron star and its immediate environment (usually leading to high-frequency components of the signal), while others depend on the convective zone behind the stalled shock front (giving rise to a low-frequency signal). The distinction is important— the gravitational-wave signal may provide a unique probe of the multidimensional flow during the first second or so of core-collapse supernova explosions.

The detailed physics dictates the character of the gravitational-wave signal. The electron fraction, Y_e, of the inner core determines the mass of the inner core, which, in turn, influences the emitted waves. In order to explore this issue we need to reliably account for the neutrino losses during the collapse. This is a serious computational challenge as it involve tracking multiple neutrino species. At present, large parameter studies based on multidimensional simulations are simply inconceivable. In fact, the emergence of mature (and increasingly successful) simulations of the collapse and explosion of massive stars is fairly recent (Janka *et al.*, 2016). The progress builds on decades of effort aimed at understanding the role of neutrinos (starting in the 1960s, Colgate and White (1966)) and the interplay with hydrodynamical instabilities.

A typical supernova explosion releases of the order of 10^{53} erg of energy. As it is a violent phenomenon, naturally involving some degree of asymmetry, one might expect a significant amount of this energy to be released as gravitational waves. In fact, in the early days of efforts to detect gravitational radiation, supernovae were considered prime source targets. However, as the understanding of the supernova mechanism(s) improved and multidimensional simulations became possible, the promise faded. We now know that most of the energy (99%) is released through neutrinos. A tiny amount (likely less than the equivalence of $10^{-6}M_\odot$) is channeled through gravitational waves (Müller *et al.*, 2013). This is yet another example of how the devil is in the detail. The gravitational-wave emission depends not on how much energy may be available, but rather on how asymmetric the process is. In essence, we should not expect to detect supernova signals from distant galaxies. However, Advanced LIGO should be able to see events throughout the Milky Way out to the Magellanic Clouds and perhaps the local group of galaxies (Powell *et al.*, 2016). The event rate may be low, but a detection would be richly rewarded as the gravitational-wave signal brings detailed information about the explosion mechanism—information that is otherwise hidden from view.

Despite the caveats, observations and theory have shown that core-collapse supernova explosions exhibit strong asymmetries. Observed supernova remnants have inhomo- geneities that may be a smoking gun for an asymmetric explosion. Inferred kick velocities are also suggestive. For slowly rotating progenitors, asymmetries arise early on as neutrino heating drives convective overturns behind the shock or from the large-scale 'standing accretion shock instability' (SASI, Blondin and Mezzacappa (2006)). This large-scale asymmetry is illustrated in Figure 20.8. The SASI plays an important role in the neutrino-driven explosion mechanism. Basically, a small fraction (5–10%) of the outgoing neutrino luminosity is deposited behind the stalled shock. This drives turbulence and increases the thermal pressure. These effects may revive the shock. The mechanism has been successfully demonstrated in simulations (Janka *et al.*, 2016), and seems to explain the vast majority of core-collapse supernovae.

In the case of convection and the SASI in neutrino-driven explosions, asymmetric mass motion in the neutrino heating layer leads to gravitational-wave emission from the post-bounce phase due to time variations in the mass quadrupole moment (Murphy *et al.*, 2009). Simulations typically show several distinct phases; see Figure 20.9. Shock ringing after prompt convection leads to a low-frequency signal around 100 Hz lasting about 50 ms. This is followed by a signal at several hundred Hz, with stochastic amplitude modulations. There may also be a 'tail' of radiation from an asymmetric shock expansion in the explosion phase. Simulations in 2D and 3D differ in the predicted gravitational- wave amplitude by as much as a factor of 10 (with the 3D models being less 'optimistic', Janka *et al.* (2016)), so further progress is needed to make our understanding quantitative.

Due to the stochastic character, it may be far from easy to infer physics from this kind of signal. Without a clear relation between the physical parameters that influence the convection and the accretion flow it would also be difficult to create a reliable theoretical template to facilitate searches. Realistically, gravitational-wave searches will have to rely

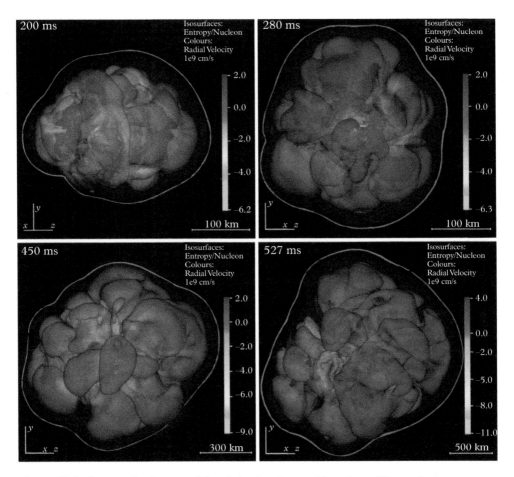

Figure 20.8 *An example of a successful explosion based on a self-consistent 3D neutrino-hydrodynamics simulation. The panels show isoentropy surfaces of neutrino-heated, buoyant matter for a 20$M_\odot$ progenitor at different times of evolution. The supernova shock is indicated as a blue, enveloping surface. During the evolution large convective plumes push out neutron-rich material at high velocities. In this particular case, strong SASI activity is observed between around 120 and 280 ms. (Reproduced from Melson et al. (2015) by permission of the AAS.)*

on generic burst algorithms. However, we can still get an intuitive understanding of the signal. For excitation by convection, we can estimate the maximum amplitude around the onset of the explosion by expressing the involved energy, E_{kin}, in terms of the mass of the gain region, M_{gain}, and the typical velocity involved, v. First we express the post-shock sound speed c_{s} in terms of the shock radius r_{sh} as (Müller, 2017)

$$c_{\text{s}} = \left(\frac{GM}{3r_{\text{sh}}} \right)^{1/2}.$$

(20.80)

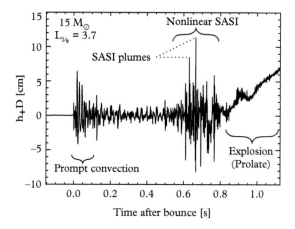

Figure 20.9 *A sample of gravitational-wave strain (h₊) times the distance, D, vs. time after core bounce. The signal was extracted from a simulation using a $15M_\odot$ progenitor model and an electron-type neutrino luminosity of $L_{\nu_e} = 3.7 \times 10^{52}$ erg s^{-1}. Prompt convection, resulting from a negative entropy gradient left by the stalling shock, leaves a distinctive feature in the signal lasting up to $\sim$50 ms after bounce. From $\sim$50 to $\sim$550 ms post bounce, the signal is dominated by convection. Afterwards and until the onset of explosion ($\sim$800 ms), strong nonlinear SASI motion dominates the signal. The most distinctive features are spikes that correlate with dense and narrow down-flowing plumes striking the 'proto-neutron star' surface ($\sim$50 km). After $\sim$800 ms the model starts to explode. (Reproduced from Murphy et al. (2009) by permission of the AAS.)*

Then we make use of the quadrupole formula (see Chapter 3)

$$h_{ij}^{TT} \sim \frac{2\alpha GE_{\text{kin}}}{dc^4},$$ (20.81)

where α represents the unknown overlap with the quadrupole motion, to estimate the gravitational-wave amplitude. This way we arrive at

$$\left(\frac{c^4 d}{G}\right) h_{\text{max}} \sim \alpha M_{\text{gain}} v^2 \text{Ma} \sim \alpha \frac{GMM_{\text{gain}}}{3r_{\text{sh}}} \text{Ma}^3,$$ (20.82)

where $\text{Ma} = (v/c_s)^2$ is the Mach number in the convective region. Finally, relating the mass in the gain region to the explosion energy, E_{expl}, via a residual recombination energy ϵ_{rec} per baryon of around 5–6 MeV, we arrive at

$$\left(\frac{c^4 d}{G}\right) h_{\text{max}} \sim \alpha \frac{E_{\text{expl}}}{\epsilon_{\text{rec}}} \frac{GM}{3r_{\text{sh}}} \text{Ma}^3,$$ (20.83)

Explicitly using $r_{\text{sh}} \approx 200\,\text{km}$ and a typical value of $\text{Ma}^2 = 0.3$ at shock revival, this suggests that (Müller, 2017)

$$h_{\text{max}} \sim 3 \times 10^{-22} \alpha \left(\frac{E_{\text{expl}}}{10^{51}\,\text{erg}}\right)\left(\frac{10\,\text{kpc}}{d}\right), \qquad (20.84)$$

which accords fairly well with the results of numerical simulations. The estimate also demonstrates the trend towards stronger signals from more energetic explosions. Basically, one would expect the observed spread in supernova explosion energies to lead to a range of gravitational-wave amplitudes.

In contrast, in the case of rotating collapse the gravitational-wave signal is primarily determined by the mass and the ratio of rotational kinetic energy to gravitational energy $(T/|W|$; see Chapter 6) of the inner core at bounce (Dimmelmeier *et al.*, 2008; Richers *et al.*, 2017). The nuclear equation of state comes into play through its effect on the mass of the inner core at bounce and the central density of the post-bounce proto-neutron star. However, as the gravitational-wave signal depends on the interplay of different nuclear physics aspects, it may be difficult for instruments like Advanced LIGO to distinguish between theoretical matter models. Moreover, rotating core collapse does not probe physics above about twice nuclear density, so very little exotic physics (like deconfined quarks) can be probed by gravitational-wave observations. These conclusions are illustrated by results like those shown in Figure 20.10 (assuming the j-constant rotation law; see Chapter 12).

Simple estimates suggest a linear relationship between the bounce amplitude and $T/|W|$ of the inner core. As usual, the gravitational-wave amplitude depends on the second time derivative of the mass quadrupole moment, $I \sim M(x^2 - z^2)$, where M is the mass of the oscillating inner core and x and z are the equatorial and polar equilibrium radii, respectively. Treating the inner core as an oblate sphere, we can take the radius of the inner core in the polar direction to be $z = R$ and the larger radius in the equatorial direction (due to the centrifugal support) $x = R + \delta R$. To first order in δR, the mass quadrupole moment then becomes

$$I \sim M[(R+\delta R)^2 - R^2] \sim MR\delta R. \qquad (20.85)$$

Adapting the argument that led to the Kepler limit in Chapter 6, one may use

$$(R+\delta R)^2 \Omega^2 + \frac{GM}{(R+\delta R)} = \frac{GM}{R}, \qquad (20.86)$$

where Ω is the rotation rate. Assuming δR is small, we then find that

$$\delta R \sim \Omega^2 R^4 / GM. \qquad (20.87)$$

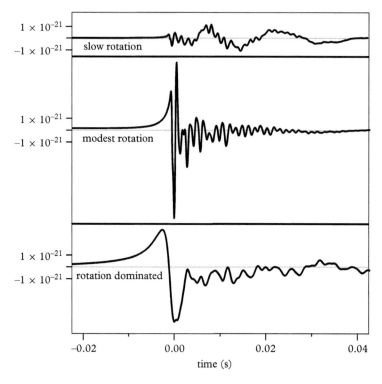

Figure 20.10 *Sample waveforms of the core bounce signal for three models with varying initial rotation states (with a fixed progenitor mass and equation of state). Note the relatively small signal peak at the time of core bounce and the significant late-time contribution from post-bounce convection for the slowly rotating model (top), and the overall lower signal frequency for the rotation-dominated and centrifugally bouncing model (bottom). The three signals cover the waveform morphology of the simulations by Dimmelmeier et al. (2008). (Adapted from Röver et al. (2009), copyright (2009) by the American Physical Society.)*

As we are only looking for an order-of-magnitude estimate, we can replace the time derivatives in the quadrupole formulate with the inverse of the timescale of core bounce

$$t_{\text{dyn}}^{-2} \sim \frac{GM}{R^3}.$$

(20.88)

We can also approximate (see Chapter 6)

$$T/|W| \sim \frac{R^3 \Omega^2}{GM}.$$

(20.89)

This way we arrive at

$$h \sim \frac{GM\Omega^2 R^2}{c^4 d} \sim \frac{T}{|W|}\frac{(GM)^2}{Rc^4 d},\tag{20.90}$$

or

$$h \sim 10^{-19}\left(\frac{T/|W|}{0.1}\right)\left(\frac{10\ \mathrm{kpc}}{d}\right).\tag{20.91}$$

The linear dependence of the bounce signal amplitude on $T/|W|$ is brought out by simulations; see for example Figure 20.11. In this case, the different simulations were initiated by imposing rotating initial conditions for the same $12M_\odot$ progenitor model, assuming a pre-collapse rotation profile

$$\Omega(\varpi) = \Omega_0\left[1 + \left(\frac{\varpi}{A}\right)^2\right]^{-1},\tag{20.92}$$

where ϖ is the cylindrical radius and models A1–5 shown in Figure 20.11 represent increasing values of A.

As discussed in Chapter 13, the bar-mode instability operates at very high spin rates (in terms $\beta = T/|W|$ the limit is about 0.25 in general relativity), which are not expected

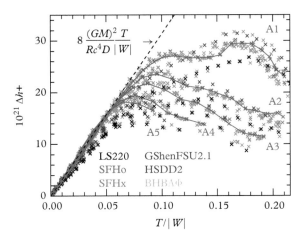

Figure 20.11 *The difference between the maximum and minimum gravitational-wave strain Δh_+, assuming a distance of $10\,kpc$, as a function of the ratio of rotational to gravitational energy $T/|W|$ of the inner core at bounce. Each (2D) simulation is represented a single point. Simulations for different equations of state and rates of differential rotation (set by the value of A) behave similarly for $T/|W| \lesssim 0.06$, but branch out when the rotation becomes dynamically important. (Reproduced from Richers et al. (2017), copyright (2017) by the American Physical Society.)*

even in more extreme collapse scenarios. Moreover, as we have already discussed, once it is active the instability may not persist for long due to nonlinear mode coupling. However, if the proto-neutron star is sufficiently differentially rotating, the low $T/|W|$ instability (Centrella *et al.*, 2001) may be active. This could lead to substantial deformations and a stronger signal. The associated signal-to-noise ratio depends (essentially) on the number of cycles and the saturation level of the instability. A typical value for the maximal strain obtained in numerical simulations is 10^{-21} at 10 kpc with a frequency of $400 - 900\,\text{Hz}$ (Ott *et al.*, 2007).

To summarize the conclusions from available core-collapse simulations: for slowly rotating iron cores, bounce and initial ringdown are expected to lead to signals with peak frequencies in the range of $700 - 900\,\text{Hz}$ and dimensionless strain amplitudes of less than 5×10^{-22} at a distance of 10 kpc, i.e. within our Galaxy. Fast rotation amplifies the bounce signal. If the iron core has moderate rotation, the peak frequencies span the larger range of $400 - 800\,\text{Hz}$ with amplitudes of 5×10^{-22} up to 10^{-20}. Very rapid rotation leads to bounce at subnuclear densities, and gravitational-wave signals in a lower frequency band of hundreds of hertz with strains around 5×10^{-21} at 10 kpc.

Prompt convection occuring shortly after core bounce due to negative lepton gradients leads to galactic signal amplitudes in the range 10^{-23} to 10^{-21} at frequencies of $50 - 1000\,\text{Hz}$, whereas signals of convection in the proto-neutron star may have strains of up to 5×10^{-23} for a somewhat wider range of frequencies. Neutrino-driven convection and the instability of the accretion-shock (SASI) could be relevant sources as well, with strain amplitudes up to 10^{-22} at $100 - 800\,\text{Hz}$. In addition, an acoustic mechanism has been proposed for supernova explosions (Burrows *et al.*, 2006). This mechanism is connected with low-order g-mode oscillations in the proto-neutron star. If this mechanism is active, large strain amplitudes of up to 5×10^{-20} at 10 kpc could be reached in extreme cases.

The different emission mechanisms have characteristic signatures, but the estimated signal-to-noise ratios make a detection of an extragalactic core-collapse supernova from a slowly rotating (canonical) iron core seem unlikely with the current generation of instruments. Even if fast core rotation rates are assumed, detections will be possible from a distance of at most $1 - 2\,\text{Mpc}$. Since the rate of (successful) supernovae is known from observations, we know that galactic events happen every $30 - 100$ years. Even at a distance of 1 Mpc—that is, for very optimistic detection estimates—the event rate will be low enough that we have to be lucky to see a single event during the operation of Advanced LIGO. However, at a distance of $3 - 5$ Mpc, a range which could admit a detectable signal in a third-generation detector like the Einstein Telescope, the event rate could be a few per year.

A major uncertainty connected with supernova models is the initial state, in particular the angular momentum distribution in the evolved iron core. Expectations from stellar evolution calculations suggest a very slowly rotating core would be typical (Heger *et al.*, 2005). This is further supported by the observation that neutron stars seem to be born with comparatively low rotation rates, and by recent evidence for loss of angular momentum of stars before the white-dwarf stage. Strong gravitational-wave signals can only be obtained by invoking processes which break the approximate spherical symmetry of the system. If the core rotates faster than expected, for example for collapsing stars that lead to gamma-ray bursts, then rotational instabilities may become relevant sources of gravitational waves. Of course, these instabilities (or magnetic fields) must be efficient enough to spin down young neutron stars after birth, in order to reconcile this scenario with the observed neutron-star spin distribution. Also, magnetic wind-up may in some cases open channels to magnetically driven explosions, which could give rise to detectable signals (Mösta *et al.*, 2014). These cases may be rare, but they are nevertheless very interesting, especially since they may have an electromagnetic counterpart.

20.9   Hypernovae

Supernovae are the most powerful explosions in the Universe, but (as we have seen) most of the energy is released in neutrinos. These neutrinos are not likely to be observed unless we are lucky enough to have a nearby event. The brightest cosmic explosions, in terms of electromagnetic radiation, are the gamma-ray bursts. Some observed bursts are bright enough to convert the rest mass of the Sun (2×10^{54} erg) into gamma-rays in less than 10 seconds. One of the main challenges in understanding these enigmatic bursts involves the central engine. What astrophysical mechanism(s) can possibly accelerate matter up to ultra-relativistic speeds and at the same time collimate the emission into a powerful jet?

The answer may involve an alternative supernova explosion mechanism (MacFadyen and Woosley, 1999; Woosley and Bloom, 2006). In the magnetorotational mechanism (Mösta *et al.*, 2014), rapid rotation and a strong magnetic field conspire to generate jet-like outflows (naturally asymmetric) that explode the star. This mechanism can, at least in principle, drive explosions up to ten times more energetic than regular supernovae. This may explain the observed gamma-ray bursts. These so-called hypernovae (Paczyński, 1998; Hartmann and MacFadyen, 2000) are expected to be rare, but they could make up about 1% of core-collapse events.

The key issue for this mechanism is the required fast core spin, leading to the formation of a proto-neutron star (possibly a magnetar) spinning with a period of about a millisecond. This is problematic as most massive progenitor stars are thought to have slowly spinning cores (Heger *et al.*, 2005).

Observed gamma-ray bursts exhibit a diverse phenomenology (Piran, 2004). The burst duration ranges from a millisecond to a thousand seconds, with a (very roughly) bimodal distribution of long bursts (lasting longer than ~ 2 s) and short bursts; see Figure 21.16. The radiation is beamed into a narrow solid angle, increasing the intensity of the emission. The (geometrically corrected) energy of long gamma-ray bursts typically lies in the range $E \sim 10^{50-52}$ erg, much smaller than the isotropic energy $E_{\mathrm{iso}} \sim 10^{52-54}$ ergs and more in line with the energy associated with core-collapse supernovae. The required energy budget and the timescales involved suggest that long gamma-ray bursts involve the formation of a black hole. In the case of rotating collapse, the newly born black hole is likely to be surrounded by a massive disk of material. The combined effects of rotation and the remnant magnetic field may collimate and launch the observed jet (Mösta *et al.*, 2014).

The collapse scenario has been confirmed by the discovery of supernovae associated with gamma-ray bursts. Some of the associated supernovae show evidence for broad emission lines indicating high-velocity ejecta and an inferred explosion energy of order 10^{52} ergs. The connection between a subclass of supernovae (type Ic) and long gamma-ray burst was first established in the case of GRB 980425 and SN 1998bw. Other events, like GRB 030329/SN 2003dh and GRB 031203/SN 2003lw, strengthen the connection (Nakamura *et al.*, 2001; Nomoto *et al.*, 2007).

The hypernova scenario adds complexity to the—already challenging—problem of modelling supernovae. In addition to the original explosion, a viable gamma-ray burst model must deliver a powerful focused jet. The jet will typically have an opening angle of about 0.1 radians and involve a power of $\sim 10^{50}$ erg/s. In addition, the model must (at least in some instances) deliver $\sim 10^{52}$ erg of kinetic energy to a larger solid angle (up to about 1 radian) to produce supernovae like SN 1998bw. This is ten times the level of emission from an ordinary supernova.

The phenomenology suggests that long gamma-ray bursts originate from metal-poor progenitors with degenerate iron cores, ending their lives in type Ibc core-collapse supernovae (Nakamura *et al.*, 2001; Nomoto *et al.*, 2007). In some cases, the supernova explosion may fail, leading to the formation of a black hole rapidly accreting fallback material. The engine that converts energy from the accretion may power a jet in excavated polar regions. Depending on the black-hole mass and the angular momentum of the collapsing envelope, this central engine may operate for several seconds—enough to explain a long gamma-ray burst. Parts of this story are supported by numerical simulations, but some pieces are still missing. In particular, the actual launch of the jet presents a problem.

Not surprisingly, given the intimate connection with the core-collapse problem, the level of gravitational-wave emission associated with the hypernova scenario is 'disappointing'. An intermittent period of turbulent, low-amplitude emission ends, leading to a pronounced spike in the waveform associated with the formation of the black hole (Ott *et al.*, 2011). The collapse signal then evolves into the familiar black-hole ringdown. However, the emission is quenched as the axisymmetric accretion flow does not excite the quasinormal modes to a significant amplitude. In essence, Advanced LIGO should be able to see gravitational waves from this scenario at a galactic distance, but the signal will not be observable from the cosmological distances where these events regularly occur.

21

Anatomy of a merger

Neutron star mergers provide rich cosmic laboratories which can be used test our understanding of extreme physics. The wealth of information one may glean from these events was clearly illustrated by the first detection on 17 August 2017 (Abbott *et al.*, 2017*i*). With a signal that lasted over 100 seconds in the detector sensitivity band, GW170817 was the largest signal-to-noise ratio event detected and the precise sky localization allowed rapid follow-up searches with a range of telescopes. When the event was found to be shining across the electromagnetic spectrum it was clear that we were witnessing the birth of multi-messenger astronomy involving gravitational waves (Abbott *et al.*, 2017*f*; Abbott *et al.*, 2017*j*; Troja *et al.*, 2017; Cowperthwaite *et al.*, 2017).

21.1   GW170817

The first binary neutron star event was detected with a (combined) signal-to-noise ratio of just over 32 and a false-alarm-rate estimate of less than one event every 80,000 years. (Abbott *et al.*, 2017*i*) The source was localized within a sky region of about 30 square degrees and had a luminosity distance of 40 Mpc, making it the closest and most precisely localized gravitational-wave signal in the first two Advanced LIGO observing runs. The detection enabled a swift electromagnetic follow-up campaign that identified, first of all, a gamma-ray flash just under two seconds after the merger (Troja *et al.*, 2017; Kasliwal *et al.*, 2017; Goldstein *et al.*, 2017). This was followed by an optical counterpart near the galaxy NGC 4993, consistent with the position and distance inferred from the gravitational-wave data (Hjorth *et al.*, 2017; Coulter *et al.*, 2017); see Figure 21.15. The event was well inside the detector horizon for binary neutron stars—the maximum distances from which the LIGO-Livingston and LIGO-Hanford detectors would have been able detect such a system (with signal-to-noise-ratio of 8), were 218 Mpc and 107 Mpc, respectively. At the same time, the horizon for the Virgo instrument was 58 Mpc. The relative weakness of the signal in the Virgo detector helped pinpoint the source location, as it is had to be in one of the detector's dark spots.

From the roughly 3,000 gravitational-wave cycles in the observed frequency range, the chirp mass in the detector frame was precisely constrained to $\mathcal{M}^{\text{det}} \approx 1.1977 M_{\odot}$. The mass in the detector frame is related to the rest-frame mass of the source by the redshift

Gravitational-Wave Astronomy: Exploring the Dark Side of the Universe. Nils Andersson, Oxford University Press (2020).
© Nils Andersson. DOI: 10.1093/oso/9780198568032.001.0001

z (see Chapter 22) as $\mathcal{M}^{\text{det}} = \mathcal{M}(1+z)$. Assuming the standard ΛCDM cosmology, the gravitational-wave distance measurement implied a redshift of $z \approx 0.008$, consistent with that of NGC 4993, strengthening the association with this galaxy. The source-frame chirp mass is thus found to be $\mathcal{M} \approx 1.188 M_\odot$ (with slightly larger error bars due to uncertainties in the redshift). This is the most accurately measured combination of the two masses. The individual mass estimates are affected by a degeneracy between the mass ratio and the aligned spin components. As a result, estimates of the mass ratio and the component masses depend on assumptions made about the spins. This is apparent from the probability distributions in Figure 21.1. However, by extrapolating the spins of known binary neutron stars, one may assume that the merging stars were unlikely to be rapidly spinning. This leads to tighter constraints on the masses. Restricting the spin to be consistent with the observed population ($\chi = a/M \leq 0.05$ in Figure 21.1), one arrives at the mass ratio $0.7 \leq q \leq 1.0$ and component masses $1.36 \leq m_1/M_\odot \leq 1.60$ and $1.17 \leq m_2/M_\odot \leq 1.36$.

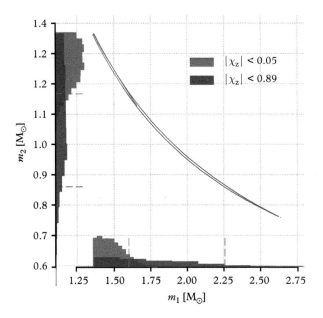

Figure 21.1 *The posterior distribution for the two component masses (m_1 and m_2) in the rest frame of the GW170817 merger for a low-spin scenario ($|\chi| < 0.05$, blue) and a (less likely based on the known radio pulsar population) high-spin scenario ($|\chi| < 0.89$, red). The coloured contours enclose 90% of the probability from the joint posterior probability density function for the masses. The shape of the two-dimensional posterior is determined by a line of constant total mass M and its width is determined by the uncertainty in this quantity. The widths of the marginal distributions (shown on the individual axes as dashed lines enclosing 90% probability away from equal mass of $1.36 M_\odot$) are strongly affected by the choice of spin priors. The result for the low-spin prior (blue) is consistent with the masses of all known binary neutron star systems. (Reproduced from Abbott et al. (2017i), Creative Commons Attribution 4.0 licence.)*

21.2 Tidal deformation

As an inspiralling binary gets tighter and the gravitational-wave frequency increases, the stars' internal structure becomes increasingly important (see Figure 21.2). For neutron stars, the tidal field of the companion introduces an additional quadrupole moment, which accelerates the inspiral.

This is an extreme version of a familiar phenomenon. When a body is exposed to an external gravitational field it responds by changing shape. This is most easily understood by considering the gravitational effect the Moon has on the Earth. The oceans move to reach equilibrium as the Moon orbits, leading to the observed tides. The effect also deforms the Earth's elastic crust, again to reach an equilibrium with the passing body (although this effect is much less pronounced). The tidal deformation can be expressed in terms of dimensionless quantities known as the Love numbers (Flanagan and Hinderer, 2008; Hinderer, 2008). The effect may leave an observational imprint on the gravitational-wave signal from a neutron star in a close binary system, in principle encoding information about the supranuclear equation of state (Hinderer *et al.*, 2010).

The tide raised by a binary companion (here treated as a point particle, which should be a good approximation) induces a linear response in the primary. In order to quantify this response we need to solve the linearized fluid equations, with a driving force given by the tidal potential (χ) associated with the companion star. In the Newtonian case, this

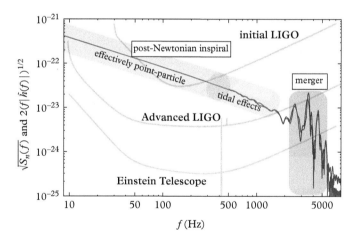

Figure 21.2 *An illustration of the late stages of binary neutron star inspiral. When the system enters the sensitivity band of ground based detectors, it can be accurately represented by post-Newtonian theory, but as the orbital separation decreases finite size effects become increasingly important. When the signal reaches a few 100 Hz we must account for the tidal deformation of the two bodies. The violent dynamics of the final merger (leading to a signal above 1 − 2 kHz) must be modelled using nonlinear numerical simulations. (Adapted from an original figure by J. Read based on data from Read et al. (2013).)*

tidal potential is given by a solution to $\nabla^2\chi = 0$, and (in a coordinate system centred on the primary) we have (Press and Teukolsky, 1977; Ho and Lai, 1999)

$$\chi = -\frac{GM_2}{|\boldsymbol{r} - \boldsymbol{a}(t)|} = -GM_2 \sum_{l\geq 2}\sum_{m=-l}^{l} \frac{W_{lm}r^l}{a^{l+1}(t)} Y_{lm} e^{-im\psi(t)}. \tag{21.1}$$

The orbit of the companion (with mass M_2) is taken to be in the equatorial plane with a the binary separation and ψ the orbital phase. For $l = 2$, which makes the dominant contribution to the gravitational-wave signal, we have the coefficients (Press and Teukolsky, 1977)

$$W_{20} = -\sqrt{\pi/5}, \qquad W_{2\pm2} = \sqrt{3\pi/10}, \qquad \text{and} \qquad W_{2\pm1} = 0. \tag{21.2}$$

The last result follows from the symmetry of the tidal potential; W_{lm} must vanish for all odd $l + m$.

If we assume that the inspiral, driven by gravitational-wave emission, is adiabatic then the tidal deformability is obtained from the 'static' part of the tidal potential (21.1). For a system in circular orbit, the distance a is constant and χ is time independent. The induced response is obviously time independent, as well, so we only need the zero-frequency solution to the perturbation problem. That is, we need to solve

$$\frac{1}{\rho}\nabla\delta p - \frac{1}{\rho^2}\delta\rho\nabla p + \nabla\delta\Phi = -\nabla\chi, \tag{21.3}$$

along with

$$\nabla^2\delta\Phi = 4\pi G\delta\rho. \tag{21.4}$$

We also know that the equation of state will (effectively) be barotropic (as long as the star is assumed to have time to reach chemical equilibrium). see Andersson and Pnigouras (2019)). This means that

$$\delta p = c_s^2\delta\rho, \tag{21.5}$$

where c_s^2 is the adiabatic sound speed. Finally, the background is such that

$$\nabla p = -\rho\nabla\Phi \quad\longrightarrow\quad p' = -\rho\Phi' = -\rho g, \tag{21.6}$$

where a prime indicates a radial derivative and we have introduced the gravitational acceleration g.

As in the f-mode problem (see Chapter 13), we expand in spherical harmonics, such that (for $m = 0$)

$$\delta p = \sum_l p_l Y_{l0}, \tag{21.7}$$

and similar for the other variables, and introduce

$$U_l = \Phi_l + \chi_l.$$ (21.8)

This leads to the radial component of (21.3)

$$p_l' - \frac{p'}{\rho}\rho_l = -\rho U_l',$$ (21.9)

the angular part

$$p_l = -\rho U_l,$$ (21.10)

and the Poisson equation

$$r^2 U_l'' + 2r U_l' - l(l+1)U_l = 4\pi G r^2 \rho_l,$$ (21.11)

where we have used the fact that χ_l solves the corresponding homogeneous equation.

It is worth noting that the problem seems to be overdetermined. We appear to have too many equations for the number of variables. However, taking a radial derivative of (21.10), we get

$$p_l' = -\rho' U_l - \rho U_l' = \frac{\rho'}{\rho}p_l - \rho U_l'.$$ (21.12)

Using this in (21.9) we have

$$\frac{\rho'}{\rho}p_l - \frac{p'}{\rho}\rho_l = 0 \longrightarrow \rho' p_l = p' \rho_l,$$ (21.13)

which is consistent with (21.5). This reduces the number of equations, so the problem is well posed, after all.

Now we have

$$r^2 U_l'' + 2r U_l' + \left[\frac{4\pi G \rho r^2}{c_s^2} - l(l+1)\right]U_l = 0.$$ (21.14)

This equation is solved by integrating from the centre to the surface of the star. At the surface we match to the exterior potential. In general, this provides the multipole moments of the body, I_l, according to (as we have $m = 0$)

$$\Phi_l = -\frac{4\pi G}{2l+1}\frac{I_l}{r^{l+1}}.$$ (21.15)

We also know that

$$\chi_l = \frac{4\pi}{2l+1} d_l r^l,$$ (21.16)

where d_2 can be read off from (21.2).

At the surface of the star we match U_l and its derivative to the exterior solution. This matching allows us to extract the Love number, k_l, which is defined by the relation

$$Gl_l = -2k_l R^{2l+1} d_l.$$ (21.17)

Taking a simple $n = 1$ polytrope as an example, we find that $k_2 \approx 0.26$ (Poisson and Will, 2014). The Love number measures how easy it is to deform the bulk of the matter in the star. If most of the star's mass is concentrated at the centre then the tidal deformation will be smaller. For polytropes, matter with a larger polytropic index n is softer and more compressible, so these polytropes are more centrally condensed. As a result, k_2 decreases as n increases. For example, for $n = 2$ we get $k_2 \approx 0.07$. For realistic equations of state and in general relativity (see later), the value of k_2 tends to lie in the range $0.05 - 0.15$. The Love number also decreases with increasing compactness. This explains the features in Figure 21.4.

In order to illustrate the impact of the tidal deformability on the gravitational-wave signal, let us focus on the phasing. The idea is simple; once an additional mechanism leads to an overall shift of about half a cycle in the waveform there would be no further accumulation of signal-to-noise in a matched filter search (see Chapter 8 and Flanagan and Hinderer (2008)). Hence, if the total number of cycles is $\mathcal{N}(f)$ (where f is the gravitational wave frequency) then a shift $\Delta\mathcal{N} > 0.5$ (or equivalently, a phase shift $\Delta\Phi = 2\pi\Delta\mathcal{N}$ of order a few radians) would mean that the effect could be distinguished. This rough criterion will be sufficient for now. Basically, we assume that an additional dynamical effect would i) suppress detectability and impact on parameter extraction with a given search template (that does not account for the effect) if $\Delta\mathcal{N} > 0.5$, but it should be safe to assume that, ii) the effect will not be distinguishable if $\Delta\mathcal{N} \ll 1$.

We take the leading-order gravitational radiation reaction from Chapter 5 as our starting point. That is, we assume that gravitational-wave emission drains energy from the orbit at a rate

$$\dot{E}_{gw} = -\frac{32\mathcal{M}\Omega}{5c^5}(G\mathcal{M}\Omega)^{7/3},$$ (21.18)

where the chirp mass is given by

$$\mathcal{M} = \mu^{3/5} M^{2/5} = M_1 \left(\frac{q^3}{1+q}\right)^{1/5},$$ (21.19)

with the total mass $M = M_1 + M_2$, reduced mass $\mu = M_1 M_2 / M$ and mass ratio $q = M_2/M_1$. In the case of a pair of $1.4 M_\odot$ neutron stars (which we take as our canonical example) we have $\mathcal{M} = 1.2 M_\odot$, close to the observed result for GW170817.

As we are interested in using observations to constrain neutron star physics, it is important establish to what extent the various parameters are already known. For the mass ratio q, we know from radio observations that double neutron star systems may be asymmetric, as in the case of PSR J0453+1559 where the two masses are $1.174 M_\odot$ and $1.559 M_\odot$; see table 9.2. Given this, the inferred range for the mass ratio in GW170817 (taking the primary to be the heavier companion) $0.7 \leq q \leq 1$, is not too surprising.

As usual, the orbital frequency Ω follows from Kepler's law

$$\Omega^2 = \frac{GM}{a^3}, \tag{21.20}$$

which links the observed gravitational-wave frequency $f = \Omega/\pi$ to the orbital separation a. Given the Newtonian orbital energy

$$E_{\text{orb}} = E_N = -\frac{GM_1 M_2}{2a} = -\frac{\mathcal{M}}{2}(G\mathcal{M}\Omega)^{2/3}, \tag{21.21}$$

it follows that the orbit evolves in such a way that

$$\frac{\dot{\Omega}}{\Omega} = -\frac{3}{2}\frac{\dot{a}}{a} = \frac{3}{2}\frac{\dot{E}_{\text{orb}}}{E_{\text{orb}}} \approx \frac{3}{2}\frac{\dot{E}_{\text{gw}}}{E_N} = \frac{96}{5c^5}(G\mathcal{M}\Omega)^{5/3}\Omega \equiv \frac{1}{t_D} \tag{21.22}$$

defines the inspiral timescale t_D (as in Chapter 5). That is, we have

$$t_D \approx 140 \left(\frac{\mathcal{M}}{1.2 M_\odot}\right)^{-5/3} \left(\frac{f}{30 \text{ Hz}}\right)^{-8/3} \text{ s.} \tag{21.23}$$

The two neutron stars merge about 2 minutes after the system enters the frequency range above 30 Hz. The result also manifests the well-known fact that the leading order gravitational-wave signal only encodes the chirp mass, which is why this quantity was better constrained than any other combination of the masses in the case of GW170817.

A determination of the individual masses requires higher order post-Newtonian corrections. These effects are, of course, subtle and a key question concerns to what extent unmodelled features may limit the precision of the parameter extraction. It is important to keep in mind that, while one may expect to obtain fairly good estimates for the masses (as in Figure 21.1), it will be more difficult to infer the spin rates (the spin–spin and spin–orbit coupling effects are likely to be weak).

As long as it is safe to ignore other aspects, the binary signal is associated with a total number of cycles

$$N_{\text{gw}} = \int_{t_a}^{t_b} f\,dt = \int_{f_a}^{f_b} \frac{f}{\dot f}\,df = \int_{f_a}^{f_b} t_D\,df$$

$$= \frac{c^5}{32\pi\,(GM\pi f_a)^{5/3}} \left[1 - \left(\frac{f_a}{f_b}\right)^{5/3} \right]. \tag{21.24}$$

As an example, consider a signal between a frequency $f = f_a$, when the signal first enters the detector band, and f_b, above which it is suppressed by the detector noise. Inspired by the case of GW170817, we may use the (somewhat conservative) frequency range from $f_a \approx 30$ Hz to $f_b \approx 300$ Hz for which the total number of cycles would be $N_{\text{gw}} \approx 2{,}500$.

Let us now consider the possibility that tidal dynamics introduces an additional change of orbital energy, say at a rate $\dot E_{\text{tide}}$, leading to a change in the number of wave cycles in the observed frequency range. With

$$\dot E_{\text{orb}} = \dot E_{\text{gw}} + \dot E_{\text{tide}}, \tag{21.25}$$

we have

$$N = \frac{2}{3} \int_{f_a}^{f_b} \frac{E_{\text{orb}}}{\dot E_{\text{orb}}}\,df \approx \int_{f_a}^{f_b} t_D \left(1 - \frac{\dot E_{\text{tide}}}{\dot E_{\text{gw}}} \right) df = N_{\text{gw}} + \Delta N, \tag{21.26}$$

where the last step should be a good approximation as long as $\dot E_{\text{tide}} \ll \dot E_{\text{gw}}$. We see that the additional torque leads to a contribution

$$\Delta N = - \int_{f_a}^{f_b} t_D \left(\frac{\dot E_{\text{tide}}}{\dot E_{\text{gw}}} \right) df, \tag{21.27}$$

allowing us to estimate the relevance of any mechanism that is active in the observed frequency range.

However, the problem is a little bit more complicated. In general, we also need to account for changes to the orbital energy associated with the tidal effect. If we do this by letting

$$E_{\text{orb}} = E_{\text{N}} + E_{\text{t}}, \tag{21.28}$$

then we arrive at

$$N = \frac{2}{3} \int_{f_a}^{f_b} \frac{E_{\text{orb}}}{\dot E_{\text{orb}}}\,df \approx \int_{f_a}^{f_b} t_D \left(1 + \frac{E_{\text{t}}}{E_{\text{N}}} - \frac{\dot E_{\text{tide}}}{\dot E_{\text{gw}}} \right) df. \tag{21.29}$$

In order to progress, let us assume that we are in the adiabatic regime where the orbital evolution can be estimated from the energy of the system and the rate at which gravitational waves are emitted. The tidal contribution to the energy E and luminosity

dE/dt for a quasi-circular inspiral, which adds the following leading-order terms to the post-Newtonian point-particle corrections (pN), are (Flanagan and Hinderer, 2008)

$$E(x) = -\frac{1}{2}M\eta x\left[1 + [\text{pN}] - 9\frac{M_2}{M_1}\frac{\lambda_1}{M^5}x^5 + 1 \leftrightarrow 2\right], \tag{21.30}$$

$$\dot{E}(x) = -\frac{32}{5}\eta^2 x^5\left[1 + [\text{pN}] + 6\frac{M_1+3M_2}{M_1}\frac{\lambda_1}{M^5}x^5 + 1 \leftrightarrow 2\right]. \tag{21.31}$$

In these relations $\lambda_1 = \lambda(M_1)$ and $\lambda_2 = \lambda(M_2)$ are the tidal deformabilities of stars 1 and 2, respectively. The parameter $\eta = M_1 M_2/M^2$ is the dimensionless reduced mass, and x is the (dimensionless) post-Newtonian parameter $x = (\Omega M)^{2/3}$ (see Chapter 11). Notably, the tidal deformability enters (21.30) and (21.31) only through the combination

$$\frac{\lambda}{M^5} = \frac{2}{3}k_2\left(\frac{R}{M}\right)^5 \sim 10^2 - 10^5. \tag{21.32}$$

Note that, even though the tidal effect formally enters at 5th post-Newtonian (order x^5 in (21.31)), the numerical prefactor is large. In principle, this is a reflection of the fact that the internal dynamics of the binary companions do not obey the post-Newtonian ordering of the binary motion—there is no reason why the tidal deformability should fit neatly into the usual scheme.

For a given equation of state, these results allow us to predict the tidal phase contribution for a given binary system. Since both stars will be deformed, it is natural to discuss the problem in terms of the weighted average (Hinderer *et al.*, 2010)

$$\tilde{\lambda} = \frac{1}{26}\left[\frac{M_1+12M_2}{M_1}\lambda(M_1) + \frac{M_2+12M_1}{M_2}\lambda(M_2)\right], \tag{21.33}$$

which reduces to λ in the equal-mass case. Making contact with (21.29), we find that

$$\dot{E}_{\text{tide}} = \dot{E}_{\text{gw}}\frac{4(1+3q)}{(1+q)^{5/3}}\left(\frac{\pi f}{\Omega_0}\right)^{10/3}k_2, \tag{21.34}$$

with

$$\Omega_0 = \left(\frac{GM_1}{R_1^3}\right)^{1/2} \approx 2\pi \times 2{,}200\text{ Hz}\left(\frac{M_1}{1.4M_\odot}\right)^{1/2}\left(\frac{R_1}{10\text{ km}}\right)^{-3/2}, \tag{21.35}$$

where R_1 is the radius of the neutron star, together with a similar expression for the tidal change to the orbital energy. The tidal contribution to the number of gravitational-wave cycles (from one of the stars) then follows from

$$2\pi\,\Delta\mathcal{N} \approx -\frac{13}{2}\frac{1}{q(1+q)^{4/3}}\left(\frac{c^2 R_1}{GM_1}\right)^{5/2}\left(\frac{\pi f}{\Omega_0}\right)^{5/3}\tilde{k}_2, \qquad (21.36)$$

where q is the mass ratio and $\tilde{k}_2$ is given by

$$\tilde{k}_2 = \frac{1}{26}(1+12q)k_2. \qquad (21.37)$$

Typical estimates, for an equal mass binary of Newtonian $n = 1$ polytropes, are shown in Figure 21.3. As expected, the tidal deformability comes into play at late stages of inspiral. In the illustrated example the effect would not be 'detectable' below $f \approx 400$ Hz or so. However, this is only an indication that this tidal imprint could be relevant for real systems. In order to make the argument quantitative, we need to consider more realistic neutron star models.

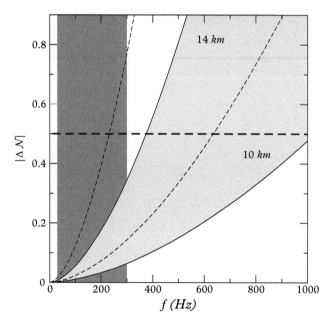

Figure 21.3 *A schematic illustration of the impact of tidal deformability on a binary neutron star signal. We indicate the estimated shift in the number of gravitational wave cycles $|\Delta\mathcal{N}|$ as a function of the gravitational-wave frequency f. The grey band follows from (21.36) if we assume a Newtonian $n = 1$ polytrope (for which $k_2 \approx 0.26$), two equal $1.4M_\odot$ neutron stars (thus doubling the value of $\Delta\mathcal{N}$) and the 'reasonable' range of radii $10 - 14$ km. The dashed curves show how this band shifts if we consider the (likely unrealistic) case of two $1.1M_\odot$ stars. The dashed horizontal line represents the indicative level of $|\Delta\mathcal{N}| \approx 0.5$ above which the effect should leave an imprint in a matched filter search, and the vertical shaded region represents an example observed frequency range between 30 and 300 Hz (similar to the actual range for GW170817). (Reproduced from Andersson and Ho (2018), copyright (2018) by the American Physical Society.)*

21.3   The relativistic Love number

Our estimates suggest that observational constraints on the tidal deformability may allow us—assuming that the individual masses can be inferred from the inspiral waveform—to infer the neutron star radius. Let us try to quantify how well one would expect to be able to do this in reality. First of all, we need to formulate the problem in general relativity. This is crucial if we want to confront realistic matter equations of state with observational data.

As in the Newtonian case, we consider a non-rotating spherically symmetric star in a static external (quadrupolar) tidal field $\mathcal{E}_{ij}$. The star responds to the tidal interaction by developing a quadrupole moment Q_{ij}. At large distances, r in the star's local asymptotic rest frame, the metric coefficient g_{tt} is then given by (Hinderer, 2008)

$$\frac{1 - g_{tt}}{2} = -\frac{M}{r} - \frac{3Q_{ij}}{2r^3}\left(\frac{x^i x^j}{r^2} - \frac{1}{3}\delta^{ij}\right) + \frac{1}{2}\mathcal{E}_{ij}x^i x^j, \tag{21.38}$$

where we have dropped terms of order $O(1/r^3)$ and $O(r^3)$. To linear order we relate the quadrupole moment to the tidal moment and introduce the Love number as

$$Q_{ij} = -\lambda \mathcal{E}_{ij} = -\frac{2k_2}{3G}R^5 \mathcal{E}_{ij}. \tag{21.39}$$

In essence, we extract the Love number, k_2, from the asymptotic behaviour of the gravitational field of the tidally deformed body. As we have assumed the star to be non-rotating, this involves solving the problem of static perturbations (the time-independent version of Eqs. (18.9)–(18.15) from Chapter 18), and connecting the two problems via

$$g_{tt} = -\left(1 - \frac{2M}{r}\right)(1 - H_0 Y_{lm}) . \tag{21.40}$$

Outside the star, the static perturbation problem leads to a single equation for H_0

$$H_0'' + \left(\frac{2}{r} - \lambda'\right)H_0' - \left[\frac{l(l+1)e^\lambda}{r^2} - (\lambda')^2\right]H_0 = 0, \tag{21.41}$$

where we have used $\nu = -\lambda$, as appropriate for the Schwarzschild solution (see Chapter 4). This equation may be solved in terms of associated Legendre polynomials, leading to

$$H(r) = a_P P_{l2}\left(\frac{r}{M} - 1\right) + a_Q Q_{l2}\left(\frac{r}{M} - 1\right), \tag{21.42}$$

where we have kept both the decreasing (P_{lm}) and growing solution (Q_{lm}).

Inside the star, the static perturbation equations reduce to (the relativistic analogue of (21.11))

$$r^2 H_0'' + \left[\frac{2}{r} + \frac{1}{2}(v' - \lambda') \right] r^2 H_0'$$

$$+ \left[2(1 - e^\lambda) - l(l+1)e^\lambda + 2r(2v' + \lambda') - r^2 (v')^2 \right] H_0$$

$$= -8\pi r^2 e^\lambda (\delta p + \delta \varepsilon) . \tag{21.43}$$

and

$$8\pi r^2 \delta p = \frac{1}{2} r e^{-\lambda} (v' + \lambda') H_0. \tag{21.44}$$

Once we provide an equation of state $\delta p = c_s^2 \delta \varepsilon$, where c_s^2 is the sound speed, we can solve this equation for H_0 to obtain the coefficients a_P and a_Q by matching to the exterior solution at the star's surface. Since the problem is studied within perturbation theory the overall amplitude of the solution is arbitrary, so we only need the ratio $a_l = a_Q/a_P$. In the case of quadrupole deformations ($l = 2$), we then have

$$k_2 = \frac{4G}{15} \left(\frac{M}{R} \right)^5 a_2. \tag{21.45}$$

For a given equation of state, this procedure leads to results similar to those shown in Figure 21.4.

The question is how accurately one would expect to be able to infer the tidal deformability from an observed signal. In order to consider this issue, we first need to work out λ (or k_2) for a range of equations of state. Typical results from this exercise show that the value of λ may span about an order of magnitude; see Figure 21.5. It should be noted that the values of λ for a given set of equations of state vary over a much wider range than the corresponding results for k_2 simply because of the factor of R^5 in the relation between the two quantities.

There are two aspects to the problem of detecting the tidal imprint. First of all, we may consider whether we will be able to extract the information from a single event. We can bring the Fisher-matrix approach (see Chapter 8) to bear on this question (Hinderer *et al.*, 2010). This leads to the constraints in Figure 21.5. The measurement error in λ generally increases with the total mass of the binary. By comparing the predicted errors to the expected range of values for λ, one finds that Advanced LIGO observations of binaries at a distance of 100 Mpc will be able to probe only unusually stiff equations of state. However, a third-generation instrument like the Einstein Telescope should be able to see a clean tidal signature. The tenfold increase in sensitivity allows a more precise discrimination between equations of state leading to the tidal signature being detectable across the expected neutron star mass range.

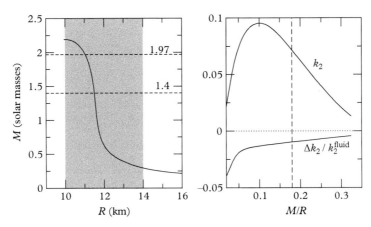

Figure 21.4 *Left: Mass–radius relation for a sequence of stellar models determined form the APR equation of state (Akmal et al., 1998), demonstrating consistency with the observational constraint on the maximum mass (upper dashed horizontal line, representing the lower limit of the mass range for the most massive known pulsar, PSR J0348+0432) and the expected radius range $10 - 14$ km (grey region). Right: The Love number k_2 (upper curve) as a function of the stellar compactness M/R. We also show the relative influence of the crust on the tidal deformability, represented by $\Delta k_2 / k_2^{fluid}$ (lower curve). As expected, the crust has a small effect on the tidal signal. This contribution can probably be ignored, although it introduces a systematic effect that limits the ultimate precision of the parameter extraction. (Reproduced from Penner et al. (2012) by permission of the AAS.)*

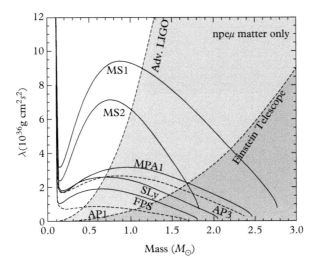

Figure 21.5 *Tidal deformability, in terms of λ, of a single neutron star as a function of mass for a range of realistic equations of state (including only $npe\mu$ matter models). The dashed lines between the various shaded regions represent the expected uncertainties in measuring λ for an equal-mass binary inspiral at a distance of $d = 100$ Mpc as it passes through the frequency range 10–450 Hz. This leaves out the last 20 or so cycles in the gravitational-wave phase (which are expected to be significantly affected by nonlinear effects). Observations of single events with Advanced LIGO will be sensitive to λ in the unshaded region, while the Einstein Telescope will be able to measure λ in the unshaded and light shaded regions. (Reproduced from Hinderer et al. (2010), copyright (2010) by the American Physical Society.)*

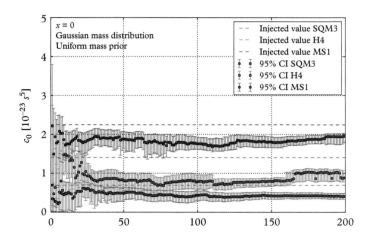

Figure 21.6 *Representative results from a Bayesian analysis of the tidal deformation detectability problem, showing the evolution of the medians and 95% confidence intervals in the measurement of $c_0 = \lambda(m_0)$, the tidal deformability at the reference mass $m_0 = 1.35\,M_\odot$, for three injected equations of state. The neutron star spins are set to 0 both in the injections and the templates and the injected masses are drawn from a peaked Gaussian distribution. (Reproduced from Agathos et al. (2015), copyright (2015) by the American Physical Society.)*

The Fisher-matrix analysis is, however, not particularly reliable for low signal-to-noise ratios. Hence, it makes sense to consider the problem from a Bayesian point of view. This can be done through numerical experiments which simulate a realistic data analysis setting (Agathos *et al.*, 2015). Representative results from such an effort are provided in Figure 21.6. The analysis coherently adds the binary signal to simulated stationary Gaussian noise following the predicted Advanced LIGO and Virgo design sensitivities. The neutron star masses are drawn from a peaked distribution and the sky positions, inclinations and polarizations of the sources are taken to be uniform on the sphere. In the illustrated case it is assumed that the stars are not spinning. Sources are distributed uniformly in co-moving volume, with luminosity distances between 100 and 250 Mpc, so that the majority of events will be near the threshold of detectability, chosen at a network signal-to-noise ratio of 8. The results suggest that the detection of the tidal imprint on the signal, to sufficient accuracy to distinguish the equation of state, will require the stacking of several tens of observations.

But sometimes you get lucky... Since the GW170817 event was closer than we assumed in the various estimates (at a distance of about 40 Mpc), we can make progress on constraining the matter. Results from such an analysis are provided in Figure 21.7 (assuming slowly spinning binary components). The shaded regions represent the values of the tidal deformabilities, λ_1 and λ_2, for models consistent with the inferred masses (from Figure 21.1). The results do not favour equations of state that predict less compact stars, which is consistent with radii inferred from X-ray observations (see Chapter 12). The best constraints on the neutron star radius (Abbott *et al.*, 2018*d*) are obtained

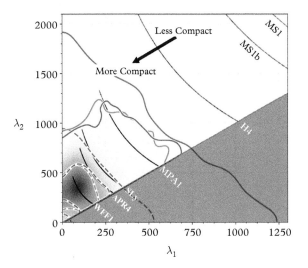

Figure 21.7 *Marginalized posterior for the tidal deformabilities of the two components of GW170817. The green shading shows the posterior obtained assuming a common equation of state for the two bodies, while the green, blue, and orange lines denote 50% (dashed) and 90% (solid) credible levels for the posteriors obtained using equation-of-state insensitive relations, a parameterized equation of state without a maximum mass requirement, and independent equations of state, respectively. Assuming a common equation of state shrinks the uncertainty region by about a factor of 3. The grey shading corresponds to the unphysical region $\lambda_2 < \lambda_1$, while the diagonal line represents $\lambda_1 = \lambda_2$. As a comparison, results for a set of representative equations of state are shown (as shaded filled regions). Some of these models are clearly not favoured. (Adapted from Abbott et al. (2018d).)*

if one assumes that both stars are determined by the same equation of state (which seems eminently reasonable) and that they spin slowly (in line with all known galactic binary neutron stars). If one also accounts for the need for the equation of state to allow neutron stars with a mass above $1.97 M_\odot$ (the lower limit of the mass range for the most massive known pulsar, PSR J0348+0432, see Figure 21.5), then the neutron star radius is constrained to the range $10.5 - 13.3$ km (at 90% credible level).

Before we move on, it is also worth pointing out that the Love number is closely related to both the moment of inertia and the star's quadrupole moment. One can identify robust (seemingly universal) relations between these quantities, similar to the empirical relations for the oscillation modes discussed in Chapter 18. These so-called I-Love-Q relations may have important implications for gravitational-wave observations (Yagi and Yunes, 2013b). In particular, they may help break the degeneracy between the spins and the quadrupole moment in non-precessing binaries (Yagi and Yunes, 2013a). The neutron star quadrupole moment is degenerate with the stars' individual spins, because there is a spin-spin interaction term in the gravitational-wave phase that enters at the same order in v/c as the quadrupole one. This kind of degeneracy may prevent simultaneous extraction of the quadrupole moment and the spins from a detected signal. However, one can use the Love-Q relation to break this degeneracy by rewriting the quadrupole moment as a function of the Love number. If the Love number can be extracted from the inspiral signal, then one can also 'measure' the spins.

21.4   Dynamical tides: resonances

In addition to the static deformation, the tidal field (21.1) has time-dependent components. This drives motion in the stellar fluid, which is interesting because it may lead to resonances between the orbit and the various oscillation modes in the star (Reisenegger and Goldreich, 1994; Kokkotas and Schafer, 1995; Ho and Lai, 1999; Andersson and Ho, 2018).

In order to quantify this effect, we add the tidal potential to the Lagrangian perturbation problem for a non-rotating star (see Chapter 13). Thus, we have

$$\rho \partial_t^2 \xi + C\xi = -\rho \nabla \chi, \tag{21.46}$$

where $C\xi$ is a messy expression. We want to examine the driven response of the star's fluid in terms of a set of normal modes, corresponding to solutions ξ_α (where α is a label that identifies the modes, say in terms of the number of nodes in the radial eigenfunction and the corresponding spherical harmonics). Letting the (real) mode-frequency be ω_α, we have

$$\xi = \sum_\alpha a_\alpha(t) e^{i\omega_\alpha t} \xi_\alpha(\mathbf{r}). \tag{21.47}$$

The individual modes then satisfy

$$-\rho \omega_\alpha^2 \xi_\alpha + C\xi_\alpha = 0. \tag{21.48}$$

We normalize the modes using the inner product from Chapter 13, such that

$$\langle \xi_{\alpha'}, \rho \xi_\alpha \rangle = \int \rho \xi_{\alpha'}^* \xi_\alpha d^3 x = \delta_{\alpha\alpha'}. \tag{21.49}$$

We can use this orthogonality to rewrite (21.46) as an evolution equation for the amplitudes b_α. This leads to

$$\ddot{b}_\alpha + \omega_\alpha^2 b_\alpha = -\langle \xi_\alpha, \rho \nabla \chi \rangle. \tag{21.50}$$

Finally, making use of the perturbed continuity equation

$$\delta \rho_\alpha = -\nabla \cdot (\rho \xi_\alpha), \tag{21.51}$$

and integrating by parts, we have

$$\langle \xi_\alpha, \rho \nabla \chi \rangle = \int \rho \xi_\alpha^* \nabla \chi \, d^3 x = - \int \chi \nabla \cdot (\rho \xi_\alpha^*) d^3 x = \int \chi \delta \rho_\alpha^* d^3 x. \tag{21.52}$$

In general, e.g. when the star is spinning, it may be practical to express the stellar perturbations with respect to a different set of spherical harmonics, perhaps symmetric with respect to the spin axis rather than the axis pointing toward the binary partner. In order to allow for this, we note the general transformation

$$Y_{lm}(\theta, \varphi) = \sum_{m'} \mathcal{D}_{mm'}^{(l)} Y_{lm'}(\theta', \varphi'), \tag{21.53}$$

where the Wigner D-function $\mathcal{D}_{mm'}^{(l)}$ is given by, for example, Ho and Lai (1999). In effect, we then have the final equation for the driven modes (Lai, 1994)

$$\ddot{b}_\alpha + \omega_\alpha^2 b_\alpha = \sum_{lmm'} \frac{GM'}{a^{l+1}(t)} W_{lm} \mathcal{D}_{mm'}^{(l)} Q_{\alpha, lm'} e^{-im\psi(t)}, \tag{21.54}$$

where we have introduced the 'overlap integral'

$$Q_{\alpha, lm'} = \int \delta \rho_\alpha^* r'^l Y_{lm'}(\theta', \varphi') d^3 x'. \tag{21.55}$$

This is the main result. In general, we have a driven set of oscillations that become resonant when (since $\psi = \Omega t$)

$$|m| \Omega = |\omega_\alpha|. \tag{21.56}$$

However, in order for this resonance to be relevant, we must have $l + m =$ even (otherwise $W_{lm} = 0$ due to the symmetry of the tidal potential). We must also have

$$D_{mm'}^{(l)} Q_{\alpha, lm'} \neq 0. \tag{21.57}$$

Now, if we assume that the resonant mode is associated with its own spherical harmonics, in such a way that $\delta \rho = \delta \bar{\rho} Y_{jk}$ then we must have $m' = k$ otherwise $Q_{\alpha, lm'} = 0$. There will also be a constraint associated with the symmetry of the mode. For example, for (spheroidal) modes like the f-modes, we much have $l + j =$ even. Finally, we see from (21.55) that the eigenfunction of the mode should ideally be as 'similar' to r^l as possible. When the mode has a number of radial nodes (as in the case of the p- and g-modes from Chapter 13) there are cancellations in the integral and the effective overlap with the tidal driving force will be weaker (Lai, 1994; Andersson and Ho, 2018). In order to quantify the impact of a resonance, we need

$$\Delta \mathcal{N} \approx - \int_{f_a}^{f_b} f \frac{dE_{tide}}{da} \frac{1}{\dot{E}_{orb}} da. \tag{21.58}$$

After integration, this leads to

$$\Delta \mathcal{N} \approx - \left(\frac{f \Delta E_{tide}}{\dot{E}_{orb}} \right)_{f=f_a} \approx - \left(\frac{3}{2} \frac{f t_D \Delta E_{tide}}{E_N} \right)_{f=f_a}, \tag{21.59}$$

where ΔE_{tide} is the total energy transferred from the orbit to the resonant mode, and the expression should be evaluated at the resonance frequency, $f = f_a$.

Let us first consider the case of a non-rotating star, for which the different m harmonics are degenerate. In this case, we have (Lai, 1994)

$$\Delta E_{tide} \approx - \frac{\pi^2}{512} \left(\frac{GM_1^2}{R_1} \right) \hat{\omega}_\alpha^{1/3} Q_\alpha^2 \left(\frac{R_1 c^2}{GM_1} \right)^{5/2} q \left(\frac{2}{1+q} \right)^{5/3}, \tag{21.60}$$

suppressing the harmonic indices on the overlap integral and introducing the dimensionless mode frequency $\hat{\omega}_\alpha$ through

$$\omega_\alpha = \hat{\omega}_\alpha \Omega_0. \tag{21.61}$$

We also have the resonance condition

$$\omega_\alpha = 2\pi f_\alpha = 2\Omega = 2\pi f. \tag{21.62}$$

It is important to note that, in the quadrupole case, the oscillation frequency of the resonant mode (f_α) is equal to the observed gravitational-wave frequency (f).

Quantifying the impact of a given mode at the corresponding resonance radius (Ho and Lai, 1999),

$$a_\alpha = \left[\frac{4GM_1(1+q)}{\omega_\alpha^2} \right]^{1/3}, \tag{21.63}$$

we find that

$$E_N = - \frac{1}{2^{5/3}} \left(\frac{GM_1^2}{R_1} \right) \hat{\omega}_\alpha^{2/3} \frac{q}{(1+q)^{1/3}}, \tag{21.64}$$

and

$$\frac{\Delta E_{\text{tide}}}{E_N} \approx \frac{\pi^2}{128 \times 2^{1/3}} (\pi \hat{f}_\alpha)^{-1/3} Q_\alpha^2 \left(\frac{R_1 c^2}{GM_1}\right)^{5/2} \left(\frac{2}{1+q}\right)^{4/3}, \tag{21.65}$$

where $\hat{f}_\alpha = \hat{\omega}_\alpha / 2\pi$. At resonance, we also have

$$f_{tD} = \frac{5}{96\pi} (\pi \hat{f}_\alpha)^{-5/3} \left(\frac{c^2 R_1}{GM_1}\right)^{5/2} \frac{(1+q)^{1/3}}{q}, \tag{21.66}$$

and it follows from (21.59) that

$$\Delta \mathcal{N} \approx -4 \times 10^{-4} \hat{f}_\alpha^{-2} Q_\alpha^2 \left(\frac{c^2 R_1}{GM_1}\right)^5 \frac{1}{q(1+q)}. \tag{21.67}$$

As one might have expected, this is a small effect. Nevertheless, it is instructive to consider to what extent the different contributions can be considered known. We have some handle on the range for the mass ratio q from radio pulsar observations (see Table 9.2. The star's compactness is also constrained by observations. From X-ray observations of accreting neutron stars (and the GW170817 results!) one would expect the radius of a $1.4M_\odot$ star to lie in the range $10 - 14$ km (see Figure 12.2). As the mass–radius curve tends to rise steeply in the relevant mass range (for a typical equation of state) we might assume the radius to be inside this range for all plausible binary masses.[1] This constrains the compactness to the range

$$0.12 \le \frac{GM_1}{c^2 R_1} \le 0.24, \tag{21.68}$$

which introduces an uncertainty of about a factor of 30 in the estimate for $\Delta \mathcal{N}$, illustrating the importance of obtaining tighter constraints on the neutron star radius. This is, of course, one of the main targets of the observations in the first place. One would hope to (eventually) get a tighter radius constraint from the tidal imprint. In addition, a measurement of the neutron star radius to within 5% is a key science aim of the NICER mission which is currently flying on the International Space Station (Gendreau *et al.*, 2016).

With a narrower region of uncertainty for the stellar compactness, one may be able to use observed deviations from a pure radiation reaction inspiral to constrain the value of Q_α for any resonant mode in a given frequency range. We illustrate this idea in Figure 21.8. Imagine that one sets an upper limit on the deviation from a post-Newtonian radiation reaction inspiral of order $\Delta \mathcal{N} \le 0.5$ in a given frequency range,

---

[1] Note that this argument does not account for the softening effect of internal phase transitions.

say $f = 100 - 150$ Hz. Then, we know from (21.67) (assuming canonical neutron star parameters) that

$$Q_\alpha \leq 10^{-2} \left(\frac{f}{100 \text{ Hz}} \right) |\Delta \mathcal{N}|^{1/2}. \tag{21.69}$$

This constraint is shown in Figure 21.8. We see that the chances of observing the imprint of a tidal resonance is better at frequencies below a few tens of hertz, where the detectors start to become less sensitive. Moreover, given the dependence on the stellar compactness, the effect would be more prominent for a larger neutron star radius. In fact, given a reliable theoretical calculation for Q_α one can turn this argument into a constraint on the radius.

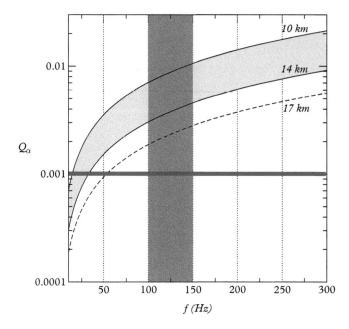

Figure 21.8 *Constraints on Q_α if a limit $|\Delta \mathcal{N}| \leq 0.5$ were to be inferred from inspiral data. The thin black lines represent equal mass $1.4 M_\odot$ binaries with neutron star radius 10 km (upper curve) and 14 km (lower curve). The grey region represents the expected radius range from X-ray observations. As an indication, the thick horizontal (red) line represents the largest values of Q_α expected for the g-modes of a non-rotating star (Lai, 1994). This should be taken as indicative of what is expected from theory. Finally, the shaded vertical region relates to an example where the observational constraint is obtained for a distinct frequency band (here taken to be 100–150 Hz). This figure illustrates that the resonant modes of a non-rotating star may be difficult to detect, but there could be a relevant effect below 50 Hz or so, if the neutron star radius were to be surprisingly large (the dashed curve shows the result for a radius of 17 km). One should also keep in mind that rotation may lead to slightly larger values of Q_α, in which case the chance of detection would improve. (Reproduced from Andersson and Ho (2018), copyright (2018) by the American Physical Society.)*

In order to understand the wider implications of this kind of constraint for neutron star physics, we need to consider the nature of specific oscillation modes. For non-rotating stars, the most likely set of modes to exhibit tidal resonance are the gravity g-modes (Lai, 1994). In a mature (cold) neutron star, these modes are associated with internal composition stratification (see Chapter 13). If the motion of a moving fluid element is faster than the nuclear reactions that would equilibrate the fluid to its new surroundings, then the chemical differences lead to buoyancy that provides the relevant restoring force. In the simplest models, the g-modes are associated with a varying proton fraction (Reisenegger and Goldreich, 1992). This typically leads to mode frequencies below a few 100 Hz and a dense spectrum of high overtone modes at lower frequencies. The lowest order (highest frequency) mode couples the strongest to the tide, with a typical value of the coupling constant $Q_\alpha \approx 10^{-4} - 10^{-3}$ (Lai, 1994; Kokkotas and Schafer, 1995). Most likely, this makes the effect too weak to be detected by the current generation of instruments; see Figure 21.8.

The problem is nevertheless interesting. The g-modes rely on physics beyond the bulk properties of the star, reflecting how the strong interaction determines the composition of matter at high densities. The state of matter is also important. For example, if the star's core contains a superfluid, then the charged components (in the simplest case, protons and electrons) can move relative to the neutrons. As a result, as long as we assume that the electrons and protons are electromagnetically coupled, the origin of the buoyancy is removed and there will no longer be any g-modes (Lee, 1995; Andersson and Comer, 2001a). This would obviously remove any related resonances, as well. However, there are twists to this story. The composition of a neutron star core is more complex than pure npe matter. Close to the nuclear saturation density the formation of muons becomes energetically favourable. This leads to stratification (now associated with the electron–muon fraction) also in a superfluid star, which reinstates the composition g-modes (Gusakov and Kantor, 2013; Passamonti *et al.*, 2016). These new g-modes are expected to have higher frequencies, perhaps by a factor of a few, which means that resonances become relevant at later stages of the inspiral. The first estimates of the tidal coupling for these modes suggest that they may be associated with an increased transfer of energy but this is compensated for by the fact that the inspiral is accelerated at the higher frequencies (Yu and Weinberg, 2017). As a result, the estimated values of Q_α are similar to those of the original g-modes. The example in Figure 21.8 then suggests that the higher frequency g-mode resonances are likely to leave a weaker imprint on the gravitational-wave signal.

In principle, the star's fundamental f-mode provides the most promising overlap with the tidal potential. However, since the f-mode frequency tends to be relatively high, this resonance is unlikely to happen before merger (Kokkotas and Schafer, 1995). Nevertheless, the interaction with the f-mode causes an increasing (dynamical) correction to the tide (Hinderer *et al.*, 2016; Steinhoff *et al.*, 2016). This effect can be significant, and should be included in any robust gravitational-wave template; see Figure 21.9. As a result, one would end up with an effective description that encodes both the (static) tidal deformability and the f-mode frequency. It is important to note that the relevant mode frequency is not the redshifted one measured by a distant observer.

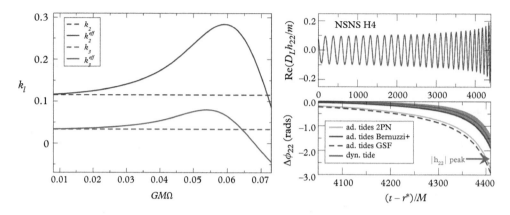

Figure 21.9 *Left: Dimensionless effective tidal deformability from a two timescale approximation (for the H4 equation of state and a neutron star mass of $1.35M_\odot$). The index l refers to the multipoles, with k_2 is the quadrupolar tidal deformability and k_3 the corresponding octupolar one. Right: The $l = m = 2$ mode gravitational waveform for a merging neutron star binary and an effective-one-body model including the tidal corrections (red curve) as well as numerical relativity simulations (blue curve), for the neutron star model used in the left panel. The lower panel shows the phase difference between the numerical simulation and different template models. (Reproduced from Steinhoff et al. (2016) and Hinderer et al. (2016), respectively. Copyright (2016) by the American Physical Society.)*

Instead, we need the mode in the rest frame of the star. Similarly, from the point of view of the star, the frequency of the driving tidal force is higher than inferred by an asymptotic observer. This enhances the dynamical contribution to the tide by shifting the f-mode resonance to a lower orbital frequency. This means that the system spends more time close to resonance and, hence, transfers more energy from the orbit to the tide.

The resonance problem also becomes more intricate for rotating stars. First of all, the resonance condition involves the mode frequency in the inertial frame. That is, we have (again for the quadrupole tide)

$$\omega_\alpha^{(i)} = 2\Omega_s. \tag{21.70}$$

Secondly, rotation breaks the degeneracy associated with the azimuthal angle and modes associated with different values of m become distinct (i.e. we need to separately consider $m = \pm 2$). To leading order, the mode frequency is then given by (as before)

$$\omega_\alpha^{(i)} = \omega_\alpha^{(r)} - m\Omega_s, \tag{21.71}$$

where $\omega_\alpha^{(r)}$ is the frequency of the mode in the rotating frame and Ω_s is the spin of the star. If the star rotates rapidly then the change in shape due to the centrifugal force provides an additional correction, but (as we have already suggested) it seems reasonable to argue that most binary systems will be old enough that the neutron stars will have slowed down by the time they become detectable with ground-based interferometers. Nevertheless, it is clear from (21.71) that low-frequency modes may be significantly affected by the rotation. This adds parameters to the problem (and the individual spin rates may be difficult to extract from the inspiral waveform) which complicates any effort to use an observed signal to constrain the physics.

The rotation-induced shift of the mode frequency (21.71) may also make a given mode susceptible to the CFS instability (discussed in Chapter 13), where the oscillation is driven unstable by the emission of gravitational waves. The instability sets in when a retrograde mode in the rotating frame becomes prograde in the inertial frame. Effectively, the mode energy then becomes negative. In an isolated star one would not expect the instability of g-modes to be particularly relevant because these modes are not efficiently emitting gravitational waves and the instability does not overcome viscous damping (Lai, 1999). The case of tidal driving is different. The main impact of a mode being unstable is that the growth of the mode pumps energy back into the orbit. This would alter the sign of $\dot{E}_{\text{tide}}$, slow down the inspiral and lead to an increase in $\mathcal{N}$, rather than a decrease. This is an important feature, which if observed would provide exciting insight into the stellar dynamics.

We have also seen that (see Chapter 13), a spinning star has a richer spectrum of oscillation modes. In particular, the Coriolis force provides an additional restoring force, which brings new sets of modes into existence. These inertial modes scale with the rotation frequency. This means that one may confuse the identification of an observed resonance. However, if the inspiral signal provides an independent constraint on the star's spin then one could potentially rule out inertial modes as they have to lie in the range $-2\Omega_s \le \omega_\alpha^{(r)} \le 2\Omega_s$. The first estimates of tidal inertial-mode excitation suggested that the impact would be minor, but the gravito-magnetic coupling enhances the importance of the inertial modes (Flanagan and Racine, 2007). In the specific case of the r-modes, the phase shift may amount to

$$\Delta\mathcal{N} \approx \frac{0.1}{2\pi} \left(\frac{R_1}{10 \text{ km}}\right)^4 \left(\frac{M_1}{1.4M_\odot}\right)^{-10/3} \left(\frac{f_s}{100 \text{ Hz}}\right)^{2/3}, \qquad (21.72)$$

where $f_s = \Omega_s/2\pi$. Based on the available results, this could be the strongest relevant mode resonance. As the r-modes are generically unstable to gravitational-wave emission (obviously assuming that the star spins fast enough that the system is above the relevant instability curve; see Chapter 15), the main resonance effect may then tend to slow down the inspiral. Moreover, as the r-mode frequency is (not accounting for relativistic effects; see Chapter 18) given by $f = 4f_s/3$, an observational constraint on the star's spin would directly indicate the frequency of the associated resonance.

21.5　Shattering the crust

In addition to affecting the gravitational-wave phasing during inspiral, the tidal interaction (and related mode resonances) may trigger an electromagnetic counterpart signal.

A particularly interesting scenario involves releasing energy stored in the star's crust. If the crust were to shatter during the inspiral, it may lead to an observable precursor to the merger event (Troja *et al.*, 2010; Tsang *et al.*, 2012; Penner *et al.*, 2012). In order to estimate the likelihood of this happening, we revisit the neutron star mountain problem from Chapter 6, but from a different point of view.

Let us consider a star of mass M_1, radius R_1, a distance a from a binary companion of mass M_2. We focus on the situation where the star is axisymmetric, deformed away from sphericity such that its quadrupole moment is changed by a fractional amount ϵ. The difference from the discussion in Chapter 6 is that the star is now deformed due to the tidal field of the companion. The star's energy can then be written

$$E(\epsilon) = E_0 + E_{\text{gravity}} + E_{\text{tidal}} + E_{\text{elastic}}, \tag{21.73}$$

where E_0 denotes the energy of the equivalent fluid star in the absence of a companion, E_{gravity} the perturbation in the star's self-gravity, E_{tidal} the energy perturbation due to the tidal interaction, and E_{elastic} is the elastic energy. Since the self-gravity is minimized for a spherical configuration, we have $E_{\text{gravity}} \sim A\epsilon^2$, where $A \sim GM^2/R$. Assuming that the elastic energy is minimized when the star is spherical (which would make sense for an inspiralling binary), we have $E_{\text{elastic}} \sim B\epsilon^2$, where B is of order the Coulomb binding energy of the solid phase (as before). The gravitational field of the binary partner is, of course, $-GM_2/r_c$, where $r_c = 0$ is the centre of mass of the companion. Expanding this about the centre of mass of the star of mass M_1, defined by $r = 0$, leads to the tidal field

$$\Phi_{\text{tidal}}(r) = \frac{1}{2}r^2\frac{GM_2}{a^3}[\delta_{ij} - 3n_in_j]\hat{r}_i\hat{r}_j, \tag{21.74}$$

where n_i is a unit vector along the line separating the two stars. The corresponding energy perturbation is

$$E_{\text{tidal}} \sim \Phi_{\text{tidal}}(\epsilon M) \sim A\frac{M_2}{M_1}\left(\frac{R_1}{a}\right)^3\epsilon. \tag{21.75}$$

The perturbation in the star's quadrupole moment is simply found by multiplying by the moment of inertia I. Thus, we have

$$Q_{\text{tidal}} = -\frac{1}{2}\frac{M_2}{M_1}\left(\frac{R_1}{a}\right)^3 I. \tag{21.76}$$

It follows that strain builds in the star's crust in such a way that

$$\sigma_{\text{tidal}} \sim \epsilon_{\text{tidal}} \sim Q_{\text{tidal}}/I. \tag{21.77}$$

This estimate shows that the crust elasticity will not have a significant effect on the tidal problem (as, indeed, demonstrated by the results in Figure 21.4). However, we can now estimate at what point during inspiral the crust fails. Combining (21.76) and (21.77) with Kepler's law, we find that the crust fails ($\sigma_{\text{tidal}} = \sigma_{\text{br}}$) when

$$f_{\text{GW}}^{\text{break}} \approx 3\,\text{kHz} \left(\frac{M_1}{1.4M_\odot}\right)^{1/2} \left(\frac{10^6\,\text{cm}}{R_1}\right)^{3/2} \left(\frac{\sigma_{\text{br}}}{0.1}\right)^{1/2}, \qquad (21.78)$$

for an equal mass system. We have scaled the order-of-magnitude result to the estimated breaking strain $\sigma_{\text{br}} \approx 0.1$ from molecular dynamics simulations (Horowitz and Kadau, 2009). Comparing to the estimated gravitational-wave frequency at the innermost stable orbit

$$f_{\text{GW}}^{\text{isco}} \approx \frac{c^3}{\pi 6^{3/2} GM_{\text{total}}} = 1.6\,\text{kHz} \left(\frac{1.4M_\odot}{M}\right), \qquad (21.79)$$

(again, for an equal-mass binary) we see that crust failure should not be expected significantly before coalescence. This expectation is brought out by detailed (fully relativistic) calculations (Penner *et al.*, 2012).

What happens if the crust fails? Again, key insights are provided by the molecular dynamics simulations. The indications are that when the critical strain is reached, there is a catastrophic failure, with energy released throughout the strained volume, rather than the formation of a lower-dimensionality crack. What happens next is less clear. Two extreme scenarios can be envisaged. The relieved strain is either dissipated locally as heat, or converted into phonons/seismic waves and transported throughout the star prior to dissipation. An interesting question is whether the relieved strain is capable of melting the crust, but estimates suggest that this is not likely to happen (Penner *et al.*, 2012).

The crust may not melt, but a significant amount of energy is still being released. Could this have observable consequences? Given the importance of neutron star binary mergers for gravitational-wave astronomy, and the association of such events with short gamma-ray bursts (see later), a precursor signal would be interesting. An interesting argument (Tsang *et al.*, 2012) involves the energy being released into seismic waves that generate Alfvén waves in the magnetosphere, eventually leading to a gamma-ray signal. In principle, such an event could be as energetic as the largest observed magnetar flare (the 27 December 2004 event in SGR 1806-20), as long as all the available strain energy is transferred to the magnetosphere. Such a signal—which would precede the merger itself by a fraction of a second—could be observable from a distance of 100 Mpc and the corresponding merger signal would be comfortably detectable by the Advanced LIGO–Virgo detector network.

21.6   Merger dynamics

The gradual inspiral, moderated by tidal effects, finally gives way to the dramatic dynamics of the merger. This leads to a complex, high-frequency gravitational-wave signal (see Figure 21.2) encoding the transition from binary system to either a solar-mass black hole or a more massive neutron star. Regardless of the eventual outcome, the merger is likely to initially lead to the formation of a hypermassive neutron star supported by thermal pressure and rotation (Baiotti and Rezzolla, 2017). The gravitational-wave energy released during the first $\sim 10\,\text{ms}$ of the life of this object is about twice the energy emitted over the entire inspiral history, comparable to the emission from a black-hole merger.

We have already discussed the different inspiral phases of the gravitational-wave signal. As the binary orbit shrinks due to the energy lost to radiation, the gravitational-wave amplitude rises and the frequency increases as well. We have seen that the inspiral chirp is well modelled by post-Newtonian methods (see Chapter 11), as it does not depend (much) on the physics of the compact objects involved. In contrast, the merger encodes a rich amount of interesting information. However, in order to catch this part of the signal, we need the detectors to be sensitive at higher frequencies. The tidal disruption that precedes the merger occurs above 600 Hz or so (Read *et al.*, 2013) and the oscillations of the remnant typically leads to a signal at several kHz. Simulations show that $1-3\%$ of the original binary's mass-energy is released at these frequencies. This is a lot of energy, but we may nevertheless need third-generation instruments to detect this signal (Sathyaprakash *et al.*, 2012; Clark *et al.*, 2016). However, the information it encodes is extremely valuable. In particular, the merger signal should tell us whether a massive neutron star or a black hole is formed, placing constraints on the (hot) supranuclear equation of state.

Increasingly precise numerical simulations illustrate the complexity of the merger (see Baiotti and Rezzolla (2017) for a recent review). This complexity is 'problematic' because, given the amount of 'unknown' physics involved, we are unlikely to reach a stage where we have truly reliable signal templates. Nevertheless, we can make progress by identifying 'robust' characteristics of the signal.

On a dynamical timescale (on the order of a millisecond) the two stars form a single, massive, differentially rotating object (see Figure 21.10). As the merger proceeds, streams of matter are squeezed out of the interface between the two stars. Part of this material becomes unbound while the rest forms a thick torus around the merger remnant. The nonlinear shock associated with the matter interface ramps up the temperature to levels as high as $\sim 50\,\text{MeV}$—hotter than a neutron star born in core collapse. Combined with the rotational support, the thermal pressure may prevent gravitational collapse even if the total binary mass significantly exceeds the maximum mass of non-rotating neutron stars. In fact, simulations show that prompt collapse only takes place for very massive binaries. The most likely (immediate) remnant is thus a massive, hot, differentially rotating 'neutron star'.

Large amplitude oscillations are induced by the merger, and the impact of the matter equation of state is apparent in the post-merger dynamics. In particular, the fundamental mode of the remnant tends to dominate the post-merger gravitational signal (Bauswein and Janka, 2012; Bauswein *et al.*, 2014; Takami *et al.*, 2014; Bose *et al.*, 2018). The reason for this is easily understood from the sequence of snapshots in Figure 21.10. The initial 'shape' of the merger remnant—basically a dumbbell—resembles a large amplitude

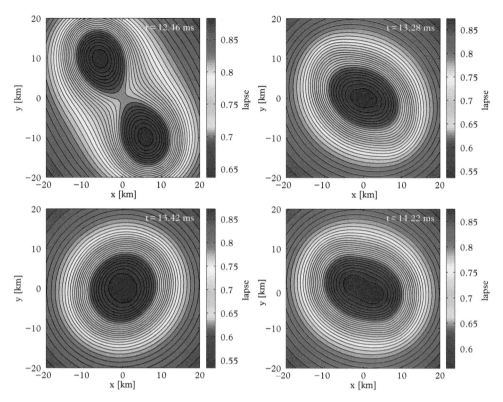

Figure 21.10 *Evolution of the lapse function, which can be seen as a relativistic proxy for the gravitational potential, of a 1.35–1.35 $M_\odot$ neutron star merger (with the DD2 equation of state from Hempel and Schaffner-Bielich (2010) and Typel et al. (2010)) in the equatorial plane. (Reproduced from Bauswein et al. (2016) with kind permission from the European Physics Journal (EPJ).)*

quadrupole f-mode. The overall impact of the equation of state is also readily explained. Stiff equations of state lead to large neutron stars. These are more easily deformed by the tidal field; see Figure 21.5. Consequently, finite size effects set in at a larger orbital separation and the stars merge at a lower orbital frequency. In contrast, soft equations of state yield more compact neutron stars, which are more difficult to deform and reach higher orbital frequencies before they merge. This implies a higher impact velocity. The stiffness also affects the dynamics of the post-merger phase and, in particular, the frequencies of the excited oscillation modes (just as in the case of mature neutron stars; see Chapter 13). A stiff equation of state leads to a relatively large merger remnant, with lower frequency oscillations (scaling approximately with the average density). In the case of a soft equation of state, the merger remnant is more compact and thus oscillates at higher frequencies. In addition, the higher impact velocity during merger leads to a stronger excitation of the quasi-radial oscillation mode—essentially inducing a large amplitude 'breathing' of the remnant.

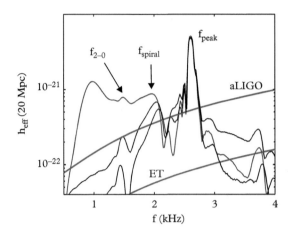

Figure 21.11 *Gravitational-wave spectrum for a 1.35–1.35 $M_\odot$ neutron star merger (for the equation of state used in Figure 21.10) along the polar direction at a distance of 20 Mpc (blue solid curve). The vertical axis shows $h_{eff} = f\tilde{h}(f)$ with the Fourier transform of the waveform and frequency f. The distinct peaks at $f_{peak}, f_{spiral},$ and f_{2-0} are features of the post-merger phase, which can be associated with specific features of the remnant. The expected sensitivity of Advanced LIGO and the Einstein Telescope are shown as red curves. (Reproduced from Bauswein et al. (2016) with kind permission from the European Physics Journal (EPJ).)*

These are the stand-out features of the gravitational-wave spectrum; see Figure 21.11. For a fixed total binary mass there appears to be a tight correlation between the dominant post-merger oscillation frequency (f_{peak}) and the radii of non-rotating neutron stars (Bauswein *et al.*, 2016). This suggests that one may be able to infer the neutron star radius from an observed merger signal (perhaps with an accuracy of a few hundred meters (Clark *et al.*, 2016). Additional features in the spectrum relate more specifically to the merger dynamics. These features tend to evolve during the post-merger phase, cf. Figure 21.12. A determination of the frequencies of the quadrupole f-mode (f_{peak}) and the radial mode (f_0) suggests that a secondary peak should be expected (due to nonlinear mode-coupling) at $f_{peak} - f_0$. This is, indeed, seen in Figure 21.11 (where this feature is labelled f_{2-0}). The final feature (f_{spiral}) in the spectrum shown in Figure 21.11 is generated by a spiral deformation created during merger. This deformation cannot follow the faster rotation of the inner remnant. Instead, the deformation forms antipodal bulges, which orbit around the central part of the remnant for several milliseconds (Bauswein *et al.*, 2016). Being associated with a non-axisymmetric deformation, the antipodal bulges generate a gravitational-wave signal at twice their orbital frequency.

One can use time-frequency maps of the gravitational-wave signal (like that in Figure 21.12) to analyse the post-merger dynamics (Clark *et al.*, 2016). The spectral features are broad, reflecting the nontrivial time-evolution of the frequencies. In terms of detectability, the peak frequency should be measurable by a coherent burst search analysis. A Fisher-matrix study provides an estimate of the typical uncertainty in the

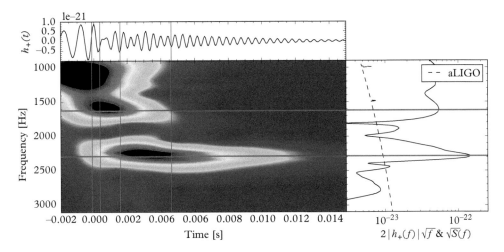

Figure 21.12 *Time-frequency analysis for a 1.35–1.35 $M_\odot$ neutron star merger waveform (for an optimally-oriented source at 50 Mpc). The top and right panels show the time-domain waveform-component h_+ and its Fourier spectrum, respectively. The time-frequency map is constructed from the magnitudes of the coefficients of a continuous wavelet transform. Horizontal red lines emphasize the locations of the peak frequency f_{peak} and the secondary peak which, in this case, corresponds to f_{spiral}. (Reproduced from Clark et al. (2016).)*

determination of the frequency, $\delta f_{\text{peak}} \sim 50$ Hz. Making use of the correlation between the frequency and the neutron star radius, this suggests a possible constraint on the radius of a fiducial neutron star of ~ 200 m. However, such measurements may only be possible for nearby (~ 30 Mpc) sources with Advanced LIGO. Such precision measurements may require future instruments—although it is worth noting that the sensitivity of the detectors during the O2 run was not too far away from being able to distinguish the GW170817 merger; see Figure 21.13. An identification of this part of the signal may eventually lead to independent constraints on the equation of state, complementing those from the inspiral phase.

The nature of the GW170817 remnant has not (yet) been established observationally. A LIGO search for short ($\lesssim 1$ s) and intermediate duration ($\lesssim 500$ s) emission did not identify a signal in the data (Abbott *et al.*, 2019a). However, as is evident from Figure 21.13, the upper limits achieved in the relevant frequency range were about one order of magnitude above the amplitudes expected from simulations.

The eventual fate of the remnant depends on a range of factors. The total mass is obviously of paramount importance, but factors like the mass ratio and the nature of the differential rotation induced in the remnant are also relevant. The former determines how much mass is ejected from the system and the latter helps support an object that is too massive to survive as a uniformly rotating star. Microphysics also comes into play. The efficiency of energy transport (mainly due to neutrinos) and the redistribution of angular momentum (likely due to the internal magnetic field and the magneto-rotational

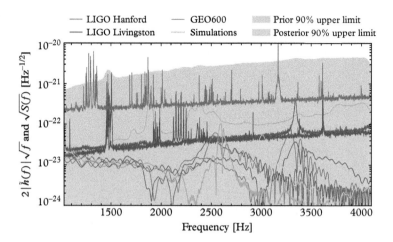

Figure 21.13 *The 90% credible upper limits on the gravitational-wave strain from the GW170817 merger signal for the Hanford detector. The noise for each instrument is shown for comparison, and indicative results from numerical simulations are also shown. The analysis suggests that, at this level the signal-to-noise associated with the merger signal is about 0.5. This provides an indication of the improvement in detector sensitivity required to detect post-merger dynamics. (Reproduced from Abbott et al. (2019a).)*

instability) dictates the evolution of the object. The problem is messy, but a clever argument allows us to constrain the maximum neutron star mass. The key part of this argument is a robust relation between the maximum mass of a non-rotating star (M_0, determined by the Tolman–Oppenheimer–Volkoff equations) and the largest mass that can be supported by uniform rotation (M_{max}; see Breu and Rezzolla (2016))

$$M_{\mathrm{max}} \approx 1.20^{+0.02}_{-0.05} M_0. \tag{21.80}$$

That is, a rotating star can support about 20% more mass than a non-spinning one. In the case of GW170817 we have some idea of the amount of mass ejection, and it seems reasonable (based on what we think we know) to assume that the emergence of a gamma-ray burst within 2 seconds of the merger indicates a fairly prompt collapse to a black hole. Still, in order to produce ejected material with the relatively high electron fraction indicated by the electromagnetic signal, the merged object needs to survive for some time. Taken together, these arguments point to a collapse close to M_{max}. This then allows us to estimate the maximum mass of a non-rotating neutron star sequence (Rezzolla *et al.*, 2018)

$$M_0 \approx 2.16^{+0.17}_{-0.15} M_\odot. \tag{21.81}$$

With future gravitational-wave observations we should be able to refine this argument.

Finally, there appears to be a strong correlation between the pre- and post-merger signals (Bernuzzi *et al.*, 2015). Specifically, one can show that f_{peak} is correlated with a coefficent, κ_2^T, that characterizes the tidal interaction during the late stages of inspiral

$$\kappa_2^T = \frac{2}{M}\left[\frac{M_2}{M_1}k_2^1 R_1^5 + \frac{M_1}{M_2}k_2^2 R_2^5\right] \propto \frac{1}{M}\tilde{\lambda}, \qquad (21.82)$$

with $\tilde{\lambda}$ given by (21.33) (and we have assumed that the two masses are similar). The correlation is robust, and depends weakly on the binary total mass, the mass ratio, and the equation of state; see Figure 21.14. This is an interesting result, indicating that measurements of the inspiral signal (which determine κ_2^T) could be used to constrain f_{peak} (and vice versa).

These conclusions obviously come with caveats. The post-merger signal may be affected by thermal effects, magnetohydrodynamics, various instabilities and dissipation channels (like neutrino emission). State-of-the-art simulations tend to include some (not all) of these aspect at different levels of realism. One may argue that the timescale of gravitational-wave emission is so short that only hydrodynamics and shock-heating are likely to affect it. Neutrino cooling would be inefficient and magnetic field effects may be too subtle. However, these aspects may significantly contribute to the overall picture. Magnetic field features (like the MRI and its ability to redistribute angular momentum) may drive the hypermassive neutron star towards collapse. The dynamics of the magnetic field is also key to the expected electromagnetic counterpart signals.

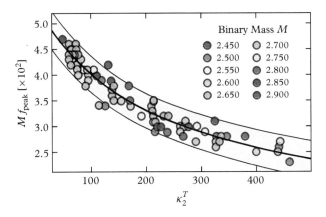

Figure 21.14 *As an illustration of the robust correlations between the pre- and post-merger signals for a neutron star binary, the dimensionless frequency Mf_{peak} is shown as a function of the tidal coupling constant κ_2^T. The colour code indicates different values of the total binary mass. The black solid line is a functional fit, with the grey area marking the 95% confidence interval. (Adapted from Bernuzzi et al. (2015), copyright (2015) by the American Physical Society.)*

21.7   Gamma-ray bursts

The slow crescendo of the inspiral cues a spectacular fireworks display. Just under two seconds after the GW170817 merger time, NASA's orbiting Fermi Gamma-ray Space Telescope detected a short pulse of gamma rays (GRB 170817A; see Abbott *et al.* (2017*f*), Goldstein *et al.* (2017), and Kasliwal *et al.* (2017)). Other telescopes immediately took aim at the suggested position in the sky (see Figure 21.15). Within 11 hours, teams of optical and infrared astronomers had found a bright new beacon on the edge of the galaxy NGC 4993 (Abbott *et al.*, 2017*j*). This source faded over several days from bright blue to dimmer red. After 9 days it began to glow in X-rays (Troja *et al.*, 2017). Finally, just over two weeks after the merger, the radio astronomers were able to join the party (Alexander *et al.*, 2017).

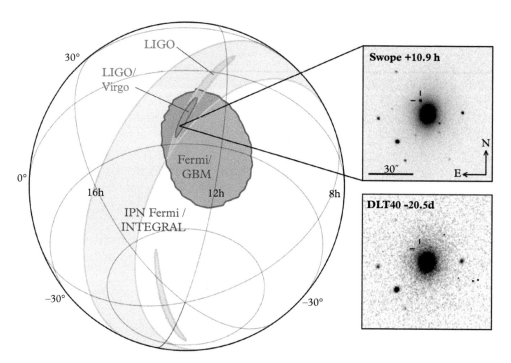

Figure 21.15 *Sky localization of the gravitational-wave, gamma-ray, and optical signals associated with the GW170817 neutron star merger. The left panel shows an orthographic projection of the 90% credible regions from LIGO (190 deg^2; light green), the initial LIGO–Virgo localization (31 deg^2; dark green), triangulation from the time delay between Fermi and INTEGRAL (light blue), and Fermi–GBM (dark blue). The inset shows the location of the apparent host galaxy NGC 4993 in the Swope optical discovery image at 10.9 hr after the merger (top right) and the DLT40 pre-discovery image from 20.5 days prior to merger (bottom right). The position of the transient is indicated in both images. (Reproduced from Abbott et al. (2017j).)*

This spectacular cosmic event was a virtual gold mine for modern astronomy. It has long been known that the detection of electromagnetic counterpart signals is essential for a full exploration of the merger physics. It can provide a precise location by pinpointing the event to a specific galaxy (and hence provide a measured redshift). This may also associate the event with a specific stellar population (old and evolved or relatively young in an active star-forming region). Electromagnetic signals may probe the behaviour of matter following the merger, including the formation of relativistic jets and outflows. This may, in turn, address the question of whether mergers are responsible for r-process nucleosynthesis, and shed light on the nature and evolution of the merger remnant.

Before 17 August 2017 we had circumstantial evidence for some of these features. Following the event, we have convincing and detailed insight. In particular, the extremely low probability of the near-simultaneous temporal and spatial observation of GRB 170817A and GW170817 occurring by chance (5.0×10^{-8}) confirms neutron star mergers as a progenitor of short gamma-ray bursts (Abbott *et al.*, 2017*f*). The arguments in favour of this explanation had been building for decades (Piran, 2004; Fernández and Metzger, 2016), but it was expected that one might need third-generation gravitational-wave detectors to confirm the connection. Basically, the bulk of observed short gamma-ray bursts have been found at redshifts $z > 0.1$, at luminosity distances $d_L > 460$ Mpc (assuming standard ΛCDM cosmology) beyond the Advanced LIGO design sensitivity horizon (~ 200 Mpc) for neutron star mergers. Moreover, estimated event rates suggest a few per year in the Advanced LIGO era. A solid identification of a short gamma-ray burst with a gravitational-wave signal would require either years of observation or a nearby event. Sometimes you just get lucky...

The gamma-ray flash GRB 170817A lasted roughly half a second; see Figure 1.5. It was well described by a single pulse and consistent with a single intrinsic emission episode, with no evidence for significant substructure. In terms of the burst duration, the event was towards the long end of bursts typically characterized as short (usually taken to mean $T_{90} < 2$ s; see Figure 21.16). The observation accorded with expectations, but the event was nevertheless rather unusual. In particular, the gamma-ray emission was faint. In terms of isotropic emission—an upper bound on the true energetics, given that a typical gamma-ray burst is observed within the beaming angle of the brightest part of the jet (Piran, 2004)—the energy released in gamma-rays was estimated to be $E_{\text{iso}} \approx 10^{46}$ erg. This makes the event a distinct outlier among observed bursts with measured redshifts; see Figure 21.17.

The detailed dynamics of the engine that drives these cosmic explosions remains poorly understood (Paschalidis, 2017). From both duration and variability of the gamma-ray signals, we know that the emission region is confined to volumes tens of kilometers across. Since this region will always be hidden from direct electromagnetic observations, we need gravitational-wave detections to provide the evidence. In particular, there will be a clear difference between bursts originating from gravitational collapse (see Chapter 20) and those from mergers. The latter scenario is much brighter in terms of gravitational-wave emission—and it is, obviously, preceded by the inspiral phase. Still, in all cases, the formation of a massive accretion disk is thought to be crucial for launching the gamma-ray emission. Such a disk may form via tidal disruption of a

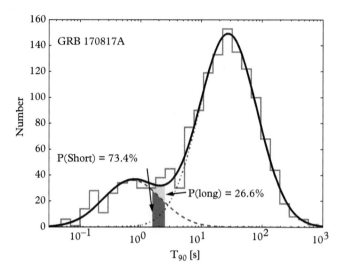

Figure 21.16 *The Fermi–GBM T_{90} (the time interval over which 90% of the total background-subtracted photon counts are observed) distribution fit with two log-normal distributions. The 1-s confidence interval for GRB 170817A is shaded below the summed curve. The red region is the probability that the event belongs to the short class, while the light blue is the probability that it belongs to the long class. (Reproduced from Kasliwal et al. (2017) by permission of the AAS.)*

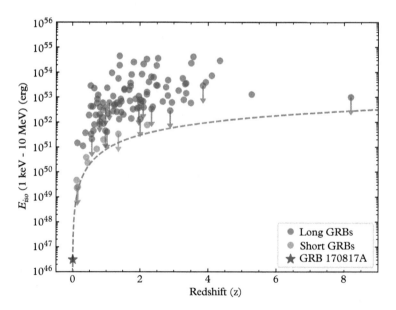

Figure 21.17 *GRB 170817A is a dim outlier in the distributions of isotropic energy emission (E_{iso}, shown as a function of redshift for all GBM-detected GRBs with measured redshifts). Short- and long-duration GRBs are separated by the standard $T_{90} = 2$ s threshold. The green curve demonstrates how the (approximate) GBM detection threshold varies with the redshift. (Reproduced from Abbott et al. (2017f), Creative Commons Attribution 3.0 licence.)*

neutron star during the merger with a black hole—although we know from the discussion in Chapter 20 that this only happens for modest mass ratios. For heavy black holes, there would be a direct plunge, no significant remnant disk and most likely no gamma-ray burst. In the case of binary neutron stars, material with centrifugal support can be left outside the newly formed remnant black hole. Whether such a disk forms or not depends on the binary properties (mass, spin, etc.), as well as the matter equation of state (Sekiguchi *et al.*, 2016). It is generally expected that a disk mass of $\sim 0.01\,M_\odot$ would be sufficient to supply the energy for the creation of a short gamma-ray burst. A comparison between numerical simulations and observations, indeed, indicate disk masses $\lesssim 0.01\,M_\odot$ (Fernández and Metzger, 2016). This favours 'high-mass' neutron star mergers ($M \gtrsim 3\,M_\odot$). The black hole that eventually forms from a hypermassive merger remnant (which may survive for $\gtrsim 10\,\text{ms}$) will be limited to dimensionless spins $\lesssim 0.7$. Larger spins can be obtained only if the remnant collapses promptly. In the case of mixed binaries, one may need a rapidly spinning black hole (with spin $\gtrsim 0.9$). Since both avenues seem viable, it is important to note that their gravitational-wave signature would be distinct; see Figure 21.18.

After about two decades of simulations of binary systems in full general relativity we are beginning to understand the ingredients that are necessary for these systems to launch jets (Rezzolla *et al.*, 2011; Ruiz and Shapiro, 2017), but we are still far from simulating gamma-ray bursts, starting from inspiral and merger all the way to jet acceleration and the emergence of the gamma rays. Simulating the launch of the jet is particularly

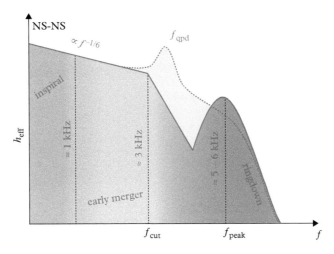

Figure 21.18 *Schematic spectrum of effective amplitude h_{eff} for binary neutron star coalescences. During the inspiral phase, up to $\approx 1\,kHz$ and the early-merger phase up to $f_{\text{cut}} \sim 3\,kHz$, the system retains its binary structure and h_{eff} scales as $f^{-1/6}$. If a black hole is promptly formed, matter quickly falls in the black hole, losing angular momentum through emitting gravitational waves around a peak frequency $f_{\text{peak}} \sim 5 - 6\,kHz$. If a proto-neutron star is formed, it may radiate gravitational waves through its quasiperiodic rotation at $f_{\text{qpd}} \sim 2 - 4\,kHz$. Once matter falls into the black hole, it rings down, emitting gravitational waves at $\approx 6.5 - 7\,kHz$ with exponentially decaying amplitude. (Reproduced from Bartos et al. (2013).)*

problematic. It is easy to understand why this is so. First of all, we need the jet to launch sharply after merger, given the limited simulation time. We simply can not afford to track systems where the hypermassive neutron star survives for an extended period. This is unfortunate, given the indications that the hypermassive remnant may last for a considerable time (Fernández and Metzger, 2016). A significant fraction (perhaps a quarter) of short gamma-ray bursts are followed by extended X-ray emission, which lasts for a minute or longer after the initial burst. The variability of this emission suggests that it is powered by continued activity from the central engine (Piran, 2004). This is in contrast to any gamma-ray burst afterglow, which originates at much larger radii. The total isotropic energy of the emission in some cases exceeds that of the initial gamma-ray burst itself.

Moreover, even if a jet emerges shortly after merger, tracking its evolution until it reaches the ultimate Lorentz factor involves following the outflow to distances of several hundreds of thousands to millions of kilometers away from the central engine (Uzdensky and MacFadyen, 2007; Xie *et al.*, 2018). At the same time, we need to resolve small scale magnetohydrodynamics features (like the MRI) which may be a pre-requisite for launching the emission in the first place. The vast difference in scales illustrates why the problem is so challenging. State-of-the-art numerical relativity simulations capture the merger through to collapse. Within the MHD approximation (typically with the magnetic field confined to the stellar interior, and sometimes including resistivity; see Dionysopoulou *et al.* (2015)), some evidence for relatively empty polar region and the early alignment of what may develop into a jet have been seen (see Figure 21.19). The involved Lorentz factors are, however, typically much lower than required. Moreover, efforts designed to explore the jet dynamics make use of additional approximations, like a fixed black-hole spacetime. Further progress is needed before we can compare these models to detailed observations.

Nevertheless, the available hydrodynamics simulations provide qualitative insight into the phenomenon and we can meaningfully compare the GW170817 event to the emerging understanding. Immediately following the observations, several models seemed viable. One possible explanation for the dim nature of the observed gamma-ray burst would be that we are witnessing a standard paradigm explosion, but we are not looking down the barrel of the jet (Kasliwal *et al.*, 2017). Standard gamma-ray burst models would involve a narrow (opening angle $\theta_{\rm jet} \sim 10°$) and ultra-relativistic (Lorentz factor $\Gamma > 100$) jet in the line-of-sight of the observer, but such event would have to be much brighter in order to explain the bulk of the observations in Figure 21.17. The isotropic electromagnetic emission level from GW170817 is about four orders of magnitude lower. Of course, it could be that we are simply seeing an extremely weak version of this scenario. The successful breakout of a narrow, ultra-relativistic jet would only require $<3\times10^{-6}$ $M_\odot$ of material. If the jet opening angle were wider, it would need to involve even less material to successfully break out. However, such a low ejecta mass is in contradiction with the observed bright UV-Optical-Infrared (UVOIR in Figure 21.20) counterparts, which hints at ≈ 0.05 $M_\odot$ of ejecta. This scenario also fails to account for the delayed onset of X-ray and radio emission. However, the case of an off-axis observation of a standard event is also problematic. In particular, given the sharp drop in observed

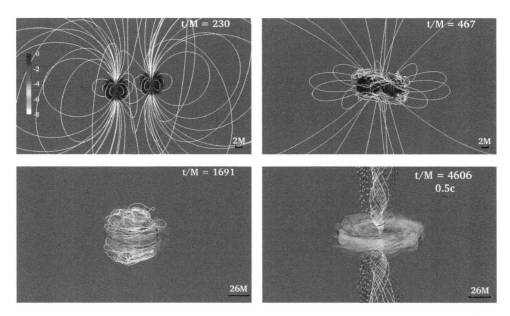

Figure 21.19 *Snapshots of the rest-mass density throughout a magnetized neutron star merger. Green arrows indicate plasma velocities and white lines show the magnetic-field structure. The final panel suggests that a jet may be launched. (Reproduced from Ruiz et al. (2016) by permission of the AAS.)*

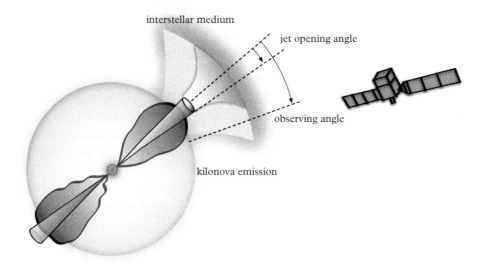

Figure 21.20 *An illustration of the most likely scenario for the gamma-ray burst associated with GW170817—a wide-angle, mildly relativistic, weak cocoon with a successful off-axis jet. (Based on the discussions of Kasliwal et al. (2017) and Mooley et al. (2018).)*

luminosity with observing angle, this scenario would involve fine-tuning. One would also expect a bright afterglow at all wavelengths roughly one day after the merger, when the external shock decelerates to $\Gamma \sim 10$, and this was not seen.

Based on the electromagnetic observations, one can estimate that a few hundredths of a solar mass of ejecta were propelled into the surrounding medium with velocities spanning a few tenths of the speed of light. If the jet were launched with a slight delay—perhaps representing the brief lifetime of the hypermassive remnant—then it would have to drill through the ejecta. The material enveloping the jet may then inflate to form a pressurized cocoon that expands outwards at a mildly relativistic speed. This leads to two possibilities. If the jet has a wide angle ($\approx 30°$), it will choke and fail to drill through the enveloping cocoon. If, on the other hand, the jet is narrow ($\approx 10°$) and long-lived, it may penetrate the ejecta and look like a classical short gamma-ray burst to an on-axis observer. Initially, both scenarios seemed consistent with the features of the GW170817 event.

Continued monitoring with very long baseline radio interferometry (Mooley *et al.*, 2018) provides a possible resolution, with evidence of superluminal apparent motion between 75 and 230 days after the GW170817 event. This helps break the degeneracy between the choked- and successful-jet cocoon models and suggests that the late-time emission was most likely dominated by an energetic and narrowly collimated jet (with opening angle of less than 5 degrees), observed from a viewing angle of about 20 degrees. This leaves us with the situation illustrated in Figure 21.20 as the most likely scenario.

21.8   The signature of a kilonova

The precise localization of the electromagnetic counterpart to GW170817 opened the floodgates, quickly making it one of the most closely studied astronomical events in history. The obtained UV/optical/IR light curves and spectra of the transient are unlike any previously seen, exhibiting a rapid decline in brightness and a transition of the spectral peak from UV to IR (Cowperthwaite *et al.*, 2017; Villar *et al.*, 2017). Spectra in different bands show a similar evolution from an initial featureless (black-body) shape, peaking in the UV about 1 day after merger, to an IR-dominated spectrum with broad absorption features only a few days later; see Figure 21.21. The observations confirm the hypothesis that neutron-star mergers produce short gamma ray bursts and bolstered the kilonova model in which neutron-rich matter flung into space by colliding neutron stars hosts a chain of nuclear interactions known as the r-process (Metzger, 2017). This process is thought to produce half the elements heavier than iron. The heaviest of these elements would soak up blue light, tinting the glowing radioactive cloud red.

The light curves and spectra observed for the GW170817 counterpart closely resemble theoretical kilonova predictions (Evans *et al.*, 2017; Cowperthwaite *et al.*, 2017; Tanaka *et al.*, 2018), making the event another first in astronomy: A clear demonstration that r-process nucleosynthesis occurs in neutron star binary mergers (Lattimer and Schramm, 1974). The observed slow, red evolution is a generic feature of all kilonova

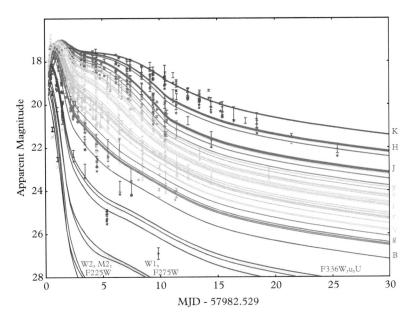

Figure 21.21 *Observed light curves for the optical counterpart to the GW170817 merger event. UV, optical, and near-IR data points are compared to a thee-component model for r-process heating and opacities (shown as solid lines). (Reproduced from Villar et al. (2017) by permission of the AAS.)*

models regardless of matter dynamics, outflow geometry, nuclear heating, opacities and radiation transfer. Meanwhile, the observed luminosity, temperature and time evolution roughly matches predictions for an ejecta mass of ~ 0.05 solar masses ($M_\odot$) and velocities of $\sim 0.1c$ The presence of transient UV emission, followed by longer term IR emission hints at two ejecta components: a lower-mass, high-velocity component (with a low lanthanide—heavier elements with $A \gtrsim 140$—abundance), and a slower, higher-mass component (rich in lanthanides). One may interpret the emission as arising from distinct physical regions. The high velocity ($v \approx 0.3c$) of the blue ejecta suggests that it originates from the shock-heated polar region created when the neutron stars collide. Meanwhile, the low-velocity ($v \approx 0.1c$) red component may originate from the dynamically ejected tidal tails in the equatorial plane of the binary, in which case the relatively high ejecta mass $\gg 0.01 M_\odot$ suggests an asymmetric mass ratio of the merging binary ($q \lesssim 0.8$, consistent with the possible mass ratio for GW170817).

The combination of high inferred velocities, rapid optical decline, slow infrared evolution and broad peaks in the infrared spectra accord with theory, which hold that about 2% of the combined mass of the stars would escape the fate of the rest. Within one second of the collision, this material expands to a cloud tens of thousands of kilometres across, but still about as dense as the Sun. In this cauldron, protons and neutrons clump together to form neutron-rich nuclei, which then start to decay radioactively. This radioactivity keeps the cloud glowing hot for several days, even as it reaches the size of our Solar system. Within a million years, the ejected material spreads across an entire galaxy.

The previous evidence was not entirely convincing. A single data point hinting at a late-time infrared excess associated with GRB 130603B had been interpreted as a signature of r-process nucleosynthesis (Tanvir *et al.*, 2013), and there were a few similar candidates events. The August 2017 event was a game changer. The origin of heavy elements like gold, platinum, and uranium is no longer a mystery—they are synthesized by the rapid capture of neutrons in matter ejected from neutron star mergers.

The ejecta mass inferred for the GW170817 counterpart and the expected merger rate suggest that similar neutron star events could be the dominant r-process site. The solar abundance pattern shows that the first of three r-process peaks accounts for about 80% of the total abundance. To account for the observed solar abundance in all three r-process peaks with neutron star mergers, one would need a rate of $500\,\mathrm{Gpc}^{-3}\mathrm{yr}^{-1}\,(M_{\mathrm{ej}}/0.05M_{\odot})^{-1}$. In order to explain the observed abundance in the two heavier r-process peaks with mergers, the rate would only have to be something like $100\,\mathrm{Gpc}^{-3}\mathrm{yr}^{-1}$. It makes sense to compare this requirement to the rate inferred from gravitational-wave searches. Based on the detection of GW170817, one arrives at a neutron star merger rate of 320–$4{,}740\,\mathrm{Gpc}^{-3}\,\mathrm{yr}^{-1}$ (Abbott *et al.*, 2017*i*). This is larger than the beaming-corrected rate for short gamma-ray bursts and also larger than the merger rate estimated from the galactic population of neutron star binaries (see Chapter 9). The picture seems consistent. The large ejecta mass and the high rate estimates from GW170817 are consistent with the scenario that mergers are the main production sites of r-process elements of the Milky Way (Lattimer and Schramm, 1974).

22

Whispers from the Big Bang

The Universe is a bit of a mess. We have to deal with things we know and 'understand'; planets, stars, galaxies, galaxy clusters, voids of empty space, and things we know we do not understand particularly well (if at all), like dark matter and dark energy. In order to describe the Universe at large, we need to account for all different 'energy' contributions and how they combine and interact with each other (Frieman *et al.*, 2008). This is an extraordinarily challenging problem.[1] In many ways it is remarkable that we know as much as (we think) we do. As our observational capabilities continue to improve, we should be able to advance this understanding. Gravitational-wave searches may play an important role, complementing the electromagnetic data. The detection of cosmological signals may allow us to probe aspects of the Universe that are beyond the reach of traditional astronomy.

Gravitational-wave cosmology involves a range of issues. First of all, we need to consider how the evolution of the Universe affects gravitational waves that reach our detectors from large distances. This involves accounting for the expansion of the Universe when working out waveforms. We also need to consider effects like gravitational lensing, which may affect the signal strength and confuse a detection algorithm. As we start to see sources from cosmological distances, the detection volume obviously increases. Along with this, the number of potential events goes up and we may have to deal with an unresolved gravitational-wave background composed of the difference source categories we have already considered. The detection of such a stochastic background brings its own specific challenges. Finally, we have new classes of sources, which owe their existence to the evolution of the Universe. We know, for example, that density fluctuations in the early Universe should generate gravitational waves (at some level). The end of the inflationary era, during which the Universe expanded exponentially, marks the beginning of the first time when detectable gravitational radiation may have been generated. Such waves would originate when the Universe was something like 10^{-40} s old. This is is sharp contrast with the information contained in the cosmic microwave background (CMB), which

---

[1] This chapter is, quite naturally, focussed on the gravitational-wave aspects of cosmology. The wider scope is covered in many books, for example, Liddle (2003) and Weinberg (2008). A number of additional gravitational-wave aspects are analysed in detail by Maggiore (2018).

Gravitational-Wave Astronomy: Exploring the Dark Side of the Universe. Nils Andersson, Oxford University Press (2020).
© Nils Andersson. DOI: 10.1093/oso/9780198568032.001.0001

describes a Universe some 50 orders of magnitude older. As these gravitational waves will (essentially) not have interacted with matter since they were generated, they carry pristine information about the Universe as it was being born. The existence of such primordial gravitational waves is predicted by some of the simplest and most compelling models of the early Universe, leading to a stochastic signal with a characteristic signature in the so-called tensor modes of the CMB. This would tell us what these waves looked like at the time of decoupling, on scales comparable to the corresponding horizon size. If we want information about shorter wavelengths, as required to probe earlier times, we have to examine the spectrum at higher frequencies today.

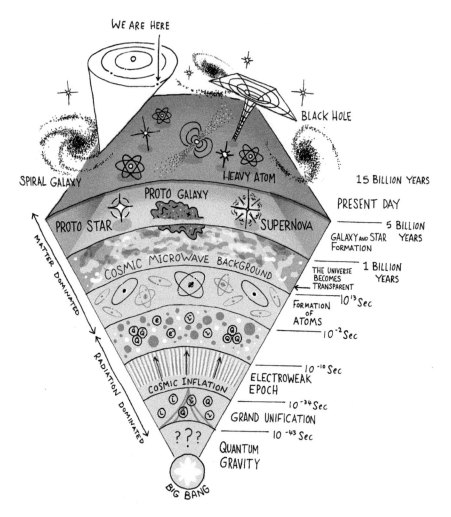

Figure 22.1 *Artist's impression of the evolution of the Universe, indicating the key stages. (Illustration by O. Dean.)*

The first stars were formed during an era lasting about 1 billion years, known as the cosmic dawn (see Figure 22.1). Ultraviolet emission from the first galaxies ionized the cosmos and the nature of the first generation of stars (Abel *et al.*, 1998*a*; Abel *et al.*, 1998*b*) impact on many of the astrophysical scenarios we have discussed. The basic story of how primordial gas collects in dark matter mini halos to form the first stars is well known (Tegmark *et al.*, 1997), but the subsequent stages introduce uncertainties (Norman *et al.*, 2018). If the primordial stars are massive, they will be luminous and short lived (a few million years). Their death is supernovae chemically enriches the Universe and leave behind stellar remnants (neutron stars and black holes; see Heger and Woosley (2002) and Chapter 9). This is the standard scenario. Alternatives involve low-mass primordial stars, but at least some of these must have been massive enough to create the first heavy elements of the periodic table (at this stage there is no other mechanism available). This should also leave a generation of compact objects, but they may be less common. In the not too distant future, when gravitational-wave observations allow us to consider populations, we should be able to constrain the different scenarios.

22.1 The standard model of cosmology

We have come a long way since 1929, when Edwin Hubble first noted that the light from distant galaxies is redshifted—they tend to move away from us (Hubble, 1929). This was the trigger that led to the development of the Big Bang theory and a relativistic Universe described the Friedmann equations (from Chapter 4). The serendipitous detection of the cosmic microwave background (Penzias and Wilson, 1965) provided the smoking gun for the model and a sequence of increasingly precise CMB experiments—from COBE in the 1990s to WMAP in the 2000s and most recently Planck[2]—have provided exquisite information about the Universe at an age of a few hundred thousand years, when the radiation decoupled from matter.

Compared to the situation today, the early Universe was simple—a homogeneous and isotropic soup of particles, where the temperature and density were nearly identical from one place to the next at any given time. The CMB provides a dramatic demonstration of this level of homogeneity. When the signal was imprinted, density and temperature varied from one point to another by about one part in 10,000. The statistical properties of these minuscule cosmological variations encode information about the very early Universe, pretty much immediately after the Big Bang. If we consider these variations as perturbations away from perfect homogeneity, then they belong to one of three different kinds: scalar, vector, and tensor perturbations. Scalar perturbations (e.g. variations in the energy density) are the easiest to measure, and we know a lot about them from the fluctuations in the CMB (see Figure 22.2). The anisotropies can be treated as small deviations from the Friedmann–Robertson–Walker metric, the evolution of which

---

[2] http://sci.esa.int/planck/

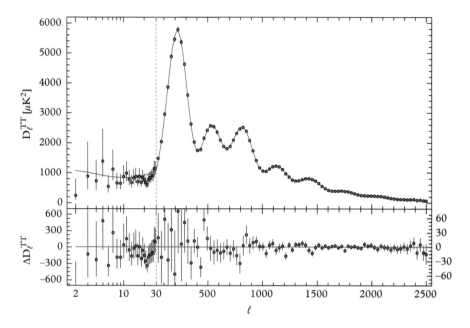

Figure 22.2 *Temperature power spectrum (with foreground and other parameters fixed to their best-fit values) from the 2015 data release of the Planck experiment. The top panel shows the power spectrum multipole-by-multipole and the red line shows the temperature spectrum for the best-fit ΛCDM cosmology. The error bars show $\pm 1\sigma$ uncertainties. The lower panel provides the power spectrum residuals with respect to this model. The first three peaks in the spectrum determine the relative contribution from dark energy, matter, and dark matter. (Reproduced from Adam et al. (2016) with permission from Astronomy and Astrophysics. Copyright ESO.)*

is described by general relativity (Chapter 4). Combined with observations this has helped establish the standard ΛCDM model of cosmology (Frieman *et al.*, 2008) and the understanding that inflation played a crucial early role. The scale of the fluctuations indicate that the Universe is (very close to) flat (with $k = 0$) and the first three peaks in the CMB spectrum (see Figure 22.2) give the relative contribution from dark energy, matter, and dark matter (Adam *et al.*, 2016). An early epoch of exponential expansion is needed to explain both large-scale homogeneity and isotropy, as well as the structure that has evolved on smaller scales.

In essence, we live in a Universe that expands monotonically, approaching the simple de Sitter model at late times (when the cosmological constant/dark energy dominates). The expansion decelerated at early times, but will accelerate at late times and the Universe eventually ends in a 'Big Chill'.

The ΛCDM model is the simplest parameterization of a Big Bang cosmology that is (broadly) consistent with observations. The model has two principal ingredients: Λ refers to the cosmological constant (often described as the energy density of the vacuum, the 'dark energy'), and CDM stands for cold dark matter (Spergel and Steinhardt,

2000), an invisible matter contribution required to explain observed galaxy rotation curves and gravitational lensing. Since it does not emit or absorb light like ordinary matter, its distribution must be inferred from gravitational effects. That the dark matter component plays a crucial role in structure formation is brought out by large scale numerical simulations (Springel *et al.*, 2005). The model also assumes that there was an epoch of inflation shortly after the Big Bang.

Although this is the 'standard model of cosmology' it is more at the level of a paradigm than a complete theory. The two main ingredients are poorly understood. No first principle calculations predict the magnitude of the dark energy contribution and, although there are many competing ideas for the composition and properties of dark matter none of the associated particles have yet been detected (Bertone *et al.*, 2005). Still, it is generally accepted that cosmological observations are most simply explained by assuming that dark matter is 'cold', i.e. that the involved particles move slowly.

The picture that emerges is that of a Universe which began in a Big Bang explosion some 13.7 billion years ago. For a brief time, it expanded rapidly (inflation) and it has continued expanding ever since. The structures we observe, galaxies and clusters, developed as a result of gravitational interactions involving dark matter and the ordinary (baryonic) matter we see with our telescopes. An example of how the different contributions to the Universe are constrained by modern datasets is provided in Figure 22.3.

As an impression of the quantitative picture, let us consider the results from the 2015 data release from the Planck experiment (Adam *et al.*, 2016). The high-multipole peaks in the obtained CMB spectrum (see Figure 22.2), are extremely well extremely

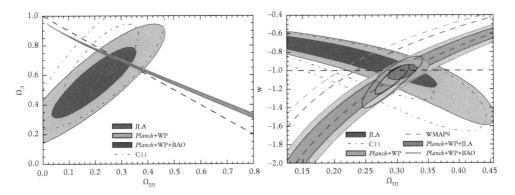

Figure 22.3 *Left: An illustration of inferred contributions from Ω_m and Ω_Λ (showing 68% and 95% confidence contours) for the standard ΛCDM model. Labels for various data sets correspond to the supernova compilation from Figure 22.4 (JLA), an earlier SN Ia compilation (C11), the combination of Planck temperature and WMAP polarization measurements of the CMB fluctuations (Planck+WP), and a combination of measurements of the baryon acoustic oscillation scale (BAO). The black dashed line corresponds to a flat Universe. Right: Confidence contours at 68% and 95% for Ω_m and w (the dark energy 'equation state' parameter; see Chapter 4) for the flat ΛCDM model. The black dashed line corresponds to the cosmological constant hypothesis. (Reproduced from Betoule et al. (2014) with permission from Astronomy and Astrophysics. Copyright ESO.)*

well described by the ΛCDM model with a power-law spectrum of adiabatic scalar perturbations. Within the context of this model, one can then infer the relevant cosmological parameters. First of all, the data is consistent with spatial flatness to percent level precision, so the $k = 0$ assumption should be accurate (see Chapter 4). Regarding the Hubble constant, there is some tension between the Planck result

$$H_0 = 67.4 \pm 1.4 \text{ kms}^{-1}\text{Mpc}^{-1}, \qquad (22.1)$$

and other experiments (see Figure 22.5), but the difference is now smaller than the discrepancy that plagued cosmology for decades. Still, given the uncertainty it is useful to introduce a parameter h such that[3]

$$H_0 = 100h_{100} \text{ kms}^{-1} \text{ Mpc}^{-1}, \qquad (22.2)$$

in order to express the different energy contributions. We then find that the Universe is dominated by the dark energy (which tends to make gravity push instead of pull, as required to drive the accelerated expansion)

$$\Omega_\Lambda = 0.686 \pm 0.020. \qquad (22.3)$$

The total matter contribution is

$$\Omega_m h_{100}^2 = 0.1423 \pm 0.0029, \qquad (22.4)$$

while the baryon contribution (mainly Hydrogen and Helium) is

$$\Omega_b h_{100}^2 = 0.02207 \pm 0.0033. \qquad (22.5)$$

We also learn that the dark energy component is such that

$$w = -1.13^{+0.13}_{-0.08}, \qquad (22.6)$$

consistent with a cosmological constant, for which we would have $w = -1$.

The overall conclusions are corroborated by a range of experiments, as illustrated in Figure 22.3. In essence, the evolution of the Universe can be described by a six-parameter model. But there are open questions. For example, it is apparent from Figure 22.2 that the model does not provide a good fit to the temperature power spectrum at low multipoles. At a conceptual level, we also need to be mindful of the fact that something like 95% of the energy content of the Universe remains a mystery. There are many proposed dark

---

[3] Here, and in the following, h_{100} is the scaled Hubble parameter and should not be confused with a gravitational-wave strain. The meaning should be clear from the context.

matter scenarios, but we need observations to make progress. When it comes to the dark energy... just because we have given something a name, it does not mean we understand it.

22.2   The cosmological redshift

The cosmological redshift is the key measure of distance in the Universe. As distant galaxies move away from us, the light they emit (as well as any gravitational waves) is redshifted. As this will impact on the detailed nature of any signals that arrive at our detectors, we need to quantify the effect. In order to do this, let us focus on the Friedmann–Robertson–Walker model from Chapter 4. That is, we have

$$ds^2 = -dt^2 + [R(t)]^2 \left[\frac{dr^2}{1 - kr^2} + r^2 \left(d\theta^2 + \sin^2 \theta \, d\varphi^2 \right) \right], \qquad (22.7)$$

where R is the cosmological scale-factor. In a flat Universe, with $k = 0$, it follows from the equation for radial (null) geodesics that

$$\frac{d|v|}{dt} = -\frac{\dot{R}}{R}|v|, \qquad (22.8)$$

where

$$v = \frac{dr}{dt}. \qquad (22.9)$$

We see that, in the frame of the chosen observer, $|v|$ remains zero if it vanished initially. In essence, this is the statement that the coordinates are co-moving and stretch along with the cosmological expansion. For a given gravitational-wave signal, e.g. from a binary, the implication is that the coordinate distance will remain the same even though the Universe expands. This is important because it means that much of our previous analysis will remain unchanged. Of course, we know that coordinate distances have no real meaning. We should be using the proper (spatial) distance

$$dr_p^2 = g_{ij} dx^i dx^j. \qquad (22.10)$$

In the general case, the distance from the origin to a radial location r is given by

$$dr_p^2 = \frac{R^2 dr^2}{1 - kr^2} \qquad \longrightarrow \qquad r_p = R(t) \int_0^r \frac{dr}{(1 - kr^2)^{1/2}}, \qquad (22.11)$$

and in the case of a flat Universe we simply have $r_p = R(t)r$.

Now consider a source at a distance r as seen by an observer at the origin. The signal travels along null geodesics. That is, we have $ds^2 = 0$, which leads to

$$\int_{t_s}^{t_o} \frac{c\,dt}{R} = \int_0^r \frac{dr}{(1 - kr^2)^{1/2}}, \tag{22.12}$$

where t_s and t_o are the time at which the signal is emitted by the source and observed, respectively. Noting that the right-hand side remains the same for a pulse emitted a small time interval Δt later, we have

$$\int_{t_s}^{t_o} \frac{c\,dt}{R} = \int_{t_s + \Delta t_s}^{t_o + \Delta t_o} \frac{c\,dt}{R}, \tag{22.13}$$

which, after linearizing, leads to

$$\Delta t_o = \frac{R(t_o)}{R(t_s)} \Delta t_s. \tag{22.14}$$

As the observer and source clocks are related by

$$dt_o = (1 + z)dt_s, \tag{22.15}$$

it follows that the cosmological redshift is given by

$$1 + z = \frac{R(t_o)}{R(t_s)}, \tag{22.16}$$

leading to the frequency relation

$$f_o = \frac{f_s}{1 + z}. \tag{22.17}$$

Let us now make the connection between these results and a given gravitational-wave signal. For a source emitting radiation with a given luminosity we have

$$L = \frac{dE_s}{dt_s}, \tag{22.18}$$

in the rest-frame of the source. In the typical case of a binary inspiral, this would correspond to the luminosity we worked out back in Chapter 5. This leads to the energy flux (the energy per unit time and area)

$$F = \frac{L}{4\pi d_l^2}, \tag{22.19}$$

which defines the luminosity distance d_l. Of course, in our expanding Universe we have (since energy if proportional to frequency, it is redshifted by the same factor)

$$E_o = \frac{E_s}{1+z},$$ (22.20)

so

$$\frac{dE_o}{dt_o} = \frac{1}{(1+z)^2} \frac{dE_s}{dt_s}.$$ (22.21)

Noting that the surface of a sphere with co-moving radius r has area $4\pi R^2(t_o)r^2$ we have

$$F = \frac{L}{4\pi R^2(t_o)r^2(1+z)^2} \quad\longrightarrow\quad d_l = (1+z)R(t_o)r,$$ (22.22)

where we should take t_o to be the present time.

Finally, for small redshifts we can make a connection with the cosmological parameters (introduced in Chapter 4) by Taylor expanding

$$\frac{R(t)}{R(t_0)} = 1 + z \approx 1 + H_0(t - t_0) - \frac{1}{2}q_0 H_0^2(t - t_0)^2.$$ (22.23)

After a bit of algebra this leads to

$$\frac{H_0 d_l}{c} = z + \frac{1}{2}(1 - q_0)z^2,$$ (22.24)

which shows why it is natural to work with the luminosity distance, $d_l(z)$, for large redshifts.

22.3 Scaling the distance ladder

Having outlined the link between cosmic distance and the main parameters, let us consider how we best make use of the idea. For nearby objects we can obtain an accurate distance using parallax measurements (observing how an object's position in the sky changes at two opposing points of the Earth's orbit). However, this only allows us to measure distances out to (perhaps) a few kpc—we only measure distances within the Galaxy. In some cases, we may also be able to directly measure the redshift of spectral lines. In order to reach further we have to make use of the theory. This inevitably involves additional assumptions.

Traditionally, the next rung on the cosmic ladder involves the characteristic variability of a particular class of white dwarfs, the Cepheids (Liddle, 2003). In essence, the timescale of pulsation is thought to be linked to the star's brightness. A comparison of the inferred (true) brightness to the observed (apparent) value, reveals the distance. The systems act as 'standard candles'—providing both the observed flux F and the intrinsic luminosity L—from which we can work out the luminosity distance, d_l. The measurements are calibrated using direct observations at low redshifts. This way we may reach distances of a few Mpc, a bit beyond the Andromeda galaxy.

At larger distances, we need more powerful events. The next level is reached via type Ia supernovae, associated with explosions of accreting white dwarfs that reach the critical mass (and hence would have an internal calibration, as well). Such supernova data provided the first evidence (in the early 2000s) that we live in an accelerating Universe, leading to the introduction of the vexing dark energy (Schmidt *et al.*, 1998; Riess *et al.*, 1998; Perlmutter *et al.*, 1999). Direct redshift measurements may also be possible for remote galaxies, but this will only yield distance estimates within an assumed theory.

The results in Figure 22.4 illustrate the current data (Betoule *et al.*, 2014). The figure combines cosmological constraints from a joint analysis of type Ia supernova

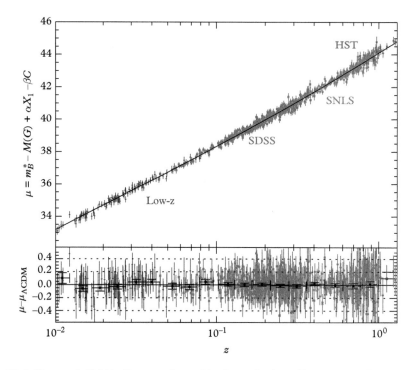

Figure 22.4 *Top panel: Hubble diagram of a combined sample of type Ia supernova observations from the Sloan Digital Sky Survey (SDSS-II) and the Supernova Legacy Survey (SNLS) collaborations. The distance modulus redshift relation of the best-fit ΛCDM cosmology for a fixed $H_0 = 70$ km s^{-1} Mpc^{-1} is shown as the black line. Bottom panel: Residuals from the best-fit ΛCDM cosmology as a function of redshift. (Reproduced from Betoule et al. (2014) with permission from Astronomy and Astrophysics. Copyright ESO.)*

observations obtained by the Sloan Digital Sky Survey (SDSS-II) and the Supernova Legacy Survey (SNLS) collaborations. The data set includes low-redshift samples ($z <$ 0.1), all three seasons from the SDSS-II ($0.05 < z < 0.4$), and three years from SNLS ($0.2 < z < 1$), involving a total of 740 confirmed type Ia supernovae with high quality light curves.

22.4   Standard sirens

Interestingly, we can use gravitational-wave observations to measure cosmological distance, as well (Schutz, 1986). Such measurements would be extremely valuable as they are independent of the electromagnetic distance ladder. They also involve a totally different internal 'calibration'.

In order to outline the argument, consider the problem of inspiralling binaries in light of what we have learned. We know that, in a curved spacetime, the time measured by a clock at the source and one at the observer location will be different. When we derived the quadrupole formula (in Chapter 3), we worked in a flat background so this difference did not enter the discussion. Neither does it impact on the result if we imagine a 'local' wave zone relatively close to the source. The results still hold. However, they will change when we consider cosmological distances. To make progress, we need to figure out how the cosmology enters the problem.

Let us focus on the case of a spatially flat Universe, with $k = 0$ (an assumption supported by observations). Then we know, from the above discussion, that we should first of all replace $r \to R(t_o)r$, with t_o the time at the observer location. Later we will show that the two polarizations decouple (at least as long as the geometrical optics approximation is valid) as the waves propagate across the Universe, so let us not worry about this issue for the moment. We do, however, need to note that the frequency of the wave f_s^{gw} is measured in the retarded time of the source t_s^{ret}. That is, we have the characteristic amplitude (see Chapter 5)

$$h_c = \frac{4}{rR(t_o)} \left(\frac{G\mathcal{M}}{c^2} \right)^{5/3} \left(\frac{\pi f_s^{\text{gw}}(t_s^{\text{ret}})}{c} \right)^{2/3}. \tag{22.25}$$

However, from (22.17) it follows that

$$\int^{t_s^{\text{ret}}} f_s^{\text{gw}} \, dt_s = \int^{t_o^{\text{ret}}} f_o^{\text{gw}} \, dt_o, \tag{22.26}$$

which means that the redshift factors cancel! At the observer's location the frequency is redshifted but the time interval has changed by the same amount so the two effects compensate for each other. This is important, because it means that the gravitational-wave phasing does not change.

We also have

$$rR(t_0) = \frac{d_l}{1+z},$$
(22.27)

so

$$h_c = \frac{4}{d_l}(1+z)^{5/3}\left(\frac{G\mathcal{M}}{c^2}\right)^{5/3}\left[\frac{\pi f_0^{\mathrm{gw}}(t_0^{\mathrm{ret}})}{c}\right]^{2/3}.$$
(22.28)

Introducing a redshifted chirp mass $\mathcal{M}_z = (1+z)\mathcal{M}$ (which makes intuitive sense if we recall the equivalence between mass and energy), we have the final result

$$h_c = \frac{4}{d_l}\left(\frac{G\mathcal{M}_z}{c^2}\right)^{5/3}\left[\frac{\pi f_0^{\mathrm{gw}}(t_0^{\mathrm{ret}})}{c}\right]^{2/3}.$$
(22.29)

At the end of the day, we have (formally) the 'same' result as in the flat spacetime case (as long as we replace $r \to d_l$ and $\mathcal{M} \to \mathcal{M}_z$). Of course, searching for the signal we need to keep in mind that the frequency is redshifted. The redshift of the chirp mass makes binaries seem more massive, which leads to a louder gravitational-wave signal and therefore the ability to see sources at larger distances.

We now have the results we need to explain how binary inspirals can be used as 'standard sirens' (Schutz, 1986; Holz and Hughes, 2005). First of all, if we detect the two polarizations and the rate of change of the frequency $\dot{f}_{\mathrm{gw}}$, then we can infer the orbital inclination from the relation between h_+ and $h_\times$. We get $\mathcal{M}_z$ from $\dot{f}_{\mathrm{gw}}$, as in the flat spacetime case. This fixes everything in (22.29), apart from the luminosity distance d_l, which is therefore determined by the observation. If we want to make contact with a cosmological model, then we need an independent (electromagnetic, say) measurement of z. Combining such results we can work out $H_0(z)$. This is a robust argument, although it obviously requires an electromagnetic counterpart (Nissanke *et al.*, 2010).

Luckily we now have such an event. The proposed strategy was executed following the GW170817 neutron star merger (Abbott *et al.*, 2017a). In this case, analysis of the gravitational waveform yielded a distance estimate of about 44 Mpc assuming that the sky position of the event was exactly coincident with the optical counterparts. This distance estimate comes with an uncertainty of about 15%, due to a combination of instrumental noise and the fact that we do not precisely know the inclination of the orbital plane of the binary system. In order to estimate H_0 one also has to combine the gravitational-wave distance with the galaxy's radial velocity, keeping in mind that the expansion of the Universe is not uniform on the relevant scale. We need to account for 'lumpiness' which distorts the Hubble expansion on smaller scales—galaxies have additional motion ('peculiar velocities') due to the gravitational attraction of other galaxies, clusters and dark matter.

The initial analysis, based on the gravitational-wave data, led to a value for the Hubble constant consistent with other measurements

$$H_0 = 74^{+16}_{-8} \text{ km s}^{-1}\text{Mpc}^{-1}. \tag{22.30}$$

The precision of the result is not spectacular, but it was an important first—future observations will improve on this. In fact, continued radio monitoring (Mooley *et al.*, 2018) has helped break the degeneracy between the source distance and the viewing angle, which dominated the uncertainty. This leads to a significantly improved measurement (see Figure 22.5 and Hotokezaka *et al.* (2019))

$$H_0 = 68.9^{+4.7}_{-4.6} \text{ km s}^{-1}\text{Mpc}^{-1}. \tag{22.31}$$

Estimates suggests that 15 (or so) similar events events could bring resolution to the tension between the Planck (Ade *et al.*, 2016*a*) and SHoES (Riess *et al.*, 2016) measurements.

Interestingly, we can also hope to make progress without optical counterparts. By combining information from ~ 100 independent gravitational-wave detections, each with a set of potential host galaxies, a $\sim 5\%$ estimate of H_0 can be obtained even without

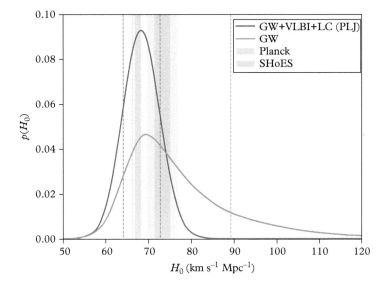

Figure 22.5 *Posterior distributions for H_0. The results of the pure gravitational-wave analysis and the first combined analysis with electromagnetic data are shown (vertical dashed lines show symmetric 68% credible interval for each model). The 1 and 2σ regions determined by Planck (green, see Ade et al. (2016a)) and SHoES (orange, see Riess et al. (2016)) are also shown as vertical bands. It is worth noting the tension between these results and the simple fact that the single gravitational-wave event is consistent with both. (Reproduced from Hotokezaka et al. (2019).)*

the detection of any transient counterparts (Del Pozzo *et al.*, 2011). This is relevant given that we expect to catch many binary black-hole mergers in the next few years.

22.5   Geometrical optics and lensing

We have learned that, when we consider the Universe at large, we need to account for the effect of the cosmology on any given gravitational-wave signal. We also need to understand to what extent the signal changes as it propagates from the source to the detector. We know that gravitational waves interact very weakly with matter so we do not have to worry too much about the signal deteriorating (in some sense). However, we still need to account for effects like gravitational lensing, just as we do for electromagnetic signals (Schneider *et al.*, 1992). In essence, this turns out to be a two-stage argument. First we show that gravitational waves are lensed in the same way as light, then we worry about the effect this may have on our signal searches.

Let us start by extending our description of gravitational-wave propagation. Mathematically, the problem boils down to an expansion using the ratio of the reduced wavelength $\lambdabar$ to the average spacetime curvature R as a small parameter, as in the discussion of general perturbations in Chapter 3. For obvious reasons, this is often referred to as the short-wavelength approximation. We now add the assumption that the wavefront has curvature $\mathcal{L}$ and require that $\lambdabar \ll \mathcal{L}$. In essence, we are considering the region far away from the source.

Under these conditions we are (to a good approximation) dealing with plane waves. Then we can use the geometrical optics approximation, where (Isaacson, 1968*a*)

$$h_{ab}^{\mathrm{TT}} = \mathrm{Re}\left[A_{ab}e^{i\phi}\right].$$

(22.32)

The amplitude A_{ab} is slowly varying, on the scale $\mathcal{L}$ or R (whichever is the shortest), while the phase $\phi = k_a x^a$, where k_a is the wave vector, varies rapidly on the scale $\lambdabar$.

In order to understand the results it is helpful to consider a monochromatic wave propagating in the z-direction of flat space. Then $\phi = \omega(z - t)$, so

$$\partial_z\phi = \omega \quad \text{and} \quad \partial_t\phi = -\omega.$$

(22.33)

In terms of the wave vector we have

$$k_a = \partial_a\phi \rightarrow k_t = k_z = \omega.$$

(22.34)

Motivated by this, we return to the gravitational-wave problem and define

$$k_a \equiv \phi_{|a},$$

(22.35)

where the | represents the covariant derivative with respect to the (presumably) curved background (as in Chapter 3). Then the Lorenz gauge condition (3.110) implies that

$$A_{ab}k^b = 0, \tag{22.36}$$

which means that the waves are transverse. Meanwhile, expanding in powers of $\hbar$ over $\mathcal{L}$ or R, the wave equation (3.111) leads to the two equations

$$k_a k^a = 0, \tag{22.37}$$

$$A_{ab|c}k^c = -\frac{1}{2}A_{ab}k_c{}^{|c}. \tag{22.38}$$

The first of these equations shows that the wave vector is null, and since $\phi_{|ab} = \phi_{|ba}$ one readily finds that

$$0 = (k_a k^a)_{|b} = 2k^a k_{a|b} = 2k^a k_{b|a}. \tag{22.39}$$

In other words, k_a is tangent to a null geodesic of the background (often called a ray in this context).

The second equation shows that

$$k^c A_{ab|c} = -\frac{1}{2}(k^c{}_{|c})A_{ab}, \tag{22.40}$$

the interpretation of which is that the change in amplitude is governed by the divergence of the wave vector.

These results confirm what we already knew—gravitational waves move along geodesics at the speed of light. We should encounter the same lensing effects as for light. This is important. Gravitational lensing has developed into an powerful tool for astrophysics and cosmology, e.g. shedding light on the dark matter component of the Universe (Schneider *et al.*, 1992). At the same time, lensing distorts images of distant sources, so it can have a detrimental effect as well. Having shown that gravitational waves will be lensed in the same way as electromagnetic waves we can see how this can be an advantage, e.g. by focussing the waves to enhance a distant signal, and a problem, e.g by distorting the wave phasing in such a way that our search templates no longer match the signal. It would be particularly concerning if lensing were to mix up the wave polarizations. However, as we will now show, we do not need to worry (too much) about this.

In order to discuss the implications of the geometric optics results, let us first of all note that, if we consider the gravitational wave as a deviation from an otherwise flat spacetime (as in the first part of Chapter 3) then it follows from (22.34) that

$$h_{ab}(t,\boldsymbol{x}) \sim h_{ab}(t - \hat{k} \times \boldsymbol{x}), \tag{22.41}$$

for a wave moving in the $\hat{k}$ direction. Moreover, working in the TT-gauge, it is natural to introduce a basis $[\hat{m}, \hat{n}]$ in the plane orthogonal to the direction of propagation (as in Figure 22.8). We can then make use of the polarization tensors

$$e_{ij}^+ = \hat{m}_i \hat{m}_j - \hat{n}_i \hat{n}_j, \tag{22.42}$$

and

$$e_{ij}^\times = \hat{m}_i \hat{n}_j + \hat{m}_j \hat{n}_i, \tag{22.43}$$

in terms of which we have

$$h_{ij} = h_+ e_{ij}^+ + h_\times e_{ij}^\times = \sum_{P=+,\times} h_P e_{ij}^P. \tag{22.44}$$

It is also worth noting that

$$\sum_P e_{ij}^P e_{ij}^P = 4, \tag{22.45}$$

a result we will make use of later. Finally, one would often opt to work in the frequency domain. That is, we express the signal in terms of the Fourier transform

$$h_{ij}(t, \boldsymbol{x}) = \sum_{P=+,\times} \int \tilde{h}_P(f, \boldsymbol{x}) e_{ij}^P(\hat{k}) \exp\left[-2\pi i f\left(t - \hat{k} \cdot \boldsymbol{x}\right)\right] df. \tag{22.46}$$

Returning to the discussion of geometrical optics, we can (obviously) write

$$k^c h_{ab|c} = k^c (h_+ e_{ab}^+ + h_\times e_{ab}^\times)_{|c} = 0 \quad \longrightarrow \quad e^{i\phi} k^c A_{ab|c} = 0. \tag{22.47}$$

It is natural to assume that the polarization tensors are parallel transported along geodesics, i.e. to take

$$k^c e_{ab|c}^P = 0, \tag{22.48}$$

which means that we have, cf. Eq. (22.40),

$$k^c h_{P|c} = -\frac{1}{2}(k^c{}_{|c}) h_P. \tag{22.49}$$

This can be rewritten as

$$\left(h_P^2 k^a\right)_{|a} = 0, \tag{22.50}$$

i.e. as a conservation law. In a local inertial frame we have

$$\frac{\partial}{\partial t}(h_{\mathrm{p}}^2 k^0) + \nabla_j(h_{\mathrm{p}}^2 k^j) = 0, \tag{22.51}$$

which we can compare to, for example, the continuity equation from fluid dynamics from Chapter 4. The result suggests that it would be natural to introduce a 'number flux four vector for gravitons' as

$$N_{\mathrm{P}}^a = h_{\mathrm{p}}^2 k^a, \tag{22.52}$$

the interpretation of which would be

$$N_{\mathrm{P}}^0 = h_{\mathrm{p}}^2 k^0 \qquad \text{graviton number density,} \tag{22.53}$$
$$N_{\mathrm{P}}^j = h_{\mathrm{p}}^2 k^j \qquad \text{graviton flux .} \tag{22.54}$$

One can also show that

$$h_{\mathrm{p}}^2 \mathcal{A} = \text{constant along a given ray,} \tag{22.55}$$

where $\mathcal{A}$ is the cross section area of a particular bundle of rays. This shows that there is no mixing of the polarizations and allows us to discuss the effects of gravitational-wave lensing for cosmological sources.

Consider, for example, the situation illustrated in Figure 22.6. Let us assume that the problem is stationary and that the impinging gravitational waves have pure plus-polarization. Then we have

$$(h_+^{\mathrm{f}})^2 \mathcal{A}_{\mathrm{f}} = (h_+^{\mathrm{i}})^2 \mathcal{A}_{\mathrm{i}}, \tag{22.56}$$

or

$$h_+^{\mathrm{f}} = \sqrt{\frac{\mathcal{A}_{\mathrm{i}}}{\mathcal{A}_{\mathrm{f}}}} h_+^{\mathrm{i}}. \tag{22.57}$$

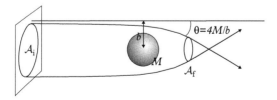

Figure 22.6 *A schematic illustration of the effect of gravitational-wave lensing by a gravitating centre, a star like the Sun or a distant galaxy.*

For a strong lens (not the Sun!) there may be a significant amplification of the gravitational-wave amplitude in cases that lead to the focussing of rays. At a focal point the amplitude formally diverges. The associated singularity is known as a caustic, and its presence signals the breakdown of the geometric optics approximation. In the neighbourhood of a caustic the amplitude changes significantly over a wavelength, introducing diffraction phenomena and possible constructive/destructive interference (Takahashi and Nakamura, 2003).

Lensing may have significant impact on the gravitational waves that reach us from cosmological sources, e.g. from coalescing black-hole binaries. In order to quantify the effect, we need to i) understand the matter distribution in the Universe, and ii) work out how the signal would be affected. Somewhat simplistically, lensing may make intrinsically too faint mergers observable. Strong lensing may also confuse parameter extraction—as the actual source is further away than it appears the masses in the source frame may be reduced by a factor of $1 + z$ (Sereno *et al.*, 2011). Lensing effects should become more important as the sensitivity of the detectors improves and signals from greater distances come within reach. In practice, this may require third-generation ground based instruments or space-based observations. It has, for example, been suggested that the Einstein Telescope may register as many as 50–100 lensed events each year (Biesiada *et al.*, 2014). In the case of LISA, which may detect massive black hole signals out to a redshift of 10–15, the impact of lensing may be significant (Sereno *et al.*, 2011). It could, in turn, shed light on the formation history of the merging binaries as lensing amplification might help identify the host galaxies.

22.6   Astrophysical backgrounds

As the gravitational-wave instruments become more sensitive we expect to probe events throughout a significant fraction of the Universe. In particular, a space bourne interferometer like LISA should be able to detect the merger of supermassive black holes (with masses in the range $10^3 - 10^6 M_\odot$) for much of cosmic history; see Figure 22.7 (Amaro-Seoane *et al.*, 2017). Meanwhile, third-generation ground-based detectors like ET will be able to catch lower mass systems out to a redshift of several (Sathyaprakash *et al.*, 2012). As the number of detectable systems increases, it becomes increasingly likely that we will not be able to resolve individual systems. We need to worry about the overall population and—if we reach deep enough—we will be dealing with a stochastic background.

Many classes of astrophysical sources may lead to the presence of an unresolvable background. However, in order to provide a concrete example, let us focus on binary mergers. We will start by demonstrating a simple relationship between the spectrum of the gravitational-wave background produced by a cosmological distribution of discrete sources, the total time-integrated energy spectrum of an individual source, and the present-day co-moving number density of remnants (Phinney, 2001; Sesana *et al.*, 2008). This is useful because it shows that the background is (essentially) independent of the cosmology and only weakly dependent on the evolutionary history of the sources.

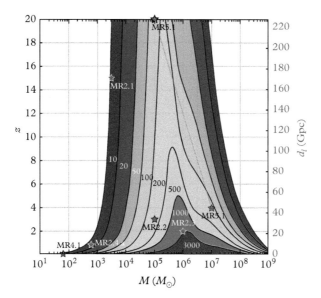

Figure 22.7 *Massive black-hole binary coalescences detectable with the LISA mission. Contours of constant signal-to-noise for the baseline observatory in the plane of total source-frame mass, M, and redshift, z (left axis, assuming standard ΛCDM cosmology), and luminosity distance, d_l (right axis), for binaries with constant mass ratio q = 0.2. Overlaid are the positions of threshold binaries used to define the mission requirements. (Reproduced from Amaro-Seoane et al. (2017).)*

First of all, we let f_s be the frequency of gravitational waves in the rest frame of the source (as before), while f is the frequency of the waves observed today on Earth. This means that we have $f_s = (1 + z)f$. We assume that the energy emitted in gravitational waves in the frequency interval between f_s and $f_s + df_s$ is

$$\frac{dE_{gw}}{df_s} df_s. \tag{22.58}$$

This energy, like the frequency f_s, is measured in the source frame. It is also integrated over all angles and the lifetime of the source.

Secondly, we take the number of events per unit co-moving volume between redshift z and $z + dz$ to be $N(z)dz$. As in Chapter 8, we define $\Omega_{gw}(f)$ to be the present-day energy density per logarithmic frequency interval, divided by the energy density (ρ_c) required to close the Universe (see Chapter 4). With these definitions, the total (present day) energy density in gravitational radiation is

$$\mathcal{E}_{gw} \equiv \int_0^\infty \rho_c c^2 \Omega_{gw}(f) \frac{df}{f} \equiv \frac{\pi}{4} \frac{c^2}{G} \int_0^\infty f^2 h_c^2(f) \frac{df}{f}, \tag{22.59}$$

where h_c is the characteristic amplitude of the gravitational-wave spectrum over a logarithmic frequency interval.

It is easy to see that, in a homogeneous and isotropic Universe, the present day energy density $\mathcal{E}_{gw}$ should follow from the sum of the energy densities radiated at each redshift, divided by $(1+z)$ to account for the overall effect. This means that we have

$$\mathcal{E}_{gw} \equiv \int_0^\infty \int_0^\infty \frac{N(z)}{1+z} \frac{dE_{gw}}{df_s} f_s \frac{df_s}{f_s} dz = \int_0^\infty \int_0^\infty \frac{N(z)}{1+z} f_s \frac{dE_{gw}}{df_s} dz \frac{df}{f} . \tag{22.60}$$

Equating the expressions from (22.59) and (22.60) for $\mathcal{E}_{gw}$, frequency by frequency, we arrive at (Phinney, 2001)

$$\rho_c c^2 \Omega_{gw}(f) = \frac{\pi}{4} \frac{c^2}{G} f^2 h_c^2(f) = \int_0^\infty \frac{N(z)}{1+z} \left(f_s \frac{dE_{gw}}{df_s} \right) \Bigg|_{f_s=f(1+z)} dz . \tag{22.61}$$

This is the main result. The interpretation is that the energy density in gravitational waves per logarithmic frequency interval is equal to the co-moving number density of events, multiplied by the (redshifted) energy each event produced (again, per logarithmic frequency interval). Notably, the result does not depend on the cosmology (apart from the assumption of a homogeneous and isotropic Universe). Moreover, it only involves the time-integrated energy spectrum of the sources, as long as they are randomly oriented with respect to the detector. If more than one type of source is of interest, the right-hand side of Eq. (22.61) is simply summed over the different source categories.

As an application of the result, let us estimate the low-frequency background from a cosmic ensemble of (adiabatically inspiralling) massive binaries. We then know from the quadrupole formula that (see Chapter 3)

$$\frac{dE_{gw}}{df_s} = \frac{\pi}{3G} \frac{(G\mathcal{M})^{5/3}}{(\pi f_s)^{1/3}} \quad \text{for } f_{min} < f_s < f_{max} , \tag{22.62}$$

where $\mathcal{M}$ is the chirp mass.

The low-frequency cut-off, f_{min}, is determined by the separation of the system at birth (or circularization, whichever comes first), and we obviously consider only systems with small enough initial separation that their merger time is shorter than the Hubble time. Over their lifetime, such systems radiate a broad spectrum of frequencies, up to an upper limit, f_{max}, set by the frequency at which the two bodies either 'come into contact' or plunge. It is natural to assume that this frequency is associated with the last stable orbit (the ISCO; see Chapter 10), in which case we have

$$f_{max} \approx 4 \times 10^{-4} \frac{1}{1+z} \left(\frac{\mathcal{M}}{10^9 M_\odot} \right)^{-1} \text{Hz.} \tag{22.63}$$

Combining Eq. (22.62) with (22.61) we obtain the gravitational-wave background (at $f < f_{max}$) from such a binary population (Phinney, 2001)

$$\Omega_{gw}(f) = \frac{8\pi^{5/3}}{9} \frac{(G\mathcal{M})^{5/3} f^{2/3}}{c^2 H_0^2} N_0 \langle (1+z)^{-1/3} \rangle, \tag{22.64}$$

where

$$N_0 = \int_0^\infty N(z) \, dz, \tag{22.65}$$

is the present-day co-moving number density of merger remnants, and

$$\langle (1+z)^{-1/3} \rangle = \frac{1}{N_0} \int_{z_{min}}^{z_{max}} \frac{N(z)}{(1+z)^{1/3}} \, dz. \tag{22.66}$$

The limits of integration

$$z_{min} = \max[0, f_{min}/f - 1] \tag{22.67}$$

and

$$z_{max} = f_{max}/f - 1, \tag{22.68}$$

can (effectively) be set to 0 and ∞, respectively, except for frequencies just below f_{min} or f_{max}. In practice, the value of $\langle (1+z)^{-1/3} \rangle$ is also not very sensitive to the details of $N(z)$ (for a flat Universe one would expect $\langle (1+z)^{-1/3} \rangle \approx 0.74$; see Phinney (2001)).

The main conclusion from this exercise is that one would expect the effective amplitude to scale as

$$h_c \sim \frac{1}{f} \Omega_{gw}^{1/2} \sim f^{-2/3}. \tag{22.69}$$

In order to complete the picture, we need to combine (22.64) with some model for the distribution, $N(z)$. For binaries, this boils down to providing $dN/d\mathcal{M}$ for a given cosmology, together with some mechanism for assembling massive black holes. Starting from some (model-dependent) prescription for the history of massive black-hole growth, one can generate a population of massive binaries using Monte-Carlo sampling. In essence, one has to work out the integral in Eq. (22.64). Typical results suggest that the result is not very sensitive to the nature of the massive black hole seeds. The main uncertainties, up to a factor of (perhaps) a few, are associated with the massive black-hole mass function at low redshifts, the variation in the halo merger rate and the accretion model (Sesana *et al.*, 2008).

Based on simulations (Rajagopal and Romani, 1995) one might expect a typical value for the comoving density of merger remnants to be $N_0 \approx 10^{-4}$ Mpc^3. For equal-mass binaries, with mass $10^9 M_\odot$, one then arrives at

$$h_c \approx 2 \times 10^{-16} \left(\frac{f}{\text{yr}^{-1}} \right)^{-2/3}. \tag{22.70}$$

This is indicative of what one might expect from a cosmological background of massive binaries.

22.7 Pulsar timing arrays

We know that supermassive black holes reside at the heart of massive galaxies. As part of cosmic evolution, galaxy mergers should form massive binary systems, which (eventually) emit gravitational waves and merge (Rees, 1984). Such mergers are a key part of hierarchical assembly scenarios, the backbone of modern structure formation models. Binaries with masses in the range $10^4 - 10^{10} M_\odot$ generate low-frequency signals that may be detectable with LISA at the low end and pulsar timing arrays at the high end of the mass-range. We have already argued that LISA will be able to detect individual events throughout cosmic history (see Figure 22.7). Meanwhile, pulsar timing experiments are most likely to detect the background generated by an incoherent superposition of the cosmic population of supermassive binaries (Manchester *et al.*, 2013).

In order to get an idea of how the pulsar timing array setup works, we first of all recall that each individual pulsar is a regular clock (with frequency ν_0, say). A passing gravitational wave impacts on the effective travel time of the radio pulses (Detweiler, 1979). The radio signal moves along null geodesics of a perturbed spacetime and if a gravitational wave crosses the line of sight the pulses arrive earlier/later than expected. When the signal arrives on Earth, it has accumulated a frequency redshift. That is, even if the pulsar emission is stable, we would observe a time-varying frequency such that

$$z(t) = \frac{\nu(t) - \nu_0}{\nu_0} = \frac{\delta\nu}{\nu_0}, \tag{22.71}$$

where $\nu(t)$ is the observed frequency. By comparing the arrival time of the pulses to predictions, we infer a timing residual which may—in addition to variations in the interstellar medium, the effects of the solar wind, and so on—encode the effects of a passing gravitational wave. In order to identify a gravitational-wave background, we can correlate data from an array of pulsars across the sky and exploit the fact that a gravitational wave would affect all pulsars in the same way, while intrinsic timing noise does not.

The strain sensitivity of a pulsar timing array depends (roughly) on the timing residuals, $\delta t \sim 1/\delta\nu$, and the observation time, T, as

$$h \sim \frac{\delta t}{T}. \tag{22.72}$$

From the discussion (leading to (22.70)) of the stochastic background from supermassive black-hole binaries, we know that $h_c \sim 10^{-15}$ for $f \sim 1$ nHz. In order to detect such a signal, we need a set of pulsars with timing residuals of order

$$\delta t \sim 3.2 \times 10^{-7} \left(\frac{h_c}{10^{-15}} \right) \left(\frac{T}{10 \text{ yr}} \right) \text{ s.} \tag{22.73}$$

The road to detection is now (relatively) clear. We need to accumulate longer and more accurate timing data for large arrays of monitored pulsars. The first requires patience and the second involves an element of luck, as we need to find suitably stable pulsars. Over the last decades, different international pulsar timing collaborations have made impressive progress (see Manchester *et al.* (2013)). For example, the recently released dataset from the NANOGrav collaboration involves 45 pulsars timed with the Green Bank Telescope over a period of 11.4 years (Arzoumanian *et al.*, 2018). This has led to results like those shown in Figure 22.10, setting upper limits close to the predicted level for the gravitational-wave background.

The analysis of the timing problem proceeds in the same way as spacecraft Doppler tracking (see Chapter 7). One would typically analyse the problem with the solar system barycentre representing the 'detector frame' $[\hat{x}, \hat{y}, \hat{z}]$; see Figure 22.8. For a wave $h_{ij}(t, \boldsymbol{x})$ moving in the direction $\hat{k}$ incident on a pulsar located in direction $\hat{p}$ a distance L away (as in Figure 22.8), the observed redshift is then

$$z(t) = \frac{1}{2} \hat{p}^i \hat{p}^j \int_{t_p}^{t} dt' \, \frac{\partial}{\partial t'} h_{ij}(t', \boldsymbol{x}). \tag{22.74}$$

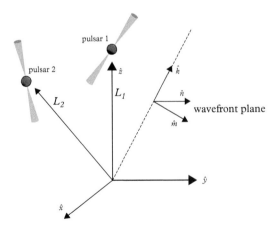

Figure 22.8 *The geometry assumed in the analysis of the impact of a gravitational wave on a pulsar timing array. We distinguish the 'detector frame' $[\hat{x}, \hat{y}, \hat{z}]$ from the wave frame with propagation direction $\hat{k}$ and wavefronts in the plane $[\hat{m}, \hat{n}]$.*

For a plane gravitational-wave (an appropriate assumption for sources at a cosmological distance), the redshift follows from

$$z(t, \hat{k}) = \frac{1}{2} \frac{\hat{p}^i \hat{p}^j}{1 + \hat{k} \cdot \hat{p}} \left[h_{ij}(t, \hat{k}) - h_{ij}\left(\tau, \hat{k}\right) \right] = \frac{1}{2} \frac{\hat{p}^i \hat{p}^j}{1 + \hat{k} \cdot \hat{p}} \Delta h_{ij}(t, \hat{k}). \tag{22.75}$$

In general, the redshift is given by the difference between the wave at Earth and at the pulsar.

For the configuration in Figure 22.8, the pulsar location is such that

$$t_p = t - L \qquad \boldsymbol{x}_p = L\hat{p}, \tag{22.76}$$

and it follows from (22.46) that

$$\tau = t_p - \hat{k} \cdot \boldsymbol{x}_p = t - L(1 + \hat{k} \cdot \hat{p}). \tag{22.77}$$

The expression for the redshift (22.75) then shows that, if the gravitational-wave source is located right behind the pulsar we have $\tau = t$ and the effect vanishes. This should be no surprise—it is a reflection of the fact that the waves are transverse. In practice, the second term in (22.75)—the pulsar term—can be considered as an extra noise term, which should vanish after averaging over a set of pulsars.

To be specific, let us focus on stochastic gravitational waves from sources distributed across the sky. That is, we consider the statistical distribution of signals from an unresolvable collection of events. Moreover, we assume that the sources are likely to be randomly distributed throughout the Universe. As long as the Universe is homogeneous and isotropic on large scales, the gravitational-wave background is then likely to be isotropic, as well. In this case, we effectively need the frequency shift over the whole sky, which follows from the integral

$$z(t) = \int z(t, \hat{k}) d\hat{k}. \tag{22.78}$$

Moreover, a timing experiment would not measure the redshift, but rather the timing residual

$$R(t) = \int_0^t z(t') dt'. \tag{22.79}$$

The effect of a stochastic background is then encoded in the correlation collected from pairs of pulsars at different sky locations. To demonstrate this, first of all note that all pulsar signals, and the Earth, are affected by the same gravitational-wave induced metric perturbation.

Using the solar system barycentre as our reference and introducing the usual polarization basis, we have

$$\Delta h_{ij} = \sum_{P=+,\times} \int \tilde{h}_P(f,\hat{k}) e^P_{ij}(\hat{k}) \exp(-2\pi i f t) \left\{ 1 - \exp\left[2\pi i f L \left(1 + \hat{p}\cdot\hat{k}\right)\right]\right\} df. \quad (22.80)$$

Thus, we identify the Fourier transform of the redshift

$$\tilde{z}(f,\hat{k}) = \left\{ 1 - \exp\left[2\pi i f L\left(1 + \hat{p}\cdot\hat{k}\right)\right]\right\} \sum_{P=+,\times} \int \tilde{h}_P(f,\hat{k}) F^P(\hat{k}), \quad (22.81)$$

where

$$F^P(\hat{k}) = \frac{1}{2} \frac{1}{1 + \hat{p}\cdot\hat{k}} \left[\hat{p}^i \hat{p}^j e^P_{ij}(\hat{k}) \right]. \quad (22.82)$$

This function is, essentially, the pulsar timing equivalent of the detector antenna pattern (see Chapter 7). From our previous results, it is easy to see that we have

$$F^+ = \frac{1}{2} \frac{1}{1 + \hat{p}\cdot\hat{k}} \left[(\hat{p}\cdot\hat{m})^2 - (\hat{p}\cdot\hat{n})^2 \right], \quad (22.83)$$

and

$$F^\times = \frac{1}{1 + \hat{p}\cdot\hat{k}} (\hat{p}\cdot\hat{m})(\hat{p}\cdot\hat{n}). \quad (22.84)$$

The stochastic background can be characterized by a one-sided power spectral density S_h (see Chapter 8) through the expectation value

$$\langle \tilde{h}^*_P(f,\hat{k}) \tilde{h}_{P'}(f',\hat{k}') \rangle = \frac{1}{2} S_h \delta^{(2)}(\hat{k},\hat{k}') \delta_{PP'} \delta(f - f'), \quad (22.85)$$

where $\delta^{(2)}(\hat{k},\hat{k}')$ is the delta-function on the sphere. We can use this result to work out the expectation of the product of signals from two pulsars in directions $\hat{p}_1$ and $\hat{p}_2$. This leads to

$$\langle \tilde{z}_1(f) \tilde{z}^*_2(f') \rangle = \frac{1}{2} S_h(f) \delta(f - f') \Gamma(f), \quad (22.86)$$

where

$$\Gamma(f) = \sum_{P=+,\times} \int \int_{S^2} \left\{ 1 - \exp\left[2\pi i f L_1 \left(1 + \hat{p}_1 \cdot \hat{k} \right) \right] \right\}$$
$$\times \left\{ 1 - \exp\left[2\pi i f L_2 \left(1 + \hat{p}_2 \cdot \hat{k} \right) \right] \right\} F_1^P(\hat{k}) F_2^P(\hat{k}) d\hat{k}. \tag{22.87}$$

The result simplifies, as the distance to the pulsars is expected to be large compared to the gravitational-wave wavelength. In this case $\Gamma(f)$ limits to a constant value, and we have

$$\Gamma(f) \approx \Gamma_0 = \sum_{P=+,\times} \int \int_{S^2} F_1^P(\hat{k}) F_2^P(\hat{k}) d\hat{k}. \tag{22.88}$$

This is the frequency domain equivalent of neglecting the pulsar term in (22.75). Evaluating the integral, we arrive at an expression with depends only on the angle θ between the pulsars. This leads to the famous Hellings and Downs curve (Hellings and Downs, 1983), illustrated in Figure 22.9.

In summary, the basic technique used to search for a gravitational-wave imprint consists in correlating the timing residuals from N pulsars. In general, the residual from a given pulsar is given by

$$\delta t = \delta t_n + \delta t_h, \tag{22.89}$$

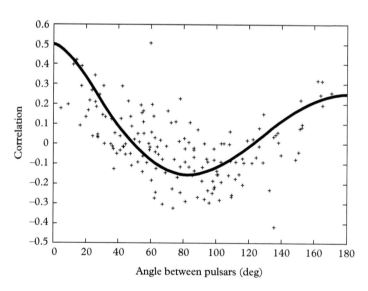

Figure 22.9 *The expected correlation in timing residuals of pairs of pulsars as a function of angular separation for an isotropic gravitational-wave background. The theoretical result is compared to a simulated background of inspiralling binaries (each '+' represents a pulsar pair of idealized noiseless data). (Reproduced from Hobbs et al. (2010a).)*

where $\delta t_h \sim h/f$ and δt_n are the gravitational-wave signal and noise contribution, respectively. Taking the case of two pulsars as an example, we have $N = 2$ and the minimum detectable stochastic signal is characterized by

$$h^2 \Omega_{\text{gw}}(f) \propto \frac{\delta t_{\text{rms}}^2 f^4}{\sqrt{T \Delta f}}, \tag{22.90}$$

where $\delta t_{\text{rms}} = \sqrt{\langle \delta t^2 \rangle}$ is the root-mean-square value of the timing residuals and Δf the bandwidth of the search. This is the pulsar-timing equivalent of the result one obtains by considering direct searches for a stochastic background using the cross-correlation of data from two interferometers (discussed in Chapter 8). If instead of two, one has many pulsars, then the optimal signal-to-noise ratio is given by the combination of all statistically independent correlations that can be formed. For N sufficiently large, this signal-to-noise ratio scales as N^2. In effect, we have

$$h^2 \Omega_{\text{gw}}(f) \propto \frac{\delta t_{\text{rms}}^2 f^4}{N \sqrt{T \Delta f}}. \tag{22.91}$$

Making contact with the discussion of astrophysical backgrounds, this leads to

$$h_c(f) \propto \frac{\delta t_{\text{rms}} f}{N^{1/2} (T \Delta f)^{1/4}}. \tag{22.92}$$

The sensitivity of a given timing array scales as $h_c(f) \propto f$ and reaches a minimum detectable frequency $f \sim 1/T$. This leads to a characteristic wedge-like sensitivity curve with a sharp low-frequency cut-off.

As an example of the state-of-the-art, let us consider the recent data release from the NANOGrav collaboration, leading to the results shown in Figure 22.10. The top panel shows 95% upper limits for free-spectrum amplitudes (jagged black line), which indicate the sensitivity of the dataset to individual monochromatic signals. The anticipated $f^{-2/3}$ power law is also shown. The coloured dashed lines and bands indicate a selection of theoretical expectations for the stochastic background. The models differ in the choice of black-hole–host-galaxy mass relationship and the impact of possible selection biases in dynamically measured masses (see Colpi and Sesana (2017) for discussion). These same results are shown in the bottom panel, expressed in terms of the stochastic energy density (per logarithmic frequency bin) in the Universe as a fraction of closure density, $\Omega_{\text{gw}}(f) h_{100}^2$, where the scaling by h_{100}^2 removes the impact of the specific value of the Hubble constant (as before). The fractional energy density scales as $\Omega_{\text{gw}} h^2 \propto f^2 h_c(f)^2$.

The results in Figure 22.10 correspond to a (95%) upper limit on the gravitational-wave strain of

$$h_c < 1.45 \times 10^{-15} \left(\frac{f}{\text{yr}^{-1}} \right)^{-2/3}. \tag{22.93}$$

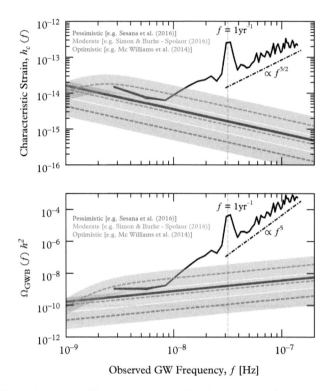

Figure 22.10 *Top: Inferred upper limit on a stochastic low-frequency gravitational-wave-amplitude (95% upper limits for an uncorrelated common process with a $f^{-2/3}$ power law (straight black line) or with independently determined free-spectrum components (jagged black line)). The dash-dotted line shows the expected sensitivity scaling behaviour for white noise. The coloured dashed lines and bands show median and 1σ ranges for the amplitudes predicted by different models (see Arzoumanian et al. (2018) for the original references). Bottom: As in the top panel, except showing the results in terms of the stochastic background energy density (per logarithmic frequency bin) in the Universe as a fraction of the closure density, $\Omega_{gw}(f)h^2_{100}$. (Reproduced from Arzoumanian et al. (2018) by permission of the AAS.)*

The timing data will improve with continued observations and we should (hopefully) soon reach the sensitivity where several scenarios (and/or parameter values within a given model) may yield a detection. If there are no detections, we may still be able to rule out proposed scenarios. In the next decade, the Square Kilometer Array (SKA) will provide a major increase in pulsar timing capability (Stappers *et al.*, 2018). Assuming that the SKA will be able to monitor 20 millisecond pulsars for 10 years at a precision level of $\delta t_{\rm rms} \sim 10$ ns, it will allow us to not only detect the stochastic background, but also study its spectrum in the frequency range $3 \times 10^{-9} \lesssim f \lesssim$ few $\times 10^{-8}$. This would provide detailed information about the formation and evolution history of supermassive black holes.

22.8 AC/DC

The intuition from linear theory suggests that gravitational waves created by an astrophysical event typically consists of a short-lived high-frequency signal. In addition, the nonlinear theory unveils a non-oscillatory low-frequency part of the signal. In the case of a binary merger, this contribution grows rapidly during the merger and—remarkably—leaves a permanent metric change once the wave burst has passed. In effect, the spacetime relaxes to a configuration that differs from the original one. This effect if known as the gravitational-wave memory (Christodoulou, 1991; Favata, 2009; Garfinkle, 2016). In an electromagnetic analogue it would represent a DC complement to the AC signals we have so far considered. This memory depends on the entire past history of the source.

As a direct illustration of the memory effect, let us consider the specific case of the black-hole merger GW150914. In essence, we want to consider the detectability of the memory with instruments like Advanced LIGO. We know that the gravitational-wave strain scales with the mass of the binary, but the energy involved only makes up a small fraction of the total energy release. This is apparent from the results shown in Figure 22.11. This example shows that the imprint of the memory on the GW150914 signal is too weak to be detected (Lasky *et al.*, 2016). However, it is not negligible. One may be able to detect it by combining data from an ensemble of events. This is promising, as black-hole mergers appear to be common. However, it could be that GW150914 represents a relatively high-mass binary compared to the overall population. If this is the case then the number of events required to detect the gravitational-wave memory obviously increases.

22.9 Astrometry

In addition to introducing a timing residual, a cosmological gravitational-wave signal may cause the apparent position of distant stars to fluctuate. It is easy to see that this should happen. The argument is a straightforward variation of the timing calculation we have already done. The angular deflection should simply be proportional to the gravitational-wave strain. These fluctuations will be tiny but they may nevertheless be detectable with accurate astrometry (Book and Flanagan, 2011; Moore *et al.*, 2017), an approach that is particularly well suited to constraining the polarization content of gravitational waves (O'Beirne and Cornish, 2018). This idea is relevant given that Gaia[4] will make high-precision observations of the positions of about a billion stars in our Galaxy and several hundred thousand distant quasars.

Gaia's sensitivity to gravitational waves relies on observing a large number of stars. Since stars are typically separated by many gravitational wavelengths, each 'star term' will be different. Just as in the pulsar timing case, this means that the signal is dominated by the 'Earth term', which is common to all stars. A gravitational-wave background will

---

[4] sci.esa.int/gaia/

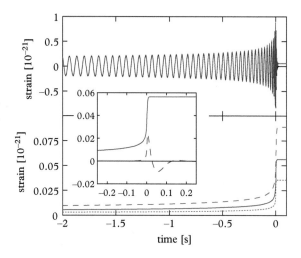

Figure 22.11 *Gravitational-wave strain for parameters consistent with GW150914. The top panel shows the strain including (blue curve) and excluding (black) the memory effect. The bottom panel shows only the memory-induced strain. The red dotted and dashed curves are binaries assumed to be at the same distance (410 Mpc) and with the same orientation, but have equal masses with $M_{1,2} = 20\,M_\odot$ and $50\,M_\odot$, respectively (compared to $65\,M_\odot$ for the blue curve). Inset: the solid blue curve shows a zoomed-in version of the blue curve from the bottom panel, while the dashed curve is after a high-pass filter to show the signal visible in an Advanced LIGO instrument. (Reproduced from Lasky et al. (2016), copyright (2016) by the American Physical Society.)*

lead to deflections that are correlated over the sky and vary randomly with time. The rms deflection is roughly given by (Book and Flanagan, 2011)

$$\delta_{\rm rms} \sim h_{\rm rms} \sim \frac{H_0}{f}\sqrt{\Omega_{\rm gw}}. \qquad (22.94)$$

Suppose we track N sources in the sky with angular precision $\Delta\theta$ for a total time T. For a single source one may detect an angular velocity of order $\sim \Delta\theta/T$ and the precision improves with the number of sources as $1/N$. The angular rms velocity is

$$\omega_{\rm rms} \sim f\delta_{\rm rms} \sim H_0\sqrt{\Omega_{\rm gw}}, \qquad (22.95)$$

and it follows that one should be able to obtain an upper limit of order

$$\Omega_{\rm gw} \sim \frac{(\Delta\theta)^2}{NT^2 H_0^2}. \qquad (22.96)$$

To be specific, by tracking the position of $N \sim 10^6$ quasars over a time of $T \sim 1$ yr with an angular resolution of $\Delta\theta \sim 10$ μas we would reach (Book and Flanagan, 2011)

$$\Omega_{gw} \sim 10^{-6}, \tag{22.97}$$

a level comparable to the limit expected from pulsar timing. Gravitational-wave astronomers should keep a keen eye on the releases of Gaia data in the early 2020s.

22.10   Detecting a primordial background

Gravitational waves decouple from matter very early in the history of the Universe, much earlier than the electromagnetic radiation (see Chapter 4). This means that one may, at least in principle, be able us to use gravitational-wave data to probe the very first stages of cosmic evolution. As this may include the inflationary era, this kind of observation would (arguably) be the most fundamental observation in all of physics. The question is if it is a realistic prospect. After all, the basic effect of inflation is to smooth things out—density variations and spatial curvature wash out. On the other hand, we know from quantum mechanics that there will always be some level of fluctuation.

In order to assess to what extent we can expect to detect primordial gravitational waves, we need to better understand the nature of such signals (see Caprini and Figueroa (2018) for a recent review). Perhaps not surprisingly, this leads to a new set of questions. First of all, the wave-generation mechanisms will (most likely) be stochastic in nature. This means that, as in the case of pulsar timing, we have to consider the statistical distribution of signals from an unresolvable collection of events. As in the case of distant astrophysical sources, we expect an isotropic gravitational-wave background. This means that we are unlikely to be able to distinguish different contributions from one another. In fact, astrophysical stochastic backgrounds may swamp the primordial signal in different frequency bands.

A number of processes in the Universe may generate gravitational waves. The most commonly considered sources are (Maggiore, 2000):

- fluctuations amplified during inflation,
- first-order phase transitions,
- cosmic strings.

We will focus on these three mechanisms, even though the list goes on (including different pre-Big-Bang scenarios, branes, and quintessence; see Maggiore (2018)). Our aim is to introduce the main detection strategy and provide a rough idea of the level of detectability. The same logic applies to other mechanisms.

The gravitational-wave spectrum for any given mechanism typically covers a range of frequencies and scales. The lowest possible frequency will always correspond to the largest wavelength of oscillation—bounded by the scale of the Universe, this should be on the order of Hubble radius at any given time. That is, we have $\lambda \sim H_0^{-1}$, leading to the lowest frequency today of $f_{gw} \sim 10^{-18}$ Hz. Meanwhile, the highest frequencies correspond to the highest temperature of the primordial Universe. Taking this to be the Planck scale $T \approx 10^{32}$ K (above this temperature quantum gravity effects would be important), the high end of the frequency band today is on the order of 10^{12} Hz (taking into account the redshift in an expanding Universe). The main message is that we should expect stochastic gravitational waves across up to 30 decades in frequency.

As discussed in Chapters 4 and 8, it is useful to express any gravitational-wave estimates in terms of the corresponding energy spectrum and compare the result to the standard critical density ρ_c from (4.72). That is—as in (8.49)—we consider the energy spectrum

$$\Omega_{gw} = \frac{1}{\rho_c} \frac{d\rho_{gw}}{d \log f} = \frac{f}{\rho_c} \frac{d\rho_{gw}}{df}. \tag{22.98}$$

In order for this to be useful we need to make contact with the gravitational-wave amplitude. To do this, we recall that a stochastic background can be thought of as a superposition of waves coming from all angles and with all possible frequencies. Building on the previous plane-wave results, we express such signals as

$$h_{ij}(t, \boldsymbol{x}) = \sum_{P=+,\times} \int_{-\infty}^{+\infty} df \int d\hat{k}\, h_P(f, \hat{k}) e^{-2\pi i f t} \epsilon_{ij}^P(\hat{k}) e^{i\boldsymbol{k}\cdot\boldsymbol{x}} + \text{c.c.}, \tag{22.99}$$

where $\hat{k}$ is the incoming unit vector of the propagation and the polarization tensors are defined to be normal to this vector. As before, the key information is encoded in the Fourier transform $h_P(f, \hat{k})$. Moreover, in the isotropic and unpolarized case, we simply have h_k (in terms of $k = 2\pi f$), given by

$$h_{ij}(t, \boldsymbol{x}) = \int \frac{d^3 k}{(2\pi)^{3/2}} h_k e^{i\boldsymbol{k}\cdot\boldsymbol{x}} \sum_P \epsilon_{ij}^P(\hat{k}) + \text{c.c.}. \tag{22.100}$$

This gives us a spectrum of gravitational waves, expressed in terms of h_k. The final connection is made via the expression for the gravitational-wave energy

$$\rho_{gw} = \frac{1}{32\pi G} \langle \dot{h}_{ij} \dot{h}^{ij} \rangle, \tag{22.101}$$

which involves averaging over a range of wavelengths. As we are dealing with a stochastic signal this amounts to taking an ensemble average. In order to complete the model, and work out the abundance of gravitational waves in the present Universe, we need to be able to (for each given mechanism) track the evolution of h_k back to the source.

22.11 Parametric amplification of quantum fluctuations

As an example of a robust mechanism for primordial gravitational waves, let us consider the amplification of quantum fluctuations during the inflationary epoch (Grishchuk, 1975; Grishchuk, 1993). The idea behind the mechanism is simple. The very early Universe deviated from absolute homogeneity due to quantum fluctuations. As the Universe expanded, some of these fluctuations were amplified. After the end of inflation, the amplified perturbations continued to propagate as waves, redshifted in the usual way.

In order to illustrate the mechanism, let us consider small deviations away from the standard Friedmann–Robertson–Walker model (see Chapter 4). The background metric is then given by (22.7) (with $k = 0$, not to be confused with the wave number in the following) and one can show that the metric perturbations are governed by the wave equation

$$\Box h_{ij}(t, \boldsymbol{x}) = 0. \tag{22.102}$$

In this case we have $\sqrt{-g} = R^3(t)$ and it follows that

$$\left[-\partial_t^2 - 3\frac{\dot{R}}{R}\partial_t + \frac{1}{R^2}\nabla^2 \right] h_{ij}(t, \boldsymbol{x}) = 0. \tag{22.103}$$

If we assume that the signal is isotropic (for simplicity), we have

$$\left[\partial_t^2 + 3\frac{\dot{R}}{R}\partial_t + \frac{k^2}{R^2} \right] h_k(t) = 0. \tag{22.104}$$

At this point it is useful to introduce a new (conformal) time coordinate η, such that $dt = ad\eta$. This leads to

$$h_k''(\eta) + 2\frac{R'}{R}h_k'(\eta) + k^2 h_k(\eta) = 0, \tag{22.105}$$

where

$$h_k' = \frac{dh_k}{d\eta} = R\dot{h}_k. \tag{22.106}$$

Moreover, we can solve the equation by changing the dependent variable to

$$\psi_k = a(\eta)h_k(\eta). \tag{22.107}$$

This leads to the final equation

$$\psi_k'' + \left(k^2 - \frac{R'}{R}\right)\psi_k = 0. \tag{22.108}$$

This is the equation for a parametric oscillator, with frequency depending on time. The time-dependent part depends on the cosmology.

We get an intuitive picture of what is going on by considering two extreme limits. In the first limit, let $k^2 \gg R'/R$. In this regime—representing what we will call sub-Hubble modes, since the wavelengths are much shorter than the Hubble scale—we simply have a harmonic oscillator, so the solution is

$$\psi_k \propto \frac{1}{\sqrt{2k}} e^{\pm ik\eta}. \tag{22.109}$$

In the opposite limit, when $k^2 \ll R'/R$, the evolution of the (super-Hubble) modes is such that

$$\psi_k \propto R\left(C_k + D_k \int \frac{d\eta}{R^2}\right), \tag{22.110}$$

where C_k and D_k are integration constants. The two solutions should be matched in the intermediate regime, but as our discussion is qualitative we do not need to work out the details. In terms of the gravitational-wave amplitude, we have

$$h_k \propto \frac{1}{\sqrt{2k}R} e^{\pm ik\eta}, \tag{22.111}$$

and

$$h_k \propto C_k + D_k \int \frac{d\eta}{R^2}. \tag{22.112}$$

We see that the sub-Hubble modes are suppressed as the Universe expands while the amplitude of the super-Hubble modes is constant. Effectively, modes that remain outside the Hubble radius are amplified relative to modes inside. However, as the Universe expands, the amplified modes may re-enter the Hubble radius (as sketched in Figure 22.12). Once they do, they turn into propagating gravitational waves

$$\psi_k = A_k e^{+ik\eta} + B_k e^{-ik\eta}. \tag{22.113}$$

Finally, we make contact with (22.101). Basically, we need

$$\dot{h} = \frac{\psi'}{R^2} - \frac{\psi}{R^2}\frac{R'}{R}, \tag{22.114}$$

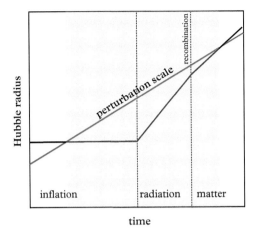

Figure 22.12 *A schematic illustration of how perturbation modes of different wavelengths exit and eventually re-enter the Hubble radius in an inflationary Universe.*

leading to

$$\langle \dot{h}_{ij}\dot{h}^{ij}\rangle = \frac{1}{R^4}\langle \psi'_{ij}\psi'^{ij} - 2H\psi^{ij}\psi'_{ij} + H^2\psi_{ij}\psi^{ij}\rangle \approx \frac{1}{R^4}\langle \psi'_{ij}\psi'^{ij}\rangle. \tag{22.115}$$

For the sub-Hubble modes we get

$$\langle \psi'_{ij}\psi'^{ij}\rangle V = \frac{16\pi}{V}\int_0^{+\infty} k^2 \psi'_k\psi'^*_k dk, \tag{22.116}$$

where we have used $\epsilon^{\rm P}_{ij}\epsilon^{ij}_{\rm P} = 4$ (from before) and $\int d\hat{k} = 4\pi$. In terms of our wave solution, and after averaging, we get

$$\langle \psi'_k\psi'^*_k\rangle = \frac{k^2}{2}\left(|A_k^2| + |B_k^2|\right), \tag{22.117}$$

so

$$\rho_{\rm gw} = \frac{1}{4GR^4V}\int k^4\left(|A_k^2| + |B_k^2|\right)dk, \tag{22.118}$$

and

$$\Omega_{\rm gw} \propto k^3\langle \psi'_k\psi'^*_k\rangle. \tag{22.119}$$

However, we are not quite done. We still need to work out the amplitudes A_k and B_k for a given scenario. This turns out to be the messy part of the problem, as the result is sensitive to the assumed inflationary model. In the simple, but perhaps unrealistic, model of deSitter expansion the Hubble parameter stays constant ($= H_I$) and we have

$$R = -\frac{1}{H_I \eta} \quad \text{for} \quad -\infty < \eta < \eta_I, \qquad (22.120)$$

where η_I represents the end of inflation. In this case one can find an analytic solution for the amplitudes we are interested in

$$A_k = \frac{1}{2\sqrt{2}k^{5/2}\eta_I^2}, \qquad (22.121)$$

and

$$B_k = \frac{1}{\sqrt{2k}}\left(1 - \frac{i}{k\eta_I} - \frac{1}{2k^2\eta_I^2}\right). \qquad (22.122)$$

This leads to a flat (k-independent) spectrum with

$$h^2\Omega_{\text{gw}} \approx 4 \times 10^{-14}\left(\frac{H_I}{6 \times 10^{-5}M_{\text{Pl}}}\right)^2, \qquad (22.123)$$

where M_{Pl} is the Planck mass

$$M_{\text{Pl}} = \left(\frac{\hbar c}{G}\right)^{1/2}. \qquad (22.124)$$

A more realistic model for inflation is the so-called slow-roll model, in which the expansion is driven by a scalar field ϕ governed by a potential $V(\phi)$ that satisfies specific conditions (Liddle, 2003). The main difference between this model and the deSitter case is that the Hubble radius varies during the inflationary epoch (as indicated in Figure 22.12). As ϕ rolls down the potential towards the minimum, the Hubble radius decreases. As a result the model is not scale invariant—the spectrum depends on k and acquires a small tilt. In principle, this could be good news for the prospects of detection, but detailed models suggest that the tilt tends to make the situation worse (see Figure 22.16).

22.12 Phase transitions

The early Universe cooled as it expanded. As the temperature dropped below critical values (typically associated with some mass scale, like that of the Higgs boson for the electroweak transition), there should have been a series of phase transitions. If these phase transitions were first order they may have generated gravitational waves (Caprini *et al.,* 2009; Binétruy *et al.,* 2012).

The basic idea is that, at high temperatures the Universe is in a metastable vacuum phase and—as the temperature drops—a new 'true' vacuum emerges, separated from the old 'false' vacuum by a potential barrier. This potential barrier prevents the transition from being instantaneous. Instead, the transition proceeds via quantum tunneling driven by random fluctuations. As bubbles of the new phase expand into the old phase they may collide and generate gravitational waves. The process also leads to heating, which may trigger turbulence and additional gravitational-wave emission.

The calculation of the gravitational-wave signature associated with phase transitions is complex, but one typically finds that it is characterized by a peak frequency which depends on the temperature at with the transition happened. In the case of the electroweak transition, which happened at $k_B T \approx 100$ GeV, the peak frequency would be $f_{peak} \sim 4 \times 10^{-3}$ Hz (Apreda *et al.,* 2002). Basically, this phase transition might leave a signature in the sensitivity range for space interferometers like LISA. Detailed calculations also show that the spectrum rises as f^3 towards the peak and then drops off as f^{-1}.

The amplitude estimated for bubble collisions suggests that the gravitational-wave signature may be within reach of future detectors, but the scenario is speculative. There are two well-known phase transitions that should have taken place in the early Universe, the QCD phase transition and the electroweak one. The first of these (when baryonic matter went from a quark-gluon plasma to the confined state, with neutrons and protons, we see today) was a smooth crossover rather than a first-order transition. Within the Standard Model, as similar crossover is predicted for the electroweak transition. These would not be very promising in terms of gravitational-wave production. Supersymmetric models may lead to first-order transitions, but at the moment there is little observational support for these suggestions.

22.13   Cosmic strings

Phase transitions associated spontaneous symmetry breaking may also generate a network of cosmic strings (topological defects) through the so-called Kibble mechanism (Hindmarsh and Kibble, 1995). The dynamics of these cosmic strings may lead to an observational gravitational-wave signature, opening up a fascinating window to fundamental physics at very high energies.

Cosmic strings are, essentially, one-dimensional massive objects with extraordinary density. The relevant quantity is the mass per unit length, μ, which can reach values as high as 10^{22} g/cm if the string formation takes place at scales above 10^{16} GeV. A network of such strings may lead to intersections, the formation of kinks and small loops being detached from the main body of each string. The dynamics shares many aspects familiar from superfluid turbulence in the laboratory setting (Schwarz, 1982; Tsubota *et al.,* 2003). As the tension is remarkably high, similar to the mass per unit length, the interactions will lead to the string vibrating at relativistic speeds. The vibrations would be damped by gravitational radiation. Small loops decay the fastest, roughly on the timescale needed for light to travel across their diameter. It is easy to understand how the string vibrations lead to quadrupole variations and the emission of gravitational waves. Unfortunately, the signal is strongly model dependent (Allen and Shellard, 1992; Caldwell *et al.*, 1996; Damour and Vilenkin, 2000).

Considering the statistics of such a cosmic string network one may argue that, at any given time, a Hubble-sized volume contains a constant number of strings passing through it and a large number of small strings that are decaying and replaced by new ones. Loops of a variety of sizes are formed and radiate at different frequencies as they shrink. This would lead to a relatively flat gravitational-wave spectrum, with a small bump at low frequencies. However, many of the proposed cosmic strings scenarios are already ruled out by astrophysical constraints; see Figure 22.13.

About 50 days of coincident data from the two LIGO detectors, collected during the O1 run in 2016, were used to constrain the string tension for different models (Abbott *et al.*, 2018a). Focusing on the classic scenario (Damour and Vilenkin, 2000; Damour and Vilenkin, 2005), where all loops chopped off the infinite string network are formed with the same relative size, the limits for stochastic and burst signals are shown in Figure 22.13. The obtained limits are not as strict as those obtained from pulsar timing,

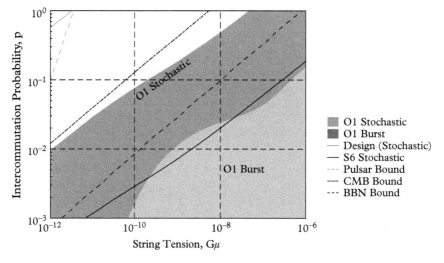

Figure 22.13 *Exclusion regions (at 90% confidence) for the classic cosmic string model where all loops are formed with the same relative size. Shaded regions are excluded by the (O1) Advanced LIGO stochastic and burst measurements. Bounds from the S6 LIGO–Virgo stochastic measurement are also shown, along with constraints from the indirect Big Bang Nucleosynthesis, CMB bounds and pulsar timing array measurements. The anticipated design sensitivity of Advanced LIGO–Virgo detectors is also indicated (Design, Stochastic, in the upper left corner of the figure). The excluded regions are below the respective curves. (Reproduced from Abbott et al. (2018a), copyright (2018) by the American Physical Society.)*

which correspond to $G\mu < 3.8 \times 10^{-12}$ for this specific model. Future detector upgrades are expected to make the interferometer constraint competitive.

22.14 E/B-modes

Gravitational waves from the very early Universe may leave an imprint on the cosmic microwave background (Kamionkowski *et al.*, 1997). However, in order to extract this signature we need to consider the fineprint of the CMB. As gravitational waves are transverse and area-conserving (stretch one way—squeeze the other) they do not produce density or temperature variations. Rather, they leave an imprint in the polarization. Detecting this imprint is far from easy and, in addition, cosmic magnetic fields and foreground dust may affect the polarization.

Let us, nevertheless, consider the expected gravitational-wave signature. A gravitational wave moving directly towards the Earth pushes the particles in the plasma apart in one direction and together in the perpendicular direction. This produces a bias in the polarization of the scattered photons and leads to a slight excess along directions in which the scatterers are being pushed. Since the gravitational waves are themselves polarized, this leads to a rotation of the polarization map vectors (which will be arbitrary). A map of the polarization takes the form of line segments on the sky, representing the direction of the net oscillation in the electric field. With a single point, this is all the information we have, but with a map of some area, one can decompose the polarization into E-modes (due the scalar density variations) and B-modes (due to tensor perturbations); see Figure 22.14. The main difference is that B-modes have a net twist as you travel around in a circle.

One would typically quantify cosmological perturbations by two numbers: the overall amplitude, A, and the 'spectral tilt', n, which tells you how the perturbations vary from large wavelengths to shorter ones. For density perturbations, we have a reasonably good idea of what these numbers are. The amplitude is about 10^{-5}; see Figure 22.14, and the tilt is $n \approx 0.96$ (Ade *et al.*, 2016*b*). For historical reasons, scalar (density) perturbations that are the same on all wavelengths are taken to have $n_S = 1$, while tensor (gravitational-wave) perturbations that are the same on all wavelengths have $n_T = 0$. It is common to relate the different perturbations by giving the amplitude ratio $r = A_T / A_S$.

In early 2014 the BICEP2 team announced the discovery of B-modes (on an angular scale of a few degrees; see Ade *et al.* (2014)). They suggested that the power in the B-modes corresponded to a tensor-to-scalar ratio $r \approx 0.2$, compared to the value $r = 0.13$ expected from simple (single-field) slow-roll inflation. However, the analysis assumed dust contamination was negligible. The immediate question was to what extent foreground dust might corrupt the conclusions. This is an important issue because (at the relevant frequency) polarized emission in the sky is expected to be dominated by dust. Indeed, a closer inspection demonstrates that the uncertainties are significant enough that a detection can not be claimed (Ade *et al.*, 2015; Gott and Colley, 2017). By correlating the BICEP2 polarization maps to those from the Planck experiment one

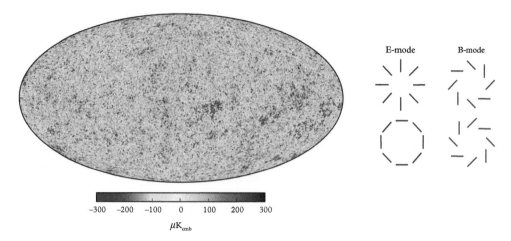

Figure 22.14 *Left: The Planck data for the cosmological background. Maps like this show differences in temperature from point to point in the sky. It is these tiny differences (one part in 10^5) that grow into stars, galaxies, and clusters as the Universe evolves. Right: The different nature of scalar density variations (E-modes) and gravitational-wave tensor perturbations (B-modes). We can (at least in principle) distinguish the different modes from the shape of the polarization pattern in the sky. (Left panel reproduced from Adam et al. (2016) with permission from Astronomy and Astrophysics. Copyright ESO.)*

finds that $r = 0.09^{+0.06}_{-0.04}$, suggesting that the gravitational-wave contribution is much less significant than initially claimed. The search for primordial gravitational waves goes on.

22.15   Twenty-nine decades of frequency

We have learned that gravitational-wave astronomy spans a frequency range of many decades. We have seen how astrophysical systems radiate at specific frequencies and how different frequency ranges—from fluctuations in the CMB at the very lowest frequencies to the high-frequency oscillations of neutron stars—are probed by distinct observation strategies. The different observations pose (to some extent) unique data analysis challenges, but there are overlaps. For example, systems like the black-hole binary GW150914 detected by LIGO could also have been seen (at an earlier stage of evolution) by a space interferometer like LISA. We are searching for gravitational waves from rotating neutron stars, but at the same time we are trying to use pulsar timing to detect cosmological gravitational waves. Still, there is only one problem that brings together results from all available observation channels—the stochastic background, which can be constrained across about 29 decades of frequency.

As often is the case, electromagnetic data allow us to rule out the most 'ambitious' gravitational-wave predictions. We have already considered indirect bounds obtained from pulsar timing and CMB measurements. We also gain insight from Big Bang

Nucleosynthesis models. Detailed calculations predict the abundances of light elements in the Universe. The bulk of today's deuterium, the helium isotopes ^{3}He and ^{4}He, and ^{7}Li was created during the primordial nucleosynthesis, the outcome of which was sensitive to the coupling constants of the fundamental interactions and the expansion rate of the Universe (Pagano *et al.*, 2016). Basically, a change in H_0 alters the freeze-out temperature at which nucleosynthesis takes place, affecting the ratio of proton and neutron production and the light-element abundances.

It is relatively straightforward to account for the impact of gravitational waves on the nucleosynthesis. We know that the expansion of the Universe relates to the energy density through the Friedmann equations (see Chapter 4). If the total energy density of gravitational waves, Ω_{gw}, is too large at the time of nucleosynthesis, then the temperature at freeze out would be too high. If we assume that the gravitational-wave spectrum is flat, this leads to the constraint

$$h_{100}^2 \Omega_{gw} \leq 5 \times 10^{-6}. \tag{22.125}$$

It is instructive to compare this limit to the different upper limits we have discussed. We may also add the most recent constraint from ground-based detectors. This is relevant, as we know from the successful detections during the LIGO O1-2 runs that the rate and masses of coalescing binary black holes may be greater than expected. It could perhaps be that the stochastic background from unresolved compact binary coalescences is also surprisingly loud.

As we have seen, one may represent many of the possible sources for a stochastic background in terms of a simple power-law spectrum such that

$$\Omega_{gw} = \Omega_\alpha \left(\frac{f}{f_{ref}} \right)^\alpha, \tag{22.126}$$

where we expect $\alpha \approx 0$ for an inflationary model and $\alpha = 2/3$ for a background of inspiralling binaries. An analysis of the data from the O1 run did not display evidence of a stochastic gravitational-wave signal (Abbott *et al.*, 2017*m*), but the observations improves on previous (direct) constraints on energy density of gravitational waves by about a factor of 30. The new limit corresponds to

$$\Omega_0 < 1.7 \times 10^{-7}, \tag{22.127}$$

with f_{ref} in the most sensitive part of the LIGO band (20–86 Hz). The constraints obtained for a range of values of α are shown Figure 22.15. In the case of binary systems, the dominant contribution to the background comes from the inspiral phase ($\alpha = 2/3$) and it is worth noting that systems beyond a redshift of 5 or so contribute little to the overall integral due to the small number of stars formed at such high redshifts.

The O1 results are summarized in Figure 22.16 and compared to upper limits from the range of observations we have discussed. The comparison highlights why it may be difficult to detected a primordial background with either ground- or space-

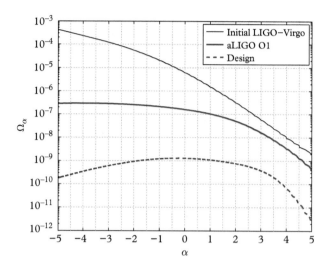

Figure 22.15 *Observational limits for a stochastic background in the $\Omega_\alpha - \alpha$ plane. The region above each curve is excluded at 95% confidence. The different curves represent constraints from the final science run of initial LIGO-Virgo detectors and from the first Advanced LIGO detector run (O1). The projected design sensitivity limit for Advanced LIGO–Virgo is also shown. (Reproduced from Abbott et al. (2017m), copyright (2017) by The American Physical Society.)*

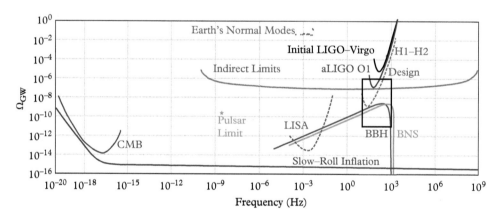

Figure 22.16 *Constraints on the stochastic gravitational-wave background across 29 decades of frequency. The limits from the final science run of initial LIGO–Virgo, the co-located detectors at Hanford (H1–H2), Advanced LIGO (aLIGO) O1, and the projected design sensitivity of the advanced detector network assuming two years of coincident data, are compared to constraints from other observations. These include CMB measurements, Big Bang nucleosynthesis results, pulsar timing, and the ringing of the Earth's normal modes. The figure also shows projected limits from a space-based detector like LISA. In addition to an estimate background level from slow-roll inflation, the estimated contribution from compact binaries (both neutron star and black-hole binaries), extended using the expected $\alpha = 2/3$ power-law to low frequencies, is indicated, with a cut-off imposed where the inspiral timescale is of the order of the Hubble scale. (Reproduced from Abbott et al. (2017m), copyright (2017) by the American Physical Society.)*

based interferometers. The expected signal may simply be overwhelmed by astrophysical counterparts. A stiff equation of state may lead to the primordial spectrum rising fast enough for the signal to become relevant, but it is not clear how seriously this suggestion should be taken. If our main interest is in unveiling the details of the origin of the Universe, this is bad news. However, in terms of detecting stochastic signals in general, there is a glimpse of promise. As the advanced detectors reach design sensitivity, there is a reasonable chance that we will detect the background due to binary black holes. The situation should further improve with third-generation detectors like the Einstein Telescope or the Cosmic Explorer, and the launch of LISA in the 2030s could help unveil the spectral shape of the gravitational-wave background. This would provide insights into the formation of these systems and may also help constrain cosmological parameters. The measurement of tiny spacetime wiggles may not, in the end, help us probe the very first moments of the Universe but we should still learn a lot about its subsequent evolution.

Apologies and thanks

The effort to understand, catch, and explore gravitational waves stretches over more than a century. The short version is simple: 50 years of confusion, 50 years of frustration, and a few years of celebration. The real story is much more complicated, with many individuals making key contributions along the way. It is impossible to credit them all. It is also impossible to include—or give justice to—all the different ideas in such a narrative. There are choices to make and this is far from easy. At the end of the day, many decisions will be subjective. So it is with this book. This is my version of the story, based on my particular experience and interests. I have tried to give credit where it is due, but I am (painfully) aware that I have not managed to include all original work in the reference list. The literature is simply too vast... so I have to apologize to anyone who feels left out.

I should also take the opportunity to apologize for mistakes and misunderstandings that have made their way (sneakily) through the many edits. If I knew where they were I could do something about them, but I don't so I can't. Maybe next time.

In a more positive vein, I want to thank everyone that helped out along the way. Many friends and colleagues have contributed to my understanding (as it is) of gravity over the years. I would like to express my heartfelt gratitude to all of you! I am not going to attempt making a list—you know who you are. I will, however, single out three people without whom this book would certainly not have been written. First of all, Bernard Schutz, who gave me the opportunity to join his group all those years ago and pushed me in all sorts of interesting directions. Secondly, Kostas Kokkotas, who introduced me to many aspect of gravitational-wave physics and continues to be a great collaborator. Finally, Greg Comer, who keeps trying to teach me stuff. This book is for all of you—I am honoured to be your friend.

References

Aasi, J. *et al.* (2014). Gravitational waves from known pulsars: Results from the initial detector era. *Astrophysical Journal*, **785**, 119.

Aasi, J. *et al.* (2015*a*). Characterization of the LIGO detectors during their sixth science run. *Classical and Quantum Gravity*, **32**, 115012.

Aasi, J. *et al.* (2015*b*). Searches for continuous gravitational waves from nine young supernova remnants. *Astrophysical Journal*, **813**, 39.

Abadie, J. *et al.* (2010). First search for gravitational waves from the youngest known neutron star. *Astrophysical Journal*, **722**, 1504.

Abadie, J. *et al.* (2010). Predictions for the rates of compact binary coalescences observable by ground-based gravitational-wave detectors. *Classical and Quantum Gravity*, **27**, 173001.

Abadie, J. *et al.* (2011*a*). Beating the spin-down limit on gravitational wave emission from the Vela Pulsar. *Astrophysical Journal*, **737**, 93.

Abadie, J. *et al.* (2011*b*). Search for gravitational wave bursts from six magnetars. *Astrophysical Journal Letters*, **734**, L35.

Abadie, J. *et al.* (2012). All-sky search for gravitational-wave bursts in the second joint LIGO-Virgo run. *Physical Review D*, **85**, 122007.

Abbott, B. *et al.* (2009*a*). Einstein@Home search for periodic gravitational waves in LIGO S4 data. *Physical Review D*, **79**, 022001.

Abbott, B. P. *et al.* (2007). Upper limits on gravitational-wave emission from 78 radio pulsars. *Physical Review D*, **76**.

Abbott, B. P. *et al.* (2008*a*). Beating the spin-down limit on gravitational wave emission from the Crab Pulsar. *Astrophysical Journal Letters*, **683**, L45.

Abbott, B. P. *et al.* (2008*b*). Search for gravitational-wave bursts from soft gamma repeaters. *Physical Review Letters*, **101**, 211102.

Abbott, B. P. *et al.* (2009*b*). Stacked search for gravitational waves from the 2006 SGR 1900+14 storm. *Astrophysical Journal Letters*, **701**, L68.

Abbott, B. P. *et al.* (2010). Searches for gravitational waves from known pulsars with science run 5 LIGO data. *Astrophysical Journal*, **713**, 671.

Abbott, B. P. *et al.* (2016*a*). Binary black-hole mergers in the first Advanced LIGO observing run. *Physical Review X*, **6**, 041015.

Abbott, B. P. *et al.* (2016*b*). Observation of gravitational waves from a binary black-hole merger. *Physical Review Letters*, **116**, 061102.

Abbott, B. P. *et al.* (2016*c*). Properties of the binary black-hole merger GW150914. *Physical Review Letters*, **116**, 241102.

Abbott, B. P. *et al.* (2016*d*). Tests of general relativity with GW150914. *Physical Review Letters*, **116**, 221101.

Abbott, B. P. *et al.* (2016*e*). The rate of binary black-hole mergers inferred from Advanced LIGO observations surrounding GW150914. *Astrophysical Journal Letters*, **833**, L1.

Abbott, B. P. *et al.* (2017*a*). A gravitational-wave standard siren measurement of the Hubble constant. *Nature*, **551**, 85.

Abbott, B. P. *et al.* (2017*b*). Directional limits on persistent gravitational waves from Advanced LIGO's first observing run. *Physical Review Letters*, **118**, 121102.

Abbott, B. P. *et al.* (2017c). Exploring the sensitivity of next generation gravitational-wave detectors. *Classical and Quantum Gravity*, **34**, 044001.

Abbott, B. P. *et al.* (2017d). First low-frequency Einstein@Home all-sky search for continuous gravitational waves in Advanced LIGO data. *Physical Review D*, **96**, 122004.

Abbott, B. P. *et al.* (2017e). First search for gravitational waves from known pulsars with Advanced LIGO. *Astrophysical Journal*, **839**, 12.

Abbott, B. P. *et al.* (2017f). Gravitational waves and gamma-rays from a binary neutron star merger: GW170817 and GRB 170817A. *Astrophysical Journal Letters*, **848**, L13.

Abbott, B. P. *et al.* (2017g). GW170104: Observation of a 50-solar-mass binary black hole coalescence at redshift 0.2. *Physical Review Letters*, **118**, 221101.

Abbott, B. P. *et al.* (2017h). GW170814: A three-detector observation of gravitational waves from a binary black hole coalescence. *Physical Review Letters*, **119**, 141101.

Abbott, B. P. *et al.* (2017i). GW170817: Observation of gravitational waves from a binary neutron star inspiral. *Physical Review Letters*, **119**, 161101.

Abbott, B. P. *et al.* (2017j). Multi-messenger observations of a binary neutron star merger. *Astrophysical Journal Letters*, **848**, L12.

Abbott, B. P. *et al.* (2017k). Search for gravitational waves from Scorpius X-1 in the first Advanced LIGO observing run with a hidden Markov model. *Physical Review D*, **95**, 122003.

Abbott, B. P. *et al.* (2017l). Upper limits on gravitational waves from Scorpius X-1 from a model-based cross-correlation search in Advanced LIGO data. *Astrophysical Journal*, **847**, 47.

Abbott, B. P. *et al.* (2017m). Upper limits on the stochastic gravitational-wave background from Advanced LIGO's first observing run. *Physical Review Letters*, **118**, 121101.

Abbott, B. P. *et al.* (2018a). Constraints on cosmic strings using data from the first Advanced LIGO observing run. *Physical Review D*, **97**, 102002.

Abbott, B. P. *et al.* (2018b). First search for non-tensorial gravitational waves from known pulsars. *Physical Review Letters*, **120**, 031104.

Abbott, B. P. *et al.* (2018c). Full band all-sky search for periodic gravitational waves in the O1 LIGO data. *Physical Review D*, **97**, 102003.

Abbott, B. P. *et al.* (2018d). GW170817: Measurements of neutron star radii and equation of state. *Physical Review Letters*, **121**, 161101.

Abbott, B. P. *et al.* (2019a). Properties of the binary neutron star merger GW170817. *Physical Review X*, **9**, 011001.

Abbott, B. P. *et al.* (2019b). Searches for continuous gravitational waves from fifteen supernova remnants and Fomalhaut b with Advanced LIGO. *Astrophysical Journal*, **875**, 122.

Abbott, B. P. *et al.* (2019c). GWTC-1: A gravitational-wave transient catalog of compact binary mergers observed by LIGO and Virgo during the first and second observing runs. *Physical Review X*, **9**, 031040.

Abedi, J., Dykaar, H., and Afshordi, N. (2017). Echoes from the abyss: Tentative evidence for Planck-scale structure at black-hole horizons. *Physical Review D*, **96**, 082004.

Abel, T. *et al.* (1998a). First structure formation. I. Primordial star-forming regions in hierarchical models. *Astrophysical Journal*, **508**, 518.

Abel, T. *et al.* (1998b). First structure formation. II. Cosmic string plus hot dark matter models. *Astrophysical Journal*, **508**, 530.

Abramovici, A. *et al.* (1992). Ligo: The laser interferometer gravitational wave observatory. *Science*, **256**, 325.

Abuter, R. *et al.* (2018). Detection of orbital motions near the last stable circular orbit of the massive black hole SgrA★. *Astronomy and Astrophysics*, **618**, L10.

Acernese, F. *et al.* (2015). The Advanced Virgo detector. *Journal of Physics Conference Series*, **610**, 012014.

Acero, F. *et al.* (2015). Fermi Large Area Telescope third source catalog. *Astrophysical Journal Supplement*, **218**, 23.

Adam, R. *et al.* (2016). Planck 2015 results. I. Overview of products and scientific results. *Astronomy and Astrophysics*, **594**, A1.

Ade, P. A. R. *et al.* (2014). Detection of B-mode polarization at degree angular scales by BICEP2. *Physical Review Letters*, **112**, 241101.

Ade, P. A. R. *et al.* (2015). Joint analysis of BICEP2/Keck array and Planck data. *Physical Review Letters*, **114**, 101301.

Ade, P. A. R. *et al.* (2016*a*). Planck 2015 results. XIII. Cosmological parameters. *Astronomy and Astrophysics*, **594**, A13.

Ade, P. A. R. *et al.* (2016*b*). Planck 2015 results. XX. Constraints on inflation. *Astronomy and Astrophysics*, **594**, A20.

Aerts, C., Christensen-Dalsgaard, J., and Kurtz, D. W. (2010). *Asteroseismology*. Springer, Heidelberg.

Agathos, M. *et al.* (2015). Constraining the neutron star equation of state with gravitational wave signals from coalescing binary neutron stars. *Physical Review D*, **92**, 023012.

Akiyama, K. *et al.* (2019). First M87 Event Horizon Telescope results. I. The shadow of the supermassive black hole. *Astrophysical Journal Letters*, **875**, L1.

Akmal, A., Pandharipande, V. R., and Ravenhall, D. G. (1998). Equation of state of nucleon matter and neutron star structure. *Physical Review C*, **58**, 1804.

Akutsu, T. *et al.* (2017). The status of KAGRA underground cryogenic gravitational wave telescope. In *Proceedings for XV International Conference on Topics in Astroparticle and Underground Physics (TAUP2017)*.

Alcock, C., Farhi, E., and Olinto, A. (1986). Strange stars. *Astrophysical Journal*, **310**, 261.

Alcubierre, M. (2003). Hyperbolic slicings of spacetime: Singularity avoidance and gauge shocks. *Classical and Quantum Gravity*, **20**, 607.

Alcubierre, M. (2008). *Introduction to 3+1 numerical relativity*. Oxford University Press, Oxford.

Alcubierre, M. *et al.* (2001). Black hole excision for dynamic black holes. *Physical Review D*, **64**, 061501.

Alcubierre, M. *et al.* (2003). Gauge conditions for long-term numerical black-hole evolutions without excision. *Physical Review D*, **67**, 084023.

Alexander, K. D. *et al.* (2017). The electromagnetic counterpart of the binary neutron star merger LIGO/Virgo GW170817. VI. Radio constraints on a relativistic jet and predictions for late-time emission from the kilonova ejecta. *Astrophysical Journal Letters*, **848**, L21.

Alford, M. *et al.* (2005). Hybrid stars that masquerade as neutron stars. *Astrophysical Journal*, **629**, 969.

Alford, M. G. *et al.* (2018). Viscous dissipation and heat conduction in binary neutron-star mergers. *Physical Review Letters*, **120**, 041101.

Allen, B. (2005). χ^2 time-frequency discriminator for gravitational wave detection. *Physical Review D*, **71**, 062001.

Allen, B. and Shellard, E. P. S. (1992). Gravitational radiation from cosmic strings. *Physical Review D*, **45**, 1898.

Allen, G. *et al.* (1998). Gravitational waves from pulsating stars: Evolving the perturbation equations for a relativistic star. *Physical Review D*, **58**, 124012.

Alpar, M. A. *et al.* (1982). A new class of radio pulsars. *Nature*, **300**, 728.

Alpar, M. A. *et al.* (1984*a*). Vortex creep and the internal temperature of neutron stars. I- General theory. *Astrophysical Journal*, **276**, 325.

Alpar, M. A., Langer, S. A., and Sauls, J. A. (1984*b*). Rapid post-glitch spin-up of the superfluid core in pulsars. *Astrophysical Journal*, **282**, 533.

Amaro-Seoane, P. *et al.* (2013). eLISA: Astrophysics and cosmology in the millihertz regime. *GW Notes*, **6**, 4.

Amaro-Seoane, P. *et al.* (2017). Laser interferometer space antenna. *preprint arXiv:1702.00786.*

Anderson, P. W. and Itoh, N. (1975). Pulsar glitches and restlessness as a hard superfluidity phenomenon. *Nature*, **256**, 25.

Andersson, N. (1992). A numerically accurate investigation of black-hole normal modes. *Proceedings of the Royal Society of London Series A*, **439**, 47.

Andersson, N. (1995). Excitation of Schwarzschild black-hole quasinormal modes. *Physical Review D*, **51**, 353.

Andersson, N. (1997). Evolving test fields in a black-hole geometry. *Physical Review D*, **55**, 468.

Andersson, N. (1998). A new class of unstable modes of rotating relativistic stars. *Astrophysical Journal*, **502**, 708.

Andersson, N. (2003). Gravitational waves from instabilities in relativistic stars. *Classical and Quantum Gravity*, **20**, R105.

Andersson, N. *et al.* (2000). R-mode runaway and rapidly rotating neutron stars. *Astrophysical Journal Letters*, **534**, L75.

Andersson, N. *et al.* (2005*a*). Modelling the spin equilibrium of neutron stars in low-mass X-ray binaries without gravitational radiation. *Monthly Notices of the Royal Astronomical Society*, **361**, 1153.

Andersson, N. *et al.* (2011). Gravitational waves from neutron stars: Promises and challenges. *General Relativity and Gravitation*, **43**, 409.

Andersson, N. *et al.* (2012). Pulsar glitches: the crust is not enough. *Physical Review Letters*, **109**, 241103.

Andersson, N. *et al.* (2017). Beyond ideal magnetohydrodynamics: From fibration to 3+1 foliation. *Classical and Quantum Gravity*, **34**, 125003.

Andersson, N. *et al.* (2018). The enigmatic spin evolution of PSR J0537-6910: r-modes, gravitational waves, and the case for continued timing. *Astrophysical Journal*, **864**, 137.

Andersson, N. and Comer, G. L. (2001*a*). On the dynamics of superfluid neutron star cores. *Monthly Notices of the Royal Astronomical Society*, **328**, 1129.

Andersson, N. and Comer, G. L. (2001*b*). Probing neutron-star superfluidity with gravitational-wave data. *Physical Review Letters*, **87**, 241101.

Andersson, N. and Comer, G. L. (2006). A flux-conservative formalism for convective and dissipative multi-fluid systems, with application to Newtonian superfluid neutron stars. *Classical and Quantum Gravity*, **23**, 5505.

Andersson, N. and Comer, G. L. (2007). Relativistic fluid dynamics: physics for many different scales. *Living Reviews in Relativity*, **10**, 1.

Andersson, N., Comer, G. L., and Glampedakis, K. (2005*b*). How viscous is a superfluid neutron star core? *Nuclear Physics A*, **763**, 212.

Andersson, N. and Ho, W. C. G. (2018). Using gravitational-wave data to constrain dynamical tides in neutron star binaries. *Physical Review D*, **97**, 023016.

Andersson, N., Jones, D.I., and Ho, W. C. G. (2014). Implications of an r-mode in XTE J1751−305: Mass, radius and spin evolution. *Monthly Notices of the Royal Astronomical Society*, **442**, 1786.

Andersson, N., Jones, D. I., and Kokkotas, K. D. (2002). Strange stars as persistent sources of gravitational waves. *Monthly Notices of the Royal Astronomical Society*, **337**, 1224.

Andersson, N., Kojima, Y., and Kokkotas, K. D. (1996*a*). On the oscillation spectra of ultracompact stars: an extensive survey of gravitational-wave modes. *Astrophysical Journal*, **462**, 855.

Andersson, N., Kokkotas, K., and Schutz, B. F. (1999*a*). Gravitational radiation limit on the spin of young neutron stars. *Astrophysical Journal*, **510**, 846.

Andersson, N. and Kokkotas, K. D. (1998). Towards gravitational-wave asteroseismology. *Monthly Notices of the Royal Astronomical Society*, **299**, 1059.

Andersson, N. and Kokkotas, K. D. (2001). The r-mode instability in rotating neutron stars. *International Journal of Modern Physics D*, **10**, 381.

Andersson, N., Kokkotas, K. D., and Schutz, B. F. (1995). A new numerical approach to the oscillation modes of relativistic stars. *Monthly Notices of the Royal Astronomical Society*, **274**, 1039.

Andersson, N., Kokkotas, K. D., and Schutz, B. F. (1996*b*). Space-time modes of relativistic stars. *Monthly Notices of the Royal Astronomical Society*, **280**, 1230.

Andersson, N., Kokkotas, K. D., and Stergioulas, N. (1999*b*). On the relevance of the r-mode instability for accreting neutron stars and white dwarfs. *Astrophysical Journal*, **516**, 307.

Andersson, N. and Pnigouras, P. (2019). The g-mode spectrum of reactive neutron star cores. *preprint arXiv:1905.00010*.

Andersson, N., Sidery, T., and Comer, G. L. (2006). Mutual friction in superfluid neutron stars. *Monthly Notices of the Royal Astronomical Society*, **368**, 162.

Ando, S., Beacom, J. F., and Yüksel, H. (2005). Detection of neutrinos from supernovae in nearby galaxies. *Physical Review Letters*, **95**, 171101.

Anninos, P. *et al.* (1993). Collision of two black holes. *Physical Review Letters*, **71**, 2851.

Anninos, P. *et al.* (1995). Head-on collision of two equal mass black holes. *Physical Review D*, **52**, 2044.

Anninos, P. and Brandt, S. (1998). Head-on collision of two unequal mass black holes. *Physical Review Letters*, **81**, 508.

Ansorg, M., Kleinwächter, A., and Meinel, R. (2002). Highly accurate calculation of rotating neutron stars. *Astronomy and Astrophysics*, **381**, L49.

Antoniadis, J. *et al.* (2013). A massive pulsar in a compact relativistic binary. *Science*, **340**, 448.

Antonopoulou, D. *et al.* (2018). Pulsar spin-down: The glitch-dominated rotation of PSR J0537−6910. *Monthly Notices of the Royal Astronomical Society*, **473**, 1644.

Apreda, R. *et al.* (2002). Gravitational waves from electroweak phase transitions. *Nuclear Physics B*, **631**, 342.

Archibald, A. M. *et al.* (2010). X-ray variability and evidence for pulsations from the unique radio pulsar/X-ray binary transition object FIRST J102347.6+003841. *Astrophysical Journal*, **722**, 88.

Armano, M. *et al.* (2018). Beyond the required LISA free-fall performance: new LISA Pathfinder results down to 20 μHz. *Physical Review Letters*, **120**, 061101.

Armstrong, J. W. (2006). Low-frequency gravitational wave searches using spacecraft Doppler tracking. *Living Reviews in Relativity*, **9**, 1.

Arnowitt, R., Deser, S., and Misner, C. W. (2008). Republication of: the dynamics of general relativity. *General Relativity and Gravitation*, **40**, 1997.

Arras, P. *et al.* (2003). Saturation of the r-mode instability. *Astrophysical Journal*, **591**, 1129.

Arzoumanian, Z. *et al.* (2018). The NANOGrav 11 year data set: pulsar-timing constraints on the stochastic gravitational-wave background. *Astrophysical Journal*, **859**, 47.

Ashby, Neil (2003). Relativity in the global positioning system. *Living Reviews in Relativity*, **6**, 1.

Ashtekar, A. and Krishnan, B. (2004). Isolated and dynamical horizons and their applications. *Living Reviews in Relativity*, **7**, 10.

Ashtekar, A. and Pullin, J. (2017). *Loop quantum gravity: The first 30 years.* World Scientific, Singapore.

Ashton, G., Jones, D.I., and Prix, R. (2016). Comparing models of the periodic variations in spin-down and beamwidth for PSR B1828-11. *Monthly Notices of the Royal Astronomical Society*, **458**, 881.

Ashton, G., Prix, R., and Jones, I. (2018). A semicoherent glitch-robust continuous gravitational-wave search. *Physical Review D*, **98**, 063011.

Baade, W. and Zwicky, F. (1934). On supernovae. *Proceedings of the National Academy of Science*, **20**, 254.

Baggio, L. *et al.* (2005). Upper limits on gravitational-wave emission in association with the 27 Dec 2004 giant flare of SGR1806−20. *Physical Review Letters*, **95**, 081103.

Baibhav, V. *et al.* (2018). Black hole spectroscopy: systematic errors and ringdown energy estimates. *Physical Review D*, **97**, 044048.

Baiotti, L. *et al.* (2007). Accurate simulations of the dynamical bar-mode instability in full general relativity. *Physical Review D*, **75**, 044023.

Baiotti, L. and Rezzolla, L. (2017). Binary neutron star mergers: A review of Einstein's richest laboratory. *Reports on Progress in Physics*, **80**, 096901.

Baker, J. *et al.* (1997). Collision of boosted black holes. *Physical Review D*, **55**, 829.

Baker, J.G. *et al.* (2006a). Getting a kick out of numerical relativity. *Astrophysical Journal Letters*, **653**, L93.

Baker, J. G. *et al.* (2006b). Gravitational-wave extraction from an inspiraling configuration of merging black holes. *Physical Review Letters*, **96**, 111102.

Balbus, S. A. and Hawley, J. F. (1991). A powerful local shear instability in weakly magnetized disks. I. Linear analysis. *Astrophysical Journal*, **376**, 214.

Baldo, M. *et al.* (1992). Proton and neutron superfluidity in neutron star matter. *Nuclear Physics A*, **536**, 349.

Barack, L. (2009). Gravitational self-force in extreme mass-ratio inspirals. *Classical and Quantum Gravity*, **26**, 213001.

Barack, L. *et al.* (2002). Calculating the gravitational self-force in Schwarzschild spacetime. *Physical Review Letters*, **88**, 091101.

Barack, L. and Ori, A. (1999). Late-time decay of scalar perturbations outside rotating black holes. *Physical Review Letters*, **82**, 4388.

Barack, L. and Pound, A. (2019). Self-force and radiation reaction in general relativity. *Reports on Progress in Physics*, **82**, 016904.

Barack, L. and Sago, N. (2009). Gravitational self-force correction to the innermost stable circular orbit of a Schwarzschild black hole. *Physical Review Letters*, **102**, 191101.

Barat, C. *et al.* (1983). Fine time structure in the 1979 March 5 gamma ray burst. *Astronomy and Astrophysics*, **126**, 400.

Bardeen, J. M., Press, W. H., and Teukolsky, S. A. (1972). Rotating black holes: Locally nonrotating frames, energy extraction, and scalar synchrotron radiation. *Astrophysical Journal*, **178**, 347.

Bartos, I., Brady, P., and Márka, S. (2013). How gravitational-wave observations can shape the gamma-ray burst paradigm. *Classical and Quantum Gravity*, **30**, 123001.

Baumgarte, T. W. and Shapiro, S. L. (1999). Numerical integration of Einstein's field equations. *Physical Review D*, **59**, 024007.

Baumgarte, T. W. and Shapiro, S. L. (2010). *Numerical relativity: Solving Einstein's equations on the computer*. Cambridge University Press, New York.

Bauswein, A. and Janka, H.-T. (2012). Measuring neutron-star properties via gravitational waves from neutron-star mergers. *Physical Review Letters*, **108**, 011101.

Bauswein, A., Stergioulas, N., and Janka, H.-T. (2014). Revealing the high-density equation of state through binary neutron star mergers. *Physical Review D*, **90**, 023002.

Bauswein, A., Stergioulas, N., and Janka, H.-T. (2016). Exploring properties of high-density matter through remnants of neutron-star mergers. *European Physical Journal A*, **52**, 56.

Baym, G. *et al.* (1969). Spin up in neutron stars: The future of the Vela Pulsar. *Nature*, **224**, 872.

Baym, G. and Pines, D. (1971). Neutron starquakes and pulsar speedup. *Annals of Physics*, **66**, 816.

Begelman, M. C., Volonteri, M., and Rees, M. J. (2006). Formation of supermassive black holes by direct collapse in pre-galactic haloes. *Monthly Notices of the Royal Astronomical Society*, **370**, 289.

Benhar, O., Ferrari, V., and Gualtieri, L. (2004). Gravitational-wave asteroseismology reexamined. *Physical Review D*, **70**, 124015.

Bernuzzi, S., Dietrich, T., and Nagar, A. (2015). Modeling the complete gravitational wave spectrum of neutron star mergers. *Physical Review Letters*, **115**, 091101.

Berti, E. *et al.* (2015). Testing general relativity with present and future astrophysical observations. *Classical and Quantum Gravity*, **32**.

Berti, E. *et al.* (2016). Spectroscopy of Kerr black holes with Earth- and space-based interferometers. *Physical Review Letters*, **117**, 101102.

Berti, E., Cardoso, V., and Starinets, A. O. (2009). Quasinormal modes of black holes and black branes. *Classical and Quantum Gravity*, **26**, 163001.

Bertone, G., Hooper, D., and Silk, J. (2005). Particle dark matter: Evidence, candidates and constraints. *Physics Reports*, **405**, 279.

Bertotti, B., Iess, L., and Tortora, P. (2003). A test of general relativity using radio links with the Cassini spacecraft. *Nature*, **425**, 374.

Betoule, M. *et al.* (2014). Improved cosmological constraints from a joint analysis of the SDSS-II and SNLS supernova samples. *Astronomy and Astrophysics*, **568**, A22.

Beyer, H. R. (2001). On the stability of the Kerr metric. *Communications in Mathematical Physics*, **221**, 659.

Biesiada, M. *et al.* (2014). Strong gravitational lensing of gravitational waves from double compact binaries—perspectives for the Einstein Telescope. *Journal of Cosmology and Astroparticle Physics*, **10**, 080.

Bildsten, L. (1998). Gravitational radiation and rotation of accreting neutron stars. *Astrophysical Journal Letters*, **501**, L89.

Bildsten, L. and Ushomirsky, G. (2000). Viscous boundary-layer damping of r-modes in neutron stars. *Astrophysical Journal Letters*, **529**, L33.

Binétruy, P. *et al.* (2012). Cosmological backgrounds of gravitational waves and eLISA/NGO: Phase transitions, cosmic strings and other sources. *Journal of Cosmology and Astroparticle Physics*, **6**, 027.

Bishop, N. T. and Rezzolla, L. (2016). Extraction of gravitational waves in numerical relativity. *Living Reviews in Relativity*, **19**, 2.

Blanchet, L. (2006). Gravitational radiation from post-Newtonian sources and inspiralling compact binaries. *Living Reviews in Relativity*, **9**, 4.

Blanchet, L. *et al.* (2010). High-order post-Newtonian fit of the gravitational self-force for circular orbits in the Schwarzschild geometry. *Physical Review D*, **81**, 084033.

Blandford, R. D. and Znajek, R. L. (1977). Electromagnetic extraction of energy from Kerr black holes. *Monthly Notices of the Royal Astronomical Society*, **179**, 433.

Blondin, J. M. and Mezzacappa, A. (2006). The spherical accretion shock instability in the linear regime. *Astrophysical Journal*, **642**, 401.

Bonazzola, S. *et al.* (1993). Axisymmetric rotating relativistic bodies: A new numerical approach for 'exact' solutions. *Astronomy and Astrophysics*, **278**, 421.

Bonazzola, S., Gourgoulhon, E., and Marck, J.-A. (1997). Relativistic formalism to compute quasiequilibrium configurations of nonsynchronized neutron star binaries. *Physical Review D*, **56**, 7740.

Bond, C. *et al.* (2017). Interferometer techniques for gravitational-wave detection. *Living Reviews in Relativity*, **19**, 3.

Bondarescu, R., Teukolsky, S. A., and Wasserman, I. (2007). Spin evolution of accreting neutron stars: Nonlinear development of the r-mode instability. *Physical Review D*, **76**, 064019.

Bondarescu, R., Teukolsky, S. A., and Wasserman, I. (2009). Spinning down newborn neutron stars: Nonlinear development of the r-mode instability. *Physical Review D*, **79**, 104003.

Bondi, H. (1960). Gravitational waves in general relativity. *Nature*, **186**, 535.

Bondi, H., van der Burg, M. G. J., and Metzner, A. W. K. (1962). Gravitational waves in general relativity. VII. Waves from axi-symmetric isolated systems. *Proceedings of the Royal Society of London Series A*, **269**, 21.

Book, L. G. and Flanagan, É. É. (2011). Astrometric effects of a stochastic gravitational wave background. *Physical Review D*, **83**, 024024.

Bose, S. *et al.* (2018). Neutron-star radius from a population of binary neutron star mergers. *Physical Review Letters*, **120**, 031102.

Bowen, J. M. and York, Jr., J. W. (1980). Time-asymmetric initial data for black holes and black-hole collisions. *Physical Review D*, **21**, 2047.

Boyer, R. H. and Lindquist, R. W. (1967). Maximal analytic extension of the Kerr metric. *Journal of Mathematical Physics*, **8**, 265.

Brandt, S. *et al.* (2000). Grazing collisions of black holes via the excision of singularities. *Physical Review Letters*, **85**, 5496.

Brandt, S. and Brügmann, B. (1997). A simple construction of initial data for multiple black holes. *Physical Review Letters*, **78**, 3606.

Brans, C. and Dicke, R. H. (1961). Mach's principle and a relativistic theory of gravitation. *Physical Review*, **124**, 925.

Brault, J. W. (1962). *The gravitational red shift in the solar spectrum*. Ph.D. thesis, Princeton University.

Breu, C. and Rezzolla, L. (2016). Maximum mass, moment of inertia and compactness of relativistic stars. *Monthly Notices of the Royal Astronomical Society*, **459**, 646.

Brill, D. R. and Lindquist, R. W. (1963). Interaction energy in geometrostatics. *Physical Review*, **131**, 471.

Brink, J., Teukolsky, S. A., and Wasserman, I. (2005). Nonlinear coupling network to simulate the development of the r mode instability in neutron stars. II. Dynamics. *Physical Review D*, **71**, 064029.

Brito, R. *et al.* (2017). Gravitational wave searches for ultralight bosons with LIGO and LISA. *Physical Review D*, **96**, 064050.

Brown, E. F. and Bildsten, L. (1998). The ocean and crust of a rapidly accreting neutron star: Implications for magnetic-field evolution and thermonuclear flashes. *Astrophysical Journal*, **496**, 915.

Brügmann, B., Tichy, W., and Jansen, N. (2004). Numerical simulation of orbiting black holes. *Physical Review Letters*, **92**, 211101.

Buonanno, A. *et al.* (2009*a*). Comparison of post-Newtonian templates for compact binary inspiral signals in gravitational-wave detectors. *Physical Review D*, **80**, 084043.

Buonanno, A. *et al.* (2009*b*). Effective-one-body waveforms calibrated to numerical relativity simulations: Coalescence of non-spinning, equal-mass black holes. *Physical Review D*, **79**, 124028.

Buonanno, A. and Damour, T. (1999). Effective-one-body approach to general-relativistic two-body dynamics. *Physical Review D*, **59**, 084006.

Burke, W. L. and Thorne, K. S. (1970). Gravitational radiation damping. In *Relativity* (ed. M. Carmeli, S. I. Fickler, and L. Witten), p. 209.

Burrows, A. *et al.* (2006). A new mechanism for core-collapse supernova explosions. *Astrophysical Journal*, **640**, 878.

Burrows, A. *et al.* (2007). Simulations of magnetically driven supernova and hypernova explosions in the context of rapid rotation. *Astrophysical Journal*, **664**, 416.

Burrows, A. and Lattimer, J. M. (1986). The birth of neutron stars. *Astrophysical Journal*, **307**, 178.

Cacciapuoti, L., Dimarcq, N., and Salomon, C. (2017). The ACES mission: Scientific objectives and present status. In *Society of Photo-Optical Instrumentation Engineers (SPIE) Conference Series*, Volume 10567, p. 105673Y.

Caldwell, R. R., Battye, R. A., and Shellard, E. P. S. (1996). Relic gravitational waves from cosmic strings: Updated constraints and opportunities for detection. *Physical Review D*, **54**, 7146.

Caldwell, R. R. and Friedman, J. L. (1991). Evidence against a strange ground state for baryons. *Physics Letters B*, **264**, 143.

Campanelli, M. *et al.* (2006). Accurate evolutions of orbiting black-hole binaries without excision. *Physical Review Letters*, **96**, 111101.

Caprini, C. *et al.* (2009). General properties of the gravitational-wave spectrum from phase transitions. *Physical Review D*, **79**, 083519.

Caprini, C. and Figueroa, D. G. (2018). Cosmological backgrounds of gravitational waves. *Classical and Quantum Gravity*, **35**, 163001.

Cardoso, V. *et al.* (2016). Gravitational-wave signatures of exotic compact objects and of quantum corrections at the horizon scale. *Physical Review D*, **94**, 084031.

Carroll, S. M. (2004). *Spacetime and geometry. An introduction to general relativity*. Addison Wesley, San Fransisco.

Carter, B. (1968). Hamilton-Jacobi and Schrödinger separable solutions of Einstein's equations. *Communications in Mathematical Physics*, **10**, 280.

Casares, J., Jonker, P. G., and Israelian, G. (2017). X-ray binaries. In *Handbook of Supernovae* (ed. A. Alsabti and P. Murdin).

Centrella, J. M. *et al.* (2001). Dynamical rotational instability at low T/W. *Astrophysical Journal Letters*, **550**, L193.

Chakrabarty, D. *et al.* (2003). Nuclear-powered millisecond pulsars and the maximum spin frequency of neutron stars. *Nature*, **424**, 42.

Chamel, N. (2012). Neutron conduction in the inner crust of a neutron star in the framework of the band theory of solids. *Physical Review C*, **85**, 035801.

Chamel, N. *et al.* (2011). Masses of neutron stars and nuclei. *Physical Review C*, **84**, 062802.

Chamel, N. and Haensel, P. (2008). Physics of neutron star crusts. *Living Reviews in Relativity*, **11**, 10.

Chandrasekhar, S. (1931). The maximum mass of ideal white dwarfs. *Astrophysical Journal*, **74**, 81.

Chandrasekhar, S. (1964). The dynamical instability of gaseous masses approaching the Schwarzschild limit in general relativity. *Astrophysical Journal*, **140**, 417.

Chandrasekhar, S. (1970). The effect of gravitational radiation on the secular stability of the Maclaurin spheroid. *Astrophysical Journal*, **161**, 561.

Chandrasekhar, S. (1973). *Ellipsoidal figures of equilibrium*. Dover publications.

Chandrasekhar, S. (1992). *The mathematical theory of black holes*. Oxford University Press, New York.

Chandrasekhar, S. and Ferrari, V. (1991). On the non-radial oscillations of a star. III. A reconsideration of the axial modes. *Proceedings of the Royal Society of London Series A*, **434**, 449.

Chen, H.-Y. and Holz, D. E. (2013). Gamma-ray-burst beaming and gravitational-wave observations. *Physical Review Letters*, **111**, 181101.

Cheng, K. S. and Dai, Z. G. (1998). Gravitational waves from phase transitions of accreting neutron stars. *Astrophysical Journal*, **492**, 281.

Ching, E. S. C. *et al.* (1995). Wave propagation in gravitational systems: Late-time behavior. *Physical Review D*, **52**, 2118.

Christodoulou, D. (1991). Nonlinear nature of gravitation and gravitational-wave experiments. *Physical Review Letters*, **67**, 1486.

Chrzanowski, P. L. (1975). Vector potential and metric perturbations of a rotating black hole. *Physical Review D*, **11**, 2042.

Chugunov, A. I. and Horowitz, C. J. (2010). Breaking stress of neutron star crust. *Monthly Notices of the Royal Astronomical Society*, **407**, L54.

Clark, C. J. *et al.* (2018). Einstein@Home discovers a radio-quiet gamma-ray millisecond pulsar. *Science Advances*, **4**, eaao7228.

Clark, J. A. *et al.* (2016). Observing gravitational waves from the post-merger phase of binary neutron star coalescence. *Classical and Quantum Gravity*, **33**, 085003.

Colaiuda, A. and Kokkotas, K. D. (2012). Coupled polar-axial magnetar oscillations. *Monthly Notices of the Royal Astronomical Society*, **423**, 811.

Colgate, S. A. and White, R. H. (1966). The hydrodynamic behavior of supernovae explosions. *Astrophysical Journal*, **143**, 626.

Colpi, M. and Sesana, A. (2017). Gravitational-wave sources in the era of multi-band gravitational-wave astronomy. In *An overview of gravitational waves: Theory, sources and detection* (ed. G. Augar and E. Plagnol), pp. 43–140. World Scientific, Singapore.

Comer, G. L., Langlois, D., and Lin, L. M. (1999). Quasinormal modes of general relativistic superfluid neutron stars. *Physical Review D*, **60**, 104025.

Cook, G. B. (2000). Initial data for numerical relativity. *Living Reviews in Relativity*, **3**, 5.

Cook, G. B., Shapiro, S. L., and Teukolsky, S. A. (1994). Rapidly rotating neutron stars in general relativity: Realistic equations of state. *Astrophysical Journal*, **424**, 823.

Cornish, N. and Robson, T. (2017). Galactic binary science with the new LISA design. In *Journal of Physics Conference Series*, Volume 840, p. 012024.

Corsi, A. and Owen, B. J. (2011). Maximum gravitational-wave energy emissible in magnetar flares. *Physical Review D*, **83**, 104014.

Coulter, D. A. *et al.* (2017). Swope supernova survey 2017a (SSS17a), the optical counterpart to a gravitational wave source. *Science*, **358**, 1556.

Cowling, T. G. (1941). The non-radial oscillations of polytropic stars. *Monthly Notices of the Royal Astronomical Society*, **101**, 367.

Cowperthwaite, P. S. *et al.* (2017). The electromagnetic counterpart of the binary neutron star merger LIGO/Virgo GW170817. II. UV, optical, and near-infrared light curves and comparison to kilonova models. *Astrophysical Journal Letters*, **848**, L17.

Creighton, J. and Anderson, W. (2011). *Gravitational-wave physics and astronomy: An introduction to theory, experiment and data analysis*. Wiley, Weinheim.

Cutler, C. (2002). Gravitational waves from neutron stars with large toroidal B fields. *Physical Review D*, **66**, 084025.

Cutler, C. *et al.* (1993). The last three minutes—issues in gravitational-wave measurements of coalescing compact binaries. *Physical Review Letters*, **70**, 2984.

Cutler, C. and Flanagan, E.E. (1994). Gravitational waves from merging compact binaries: How accurately can one extract the binary's parameters from the inspiral waveform? *Physical Review D*, **49**, 2658.

Cutler, C. and Jones, D. I. (2001). Gravitational-wave damping of neutron star wobble. *Physical Review D*, **63**, 024002.

Cutler, C. and Lindblom, L. (1987). The effect of viscosity on neutron star oscillations. *Astrophysical Journal*, **314**, 234.

Cutler, C., Ushomirsky, G., and Link, B. (2003). The crustal rigidity of a neutron star and implications for PSR B1828-11 and other precession candidates. *Astrophysical Journal*, **588**, 975.

Dall'Osso, S., Stella, L., and Palomba, C. (2018). Neutron star bulk viscosity, 'spin-flip' and GW emission of newly born magnetars. *Monthly Notices of the Royal Astronomical Society*, **480**, 1353.

Damour, T. and Deruelle, N. (1981). Radiation reaction and angular momentum loss in small angle gravitational scattering. *Physics Letters A*, **87**, 81.

Damour, T., Iyer, B. R., and Sathyaprakash, B. S. (2001). Comparison of search templates for gravitational waves from binary inspiral. *Physical Review D*, **63**, 044023.

Damour, T. and Nagar, A. (2016). The effective-one-body approach to the general relativistic two-body problem. In *Lecture Notes in Physics* (ed. F. Haardt, V. Gorini, U. Moschella, A. Treves, and M. Colpi), Volume 905, p. 273. Springer, Berlin.

Damour, T. and Taylor, J. H. (1991). On the orbital period change of the binary pulsar PSR 1913+16. *Astrophysical Journal*, **366**, 501.

Damour, T. and Vilenkin, A. (2000). Gravitational-wave bursts from cosmic strings. *Physical Review Letters*, **85**, 3761.

Damour, T. and Vilenkin, A. (2005). Gravitational radiation from cosmic (super)strings: Bursts, stochastic background, and observational windows. *Physical Review D*, **71**, 063510.

Davis, M. *et al.* (1971). Gravitational radiation from a particle falling radially into a Schwarzschild black hole. *Physical Review Letters*, **27**, 1466.

Del Pozzo, W., Veitch, J., and Vecchio, A. (2011). Testing general relativity using Bayesian model selection: Applications to observations of gravitational waves from compact binary systems. *Physical Review D*, **83**, 082002.

Demorest, P.B. *et al.* (2010). A two-solar-mass neutron star measured using Shapiro delay. *Nature*, **467**, 1081.

Detweiler, S. (1979). Pulsar timing measurements and the search for gravitational waves. *Astrophysical Journal*, **234**, 1100.

Detweiler, S. and Lindblom, L. (1985). On the nonradial pulsations of general relativistic stellar models. *Astrophysical Journal*, **292**, 12.

Detweiler, S. and Whiting, B. F. (2003). Self-force via a Green's function decomposition. *Physical Review D*, **67**, 024025.

Detweiler, S. L. (1975). A variational calculation of the fundamental frequencies of quadrupole pulsation of fluid spheres in general relativity. *Astrophysical Journal*, **197**, 203.

Dhanpal, S. *et al.* (2018). A 'no-hair' test for binary black holes. *Physical Review D*, **99**, 104056.

Dietrich, T. and Bernuzzi, S. (2015). Simulations of rotating neutron star collapse with the puncture gauge: End state and gravitational waveforms. *Physical Review D*, **91**, 044039.

Dimmelmeier, H. *et al.* (2008). Gravitational-wave burst signal from core collapse of rotating stars. *Physical Review D*, **78**, 064056.

Dionysopoulou, K. *et al.* (2013). General-relativistic resistive magnetohydrodynamics in three dimensions: Formulation and tests. *Physical Review D*, **88**, 044020.

Dionysopoulou, K., Alic, D., and Rezzolla, L. (2015). General-relativistic resistive-magnetohydrodynamic simulations of binary neutron stars. *Physical Review D*, **92**, 084064.

Dodson, R. G., McCulloch, P. M., and Lewis, D. R. (2002). High time resolution observations of the January 2000 glitch in the Vela Pulsar. *Astrophysical Journal Letters*, **564**, L85.

Doneva, D.D., Kokkotas, K.D., and Pnigouras, P. (2015). Gravitational wave afterglow in binary neutron star mergers. *Physical Review D*, **92**, 104040.

Doneva, D. D. *et al.* (2013). Gravitational-wave asteroseismology of fast rotating neutron stars with realistic equations of state. *Physical Review D*, **88**, 044052.

Douchin, F., Haensel, P., and Meyer, J. (2000). Nuclear surface and curvature properties for SLy Skyrme forces and nuclei in the inner neutron-star crust. *Nuclear Physics A*, **665**, 419.

Duncan, R. C. (1998). Global seismic oscillations in soft gamma repeaters. *Astrophysical Journal Letters*, **498**, L45.

Dyson, F. W., Eddington, A. S., and Davidson, C. (1920). A Determination of the deflection of light by the Sun's gravitational field, from observations made at the total eclipse of May 29, 1919. *Philosophical Transactions of the Royal Society of London Series A*, **220**, 291.

Echeverria, F. (1989). Gravitational-wave measurements of the mass and angular momentum of a black hole. *Physical Review D*, **40**, 3194.

Einstein, A. (1915). Die feldgleichungen der gravitation. *Sitzungsberichte der Königlich Preussischen Akademie der Wissenschaften (Berlin)*, 844.

Einstein, A. (1916). Näherungsweise integration der feldgleichungen der gravitation. *Sitzungsberichte der Königlich Preussischen Akademie der Wissenschaften (Berlin)*, 688.

Ellis, G. F. R. and van Elst, H. (1999). Cosmological Models (Cargèse lectures 1998). In *NATO Advanced Science Institutes (ASI) Series C*, Volume 541, NATO Advanced Science Institutes (ASI) Series C.

Elvis, M. (2000). A structure for quasars. *Astrophysical Journal*, **545**, 63.

Elvis, M. *et al.* (1975). Discovery of powerful transient X-ray source A0620-00 with Ariel V Sky Survey Experiment. *Nature*, **257**, 656.

Epelbaum, E., Hammer, H.-W., and Meissner, U.-G. (2009). Modern theory of nuclear forces. *Reviews of Modern Physics*, **81**, 1773.

Epstein, R. I. and Baym, G. (1992). Vortex drag and the spin-up time scale for pulsar glitches. *Astrophysical Journal*, **387**, 276.

Espinoza, C. M. *et al.* (2011a). A study of 315 glitches in the rotation of 102 pulsars. *Monthly Notices of the Royal Astronomical Society*, **414**, 1679.

Espinoza, C. M. *et al.* (2011*b*). The braking index of PSR J1734-3333 and the magnetar population. *Astrophysical Journal Letters*, **741**, L13.

Etienne, Z. B. *et al.* (2008). Fully general relativistic simulations of black hole-neutron star mergers. *Physical Review D*, **77**, 084002.

Evans, P. A. *et al.* (2017). Swift and NuSTAR observations of GW170817: Detection of a blue kilonova. *Science*, **358**, 1565.

Everitt, C.W.F. *et al.* (2011). Gravity Probe B: Final results of a space experiment to test general relativity. *Physical Review Letters*, **106**, 221101.

Fabian, A. C. (1975). UHURU—the first X-ray astronomy satellite. *Journal of the British Interplanetary Society*, **28**, 343.

Fafone, V. (2004). Resonant-mass detectors: Status and perspectives. *Classical and Quantum Gravity*, **21**, S377.

Faller, J. E. *et al.* (1985). Space antenna for gravitational-wave astronomy. In *Kilometric optical arrays in space* (ed. N. Longdon and O. Melita), Volume 226, ESA Special Publication.

Faulkner, J. (1971). Ultrashort-period binaries, gravitational radiation, and mass transfer. I. The standard model, with applications to WZ Sagittae and Z Camelopardalis. *Astrophysical Journal Letters*, **170**, L99.

Favata, M. (2009). Nonlinear gravitational-wave memory from binary black hole mergers. *Astrophysical Journal Letters*, **696**, L159.

Favata, M. (2011). Conservative self-force correction to the innermost stable circular orbit: Comparison with multiple post-Newtonian-based methods. *Physical Review D*, **83**, 024027.

Fernández, R. and Metzger, B. D. (2016). Electromagnetic signatures of neutron star mergers in the Advanced LIGO era. *Annual Review of Nuclear and Particle Science*, **66**, 23.

Ferrari, V., Miniutti, G., and Pons, J.A. (2003). Gravitational waves from newly born, hot neutron stars. *Monthly Notices of the Royal Astronomical Society*, **342**, 629.

Finn, L. S. (1988). Relativistic stellar pulsations in the Cowling approximation. *Monthly Notices of the Royal Astronomical Society*, **232**, 259.

Finn, L. S. (1992). Detection, measurement, and gravitational radiation. *Physical Review D*, **46**, 5236.

Flanagan, É. É. and Hinderer, T. (2008). Constraining neutron-star tidal Love numbers with gravitational-wave detectors. *Physical Review D*, **77**, 021502.

Flanagan, É. É. and Racine, É. (2007). Gravitomagnetic resonant excitation of Rossby modes in coalescing neutron-star binaries. *Physical Review D*, **75**, 044001.

Flowers, E. and Itoh, N. (1976). Transport properties of dense matter. *Astrophysical Journal*, **206**, 218.

Flowers, E. and Itoh, N. (1979). Transport properties of dense matter. II. *Astrophysical Journal*, **230**, 847.

Fonseca, E. *et al.* (2016). The NANOGrav nine-year data set: Mass and geometric measurements of binary millisecond pulsars. *Astrophysical Journal*, **832**, 167.

Font, J. A. (2008). Numerical hydrodynamics and magnetohydrodynamics in general relativity. *Living Reviews in Relativity*, **11**, 7.

Font, J. A., Stergioulas, N., and Kokkotas, K. D. (2000). Non-linear hydrodynamical evolution of rotating relativistic stars: Numerical methods and code tests. *Monthly Notices of the Royal Astronomical Society*, **313**, 678.

Fowler, R. H. (1926). On dense matter. *Monthly Notices of the Royal Astronomical Society*, **87**, 114.

Friedman, J. L. *et al.* (2017). Limits on magnetic field amplification from the r-mode instability. *Physical Review D*, **96**, 124008.

Friedman, J. L., Ipser, J. R., and Parker, L. (1989). Implications of a half-millisecond pulsar. *Physical Review Letters*, **62**, 3015.

Friedman, J. L. and Morsink, S. M. (1998). Axial instability of rotating relativistic stars. *Astrophysical Journal*, **502**, 714.

Friedman, J. L. and Schutz, B. F. (1978*a*). Lagrangian perturbation theory of nonrelativistic fluids. *Astrophysical Journal*, **221**, 937.

Friedman, J. L. and Schutz, B. F. (1978*b*). Secular instability of rotating Newtonian stars. *Astrophysical Journal*, **222**, 281.

Friedman, J. L. and Stergioulas, N. (2013). *Rotating relativistic stars*. Cambridge University Press, Cambridge.

Frieman, J. A., Turner, M. S., and Huterer, D. (2008). Dark energy and the accelerating Universe. *Annual Review Astronomy and Astrophysics*, **46**, 385.

Fryer, C. L. and New, K. C. B. (2011). Gravitational waves from gravitational collapse. *Living Reviews in Relativity*, **14**, 1.

Fujibayashi, S. *et al.* (2018). Mass ejection from the remnant of a binary neutron star merger: Viscous-radiation hydrodynamics study. *Astrophysical Journal*, **860**, 64.

Gabler, M. *et al.* (2012). Magnetoelastic oscillations of neutron stars with dipolar magnetic fields. *Monthly Notices of the Royal Astronomical Society*, **421**, 2054.

Gaertig, E. *et al.* (2011). f-mode instability in relativistic neutron stars. *Physical Review Letters*, **107**, 101102.

Gaertig, E. and Kokkotas, K.D. (2011). Gravitational-wave asteroseismology with fast rotating neutron stars. *Physical Review D*, **83**, 064031.

Gaertig, E. and Kokkotas, K. D. (2008). Oscillations of rapidly rotating relativistic stars. *Physical Review D*, **78**, 064063.

Gaertig, E. and Kokkotas, K. D. (2009). Relativistic g-modes in rapidly rotating neutron stars. *Physical Review D*, **80**, 064026.

Galloway, D. K. *et al.* (2005). Discovery of the accretion-powered millisecond X-ray pulsar IGR J00291+5934. *Astrophysical Journal Letters*, **622**, L45.

Gamow, G. (1939). Physical possibilities of stellar evolution. *Physical Review*, **55**, 718.

Gandolfi, S., Carlson, J., and Reddy, S. (2012). Maximum mass and radius of neutron stars, and the nuclear symmetry energy. *Physical Review C*, **85**, 032801.

Garfinkle, D. (2016). A simple estimate of gravitational-wave memory in binary black-hole systems. *Classical and Quantum Gravity*, **33**, 177001.

Gehrels, N. and Mészáros, P. (2012). Gamma-ray bursts. *Science*, **337**, 932.

Gendreau, K. C. *et al.* (2016). The Neutron star Interior Composition Explorer (NICER): design and development. In *Space telescopes and instrumentation 2016: Ultraviolet to gamma ray*, Volume 9905, Proc. SPIE, p. 99051H.

Geppert, U., Page, D., and Zannias, T. (1999). Submergence and re-diffusion of the neutron-star magnetic field after the supernova. *Astronomy and Astrophysics*, **345**, 847.

Gerlach, U. H. and Sengupta, U. K. (1979). Gauge-invariant perturbations on most general spherically symmetric space-times. *Physical Review D*, **19**, 2268.

Ghosh, P. and Lamb, F. K. (1978). Disk accretion by magnetic neutron stars. *Astrophysical Journal Letters*, **223**, L83.

Ghosh, P. and Lamb, F. K. (1979). Accretion by rotating magnetic neutron stars. III. Accretion torques and period changes in pulsating X-ray sources. *Astrophysical Journal*, **234**, 296.

Gillessen, S. *et al.* (2009). Monitoring stellar orbits around the massive black hole in the Galactic center. *Astrophysical Journal*, **692**, 1075.

Glampedakis, K. and Andersson, N. (2001). Late-time dynamics of rapidly rotating black holes. *Physical Review D*, **64**, 104021.

Glampedakis, K. and Andersson, N. (2003). 'Quick and dirty' methods for studying black-hole resonances. *Classical and Quantum Gravity*, **20**, 3441.

Glampedakis, K., Andersson, N., and Samuelsson, L. (2011). Magnetohydrodynamics of superfluid and superconducting neutron star cores. *Monthly Notices of the Royal Astronomical Society*, **410**, 805.

Glampedakis, K. and Kennefick, D. (2002). Zoom and whirl: Eccentric equatorial orbits around spinning black holes and their evolution under gravitational radiation reaction. *Physical Review D*, **66**, 044002.

Glampedakis, K., Lander, S.K., and Andersson, N. (2014). The inside-out view on neutron-star magnetospheres. *Monthly Notices of the Royal Astronomical Society*, **437**, 2.

Glampedakis, K. and Lasky, P. D. (2016). The freedom to choose neutron star magnetic field equilibria. *Monthly Notices of the Royal Astronomical Society*, **463**, 2542.

Glendenning, N. K. (1996). *Compact stars*. Springer, New York.

Gold, T. (1969). Rotating neutron stars and the nature of pulsars. *Nature*, **221**, 25.

Goldstein, A. *et al.* (2017). An ordinary short gamma-ray burst with extraordinary implications: Fermi-GBM detection of GRB 170817A. *Astrophysical Journal Letters*, **848**, L14.

González, J. A. *et al.* (2007). Maximum kick from nonspinning black-hole binary inspiral. *Physical Review Letters*, **98**, 091101.

Gott, III, J. R. and Colley, W. N. (2017). Reanalysis of the BICEP2, Keck and Planck data: No evidence for gravitational radiation. *preprint arXiv:1707.06755*.

Graber, V., Andersson, N., and Hogg, M. (2017). Neutron stars in the laboratory. *International Journal of Modern Physics D*, **26**, 1730015–347.

Green, M. B., Schwarz, J. H., and Witten, E. (1987). *Superstring theory. Volume 1. Introduction.* Cambridge University Press, Cambridge and New York.

Grishchuk, L. P. (1975). Amplification of gravitational waves in an isotropic universe. *Soviet Journal of Experimental and Theoretical Physics*, **40**, 409.

Grishchuk, L. P. (1993). Quantum effects in cosmology. *Classical and Quantum Gravity*, **10**, 2449.

Gudmundsson, E. H., Pethick, C. J., and Epstein, R. I. (1983). Structure of neutron star envelopes. *Astrophysical Journal*, **272**, 286.

Gundlach, C. and Martín-García, J. M. (2000). Gauge-invariant and coordinate-independent perturbations of stellar collapse: The interior. *Physical Review D*, **61**, 084024.

Gundlach, C. and Martín-García, J. M. (2006). Well-posedness of formulations of the Einstein equations with dynamical lapse and shift conditions. *Physical Review D*, **74**, 024016.

Gundlach, C., Price, R. H., and Pullin, J. (1994). Late-time behavior of stellar collapse and explosions. I. Linearized perturbations. *Physical Review D*, **49**, 883.

Gusakov, M. E., Chugunov, A. I., and Kantor, E. M. (2014). Explaining observations of rapidly rotating neutron stars in low-mass X-ray binaries. *Physical Review D*, **90**, 063001.

Gusakov, M. E. and Kantor, E. M. (2013). Thermal g-modes and unexpected convection in superfluid neutron stars. *Physical Review D*, **88**, 101302.

Haensel, P. (1997). Solid interiors of neutron stars and gravitational radiation. In *Relativistic Gravitation and Gravitational Radiation* (ed. J.-A. Marck and J.-P. Lasota), p. 129.

Haensel, P., Salgado, M., and Bonazzola, S. (1995). Equation of state of dense matter and maximum rotation frequency of neutron stars. *Astronomy and Astrophysics*, **296**, 745.

Haensel, P. and Zdunik, J. L. (1989). A submillisecond pulsar and the equation of state of dense matter. *Nature*, **340**, 617.

Haensel, P. and Zdunik, J. L. (2003). Nuclear composition and heating in accreting neutron-star crusts. *Astronomy and Astrophysics*, **404**, L33.

Hafele, J. C. and Keating, R. E. (1972). Around-the-world atomic clocks: Observed relativistic time gains. *Science*, **177**, 168.

Hall, Evan D and Evans, Matthew (2019). Metrics for next-generation gravitational-wave detectors. *preprint arXiv:1902.09485*.

Hall, H. E. and Vinen, W. F. (1956). The rotation of liquid helium II. II. The theory of mutual friction in uniformly rotating helium II. *Proceedings of the Royal Society of London Series A*, **238**, 215.

Harrison, B. K. *et al.* (1965). *Gravitation theory and gravitational collapse*. University of Chicago Press, Chicago.

Hartle, J. B. (2003). *Gravity: An introduction to Einstein's general relativity*. Addison Wesley, San Francisco.

Hartle, J. B. and Thorne, K. S. (1968). Slowly rotating relativistic stars. II. Models for neutron stars and supermassive stars. *Astrophysical Journal*, **153**, 807.

Hartmann, D. H. and MacFadyen, A. I. (2000). Hypernovae, collapsars, and gamma-ray bursts. *Nuclear Physics B Proceedings Supplements*, **80**, 135.

Haskell, B. *et al.* (2008). Modelling magnetically deformed neutron stars. *Monthly Notices of the Royal Astronomical Society*, **385**, 531.

Haskell, B. *et al.* (2015). Detecting gravitational waves from mountains on neutron stars in the advanced detector era. *Monthly Notices of the Royal Astronomical Society*, **450**, 2393.

Haskell, B. and Andersson, N. (2010). Superfluid hyperon bulk viscosity and the r-mode instability of rotating neutron stars. *Monthly Notices of the Royal Astronomical Society*, **408**, 1897.

Haskell, B., Andersson, N., and Comer, G. L. (2012). Dynamics of dissipative multifluid neutron star cores. *Physical Review D*, **86**, 063002.

Haskell, B., Andersson, N., and Passamonti, A. (2009). r modes and mutual friction in rapidly rotating superfluid neutron stars. *Monthly Notices of the Royal Astronomical Society*, **397**, 1464.

Haskell, B., Glampedakis, K., and Andersson, N. (2014). A new mechanism for saturating unstable r modes in neutron stars. *Monthly Notices of the Royal Astronomical Society*, **441**, 1662.

Haskell, B., Jones, D. I., and Andersson, N. (2006). Mountains on neutron stars: Accreted versus non-accreted crusts. *Monthly Notices of the Royal Astronomical Society*, **373**, 1423.

Haskell, B. and Patruno, A. (2017). Are gravitational waves spinning down PSR J1023+0038? *Physical Review Letters*, **119**, 161103.

Hawking, S. W. and Israel, W. (1989). *Three hundred years of gravitation*. Cambridge University Press, Cambridge.

Hebeler, K. *et al.* (2010). Constraints on neutron star radii based on chiral effective field theory interactions. *Physical Review Letters*, **105**, 161102.

Hebeler, K. *et al.* (2013). Equation of state and neutron star properties constrained by nuclear physics and observation. *Astrophysical Journal*, **773**, 11.

Heger, A. and Woosley, S. E. (2002). The nucleosynthetic signature of population III. *Astrophysical Journal*, **567**, 532.

Heger, A., Woosley, S. E., and Spruit, H. C. (2005). Presupernova evolution of differentially rotating massive stars including magnetic fields. *Astrophysical Journal*, **626**, 350.

Hellings, R. W. and Downs, G. S. (1983). Upper limits on the isotropic gravitational radiation background from pulsar timing analysis. *Astrophysical Journal Letters*, **265**, L39.

Hempel, M. and Schaffner-Bielich, J. (2010). A statistical model for a complete supernova equation of state. *Nuclear Physics A*, **837**, 210.

Hessels, J. W. T. *et al.* (2006). A radio pulsar spinning at 716 Hz. *Science*, **311**, 1901.

Hewish, A. *et al.* (1968). Observation of a rapidly pulsating radio source. *Nature*, **217**, 709.

Heyl, J. S. (2002). Low-mass X-ray binaries may be important Laser Interferometer Gravitational-wave Observatory sources after all. *Astrophysical Journal Letters*, **574**, L57.

Hinderer, T. (2008). Tidal Love numbers of neutron stars. *Astrophysical Journal*, **677**, 1216.

Hinderer, T. *et al.* (2010). Tidal deformability of neutron stars with realistic equations of state and their gravitational-wave signatures in binary inspiral. *Physical Review D*, **81**, 123016.

Hinderer, T. *et al.* (2016). Effects of neutron-star dynamic tides on gravitational waveforms within the effective-one-body approach. *Physical Review Letters*, **116**, 181101.

Hindmarsh, M. B. and Kibble, T. W. B. (1995). Cosmic strings. *Reports on Progress in Physics*, **58**, 477.

Hjorth, J. *et al.* (2017). The distance to NGC 4993: The host galaxy of the gravitational-wave event GW170817. *Astrophysical Journal Letters*, **848**, L31.

Ho, W. C. G. (2015). Magnetic field growth in young glitching pulsars with a braking index. *Monthly Notices of the Royal Astronomical Society*, **452**, 845.

Ho, W. C. G. (2016). Gravitational waves within the magnetar model of superluminous supernovae and gamma-ray bursts. *Monthly Notices of the Royal Astronomical Society*, **463**, 489.

Ho, W. C. G. *et al.* (2014). Equilibrium spin pulsars unite neutron star populations. *Monthly Notices of the Royal Astronomical Society*, **437**, 3664.

Ho, W. C. G., Andersson, N., and Haskell, B. (2011*a*). Revealing the physics of r modes in low-mass X-Ray binaries. *Physical Review Letters*, **107**, 101101.

Ho, W. C. G., Glampedakis, K., and Andersson, N. (2012). Magnetars: super(ficially) hot and super(fluid) cool. *Monthly Notices of the Royal Astronomical Society*, **422**, 2632.

Ho, W. C. G. and Lai, D. (1999). Resonant tidal excitations of rotating neutron stars in coalescing binaries. *Monthly Notices of the Royal Astronomical Society*, **308**, 153.

Ho, W. C. G. and Lai, D. (2000). r-mode oscillations and spin-down of young rotating magnetic neutron stars. *Astrophysical Journal*, **543**, 386.

Ho, W. C. G., Maccarone, T.J., and Andersson, N. (2011*b*). Cosmic recycling of millisecond pulsars. *Astrophysical Journal Letters*, **730**, L36.

Hobbs, G. *et al.* (2010*a*). The International Pulsar Timing Array project: Using pulsars as a gravitational-wave detector. *Classical and Quantum Gravity*, **27**, 084013.

Hobbs, G., Lyne, A. G., and Kramer, M. (2010*b*). An analysis of the timing irregularities for 366 pulsars. *Monthly Notices of the Royal Astronomical Society*, **402**, 1027.

Holz, D. E. and Hughes, S. A. (2005). Using gravitational-wave standard sirens. *Astrophysical Journal*, **629**, 15.

Horowitz, C. J. and Hughto, J. (2008). Molecular dynamics simulation of shear moduli for Coulomb crystals. *preprint arXiv:0812.2650*.

Horowitz, C. J. and Kadau, K. (2009). Breaking strain of neutron star crust and gravitational waves. *Physical Review Letters*, **102**, 191102.

Hosokawa, T. *et al.* (2013). Formation of primordial supermassive stars by rapid mass accretion. *Astrophysical Journal*, **778**, 178.

Hotokezaka, K. *et al.* (2019). A Hubble constant measurement from superluminal motion of the jet in GW170817. *Nature Astronomy*.

Hubble, E. (1929). A relation between distance and radial velocity among extra-galactic nebulae. *Proceedings of the National Academy of Science*, **15**, 168.

Hulse, R. A. and Taylor, J. H. (1975). Discovery of a pulsar in a binary system. *Astrophysical Journal Letters*, **195**, L51.

Idrisy, A., Owen, B. J., and Jones, D. I. (2015). R-mode frequencies of slowly rotating relativistic neutron stars with realistic equations of state. *Physical Review D*, **91**, 024001.

Iida, K., Watanabe, G., and Sato, K. (2001). Formation of nuclear 'pasta' in cold neutron star matter. *Progress of Theoretical Physics*, **106**, 551.

Ioka, K. (2001). Magnetic deformation of magnetars for the giant flares of the soft gamma-ray repeaters. *Monthly Notices of the Royal Astronomical Society*, **327**, 639.

Ipser, J. R. and Lindblom, L. (1991). The oscillations of rapidly rotating Newtonian stellar models. II. Dissipative effects. *Astrophysical Journal*, **373**, 213.

Isaacson, R. A. (1968*a*). Gravitational radiation in the limit of high frequency. I. The linear approximation and geometrical optics. *Physical Review*, **166**, 1263.

Isaacson, R. A. (1968*b*). Gravitational radiation in the limit of high frequency. II. Nonlinear terms and the effective stress tensor. *Physical Review*, **166**, 1272.

Israel, G. L. *et al.* (2005). The discovery of rapid X-ray oscillations in the tail of the SGR 1806−20 hyperflare. *Astrophysical Journal Letters*, **628**, L53.

Iyer, B. R. and Will, C. M. (1995). Post-Newtonian gravitational radiation reaction for two-body systems: Nonspinning bodies. *Physical Review D*, **52**, 6882.

Jani, K. *et al.* (2016). Georgia tech catalog of gravitational waveforms. *Classical and Quantum Gravity*, **33**, 204001.

Janka, H.-T. (2012). Explosion mechanisms of core-collapse supernovae. *Annual Review of Nuclear and Particle Science*, **62**, 407.

Janka, H.-T., Melson, T., and Summa, A. (2016). Physics of core-collapse supernovae in three dimensions: a sneak preview. *Annual Review of Nuclear and Particle Science*, **66**, 341.

Jaranowski, P., Królak, A., and Schutz, B. F. (1998). Data analysis of gravitational-wave signals from spinning neutron stars: The signal and its detection. *Physical Review D*, **58**, 063001.

Jaranowski, P. and Schäfer, G. (1997). Radiative 3.5 post-Newtonian ADM Hamiltonian for many-body point-mass systems. *Physical Review D*, **55**, 4712.

Johnson-McDaniel, N. K. and Owen, B. J. (2013). Maximum elastic deformations of relativistic stars. *Physical Review D*, **88**, 044004.

Jones, D. I. (2010). Gravitational-wave emission from rotating superfluid neutron stars. *Monthly Notices of the Royal Astronomical Society*, **402**, 2503.

Jones, D. I. and Andersson, N. (2001). Freely precessing neutron stars: Model and observations. *Monthly Notices of the Royal Astronomical Society*, **324**, 811.

Jones, D. I. and Andersson, N. (2002). Gravitational waves from freely precessing neutron stars. *Monthly Notices of the Royal Astronomical Society*, **331**, 203.

Jones, P. B. (1976). Orientation of pulsar magnetic dipole moments. *Nature*, **262**, 120.

Kalogera, V. *et al.* (2004). The cosmic coalescence rates for double neutron star binaries. *Astrophysical Journal Letters*, **601**, L179.

Kamionkowski, M., Kosowsky, A., and Stebbins, A. (1997). A probe of primordial gravity waves and vorticity. *Physical Review Letters*, **78**, 2058.

Kasen, D. and Bildsten, L. (2010). Supernova light curves powered by young magnetars. *Astrophysical Journal*, **717**, 245.

Kasliwal, M. M. *et al.* (2017). Illuminating gravitational waves: A concordant picture of photons from a neutron star merger. *Science*, **358**, 1559.

Kaspi, V. M. and Beloborodov, A. M. (2017). Magnetars. *Annual Review of Astronomy and Astrophysics*, **55**, 261.

Kay, B. S. and Wald, R. M. (1987). Linear stability of Schwarzschild under perturbations which are non-vanishing on the bifurcation 2-sphere. *Classical and Quantum Gravity*, **4**, 893.

Kennefick, D. (2007). *Traveling at the speed of thought. Einstein and the Quest for Gravitational Waves.* Princeton University Press, Princeton, New Jersey.

Kerr, R. P. (1963). Gravitational field of a spinning mass as an example of algebraically special metrics. *Physical Review Letters*, **11**, 237.

Kidder, L. E. (1995). Coalescing binary systems of compact objects to (post)$^{5/2}$-Newtonian order. V. Spin effects. *Physical Review D*, **52**, 821.

Kidder, L. E., Will, C. M., and Wiseman, A. G. (1993*a*). Coalescing binary systems of compact objects to (post)$^{5/2}$-Newtonian order. III. Transition from inspiral to plunge. *Physical Review D*, **47**, 3281.

Kidder, L. E., Will, C. M., and Wiseman, A. G. (1993*b*). Spin effects in the inspiral of coalescing compact binaries. *Physical Review D*, **47**, R4183.

Kilic, M. *et al.* (2018). Gaia reveals evidence for merged white dwarfs. *Monthly Notices of the Royal Astronomical Society*, **479**, L113.

Kinnersley, W. (1969). Type D vacuum metrics. *Journal of Mathematical Physics*, **10**, 1195.

Kiuchi, K. *et al.* (2015). High resolution magnetohydrodynamic simulation of black hole-neutron star merger: Mass ejection and short gamma ray bursts. *Physical Review D*, **92**, 064034.

Kiuchi, K. others (2018). Global simulations of strongly magnetized remnant massive neutron stars formed in binary neutron star mergers. *Physical Review D*, **97**, 124039.

Kobayashi, S. *et al.* (2004). Gravitational waves and X-ray signals from stellar disruption by a massive black hole. *Astrophysical Journal*, **615**, 855.

Kojima, Y. (1992). Equations governing the nonradial oscillations of a slowly rotating relativistic star. *Physical Review D*, **46**, 4289.

Kojima, Y. and Nakamura, T. (1983). Gravitational radiation from a particle with zero orbital angular momentum plunging into a Kerr black hole. *Physics Letters A*, **96**, 335.

Kojima, Y. and Nakamura, T. (1984). Gravitational radiation from a particle scattered by a Kerr black hole. *Progress of Theoretical Physics*, **72**, 494.

Kokkotas, K. D., Apostolatos, T. A., and Andersson, N. (2001). The inverse problem for pulsating neutron stars: a 'fingerprint analysis' for the supranuclear equation of state. *Monthly Notices of the Royal Astronomical Society*, **320**, 307.

Kokkotas, K. D. and Schafer, G. (1995). Tidal and tidal-resonant effects in coalescing binaries. *Monthly Notices of the Royal Astronomical Society*, **275**, 301.

Kokkotas, K. D. and Schmidt, B. G. (1999). Quasi-normal modes of stars and black holes. *Living Reviews in Relativity*, **2**, 2.

Kokkotas, K. D. and Schutz, B. F. (1992). W-modes. A new family of normal modes of pulsating relativistic stars. *Monthly Notices of the Royal Astronomical Society*, **255**, 119.

Kopparapu, R. K. *et al.* (2008). Host galaxies catalog used in LIGO searches for compact binary coalescence events. *Astrophysical Journal*, **675**, 1459.

Koranda, S., Stergioulas, N., and Friedman, J. L. (1997). Upper limits set by causality on the rotation and mass of uniformly rotating relativistic stars. *Astrophysical Journal*, **488**, 799.

Kormendy, J. and Richstone, D. (1995). Inward bound—the search for supermassive black holes in Galactic nuclei. *Annual Review in Astronomy and Astrophysics*, **33**, 581.

Kramer, M. *et al.* (2006). A periodically active pulsar giving insight into magnetospheric physics. *Science*, **312**, 549.

Kramer, M. *et al.* (2006). Tests of general relativity from timing the Double Pulsar. *Science*, **314**, 97.

Krüger, C., Gaertig, E., and Kokkotas, K. D. (2010). Oscillations and instabilities of fast and differentially rotating relativistic stars. *Physical Review D*, **81**, 084019.

Krüger, C. J., Ho, W. C. G., and Andersson, N. (2015). Seismology of adolescent neutron stars: Accounting for thermal effects and crust elasticity. *Physical Review D*, **92**, 063009.

Kruskal, M. D. (1960). Maximal extension of schwarzschild metric. *Physical Review*, **119**, 1743.

Kyutoku, K. *et al.* (2011). Gravitational waves from spinning black hole-neutron star binaries: Dependence on black hole spins and on neutron star equations of state. *Physical Review D*, **84**, 064018.

Lai, D. (1994). Resonant oscillations and tidal heating in coalescing binary neutron stars. *Monthly Notices of the Royal Astronomical Society*, **270**, 611.

Lai, D. (1999). Secular instability of g-modes in rotating neutron stars. *Monthly Notices of the Royal Astronomical Society*, **307**, 1001.

Lai, D., Rasio, F. A., and Shapiro, S. L. (1993). Ellipsoidal figures of equilibrium. Compressible models. *Astrophysical Journal Supplement*, **88**, 205.

Lai, D. and Shapiro, S. L. (1995). Gravitational radiation from rapidly rotating nascent neutron stars. *Astrophysical Journal*, **442**, 259.

Landau, L. (1938). Origin of stellar energy. *Nature*, **141**, 333.

Lander, S. K. and Jones, D. I. (2009). Magnetic fields in axisymmetric neutron stars. *Monthly Notices of the Royal Astronomical Society*, **395**, 2162.

Lander, S. K. and Jones, D. I. (2012). Are there any stable magnetic fields in barotropic stars? *Monthly Notices of the Royal Astronomical Society*, **424**, 482.

Lasky, P. D. *et al.* (2016). Detecting gravitational-wave memory with LIGO: Implications of GW150914. *Physical Review Letters*, **117**, 061102.

Lasky, P. D. and Glampedakis, K. (2016). Observationally constraining gravitational-wave emission from short gamma-ray burst remnants. *Monthly Notices of the Royal Astronomical Society*, **458**, 1660.

Lattimer, J. M. (2014). Symmetry energy in nuclei and neutron stars. *Nuclear Physics A*, **928**, 276.

Lattimer, J. M. *et al.* (1991). Direct Urca process in neutron stars. *Physical Review Letters*, **66**, 2701.

Lattimer, J. M. and Schramm, D. N. (1974). Black-hole-neutron-star collisions. *Astrophysical Journal Letters*, **192**, L145.

Lau, H. K., Leung, P. T., and Lin, L. M. (2010). Inferring physical parameters of compact stars from their f-mode gravitational-wave signals. *Astrophysical Journal*, **714**, 1234.

Lazarus, P. *et al.* (2016). Einstein@Home Discovery of a Double Neutron Star Binary in the PALFA Survey. *Astrophysical Journal*, **831**, 150.

Le Tiec, A. *et al.* (2011). Periastron advance in black-hole binaries. *Physical Review Letters*, **107**, 141101.

Leaver, E. W. (1985). An analytic representation for the quasi-normal modes of Kerr black holes. *Proceedings of the Royal Society of London Series A*, **402**, 285.

Leaver, E. W. (1986). Spectral decomposition of the perturbation response of the Schwarzschild geometry. *Physical Review D*, **34**, 384.

Lee, U. (1995). Nonradial oscillations of neutron stars with the superfluid core. *Astronomy and Astrophysics*, **303**, 515.

Lee, U. (2014). Excitation of a non-radial mode in a millisecond X-ray pulsar XTE J1751-305. *Monthly Notices of the Royal Astronomical Society*, **442**, 3037.

Levin, Y. (1999). Runaway heating by r-modes of neutron stars in low-mass X-ray binaries. *Astrophysical Journal*, **517**, 328.

Levin, Y. (2007). On the theory of magnetar QPOs. *Monthly Notices of the Royal Astronomical Society*, **377**, 159.

Levin, Y. and Ushomirsky, G. (2001). Crust-core coupling and r-mode damping in neutron stars: A toy model. *Monthly Notices of the Royal Astronomical Society*, **324**, 917.

Liddle, A. (2003). *An Introduction to modern cosmology*. Wiley.

Lin, L.-M., Andersson, N., and Comer, G. L. (2008). Oscillations of general relativistic multi-fluid/multilayer compact stars. *Physical Review D*, **78**, 083008.

Lin, L.-M. and Suen, W.-M. (2006). Non-linear r modes in neutron stars: A hydrodynamical limitation on r-mode amplitudes. *Monthly Notices of the Royal Astronomical Society*, **370**, 1295.

Lindblom, L. and Cutler, C. (2016). Model waveform accuracy requirements for the Allen χ^2 discriminator. *Physical Review D*, **94**, 124030.

Lindblom, L. and Detweiler, S. L. (1983). The quadrupole oscillations of neutron stars. *Astrophysical Journal Supplement*, **53**, 73.

Lindblom, L. and Mendell, G. (1995). Does gravitational radiation limit the angular velocities of superfluid neutron stars. *Astrophysical Journal*, **444**, 804.

Lindblom, L. and Mendell, G. (2000). r-modes in superfluid neutron stars. *Physical Review D*, **61**, 104003.

Lindblom, L., Mendell, G., and Owen, B. J. (1999). Second-order rotational effects on the r-modes of neutron stars. *Physical Review D*, **60**, 064006.

Lindblom, L. and Owen, B. J. (2002). Effect of hyperon bulk viscosity on neutron-star r-modes. *Physical Review D*, **65**, 063006.

Lindblom, L., Owen, B. J., and Morsink, S. M. (1998). Gravitational radiation instability in hot young neutron stars. *Physical Review Letters*, **80**, 4843.

Lindblom, L., Tohline, J. E., and Vallisneri, M. (2001). Nonlinear evolution of the r-modes in neutron stars. *Physical Review Letters*, **86**, 1152.

Lindblom, L., Tohline, J. E., and Vallisneri, M. (2002). Numerical evolutions of nonlinear r-modes in neutron stars. *Physical Review D*, **65**, 084039.

Link, B. (2003). Constraining hadronic superfluidity with neutron star precession. *Physical Review Letters*, **91**, 101101.

Link, B. and Epstein, R. I. (2001). Precession interpretation of the isolated pulsar PSR B1828-11. *Astrophysical Journal*, **556**, 392.

Lobo, J. A. (2000). Multimode gravitational wave detection: the spherical detector theory. *preprint gr-qc/0006055*.

Lockitch, K.H., Andersson, N., and Friedman, J.L. (2001). Rotational modes of relativistic stars: Analytic results. *Physical Review D*, **63**, 024019.

Lockitch, K.H., Friedman, J.L., and Andersson, N. (2003). Rotational modes of relativistic stars: Numerical results. *Physical Review D*, **68**, 124010.

Lockitch, K. H. and Friedman, J. L. (1999). Where are the r-modes of isentropic stars? *Astrophysical Journal*, **521**, 764.

Lorimer, D. R. *et al.* (2007). A bright millisecond radio burst of extragalactic origin. *Science*, **318**, 777.

Lorimer, D. R. and Kramer, M. (2012). *Handbook of pulsar astronomy*. Cambridge University Press, Cambridge.

Lu, J. R. *et al.* (2009). A disk of young stars at the Galactic Center as determined by individual stellar orbits. *Astrophysical Journal*, **690**, 1463.

Lyne, A. G. *et al.* (2004). A double-pulsar system: A rare laboratory for relativistic gravity and plasma physics. *Science*, **303**, 1153.

Lyne, A. G. *et al.* (2010). Switched magnetospheric regulation of pulsar spin-down. *Science*, **329**, 408.

MacFadyen, A. I. and Woosley, S. E. (1999). Collapsars: Gamma-ray bursts and explosions in 'failed supernovae'. *Astrophysical Journal*, 524, 262.

Mädler, T. and Winicour, J. (2016). Bondi-Sachs formalism. *Scholarpedia*, 11.

Madsen, J. (1998). How to identify a strange star. *Physical Review Letters*, 81, 3311.

Maggiore, M. (2000). Gravitational-wave experiments and early Universe cosmology. *Physics Reports*, 331, 283.

Maggiore, M. (2007). *Gravitational waves. Volume 1: Theory and experiment.* Oxford University Press, Oxford.

Maggiore, M. (2018). *Gravitational waves. Volume 2: Astrophysics and cosmology.* Oxford University Press, Oxford.

Manchester, R. N. *et al.* (2013). The International Pulsar Timing Array. *Classical and Quantum Gravity*, 30, 224010.

Marchant, P. *et al.* (2016). A new route towards merging massive black holes. *Astronomy and Astrophysics*, 588, A50.

Marshall, F. E. *et al.* (1998). Discovery of an ultrafast X-ray pulsar in the supernova remnant N157B. *Astrophysical Journal Letters*, 499, L179.

Martynov, D. V. *et al.* (2016). Sensitivity of the Advanced LIGO detectors at the beginning of gravitational wave astronomy. *Physical Review D*, 93, 112004.

Mastrano, A. *et al.* (2011). Gravitational-wave emission from a magnetically deformed non-barotropic neutron star. *Monthly Notices of the Royal Astronomical Society*, 417, 2288.

May, M. M. and White, R. H. (1966). Hydrodynamic calculations of general-relativistic collapse. *Physical Review*, 141, 1232.

Mazur, P. O. and Mottola, E. (2015). Surface tension and negative pressure interior of a non-singular black hole. *Classical and Quantum Gravity*, 32, 215024.

McDermott, P. N., van Horn, H. M., and Hansen, C. J. (1988). Nonradial oscillations of neutron stars. *Astrophysical Journal*, 325, 725.

Melson, T. *et al.* (2015). Neutrino-driven explosion of a 20 solar-mass star in three dimensions enabled by strange-quark contributions to neutrino-nucleon scattering. *Astrophysical Journal Letters*, 808, L42.

Mendell, G. (1991). Superfluid hydrodynamics in rotating neutron stars. II. Dissipative effects. *Astrophysical Journal*, 380, 530.

Mendell, G. (2001). Magnetic effects on the viscous boundary layer damping of the r-modes in neutron stars. *Physical Review D*, 64, 044009.

Merritt, D. *et al.* (2004). Consequences of gravitational radiation recoil. *Astrophysical Journal Letters*, 607, L9.

Messenger, C. and Patruno, A. (2015). A semi-coherent search for weak pulsations in Aquila X-1. *Astrophysical Journal*, 806, 261.

Metzger, B. D. (2017). Kilonovae. *Living Reviews in Relativity*, 20, 3.

Metzger, B. D. and Berger, E. (2012). What is the most promising electromagnetic counterpart of a neutron star binary merger? *Astrophysical Journal*, 746, 48.

Michelson, A. A. and Morley, E. W. (1966). The luminiferous ether receives a mortal blow. In *The World of the Atom, Volume 1* (ed. H. A. Boorse, L. Motz, and L. Motz), p. 369. Basic Books.

Miniutti, G., Fabian, A. C., and Miller, J. M. (2004). The relativistic Fe emission line in XTE J1650-500 with BeppoSAX: Evidence for black hole spin and light-bending effects? *Monthly Notices of the Royal Astronomical Society*, 351, 466.

Mino, Y., Sasaki, M., and Tanaka, T. (1997). Gravitational radiation reaction to a particle motion. *Physical Review D*, 55, 3457.

Misner, C. W. (1960). Wormhole initial conditions. *Physical Review*, 118, 1110.

Moncrief, V. (1974). Gravitational perturbations of spherically symmetric systems. I. The exterior problem. *Annals of Physics*, **88**, 323.

Mooley, K. P. *et al.* (2018). Superluminal motion of a relativistic jet in the neutron-star merger GW170817. *Nature*, **561**, 355.

Moore, C. J. *et al.* (2017). Astrometric search method for individually resolvable gravitational-wave sources with Gaia. *Physical Review Letters*, **119**, 261102.

Morsink, S. M. and Rezania, V. (2002). Normal modes of rotating magnetic stars. *Astrophysical Journal*, **574**, 908.

Mösta, P. *et al.* (2014). Magnetorotational core-collapse supernovae in three dimensions. *Astrophysical Journal Letters*, **785**, L29.

Müller, B. (2017). Gravitational waves from core-collapse supernovae. In *Gravitational Waves: Sources and Detection, Proceedings of the 13th International Conference on Mathematical and Numerical Aspects of Wave Propagation*.

Müller, B., Janka, H.-T., and Marek, A. (2013). A new multi-dimensional general relativistic neutrino hydrodynamics code of core-collapse supernovae. III. Gravitational wave signals from supernova explosion models. *Astrophysical Journal*, **766**, 43.

Murphy, J.W., Ott, C.D., and Burrows, A. (2009). A model for gravitational wave emission from neutrino-driven core-collapse supernovae. *Astrophysical Journal*, **707**, 1173.

Nakamura, T. *et al.* (2001). Light curve and spectral models for the hypernova SN 1998BW associated with GRB 980425. *Astrophysical Journal*, **550**, 991.

Narayan, R., Piran, T., and Shemi, A. (1991). Neutron star and black-hole binaries in the Galaxy. *Astrophysical Journal Letters*, **379**, L17.

New, K. C. B., Centrella, J. M., and Tohline, J. E. (2000). Gravitational waves from long-duration simulations of the dynamical bar instability. *Physical Review D*, **62**, 064019.

Newman, E. and Penrose, R. (1962). An approach to gravitational radiation by a method of spin coefficients. *Journal of Mathematical Physics*, **3**, 566.

Newman, E. T. and Penrose, R. (2009). Spin-coefficient formalism. *Scholarpedia*, **4**.

Newton, W. G. (2013). Neutron stars: A taste of pasta? *Nature Physics*, **9**, 396.

Nicholson, D. *et al.* (1996). Results of the first coincident observations by two laser-interferometric gravitational wave detectors. *Physics Letters A*, **218**, 175.

Nissanke, S. *et al.* (2010). Exploring short gamma-ray bursts as gravitational-wave standard sirens. *Astrophysical Journal*, **725**, 496.

Nissanke, S. and Blanchet, L. (2005). Gravitational radiation reaction in the equations of motion of compact binaries to 3.5 post-Newtonian order. *Classical and Quantum Gravity*, **22**, 1007.

Nollert, H.-P. (1999). Quasinormal modes: The characteristic 'sound' of black holes and neutron stars. *Classical and Quantum Gravity*, **16**, R159.

Nollert, H.-P. and Schmidt, B. G. (1992). Quasinormal modes of Schwarzschild black holes: Defined and calculated via Laplace transformation. *Physical Review D*, **45**, 2617.

Nomoto, K. *et al.* (2007). Hypernovae and their gamma-ray bursts connection. In *A life with stars*. New Astronomy Reviews.

Norman, M. L., Smith, B. D., and Bordner, J. (2018). Simulating the cosmic dawn with Enzo. *Frontiers in Astronomy and Space Sciences*, **5**, 34.

O'Beirne, L. and Cornish, N. J. (2018). Constraining the polarization content of gravitational waves with astrometry. *Physical Review D*, **98**, 024020.

Ogata, S. and Ichimaru, S. (1990). First-principles calculations of shear moduli for Monte Carlo-simulated Coulomb solids. *Physical Review A*, **42**, 4867.

Olausen, S. A. and Kaspi, V. M. (2014). The McGill magnetar catalog. *Astrophysical Journal Supplement*, **212**, 6.

Oppenheimer, J. R. and Snyder, H. (1939). On continued gravitational contraction. *Physical Review*, **56**, 455.

Oppenheimer, J. R. and Volkoff, G. M. (1939). On massive neutron cores. *Physical Review*, **55**, 374.

Osburn, T., Warburton, N., and Evans, C. R. (2016). Highly eccentric inspirals into a black hole. *Physical Review D*, **93**, 064024.

Ott, C.D. *et al.* (2011). Dynamics and gravitational-wave signature of collapsar formation. *Physical Review Letters*, **106**, 161103.

Ott, C. D. (2009). Probing the core-collapse supernova mechanism with gravitational waves. *Classical and Quantum Gravity*, **26**, 204015.

Ott, C. D. *et al.* (2007). 3D collapse of rotating stellar iron cores in general relativity including deleptonization and a nuclear equation of state. *Physical Review Letters*, **98**, 261101.

Ou, S. and Tohline, J. E. (2006). Unexpected dynamical instabilities in differentially rotating neutron stars. *Astrophysical Journal*, **651**, 1068.

Owen, B. J. (1996). Search templates for gravitational waves from inspiraling binaries: Choice of template spacing. *Physical Review D*, **53**, 6749.

Owen, B. J. (2010). How to adapt broad-band gravitational-wave searches for r-modes. *Physical Review D*, **82**, 104002.

Owen, B. J. *et al.* (1998). Gravitational waves from hot young rapidly rotating neutron stars. *Physical Review D*, **58**, 084020.

Owen, B. J. and Sathyaprakash, B. S. (1999). Matched filtering of gravitational waves from inspiraling compact binaries: Computational cost and template placement. *Physical Review D*, **60**, 022002.

Özel, Feryal, Baym, Gordon, and Güver, Tolga (2010). Astrophysical measurement of the equation of state of neutron star matter. *Physical Review D*, **82**, 101301.

Paczyński, B. (1971). Evolutionary processes in close binary systems. *Annual Review in Astronomy and Astrophysics*, **9**, 183.

Paczyński, B. (1998). Are gamma-ray bursts in star-forming regions? *Astrophysical Journal Letters*, **494**, L45.

Pagano, L., Salvati, L., and Melchiorri, A. (2016). New constraints on primordial gravitational waves from Planck 2015. *Physics Letters B*, **760**, 823.

Page, D. *et al.* (2004). Minimal cooling of neutron stars: A new paradigm. *Astrophysical Journal Supplement*, **155**, 623.

Page, D. *et al.* (2011). Rapid cooling of the neutron star in Cassiopeia A triggered by neutron superfluidity in dense matter. *Physical Review Letters*, **106**, 081101.

Page, D., Geppert, U., and Weber, F. (2006). The cooling of compact stars. *Nuclear Physics A*, **777**, 497.

Palfreyman, J. *et al.* (2018). Alteration of the magnetosphere of the Vela pulsar during a glitch. *Nature*, **556**, 219.

Pandharipande, V. R. and Ravenhall, D. G. (1989). Hot nuclear matter. In *NATO Advanced Science Institutes (ASI) Series B* (ed. M. Soyeur, H. Flocard, B. Tamain, and M. Porneuf), Volume 205, p. 103.

Pannarale, F. *et al.* (2015). Gravitational-wave cutoff frequencies of tidally disruptive neutron star-black hole binary mergers. *Physical Review D*, **92**, 081504.

Papaloizou, J. and Pringle, J. E. (1978). Gravitational radiation and the stability of rotating stars. *Monthly Notices of the Royal Astronomical Society*, **184**, 501.

Pardo, K. *et al.* (2018). Limits on the number of spacetime dimensions from GW170817. *Journal of Cosmology and Astroparticle Physics*, **7**, 048.

Paschalidis, V. (2017). General relativistic simulations of compact binary mergers as engines for short gamma-ray bursts. *Classical and Quantum Gravity*, **34**, 084002.

Passamonti, A. *et al.* (2009). Oscillations of rapidly rotating stratified neutron stars. *Monthly Notices of the Royal Astronomical Society*, **394**, 730.

Passamonti, A., Andersson, N., and Ho, W. C. G. (2016). Buoyancy and g-modes in young superfluid neutron stars. *Monthly Notices of the Royal Astronomical Society*, **455**, 1489.

Pati, M. E. and Will, C. M. (2000). Post-Newtonian gravitational radiation and equations of motion via direct integration of the relaxed Einstein equations: Foundations. *Physical Review D*, **62**, 124015.

Pati, M. E. and Will, C. M. (2002). Post-Newtonian gravitational radiation and equations of motion via direct integration of the relaxed Einstein equations. II. Two-body equations of motion to second post-Newtonian order, and radiation reaction to 3.5 post-Newtonian order. *Physical Review D*, **65**, 104008.

Patruno, A., Haskell, B., and Andersson, N. (2017). The spin distribution of fast-spinning neutron stars in low-mass X-ray binaries: Evidence for two subpopulations. *Astrophysical Journal*, **850**, 106.

Patruno, A. and Watts, A.L. (2012). Accreting millisecond X-ray pulsars. In *Timing neutron stars: pulsations, oscillations and explosions* (ed. C. Z. T. Belloni, M. Mendez). Springer.

Penner, A.J. *et al.* (2012). Crustal Failure during Binary Inspiral. *Astrophysical Journal Letters*, **749**, L36.

Penrose, R. (1969). Gravitational collapse: The role of general relativity. *Nuovo Cimento Rivista Serie*, **1**, 252.

Penrose, R. and Floyd, R. M. (1971). Extraction of rotational energy from a black hole. *Nature*, **229**, 177.

Penzias, A. A. and Wilson, R. W. (1965). A measurement of excess antenna temperature at 4080 Mc/s. *Astrophysical Journal*, **142**, 419.

Perlmutter, S. *et al.* (1999). Measurements of Ω and Λ from 42 high-redshift supernovae. *Astrophysical Journal*, **517**, 565.

Peters, P. C. and Mathews, J. (1963). Gravitational radiation from point masses in a Keplerian orbit. *Physical Review*, **131**, 435.

Petrillo, C. E., Dietz, A., and Cavaglià, M. (2013). Compact object coalescence rate estimation from short gamma-ray burst observations. *Astrophysical Journal*, **767**, 140.

Phinney, E. S. (1991). The rate of neutron star binary mergers in the Universe—minimal predictions for gravity-wave detectors. *Astrophysical Journal Letters*, **380**, L17.

Phinney, E. S. (2001). A practical theorem on gravitational-wave backgrounds. *preprint astro-ph/0108028*.

Piekarewicz, J. *et al.* (2012). Electric dipole polarizability and the neutron skin. *Physical Review C*, **85**, 041302.

Pines, D. and Shaham, J. (1972). The elastic energy and character of quakes in solid stars and planets. *Physics of the Earth and Planetary Interiors*, **6**, 103.

Piran, T. (2004). The physics of gamma-ray bursts. *Reviews of Modern Physics*, **76**, 1143.

Pletsch, H. J. *et al.* (2013). Einstein@Home Discovery of four young gamma-ray pulsars in Fermi LAT data. *Astrophysical Journal Letters*, **779**, L11.

Pnigouras, P. and Kokkotas, K.D. (2016). Saturation of the f-mode instability in neutron stars. II. Applications and results. *Physical Review D*, **94**, 024053.

Pnigouras, P. and Kokkotas, K. D. (2015). Saturation of the f-mode instability in neutron stars: Theoretical framework. *Physical Review D*, **92**, 084018.

Poisson, E., Pound, A., and Vega, I. (2011). The motion of point particles in curved spacetime. *Living Reviews in Relativity*, **14**, 7.

Poisson, E. and Sasaki, M. (1995). Gravitational radiation from a particle in circular orbit around a black hole. V. Black-hole absorption and tail corrections. *Physical Review D*, **51**, 5753.

Poisson, E. and Will, C. M. (2014). *Gravity*. Cambridge University Press, Cambridge.

Polchinski, J. (2005). *String Theory*. Cambridge University Press, Cambridge.

Pons, J. A. *et al.* (1999). Evolution of proto-neutron stars. *Astrophysical Journal*, **513**, 780.

Pons, J. A., Viganò, D., and Rea, N. (2013). A highly resistive layer within the crust of X-ray pulsars limits their spin periods. *Nature Physics*, **9**, 431.

Portegies Zwart, S. F. and McMillan, S. L. W. (2000). Black-hole mergers in the Universe. *Astrophysical Journal Letters*, **528**, L17.

Postnov, K. A. and Yungelson, L. R. (2014). The evolution of compact binary star systems. *Living Reviews in Relativity*, **17**, 3.

Potekhin, A. Y. *et al.* (2013). Analytical representations of unified equations of state for neutron-star matter. *Astronomy and Astrophysics*, **560**, A48.

Pound, A. (2017). Nonlinear gravitational self-force: Second-order equation of motion. *Physical Review D*, **95**, 104056.

Pound, R. V. and Rebka, G. A. (1960). Apparent weight of photons. *Physical Review Letters*, **4**, 337.

Powell, J. *et al.* (2016). Inferring the core-collapse supernova explosion mechanism with gravitational waves. *Physical Review D*, **94**, 123012.

Prakash, M. and Lattimer, J. M. (2004). A tale of two mergers: searching for strangeness in compact stars. *Journal of Physics G Nuclear Physics*, **30**, S451.

Press, W. H. (1971). Long wave trains of gravitational waves from a vibrating black hole. *Astrophysical Journal Letters*, **170**, L105.

Press, W. H. and Teukolsky, S. A. (1972). Floating orbits, superradiant scattering and the black-hole bomb. *Nature*, **238**, 211.

Press, W. H. and Teukolsky, S. A. (1977). On formation of close binaries by two-body tidal capture. *Astrophysical Journal*, **213**, 183.

Pretorius, F. (2005). Evolution of binary black-hole spacetimes. *Physical Review Letters*, **95**, 121101.

Price, R. H. (1972). Nonspherical perturbations of relativistic gravitational collapse. I. Scalar and gravitational perturbations. *Physical Review D*, **5**, 2419.

Price, R. H. and Pullin, J. (1994). Colliding black holes: The close limit. *Physical Review Letters*, **72**, 3297.

Prix, R., Comer, G. L., and Andersson, N. (2002). Slowly rotating superfluid Newtonian neutron star model with entrainment. *Astronomy and Astrophysics*, **381**, 178.

Psaltis, D. *et al.* (2015). Event Horizon Telescope evidence for alignment of the black hole in the center of the Milky Way with the inner stellar disk. *Astrophysical Journal*, **798**, 15.

Punturo, M. *et al.* (2010). The Einstein Telescope: A third-generation gravitational-wave observatory. *Classical and Quantum Gravity*, **27**, 194002.

Quinn, T. C. and Wald, R. M. (1997). Axiomatic approach to electromagnetic and gravitational radiation reaction of particles in curved spacetime. *Physical Review D*, **56**, 3381.

Racine, É. (2008). Analysis of spin precession in binary black-hole systems including quadrupole-monopole interaction. *Physical Review D*, **78**, 044021.

Radhakrishnan, V. and Manchester, R. N. (1969). Detection of a change of state in the pulsar PSR 0833−45. *Nature*, **222**, 228.

Radhakrishnan, V. and Srinivasan, G. (1982). On the origin of the recently discovered ultra-rapid pulsar. *Current Science*, **51**, 1096.

Rajagopal, M. and Romani, R. W. (1995). Ultra-low-frequency gravitational radiation from massive black-hole binaries. *Astrophysical Journal*, **446**, 543.

Randall, L. and Sundrum, R. (1999). An alternative to compactification. *Physical Review Letters*, **83**, 4690.

Rea, N. *et al.* (2012). A new low magnetic field magnetar: The 2011 outburst of Swift J1822.3-1606. *Astrophysical Journal*, **754**, 27.

Read, J. S. *et al.* (2013). Matter effects on binary neutron star waveforms. *Physical Review D*, **88**, 044042.

Rees, M. J. (1984). Black-hole models for Active Galactic Nuclei. *Annual Review in Astronomy and Astrophysics*, **22**, 471.

Regge, T. and Wheeler, J. A. (1957). Stability of a Schwarzschild singularity. *Physical Review*, **108**, 1063.

Reichley, P. E. and Downs, G. S. (1969). Observed decrease in the periods of pulsar PSR 0833-45. *Nature*, **222**, 229.

Reisenegger, A. and Goldreich, P. (1992). A new class of g-modes in neutron stars. *Astrophysical Journal*, **395**, 240.

Reisenegger, A. and Goldreich, P. (1994). Excitation of neutron star normal modes during binary inspiral. *Astrophysical Journal*, **426**, 688.

Rezzolla, L. *et al.* (2001). Properties of r modes in rotating magnetic neutron stars. I. Kinematic secular effects and magnetic evolution equations. *Physical Review D*, **64**, 104013.

Rezzolla, L. *et al.* (2011). The missing link: Merging neutron stars naturally produce jet-like structures and can power short gamma-ray bursts. *Astrophysical Journal Letters*, **732**, L6.

Rezzolla, L., Lamb, F. K., and Shapiro, S. L. (2000). R-mode oscillations in rotating magnetic neutron stars. *Astrophysical Journal Letters*, **531**, L139.

Rezzolla, L., Macedo, R.P., and Jaramillo, J.L. (2010). Understanding the 'antikick' in the merger of binary black holes. *Physical Review Letters*, **104**, 221101.

Rezzolla, L., Most, E. R., and Weih, L. R. (2018). Using gravitational-wave observations and quasi-universal relations to constrain the maximum mass of neutron stars. *Astrophysical Journal Letters*, **852**, L25.

Richers, S. *et al.* (2017). Equation of state effects on gravitational waves from rotating core collapse. *Physical Review D*, **95**, 063019.

Riess, A. G. *et al.* (1998). Observational evidence from supernovae for an accelerating universe and a cosmological constant. *Astronomical Journal*, **116**, 1009.

Riess, A. G. *et al.* (2016). A 2.4% determination of the local value of the Hubble constant. *Astrophysical Journal*, **826**, 56.

Riquelme, M. A. *et al.* (2012). Local two-dimensional particle-in-cell simulations of the collisionless magnetorotational instability. *Astrophysical Journal*, **755**, 50.

Roca-Maza, X. *et al.* (2011). Neutron skin of Pb208, nuclear symmetry energy, and the Parity Radius Experiment. *Physical Review Letters*, **106**, 252501.

Röver, C. *et al.* (2009). Bayesian reconstruction of gravitational-wave burst signals from simulations of rotating stellar core collapse and bounce. *Physical Review D*, **80**, 102004.

Ruderman, M. (1969). Neutron starquakes and pulsar periods. *Nature*, **223**, 597.

Ruderman, M., Zhu, T., and Chen, K. (1998). Neutron star magnetic field evolution, crust movement, and glitches. *Astrophysical Journal*, **492**, 267.

Ruffini, R. (1973). Gravitational radiation from a mass projected into a Schwarzschild black hole. *Physical Review D*, **7**, 972.

Ruiz, M. *et al.* (2016). Binary neutron star mergers: A jet engine for short gamma-ray bursts. *Astrophysical Journal Letters*, **824**, L6.

Ruiz, M. and Shapiro, S. L. (2017). General relativistic magnetohydrodynamics simulations of prompt-collapse neutron star mergers: The absence of jets. *Physical Review D*, **96**, 084063.

Sachs, R. K. (1962). Gravitational waves in general relativity. VIII. Waves in asymptotically flat space-time. *Proceedings of the Royal Society of London Series A*, **270**, 103.

Saio, H. (1982). R-mode oscillations in uniformly rotating stars. *Astrophysical Journal*, **256**, 717.

Samuelsson, L. and Andersson, N. (2007). Neutron star asteroseismology. Axial crust oscillations in the Cowling approximation. *Monthly Notices of the Royal Astronomical Society*, **374**, 256.

Saravani, M., Afshordi, N., and Mann, R. B. (2014). Dynamical emergence of universal horizons during the formation of black holes. *Physical Review D*, **89**, 084029.

Sasaki, M. and Nakamura, T. (1990). Gravitational radiation from an extreme Kerr black hole. *General Relativity and Gravitation*, **22**, 1351.

Sathyaprakash, B. *et al.* (2012). Scientific objectives of Einstein Telescope. *Classical and Quantum Gravity*, **29**, 124013.

Sathyaprakash, B. S. and Schutz, B. F. (2009). Physics, astrophysics and cosmology with gravitational waves. *Living Reviews in Relativity*, **12**, 2.

Saulson, P. R. (2000). Interferometric gravitational-wave detection: accomplishing the impossible. *Classical and Quantum Gravity*, **17**, 2441.

Saulson, P. R. (2005). Receiving Gravitational Waves. In *100 Years of relativity: Space-time structure—Einstein and beyond* (ed. A. Ashtekar), pp. 228–256. World Scientific Publishing Co.

Saulson, P. R. (2017). *Fundamentals of interferometric gravitational wave detectors*. World Scientific Publishing Co.

Sawyer, R. F. (1989). Bulk viscosity of hot neutron-star matter and the maximum rotation rates of neutron stars. *Physical Review D*, **39**, 3804.

Schaeffer, R., Zdunik, L., and Haensel, P. (1983). Phase transitions in stellar cores. I. Equilibrium configurations. *Astronomy and Astrophysics*, **126**, 121.

Schäfer, G. (1985). The gravitational quadrupole radiation-reaction force and the canonical formalism of ADM. *Annals of Physics*, **161**, 81.

Schenk, A. K. *et al.* (2002). Nonlinear mode coupling in rotating stars and the r-mode instability in neutron stars. *Physical Review D*, **65**, 024001.

Schmidt, B. P. *et al.* (1998). The high-z supernova search: measuring cosmic deceleration and global curvature of the Universe using type Ia supernovae. *Astrophysical Journal*, **507**, 46.

Schmidt, M. (1963). 3C 273 : A star-like object with large red-shift. *Nature*, **197**, 1040.

Schneider, P., Ehlers, J., and Falco, E. E. (1992). *Gravitational lenses*. Springer, Berlin.

Schumaker, B. L. and Thorne, K. S. (1983). Torsional oscillations of neutron stars. *Monthly Notices of the Royal Astronomical Society*, **203**, 457.

Schutz, B. F. (1986). Determining the Hubble constant from gravitational-wave observations. *Nature*, **323**, 310.

Schutz, B. F. (1996). Making the transition from Newton to Einstein: Chandrasekhar's work on the post-Newtonian approximation and radiation reaction. *Journal of Astrophysics and Astronomy*, **17**, 183.

Schutz, B. F. (2009). *A first course in general relativity*. Cambridge University Press, Cambridge.

Schutz, B. F. and Will, C. M. (1985). Black-hole normal modes—a semianalytic approach. *Astrophysical Journal Letters*, **291**, L33.

Schwarz, K. W. (1982). Generation of superfluid turbulence deduced from simple dynamical rules. *Physical Review Letters*, **49**, 283.

Schwarzschild, K. (1999). On the gravitational field of a mass point according to Einstein's theory. *preprint physics/990503*.

Sedrakian, A., Wasserman, I., and Cordes, J. M. (1999). Precession of isolated neutron stars. I. Effects of imperfect pinning. *Astrophysical Journal*, **524**, 341.

Sekiguchi, Y. *et al.* (2016). Dynamical mass ejection from the merger of asymmetric binary neutron stars: Radiation-hydrodynamics study in general relativity. *Physical Review D.*, **93**, 124046.

Sereno, M. *et al.* (2011). Cosmography with strong lensing of LISA gravitational wave sources. *Monthly Notices of the Royal Astronomical Society*, **415**, 2773.

Sesana, A., Vecchio, A., and Colacino, C. N. (2008). The stochastic gravitational-wave background from massive black hole binary systems: Implications for observations with Pulsar Timing Arrays. *Monthly Notices of the Royal Astronomical Society*, **390**, 192.

Seveso, S. *et al.* (2016). Mesoscopic pinning forces in neutron star crusts. *Monthly Notices of the Royal Astronomical Society*, **455**, 3952.

Shaham, J. (1977). Free precession of neutron stars—role of possible vortex pinning. *Astrophysical Journal*, **214**, 251.

Shakura, N. I. and Sunyaev, R. A. (1973). Black holes in binary systems. Observational appearance. *Astronomy and Astrophysics*, **24**, 337.

Shapiro, I. I. (1966). Testing general relativity with radar. *Physical Review*, **141**, 1219.

Shcherbakov, R. V. *et al.* (2013). GRB060218 as a tidal disruption of a white dwarf by an intermediate-mass black hole. *Astrophysical Journal*, **769**, 85.

Shibata, M. (2016). *Numerical relativity*. World Scientific, Singapore.

Shibata, M. *et al.* (2016). Gravitational waves from supermassive stars collapsing to a supermassive black hole. *Physical Review D*, **94**, 021501.

Shibata, M., Baumgarte, T. W., and Shapiro, S. L. (2000). The bar-mode instability in differentially rotating neutron stars: Simulations in full general relativity. *Astrophysical Journal*, **542**, 453.

Shibata, M., Karino, S., and Eriguchi, Y. (2002). Dynamical instability of differentially rotating stars. *Monthly Notices of the Royal Astronomical Society*, **334**, L27.

Shibata, M. and Nakamura, T. (1995). Evolution of three-dimensional gravitational waves: Harmonic slicing case. *Physical Review D*, **52**, 5428.

Shibata, M. and Uryu, K. (2007). Merger of black hole neutron star binaries in full general relativity. *Classical and Quantum Gravity*, **24**, S125.

Shternin, P. S. *et al.* (2011). Cooling neutron star in the Cassiopeia A supernova remnant: Evidence for superfluidity in the core. *Monthly Notices of the Royal Astronomical Society*, **412**, L108.

Sidery, T., Passamonti, A., and Andersson, N. (2010). The dynamics of pulsar glitches: Contrasting phenomenology with numerical evolutions. *Monthly Notices of the Royal Astronomical Society*, **405**, 1061.

Siegel, D. M. *et al.* (2013). Magnetorotational instability in relativistic hypermassive neutron stars. *Physical Review D*, **87**, 121302.

Sigurdsson, S. and Hernquist, L. (1992). A novel mechanism for creating double pulsars. *Astrophysical Journal Letters*, **401**, L93.

Sigurdsson, S. and Hernquist, L. (1993). Primordial black holes in globular clusters. *Nature*, **364**, 423.

Smarr, L. (1979). Gauge conditions, radiation formulae and the two black hole collision. In *Sources of Gravitational Radiation* (ed. L. L. Smarr), p. 245.

Sopuerta, C. F., Yunes, N., and Laguna, P. (2006). Gravitational recoil from binary black-hole mergers: The close-limit approximation. *Physical Review D*, 74, 124010.

Sorkin, R. D. (1982). A stability criterion for many parameter equilibrium families. *Astrophysical Journal*, 257, 847.

Spergel, D. N. and Steinhardt, P. J. (2000). Observational evidence for self-interacting cold dark matter. *Physical Review Letters*, 84, 3760.

Sperhake, U. (2015). The numerical relativity breakthrough for binary black holes. *Classical and Quantum Gravity*, 32, 124011.

Spitkovsky, A. (2006). Time-dependent force-free pulsar magnetospheres: axisymmetric and oblique rotators. *Astrophysical Journal Letters*, 648, L51.

Springel, V. *et al.* (2005). Simulations of the formation, evolution and clustering of galaxies and quasars. *Nature*, 435, 629.

Spruit, H. and Phinney, E. S. (1998). Birth kicks as the origin of pulsar rotation. *Nature*, 393, 139.

Stairs, I. H., Lyne, A. G., and Shemar, S. L. (2000). Evidence for free precession in a pulsar. *Nature*, 406, 484.

Stappers, B. W. *et al.* (2018). The prospects of pulsar timing with new-generation radio telescopes and the Square Kilometre Array. *Philosophical Transactions of the Royal Society of London Series A*, 376, 20170293.

Stark, R. F. and Piran, T. (1985). Gravitational-wave emission from rotating gravitational collapse. *Physical Review Letters*, 55, 891.

Steiner, A. W. *et al.* (2018). Constraining the mass and radius of neutron stars in globular clusters. *Monthly Notices of the Royal Astronomical Society*, 476, 421.

Steiner, A. W., Lattimer, J. M., and Brown, E. F. (2010). The equation of state from observed masses and radii of neutron stars. *Astrophysical Journal*, 722, 33.

Steinhoff, J. *et al.* (2016). Dynamical tides in general relativity: Effective action and effective-one-body Hamiltonian. *Physical Review D*, 94, 104028.

Stella, L. *et al.* (2005). Gravitational radiation from newborn magnetars in the Virgo cluster. *Astrophysical Journal Letters*, 634, L165.

Stephani, H. *et al.* (2009). *Exact solutions of Einstein's field equations*. Cambridge University Press, Cambridge.

Stergioulas, N. and Friedman, J. L. (1995). Comparing models of rapidly rotating relativistic stars constructed by two numerical methods. *Astrophysical Journal*, 444, 306.

Stergioulas, N. and Friedman, J. L. (1998). Nonaxisymmetric neutral modes in rotating relativistic stars. *Astrophysical Journal*, 492, 301.

Stroeer, A. and Vecchio, A. (2006). The LISA verification binaries. *Classical and Quantum Gravity*, 23, S809.

Strohmayer, T. *et al.* (1991). The shear modulus of the neutron star crust and non-radial oscillations of neutron stars. *Astrophysical Journal*, 375, 679.

Strohmayer, T. and Mahmoodifar, S. (2014a). A non-radial oscillation mode in an accreting millisecond pulsar? *Astrophysical Journal*, 784, 72.

Strohmayer, T. and Mahmoodifar, S. (2014b). Discovery of a neutron star oscillation mode during a superburst. *Astrophysical Journal Letters*, 793, L38.

Strohmayer, T. E. and Watts, A. L. (2005). Discovery of fast X-ray oscillations during the 1998 giant flare from SGR 1900+14. *Astrophysical Journal Letters*, 632, L111.

Strohmayer, T. E. and Watts, A. L. (2006). The 2004 hyperflare from SGR 1806−20: Further evidence for global torsional vibrations. *Astrophysical Journal*, 653, 593.

Szekeres, P. (1965). The gravitational compass. *Journal of Mathematical Physics*, **6**, 1387.

Takahashi, R. and Nakamura, T. (2003). Wave effects in the gravitational lensing of gravitational waves from chirping binaries. *Astrophysical Journal*, **595**, 1039.

Takami, K., Rezzolla, L., and Baiotti, L. (2014). Constraining the equation of state of neutron stars from binary mergers. *Physical Review Letters*, **113**, 091104.

Tamburino, L. A. and Winicour, J. H. (1966). Gravitational fields in finite and conformal Bondi frames. *Physical Review*, **150**, 1039.

Tanaka, M. *et al.* (2018). Properties of kilonovae from dynamical and post-merger ejecta of neutron star mergers. *Astrophysical Journal*, **852**, 109.

Tanvir, N. R. *et al.* (2013). A 'kilonova' associated with the short-duration γ-ray burst GRB 130603B. *Nature*, **500**, 547.

Tassoul, J.-L. (1978). *Theory of rotating stars*. Princeton University Press, Princeton.

Tauris, T.M. *et al.* (2017). Formation of double neutron star systems. *Astrophysical Journal*, **846**, 170.

Tegmark, M. *et al.* (1997). How small were the first cosmological objects? *Astrophysical Journal*, **474**, 1.

Teukolsky, S. A. (1973). Perturbations of a rotating black hole. I. Fundamental equations for gravitational, electromagnetic, and neutrino-field perturbations. *Astrophysical Journal*, **185**, 635.

Thompson, C. and Duncan, R. C. (1995). The soft gamma repeaters as very strongly magnetized neutron stars. I. Radiative mechanism for outbursts. *Monthly Notices of the Royal Astronomical Society*, **275**, 255.

Thorne, K. S. (1974a). Disk-accretion onto a black hole. II. Evolution of the hole. *Astrophysical Journal*, **191**, 507.

Thorne, K. S. (1974b). The search for black holes. *Scientific American*, **231**, 32.

Thorne, K. S. (1979). Sources of gravitational waves. *Proceedings of the Royal Society of London A*, **368**, 9.

Thorne, K. S. (1980). Multipole expansions of gravitational radiation. *Reviews of Modern Physics*, **52**, 299.

Thorne, K. S. and Blandford, R. D. (2017). *Modern classical physics: Optics, fluids, plasmas, elasticity, relativity, and statistical physics*. Princeton University Press, Princeton.

Thorne, K. S. and Campolattaro, A. (1967). Non-radial pulsation of general-relativistic stellar models. I. *Astrophysical Journal*, **149**, 591.

Thorne, K. S. and Zytkow, A. N. (1975). Red giants and supergiants with degenerate neutron cores. *Astrophysical Journal Letters*, **199**, L19.

Tingay, S. J. *et al.* (1995). Relativistic motion in a nearby bright X-ray source. *Nature*, **374**, 141.

Tinto, Massimo and Dhurandhar, Sanjeev V. (2014). Time-delay interferometry. *Living Reviews in Relativity*, **17**, 6.

Tolman, R. C. (1939). Static solutions of Einstein's field equations for spheres of fluid. *Physical Review*, **55**, 364.

Toman, J. *et al.* (1998). Nonaxisymmetric dynamic instabilities of rotating polytropes. I. The Kelvin modes. *Astrophysical Journal*, **497**, 370.

Tomimura, Y. and Eriguchi, Y. (2005). A new numerical scheme for structures of rotating magnetic stars. *Monthly Notices of the Royal Astronomical Society*, **359**, 1117.

Troja, E. *et al.* (2017). The X-ray counterpart to the gravitational-wave event GW170817. *Nature*, **551**, 71.

Troja, E., Rosswog, S., and Gehrels, N. (2010). Precursors of short gamma-ray bursts. *Astrophysical Journal*, **723**, 1711.

Tsang, D. *et al.* (2012). Resonant shattering of neutron star crusts. *Physical Review Letters*, **108**, 011102.

Tsubota, M., Araki, T., and Barenghi, C. F. (2003). Rotating superfluid turbulence. *Physical Review Letters*, **90**, 205301.

Tsui, L. K. and Leung, P. T. (2005). Universality in quasi-normal modes of neutron stars. *Monthly Notices of the Royal Astronomical Society*, **357**, 1029.

Tutukov, A. V. and Yungelson, L. R. (1993). The merger rate of neutron star and black-hole binaries. *Monthly Notices of the Royal Astronomical Society*, **260**, 675.

Typel, S. *et al.* (2010). Composition and thermodynamics of nuclear matter with light clusters. *Physical Review C*, **81**, 015803.

Tyson, A. and Angel, R. (2001). The Large-aperture Synoptic Survey Telescope. In *The New Era of Wide Field Astronomy* (ed. R. Clowes, A. Adamson, and G. Bromage), Volume 232, Astronomical Society of the Pacific Conference Series, p. 347.

Ushomirsky, G., Cutler, C., and Bildsten, L. (2000). Deformations of accreting neutron star crusts and gravitational-wave emission. *Monthly Notices of the Royal Astronomical Society*, **319**, 902.

Uzdensky, D. A. and MacFadyen, A. I. (2007). Magnetically dominated jets inside collapsing stars as a model for gamma-ray bursts and supernova explosions. *Physics of Plasmas*, **14**, 056506.

Vigelius, M. and Melatos, A. (2010). Gravitational-wave spin-down and stalling lower limits on the electrical resistivity of the accreted mountain in a millisecond pulsar. *Astrophysical Journal*, **717**, 404.

Villar, V. A. *et al.* (2017). The combined ultraviolet, optical, and near-infrared light curves of the kilonova associated with the binary neutron star merger GW170817: Unified data set, analytic models, and physical implications. *Astrophysical Journal Letters*, **851**, L21.

Visco, M. and Votano, L. (2000). Resonant bar gravitational-wave detectors. p. 288.

Vishveshwara, C. V. (1970*a*). Scattering of gravitational radiation by a Schwarzschild black hole. *Nature*, **227**, 936.

Vishveshwara, C. V. (1970*b*). Stability of the Schwarzschild metric. *Physical Review D*, **1**, 2870.

Wagoner, R. V. (1984). Gravitational radiation from accreting neutron stars. *Astrophysical Journal*, **278**, 345.

Wang, Y.-M. (1995). On the torque exerted by a magnetically threaded accretion disk. *Astrophysical Journal Letters*, **449**, L153.

Watts, A. L. *et al.* (2008). Detecting gravitational-wave emission from the known accreting neutron stars. *Monthly Notices of the Royal Astronomical Society*, **389**, 839.

Watts, A. L. *et al.* (2016). Colloquium: Measuring the neutron star equation of state using X-ray timing. *Reviews of Modern Physics*, **88**(2), 021001.

Watts, A. L. and Andersson, N. (2002). The spin evolution of nascent neutron stars. *Monthly Notices of the Royal Astronomical Society*, **333**, 943.

Watts, A. L., Andersson, N., and Jones, D. I. (2005). The nature of low $T/|W|$ dynamical instabilities in differentially rotating stars. *Astrophysical Journal Letters*, **618**, L37.

Weber, J. (1967). Gravitational radiation. *Physical Review Letters*, **18**, 498.

Weber, J. (1969). Evidence for discovery of gravitational radiation. *Physical Review Letters*, **22**, 1320.

Weinberg, S. (2008). *Cosmology*. Oxford University Press, Oxford.

Weisberg, J.M., Nice, D.J., and Taylor, J.H. (2010). Timing Measurements of the relativistic Binary Pulsar PSR B1913+16. *Astrophysical Journal*, **722**, 1030.

Westerweck, J. *et al.* (2018). Low significance of evidence for black hole echoes in gravitational-wave data. *Physical Review D*, **97**, 124037.

Wette, K. *et al.* (2008). Searching for gravitational waves from Cassiopeia A with LIGO. *Classical and Quantum Gravity*, **25**, 235011.

White, N. E. and Zhang, W. (1997). Millisecond X-ray pulsars in low-mass X-ray binaries. *Astrophysical Journal Letters*, **490**, L87.

Wijnands, R. and van der Klis, M. (1998). A millisecond pulsar in an X-ray binary system. *Nature*, **394**, 344.

Will, C. M. (1998). Bounding the mass of the graviton using gravitational-wave observations of inspiralling compact binaries. *Physical Review D*, **57**, 2061.

Will, C. M. (1999). Gravitational radiation and the validity of general relativity. *Physics Today*, **52**, 38.

Will, C. M. (2005). Was Einstein right?: Testing relativity at the centenary. In *100 Years of Relativity: Space-Time Structure - Einstein and Beyond* (ed. A. Ashtekar), pp. 205–227. World Scientific Publishing Co.

Will, C. M. (2014). The confrontation between general relativity and experiment. *Living Reviews in Relativity*.

Will, C. M. (2018). Solar system versus gravitational-wave bounds on the graviton mass. *Classical and Quantum Gravity*, **35**(17), 17LT01.

Willke, B. *et al.* (2007). GEO600: Status and plans. *Classical and Quantum Gravity*, **24**, S389.

Winicour, J. (2012). Characteristic evolution and matching. *Living Reviews in Relativity*, **15**, 2.

Wiseman, A.G. (1992). Coalescing binary systems of compact objects to (post)$^{5/2}$-Newtonian order. II. Higher-order wave forms and radiation recoil. *Physical Review D*, **46**, 1517.

Witten, E. (1984). Cosmic separation of phases. *Physical Review D*, **30**, 272.

Woosley, S. E. and Bloom, J. S. (2006). The supernova gamma-ray burst connection. *Annual Review Astronomy and Astrophysics*, **44**, 507.

Xie, X., Zrake, J., and MacFadyen, A. (2018). Numerical simulations of the jet dynamics and synchrotron radiation of binary neutron star merger event GW170817/GRB 170817A. *Astrophysical Journal*, **863**, 58.

Yagi, K. and Yunes, N. (2013a). I–Love–Q relations in neutron stars and their applications to astrophysics, gravitational waves, and fundamental physics. *Physical Review D*, **88**, 023009.

Yagi, K. and Yunes, N. (2013b). I–Love–Q: Unexpected universal relations for neutron stars and quark stars. *Science*, **341**, 365.

York, J. W. (1971). Gravitational degrees of freedom and the initial-value problem. *Physical Review Letters*, **26**, 1656.

York, Jr., J. W. (1979). Kinematics and dynamics of general relativity. In *Sources of Gravitational Radiation* (ed. L. L. Smarr), p. 83.

Yoshida, S. and Eriguchi, Y. (1999). A numerical study of normal modes of rotating neutron star models by the Cowling approximation. *Astrophysical Journal*, **515**, 414.

Yu, H. and Weinberg, N. N. (2017). Resonant tidal excitation of superfluid neutron stars in coalescing binaries. *Monthly Notices of the Royal Astronomical Society*, **464**, 2622.

Yunes, N., Yagi, K., and Pretorius, F. (2016). Theoretical physics implications of the binary black-hole mergers GW150914 and GW151226. *Physical Review D*, **94**, 084002.

Zerilli, F. J. (1970a). Effective potential for even-parity Regge–Wheeler gravitational perturbation equations. *Physical Review Letters*, **24**, 737.

Zerilli, F. J. (1970b). Gravitational field of a particle falling in a Schwarzschild geometry analyzed in tensor harmonics. *Physical Review D*, **2**, 2141.

Zimmermann, M. and Szedenits, Jr., E. (1979). Gravitational waves from rotating and precessing rigid bodies. Simple models and applications to pulsars. *Physical Review D*, **20**, 351.

Index